Strict and Facultative Anaerobes
Medical and Environmental Aspects

Edited by

Michiko M. Nakano
Peter Zuber

Oregon Health & Science University,
Beaverton, OR 97006 8921, USA

Horizon Bioscience
32 Hewitts Lane
Wymondham
Norfolk NR18 0JA
England

www.horizonbioscience.com

British Library Cataloguing-in-Publication Data

A catalogue record for this book is available from the British Library

ISBN: 0-8493-3317-2
ISBN: 1-904933-03-3

Printed and bound in Great Britain

Contents

Contributors vi
Preface x

Section 1: Overview

Chapter 1. The Phylogeny and Classification of Anaerobic Bacteria 1
Erko Stackebrandt

Chapter 2. An Overview of Anaerobic Metabolism 27
Michael J. McInerney and Lisa M. Gieg

Chapter 3. Redox (Oxygen)-Dependent Gene Regulation in Facultative Anaerobes 67
R. Gary Sawers and Michiko M. Nakano

Section 2: Medical Aspects

Chapter 4. Anaerobic Metabolism by *Pseudomonas aeruginosa* in Cystic Fibrosis Airway Biofilms: Role of Nitric Oxide, Quorum Sensing, and Alginate Production 87
Daniel J. Hassett, Sergei V. Lymar, John J. Rowe, Michael J. Schurr, Luciano Passador, Andrew B. Herr, Geoffrey L. Winsor, Fiona S. L. Brinkman, Sang Sun Yoon, Gee W. Lau, and Sung Hei Hwang

Chapter 5. Oral Microbial Communities: Genetic Analysis of Oral Biofilms 109
Howard K. Kuramitsu

Chapter 6. The Gut Microflora 125
Rodrigo Bibiloni, Jens Walter, and Gerald W. Tannock

Chapter 7. The Bacillus Holds its Breath – Latency and the Hypoxic Response of *Mycobacterium tuberculosis* 145
David Sherman and David M. Roberts

Chapter 8. Molecular Basis for Aerotolerance of the Obligately Anaerobic *Bacteroides* spp. 161
Anthony D. Baughn and Michael H. Malamy

Chapter 9. Clostridial and Bacteroides Toxins: Structure and Mode of Action 171
Michel-Robert Popoff

Chapter 10. Toxin Gene Regulation in *Clostridium* 199
Michiko M. Nakano, Peter Zuber, and Abraham L. Sonenshein

Chapter 11. Use of Anaerobic Bacteria for Cancer Therapy 211
J. Martin Brown and S. C. Liu

Section 3: Environmental/Industrial Aspects

Chapter 12. Bioremediation Processes Coupled to the Microbial Reduction of Fe(III)oxides in Sedimentary Environments 221
Robert T. Anderson

Chapter 13. Prokaryotic Arsenate and Selenate Respiration
Joanne M. Santini and John F. Stolz 239

Chapter 14. Molecular and Cellular Biology of Acetogenic Bacteria 251
Volker Müller, Frank Imkamp, Andreas Rauwolf, Kirsten Küsel, and Harold L. Drake

Chapter 15. Anaerobic Oxidation of Inorganic Nitrogen Compounds 283
Ingo Schmidt and Mike S. M. Jetten

Chapter 16. Reductive Dehalogenation of Polychlorinated Benzenes and Dioxins 303
Lorenz Adrian and Ute Lechner

Chapter 17. Biotransformation of Carbon Tetrachloride by the Facultative Anaerobic Bacterium *Pseudomonas stutzeri* 317
Andrzej Paszczynski, Jonathan Sebat, Daniel Erwin, and Ronald L. Crawford

Chapter 18. Solventogenesis by Clostridia 329
Peter Dürre

Chapter 19. The Clostridial Cellulosome 343
Anne Belaich, Chantal Tardif, Henri-Pierre Fierobe, Sandrine Pagès, and Jean-Pierre Belaich

Chapter 20. Microbial Community Structure and Functions in Methane Fermentation Technology for Wastewater Treatment 361
Yuji Sekiguchi and Yoichi Kamagata

Index 385

Books of Related Interest

- **Pathogenic Fungi: Host Interaction and Emerging Strategies for Control** (2004)
  Edited by: Gioconda San-Blas and Richard A. Calderone
- **Pathogenic Fungi: Structural Biology and Taxonomy** (2004)
  Edited by: Gioconda San-Blas and Richard A. Calderone
- **Malaria Parasites: Genomes and Molecular Biology** (2004)
  Edited by: Andrew P. Waters and Chris J. Janse
- **Peptide Nucleic Acids: Protocols and Applications,** 2nd Edition (2004)
  Edited by: Peter E. Nielsen
- **Ebola and Marburg Viruses**: Molecular and Cellular Biology (2004)
  Edited by: Hans-Dieter Klenk and Heinz Feldmann
- **Metabolic Engineering in the Post-genomic Era** (2004)
  Edited by: Boris N. Kholodenko and Hans V. Westerhoff
- **MRSA: Current Perspectives** (2003)
  Edited by: Ad C. Fluit and Franz-Josef Schmitz
- **Tuberculosis: The Microbe Host Interface** (2003)
  Edited by: Larry S. Schlesinger and Lucy DesJardin
- **Genome Mapping and Sequencing** (2003)
  Edited by: Ian Dunham
- **Regulatory Networks in Prokaryotes** (2003)
  Edited by: Peter Dürre and Bärbel Friedrich
- **Bioremediation: A Critical Review** (2003)
  Edited by: I.M. Head, I. Singleton and M. Milner
- **Bioinformatics and Genomes**: Current Perspectives (2003)
  Edited by: Miguel A. Andrade
- **Frontiers in Computional Genomics** (2003)
  Edited by Michael Y. Galperin and Eugene V. Koonin :
- **Multiple Drug Resistant Bacteria: Emerging Strategies** (2003)
  Edited by: Carlos F. Amábile-Cuevas
- **Vaccine Delivery Strategies** (2003)
  Edited by: Guido Dietrich and W. Goebel
- **Probiotics and Prebiotics: Where Are We Going?** (2002)
  Edited by: Gerald W. Tannock
- **Genomic Technologies: Present and Future** (2002)
  Edited by: David J. Galas and Stephen J. McCormack
- **Genomics of GC-Rich Gram Positive Bacteria** (2002)
  Edited by: Antoine Danchin

Full details of these and all our books at:

www.horizonpress.com

Contributors

Lorenz Adrian
Fachgebiet Technische Biochemie
Institut für Biotechnologie
Technische Universität Berlin
Sekr. GG1
13353 Berlin
Germany

Robert T. Anderson
Department of Microbiology
University of Massachusetts
Morrill Science Center IVN
Amherst
MA 01003
USA

Anthony D. Baughn
Department of Molecular Biology and Microbiology
Tufts University School of Medicine
136 Harrison Avenue
Boston
MA 02111
USA

Anne Belaich
Laboratoire de Bioenergétique et Ingénierie des Proteines
CNRS
31 chemin Joseph Aiguier
13009 Marseille
France

Jean-Pierre Belaich
Université de Provence
13003 Marseille
France

and

Laboratoire de Bioenergétique et Ingénierie des Proteines
CNRS
31 chemin Joseph Aiguier
13009 Marseille
France

Rodrigo Bibiloni
Department of Microbiology
University of Otago
PO Box 56
Dunedin
New Zealand

J. Martin Brown
Dept. of Radiation Oncology
Division of Cancer and Radiation Biology
CCSR-South
Room 1255, 269 Campus Drive
Stanford University School of Medicine
Stanford
CA 94305-5152
USA

Fiona S. L. Brinkman
Department of Molecular Biology and Biochemistry
Simon Fraser University
Burnaby
BC
Canada
V5A 1S6

Ronald L. Crawford
Environmental Biotechnology Institute
University of Idaho
Moscow
ID 83844-1052
USA

Harold L. Drake
Department of Ecological Microbiology
University of Bayreuth
95440 Bayreuth
Germany

Peter Dürre
Mikrobiologie und Biotechnologie
Universität Ulm
89069 Ulm
Germany

Daniel Erwin
Environmental Biotechnology Institute
University of Idaho
Moscow
ID 83844-1052
USA

Henri-Pierre Fierobe
Laboratoire de Bioenergétique et Ingénierie des Proteines
CNRS
31 chemin Joseph Aiguier
13009 Marseille
France

Lisa M. Gieg
Department of Botany and Microbiology
University of Oklahoma
770 Van Vleet Oval
Norman
OK 73019
USA

Daniel J. Hassett
Department of Molecular Genetics
Biochemistry and Microbiology
University of Cincinnati College of Medicine
231 Albert Sabin Way
Cincinnati
OH 45267-0524
USA

Andrew B. Herr
Department of Molecular Genetics
Biochemistry and Microbiology and Pulmonary Medicine
University of Cincinnati College of Medicine
Cincinnati
OH 45267
USA

Sung Hei Hwang
Department of Molecular Genetics
Biochemistry and Microbiology and Pulmonary Medicine
University of Cincinnati College of Medicine
Cincinnati
OH 45267
USA

Frank Imkamp
Institute of Microbiology
Johann Wolfgang Goethe University
Marie-Curie-Str. 9
60439 Frankfurt
Germany

Mike S. M. Jetten
Department of Microbiology
University of Nijmegen
Toernooiveld 1
6525 ED Nijmegen
The Netherlands

Yoichi Kamagata
Microbial and Genetic Resources Research Group
Research Institute of Biol. Resources and Functions
National Inst Advanced Industrial Sci. and Technol. (AIST)
Tsukuba
Ibaraki 305-8566
Japan

Howard K. Kuramitsu
Dept Oral Biology and Microbiology-Immunology
State University of New York at Buffalo
3455 Main St.
Buffalo
NY 14214
USA

Kirsten Küsel
Department of Ecological Microbiology
University of Bayreuth
95440 Bayreuth
Germany

Gee W. Lau
Department of Molecular Genetics
Biochemistry and Microbiology and Pulmonary Medicine
University of Cincinnati College of Medicine
Cincinnati
OH 45267
USA

Ute Lechner
Institut für Mikrobiologie
Martin-Luther Universität Halle-Wittenberg
Germany

Shie-chau Liu
Division of Radiation and Cancer Biology
Department of Radiation Oncology
Stanford University School of Medicine
Stanford
CA 94305
USA

Sergei V. Lymar
Department of Chemistry
Brookhaven National Laboratory
Upton
NY 11973-5000
USA

Michael H. Malamy
Department of Molecular Biology and Microbiology
Tufts University School of Medicine
136 Harrison Avenue
Boston
MA 02111
USA

Michael J. McInerney
Department of Botany and Microbiology
University of Oklahoma
770 Van Vleet Oval
Norman
OK 73019
USA

Volker Müller
Institute of Microbiology
Johann Wolfgang Goethe University
Marie-Curie-Str. 9
60439 Frankfurt
Germany

Michiko M. Nakano
Deptartment of Environmental and Biomolecular Systems
OGI School of Science & Engineering
Oregon Health & Science University
20000 NW Walker Road
Beaverton
Oregon 97006
USA

Sandrine Pagès
Université de Provence
13003 Marseille
France

and

Laboratoire de Bioenergétique et Ingénierie des Proteines
CNRS
31 chemin Joseph Aiguier
13009 Marseille
France

Luciano Passador
Department of Microbiology and Immunology
University of Rochester School of Medicine
Rochester
NY 14618
USA

Andrzej Paszczynski
Environmental Biotechnology Institute
University of Idaho
Moscow
ID 83844-1052
USA

Michel R. Popoff
Unité des Bactéries Anaérobies et Toxines
Institut Pasteur
28 rue du Dr Roux
75724 Paris cedex 15
France

Andreas Rauwolf
Institute of Microbiology
Johann Wolfgang Goethe University
Marie-Curie-Str. 9
60439 Frankfurt
Germany

David M. Roberts
Department of Pathobiology
University of Washington
Seattle
WA 98195
USA

John J. Rowe
Department of Biology
University of Dayton
Dayton
OH 45469
USA

Joanne M. Santini
Department of Microbiology
La Trobe University
Victoria
3086 Australia

R. Gary Sawers
Department of Molecular Microbiology
John Innes Centre
Norwich
NR4 7UH
UK

Michael J. Schurr
Department of Microbiology and Immunology
Tulane University
New Orleans
LA 70112-2699
USA

Ingo Schmidt
Department of Microbiology
University of Bayreuth
Universitaetsstrasse 30
95447 Bayreuth
Germany

Jonathan Sebat
Cold Spring Harbor laboratory
1 Bungtown Road
PO Box 100
Cold Spring Harbor
NY 11724
USA

Yuji Sekiguchi
Microbial and Genetic Resources Research Group
Research Inst Biological Resources and Functions
National Inst Advanced Industrial Sci. and Technol. (AIST)
Tsukuba
Ibaraki 305-8566
Japan

David R. Sherman
Department of Pathobiology
G-153 Health Sciences Building
University of Washington School of Public Health and Community Medicine
Seattle
WA 98195-7238
USA

Abraham L. Sonenshein
Department of Molecular Biology and Microbiology
Tufts University School of Medicine
Boston
MA 02111
USA

John F. Stolz
Department of Biological Sciences
Duquesne University
Pittsburgh
PA 15282
USA

Erko Stackebrandt
DSMZ-German Collection of Microorganisms and Cell Cultures GmbH
Mascheroder Weg 1b
38124 Braunschweig
Germany

Gerald W. Tannock
Department of Microbiology
University of Otago
PO Box 56
Dunedin
New Zealand

Chantal Tardif
Laboratoire de Bioenergétique et Ingénierie des Proteines
CNRS
31 chemin Joseph Aiguier
13009 Marseille
France

and

Université de Provence
13003 Marseille
France

Jens Walter
Department of Microbiology
University of Otago
PO Box 56
Dunedin
New Zealand

Geoffrey L. Winsor
Department of Molecular Biology and Biochemistry
Simon Fraser University
Burnaby, BC
Canada
V5A 1S6

Sang Sun Yoon
Department of Molecular Genetics
Biochemistry and Microbiology and Pulmonary Medicine
University of Cincinnati College of Medicine
Cincinnati
OH 45267
USA

Peter Zuber
Department of Environmental and Biomolecular Systems
OGI School of Science & Engineering
Oregon Health & Science University
20000 NW Walker Road
Beaverton
Oregon 97006
USA

Preface

The order *Bacteria* constitutes a collection of the most abundant life forms on earth, with new species continuously being discovered, particularly among the as-yet-uncultured members. Bacteria also represent a most ecologically and physiologically diverse order of life, exhibiting a metabolic virtuosity that continues to amaze us. The majority of bacteria are able to survive and thrive under anaerobic conditions. While some can grow only in the absence of oxygen (strict or obligate anaerobes), others are able to grow either in the presence or absence of oxygen (facultative anaerobes). Recent studies, and as summarized within the chapters of this book, have shown that some bacterial species, once classified as strict aerobes, are able to grow under anaerobic conditions and some obligate anaerobes can tolerate very low concentrations of oxygen. These new findings indicate that the anaerobic way of life is more wide-spread and more complex than was originally thought. With this in mind, it is not surprising that bacterial anaerobiosis can be observed in nearly every location on earth that is characterized by low or no oxygen concentration. Nearly every ecosystem on earth has such a habitat, including the digestive tracts of human and animals, plants, soil, the ocean, and areas affected by the anthropogenic influences associated with industrial development.

In this book, we highlight recent developments in the ongoing effort to uncover and understand the anaerobic life of microbes, and how these developments relate to issues of environmental and medical importance. As an introduction, the first three chapters focus on the taxonomy and metabolism of strict anaerobic bacteria as well as genetic regulation of anaerobic processes within facultative anaerobes. The detailed descriptions of species and their metabolism within the context of their ecological realms of influence speak to the tightly woven associations at work in the processes required to bring biochemical reaction pathways to completion. Complex consortia, not individual species, constitute catalytic units assembled through metabolic interactions between anaerobic species. This theme emerges again in subsequent chapters that focus on the individual contributions of species to the metabolic ecology of their respective habitats. The remarkable physiological traits of anaerobic bacteria, beneficial in one situation, are sometimes hazardous to human and animal health in other cases. The medical aspects include the anaerobic nature of infections and latency, anaerobic biofilms formed by pathogens, and toxin production by anaerobic microbes. Conversely, the anaerobic nature of Clostridia can be put to good use as a therapeutic agent to combat cancer. In chapters of environmental/industrial aspects, unique anaerobic metabolism is discussed in depth, including reduction of iron, selenate, and arsenate, oxidation of halogenated organics, ammonium oxidation, and acetogenesis. The application of these anaerobic processes to bioremediation and wastewater treatment is of great environmental and industrial importance. Also discussed are processes associated with the formation of industrially important "organellar" cellulolytic enzyme complexes and organic solvent production by fermentation. We hope that the book provides a glimpse of the wide-ranging capabilities of anaerobic bacteria, the study of which will be undoubtedly be one of the most important areas of microbiology, not only in terms of its relation to basic biology and chemistry, but the impact it has on the medical and environmental problems that confront us all.

We are grateful to the many scientists who contributed wonderful chapters to this book. The diversity of expertise that was assembled to write this book, we believe, resulted in extensive coverage of the subject. We would like to thank Annette Griffin, Horizon Scientific Press for her continuous help and warm encouragement.

Michiko M. Nakano
Peter Zuber

July 2004

Chapter 1

The Phylogeny and Classification of Anaerobic Bacteria

Erko Stackebrandt*

Abstract

Anaerobic prokaryotes have dominated life on the planet Earth for possibly more than a billion years before the increasing oxygen content forced them to retreat into niches, hostile for oxygenic organisms. These environments were provided by geological conditions and by oxygenic microorganisms. Eukaryotic species in concert with aerobic bacteria provided additional environmental facilities in which anaerobic forms could evolve. The diversity of morphologies, chemical and physiological traits represented among the recent anaerobic organisms are found in phylogenetic lineages that branched off early in the tree of ribosomal RNA gene sequences, which indicates that genetic diversity was already high 3.8 billion years ago. The majority of archaeal and bacterial lineages embrace anaerobic prokaryotes, though only few of them are defined solely by obligate and strictly anaerobic forms. This chapter will provide a summary on the diversity of anaerobes of the domain Bacteria, emphasising the classification at levels above the genus rank.

1. Introduction

Considering the time at which the first live forms are believed to have evolved on the planet Earth and the geological time at which oxygen evolved, it must be assumed that the first prokaryotic cell thrived by an anaerobic metabolism. Evidence for pre-Phenerozoic life originates from the presence of atmospheric molecular hydrogen that may have been as high as 1% (Walker, 1977) and from the extensive graphite $^{12}C/^{13}C$ enrichment in the oldest sedimentary rocks [>3.7 billion years [Gy], the Isua formation in West Greenland (Schidlowski *et al.*, 1979)]. Though microfossils and fossil organic compounds have been identified in formations as old as >3.4 Gy, such as in the Fig Tree Group, South Africa and in Warrawoona, Australia (Schopf, 1975), these forms cannot be affiliated to any recent life forms. Results from comparative sequence analyses of homologous and evolutionary conserved genes support the early hypothesis of Oparin (1924), postulating that an anaerobic organism, though not necessarily with a hetrotrophic, *Clostridium*-type metabolism, characterized the first living cell. Many deeply branching lineages contain autotrophic organisms, obtaining energy from a wide range of organic and inorganic compounds. The finding of an early origin of autotrophy with H_2-based metabolism (methanogens, sulfate reducers) lead to the theory of Wächtershäuser (1988; 1990; 2001), in which an early scenario for primordial life is the reaction in which ferrous sulfide and hydrogen sulfide yield positively charged pyrite and H_2. The free energy of –38 kJ/mol would have been sufficiently high to drive CO_2 fixation in a reductive citric cycle (theory of surface metabolism). The surface of pyrite could have then served as a matrix for the growing pool of negatively charged organic compounds needed for anabolic reactions. The importance of hydrogen in the evolution of living systems has also been highlighted by the theory of Martin and Müller (1998), suggesting that an endosymbiotic relationship between an autotrophic, strictly H_2-dependent archaeal organism as the host and a H_2 producing bacterium (reduction equivalents formed either by respiration or fermentation) as the endosymbiont. This theory may explain the successive merging of the genomes of both organisms into a single genome, as well as the formation of both mitochondria and hydrogenosomes. The genome plasticity of early organisms can today be unraveled by measuring the extent of lateral gene transfer among remotely related recent organisms.

Different modes of anaerobic bioenergetic processes have evolved and summarising organisms under "anaerobes" underestimates the bewildering genetic and phenotypic diversity of energy-yielding mechanisms in a reduced atmosphere. The evolution of the methanogenic and the anaerobic photosynthetic pathways, using H_2, H_2S, or sulfur as electron donors, guaranteed survival of life on Earth. If, however, heterotrophs were indeed the earliest inhabitants on this planet, autotrophs provided a new primary source of energy without which the early heterotrophic prokaryotes probably would not have survived because of depletion of energy sources. It can be assumed that for more than 1 billion years, and

*For correspondence email erko@dsmz.de

in some niches for more than 3.8 billion years, until today, organisms managed to escape the toxic effect of oxygen derivatives, mostly in concert with aerobic and microaerophilic microorganisms which lowered or reduced the oxygen content to provide appropriate conditions for the anaerobes.

The origin of oxygen-evolving organisms can be traced back to cyanobacterial-type fossils in the Bulawayan limestones, Zimbabwe (Schopf and Barghoorn, 1967) with an approximate age of 2.7-3.0 Gy. Analyses of fossil stromatolites, consisting of communities of cyanobacteria, sometimes with methanogens and chloroflexi and other bacteria, give evidence of a rich microbial diversity from about 2.5 Gy, which may have dominated the Proterozoic landscape (see Margulis, 1981). The evolution of the oxygenic photosynthesis, evidenced by formations of banded iron and other minerals, may have occurred even earlier than deduced from the first occurrence of stromatolites. Banded iron formations, in which layers of oxidized iron (ferric oxides) alternate with more reduced iron (ferrous oxides) are already present in the Isua (3.7 Gy) and Swaziland formations (3.1-3.4 Gy). It should, however, be noted, that oxidation of minerals can also be due to the activities of ancient anaerobic photosynthetic organisms, such as the ancestors of the recent ferrous-iron-oxidizing proteobacteria *Rhodovulum iodosum* and *R. robiginosum* (Straub *et al.*, 1999). The origin of anaerobic photosynthesis has not yet been determined with confidence but Blankenship (1992) postulates that the evolution of oxygen has predated that of anoxygenic photosynthesis.

Depending upon the level of oxygen tolerated, the term "anaerobic" differentiates between "strictly anaerobic" (no growth in the presence of even traces of O_2) and "obligately anaerobic" (traces of O_2 tolerated). The respective organisms did not develop mechanisms to prevent the formation or to remove toxic forms of oxygen (singlet oxygen, superoxides, hydroxyl free radicals, peroxides) by catalase, peroxidase and/or superoxide dismutase. Microaerophiles grow under reduced concentrations of molecular oxygen, e.g., in microoxic habitats (about ≤ 1 -10 μM dissolved O_2 in water), while oxygen-tolerant anaerobic prokaryotes are able to tolerate oxygen at normal levels (10-100 μM O_2); these organisms express superoxide dismutase, forming H_2O_2 from the reduction of oxygen. In facultatively anaerobic bacteria that are able to grow in the presence and at reduced levels of oxygen, O_2 triggers a regulatory switch between pathways, allowing adjustment of anabolic and catabolic pathways to varying oxygen levels. These organisms, as well as aerobic life forms, which use oxygen as an external electron acceptor, did evolve appropriate measures to cope with the toxic electronic states of oxygen. (Atlas, 1996; Madigan *et al.*, 2000). The term anaerobic respirer refers to organisms which possess a fully developed electron chain and use inorganic substrates as alternative electron acceptors in the absence of oxygen. Oxidized metals (including Mn(III) and (IV), Fe(III), Cr (VI), U (VI), fumarate, NO_3^-, CO_2, SO_4^-, or elemental sulfur (Myers and Nealson, 1990) are reduced. *Shewanella oneidensis* is an example of a bacterium with remarkably diverse respiratory capabilities (Venkateswaran *et al.*, 1999).

The presence of hydrogenase in prokaryotes, the enzyme responsible for catalysing the reactions $H_2 \leftrightarrow 2H^+ + 2e^-$, can not be used to define an anaerobic organism. Though present in predominantly anaerobic species (Schwartz and Friedrich, 2003), this enzyme is also present in many aerobic organisms, including cyanobacteria, bacilli, streptomycetes, rhizobia and thiobacilli, as well as in eukaryotic microorganisms. An excellent overview on aspects of metabolism and oxygen sensitivity has been provided by Schmitz *et al.* (2001).

This chapter will be restricted to the phylogeny of anaerobic Bacteria. Bacteria are defined as members of the domain Bacteria which, together with the organisms of the domain Archaea, represent prokaryotic organisms (Woese *et al.*, 1990). Many of the archaeal taxa are strictly anaerobes (e.g., methanotrophes, sulfur and sulfate reducing archaea), representing descendants of inhabitants of ancient environments thriving under todays' conditions which may resemble those prevailing billions of years ago. The metabolic and genomic diversity of archaeae in the environment is fully acknowledged, and as as-yet uncultured archaea are increasingly detected as commensals or symbionts of the eukaryotic cell, others are found in non-extremophilic habitats, including soil, the open sea. Restricting the chapter to members of Bacteria is justified by the finding that not a single archaeal species has been described to be pathogenic, while members of the Bacteria embrace a high number of highly pathogenic anaerobes in various main lines of descent. Also, in the context of this communication, only the strictly and obligately anaerobic fermenters of the domain Bacteria are included, while facultatively anaerobic species, microaerophilic species and, facultative anaerobic respiring organisms are not included. The large phylogenetic diversity of these three groups of organism, present in almost every taxonomic group consisting of mainly aerobic species, excludes a thorough coverage in the context of this book. Emphasis is on the placing of anaerobic species into the phylogenetic framework, referring the reader interested in the taxonomic description of the species to their original descriptions published in specialized journals such as Anaerobe (ISSN 1075-9946), the International Journal of Systematic and Evolutionary Microbiology (ISSN 1466-5026), Archives of Microbiology, Applied and Environmental Microbiology (ASM, Washington), and related journals.

2. Anaerobes and Classification

The ability to thrive in the presence or absence of oxygen is one of the key properties in the classification of prokaryotes above the species level. Except for the definition of the term "strictly anaerobic", usually applied to organisms cultivated according to techniques described by Hungate (1969) or similar highly reducing techniques (see Holdeman *et al.*, 1977; DSMZ, 2001), all other categories mentioned above are less well defined with respect to tolerance to oxygen. Organisms growing in a candle-jar or using commercial systems (e.g., Gas Generating Kit BR 038B, Oxoid, Basingstoke; or Anaerocult®, 1.101611, Merck, Darmstadt) are referred to as being obligately anaerobic. Most genera are rather homogeneous with respect to their members' relationship to oxygen though exceptions exist. To name a few:

- The anaerobic species *Bacillus infernus, B. arseniciselenatis*, and *B. selenitireductens* are classified among otherwise facultative anaerobic (*B. cereus*) and aerobic species.
- *Lactobacillus* contains obligately anaerobic as well as microaerophilic species.
- *Actinomyces* harbors oligately anaerobic, facultative anaerobic and aerobic species.
- *Pasteurella, Actinobacillus, Chromobacterium, Flavobacterium, Haemophilus, Corynebacterium*, and *Cytophaga* embrace facultative anaerobes and aerobic species.

Experience shows that organisms cultivated from an aerobic environment are usually cultivated under aerobic or microaerophilic conditions, while their capability to also thrive under much more reduced oxygen levels are rarely tested thoroughly. Microbiologists often learn about the potential to ferment substrates from the lists of metabolic properties provided in the species description. The lack of clearly defined physical boundaries with respect to tolerance to oxygen makes it difficult to evaluate the classification of species according to their relationship to oxygen.

The original finding of Pasteur (1861) that microbes can grow in the absence of oxygen was already used in the first classification scheme by subdividing some of the morphology-based larger categories on the basis of fermenting properties (Cohn, 1875), e.g, the separation of *Micrococcus* into chromogenic (pigmented), zymogenic (fermenting) and pathogenic (contagious) properties. Due to the lack of phylogenetic evidence, superficial properties continued to play the decisive role even in the 8th edition of Bergey's Manual of Determinative Bacteriology (Murray, 1984) and in the 1st edition of Bergey's Manual of Systematic Bacteriology (Holt *et al.*, 1994). For more than 100 years, and in combination with other properties, the division of the three basic classes of bioenergetic processes, i.e., photosynthesis, fermentation and respiration, have strongly influenced the general outline of prokaryotic systematics. While morphology was still considered "the first and most reliable guide" (Stanier and van Niel, 1942; Stanier, 1964), the importance of specialized physiological and biochemical traits as expression of diversity among the [eu]bacteria was stated by Stanier *et al.* (1970). The grouping of organisms according to key phenotypic properties is still well established for teaching purposes and in the clinical environment, e.g. the category of "Gram-negative, anaerobic, straight, curved and helical bacteria", the "anoxygenic phototrophic bacteria", or the "endospores-forming rods and cocci". However, during the past 25 years the natural relationships among prokaryotic organisms were unravelled and a phylogenetic system was established in which the degree of gene sequence similarity allowed the ordering of organisms according to common ancestry.

3. Early Attempts to Determine Natural Relationships

The importance of bioenergetic processes in the context of evolution and phylogenetic relationships, as well as in the framework of the history of ideas and hypotheses have been covered excellently by Broda (1975). Prior to the era of gene sequences, relationships were deduced from metabolic properties and biochemical pathways. According to the conversion hypothesis of Broda (1975) several deeply branching lineages encompassing anaerobic phototrophes and clostridia first gave rise to independently evolving lineages of aerobic prokaryotic representatives. This hypothesis stood in contrast to the segregation hypothesis (Margulis 1970), in which the low G+C and C-heterotrophic fermenting organisms first gave rise to anaerobic respiring organisms which developed the respiration chain, also used later in evolution by aerobic organisms. Indeed, in the higher classification system of Margulis (1981) the anaerobic bacteria constitute five phyla, strictly separating fermenting bacteria, from the spirochetes, the anaerobic sulfur reducing forms, methanogens and the anaerobic photosynthetic bacteria. This system was tested when the first results on the phylogenetic relationships became available through comparative analysis of homologous protein involved in housekeeping functions of the prokaryotic cells, e.g., cytochrome *c* (Ambler *et al.*, 1979; Dickerson, 1979) and ferredoxin (Schwartz and Dayhoff, 1978). Though limited by the restriction that only those organisms that contained the respective genes could be subjected to phylogenetic analysis, it appeared that a hypothetical anaerobic fermenting ancestor gave rise to anaerobic phototrophes, anaerobic organotrophes and sulfate reducers, while the respiring organisms evolved individually and in parallel from anaerobic ancestors. The phylogenetic relatedness of amino acid sequences

Table 1. Examples of frequently used molecular methods generating DNA profiles for the differentiation of closely related organisms

Method	Example	Reference
Random amplification of polymorphic DNA by polymerase chain reaction (RAPD-PCR)	Separation of *Prevotella intermedia* from *Prevotella nigrescens*	Robertson *et al.*, 1999
BOX PCR GTG_5-PCR	Differentiation of *Vibrio* species	Ben-Haim *et al.*, 2003
Amplification of 16S-23S spacer region	Characterization of *Fusobacterium* species	Claros *et al.*, 1999
16S rDNA endonuclease restriction length polymorphism	Characterization of *Bacteroides* species	Santos *et al.*, 1999
Ribotyping	Characterization of *Clostridium* strains	Spring *et al.*, 2003
DNA restriction endonuclease length polymorphism	Characterization of *strains of Actinobacillus actinomycetemcomitans*	Leite *et al.*, 1999
Amplified fragment length polymorphism (AFLP)	Differentiation of *Enterovibrio norvegicus*	Thompson *et al.*, 2002
Multiple-locus, variable-number tandem repeat analysis (MLVA)	*Francisella tularensis* typing	Farlow *et al.*, 2001
BOX-PCR	Characterization of *Bifidobacterium* strains	Gomez Zavaglia *et al.*, 2000

of conservative proteins strongly supported the scenario pictured by the conversion hypothesis.

4. The Era of Ribosomal RNA Gene Sequence Analysis

Bacterial systematics consists of a series of sequential steps of which the characterization of an isolate in the process of identification is the first one. The results of this initial phase will determine whether the isolate can be affiliated to a described taxon (identified) or whether identification is not possible. In the non-medical environment, but even here increasingly so, determination of the phylogenetic nearest neighbour by one of the many available molecular tools (Table 1) is the most rapid and reliable means to direct the identification process: the genetically closest neighbor determines the degree with which characterization needs to be performed. If sufficient differentiating properties have been identified which indicates the isolate to represent a new species this new taxon is given a name following the rules and regulations laid down in the International Code of Nomenclature (Lapage *et al.*, 1992). The purpose and efficiency of a taxonomy is tested by its use in identifying organisms (Stackebrandt *et al.*, 1999a).

Phylogenetic classification is the theory and process of ordering the organisms according to common ancestry, irrespective of metabolic, chemical or morphological properties. The framework most commonly referred to today was based on the analysis of oligonucleotides of 16S ribosomal RNA (1974-1986) (Fox *et al.*, 1977; 1980). This method was then replaced by reverse transcriptase sequence analysis of the molecule (1986-1990) and from then on by analysis of gene sequences coding for 16SrRNA. More than 70.000 sequences are available in public databases (e.g. EMBL, www.ebi.ac.uk), including those of the vast majority of type strains. This molecule is the most useful molecular chronometer as it contains information that relates to nearly the entire spectrum of bacterial taxa, from domain to species; genes coding for ribosomal RNA are ubiquitous, homologous (derived from a common ancestor), functionally equivalent, and genetically stable.

The rRNA dendrogram of relatedness displays impressively the gene sequence diversity which has been the basis of the higher classification structure of the 2nd edition of Bergey's Manual of Systematic Bacteriology (Garrity *et al.*, 2003). A different classification system has been published on the basis of a similar topology of the 16S rRNA tree (Cavalier-Smith, 2002). Classification is an evolutionary process with adjustments done when new data indicate the necessity for doing so. Shortly after the establishment of 16S rRNA gene sequence analysis in bacterial systematics, the spectrum of genes sequenced was broadened by including a wider spectrum of housekeeping genes (Schleifer and Ludwig, 1994; Gupta, 2000). These studies were accompanied by analysis of fully sequenced genomes, which allowed recognition of those genes that match the phylogeny of rRNA genes, and those that were subjected to lateral gene transfer at various epoches in the evolution of the pro- and eukaryotic cell. The interested reader is referred to excellent coverage of this topic (Nelson *et al.*, 1999; Ochman *et al.*, 2000; Woese, 2003). As ribosomal RNA genes are genetically rather stable and as the number of the other houskeeping genes sequenced for type strains so far is significantly lower, the importance of ribosomal RNA genes sequences in providing a solid skeleton for bacterial classification is unmatched in prokaryotic systematics. The reader who is not familiar with gene

Table 2. Examples of reclassified *Bacteroides* and *Clostridium* species (source DSMZ, 2003)

Original name	Reclassified as	Original name	Reclassified as
Bacteroides		*Clostridium*	
B. amylophilus	*Ruminobacter amylophilus*	*C. bryantii*	*Syntrophospora bryantii*
B. asaccharolyticus	*Porphyromonas asaccharolyticus*	*C. durum*	*Paenibacillus azotofixans*
B. buccae	*Prevotella buccae*	*C. fervidum*	*Caloramator fervidus*
B. forsythus	*Tannerella forsythus*	*C. hydroxybenzoicum*	*Sedimentibacter hydroxybenzoicus*
B. furcosus	*Anaerorhabdus furcosus*	*C. lortetii*	*Sporohalobacter lortetii*
B. gracilis	*Campylobacter gracilis*	*C. oxalicum*	*Oxalophagus oxalicum*
B. hypermegas	*Megamonas hypermegale*	*C. quercicola*	*Dendrosporobacter quercicolus*
B. multiacidus	*Mitsuokella multiacida*	*C. thermoaceticum*	*Morella thermoaceticum*
B. nodosus	*Dichelobacter nodosus*	*C. thermocopriae*	*Thermoanaerobacter thermocopriae*
B. ochraceus	*Capnocytophaga ochracea*	*C. thermosaccharolyticum*	*Thermoanaerobacteriun thermosaccharolyticum*
B. putredinis	*Alistipes putredinis*	*C. villosum*	*Filifactor villosum*
B. pneumosintes	*Dialister pneumosintes*		
B. praeacutus	*Tissierella praeacuta*		
B. succinogenes	*Fibrobacter succinogenes*		

sequence analysis should be aware of the limitations of phylogenetic dendrograms and trees. Due to treeing algorithms, the topology of trees change with the number and reference sequences included in the analyses. Most likely the introduction of novel sequences will affect the branching order of neighboring lineages at any level of relationship (Stackebrandt *et al.*, 1996). Therefore, the order should be considered tentative, and the systematic treatment of the species involved may need to be adjusted.

It should be stressed that classification is (and has always been) much more than clustering organisms according to sequence similarities. Modern systematics evolved from a 100 year long period of descriptive systematics about 30 years ago and still today many elements of the early period are still influencing present-day methods of systematics. However, the "polyphasic approach" is much more demanding than any other classification strategy used in the past, as a broad spectrum of properties and not individual traits are considered important to determine the uniqueness of an organism. These include determination of properties at the genetic level (information carried directly on nucleic and amino acid sequences), as well as at the epigenetic level (information expressed by gene products), including traditional taxonomic properties. The advantage of working with the molecular approach, which is not restricted to gene sequence determination but also, especially for the level of strains and species, to comparative analysis of DNA patterns (Table 1), is the reduction of cost and time, reproducibility, the option to build a cumulative database, and universal application.

The distribution of genomic and metabolic traits of members of Bacteria can be mapped along the topology of the phylogenetic dendrogram. For example, the formation of endospores appears to be monophyletic, present so far only in the clostridial lineage; spirochete morphology is also likely to be a monophyletic trait, giving rise to the different phenotypes in members of the phylum *Spirochaetes*; a multilayered Gram-positive cell wall is more widely distributed, such as in lineages embracing clostridia, *Actinobacteria* and *Deinococcus*. Photosynthetic organisms, including the aerobic photosynthetic organisms, on the other hand, are present in several lineages, including those of the clostridia, and different classes of *Proteobacteria*, *Chloroflexi* and *Chlorobi*. Metabolic properties, such as formation of organic acids, alcohols, and gene products coding for housekeeping functions are usually of polyphyletic origin as the same taxonomic features are found widely distributed in remotely related phylogenetic lineages. Annotation of sequences of completely sequenced genomes will uncover these pathways, providing the means to identify the genes involved. With this information at hand, properties expressed will receive new phylogenetic weight. The combination of polyphyletic chemotaxonomic traits (i.e., chemical constituents of cell walls and membranes, pigments and components of electron chains) allows taxonomists to delineate taxa at different phylogenetic levels (Stackebrandt and Schumann, 2000).

New insights into the phylogenetic position of members of a genus led to dramatic taxonomic rearrangement during the past 20 years. Some affiliations changed several times and these changes have not always been fully appreciated. For example, the pathogenic *Pseudomonas pickettii* was reclassified in 1992 as *Burkholderia pickettii* (Yabuuchi *et al.*, 1992) and in 1995, as *Ralstonia picketii* (Yabuuchi *et al.*, 1995); such changes are not always immediately accepted by clinical microbiologist who needed to adjust to the changes in their daily routine. By and large, however, the advantage of working with a phylogenetic framework-based system was accepted. The vast majority of changes, adjusting systematics to phylogenetic relatedness, makes sense in the light of overall biological properties of the organisms involved, and gained acceptance from geneticists, physiologists, and medical microbiologists in academia and industry. The recent literature lists many examples of reclassifications, here exemplified by those having occurred within the genera *Bacteroides* and *Clostridium* (Table 2), based on phylogenetic and chemotaxonomic evidence. It can be expected that the number of dramatic taxonomic changes at the genus level will be reduced markedly as gene sequences are available for nearly all type strains and new species are described based on phylogenetic relatedness. Moreover, the need has been expressed to back taxonomic decisions made solely on the basis of ribosomal gene sequences with sequences of additional genes, i.e., those coding for housekeeping functions, providing stability to the phylogenetic branching patterns (Stackebrandt *et al.*, 2002).

The nomenclature of higher taxa of anaerobic species used in this chapter, i.e., above the level of genera, has been adopted from the outline of the recent edition of Bergey's Manual of Systematic Bacteriology, 2nd Edition (Garrity *et al.*, 2003). Of this hierarchical system, only the taxa proposed in the first volume have aready been formalized. Once these additional volumes have been published, these names will be validated by addition to the Validation List, published bimonthly in the International Journal of Systematic and Evolutionary Microbiology (http://ijs.sgmjournals.org). The higher ranks will be used in this chapter - though some of them may have disappeared prior to the publication of this reference journal or of the next volumes of the 2nd edition Bergey's Manual of Systematic Bacteriology. As pointed out above, a phylogenetic dendrogram is a dynamic construct which may influence the subjective treatment of taxonomic opinions, leading to changes in the delineation of phyla, classes, orders and families. Some of the higher taxa used in this chapter may therefore not be found in the Manual but the genus name will guide the reader to the most recent taxonomic name (Nomenclature–up-to-date, 2003).

5. Cultured and as Yet Uncultured Organisms

The phylogenetic radiation of members of the domain Bacteria provided by the ARB program (ARB = arbor, Ludwig *et al.*, 2004) gives a complex picture, consisting of more than 40 deeply branching lineages, termed Phyla. While about half of these phyla contain cultured organisms, represented by the type strain and additional strains of a species, the other half embrace organisms for which no cultured representative have as yet been found. This information is available from molecular studies on populations, in which *rrn* operons of DNA, recovered from an environmental sample, are subjected to polymerase chain reaction amplification, cloned and the cloned inserts sequenced. Sequences obtained are then compared to the database of homologous sequences of cultured organisms and the extent of novelty is determined (Hugenholz *et al.*, 1998). Alternative rDNA-based methods have been described, such as gradient gel electrophoresis (McCartney, 2002), single strand polymorphism (Schwieger and Tebbe, 2000), and genomic sequence tags (Dunn *et al.*, 2002). As the 16S rRNA gene sequence is the only information of these as-yet uncultured organisms, nothing is known about their morphology, metabolism or ecological role. Large fragments of the genomes of some of the putative species are currently sequenced but information is still fragmentary. Methods to investigate the metabolic potential of communities also involves the application of radioactively labelled compounds (acetate, CO_2) to samples of a natural population, followed by microautoradiography and labelling of bacterial cells in situ with fluorescently labelled RNA probes (Nübel *et al.*, 2002). This approach aims at the identification of the metabolically active part of the population, down to the species level, depending on the specificity of the oligonucleotide probe used.

6. The Main Lines of Descent of the Domain Bacteria

Rooting the radial phylogenetic tree with members of the Archaea, a few deeply branching phyla emerge, some of which embrace anaerobic organisms (e.g., *Thermotogae*, *Dictyoglomus*, *Thermodesulfobacteria*). The exception refers to the aerobic hyperthermophilic *Aquificales* (see below). Different ARB trees, generated with changing composition of sequences of different length and origin, may differ in the order which phyla appear in the tree. It should also be noted that a 16S rRNA phylogenetic tree based on the most conserved positions determines the peptidoglycan-less *Planctomycetales* as the first branching bacterial group and not a hyperthermophilic bacterium (Brochier and Philippe, 2002), from a group of organisms which may either be ancient or whose genomes may have undergone a high rate of mutation than other bacteria (Liesack *et al.*, 1992). By and large the order of phyla within the 16S rRNA gene sequences tree is not well resolved, due to highly conserved backbone structure at all structure levels. The phylogenetic tree has a bush- or fan-like appearance. Not all anaerobic species appear to cluster together, though most of them form individual clusters (*Bacteroidetes, Fibrobacteres* and *Chlorobi*; *Spirochaetes, Aminobacteria, Deferribacteres,* and *Chrysiogenetes*; *Synergistes, Thermotogae, Coprothermobacteria, Dictyoglomus* and *Thermodesulfobacteria*), while others are more isolated (*Fusobacteria, Chloroflexi*). Besides phyla which are almost solely embracing anaerobic organisms, others contain genera defined by anaerobic and aerobic species (e.g., *Firmicutes, Actinobacteria, Acidobacteria, Proteobacteria*), yet others mainly aerobic representatives (e.g., *Planctomycetes, Cyanobacteria, Chlamydia-Verrucomicrobia, Nitrospira, Deinococcus-Thermus*). Considering the evolution of oxygen, it is not surprising to see within phyla the anaerobic members occupying the deepest branches of the taxa, given the evolution of oxygen on Earth. This is especially obvious in the branching point of the various families of the *Firmicutes*, the phototrophic members in the *Gammaproteobacteria*, and the *Coriobacteraceae* within *Actinobacteria*. On the other hand, an anaerobic metabolism is sometimes seen in close phylogenetic relationship with aerobic organisms. The neighboring genera of anaerobic *Rhodopseudomonas* spp. and aerobic *Nitrobacter* spp. (*Alphaproteobacteria*) and of anaerobic *Propionibacterium* spp. and aerobic *Microlunatus* spp. (*Actinobacteria*), as well as the families of anaerobic *Succinivibrionaceae* and aerobic *Aeromonadaceae* (*Gammaproteobacteria*) species are only three of many examples.

Calibration of sequence similarity with geological time is rather imprecise because of different rates of tempo and mode of evolution of ribosomal RNA genes in different lines of descent. Isotope discrimination in biogenic compounds (enrichment of ^{13}C versus ^{12}C and ^{34}S versus ^{32}S) on dated rocks allow a very crude estimation of the evolution of a few key metabolic properties. Fixation of CO_2 is a common property of species considered to be descendants of members of deeply branching lineages. Reduction of CO_2 with the formation of methane by archaeal methanogens (Balch *et al.*, 1979), and carbon dioxide fixation via the reductive citric acid cycle in the *Aquificales* (Deckert *et al.*, 1998) are only two of many examples of ancient metabolic pathways, possibly present on this planet for more than 3.7 Gy. The evolution of sulfur and sulfate reduction, as carried out by members of *Desulfovibrio* and relatives (*Deltaproteobacteria*), *Desulfotomaculum* (*Clostridiales*) and the sulfur dependent Archaea, can be traced through isotope discrimination in rocks as old as 3 Gy. Determination of fossilized organic compounds, such as phytane and pristane points towards the former presence of photosynthetic pigments. The origin of oxygen-evolving cyanobacteria is still under discussion as the fossil record of putative cyanobacterial morphologies (3.5 Gy) does not agree with the rather late origin of the most deeply branching cyanobacteria species in the 16S rRNA gene tree. Some possible explanations are: (i) the fossil record is interpreted erroneously as cyanobacterium-like organisms, (ii) early cyanobacteria became extinct, (iii) the more older cyanobacterial species have not yet been found, and (vi) the formation of Fe(II) was not due to oxygen but to oxidation by anaerobic phototrophic bacteria. Due to the activities of cyanobacteria and possibly eukaryotic algae, the atmospheric oxygen content was determined to be about 0.2 % (about 1.2 Gy ago). The Pasteur point of an oxygen content of 0.2% has been considered a threshold value at which the energy gained from respiration exceeds that of fermentation, pushing back the anaerobes to niches in which aerobic organisms provided the oxygen-free environment. An oxygen content of 2% was reached about 500 million years ago and indeed, the phylogenetic tree sees a significant diversification and radiation of aerobic species from anaerobic ancestors in various lines of descent, e.g., the transition from anaerobic clostridia to facultatively anaerobic lactobacilli and streptocococi to the aerobic bacilli; or from the anaerobic coriobacteria and bifidobacteria to propionibacteria to streptomycetes, micrococci and relatives; other examples are found in most classes of the phylum *Proteobacteria*.

The majority of genera embracing anaerobic species are listed according to their higher taxonomic affiliation as devised by Garrity *et al.* (2003). The reader interested in the original description of taxa at the genus and species level is referred to the homepage of Jean Euzeby (http://www.bacterio.cict.fr) providing a curated nomenclatural database

7. The Deeply Rooting Anaerobes

In contrast to the expection of anaerobes defining the most deeply branching lineages in the domain bacteria, members of the *Aquificales* (*Aquifex, Hydrogenothermus, Hydrogenobacter, Thermocrinis*) are strictly aerobic, growing preferentially under microaerophilic culture conditions. They perform the "Knallgas" reaction ($2\ H_2 + O_2 \rightarrow 2\ H_2O$), i.e., the oxidation of hydrogen and reduction of oxygen. Instead of hydrogen, most *Aquificales* species can also use thiosulfate or sulfur as single energy sources, and sulfuric acid is formed as the endproduct. Some species grow exclusively in the presence of sulfur or thiosulfate (Shima and Suzuki, 1993; Skirnisdottir *et al.*, 2001), others produce H_2S in addition to sulfuric acid in the presence of sulfur.

Aquificales resemble many of the anaerobic members of neighboring phyla in their thermophilic way of life, thriving in geothermally and volcanically heated environments. *Aquificales* represent the bacteria with the highest growth temperatures known so far (95°C being

Table 3. Genera of the phyla *Thermotogae, Thermodesulfobacteria, Chrysiogenetes and Deferribacterales,*containing strictly or obligately anaerobic members. Genera following the type genus are listed in alphabetical order

Phylum	***Thermotoga***	***Thermodesulfobacteria***	***Chrysiogenetes***	***Deferribacterales***
Order	***Thermotogales***	***Thermodesulfobacteriales***	***Chrysiogenales***	***Deferribacterales***
Family	***Thermotogaceae***	***Thermodesulfobacteriaceae***	***Chrysiogenaceae***	***Deferribacteraceae***
Genera	*Thermotoga*	*Thermodesulfobacterium*	*Chrysiogenes*	*Deferribacter*
	Fervidobacterium			*Denitrovibrio*
	Geotoga			*Flexistipes*
	Marinitoga			*Geovibrio*
	Petrotoga			*Synergistes*
	Thermosipho			

the upper limit of their growth temperature) (Huber and Eder, 2002). The topology of phylogenetic trees of the 16S rRNA gene of *Aquifex* is not always in accord with those of other gene and protein sequences of this organism (e.g., Brown and Doolittle, 1995; Baldauf *et al*., 1996; Klenk *et al*., 1999; Schütz *et al*., 2000). About 16% of the *A. aeolicus* genes (Deckert *et al*., 1998) are of archaeal origin (Aravind *et al*., 1998). The question of whether hyperthermophilic *Bacteria* and *Archaea* have exchanged genes particularly often by lateral transfer (Aravind *et al*., 1998; Nelson *et al*., 1999) or whether some of the genes of these organisms of different domains are shared ancient properties, has not yet been answered. No pathogenic representatives are known.

8. Phylum *Thermotogae*

Members of *Thermotogae* (Table 3) are widespread and cosmopolitan. They have been isolated from shallow and deep-sea marine hydrothermal systems, high-temperature marine and continental oil fields and from within continental solfatara springs of low salinity. Morphologically, thermotogae are defined by rod-shaped cells with an outer sheath-like envelope ('toga'). Together with members of the order *Aquificales* thermotogae represent the bacteria with the highest growth temperatures known so far. Pathogenic organisms have not yet been reported. Thermotogae are strictly anaerobic, organotrophic, fermentative bacteria. They stain Gram-negatively and are lysozyme sensitive, but *meso*-diaminopimelic acid, the hallmark of Gram-negative bacteria, is not present in the peptidoglycan (Huber and Hanning, 2003). Fatty acids are unusual long-chain dicarboxylic acids.

As already seen with *Aquifex aeolicus*, the phylogentic position of *T. maritima* is not stable when phylogenetic relationships are inferred from different gene sequences. The 16S rRNA gene tree is supported by comparative analysis of 23S rRNA, elongation factor Tu and G, β-subunit of the ATPase, *fus* gene and of ferredoxins (e.g., Bachleitner *et al*., 1989; Tiboni *et al*., 1991; Darimont and Sterner, 1994; Schleifer and Ludwig, 1994) and by whole genome-based phylogenetic analysis (Fitz-Gibbon and House, 1999). On the other hand, sequence comparisons of large subunits of RNA-polymerase placed *T. maritima* next to the chloroplasts, while DNA polymerase III, sequences places *T. maritima* next to *Clostridium acetobutylicum* (Palm *et al*., 1993). However, based on conserved inserts and deletions found in various proteins, Gupta and Griffith (2002) places *T. maritima* phylogenetically within the radiation of Gram-positive Bacteria.

Members of the *Thermotogae* are extremely thermophilic or hyperthermophilic (optimal growth temperature around 80°C and a maximum growth temperature of 90°C) growing optimally in the neutral pH range. They are strict organotrophs, fermenting preferentially simple and complex carbohydrates or complex organic matter. Glucose is metabolized mainly via the Embden-Meyerhof glycolytic pathway and, to a lesser extent, via the Entner-Doudoroff pathway. With glucose as growth substrate, L(+)-lactate, acetate, ethanol, L-alanine, carbon dioxide and hydrogen are formed as major final products. Hydrogen, an inhibitor of *Thermotogae* growth, can be removed by addition of sulfur, organic sulfur-containing compounds or Fe(III) (see Huber and Hanning, 2003). The 1.8 Mb genome from *T. maritima* has been sequenced (Nelson *et al*., 1999). It contains 1877 predicted coding regions, about 24% of which are of archaeal origin. About 7% of the predicted codons are involved in the metabolism of simple and complex sugars, about twice that seen in the genomes of other prokaryotic species (Nelson *et al*., 2001).

9. Phylum *Thermodesulfobacteria*

This phylum (Table 3) has been described for the single genus *Thermodesulfobacterium*, containing strictly anaerobic, nonsporeforming, thermophilic (optimum

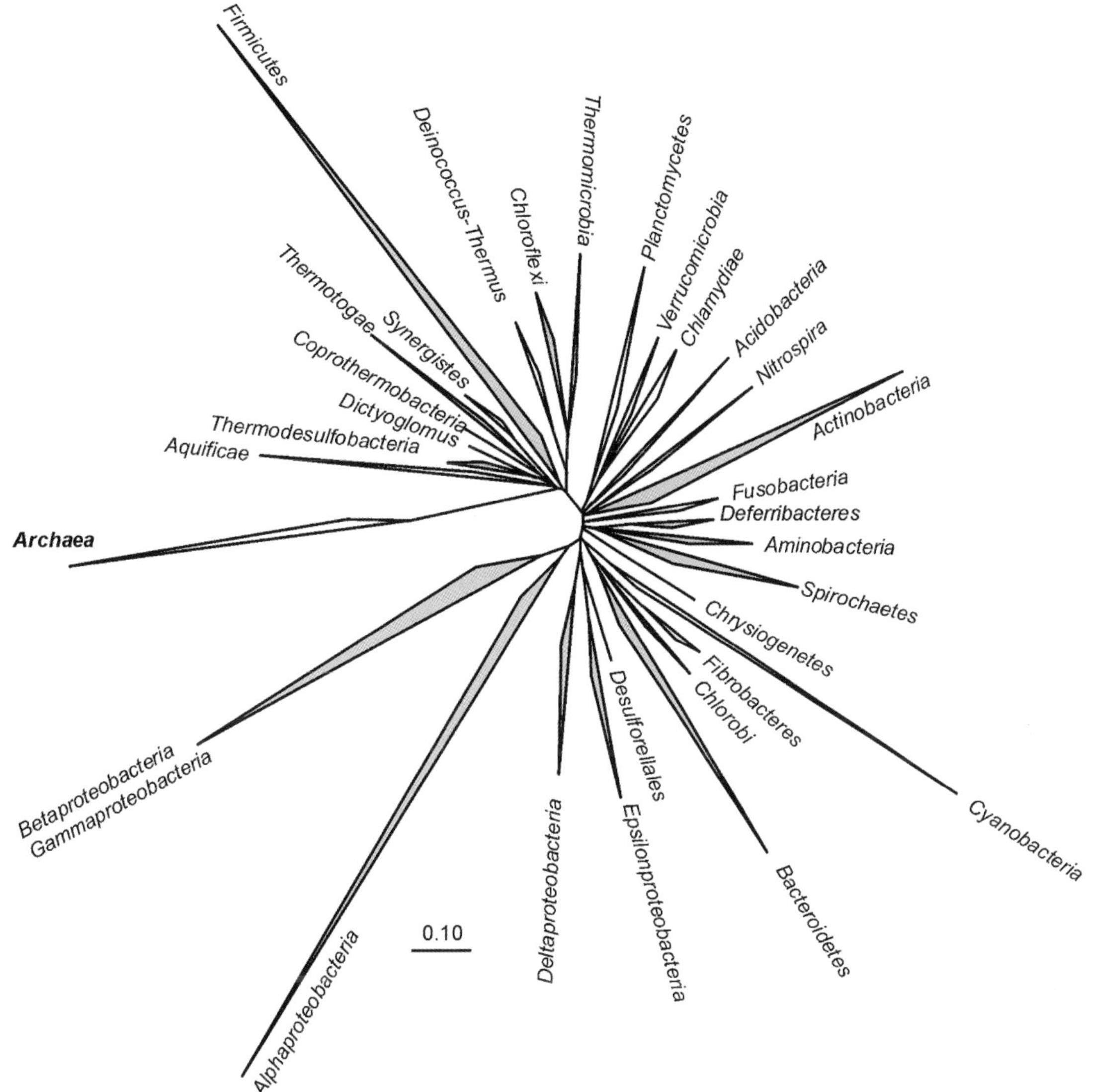

Figure 1. Radial tree displaying the 16S rRNA gene sequence similarities between the main lines of descent (Phyla) of the Domain Bacteria, using the ARB programme (Ludwig *et al.* 2002). Species of the Archaea were used as a root. The order of branching points is tentative and subject to changes based on the number and selection of sequences included in the treeting program. Triangles in grey indicate the presenc of anaerobic species. The dimension of the triangle does by and large reflect the number of sequences available in the database. In addition to the phyla indicated here, Zhang *et al.* (2003) have described the phylum *Gemmatimonas*, embracing an aerobic species. Bar scale=1% difference in sequence similarities.

65-75°C), sulfate-reducing bacteria from hot springs, hot oil reservoirs and hydrothermal vents. Species are either chemoorganotrophic or chemolithoautotrophic, in which sulfate is reduced to sulfide. Hydrogen or C1-C3 acids serve as electron donors. Chemotaxonomically they are unique as they contain non-phytanyl ether linked lipids. Desulfoviridin is absent while desulfofuscidin and cytochrome c_3 are present (Jeanthon *et al.*, 2002)

10. Phylum *Chrysiogenetes*

This monogeneric phylum (Figure 1, Table 3) contains a single species, *Chrysiogenes arsenatis* (Macy *et al.*, 1996), isolated from a goldmine in Australia. This Gram-negative, curved and motile rod thrives by arsenate respiration, reducing arsenate [As(V)] to arsenite [As(IV)], using acetate as electron donor and carbon source (see Chapter 13). Acetate can be replaced by a variety of organic acids.

11. Phylum *Deferribacteres*

Deferribacter species (Caccavo *et al.*, 1996; Greene *et al.*, 1997) (Figure 1, Table 3) are Gram-negative and non-fermenting thermophilic rods with a low DNA G+C composition (around 35 mol%) that have been isolated from an oil field in the North Sea and from surface sediments. Iron (FeIII), manganese (IV), and nitrate serve as electron acceptors, while complex organic

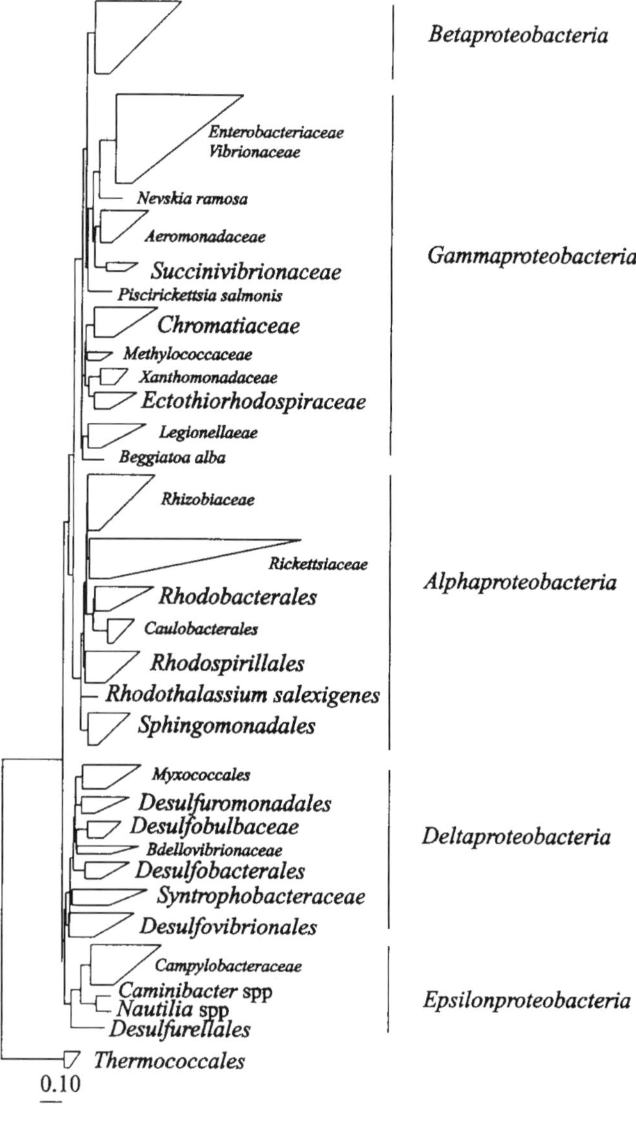

Figure 2. Dendrogram of 16S rRNA gene sequence similarities among some higher taxa containing anaerobic species within the phylum *Proteobacteria*, using the database of the ARB programme (Ludwig *et al.* 2002). Species of the archaeal *Thermococcales* were used as a root. The order of branching points is tentative and subject to changes based on the number and selection of sequences included in the treeting program. Some aerobic reference organisms and higher taxa are indicated in smaller print. The dimension of the triangle does by and large reflect the number of sequences available in the database. Bar scale=1% difference in sequence similarities.

extracts, as well as acetate and hydrogen or selected amino acids serve as electron donors. *Flexistipes sinusarabici* is a halophilic hetrotrophic organism from the Red Sea (Fiala *et al.*, 1990). The genus *Synergistes* (Allison *et al.*, 1992) is taxonomically treated as genus incertae sedis, as its phylogenetic position is not yet settled. *S. jonesii*, a strictly anaerobic heterotrophic rumen bacterium is capable of degradation of the toxic compound pyridinediol (2,3 DHP and 3,4 DHP), as well as some amino acids (arginine, histidine and glycine). Thus, *S. jonesii* is the first rumen bacterium for which the use of arginine and histidine as major energy yielding substrates has been reported (McSweeny *et al.*, 1993).

12. Phylum *Proteobacteria*

This phylum (Figure 1 and 2) encompasses an enormous genomic diversity, as reflected in the many different morphologies and physiologies. More than 40% of all published genera are members of this phylum which contain the majority of the traditional Gram-negative organisms (Kersters *et al.*, 2003). Besides the aerobic enterobacteria and rhizobia, the most well known members of this class are the anaerobic photosynthetic bacteria, previously classified as "purple nonsulfur bacteria", able to perform anoxygenic photosynthesis by means of bacteriochlorophyl *a* or *b* (see Chapter 3). In contrast to their traditional classification as members of a single order *Rhodospirillales* (Pfennig and Trüper, 1971), molecular and chemical analyses pointed towards the presence of a phylogenetically highly diverse and chemotaxonomically heterogeneous group of Gram-negative bacteria, belonging to two different classes of Proteobacteria (Woese *et al.*, 1984a; 1984b; 1985a; Gibson *et al.*, 1980; Stackebrandt *et al.*, 1988). Different absorption spectra of photosynthetic pigments are responsible for the different colors of cell suspensions (variations of beige, brown, red or pink). The metabolism of these organisms is diverse and flexible: all can grow photoorganoheterotrophically, but many species can grow photolithoautotrophically with either hydrogen, sulfide, thiosulfide, and in some cases ferrous iron, as electron donors. In the dark, anaerobic growth by either fermentation, or chemotrophic growth under microoxic (zone between oxygenic and anoxygenic phase) to oxic conditions may occur (Imhoff, 2001a; 2001b). A single term to designate all the non-photosynthetic bacteria does not make sense from a phylogenetic point of view. In fact, phototrophic, chemolithotrophic and chemoorganotrophic proteobacteria are thoroughly intermingled phylogenetically. The discovery that energy source does not serve well as the primary criterion for classification was one of the first major findings that originated from molecular phylogenetic research.

12.1. Class *Alphaproteobacteria*

The majority of "purple nonsulfur bacteria" are members of the *Alphaproteobacteria*, where they are placed in different major branches of the 16S rRNA gene tree (Kersters *et al.*, 2003; Table 4, Figure 2). Most of the traditionally defined genera have undergone major reclassification, based upon gene sequence, chemotaxonomic and ecophysiological properties (Imhoff, 2001a). Many aerobic non-phototrophic species are moderately related to and intermixed with anaerobic phototrophic bacteria and can be considered descendents

Table 4. Main groups of anaerobic phototophic taxa of the class *Alphaproteobacteria*, phylum *Proteobacteria*. Genera following the type genus are listed in alphabetical order

Order	*Rhodospirillales*	*Rhodobacterales*	*Rhizobiales*	
Family	***Rhodospirillaceae***	***Rhodobacteraceae***	***Bradyrhizobiaceae***	***Hyphomicrobiaceae***
Genera	*Rhodospirillum*	*Rhodobacter*	*Rhodoblastus*	*Blastochloris*
	Phaeospirillum	*Rhodobaca*	*Rhodopseudomonas*	*Rhodomicrobium*
	Rhodocista	*Rhodothalassium*		*Rhodoplanes*
	Rhodospira	*Rhodovulum*		
	Rhodovibrio			
	Skermanella			
	Thalassospira			

of previously phototrophic ancestors. Characteristic properties of taxonomic importance, including carbohydrate utilization, G+C mol% of DNA, vitamin requirement and aerobic growth in the dark are listed by Imhoff (2001a).

Most species of the genus *Rhodospirillum* were reclassified as species of *Rhodocista, Phaeospirillum, Rhodovibrio* and *Rhodoyclus*. The genera and a variety of other recently described or reclassified mostly spiral-shaped phototrophic genera are members of the order *Rhodospirillales*, family *Rhodospirillaceae*, formerly affiliated with the alpha −1 group in the scientific literature. This subgroup also contains several spiral non-phototrophic (e.g., *Magnetospirillum, Azospirillum*) and non-spiral (e.g., *Acidiphilium*) alphaproteobacteria. *Rhodopila* is a member of the family *Acetobacteraceae*. A second portion of the 16S rRNA gene tree (formerly alpha-2 group) is defined by members of *Rhodospeudomonas* and relatives (Table 4), and placed in the family *Bradyrhizobiaceae*, order *Rhizobiales*. *Rhodopseudomonas* had been dissected to exclude those species which were not characterized by a budding mode of reproduction and by having internal lamellae membranes that were lying parallel to the cytoplasmic membrane. This taxon contains mainly aerobic taxa (e.g., *Bradyrhizobium, Afipia, Nitrobacter*). The genera *Rhodomicrobium, Rhodoplanes,* and *Blastochloris* are members of *Hyphomicrobiaceae*, which also embraces many taxa defined by a budding mode of reproduction and aerobic metabolism (e.g., *Hyphomicrobium, Dichotomicrobium, Pedomicrobium* and *Prosthecomicrobium*).

The third group is classified in the order *Rhodobacterales* (formerly alpha-3 group), family *Rhodobacteraceae*. Several species of *Rhodobacter* were reclassified as *Rhodovulum* spp, which, together with other genera (Table 4) are characterized by the presence of carotenoids of the spheroidene series, and metabolic versatility and flexibility (Imhoff, 2001a). The internal membrane system is composed mainly of vesicles (exception: *Rhodobacter blasticus*) and cell division is by binary fission. Aerobic representatives of this phylogenetic group are among others, *Roseobacter* and its marine relatives, *Paracoccus* and *Hyphomonas*. Besides these main groups, the *Alphaproteobacteria* contains the anaerobic organisms *Rhodobium* and *Roseospirillum* (family *Rhodobiaceae*, order *Rhizobiales*) and *Aminobacter* (family *Phyllobacteriaceae*, order *Rhizobiaceae*).

12.2. Class *Betaproteobacteria*

Like their relatives in the *Alphaproteobacteria*, the "purple nonsulfur bacteria" of the *Betaproteobacteria* perform anoxygenic photosynthesis with bacteriochlorophyll *a* and carotinoids as photosynthetic pigments. The distinct phylogenetic position is supported by clear differences in chemotaxonomic properties (Hiraishi *et al.*, 1991; Imhoff, 2001b): in contrast to phototrophic alphaproteobacteria they possess ubiquinones plus menaquinones and they have a "smalltype" cytochrome *c* (Ambler *et al.*, 1979; Dickerson, 1979); their lipopolysaccharides contain unique features (Weckesser *et al.* 1995). These organisms, too, grow preferentially photohetrotrophically under anoxigenic conditions in the light, though some species can also grow photoautotrophically when hydrogen is used as electron donor. In the dark, anaerobic growth occurs by fermentation, while under microaerobic and oxic condition organisms thrive by chemotrophic growth. There is evidence that lateral gene transfer of the L and M subunits of an alphapreobacterial ancestor to the genome of an (heterotrophic?) betaproteobacterial donor cell has occurred, as sequences of these genes from *Rubrivivax gelatinosa, Rhodocyclus tenuis* and *Rhodoferax fermentans* were found in members of *Alphaproteobacteria* (Stackebrandt *et al.*, 1996).

Members of the phototrophic genera of *Betaproteobacteria* do not form a phylogenetically coherent cluster but the genus *Rhodocyclus* is well separated from the genera *Rhodoferax* and *Rubrivivax*;

Table 5. Genera containing anaerobic phototrophic species in the class *Gammaproteobacteria*, phylum *Proteobacteria*. Genera following the type genus are listed in alphabetical order

Order	***Chromatiales***		
Family	***Chromatiaceae***		***Ectothiorhodospiraceae***
Genera	*Chromatium*	*Thioalkalicoccus*	*Ectothiorhodospira*
	Allochromatium	*Thiobaca*	*Alkalilimnicola*
	Amoebobacter	*Thiocapsa*	*Alkalispirillum*
	Halochromatium	*Thiococcus*	*Halorhodospira*
	Halothiobacillus	*Thiodictyon*	*Thiorhodospira*
	Isochromatium	*Thiocystis*	
	Lamprobacter	*Thioflavicoccus*	
	Lamprocystis	*Thiohalocapsa*	
	Marichromatium	*Thiolamprovum*	
	Pfennigia	*Thiopedia*	
	Phaeospirillum	*Thiorhodococcus*	
	Rhabdochromatium	*Thiorhodovibrio*	
	Thermochromatium	*Thiospirillum*	

these latter two genera are more closely related to aerobic organisms, such as *Leptothrix* (family *Comamonadaceae*), *Hydrogenophaga*, *Aquabacterium* and the aerobic bacteriochlorophyll *a* containing *Roseateles*. *Rhodocyclus*, together with the anaerobic malate-fermenting *Propionivibrio* (Tanaka *et al.*, 1990), and *Ferribacterium*, a dissimilatory Fe(III) reducing organism (Cummings *et al.*, 1999), have been classified in the family *Rhodocyclaceae* which mainly embraces aerobic representatives (e.g., *Azoarcus* [containing some anaerobic species] and *Thauera*) as well as facultative anaerobes (e.g., *Dechloromonas, Dechlorosoma*).

Formivibrio (family *Neisseriaceae*), *Hydrogenophilus* (family *Hydrogenophilaceae*) and *Oxalobacter* (family *Oxalobacteraceae*, order *Burkholderiales*) are additional anaerobic representatives of *Betaproteobacteria*.

12.3. Class *Gammaproteobacteria*

Anaerobic bacteria previously classified as phototrophic purple sulfur bacteria are now included in two families of the order *Chromatiales*, i.e. *Chromatiaceae* and *Ectothiorhodospira* (Table 5, Figure 2). These organisms are capable of using sulfur compounds (all use sulfide and elemental sulfur, some also sulfite and thiosulfate) as electron donors, oxidizing them to metastable state sulfate as the final oxidation product (polysulfides) (Steudel *et al.*, 1990). This compound may support aerobic growth in the dark in that it serves as an electron donor for chemolitothrophic growth (Kämpf and Pfennig, 1986). Anaerobic growth in the dark is also possible, in which sulfur is reduced to sulfide during fermentation (Van Gemerden, 1974). Within *Chromatiaceae*, defined by organisms depositing elemental sulfur inside the cells (Table 5), strains inhabiting the marine environment are affiliated with *Halochromatium* and *Marichromatium*. These are phylogenetically separated from those of freshwater environments. Organisms depositing elemental sulfur globules outside the cell are members of *Ectothiorhodospiraceae* (Table 5) (*Thiorhodospira sibirica* is an exception as it deposits sulfur inside the cell). This distinction is in accord with phylogenetic analyses and chemotaxonomic properties [composition of quinones, lipids, lipopolysaccharides and fatty acids, (Weckesser *et al.*, 1979; Imhoff, 1984; Imhoff and Bias-Imhoff, 1995)]. The genera *Arhodomonas* and *Thioalkalivibrio* are non-phototrophic members branching within the radiation of ectothiorhodospiras.

A second group of obligately anaerobic *Gammaproteobacteria* belong to the family *Succinivibrionaceae* (*Succinivibrio, Ruminobacter, Anaerobiospirillum, Succinimonas*) which are able to ferment carbohydrates to acetate and succinate. These morphologically heterogeneous assemblage of species, isolated from the rumen of sheep and cattle and from feces of dogs and cats, are classified in a family on the basis of 16S rRNA gene sequence analysis (Hippe *et al.*, 1999).

The genus *Thiomicrospira* (order *Thiotrichales*) contains a single anaerobic species. A high number of facultatively anaerobic Gram-negative rods of

Table 6. Genera of families of the *Deltaproteobacteria*, containing strictly or obligately anaerobic species. Genera following the type genus are listed in alphabetical order

Order	***Desulfovibrionales***			
Family	***Desulfovibrionaceae***	***Desulfomicrobiaceae***	***Desulfohalobiaceae***	***Desulfonatronumaceae***
Genera	*Desulfovibrio*	*Desulfomicrobium*	*Desulfohalobium*	*Desulfonatrum*
	Bilophila		*Desulfomonas*	
	Lawsonia		*Desulfonatronovibrio*	
			Desulfothermus	
Order	***Desulfobacterales***			
Family	***Desulfobacteriaceae***			***Desulfobulbaceae***
Genera	*Desulfobacter*	*Desulfofaba*	*Desulfonema*	*Desulfobulbus*
	Desulfobacterium	*Desulfofrigus*	*Desulforegula*	*Desulfocapsa*
	Desulfobacula	*Desulfomusa*	*Desulfosarcina*	*Desulforhopalus*
	Desulfocella		*Desulfospira*	*Desulfotalea*
	Desulfococcus		*Desulfotignum*	*Desulfofustis*
Order	***Desulfuromonales***		***Synthrophobacterales***	
Family	***Desulfuromonaceae***	***Geobacteraceae***	***Syntrophobacteraceae***	***Syntrophaceae***
Genera	*Desulfuromonas*	*Geobacter*	*Syntrophobacter*	*Synthrophus*
	Desulfuromusa	*Trichlorobacter*	*Desulfacinum*	*Desulfobacca*
	Pelobacter		*Desulforhabdus*	*Desulfomonile*
	Malonomonas		*Desulfovirga*	*Smithella*
			Thermodesulforhabdus	

clinical significance are either members of the family *Enterobacteriaceae* (e.g., *Escherichia, Shigella, Salmonella, Klebsella, Enterobacter, Serratia, Proteus*), *Vibrionaceae* (e.g., *Vibrio*) or *Actinobacillaceae* (e.g., *Pasteurella*, and *Haemophilus*), but will not be considered in the context of obligate anaerobic bacteria.

12.4. Class *Deltaproteobacteria*

The phylogenetic composition of this class came as a surprise as it combines the fruiting bodies forming myxobacteria, the predatory bdellovibrios and the large group of morphologically and physiologically diverse Gram-negative sulfate and sulfur reducers (Figure 2). The outline of Bergey's Manual, vol 2, treats the genus *Desulfurella* and *Hippea* as a member of this phylum, though phylogenetic analysis indicates that these genera may merit a rank at the level of an individual class.

The publications by Widdel and Bak (1992), Widdel and Pfennig (1984) and Rabus *et al.* (2000) summarize recent developments of the taxonomy of these organisms, including a description of families and genera, along with an extensive coverage of metabolic properties. The sulfur- and sulfate-reducing organisms of this phylum are classified in four orders (Table 6) which contain a few genera only for which sulfur-dependent pathways have not been detected: Within the family *Desulfovibrionaceae*, these are the obligately intracellular *Lawsoni intracellularis* (McOrist *et al.*, 1995) and *Bilophila wadsworthia*, isolated from patients with gangrenous and perforated appendicitis (Baron *et al.*, 1989). Another group of anaerobic organisms of the family *Desulfuromonadaceae* that grow without sulfate or sulfur reduction are members of *Pelobacter*, which ferment either acetoin or trihydroxybenzenoids to acetate and CO_2, though some species ferment either polyethylene glycol, 2,3-butandiol, or acetylene to acetate and alcohols (Schink, 1992). *Malonomonas rubra* (Dehning and Schink, 1989) is also placed in this family. The family *Geobacteriaceae* comprises two genera. *Geobacter* contains strict anaerobic species that obtain energy for growth by coupling the oxidation of acetate to the reduction of Fe(III) (Chapter 12). Differences between the species exist in the ability to oxidize aromatic compounds and organic electron donors oxidized with Fe(III) (Coates *et al.*, 2001).

Table 7. Genera of the phyla *Chloroflexi, Chlorobi, Dictyoglomus* and *Fibrobacteres,* containing strictly or obligately anaerobic species. Genera following the type genus are listed in alphabetical order

Phylum	*Chloroflexi*	*Chlorobi*	*Dictyoglomus*	*Fibrobacteres*
Family	***Chloroflexaceae***	***Chlorobiaceae***	***Dictyoglomaceae***	***Fibrobacteraceae***
Genera	*Chloroflexus*	*Chlorobium*	*Dictyoglomus*	*Fibrobacter*
	Chloronema	*Ancalochloris*		
	Heliothrix	*Chloroherpeton*		
	Roseiflexus	*Pelodictyon*		
		Prosthecochloris		

Trichlorobacter thiogenes, isolated from a contaminated soil sample, grows anaerobically using a process in which reductive dehalogenation of trichoroacetic acid is coupled to the oxidation of acetate or acetoin via a sulfur-sulfide redox cycle (de Wever *et al.*, 2000).

12.5. Class *Epsilonproteobacteria*

The majority of species of this class are aerobes, classified as members of *Campylobacter*, *Arcobacterium*, and *Helicobacter*. About half of the *Campylobacter* species are in fact anaerobes, as is the related species *Sulfospirillum deleyianum*, a facultative anaerobic species using sulfur, nitrate and additional electron acceptors for growth on hydrogen or formate. These properties are shared with *Campylobacter* species and *Wolinella succinogenes*, a member of the *Helicobacteriaceae*. *Thiovolum*, a motile, ovoid organism that stores intracellular sulfur particles, has not yet been brought into pure culture. Recently, the order *Nautiliales* has been described (Miroshnichenko *et al*, 2003) for rod-shaped, chemolithotrophic or mixotrophic, strictly anaerobic organisms of the genera *Nautilia* and *Caminibacter* from deep see hydrothermal vents, growing with H_2 as energy source and sulfate as electron acceptor.

13. Phylum *Chloroflexi*

Two orders have been described for this phylum (Table 7), with *Herpetosiphonales* embracing gliding non-photosynthetic bacteria, while in *Chloroflexales* the gliding, multicellular filamentous and phototrophic members with bacteriochlorophyll *a* are classified. The genera within this phylum differ with respect to the presence of chlorosomes and chlorophyll *c* (present in *Chloroflexus* and *Oscillochloris*, absent in *Roseiflexus* and *Heliothrix*) and growth temperature (Hanada *et al.*, 2002). The chlorobi represent the deepest branch among phototrophic bacteria. The photosynthetic reaction center is shared with those of proteobacteria and is similar to the Photosystem II of cyanobacteria. In contrast to the 16S rRNA gene tree, which determines chloroflexi as an early emerging branch and cyanobacteria and proteobacteria evolving later, sequence analyses of the reaction centre L(light)- and M(medium)- subunits of protein sequences cluster *Chloroflexus* close to the *Proteobacteria*, while cyanobacteria are clearly separated (Blankenship, 1992). It appears that an ancestral reaction centre split to develop the D1/D2 lines of cyanobacteria and chloroplasts, and the L and M protein of the photosynthetic proteobacteria and *Chloroflexus*.

The presence of genetic, biochemical and structural similarities in the chlorosomes of chloroflexi and chlorobia has puzzled biochemist as these organisms are phylogenetically quite unrelated. Horizontal gene transfer may be an explanation for this phenomenon, but questions about the common ancestor and the recipient that contained a photosynthetic apparatus with either an iron-sulfur or a pheophytin/quinone reaction center, remains unanswered (Gibson *et al.*, 1985). As pointed out above, the solution to this problem is further complicated by the lack of information about the order of events leading to oxygen-evolving pathways.

14. Phylum *Chlorobi*

The green sulfur bacteria, characterized by chlorosomes containing either bacteriochlorophylls *c*, *d*, or *e* as accessory pigments are classified in the class *Chlorobia* (Table 7). This phylum branches adjacent to *Bacteriodetes*, which is also supported by sequence homology of the *recA* gene of both groups (Gruber *et al.*, 1998). The single family *Chloribiaceae* (Table 7) contains five genera. In addition, the invalid genus "*Clathrochloris*" should be added to the family (Witt *et al.*, 1989). While the species share most of the physiological capacities, the genera are presently differentiated by cell morphology, motility, and ability to form gas vesicles (Overmann, 2000).

As already indicated above, chlorosomes and bacteriochlorophylls *c* or *d* are also present in the multicellular filamentous gliding bacteria of the family *Chloroflexaceae*. The similarity of the chlorosome extends to the homology of the CsmA proteins of the chlorosome envelope. Differences between the chlorosome types of the two taxa have been summarized by Overmann (2000). Of these, the structure of the reaction centre is the most salient one: while chlorobi contain the iron-sulfur-type, the quinone-type is found in *Chloroflexus* and relatives. The reaction centre of chlorobi resembles that of heliobacteria (phylum *Firmicutes*) and this is functionally and structurally similar with the core proteins of reaction centre I of chloroplasts and cyanobacteria.

15. Phylum *Dictyoglomus*

The genus *Dictyoglomus* (Table 7) contains two species with a unique morphology. Large spherical bodies are formed by association of separate rods (up to 100) which are surrounded by their outer wall membrane. The organisms, isolated from hot springs, are non-motile Gram-negative fermenters of carbohydrates. The DNA base composition of 29 mol% (Saiki *et al.*, 1985) ranges at the lower end of the G+C scale.

16. Phylum *Fibrobacteres*

Fibrobacter (Table 7) is the only genus of this phylum. Its members, affiliated to two species, are Gram-negative, nonsporulating bacteria of the mammalian gastrointestinal tract, digesting cellulose with succinic and acetic acid as primary fermentation products.

17. Lineages Containing Anaerobic Gram-Positive Bacteria

The separation of organisms defined by a multilayered peptidoglycan, the Gram-positive bacteria, into *Clostridium*, *Bacillus* and relatives and in actinomycetes and relatives on the basis of the base composition of their DNA has been proven by and large valid on phylogenetic grounds. These two groups of organisms have received the rank of separate phyla, the *Firmicutes* for the Clostridia and relatives and the *Actinobacteria* for those with a DNA G+C content of higher than 55 mol% (at least in the vast majority of its members). The term Firmicutes was originally introduced by Murray (1984) for uniting Gram-positive organisms (with the exception of *Deinococcus* spp.) in a higher level rank. The rational for separating *Firmicutes* from *Actinobacteria* is based on the inability of phylogenetic analysis to clearly demonstrate a common root exclusively for these two species-rich deep lineages. If present, the unity of these two subphyla is statistically not significant. Analyses of 16S rDNA reveal that the order of divergence of the main lines of descent is not constant but changes as a function of the particular set of sequences included in the analyses (Van de Peer *et al.*, 1994). Similarly, analyses of 23S rDNA and the elongation factor Tu (EF-Tu) show the two main lineages of Gram-positive bacteria to be separated (Schleifer and Ludwig, 1994). Another argument is the finding that, the most deeply branching species within the *Clostridium* lineage are not Gram-positives but possess a Gram-negative cell wall, such as *Sporomusa, Selenomonas* and *Megasphaera* (Stackebrandt *et al.*, 1985). Some of these organisms even contain an outer membrane with LPS. Thus, the topology of the *Sporomusa* cell wall does not appear to be a reduction of the Gram-positive wall but rather an ancestral feature of a Gram-negative ancestor (Stackebrandt *et al.*, 1988). Even a photosynthetic ancestor of the Gram-positive bacteria cannot be excluded, as the heliobacteria, characterized by bacteriochlorophyll *g*, branch deeply in the phylum Firmicutes (Woese *et al.*, 1985a).

18. Phylum Actinobacteria

This group of organisms (Figure 1) was the first to be classified completely along phylogenetic lines (Stackebrandt *et al.*, 1997). However, the hierarchic system included the class as the highest rank, while the phylum level will now be proposed in Bergey's Manual (Garrity *et al.*, 2003). The vast majority of species of this phylum are strict aerobes or microaerophilic and many representatives form acids from carbohydrate under reduced oxygen levels. As judged from the branching point of the most deeply branching taxa, the actinobacteria represent a phylogenetically tight group that evolved rather late in evolution. It appears that the radiation of actinobacteria occurred in parallel with the increased level of oxygen, leading to a rapid diversification at the genetic and epigenetic level. Being metabolically rather uniform, the actinobacteria are defined by a high morphological and chemotaxonomic diversity. A few well known examples of anaerobic species have been described. Some organisms, previously classified as facultatively anaerobic clostridia were transferred to the class *Actinobacteria* in which they form the most deeply branching lineages. Today these sugar-fermenting organisms are classified in the family *Coriobacteriaceae* (*Atopobium, Slackia, Coriobacterium,* and *Colinsella*). The taxonomy of organisms classified in the families *Actinomycetaceae* (*Actinomyces, Arcanobacterium, Actinobaculum,* and *Mobiluncus=Falcivibrio*) (Schaal, 1992) in *Bifidobacteriaceae* (*Bifidobacterium, Gardnerella, Parascardovia,* and *Scardovia*) (Biavati *et al.*, 1991; Biavati and Matarelli, 2001) and Propionibacteriaceae (*Propionibacterium, Propionimicrobium*) has been covered extensively because of their importance as human and animal pathogens and as commensals of the alimentary tract, respectively.

19. Phylum Firmicutes

This phylum is without any doubt the greatest challenge for taxonomists (Cato and Stackebrandt, 1989; Collins *et al.*, 1994; Stackebrandt *et al.*, 1999b). Various lineages

Table 8. Genera of the phylum *Firmicutes* containing strictly or obligately anaerobic species. Genera following the type genus are listed in alphabetical order.

Class	***Clostridia***			
Order	***Clostridiales***			
Family	***Clostridiaceae***			***Eubacteriaceae***
Genus	*Clostridium*	*Caloranaerobacter*	*Sporobacter*	*Eubacterium*
	Acetivibrio	*Coprobacillus*	*Thermobrachium*	*Acetobacterium*
	Acidaminobacter	*Dorea*	*Thermohalobacter*	*Anaerovorax*
	Alkaliphilus	*Natronincola*	*Tindallia*	*Mogibacterium*
	Anaerobacter	*Oxobacter*		*Pseudoramibacter*
	Caloramator	*Sarcina*		
Family	***Lachnospiraceae***		***Peptostreptococcaceae***	
Genus	*Lachnospira*	*Johnsonella*		*Helcococcus*
	Acetitomaculum	*Lachnobacterium*	*Anaerococcus*	*Micromonas*
	Anaerofilum	*Pseudobutyrivibrio*	*Filifactor*	*Peptoniphilus*
	Butyrivibrio	*Roseburia*	*Finegoldia*	*Sedimentibacter*
	Catenibacterium	*Ruminococcus*	*Fusibacter*	*Sporanaerobacter*
	Catonella	*Sporobacterium*	*Gallicola*	*Tissierella*
	Coprococcus			
Family	***Peptococcaceae***			***Heliobacteriaceae***
Genus	*Peptococcus*	*Dehalobacter*	*Mitsuokella*	*Heliobacterium*
	Anaeroarcus	*Dendrosporobacter*	*Propionispira*	*Heliobacillus*
	Anaerosinus	*Desulfitobacterium*	*Succinispira*	*Heliophilum*
	Anaerovibrio	*Desulfonispora*	*Syntrophobotulus*	*Heliorestis*
	Carboxydothermus	*Desulfosporosinus*		
	Centipedia	*Desulfotomaculum*		
Family	***Acidaminococcaceae***			
Genus	*Acidaminococcus*	*Megasphaera*	*Quinella*	*Veillonella*
	Acetonema	*Papillibacter*	*Schwartzia*	*Zymophilus*
	Anaeroglobus	*Pectinatus*	*Selenomonas*	
	Anaeromusa	*Phascolarcobacterium*	*Sporomusa*	
	Dialister	*Propionispora*	*Succiniclasticum*	
Family	***Syntrophomonadaceae***			
Genus	*Syntrophomonas*	*Anaerobaculum*	*Pelospora*	*Thermohydrogenium*
	Acetogenium	*Anaerobranca*		*Thermosyntropha*
	Aminobacterium	*Caldicellulosiruptor*		
	Aminomonas	*Dethiosulfovibrio*		

Table 8, continued

Order	*Thermoanaerobacteriales*			
Family	***Thermoanaerobacteriaceae***			
Genus	*Thermoanaerobacterium*	*Gelria*	*Thermoacetogenium*	
	Ammonifex	*Moorella*	*Thermoanaerobacter*	
	Carboxydobrachium	*Sporotomaculum*	*Thermoanaerobium*	
	Coprothermobacter			
Order	***Haloanaerobiales***			
Family	***Haloanaerobiaceae***	***Halobacteroidaceae***		
Genus	*Haloanaerobium*	*Halobacteroides*	*Halonatronum*	*Selenihalanaerobacter*
	Halocella	*Acetohalobium*	*Natroniella*	*Sporohalobacter*
	Halothermothrix	*Haloanaerobacter*	*Orenia*	

*, *Thermaerobacter* is the only genus defined by memers with aerobic metabolism

have emerged during evolution in which the organisms are clustered according to their relationship to oxygen. The most deeply branching organisms are strict anaerobes; in one of these lineages the ability was acquired to cope with oxygen, leading first to facultatively anaerobic (lactobacilli, streptococci, and relatives) and then to aerobic organisms, such as listeria, bacilli, staphylococci, and mycoplasmas and others. More than fifteen deeply branching lineages have been defined by Collins *et al.* (1994) in their pioneering study on the elucidation of the phylogenetic structure of clostridia, most of which contained *Clostridium* species. This study had unraveled the enormous phylogenetic heterogeneity of many of those genera which were classified on superficial phenotypic grounds: presence or absence of spore, morphology, relationship to oxygen, ability to reduce sulfur, and the like. Though the relationship of nearly all type strains is known by now, the reclassification process is slow: this is due to the fact that, other than the situation seen among actinobacterial genera, chemotaxonomic or other taxonomic differentiating properties are lacking that would allow delineation of groups of organisms at the generic level. It is for this reason that, at this taxonomic level, the new edition of Bergey's Manual (Table 8) still follows the present classification, though the phylogenetic unrelatedness of species of certain genera has already been demonstrated. This is especially obvious in the genera *Clostridium, Desulfotomaculum* and *Eubacterium.* In order to taxonomically restructure this group of Gram-positive bacteria, the 16S rDNA phylogenetic dendrogram (Figure 3) should be reconfirmed by investigating other markers of phylogenetic significance, e.g., houskeeping genes (Stackebrandt *et al.*, 2002). Rarely have phylogenetic analyses on genes other than ribosomal RNA genes been performed on more than a few *Clostridium* species. Either confirmation of phylogenetic relationships and the emergence of a redefined pattern of relationships will facilitate reclassification so that phenotypic coherence of phylogenetic groups of organisms may be detected. Novel taxonomic properties should be uncovered and the same techniques should be applied in parallel to a broad spectrum of members of phylogenetically heterogeneous genera. Future reclassifications should also consider the medical importance of many of these clostridial species. Taxonomists have the option to overrule scientific decisions by conserving a taxonomic name (*nomen pericilosum*, Rule 56a, Lapage *et al.*, 1992), as for example seen in the conservation of the genus name *Clostridium*, which otherwise would have been changed to *Sarcina* because of publication date priority (Rule 24b, Lapage *et al.*, 1992).

A detailed description of the individual higher taxa of the phylum *Firmicutes* is beyond the scope of this chapter. The reader will find the basis for the new hierarchic system in the 2nd edition of Bergey's Manual dealing with the Firmicutes, published in the coming years. As new genera are regularly being described and 16S rRNA gene sequences are added to the database it is likely that new adjustments to the system will be done. The affiliation of genera containing anaerobic species to higher taxa as depicted in the Tables is therefore tentative but gives an impression of the rich systematic structure. In contrast to the proposed higher classification, phylogenetic analysis indicates that the genus *Coprothermobacter* (*Thermoanaerobacteriales*) (Rainey and Stackebrandt, 1993) is a distinct phylum-level lineage (Figure 1). This could also be proposed for some members of the family *Syntrophomonadaceae* (i.e., *Aminomonas, Anaerobaculum, Dethiosulfovibrio*, and *Thermanaerovibrio*).

Additional anaerobic species which are not listed in these Tables are included in the family *Erysipelothrichaceae* (*Bulleidia, Holdemania* and

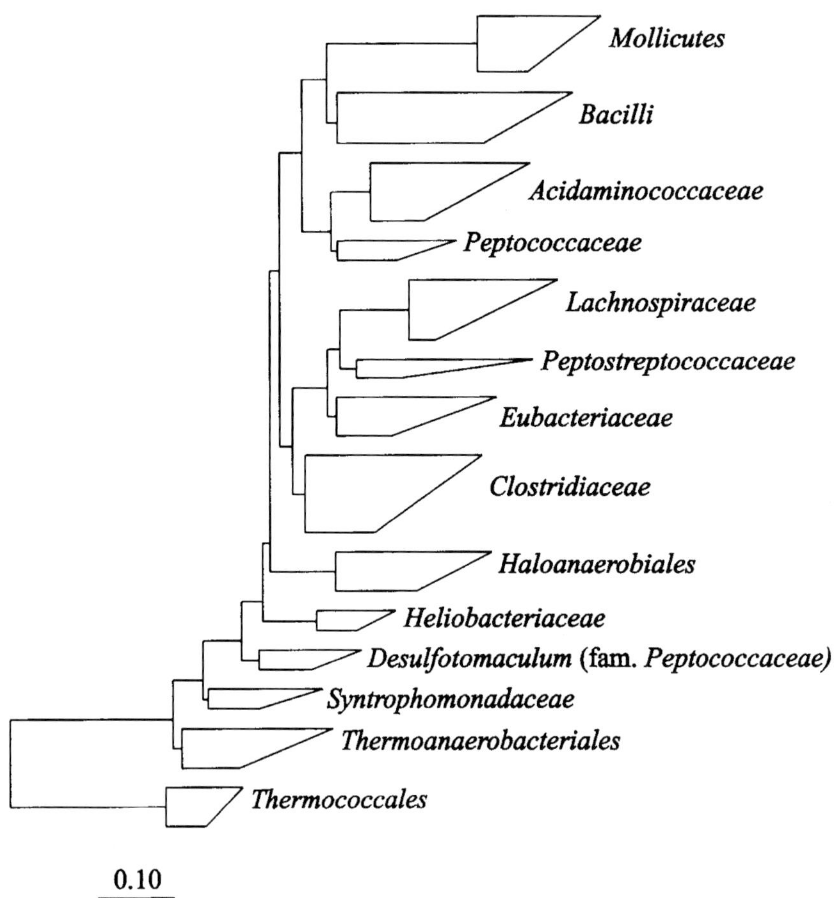

Figure 3. Dendrogram of 16S rRNA gene sequence similarities among some higher taxa containing anaerobic species within the phylum *Firmicutes*, using the database of the ARB programme (Ludwig *et al.* 2002). Species of the archaeal *Thermococcales* were used as a root. Also included are *Bacilli* and *Mollicutes* (wall-less Gram-positive bacteria). Note that non-authentic and misclassified species of e.g., *Clostridium* and *Eubacterium* are not shown. The order of branching points is tentative and subject to changes based on the number and selection of sequences included in the treeting program. The dimension of the triangle does by and large reflect the number of sequences available in the database. Bar scale= 1% difference in sequence similarities.

Solobacterium), but this family has recently been restricted to the genus *Erysipelothrix* (Verbarg *et al.*, 2003) and in family *Anaeroplasmataceae* (*Anaeroplasma*). The genus *Oxalophagus* is a member of *Paenibacillaceae*, while the order *Lactobacillales* embraces besides *Lactobacillus* several anaerobic genera such as *Gemella, Trichococcus, Abiotrophia,* and *Syntrophococcus* in different families containing otherwise less anaerobic members.

20. Phylum *Bacteroidetes*

Bacteroidetes encompasses several orders, some of which contains mainly anaerobic species. The relationship of the strictly anaerobic (*Bacteroides* and relatives) and aerobic (*Flavobacteria* and relatives) species were never considered phylogenetic neighbours prior to the introduction of molecular systematic analyses (Paster *et al.*, 1985), as these two groups have few, if any, phenotypic properties in common. *Bacteroides* had once been a dumping ground for anaerobic, fermenting and nonsporing Gram-negative rods, which produce C1-C4 acids as major end products. The range of DNA mol% G+C of 28 to 61 was a clear indication of the phylogenetic heterogeneity of the genus. Due to the massive taxonomic restructuring of the genus *Bacteroides* on the basis of biochemical and chemotaxonomic data and DNA-DNA hybridization (Shah and Collins, 1983; see Table 2), several new genera were described, most of which are distantly related to *Bacteroides* but grouped in the family *Rikenellaceae*, *Porphyromonadaceae*, and *Prevotellaceae* (Table 9). The restructuring was by and large supported by 16S rRNA gene analysis but the range of DNA G+C content is still high (28-60mol%), which indicates that the reclassification process will be continued once appropiate discriminating properties are determined. Many species of *Bacteroides* and genera resulting from reclassification of former *Bacteroides* species have been isolated from various types of human clinical specimen, causing various infections, including those originating from fecal contamination. New genera were described harboring fermenting thermophilic species from sewage sludge, and anaerobic subsurface sites. Despite the massive restructuring of *Bacteroides*, the genus is still phylogenetically heterogeneous, with some species branching outside the core species cluster of the genus (Lawson *et al.*, 2002). The genus *Alistipes* (Rautio *et al.*, 2003) is the latest addition to the family *Rikenellaceae*.

21. Phylum *Spirochaetes*

The spirochetes are one of few examples in which morphology and mode of motility are responsible for phylogenetic coherency, hence the presence of sometimes

Table 9. Genera containing strictly or obligately anaerobic species of families of the phylum *Bacteroidetes*. Genera following the type genus are listed in alphabetical order

Order	***Bacteroidetes***			
Family	***Bacteroidaceae***	***Rikenellaceae***	***Porphyromonadaceae***	***Prevotellaceae***
Genera	*Bacteroides*	*Rikenella*	*Porphyromonas*	*Prevotella*
	Acetomicrobium	*Alistipes*	*Dysgonomonas*	
	Acetothermus	*Marinilabilia*	*Tannerella*	
	Acidofilamentum			
	Anaerophaga			
	Anaerorhabdus			
	Megamonas			

more than 100 axial endoflagella wound around the cell is a monophyletic trait. At least two flagella are attached to the opposite ends if the spirochete cell. The phylum consists of genera containing anaerobic (Table 10) and aerobic species (*Leptonema, Borrelia, Leptospira*). Many spirochetes previously considered uncultivable have been cultured. Others remain uncultured, including those thriving in the intestine of dry wood eating cockroaches and termites (*Clevelandia, Diplocalyx, Hollandia, Pillotina*) or inhabiting the crystalline style or gut fluid of marine and fresh water mollusks (*Cristispira*). The recent taxonomy of the spirochetes has been reported by Olsen *et al.* (2000). Members of the anaerobic *Spirochaeta* are non-pathogenic, saccharolytic and grow in mud and sediments of aquatic environments and in sewage. Some spirochetes live in associations with animals (e.g. in the oral cavity or the gastrointestinal tract) and a few are pathogens of animals. *Treponema* species, for which physiological requirements are not well studied, are human pathogens, of which *T. pallidum*, the causative agent of syphilis, is the most well known spirochete representative. Some human parasitic *Treponema* species were first reclassified as members of *Serpulina*, then as members of *Brachyspira*. Nevertheless, two *Serpulina* species, *S. intermedia* and *S. murdochii*, are still listed as valid species.

22. Phylum *Fusobacteria*

This group of genera (Table 10) has received phylum status though in previous phylogenetic analysis individual members consistantly grouped within the phylum Firmicutes. With the database of fusobacterial 16S rRNA gene sequences increasing, a remote relationship to the *Proteobacteria* was observed. Affiliation of members of this phylum can be problematic, as *Leptothrichia buccalis, Sebaldella termitidis,* and *Strepotobacillus moniliformis* share only very low similarities with other members of the phylum, which are almost as low as those shared between members of different phyla (< 80% sequence similarity). *Fusobacterium* and *Leptotrichia* have previously been classified in the family Bacteroidaceae (Holdeman *et al.*, 1984) and can be distinguished from members of *Bacteroides* in the production of butyric acid and lactic acid, respectively as major fermentation end products. Members of the genera affiliated with the phylum resemble each other with respect to morphology and their carbohydrate and/or amino acid fermentation end products. Affiliation to the phylum and to individual genera are preferentially done by analysis of 16S rRNA gene sequences.

Table 10. Genera containing strictly or obligately anaerobic species of families of the phylum *Spirochaetes* and *Fusobacteria*. Genera following the type genus are listed in alphabetical order

Phylum	***Spirochaetes***		***Fusobacteria***	
Order	***Spirochaetales***		***Fusobacteriales***	
Family	*Spirochaeta*	*Serpulinaceae*	*Fusobacteriaceae*	
Genera	*Spirochaeta*	*Serpulina*	*Fusobacterium*	*Propionigenium*
	Treponema	*Brachyspira*	*Cetobacterium*	*Sebaldella*
			Ilyobacter	*Sneathia*
			Leptotrichia	*Streptobacillus*

23. Phylum *Acidobacteria*

This phylum contains only three monospecific genera, though based on molecular environmental studies at least for the genus *Acidobacterium* a huge diversity of putative novel species has been detected in widely different terrestrial samples. *Geothrix fermentans* is an anaerobic chemoorganotrophic rod, reducing Fe(III) with acetate as electron donor (Coates *et al.*, 1999). *Holophaga foetida* is a homoacetogenic bacterium that degrades methoxylated aromatic compounds (Liesack *et al.*, 1994).

24. Phylum Nitrospira

Thermodesulfovibrio spp. are non-sporeforming, motile thermophilic and Gram-negative curved rods that reduce sulfate, thiosulfate and sulfite, while sulfur, fumarate and nitrate are not reduced. In the presence of sulfate, growth was observed only with lactate, pyruvate, hydrogen plus acetate, or formate plus acetate. Pyruvate, oxidized to acetate, was the only compound observed to support fermentative growth (Henry *et al.*, 1994). Other members of the phylum are strict aerobes (*Nitrospira, Leptospirillum*).

References

Allison, M.J., Mayberry, W.R., McSweeney, C.S., and Stahl, D.A. 1992. *Synergistes jonesii*, gen. nov., sp. nov.: a rumen bacterium that degrades toxic pyridinediols. System. Appl. Microbiol. 15: 522-529.

Ambler, R. P., Daniel, M., Hermoso, J., Meyer, T.E., Bartsch, R.G., and Kamen, M.D. 1979. Cytochrome c_2 sequence variation among the recognised species of purple nonsulphur photosynthetic bacteria. Nature. 278: 659-660.

Aravind, L., Tatusov, R.L., Wolf, Y.I., Walker, D.R., and Koonin, E.V. 1998. Evidence for massive gene exchange between archaeal and bacterial hyperthermophiles. TIG. 14: 442-444.

Atlas, R.M. 1996. Principles of Microbiology.2nd, ed. Wm. C. Brown, Boston, Massachusetts.

Bachleitner, M., Ludwig, W., Stetter, K.O., and Schleifer, K.-H. 1989. Nucleotide sequence of the gene coding for the elongation factor Tu from the extremely thermophilic eubacterium *Thermotoga maritima*. FEMS Microbiol. Lett. 48: 115-120.

Balch, W.E., Fox, G.E., Magrum, L.J., Woese, C.R., and Wolfe, R.S. 1979. Methanogens: reevaluation of a unique biological group. Microbiol. Rev. 43: 260-296.

Baldauf, S.L., Palmer, J.D., and Doolittle, W.F. 1996. The root of the universal tree and the origin of eukaryotes based on elongation factor phylogeny. Proc. Natl. Acad. Sci. USA. 93: 7749-7754.

Baron, E.J., Summanen, P., Downes, J., Roberts, M.C., Wexler, H., and Finegold, S.M. 1989. *Bilophila wadsworthia*, gen. nov. and sp. nov., a unique Gram-negative anaerobic rod recovered from appendicitis specimens and human feces. J. Gen. Microbiol. 135: 3405-3411.

Ben-Haim, Y., Thompson, F.L., Thompson, C.C., Cnockaert, M.C., Hoste, B., Swings, J., and Rosenberg, E. 2003. *Vibrio coralliilyticus* sp. nov., a temperature-dependent pathogen of the coral *Pocillapora damicornis*. Int. J. Syst. Evol. Microbiol. 53:309-315.

Biavati, B., and Matarelli, P. 2001. The Family *Bifidobacteriaceae*. In: The Prokaryotes. An Evolving Electronic Resource for the Microbiological Community, 3rd edition, release 3.3. M. Dworkin *et al.*, eds. Springer-Verlag, New York, New York.

Biavati, B., Sgorbati, B., and Scardovi, V. 1991. The genus *Bifidobacterium*. In: The Prokaryotes, 2nd ed. A. Balows, H. G. Trüper, M. Dworkin, W. Harder, and K.-H. Schleifer, eds. Springer-Verlag, New York, New York. p. 816-833.

Blankenship, R.B. 1972. Origin and early evolution of photothynthesis. Photosyn. Res. 33: 91-111.

Blankenship, R.E. 1992. Origin and early evolution of photothynthesis. Photosyn. Res. 33: 91-111.

Brochier, C., and Philippe, H. 2002. Phylogeny: a non-hyperthermophilic ancestor for bacteria. Nature. 417: 244.

Broda, E. 1975. The Evolution of the Bioenergetic Processes. Pergamon Press, Oxford.

Brown, J.R., and Doolittle, W.F. 1995. Root of the universal tree of life based on ancient aminoacyl-tRNA synthetase gene duplications. Proc. Natl. Acad. Sci. USA. 92: 2441-2445.

Caccavo, F. Jr., Coates, J.D., Rosello-Mora, R.A., Ludwig, W., Schleifer, K.H., Lovley, D.R., and McInerney, M.J. 1996. *Geovibrio ferrireducens*, a phylogenetically distingt dissimilatory Fe(III)-reducing bacterium. Arch. Microbiol. 165: 370-376.

Cato, E.P., and Stackebrandt, E. 1989. Taxonomy and Phylogeny. In: The Genus *Clostridium*. N.P. Nigel and D.J. Clarke, eds., Biotechnology Handbooks, Plenum, New York, New York. p. 1-26.

Cavalier-Smith, T. 2002. The neomuran origin of archaebacteria, the negibacterial root of the universal tree and bacterial megaclassification. Int. J. Syst. Evol. Microbiol. 52: 7-76.

Claros, M.C., Papke, Y., Kleinkauf, N., Adler, D., Citron, D.M., Hunt-Gerardo, S., Montag, T., Goldstein, E.J.C., and Rodloff, A.C. 1999. Characteristics of *Fusobacterium ulcerans*, a new and unusual species compared with *Fusobacterium varium* and *Fusobacterium mortiferum*. Anaerobe. 5: 137-140.

Coates, J.D., Bhupathiraju, V.K., Achenbach, L.A., McInerney, M.J., and Lovley, D.R. 2001. *Geobacter hydrogenophilus, Geobacter chapellei* and *Geobacter grbiciae*, three new, strictly anaerobic, dissimilatory Fe(III)-reducers. Int. J. Syst. Evol. Microbiol. 51: 581-588.

Coates, J.D., Ellis, D.J., Gaw, C.V., and Loveley, D.R. 1999. *Geothrix fermentans* gen. nov., sp.nov., a novel Fe(III)-reducing bacterium from a hydrocoarbon-conaminated aquifer. Int. J. Syst. Evol. Microbiol. 49: 1615-1622.

Cohn, F. 1875. Untersuchungen über Bakterien II. Beitr. Biol. Pflanzen 1: 141-207.

Collins, M.D., Lawson, P.A., Willems, A., Cordoba, J.J., Fernandez-Garayzabal, J., Garcia, P., Cai, J., Hippe, H., and Farrow, J.A.E. 1994. The phylogeny of the genus *Clostridium*: proposal of five new genera and eleven new species combinations. Int. J. Syst. Bacteriol. 44: 812-826.

Cummings, D.E, Caccavo, F., Spring, S., and Rosenzweig, R.F. 1999. *Ferribacterium limneticum* gen. nov., sp. nov., an Fe(III)-reducing microorganism isolated from mining-impacted freshwater lake sediments. Arch. Microbiol. 171: 183-188.

Darimont, B., and Sterner, R. 1994. Sequence, assembly and evolution of a primordial ferredoxin from *Thermotoga maritima*. EMBO J. 13: 1772-1781.

De Weaver H., Cole, J.R., Fettig, M.R., Hogan, D.A., and Tiedje, J.M. 2000. Reductive dehalogenation of trichloroacetic acid by *Trichlorobacter thiogenes* gen. nov., sp. nov. Appl. Environ. Microbiol. 66: 2297-2301.

Deckert, G., Warren, P.V., Gaasterland, T., Young, W.G., Lenox, A.L., Graham, D.E., Overbeek, R., Snead, M.A., Keller, M., Aujay, M., Huber, R., Feldman, R.A., Short, J.M., Olsen, G.J., and Swanson, R.V. 1998. The complete genome of the hyperthermophilic bacterium *Aquifex aeolicus*. Nature. 392: 353-358.

Dehning, I., and Schink, B. 1989. *Malonomonas rubra* gen. nov., sp.nov., a microaerotolerant anaerobic bacterium growing by decarboxylation of malonate. Arch. Microbiol. 151: 427-433.

Dickerson, R.E. 1979. The cytochromes c: an exercise in scientific serendipity. UCLA Forum Med Sci. 21: 173-202

DSMZ-German Collection of Microorganisms. CaTalogue of Strains, 7th edition, 2001. Braunschweig, Germany.

Dunn, J.J., McCorkle, S.R., Praissman, L.A., Hind, G., van der Lelie, D., Bahou, W.F., Gnatenko, D.V., and Krause, M.K. 2002. Genomic signature tags (GSTs): a system for profiling genomic DNA. Genome Res. 12: 1756-1765.

Farlow, J., Smith, K.L., Wong, J., Abrams, M., Lytle, M., and Keim, P. 2001. *Francisella tularensis* strain typing using multiple-locus, variable-number tandem repeat analysis. J. Clin. Microbiol. 39: 3186-3192.

Fiala, G., Woese, C.R., Langworthy, T.A., and Stetter, K.O. 1990. *Flexistipes sinusarabic*i a novel genus and species of eubacteria occurring in the Atlantis II Deep brines of the Read Sea. Arch. Microbiol. 154: 120-126.

Fitz-Gibbon, S. T., and House, C. H. 1999. Whole genome-based phylogenetic analysis of free-living microorganisms. Nucl. Acids Res. 27: 4218-4222.

Fox, G. E., Pechman, K. R., and Woese, C. R. 1977. Comparative cataloguing of 16S ribosomal ribonucleic acid: Molecular approach to prokaryotic systematics. Int. J. Syst. Bacteriol. 27: 44-57.

Fox, G.E., Stackebrandt, E., Hespell, R.B., Gibson, J., Maniloff, J., Dyer, T.A., Wolfe, R.S., Balch, W.E., Tanner RS, Magrum LJ, Zablen LB, Blakemore R, Gupta R, Bonen L, Lewis BJ, Stahl DA, Luehrsen KR, Chen KN, and Woese CR. 1980. The phylogeny of prokaryotes. Science. 209: 457-463.

Garrity, G.M., Bell, J.A. and Lilburn, T.G. 2003. Taxonomic Outline of the Procaryotes. Bergey's Manual of Systematic Bacteriology, Second Edition., Release 4.0., Springer-Verlag, New York, New York. DOI: 10.1007/bergeysoutline200310. http://www.springerny.com/bergeysoutline/main.htm

Gibson, J., Ludwig, W., Stackebrandt, E., and Woese, C.R. 1985. The phylogeny of the green photosynthetic bacteria: Lack of a close relationship between *Chlorobium* and *Chloroflexus*. System. Appl. Microbiol. 6: 152-156.

Gibson, J., Stackebrandt, E., Zablen, L.B., Gupta, R., and Woese, C.R. 1980. A genealogical analysis of the purple photosynthetic bacteria. Curr. Microbiol. 3: 59-66.

Gomez Zavaglia, A., de Urraza, P., and De Antoni, G. 2000. Characterization of *Bifidobacterium* strains using BOX primers. Anaerobe. 6: 167-177.

Greene, A.C., Patel, B.K.C., and Sheehy, A.J. 1997. *Deferribacter thermophilus* gen. nov., sp. nov., a novel thermophilic manganese- and iron-reducing bacterium isolated from a petroleum reservoir. Int. J. Syst. Bacteriol. 47: 505-509.

Gruber, T.M., Eisen, J.A., Gish, K., and Bryant, D.A. 1998. The phylogenetic relationships of *Chlorobium tepidum* and *Chloroflexus aurantiacus* based upon their RecA sequences. FEMS Microbiol. Lett. 162: 53-60.

Gupta, R.S., and Griffiths, E. 2002. Critical issues in bacterial phylogeny. Theor. Pop. Biol. 61: 423-434.

Gupta, R.S. 2000. The phylogeny of proteobacteria: relationships to other eubacterial phyla and eukaryotes. FEMS Microbiol. Rev. 24: 367-402.

Hanada, S. Takaichi, S. Matsuura, K., and Nakamura K. 2002. *Roseiflexus castenholzii* gen. nov., sp. nov., a thermophilic, filamentous, photosynthetic bacterium that lacks chlorosomes. Int. J. Syst. Evol. Microbiol. 52: 187-193.

Henry, E.A., Devereux, R., Maki, J.S., Gilmour, C.C., Woese, C.R., Mandelco, L., Schauder, R., Remsen, C.C., and Mitchell, R. 1994. Characterization of a new thermophilic sulfate-reducing bacterium *Thermodesulfovibrio yellowstonii*, gen. nov. and sp. nov.: its phylogenetic relationship to *Thermodesulfobacterium commune* and their origins deep within the bacterial domain. Arch. Microbiol. 161: 62-69.

Hippe, H, Hagelstein, Kramer, I., Swiderski, J., and Stackebrandt, E. 1999. Phylogenetic analysis of *Formivibrio citricus*, *Propionivibrio dicarboxylicus*, *Anaerobiopspirillum thomasii*, *Succinimonas amylolytica* and *Succinivibrio dextrinisolvens* and proposal for *Succinivibrionaceae* fam. nov. Int. J. Syst. Bacteriol. 49: 779-782.

Hiraishi, A., Hoshino, Y., and Satoh, T. 1991. *Rhodoferax fermentans* gen. nov., sp. nov., a phototrophic purple nonsulfur bacterium previously referred to as the "*Rhodocyclus gelatinosus*-like" group. Arch. Microbiol. 155: 330-336.

Holdeman, L.V., Cato, E.P., and Moore, W.E.C. 1977. Anaerobe. Laboratory Manual, Virginia Polytechic Institute and State University, Blacksburg, Virginia.

Holdeman, L.V., Kelley, R., and Moore, W.E.C. 1984. *Bacteroidaceae* In Bergey's Manual of Systematic Bacteriology. N.R. Krieg, and J.G. Holt, eds. Williams and Wilkins, Baltimore, Maryland. p. 602-662.

Holt, J.G., Krieg, N.R., Sneath, P.H.A., Staley, J.T., and Williams, S.T. 1994. Bergy's Manual of Determinative Bacteriology, 9th edition. Williams and Wilkins, Baltimore, Maryland.

Huber, R., and Eder, W. 2002. *Aquificales*. In: The Prokaryotes: An Evolving Electronic Resource for the Microbiological Community, 3rd edition, release 3.14. M. Dworkin, S. Falkow, E. Rosenberg, K.-H. Schleifer and E. Stackebrandt, eds. Springer-Verlag, New York, New York.

Huber, R., and Hanning, M. 2003. *Thermotogales*. In: The Prokaryotes: An Evolving Electronic Resource for the Microbiological Community, 3rd edition, release 3.14. M. Dworkin, S. Falkow, E. Rosenberg, K.-H. Schleifer, and E. Stackebrandt, eds. Springer-Verlag, New York, New York.

Hugenholtz, P., Goebel, B.G., and Pace, N.R. 1998. Impact of culture-independent studies on the emerging phylogenetic view of bacterial diversity. J. Bacteriol. 180: 4765-4774.

Hungate, R.E. 1969. A roll tube method for cultivation if strict anaerobes. In: Methods in Microbiology. J.R. Norris, and D.W. Ribbons, eds. Academic Press, New York, New York. p.117-132.

Imhoff, J.F. 1984. Quinones of phototrophic purple bacteria. FEMS Microbiol. Lett. 25: 85-89.

Imhoff, J.F., and Bias-Imhoff, U. 1995. Lipids, quinones and fatty acids of anoxygenic phototrophic bacteria. In: Anoxygenic Photosynthetic Bacteria. R. E. Blankenship, M.T. Madigan, and C.E. Bauer, eds. Kluwer Academic Publishers, Dordrecht, The Netherlands, p. 179-205

Imhoff, J.F. 2001a. The phototrophic alpha-Proteobacteria. In: The Prokaryotes. Release 3.5 The Anaerobic Way of Life. In: The Prokaryotes: An Evolving Electronic Resource for the Microbiological Community, 3rd edition, release 3.6. M. Dworkin, S. Falkow, E. Rosenberg, K.-H. Schleifer, and E. Stackebrandt, eds. Springer-Verlag, New York, New York.

Imhoff, J.F. 2001b. The Phototrophic Beta-Proteobacteria In: The Prokaryotes. Release 3.6 The Anaerobic Way of Life. In: The Prokaryotes: An Evolving Electronic Resource for the Microbiological Community, 3rd edition, release 3.3. M. Dworkin, S. Falkow, E. Rosenberg, K.-H. Schleifer, and E. Stackebrandt, eds. Springer-Verlag, New York, New York.

Jeanthon, C., L'Haridon, S., Cueff, V., Banta, A., Reysenbach, A.-L., and Prieur, D. 2002. *Thermodesulfobacterium hydrogeniphilum* sp. nov., a thermophilic, chemolithoautotrophic, sulfate reducing bacterium isolated from a deep-sea hydrothermal vent at Guaymas Basin, and emendation of the genus *Thermodesulfobacterium* Int. J. Syst. Evol. Microbiol. 52: 765-772.

Kämpf, C., and Pfennig, N. 1986. Isolation and characterization of some chemoautotrophic *Chromatiaceae*. J. Basic Microbiol. 9: 507-515.

Kerster, K., De Vos, P., Gillis, M., Swings, J., Vandamme, P., and Stackebrandt, E. 2003. Introduction to the *Proteobacteria*. In: The Prokaryotes. Release 3.5 The Anaerobic Way of Life. In: The Prokaryotes: An Evolving Electronic Resource for the Microbiological Community, 3rd edition, release 3.12. M. Dworkin, S. Falkow, E. Rosenberg, K.-H. Schleifer, and E. Stackebrandt, eds. Springer-Verlag, New York, New York.

Klenk, H.-P., Meier, T.-D., Durovic, P., Schwass, V., Lottspeich, F., Dennis, P.P., and Zillig, W. 1999. RNA polymerase of *Aquifex pyrophilus*: Implications for the evolution of the bacterial *rpoBC* operon and extremely thermophilic bacteria. J. Molec. Evol. 48: 528-541.

Lapage, S.P., Sneath, P.H.A., Lessel, E.F., Skerman, V.B.D., Seeliger, H.P.R., and Clark, W.A. 1992. International Code of Nomenclature of Bacteria (1990 Revision). Bacteriological Code. American Society for Microbiology, Washington, D.C.

Lawson, P.A., Falsen, E., Inganäs, E., Weyant, R.S., and Collins, M.D. 2002. *Dysgonomonas mosii* sp. nov., from human sources. Syst. Appl. Microbiol. 25: 194-197.

Leite, A.A., Saddi-Ortega, L., Andrade, F.A.R.S., Macedo, A.M., Carvalho, M.A.R., Petrillo-Teixoto, M.L. and Moreira, E.S.A. 1999. Analysis of genetic polymorphism in *Actinobacillus actinomycetemcomitans*. Anaerobe. 5: 149-152

Liesack, W., Bak, F., Kreft, J.-U. and Stackebrandt, E. 1994. *Holophaga foetida* gen. nov., sp. nov., a new homoacetogenic bacterium degrading methoxylated aromatic compounds. Arch. Microbiol. 162: 85-90.

Liesack, W., Söller, R., Stewart, T., Haas, H., Giovannoni, S., and Stackebrandt, E. 1992. The influence of tachytelically evolving sequences on the topology of phylogenetic trees - intrafamily relationships and the phylogenetic position of *Planctomycetaceae* as revealed by comparative analysis of 16S ribosomal RNA sequences. Syst. Appl. Microbiol. 15: 357-362.

Ludwig, W., Strunk, O., Westram, R., Richter, L., Meier, H., *et al*. 2004. ARB: a software environment for sequence data. Nucl. Acids Res. 32: 1363-1371.

Macy, J.M., Nunan, K., Hagen, K.D., Dixon, D.R., Harbour, P.J., Cahill, M. and Sly, L.I. 1996. *Chrysiogenes arsenatis* gen. nov., sp. nov., a new arsenate-respiring bacterium isolated from gold mine wastewater. Int. J. Syst. Bacteriol. 46: 1153-1157.

Madigan, M.T., Martinko, J.M. and Parker, J. 2000. Brock Biology of Microorganisms, 9th ed. Prentice-Hall, Inc., Englewood Cliffs, New Jersey.

Margulis, L. 1970. Origin of the eukaryotic cells. Yale University Press, New Haven, Connecticut.

Margulis, L. 1981. Symbiosis in Cell Evolution. W.H. Freeman and Company. San Francisco.

Martin, W., and Mueller; M. 1998. The hydrogen hypothesis for the first eukaryote. Nature. 392: 37-41.

McCartney, A.L. 2002. Application of molecular biological methods for studying probiotics and the gut flora. Br. J. Nutr. Suppl. 1: 29-37.

McOrist, S., Gebhart, C.J., Boid, R., and Barns, S.M. 1995. Characterization of *Lawsonia intracellularis* gen. nov., sp. nov., the obligately intracellular bacterium of porcine proliferative enteropathy. Int. J. Syst. Bacteriol. 45: 820-825.

McSweeny, C.S., Allison, M.J., and Mackie, R.I. 1993. Amino acid utilization by the ruminal bacterium *Synergistes jonesii* strain 78-1. Arch. Microbiol. 159:131-135.

Mireshnichenko, M.L., L' Haridon, S., Schumann, P., Spring, S., Bonch-Osmolovskaya, E.A., Jeanthon, C., and Stackebrandt, E. 2003. *Caminibacter profundus* sp. nov., isolated from a deep-sea hydrothermal vent represents a novel thermophilic bacterium of the *Nautiliales* ord. nov. within the class Epsilonproteobacteria. Int. J. Syst. Evol. Microbiol. 53: In press.

Murray, R.G.E. 1984. The higher taxa, or, a place for everything...? In: Bergey's Manual of Systematic Bacteriology, vol. 1: N.R. Krieg and J.G. Holt, eds. The Williams and Wilkins Co., Baltimore, Maryland. p. 31-34.

Myers, C.R. and Nealson, K.H. 1990. Respiration-linked proton translocation coupled to anaerobic reduction of manganese (IV) and iron (III) in *Shewanella putrefaciens* MR-1. J. Bacteriol. 172: 6232-6238.

Nelson, K.E., Clayton, R.A., Gill, S.R., Gwinn, M.L., Dodson, R.J. *et al*. 1999. Evidence for lateral gene transfer between Archaea and Bacteria from genome sequence of *Thermotoga maritima*. Nature. 399: 323-329.

Nelson, K.E., Eisen, J.A., and Fraser, C.M. 2001. *Genome of Thermotoga maritima MSB8* In: Methods in Enzymology. M. W. W. Adams, and R. M.Kelly, eds. Academic Press, San Diego, California, p. 169-180.

Nomenclature-up-to-date. 2003. DSMZ-German collection of Microorganisms and Cell Cultures, GmbH (www.dsmz.de), bimonthly update.

Nübel, U., Bateson, M.M., Vandieken, V., Wieland, A., Kuhl, M., and Ward, D.M. 2002. Microscopic examination of distribution and phenotypic properties of phylogenetically diverse *Chloroflexaceae*-related bacteria in hot spring microbial mats. Appl. Environ. Microbiol. 68: 4593-4603.

Ochman, H., Lawrence, J.G., and Groisman, E.A. 2000. Lateral gene transfer and the nature of bacterial innovation. Nature. 405: 299-304.

Olsen, I., Paster B.J. and Dewhirst., F.E. 2000. Taxonomy of spirochetes. Anaerobe. 6: 39-57.

Oparin, A. I. 1924. *The Origin of Life,* reprinted as an appendix to Bernal, J. D. 1967. *The Origin of Life,* Weidenfeld and Nicholson, London.

Overmann, J. 2000. The Family *Chlorobiaceae*. In: The Prokaryotes: An Evolving Electronic Resource for the Microbiological Community, 3rd edition, release 3.1. M. Dworkin, S. Falkow, E. Rosenberg, K.-H. Schleifer, and E. Stackebrandt, eds. Springer-Verlag, New York, New York.

Palm, P., Schleper, C., Arnold-Ammer, I., Holz, I., Meier, T., Lottspeich, F., and Zillig, W. 1993. The DNA-dependent RNA-polymerase of *Thermotoga maritima;* characterization of the enzyme and the DNA-sequence of the genes for the large subunits. Nucl. Acids Res. 21: 4904-4908.

Paster, B.J., Ludwig, W., Weisburg, W.G., Stackebrandt, E., Reichenbach, H., Hespell, R.B., Hahn, C.M., Gibson, J., Stetter, H.O. and Woese, C.R. 1985. A phylogenetic grouping of the bacteriodes, cytophages and certain flavobacteria. System. Appl. Microbiol. 6: 34-42.

Pasteur, L. 1861. Animalcules infusoires vivant sans gaz oxygene libre et determinant des fermentations. Compt. Rend. Acad. Sci. (Paris) 52: 344-347.

Pfennig, N., and Trüper, H. G. 1971. Higher taxa of the phototrophic bacteria. Int. J. Syst. Bacteriol. 21: 17-18.

Rabus, R., Hansen, T., and Widdel, F. 2001. Dissimilatory sulfate- and sulfur-reducing prokaryotes. In: The Prokaryotes: An Evolving Electronic Resource for the Microbiological Community, 3rd edition, release 3.3. M. Dworkin, S. Falkow, E. Rosenberg, K.-H. Schleifer, and E. Stackebrandt, eds. Springer-Verlag, New York, New York.

Rainey, F.A., and Stackebrandt, E. 1993. *Thermoanaerobacter acetoethylicus* comb. nov. (*Thermobacteroides acetoethylicus* Approved List No 9, 1982) and designation of *Coprothermobacter proteolyticus* gen. nov. Int. J. Syst. Bacteriol. 43: 857-859.

Rautio, M., Eerola, E., Väisänen-Tunkelrott, M.L., Molitoris, P., Collins, M.D., and Jousimies-Somer, H. 2003. Reclassification of *Bacteroides putredinis* (Weinberg *et al.*, 1937) in a new genus *Alistipes* gen. nov., as *Alistipes putredinis* comb. nov., and description of *Alistipes finegoldii* sp. nov., from human sources. Syst. Appl. Microbiol. 26: 182-188.

Robertson, K.L., Drucker, D.B., Blinkhorn, A.S., and Davies, R.M. 1999. A comparison of techniques used to distinguish strains of *Prevotella intermedia* from *Prevotella nigrescens*. Anaerobe. 5: 119-122.

Saiki, T., Kobayashi, Y., Kawagoe, K., and Beppu, T. 1985. *Dictyoglomus thermophilum* gen. nov., sp. nov., a chemoorganotrophic, anaerobic, thermophilic bacterum. Int. J. Syst. Bacteriol. 35: 253-259.

Santos, A., Abdo, M.C.B., Anacleto, C., Farias, L.M., Carvalho, M.A.R., Moreira, E.S.A., and Petrillo-Peixoto, M.L. 1999. Diversity of 16S rDNA restriction fragment length polymorphism profiles of bacteroides species isolated from human and marmosets. Anaerobe. 5: 145-148.

Schaal, K.P. 1992. The Genera *Actinomyces, Arcanobacterium,* and *Rothia*. In: The Prokaryotes. A Handbook on the Biology of Bacteria: Ecophysiology, Isolation, Identification, Applications. Balows, A., Trüper, H.G., Dworkin, M., Harder, W. and Schleifer, K.H. Springer-Verlag, New York, p. 850-905.

Schidlowski, M., Appel, P.W.U., Eichmann, R., and Junge, C.E. 1979. Carbon isotope geochemistry of the 3.7 x 10^9 year old Isua sediments. Geochim. Cosmochim. Acta. 43: 189-199.

Schink, B. 1992. The genus *Pelobacter*. 1991. In: The Prokaryotes. A. Balows, H.G. Trüper, M. Dworkin, W. Harder and K.H. Schleifer, eds. Springer, New York. p. 3393-3399.

Schleifer, K.H., and Ludwig, W. 1994. Molecular taxonomy : classification and identification. In: Bacterial Diversity and Systematics. F.G. Priest, and M. Goodfellow, eds. Plenum Press, New York, New York. p. 1-15.

Schmitz, R.A., Daniel, R., Deppenmeier, U., and Gottschalk, G. 2001. The anaerobic way of life. In: The Prokaryotes: An Evolving Electronic Resource for the Microbiological Community, 3rd edition, release 3.5. M. Dworkin, S. Falkow, E. Rosenberg, K.-H. Schleifer, and E. Stackebrandt, eds. Springer-Verlag, New York, New York.

Schopf, J.W. 1975. Biostratigraphic usefulness of stromatolite Prekambrian microbiotas: a preliminary analysis. Precambrian Res. 5: 143-175.

Schopf, J.W., and Barghoorn, E.S. 1967. Algal-like fossils from the early Precambrian of South Africa. Science. 156: 508-512.

Schütz, M., Brugna, M., Lebrun, E., Baymann, F., Huber, and others. 2000. Early evolution of cytochrome bc complexes. J. Mol. Biol. 300: 663-675.

Schwartz, E., and Friedrich, B. 2003. The H_2-metabolizing prokaryotes. In: The Prokaryotes: An Evolving Electronic Resource for the Microbiological Community, 3rd edition, release 3.14. M. Dworkin, S. Falkow, E. Rosenberg, K.-H. Schleifer, and E. Stackebrandt, eds. New York, New York.

Schwartz, R.M. and Dayhoff, M.O. 1978. Origins of prokaryotes, eukaryotes, mitochondria and chloroplasts. Science. 199: 395-403.

Schwieger, F., and Tebbe, C.C. 2000. Effect of field inoculation with *Sinorhizobium meliloti* L33 on the composition of bacterial communities in rhizospheres of a target plant (*Medicago sativa*) and a non-target plant (*Chenopodium album*)-linking of 16S rRNA gene-based single-strand conformation polymorphism community profiles to the diversity of cultivated bacteria. Appl. Environ. Microbiol. 66: 3556-3565.

Shah, H.N., and Collins, M.D. 1983. Genus *Bacteroides*: a chemotaxonomical perspective. J. Appl. Bact. 55: 403-416.

Shima, S., and Suzuki, K.I. 1993. *Hydrogenobacter acidophilus* sp. nov., a thermoacidophilic, aerobic, hydrogen-oxidizing bacterium requiring elemental sulfur for growth. Int. J. Syst. Bacteriol. 43: 703-708.

Skirnisdottir, S., Hreggvidsson, G.O., Holst, O., and Kristjansson J.K. 2001. A new ecological adaptation to high sulfide by a *Hydrogenobacter* sp. growing on sulfur compounds but not on hydrogen. Microbiol. Res. 156: 41-47.

Spring, S. Merkhoffer, B. Weiss, N. Kroppenstedt, R.M. Hippe H., and Stackebrandt, E. 2003. Characterization of novel psychrophilic clostridia from an Antarctic microbial mat: description of *Clostridium frigoris* sp. nov., *C. lacusfryxellense* sp. nov., *C. bowmanii* sp. nov. and *C. psychrophilum* sp. nov. and reclassification of *C. laramiense* as *C. estertheticum* subsp. laramiense subsp. nov. Int. J. Syst. Evol. Microbiol. 53: 1019-1029.

Stackebrandt, E., and Schumann, P. 2000. Introduction to the taxonomy of the class *Actinobacteria*. In: The Prokaryotes. An Evolving Electronic Resource for the Microbiological Community, 3rd edition, release 3.3. Dworkin, S., Falkow, M., Rosenberg, E., Schleifer K.-H. and Stackebrandt, E. (eds). Springer-Verlag, New York, New York.

Stackebrandt, E., Embley, M., and Weckesser, J. 1988. Evolutionary, phylogenetic and taxonomic aspects of phototrophic eubacteria. In: Proceedings of the 1st Symposium on Green Bacteria. J. Olson, J.G. Ormerod, J. Amesz, E. Stackebrandt, and H.G. Trüper, eds. Plenum Press, London. p. 210-216.

Stackebrandt, E., Frederiksen, W., Garrity, G. M., Grimont, P. A. D., Kämpfer, *et al.* 2002. Report of the ad hoc committee for the re-evaluation of the species definition in bacteriology. Int. J. Syst. Evol. Microbiol. 52: 1043-1052.

Stackebrandt, E., Kramer, I., Swiderski, J., and Hippe, H. 1999b. Phylogenetic basis for a taxonomic dissection of the genus *Clostridium*. FEMS Immunol. Med. Microbiol. 24: 253-258.

Stackebrandt, E., Pohla, H., Kroppenstedt, R., Hippe H., and Woese C.R. 1985. 16S rRNA analysis of *Sporomusa*, *Selenomonas,* and *Megasphaera*: on the phylogenetic origin of Gram-positive Eubacteria. Arch. Microbiol. 143: 270-276.

Stackebrandt, E., Rainey, F.A., and Ward-Rainey, N. 1996. Anoxygenic phototrophy across the phylogenetic spectrum: current understanding and future perspectives. Arch. Microbiol. 166: 211-223.

Stackebrandt, E., Rainey, F.A., and Ward-Rainey, N. L. 1997. Proposal for a new hierarchic classification system, *Actinobacteria* classis nov. Int. J. Syst. Bacteriol. 47: 479-491.

Stackebrandt, E., Tindall, B., Ludwig, W., and Goodfellow, M. 1999a. Prokaryotic diversity and systematics. In: Biology of the Prokaryotes. J.W. Lengeler, G. Drews, and H. Schlegel, eds. Thieme, Stuttgart. p. 674-722.

Stanier, R.Y. 1964. Towards a definition of the bacteria. In: The Bacteria: A Treatise on Structure and Function. I.C. Gunsalus, and R.Y. Stanier, eds. Academic Press, New York, New York. p. 445-462

Stanier, R.Y., and Van Niel, C.B. 1942. The concept of a bacterium. Arch. Microbiol. 42: 17-35.

Stanier, R.Y., Doudoroff, M., and Adelberg, E.A. 1970. The Microbial World, 3rd ed. Prentice-Hall Engelwood Cliffs, New Jersey.

Steudel, R., Holdt, G. Visscher, P. T., and van Gemerden, H. 1990. Search for polythionates in cultures of *Chromatium vinosum* after sulfide incubation. Arch. Microbiol. 153: 432-437.

Straub, K.L., Rainey, F.A., and Widdel, F. 1999. *Rhodovulum iodosum* sp. nov. and *Rhodovulum robiginosum* sp. nov., two new marine phototrophic ferrous-iron-oxidizing purple bacteria. . Int. J. Syst. Bacteriol. 49: 729-735.

Tanaka, K., Nakamura, K., and Mikami, E. 1990. Fermentation of malate by a Gram-negative strictly anaerobic non-spore former, *Propionivibrio dicarboxylicus* gen. nov., sp. nov. Arch . Microbiol. 154: 323-328.

Thompson, F.L., Hoste, B., Thompson, C.C., Goris, J., Gomez-Gil, B., Huys, L., de Vos, P., and Swings, J. 2002. *Enterovibrio norvegicus* gen. nov., sp. nov., isolated from the gut of turbot (*Scophthalmus maximus*) larvae: a new member of the family *Vibrionaceae*. Int. J. Syst. Evo. Microbiol. 52: 2015-2022.

Tiboni, O., Cantoni, R., Creti, R., Cammarano, P. and Sanangelantoni, A.M. 1991. Phylogenetic depth of *Thermotoga maritima* inferred from analysis of the fus gene: Amino acid sequence of elongation factor G and organization of the *Thermotoga* str operon J. Mol. Evol. 33: 142-151.

Van de Peer, Y., Neefs, J.-M., de Rijk, P., De Vos, P., and de Wachter, R. 1994. About the origin of divergence of the major bacterial taxa during evolution. Syst. Appl. Microbiol. 17: 32-38.

Van Gemerden, H. 1974. Coexistence of organisms competing for the same substrate: An example among the purple sulfur bacteria. Microb. Ecol. 1: 19-23.

Venkateswaran, K., Moser, D.P., Dollhopf, M.E., Lies, D.P., Saffarini, D.A. and others. 1999. Polyphasic taxonomy of the genus *Shewanella* and description of *Shewanella oneidensis*. Int. J. Syst. Bacteriol. 49: 705-724.

Verbarg, S., Rheims, H., Emus, S., Frühling, A., Kroppenstedt, R., Stackebrandt, E. and Schumann, P. 2004. *Erysipelothrix inopinata* sp. nov., isolated in the course of sterile filtration of vegetabile peptone broth and description of *Erysipelothrichaceae* fam. nov. Int. J. Syst. Evol. Microbiol. 54: 221-225.

Wächtershauser, G. 1988. Before enzymes and templates: theory of surface metabolism. Microbiol. Rev. 52: 452-484.

Wächtershauser, G. 1990. The case for the chemo-autotrophic origin of life in an iron-sulfur world. Origins of Life and Evolution of the Biosphere. 20: 173-176.

Wächtershauser, G. 2001. Origin of Life: RNA world versus autocatalytic anabolism. In: The Prokaryotes. An Evolving Electronic Resource for the Microbiological Community, 3rd edition, release 3.8. M. Dworkin, S. Falkow, E. Rosenberg, K.-H. Schleifer, and E. Stackebrandt, eds. Springer-Verlag, New York, New York.

Walker, J.C.G. 1977. Evolution of the Atmosphere. Macmillan, New York.

Weber, A.S. 2000. Nineteenth Century Science. Broadview Press, Peterborough, Ontario, Canada.

Weckesser, J., Drews, G., and Mayer, H. 1979. Lipopolysaccharides of photosynthetic procaryotes. Ann. Rev. Microbiol. 33: 215-239.

Weckesser, J., Mayer, H., and Schulz, G. 1995. Anoxygenic phototrophic bacteria: Model organisms for studies on cell wall macromolecules. In: Anoxygenic Photosynthetic Bacteria. R.E. Blankenship, M.T. Madigan, and C.E. Bauer, eds. Kluwer Academic Publishing Dordrecht, The Netherlands. p. 207-230.

Widdel, F., and Bak, F. 1992. Gram-negative mesophilic sulfate-reducing bacteria. In: The Prokaryotes, 2nd ed. A. Balows, H. G. Trüper, M. Dworkin, W. Harder, and K.-H. Schleifer. Springer-Verlag, New York, New York. p. 3352-3378.

Widdel, F., and Pfennig, N. 1984. Dissimilatory sulfate- and sulfur-reducing bacteria. In: Bergey's Manual of Systematic Bacteriology. N.R. Krieg, and J.G. Holt, eds. Williams and Wilkins, Baltimore, Maryland. p. 663-679.

Witt, D., Bergstein-Ben Dan, T., and Stackebrandt, E. 1989. Nucleotide sequence of 16S rRNA and phylogenetic position of the green sulfur bacterium *Clathrochloris sulfurica*. Arch. Microbiol. 152: 206-208.

Woese, C.R. 2003. How we do, don't and should look at bacteria and bacteriology. In: The Prokaryotes. An Evolving Electronic Resource for the Microbiological Community, 3rd edition, release 3.14. M. Dworkin, S. Falkow, E. Rosenberg, K.-H. Schleifer, and E. Stackebrandt, eds. Springer-Verlag, New York, New York.

Woese, C.R., Debrunner-Vossbrinck, B.A., Oyaizu, H., Stackebrandt E., and Ludwig W. 1985a. Gram-positive bacteria: possible photosynthetic ancestry. Science. 229: 762-765.

Woese, C.R., Kandler, O., and Wheelis, M.L. 1990. Towards a natural system of organisms: proposal for the domains Archaea, Bacteria and Eucaryas. Proc. Natl. Acad. Sci. USA. 87: 4576-4579.

Woese, C.R., Stackebrandt, E., Weisburg, W.G., Paster, B J., Madigan, M.T., Blanz, P., Gupta, R., Fowler, V.J., Hahn, C.M., and Fox G.E. 1984a. The phylogeny of purple bacteria: The alpha subdivision. System. Appl. Microbiol. 5: 315-326.

Woese, C.R., Weisburg, W., Hahn, C.M., Paster, B.J., Zablen, L.B., Lewis, B.J., Ludwig W., and Stackebrandt, E. 1985b. The phylogeny of purple bacteria: The gamma subdivision. System. Appl. Microbiol. 6: 25-33.

Woese, C.R., Weisburg, W.G., Paster, B.J., Hahn, C.M., Koops, H.P., Harms, H., and Stackebrandt, E. 1984b. The phylogeny of purple bacteria: The beta subdivision. System. Appl. Microbiol. 5: 327-336.

Yabuuchi, E., Kosako, Y., Oyaizu, H., Yano, I, Hotta, H., Hashimoto, Y., Ezaki, T., and Arakawa, M. 1992. Proposal of *Burkholderia* gen. nov. and transfer of seven species of the genus *Pseudomonas* homology group II to the new genus, with the type species *Burkholderia cepacia* (Palleroni and Holmes 1981) comb. nov. Microbiol. Immunol. 36: 1251-1275.

Yabuuchi, E., Kosako, Y., Yano, I, Hotta, H., and Nishiuchi, Y. 1992. Transfer of two *Burkholderia* and an *Alcaligenes* species to *Ralstonia* gen. nov.: proposal of *Ralstonia pickettii* (Ralston, Palleroni and Doudoroff 1973) comb. nov., *Ralstonia solanacearum* (Smith 1896) comb. nov. and *Ralstonia eutropha* (Davis 1969) comb. nov. Microbiol. Immunol. 39: 897-904.

Zhang, H.. Sekiguchi, Y., Hanada, Y., Hugenholtz, P., Kim, H., Kamagata, Y., and Nakamura, K. 2003. *Gemmatimonas aurantiaca* gen. nov., sp. nov., a Gram-negative, aerobic, polyphosphate-accumulating micro-organism, the first cultured representative of the new bacterial phylum *Gemmatimonadetes* phyl. nov. Int. J. Syst. Evol. Microbiol. 53: 1155-1163.

Chapter 2

An Overview of Anaerobic Metabolism

Michael J. McInerney* and Lisa M. Gieg

Abstract

In many anaerobic ecosystems, a consortium of microorganisms rather than a single species is the catalytic unit responsible for biodegradation. Interspecies hydrogen and formate transfer are critical in regulating the flow of carbon and electrons in many anaerobic ecosystems. Anaerobes use novel approaches such as the addition of fumarate or a carboxyl group to activate hydrocarbons, the hydrolysis of ATP to provide low potential electrons to reduce aromatic rings, and oxidation-reduction or cofactor B_{12}-mediated reactions to generate free radicals for dehydration reactions. In addition to fermentative metabolism where ATP production occurs mainly by substrate-level phosphorylation, anaerobes can use diverse inorganic or organic compounds as electron acceptors. Anaerobic respiratory chains use redox loops, redox-driven pumps, and the separation of proton-consuming and proton-producing reactions across the membrane to generate a chemiosmotic potential. Some anaerobes can generate a chemiosmotic potential by the electrogenic efflux of compounds across the membrane. The dependence of anaerobes on low potential and free radical biochemistry makes them very sensitive to oxygen. The degree to which an anaerobe depends on oxygen-sensitive systems relative to its ability to detoxify reactive oxygen compounds determines its oxygen tolerance.

1. Introduction

In strictly anaerobic ecosystems, the mineralization of carbon substrates proceeds in a step-wise manner involving several metabolic groups of bacteria. This is in contrast to metabolism under aerobic and denitrifying conditions where a single species is usually able to mineralize a compound when the electron acceptor (e. g., oxygen or nitrate) is in excess. The metabolic interactions between anaerobic species are often tightly coupled into consortia that act as single catalytic units. The degree of mutual interdependence between the different metabolic groups depends on the genetic capabilities of the respective organisms and the constraints that kinetics and thermodynamics place on key reactions. For some interactions, energy limitations are such that neither partner can operate without the activity of the other organism. Such consortia are more than just a convenient mutualism between organisms; they are thermodynamically-based.

Given the constraints that are placed on anaerobic metabolism by the lack of a high potential electron acceptor such as oxygen, the energy yields of anaerobic microorganisms are often controlled by the mechanisms that anaerobic bacteria use to reoxidize their reduced cofactors. For example, many anaerobic bacteria reoxidize their reduced cofactors by hydrogen production. However, the production of hydrogen from pyridine nucleotide or flavin cofactors is only favorable when the hydrogen concentration is at a very low level (McInerney, 1999). Terminal electron-accepting processes such as methanogenesis or sulfate reduction function to maintain low hydrogen levels and these processes are key in controlling carbon and electron flow in anaerobic ecosystems.

As discussed in Chapter 1, the diversity of anaerobic microorganisms is overwhelming. So, it is a difficult task to provide a succinct overview of the diverse mechanisms by which these various anaerobes obtain energy for growth. It is not possible to adequately review all of the novel features of anaerobic metabolism and some topics such as anaerobic photosynthesis and denitrification have been omitted due to space limitations. We will first provide an overview of different anaerobic ecosystems and the role of thermodynamics in regulating carbon flow and community structure. The unique aspects of anaerobic cultivation and nutrition will also be discussed, emphasizing the importance of metals such as selenium, tungsten, molybdenum, iron, copper, and nickel in key anaerobic processes. The lack of oxygen presents a challenge to anaerobes in the degradation of hydrocarbons, aromatic compounds, and compounds with unactivated hydroxyl groups. We will find that anaerobes have evolved novel approaches to degrade these compounds. Anaerobes generate energy both by substrate-level phosphorylation and chemiosmotic mechanisms. Once we have reviewed the

*For correspondence email mcinerney@ou.edu.

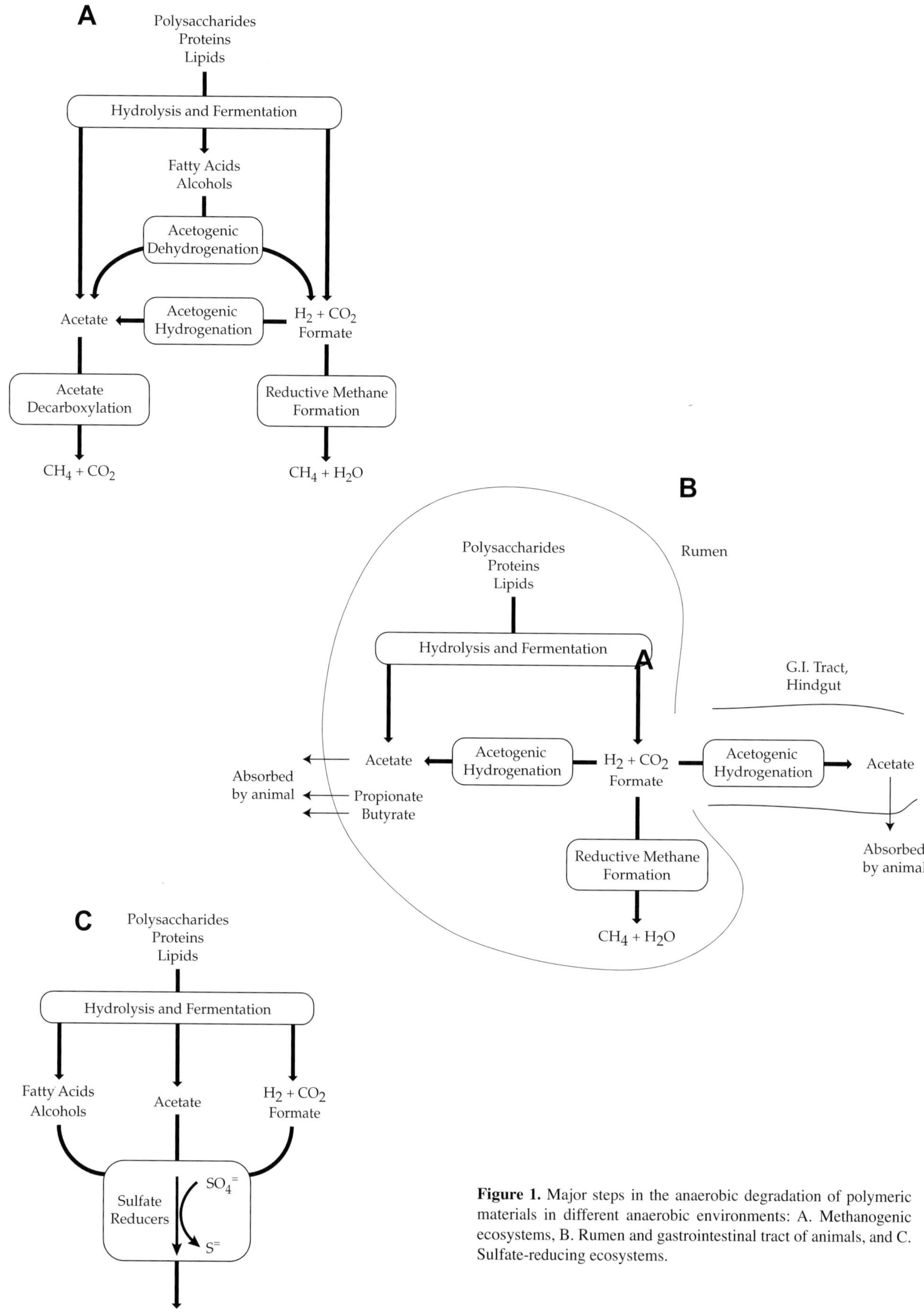

Figure 1. Major steps in the anaerobic degradation of polymeric materials in different anaerobic environments: A. Methanogenic ecosystems, B. Rumen and gastrointestinal tract of animals, and C. Sulfate-reducing ecosystems.

reactions by which anaerobes obtain their energy, we will discuss the physiological bases that make anaerobes sensitive to oxygen. For more detailed information on these topics, the reader is referred to the on-line version of *The Prokaryotes* (Dworkin, 1999) that contains a number of timely reviews on anaerobic metabolism. Unless otherwise indicated, the Gibbs free energy values and redox potentials are from Thauer *et al.* (1977), a review which also provides an excellent discussion of the mechanism by which anaerobes obtain energy for growth.

2. Carbon Flow and Its Regulation in Anaerobic Ecosystems

This section will discuss the metabolic groups involved in anaerobic degradation in several environments to illustrate the importance of hydrogen and formate in regulating anaerobic metabolism. The molecular biology and biochemistry of hydrogen use and production by hydrogenases will not be discussed, but excellent reviews of these topics are available (Wu and Mandrand, 1993; Schwartz and Friedrich, 2003).

2.1. Methanogenic Ecosystems

In methanogenic ecosystems, organic matter is degraded to CH_4 and CO_2 (McInerney, 1999). Methanogenesis predominates in environments that lack light and have limited amounts of potential terminal electron acceptors such as oxygen, ferric iron, or oxyanions of nitrogen and sulfur. Examples of such environments include freshwater and some marine sediments, flooded soils, wet wood of trees, tundra, landfills, and sewage digestors. Figure 1A illustrates the main steps involved in the conversion of complex polymeric substrates into methane and carbon dioxide (McInerney and Bryant, 1981). Fermentative bacteria hydrolyze polysaccharides such as cellulose, hemicellulose, pectin, and starch to lower molecular weight materials such as sugars and oligosaccharides (uronic acids and methanol are also formed from pectin) (Ljungdahl and Eriksson, 1985; Felix and Ljungdahl, 1993; Beguin and Aubert, 1994; Leschine, 1995; Warren, 1996; Schwarz, 2001). These compounds are then fermented to acetate and longer chain fatty acids, CO_2, formate, and H_2. Acetogeneic bacteria O-demethylate pectins and low molecular weight ligneous materials and ferment hydroxylated and methoxylated aromatic compounds with the production of acetate (Dolfing, 1988; Frazer, 1994). Proteins are hydrolyzed to peptides and amino acids that are fermented to the above products as well as to other organic compounds (McInerney, 1988). Isobutyrate, isovalerate, and D-2-methylbutyrate are produced from the branched-chain amino acids, valine, leucine, and isoleucine, respectively. Phenylacetate, phenylpropionate, indole, and indolacetate are produced from aromatic amino acids. Glycerides, phospholipids, and other fats are hydrolyzed to long-chain fatty acids, glycerol, and sugars; only the latter two are substrates for fermentative bacteria (McInerney, 1988; Mackie *et al.*, 1991).

A second group of fermentative bacteria, usually called syntrophic or proton-reducing acetogenic bacteria, degrades propionate and longer chain fatty acids, alcohols, some amino acids, hydrocarbons, and aromatic compounds to the methanogenic substrates, H_2, formate, and acetate. The degradation of these compounds with H_2 or formate production is thermodynamically unfavorable unless the H_2 and formate concentrations are kept low by H_2/formate-using bacteria such as methanogens (McInerney and Bryant, 1981). Such thermodynamically-based, microbial associations are called syntrophic associations. The final step in methanogenesis involves two different groups of methanogens, the hydrogenotrophic methanogens that use the H_2 and formate produced by other bacteria to reduce CO_2 to CH_4 and the acetoclastic methanogens that metabolize acetate to CO_2 and CH_4.

2.2. Regulation of Carbon Flow

The concentrations of H_2 and formate play an important role in regulating the kinds and amounts of products made by fermentative bacteria. Many fermentative bacteria have the ability to reoxidize NADH (nicotinamide adenine dinucleotide) by reducing protons to H_2 (equation 1) (Wolin, 1982).

$$NADH + H^+ \rightarrow NAD^+ + H_2\ ,\ \Delta G^{o\prime} = +\ 18\ kJ/mol \quad (1)$$

This reaction is favorable only when the partial pressure of H_2 is low, about 100 Pa (Wolin, 1982). The regulation of electron flow during glucose metabolism by fermentative bacteria is illustrated in Figure 2. Pyruvate: ferredoxin oxidoreductase converts pyruvate into acetyl-Coenzyme A (CoA), CO_2, and reduced ferredoxin. The activity of the enzyme is inhibited by high ratios of reduced to oxidized ferredoxin (Tewes and Thauer, 1980). NADH: ferredoxin oxidoreductase catalyzes the reduction of ferrredoxin with NADH (Thauer *et al.*, 1969; Jungermann *et al.*, 1973; Tewes and Thauer, 1980). The latter enzyme requires acetyl-CoA for activity and CoA competitively inhibits activation by acetyl-CoA. At low H_2 partial pressures, acetyl-CoA levels should be high and most of the ferredoxin should be oxidized. This will allow pyruvate to be metabolized almost exclusively to acetate, CO_2, and H_2 (Figure 2) (McInerney and Bryant, 1981). Consistent with this concept, when radioactive glucose was added to well-operated anaerobic reactors, it was almost exclusively metabolized to acetate with little or no radioactive carbon detected in other intermediates such as propionate, butyrate, or lactate (Zinder, 1986).

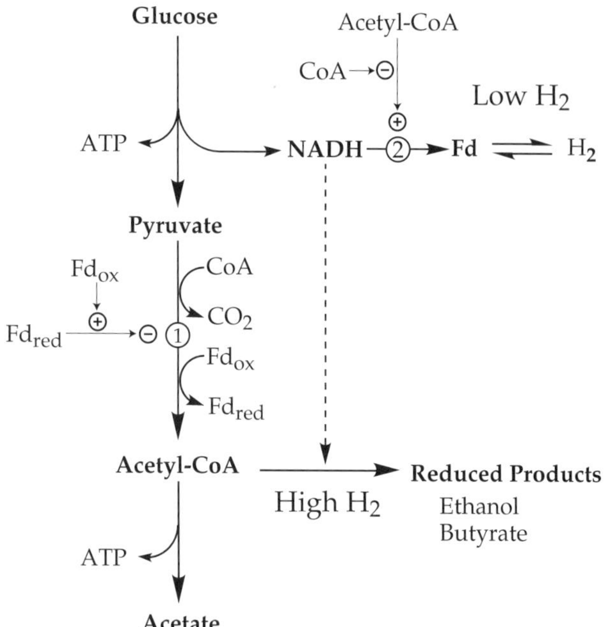

Figure 2. Regulation of end-product formation of fermentative bacteria. Symbols: minus and plus signs indicate enzyme inhibition or activation, respectively; Fd, ferredoxin. (adapted from Tewes and Thauer, 1980).

At high H_2 partial pressure, as often occurs with shock loads or when methanogenesis is inhibited (Mosey and Fernandes, 1989; Strong and Cord-Ruwisch, 1995; Cord-Ruwisch *et al.*, 1997; Guwy *et al.*, 1997), the levels of CoA and reduced ferredoxin should be high and H_2 formation from NADH will be inhibited (Figure 2). Pyruvate is then used to form reduced products such as ethanol or butyrate. Numerous studies show that little or no ethanol and lactate, much less propionate and butyrate, and more acetate and CO_2 are made when fermentative organisms are grown in association with H_2/formate-using bacteria (Bauchop and Mountfort, 1981; McInerney and Bryant, 1981; Wolin, 1982; Stams, 1994).

Some fermentative anaerobes have the ability to produce formate by coupling the reduction of CO_2 to the oxidation of NADH or reduced ferredoxin (Miller and Wolin, 1973). This reaction could be used in place of H_2 production from pyridine nucleotide-linked or ferredoxin-linked hydrogenase reactions. Here, the reducing equivalents generated in fermentative metabolism are transferred to methanogens in the form of formate rather than H_2. This would provide an ecologically relevant explanation for the widespread ability of methanogens to use formate and would allow fermentative bacteria that lack hydrogenases to optimize their energy gain from glucose.

The thermodynamic basis of syntrophic degradation of propionate, butyrate, and benzoate, important intermediates in anaerobic degradation, is illustrated in Table 1. Under standard conditions, the degradation of these compounds is endogonic. However, if the hydrogen partial pressure is low, about 1 Pa in this example, then the degradation of these compounds is exogonic. Consistent with the thermodynamic predictions, small changes in H_2 partial pressure inhibit the degradation of butyrate and benzoate by syntrophic cocultures (Ahring and Westermann, 1988; Dwyer *et al.*, 1988; Schink, 1997) and propionate degradation in methanogenic mixed cultures (Smith and McCarty, 1989).

The above examples illustrate the important role that hydrogen-using microorganisms such as methanogens play in anaerobic ecosystems. The rapid use of hydrogen by methanogens and other hydrogen-using bacteria maintains hydrogen at very low levels, even though large amounts of hydrogen are produced. The favorable thermodynamics of H_2 use by methanogens (Table 1) allows them to metabolize H_2 to very low partial pressures. Methanogens can metabolize H_2 at partial pressures ranging from 3-7 Pa (Lovley, 1985; Cord-Ruwisch *et al.*, 1988). Methanogens also have a high affinity for H_2 use, with apparent K_m values in the range of 5-13 μM (670-1700 Pa) (Robinson and Tiedje, 1984; Ward and Winfrey, 1985; Zinder, 1993). With H_2 partial pressures ranging from 2 to 1200 Pa in digestors (Mosey and Fernandes, 1989; Strong and Cord-Ruwisch, 1995; Cord-Ruwisch *et al.*, 1997; Guwy *et al.*, 1997), hydrogenotrophic methanogenesis is undersaturated with respect to H_2 and increases in H_2 production are compensated for by corresponding increases in the H_2 consumption while maintaining nearly constant hydrogen partial pressures (Shea *et al.*, 1968; Kaspar and Wuhrmann, 1978; Strayer and Tiedje, 1978; Robinson and Tiedje, 1982).

While much of the above discussion focused on the role of H_2 in syntrophic metabolism, formate is also an important interspecies intermediate in syntrophic metabolism. Interspecies formate transfer has been demonstrated in the syntrophic degradation of amino acids (Zindel *et al.*, 1988). Syntrophic propionate degradation by *Syntrophobacter fumaroxidans* (Dong *et al.*, 1994a; Dong and Stams, 1995) and butyrate degradation by *Syntrophospora bryantii* (Dong *et al.*, 1994b) required a partner that used both hydrogen and formate. Formate dehydrogenase levels were very high in both members of the syntrophic association consistent with electron flow coupled to interspecies formate transfer (de Bok *et al.*, 2002). Radioisotopic studies and thermodynamic considerations indicated that ethanol degradation involves interspecies formate transfer in methanogenic granules (Thiele and Zeikus, 1988). Analysis of flux rates of H_2 and formate indicated that the rate of H_2 diffusion was too slow to account for the rate of syntrophic propionate and butyrate degradation (Boone *et al.*, 1989; de Bok *et al.*, 2002).

Temperature and pH are other important factors that regulate carbon and electron flow in

Table 1. Reactions involved in syntrophic metabolism[a]

Reactions	ΔG°' (kJ/mol)	ΔG' [b] (kJ/mol)
Methanogenic H_2 Consumption		
$4\ H_2 + HCO_3^- + H^+ \rightarrow CH_4 + 3\ H_2O$	-135.6	-16.4
Syntrophic Metabolism		
$Propionate^- + 3\ H_2O \rightarrow Acetate^- + HCO_3^- + H^+ + 3\ H_2$	+76.1	-16.7
$Butyrate^- + 2\ H_2O \rightarrow 2\ Acetate^- + H^+ + 2\ H_2$	+48.6	-31.7
$Benzoate^- + 7\ H_2O \rightarrow 3\ Acetate^- + HCO_3^- + 3\ H^+ + 3\ H_2$	+70.1	-60.3

[a] Calculated from the data in Thauer *et al.* (1977) with the free energy of formation for benzoate given in Kaiser and Hanselmann (1982).
[b] Calculated on the basis of the following conditions observed in methanogenic ecosystems: partial pressures of H_2 of 1 Pa and of CH_4 of 50 kPa, 50 mM bicarbonate, and the concentrations of the substrates at 0.1 mM each.

methanogenic environments. In Knaack Lake (pH 6.1), hydrogenotrophic methanogenesis did not occur and all of the electron flow went through acetogenesis (Phelps and Zeikus, 1984). Temperatures below 15°C greatly reduced the rate of methanogenesis in lake sediments, rice paddies, tundra soils, and anaerobic reactors (Conrad *et al.*, 1987; Kotsyurbenko *et al.*, 1996; Nachaiyasit and Stuckey, 1997; Rebac *et al.*, 1997; van Lier *et al.*, 1997). This is in part due to the fact that the optimal temperature for methanogenesis is often in the mesophilic range (28-35°C) (Boone *et al.*, 1993), but thermodynamic considerations also come into play. Acetogenesis becomes progressively more favorable compared to hydrogenotrophic methanogenesis as the temperature declines (Conrad and Wetter, 1990; Schink, 1997). Thus, the reason why H_2 turnover in lake sediments incubated at 4°C is due to acetogenesis rather than methanogenesis is due to both physiological and thermodynamic factors (Figure 1A).

2.3. Reverse Electron Transport

Hydrogen production from substrates such as propionate and butyrate is thermodynamically difficult even at low hydrogen concentrations (Thauer and Morris, 1984; Schink, 1997). Syntrophic butyrate metabolism occurs by β-oxidation (Wofford *et al.*, 1986). The oxidation of butyryl-CoA to crotonyl-CoA generates electrons at a standard redox potential of –126 mV (Gustafson *et al.*, 1986). The reduction of protons to hydrogen ($E^{o'}$ of –414 mV) using the electrons generated from butyryl-CoA oxidation requires a hydrogen partial pressure of about 10^{-5} Pa to be thermodynamically favorable (Schink, 1997). Similar hydrogen partial pressures are needed for the oxidation of pimelyl-CoA and glutaryl-CoA, intermediates involved in syntrophic benzoate degradation (Elshahed *et al.*, 2001b). The syntrophic metabolism of propionate by *Syntrophobacter wolinii* by the methylmalonyl-CoA pathway (Houwen *et al.*, 1990) involves the oxidation of succinate to fumarate ($E^{o'}$ of +10 mV) (Thauer *et al.*, 1977). Here, much lower hydrogen partial pressures (10^{-10} Pa) are needed for this reaction to be thermodynamically favorable (Schink, 1997). Hydrogen-dependent methanogenesis reaches thermodynamic equilibrium around 0.2 Pa (Schink, 1997). Thus, proton reduction coupled to the oxidation of these intermediates can only occur if energy is expended to lower the redox potential of the electrons. The protonophore, CCCP, and the ATPase inhibitor, DCCD, inhibited hydrogen production from butyrate by *Syntrophomonas wolfei* and from benzoate by *S. buswellii* providing evidence for reverse electron transport (Wallrabenstein and Schink, 1994). Syntrophic propionate oxidation by *S. fumaroxidans* requires a very high Gibbs free energy of dissipation, 3,500 kJ mol C^{-1} of biomass [a simple thermodynamic measure for the amount of biochemical work needed to convert a carbon source into biomass (Heijnen, 1994)] in order for the theoretical yields to approach the experimentally obtained values (Scholten and Conrad, 2000). Such high values have only been obtained with nitrifiers and thiobacilli that use CO_2 as their carbon source and require reverse electron transport.

The oxidation of succinate to fumarate is also a problem for *Desulfuromonas acetoxidans*. *D. acetoxidans* grows by the oxidation of acetate to CO_2 coupled to the reduction of sulfur to sulfide. The oxidation of acetate occurs by a modified citric acid cycle (Gebhardt *et al.*, 1985). The oxidation of succinate to fumarate coupled to sulfur reduction to sulfide is endogonic ($\Delta G^{o'}$ of + 53 kJ mol^{-1}). Succinate oxidation coupled to NAD or sulfur reduction by membrane fractions was dependent on ATP and inhibited by protonophores, showing that reverse electron transport was involved in succinate oxidation (Paulsen *et al.*, 1986). Similarly, hydrogen formation from glutamate by *Acidaminococcus fermentans* (Härtel and Buckel, 1996) and glycolate by membrane vesicles of an unnamed bacterium required ATP or a proton gradient (Friedrich and Schink, 1993; 1995).

2.4. Gastrointestinal and Hindgut Fermentations

The rumen of herbivores contains about 10^5 to 10^6 protozoa and about 10^{10} bacteria per ml. These organisms degrade the main dietary components, starch, cellulose and hemicellulose. The polysaccharides are only partially converted to CO_2 and CH_4 (Wolin and Miller, 1989). Acetate and longer chain fatty acids accumulate (equation 2) and are absorbed and used by the animal as energy sources (Figure 1B) (Mackie and Bryant, 1994).

$$57.5\ C_6H_{12}O_6 \rightarrow 65\ CH_3COOH + 20\ CH_3CH_2COOH + 15\ CH_3(CH_2)_2COOH + 60\ CO_2 + 35\ CH_4 + 25\ H_2O \quad (2)$$

The metabolic groups in the rumen include fermentative bacteria and eukaryotes and H_2- and formate-using methanogens (Figure 1B). In contrast to the rumen, the human colonic fermentation involves only bacteria. Endogenous materials such as exfoliated epithelium and excretory proteins are important substrates and H_2 is a major product (Wolin and Miller, 1989). Only about 30% to 50% of the individuals that eat Western diets produce methane (Bond *et al.*, 1971). The ratio of acetate, propionate, and butyrate formed in methanogenic and non-methanogenic individuals is similar (equations 3 and 4, respectively):

$$57.5\ C_6H_{12}O_6 \rightarrow 56\ CH_3COOH + 21\ CH_3CH_2COOH + 19\ CH_3(CH_2)_2COOH + 61.75\ CO_2 + 32.25\ CH_4 + 29.5\ H_2O \quad (3)$$

$$54\ C_6H_{12}O_6 \rightarrow 55\ CH_3COOH + 22\ CH_3CH_2COOH + 23\ CH_3(CH_2)_2COOH + 56\ CO_2 + 44\ H_2 + 12\ H_2O \quad (4)$$

The only way to account for the large amount of acetate made in the absence of methane production is by reducing carbon dioxide to acetate (Wolin and Miller, 1994). Lajoie *et al.* (1988) found that labeled acetate from $^{14}CO_2$ was produced during the fermentation of vegetable polysaccharides by human fecal samples. The reduction of CO_2 to acetate by acetogenic bacteria is also the dominant electron-accepting reaction in termite guts (Breznak, 1994) and the cecum of rodents (Prins and Lankhorst, 1977). In these fermentations, the community structure appears to be similar to that of the rumen, except that H_2-using acetogenic bacteria replace the H_2-using methanogens as the terminal electron-accepting group (Figure 1B). However, mixotrophy (simultaneous utilization of H_2/CO_2 and organic substrates) may also be involved.

2.5. Other Environments

In ecosystems with high levels of sulfate, such as marine and estuarine sediments and petroleum reservoirs, sulfate reduction rather than methanogenesis is the terminal electron-accepting process. Organic matter is completely oxidized to CO_2 with the reduction of sulfate to sulfide. Again, the process involves the concerted efforts of several metabolic groups of bacteria with the sulfate-reducing bacteria apparently performing the functions of the syntrophic metabolizers along with hydrogenotrophic and acetoclastic methanogens (Figure 1C) (McInerney, 1986; Oude Elferink *et al.*, 1994; Stams, 1994). The degradation of propionate and longer chain fatty acids to CO_2 in marine sediments and toluene in sulfate-containing aquifer sediments does not require interspecies H_2 transfer (Banat and Nedwell, 1983; Elshahed and McInerney, 2001b). However, it is likely that the use of H_2 by sulfate reducers influences product formation of fermentative bacteria in a manner analogous to that found in methanogenic environments.

In environments where dissimilatory Fe(III) or Mn(IV) reduction is the terminal electron-accepting process, fermentative bacteria hydrolyze complex organic material and ferment sugars and amino acids (Lovley, 1991). These organisms may reduce small amounts of Fe(III), but their main metabolic function is fermentative. Dissimilatory metal-reducing bacteria then use acetate and longer chain fatty acids, aromatic compounds, or hydrogen to reduce Fe(III) or Mn(IV) to Fe(II) and Mn(II), respectively.

3. Cascade of Terminal Electron-Accepting Processes

Work in the last decade has shown that there is a remarkable diversity in the types of compounds that serve as electron acceptors. In addition to oxyanions of nitrogen and sulfur, anaerobic microorganisms can use a variety of metals and metalloids as terminal electron acceptors (Barton *et al.*, 2003). A number of these compounds can support growth such as chlorate, perchlorate, chromate, Mn(IV), Fe(III), Co(III), arsenate, selenate, and uranium. In addition, many anaerobic microorganisms can use multiple electron acceptors. For example, *Shewanella alga* strain BrY uses Fe(III), Mn(IV), U(VI), O_2, thiosulfate, fumarate, and trimethylamine-N-oxide (Caccavo *et al.*, 1992; 1994). The use of multiple electron acceptors provides an adaptive mechanism to survive in habitats where the local availability of a particular electron acceptor may be limited. Also, the use of multiple electron acceptors may facilitate the survival and dispersal of microorganisms by allowing these organisms to grow and metabolize in different habitats.

Equilibrium considerations are often used to predict or to understand observed redox processes in natural ecosystems (Zehnder and Stumm, 1988; Nealson and Saffarini, 1994). However, because of high reaction rates and diffusion limitations, redox conditions may be established within a microenvironment that do not coincide with those of the macroscopic environment.

Table 2. Comparison of different terminal electron-accepting processes

Terminal electron-accepting process	$p\varepsilon^{o\ a}$ (pH 7.0; 25°C)
Aerobic respiration [$O_2 \rightarrow H_2O$]	+ 13.75
Denitrification [$NO_3^- \rightarrow N_2$]	+ 12.65
Manganese reduction [Mn(IV) → Mn(II)]	+ 8.9
Dissimilatory nitrate reduction [$NO_3^- \rightarrow NH_4^+$]	+ 6.15
Iron reduction [Fe(III) → Fe(II)]	- 0.8
Sulfate reduction [$SO_4^= \rightarrow HS^-$]	- 3.5
Methanogenesis [$CO_2 \rightarrow CH_4$]	- 4.13

[a] $p\varepsilon^o$ gives the hypothetical electron activity at equilibrium and is a measure of the tendency of compound to accept or donate electrons. Data are taken from Zehnder and Stumm (1988).

Thus, a quantitative understanding of the rates at which redox reactions occur as well as equilibrium considerations are needed. Given this caveat, equilibrium conditions are useful in predicting the processes that will be operative under a given set of environmental conditions. The redox potentials of common terminal electron-accepting processes are given in Table 2. This information predicts that the sequence of use of terminal electron acceptors with depth would be oxygen, nitrate, manganese, iron, sulfate, and finally CO_2 (methanogenesis). This sequence is consistent with the profiles observed in many anaerobic environments (Zehnder and Stumm, 1988; Nealson and Saffarini, 1994). In cases where the electron acceptor is not abundant, the process could still be important if rapid recycling of the electron acceptor occurs (Nealson and Saffarini, 1994). For example, in coastal sediments off Denmark, recycling of iron and manganese was rapid and each molecule probably turned over about 100 to 300 times before being buried.

The distribution of terminal electron-accepting reactions in aquifers depends on the relative abundance of the electron acceptors, the amount and availability of the electron donor, and the rate of groundwater flow (Smith, 2002). In pristine aquifers, naturally occurring organic matter is limited. The sequence of terminal electron acceptors occurs in the order given in Table 2 with oxygen reduction the closest and methanogenesis the furthest from the source of carbon. The thickness of each redox zone in the direction of groundwater flow depends on the abundance of the electron acceptor, but the zones may be kilometers long. In fact, the aquifer may discharge to a surface water system before oxygen is depleted. When the amount of electron donor is in abundance, often the result of organic contamination, all available electron acceptors are rapidly used, creating a central methanogenic zone. As the organic compounds are consumed and diluted by downgradient transport, points are reached where sulfate reduction, Fe(III) reduction, etc. predominate. Thus, as one moves from upgradient to downgradient of the source of the organic material, the progression of electron-accepting processes is first in the order listed in Table 2 and then in the reverse order.

Since hydrogen is an important intermediate in most anaerobic food chains, it can be used to indicate the terminal electron-accepting process operative at a given site (Lovley and Goodwin, 1988). The steady state hydrogen concentration should be dictated by the thermodynamics of the terminal electron-accepting reaction and not by kinetic considerations. Electron-accepting processes that result in the most negative free energy change will maintain the lowest steady state hydrogen concentrations. Low hydrogen levels will prevent less favorable electron-accepting processes from using hydrogen. The expected steady state hydrogen concentrations are > 5 nM for methanogenesis, 1-4 nM for sulfate reduction, 0.2-0.6 nM for Fe(III) reduction, and < 0.1 nM for oxygen and nitrate reduction.

In microorganisms that use multiple electron acceptors, there is a hierarchical sequence of electron acceptor use (Gunsalus, 1992; Unden *et al.*, 1994). Distinct regulatory circuits controlled by oxygen and nitrate such as ArcA/B, NarX/L (NarQ/P), FhlA, and FNR regulate the synthesis of enzymes involved in aerobic and anaerobic metabolism in response to environmental signals. This topic will be discussed more fully in Chapter 3 of this book.

4. Thermodynamic Explanation for Anaerobic Community Structure

Equation 5 describes biomass formation in relation to the degree of reduction of the electron donor and acceptor (Heijnen, 1994; 1999):

$$(-1/Y_{sx})\ \text{donor} - [(\gamma_D / Y_{sx}) - \gamma_D][1/ -\gamma_A]\ \text{acceptor} + 1\text{C-mol biomass} = 0 \qquad (5)$$

where Y_{sx} is the yield of biomass of a substrate or electron donor (C-mol biomass C-mol donor^{-1}) and γ is

Table 3. Gibbs free energy and enthalpy changes for idealized reactions involved in anaerobic degradation of glucose by fermentation or respiration[a]

Reaction	ΔG°' (kJ/electron)	ΔH° (kJ/electron)
1. Glucose + 6 O_2 → 6 HCO_3^- + 6 H^+	-118.5	-120.1
2. Glucose + 4.8 NO_3^- → 6 HCO_3^- + 2.4 N_2 + 2.4 H_2O + 1.2 H^+	-111.9	-107.4
3. Glucose + 3 $SO_4^=$ → 6 HCO_3^- + 3 HS^- + 3 H^+	-18.9	-8.7
4. Glucose + 3 H_2O → 3 HCO_3^- + 3 CH_4 + 3 H^+	-16.8	-7.3
5. Glucose + H_2O → Propionate⁻ + Acetate⁻ + HCO_3^- + H_2 + 3 H^+	-47.1	-22.9
6. Glucose → Succinate⁼ + Acetate⁻ + H_2 + 3 H^+	-43.7	-21.7
7. Glucose + 2.6 H_2O → 0.7 Butyrate⁻ + 0.6 Acetate⁻ + 2 HCO_3^- + 2.6 H_2 + 3.3 H^+	-30.0	-5.1
8. Glucose + 2 H_2O → Ethanol + 2 HCO_3^- + 2 H^+	-56.5	-25.1
9. Glucose → 2 Lactate⁻ + 2 H^+	-49.6	-27.3
10. Glucose + $SO_4^=$ → 2 Acetate⁻ + 2 HCO_3^- + HS^- + 3 H^+	-44.8	-29.3
11. Glucose + H_2O → 2 Acetate⁻ + HCO_3^- + CH_4 + 3 H^+	-38.0	-23.4
12. Glucose → 3 Acetate⁻ + 3 H^+	-39.8	-24.1

[a] Modified from McInerney and Beaty (1988).

the degree of reduction of a chemical compound [donor (D), acceptor (A) or biomass (X)]. The empirical formula for a microbial cell ($CH_{1.8}O_{0.5}N_{0.2}$) equivalent to about 25 g dry weight of biomass is used in calculations. The degree of reduction is a stoichiometric property of a chemical compound that describes its electron content (mol electrons compound^{-1}).

The overall Gibbs energy balance for the above macrochemical equation can be written as follows (equation 6):

$$0 = \Delta G(\text{donor}) + \Delta G(\text{acceptor}) + \Delta G(\text{biomass}) + \text{Dissipation}$$

$$0 = (-1/Y_{sx})[\gamma_D \Delta G_{eD}^{01}] - [(\gamma_D/Y_{sx}) - \gamma_x][1/-\gamma_A][\gamma_A \Delta G_{eA}^{01}] + [\gamma_x \Delta G_{eD}^{01}] + D_s^{01}/r_{Ax} \qquad (6)$$

where ΔG_{eD}^{01} and ΔG_{eA}^{01} are the Gibbs energies of formation per electron in donor (kJ mol electron of donor^{-1}) and acceptor (kJ mol electron of donor^{-1}), respectively, D_s^{01} is the Gibbs energy dissipation (kJ $m^{3\ -1}$ h^{-1}), and r_{Ax} is the net growth rate (mol $m^{3\ -1}$ h^{-1}). Solving the above equation for Y_{sx} gives an equation that shows how Y_{sx} depends on D_s^{01}/r_{Ax} (equation 7):

$$Y_{sx} = [\gamma_D (\Delta G_{eD}^{01} - \Delta G_{eA}^{01})] \ / \ [(D_s^{01}/r_{Ax}) + \gamma_x(\Delta G_{eD}^{01} - \Delta G_{eA}^{01})] \qquad (7)$$

where $\Delta G_{eD}^{01} - \Delta G_{eA}^{01}$ is the Gibbs free energy released in the catabolic equation per electron. Equation 7 shows that the yield of cells (Y_{sx}) decreases at high dissipation values, is lower for electron donors with fewer electrons (γ_D is smaller) and is a hyperbolic function of $\Delta G_{eD}^{01} - \Delta G_{eA}^{01}$. For aerobic growth, $\Delta G_{eD}^{01} - \Delta G_{eA}^{01}$ is nearly constant, about 110 kJ e-mol^{-1}. For anaerobic growth, this value varies (2 to 25 kJ e-mol^{-1}). In general, catabolic reactions with greater the $\Delta G_{eD}^{01} - \Delta G_{eA}^{01}$ values will result in higher yields.

Equation 7 helps explain some of the interesting features of anaerobic metabolism discussed in the previous section. Since the value of $\Delta G_{eD}^{01} - \Delta G_{eA}^{01}$ is small compared to aerobic metabolism, there is a strong effect of concentration (particularly pH) on the Gibbs free energy of the catabolic reaction of some anaerobic processes. We have seen how the hydrogen partial pressure affects syntrophic metabolism and pH affects hydrogenotrophic methanogenesis. The formation of precipitates (sulfide minerals) also can have a strong effect on the ΔG of the reaction. The formation of insoluble sulfide precipitates is the driving force for some sulfur disproportionations (Thamdrup *et al.*, 1993). Depending on the entropy (ΔS) of the catabolic equation, an increase in temperature can either increase (e. g., syntrophic acetate oxidation) or decrease (e. g., hydrogenotrophic methanogenesis) the Gibbs free energy change. This may explain the occurrence of syntrophic acetate oxidation at thermophilic temperatures (Figure 1A) (Schink, 1997).

Table 3 shows the enthalpy (ΔH°) and Gibbs free energy changes for the mineralization of glucose under aerobic and various anaerobic conditions. For mineralization reactions under aerobic and denitrifying conditions, the Gibbs free energy change per mole of transferred electrons is large and almost equivalent to the enthalpy change. This implies that the contribution of heat to the overall energy flow is of overwhelming importance. The situation is much different for anaerobic fermentations and other anaerobic respirations. For these latter reactions, the Gibbs free energy change is larger than the enthalpy change, implicating the importance of the chemical entropy in these reactions. Thus, while aerobic and denitrifying processes are mainly heat-driven, other anaerobic processes are driven by the difference in the entropy of chemical substances exchanged between the cell and the environment.

This analysis offers a possible explanation why complete mineralization of glucose (as well as other compounds) requires a consortium of organisms under sulfate-reducing and methanogenic conditions (McInerney and Beaty, 1988). Since anaerobic processes other than denitrification rely on entropy created by the exchange or flux of materials with the environment, glucose and other organic compounds tend to be incompletely degraded by fermentative bacteria. Thus, niches are available for other organisms to use the end products produced by fermentative bacteria as energy sources for growth.

Another way to explain the concept is to compare the Gibbs free energy values per electron for the different processes. McCarty (1971) found that the free energy released per electron is a major factor in determining whether an organism will reproduce fast enough to be maintained in anaerobic digestors operated at high dilution rates. The free energy change per electron for the complete mineralization of glucose coupled to sulfate reduction and methanogenesis (reactions 3 and 4, Table 3) is much smaller than that for aerobic and denitrifying conditions (reactions 1 and 2, Table 3). This suggests that if an organism could completely mineralize glucose by sulfate reduction or methane production, its growth rate would probably be too slow to be competitive in most ecosystems. Pfeiffer *et al.* (2001) argue that when resources (e. g., substrate) are limiting, organisms that produce ATP at high rates but low yields (fermentative organisms) will be favored over those that produce ATP at high yields but at low rates (respiratory organisms). Fermentation of glucose with the production of propionate, succinate, butyrate, ethanol, or lactate (reactions 5 to 9, Table 3) releases large amounts of free energy per electron, suggesting that these process should support growth, which indeed they do. The degradation of glucose to acetate and HCO_3^- coupled to sulfate reduction, acetogenesis, or methane production (reactions 10-12, Table 3) also release large amounts of free energy per electron. Homoacetogenic fermentors are well known (Ljungdahl, 1986) and some sulfate reducers are known to metabolize fructose to acetate and HCO_3^- (Cord-Ruwisch *et al.*, 1986). Methanogens are not known to use glucose, although some store polysaccharide reserve materials (König *et al.*, 1985; Murray and Zinder, 1987) that can be degraded with the production of methane and possibly acetate under starvation conditions.

The fermentation of glucose coupled to dissimilatory iron reduction offered a direct test of the above hypothesis (Lovley and Phillips, 1989). Iron-reducing, freshwater anaerobic sediments metabolized glucose to a mixture of labeled acetate, propionate, and lactate. This pattern was similar to that observed in methanogenic sediments. Thus, the thermodynamic explanation for anaerobic community structure predicted a priori the fate of glucose carbon under a newly discovered terminal electron-accepting process.

Zwolinski *et al.* (2000) argue that an organism is unlikely to release an intermediate in which it has a significant energy investment. Terminal electron-accepting processes with large free energy changes (e. g., denitrification) have a number of steps where the release of an intermediate would be energetically favorable. On the other hand, processes such as methanogenesis are much less favorable and there would be fewer places where the free energy differences between the substrate and the intermediate are large enough to allow for the release of intermediates. Thus, one would expect a more complex community under denitrifying as compared to methanogenic conditions. With the advent of molecular tools to dissect the community structure of anaerobic ecosystems, it is now possible to test these two hypotheses.

5. Nutrition

In general, there is a large degree of unity in the biosynthetic pathways between aerobic and anaerobic microorganisms. However, there are some unique features concerning anaerobic nutrition that are important to discuss. Obligate anaerobes require a highly reduced medium for cultivation. Often, the simple removal of oxygen is not sufficient to poise the redox conditions of the medium low enough to allow growth. Usually, a reducing solution containing both cysteine and sulfide is used. Sulfide alone or palladium chloride can be used if one needs to avoid the inclusion of organic compounds in the medium. Since most anaerobes require a reduced form of sulfur as their sulfur source, cysteine and/or sulfide serve both as reducing agents and the main sulfur source for growth (Bryant, 1974; Varel and Bryant, 1974). Many anaerobic bacteria have an absolute requirement for CO_2 (Dehority, 1971). This is especially true for propionate- and succinate-producing species. Most anaerobic media contain bicarbonate and have a CO_2 gas phase to meet this nutritional requirement as well as to buffer the medium. The inclusion of bicarbonate/CO_2 in the medium probably replaces the need for serum. Most anaerobes use ammonia as their main nitrogen source (Bryant and Robinson, 1962; Herbeck and Bryant, 1974; Varel and Bryant, 1974). *Bacteriodes ruminicola* is interesting in that it uses oligopeptides as a nitrogen source (Pittman *et al.*, 1967).

Acetate is often required or highly stimulatory for the growth of many anaerobes (Bryant and Robinson, 1962; Bryant, 1974). Some gastrointestinal tract bacteria require one or more of the following volatile acids for growth, *n*-valeric, isobutyric, 2-methylbutyric, and isovaleric acids (Bryant and Robinson, 1962; Herbeck and Bryant, 1974; Whitman *et al.*, 1982; Tanner and Wolfe, 1988). Volatile acids are used to synthesize long-chain saturated fatty acids and in some cases provide the carbon skeletons for the synthesis of branched-chain amino acids (Allison *et al.*, 1962a; 1962b). *Syntrophococcus sucromutans,* however, requires octadecenoic acid or complex lipids such as triglycerides or phospholipids for growth (Doré and Bryant, 1989). *Ruminococcus albus* requires phenylpropionate (Hungate and Stack, 1982). Heme or a related tetrapyrrole is required for growth of *B. ruminicola* (Caldwell *et al.*, 1965). *Succinivibrio dextrinosolvens* requires 1,4-naphthoquinone in addition to other organic compounds for growth (Gomez-Alarcon *et al.*, 1982). Some methanogens have lost the ability to synthesize all or part of some of their unusual cofactors. Some strains of *Methanobrevibacter ruminantium* require coenzyme M (Balch and Wolfe, 1976); whereas *Methanomicrobium mobile* requires 7-mercaptoheptanoylthreonine phosphate (Tanner and Wolfe, 1988; Kuhner *et al.*, 1991). Evidently, these organisms evolved in environments where the loss of certain biosynthetic capabilities was not detrimental, presumably because the needed growth factors were present.

6. Importance of Metals

Nickel, selenium, tungsten, and molybdenum play important roles in catalysis in anaerobes. For this reason, it is important that these metals be added to anaerobic media. Insufficient amounts of these metals can lead to growth limitations (Costilow, 1977; Jones and Stadtman, 1977; Schönheit *et al.*, 1979; Diekert and Ritter, 1982). Molybdenum-containing enzymes fall into two broad classes (Stiefel, 1996; Kisker *et al.*, 1997). In nitrogenase, molybdenum is found in a special iron-molybdenum cofactor. The other class contains a molybdenum-pterin cofactor and can be found in enzymes such as dimethylsulfoxide reductase, nitrate reductase, formate dehydrogenase, pyrogallol-phloroglucinol transhydroxylase, and xanthine oxidase (Boyington *et al.*, 1997; Kisker *et al.*, 1997). The 4-hydroxybenzoyl-CoA reductases that remove the hydroxyl group from the benzoyl-CoA ring are structurally similar to xanthine oxidase and contain molybdenum (Gibson *et al.*, 1997; Breese and Fuchs, 1998). Most molybdenum-dependent enzymes catalyze the net transfer of a hydroxyl group to or from the substrate coupled to electron transfer between the substrate and other cofactors such as flavins, hemes, or iron-sulfur centers.

Tungsten is found in some formate dehydrogenases (Yamamoto *et al.*, 1983; Wagner and Andreesen, 1987), in aldehyde:ferredoxin oxidoreductases of several archaea, and in acetylene hydratase from *Pelobacter acetylenicus* (Rosner and Schink, 1995). Nickel is found in most hydrogenases (Wu and Mandrand, 1993; Schwartz and Friedrich, 2003), CO dehydrogenase (Ragsdale, 1991; Dobbek *et al.*, 2001; Thauer, 2001; Doukov *et al.*, 2002), and coenzyme F_{430} in methanogens (Deppenmeier, 2002b).

Selenium is found in formate dehydrogenase, some hydrogenases, and in the amino acid selenocysteine (Self, 2003). Selenium is in the form of selenocysteine and is closely associated with tungsten in *Moorella thermoacetica* (formally *Clostridium thermoaceticum)* formate dehydrogenase, which also contains molybdenum (Self, 2003). The fermentation of purines by clostridia involves molybdenum hydroxylases, such as xanthine dehydrogenase, that also contain selenium. Another selenium-containing, molybdenum hydroxylase is nicotinic acid hydroxylase where selenium is in a labile form and not as selenocysteine (Dilworth, 1983).

A number of anaerobic bacteria such as *Clostridium sporogenes*, *Clostridium sticklandii*, *Clostridium histolyticum*, *Clostridium butulinum*, *Peptococcus glycinophilus* and *Peptococcus magnus* use the Stickland reaction where one amino acid serves as the electron donor and a second amino acid such as glycine serves as the electron acceptor (Gottschalk, 1985; McInerney, 1988). Glycine reductase in these organisms catalyzes the following reaction:

$$NH_3CH_2COO^- + P_i + 2\ e^- \rightarrow CH_3COO\text{-}PO_4^= + NH_4^+ \quad (8)$$

This enzyme complex contains three chromatographically separable fractions, one of which is a selenoprotein (enzyme A) (Tanaka and Stadtman, 1979; Self, 2003). The pyruvoyl-enzyme (enzyme B) transfers glycine to the selenium anion on enzyme A to form a carboxymethylselenocysteine residue (Andereesen, 1994). This moiety is transferred to the third enzyme to form an acetylcysteine that is then cleaved by phosphate to yield acetyl phosphate. This novel use of selenium allows the formation of a "high-energy" intermediate, acetyl phosphate, which can be used to make ATP.

Nitrogen fixation is an important part of the nitrogen cycle and results in increased agricultural productivity in soils. The enzyme system involved in nitrogen fixation is oxygen-sensitive. There are some organisms that fix nitrogen while growing aerobically; however, these organisms have protective systems to prevent the inactivation of nitrogenase by oxygen.

Nitrogenase is an enzyme system that catalyzes the following reaction:

$$N_2 + 16\,ATP + 8\,e^- + 8\,H^+ \rightarrow 2\,NH_3 + 16\,ADP + 16\,P_i + H_2 \qquad (9)$$

Two proteins called Fe-protein and MoFe-protein are responsible for this reaction (Ormen-Johnson, 1992; Peters *et al.*, 1995). The Fe-protein has two subunits and, as a dimer, it contains a single 4Fe-4S cubane cluster. The Fe-protein is the site of ATP hydrolysis where two ATP molecules are hydrolyzed per electron transferred to MoFe-protein. ATP hydrolysis is required to generate very low potential electrons to initiate the reduction of the stable triple bond in N_2. The initial two-electron transfer is believed to be endergonic (about 205 kJ/mol) (Ormen-Johnson, 1992). The MoFe-protein has an $\alpha_2\beta_2$ arrangement and contains two unique metal clusters: the P cluster (8Fe-7S) and the (Mo:7Fe:9S):homocitrate: iron-molybdenum cofactor cluster (Georgiadis *et al.*, 1992). Recently, an improved resolution of the crystal structure of the MoFe-protein suggests the presence of an internal hexacoordinate light atom, probably nitrogen, within the FeMo-cofactor (Einsle *et al.*, 2002). This implicates the seven-Fe cluster as the site of nitrogen reduction. However, using model synthetic substrates, Yandulov and Schrock (2003) showed that di-nitrogen reduction to ammonia can occur at a single Mo center. Thus, after more than 50 years of research, there is still much to be learned about the mechanism of nitrogen fixation. The nitrogenase system does show how various metal centers are used to conduct difficult oxidation-reduction reactions and illustrates that many anaerobic redox processes often require the input of energy to drive the reactions.

7. One and Two Carbon Metabolism

Carbon dioxide can be fixed into organic compounds by three different pathways in anaerobic bacteria, the reductive acetyl-CoA pathway, the reductive citric acid cycle, and the 3-hydroxypropionate cycle (Evans *et al.*, 1966; Fuller, 1978; Ljungdahl, 1986; Holo, 1989; Strauss and Fuchs, 1993). Some Crenarchaeota possess ribulose-1, 5-bisphosphate carboxylase, the key enzyme in the Calvin-Bassham-Benson pathway, used by aerobic chemolithotrophs to fix carbon dioxide (Hügler *et al.*, 2003a). A ribulose-1, 5-bisphosphate carboxylase-like protein functions in the methionine salvage pathway of *Bacillus subtilis* (Ashida *et al.*, 2003), so the presence of this enzyme does not necessarily indicate that a complete functional pathway exists. Since the acetyl-CoA pathway is critical to supply carbon for biosynthesis in many anaerobes, especially methanogens, and because both the acetyl-CoA pathway and the citric acid cycle can be used to oxidize acetate, the salient features of acetogenesis, methanogenesis, and acetate degradation will be discussed in concert with carbon dioxide fixation pathways.

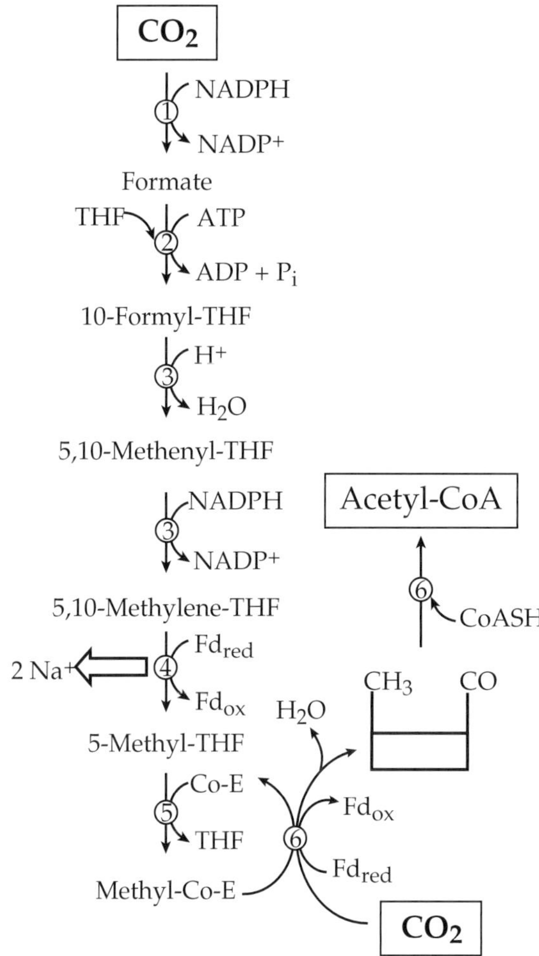

Figure 3. Pathway for acetate formation from CO_2 and H_2 by acetogenic bacteria. Symbols: block arrow, energy conservation step; box, active sites of acetyl-CoA synthetase/CO dehydrogenase; THF, tetrahydrofolate, Fd, ferredoxin, and Co-E, corrinoid protein. Enzymes: (1), formate dehydrogenase; (2), 10-formyl-tetrahydrofolate synthetase; (3), 5,10-methenyl-tetrahydrofolate cyclohydrolase and 5,10-methylene-tetrahydrofolate dehydrogenase; (4), 5,-10-methylene-tetrahydrofolate reductase; (5), methyltransferase; and (6), acetyl-CoA synthetase/CO dehydrogenase.

7.1. Acetogenesis

Many anaerobes utilize the Wood-Ljungdahl pathway for carbon dioxide fixation (Ljungdahl, 1986), which results in the non-cyclic generation of acetyl-CoA that can be used for biosynthesis (Figure 3). The bifunctional CO dehydrogenase/acetyl-CoA synthetase is an $\alpha_2\beta_2$ heterotetramer that catalyses the reduction of CO_2 to CO and the synthesis of acetyl-CoA from a methyl-group, CO, and CoA (Ragsdale *et al.*, 1983; Pezacka and Wood, 1984; Ragsdale, 1991). The β subunit contains a distorted cubane metal center called the C cluster that contains 1 nickel, 3 iron, and 4 sulfur atoms which functions to form CO from CO_2 (Doukov *et al.*, 2002). The α subunit is involved in acetyl-CoA synthesis at a metal center called the A cluster. The A cluster is unusual in that it contains an iron-sulfur cubane linked to a copper ion by a single

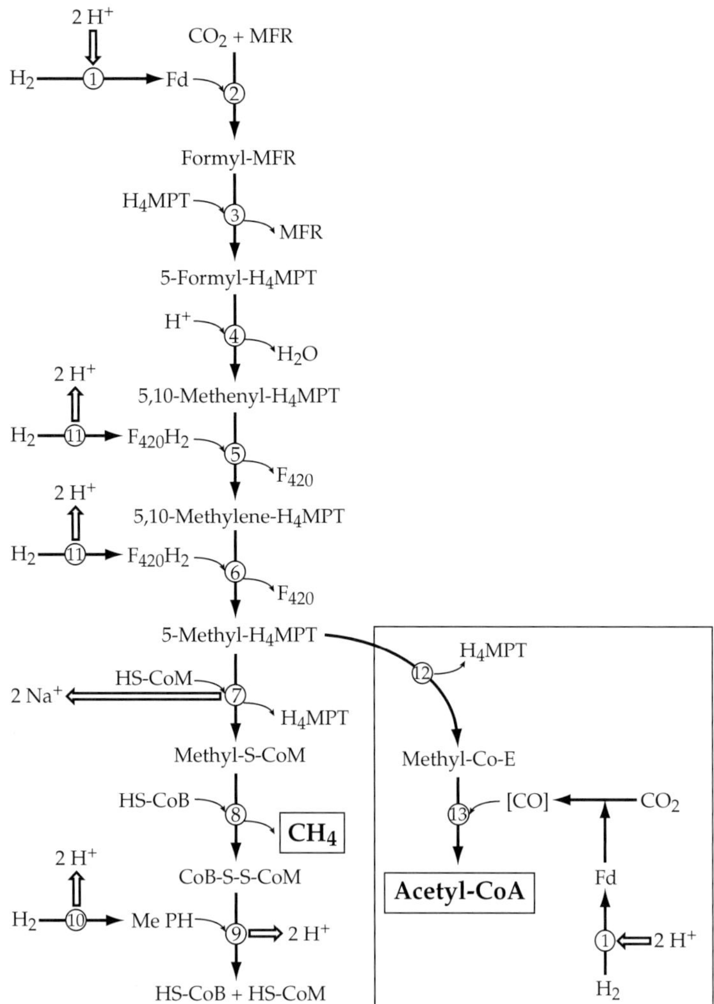

Figure 4. Pathway for methane formation and acetate synthesis by hydrogenotrophic methanogens (adapted from Deppenmeier, 2002b). Box, biosynthetic production of acetate; block arrows, sites of energy conservation (pointing outward) or energy input (pointing inward); MFR, methanofuran; H_4MPT, tetrahydromethanopterin; F_{420}, cofactor F_{420} (a deazaflavine derivative); HS-CoB, coenzyme B (N-7-mercaptoheptanoyl-L-threonine phosphate), HS-CoM, coenzyme M (2-mercaptoethane-sulfonic acid); CoB-S-S-CoM, heterodisulfide of CoM and CoB; Co-E, corrinoid protein; MePH, methanophenazine; and Fd, ferredoxin. Enzymes: (1), Ech hydrogenase (membrane-bound); (2), formate dehydrogenase; (3), formyl-MFR:H_4MPT formyltransferase; (4), methenyl-H_4MPT cyclohydrolase; (5), methylene-H_4MPT dehydrogenase; (6), methylene-H_4MPT reductase; (7), methytransferase (membrane-bound, sodium pump); (8), methyl-CoM reductase; (9), heterodisulfide reductase (membrane-bound); (10), $F_{420}H_2$- or H_2-heterodisulfide oxidoreductase system; (11), F_{420} dehydrogenase (membrane-bound); (12), methyltransferase; and (13), acetyl-CoA synthetase/CO dehydrogenase.

cysteine. The copper ion is in turn bridged to a nickel atom by two cysteines (Doukov *et al.*, 2002). The methyl group is supplied to the A cluster by a corrinoid iron-sulfur protein (Ragsdale, 1991). It is believed that CO binds to the copper ion and the methyl group binds to the nickel atom. The methyl group is transferred to the copper ion to form an acetyl-Cu intermediate before the acetyl group is transferred to CoA to form acetyl CoA. The A and C clusters are connected by a channel approximately 139 angstroms long that allows CO formed at the C cluster to migrate to the A cluster where it is incorporated into acetyl-CoA (Doukov *et al.*, 2002). The unfavorable reduction of CO_2 to CO is probably driven by a sodium gradient (Müller and Gottschalk, 1994).

In acetogenic bacteria, CO_2 is reduced to formate by a tungsten-containing, NADP-dependent, formate dehydrogenase (Figure 3) (Ljungdahl, 1986; Ragsdale, 1991). Alternatively, sugar-fermenting acetogens can use the carboxyl group of pyruvate. Formate is activated to 10-formyl-tetrahydrofolate by an ATP-dependent condensation reaction with tetrahydrofolate (THF). A bifunctional enzyme, 5, 10-methenyl-THF cyclohydrolase and 5, 10-methylene-THF dehydrogenase, catalyzes the conversion of 10-formyl-THF to 5, 10-methylene-THF. The latter reaction is NADP-dependent. 5, 10-Methylene-THF reductase converts 5, 10-methylene-THF to 5-methyl-THF. The enzyme contains an iron-sulfur cluster, zinc, and FAD and uses reduced ferredoxin or $FADH_2$, but not pyridine nucleotides, as electron donors. The reaction is exergonic and is thought to be a site of energy conservation in acetogens (Müller and Gottschalk, 1994). The formation of acetate from H_2/CO_2 or $HCHO/H_2$ but not CH_3OH/H_2 is dependent on sodium ions. Since formaldehyde enters the pathway at the level of methylene-THF, experiments indicate that some step between methylene-THF and methyl-THF is sodium-dependent. Consistent with its role in bioenergetics, methylene-THF reductase is membrane-bound. The final step in the synthesis of the methyl group of acetate is the transfer of the methyl group from 5-methyl-THF to the corrinoid iron-sulfur protein by a methyl transferase. The corrinoid iron-sulfur protein is the substrate for CO dehydrogenase/acetyl-CoA synthetase.

Desulfobacterium autotrophicum and *Desulfotomaculum acetoxidans* use the CO dehydrogenase/acetyl-CoA pathway in reverse to oxidize acetate to carbon dioxide coupled to the reduction of sulfate to sulfide (Spormann and Thauer, 1988; Länge *et al.*, 1989).

Experiments designed to mimic critical elements of ancient hydrothermal vent solution chemistry demonstrated the synthesis of acetate from methyl thiol and CO in the presence of iron and nickel sulfides and catalytic amounts of selenium (Huber and Wächtershäuser, 1997). More recently, the formation of pyruvate and peptides under the above conditions has been documented (Cody *et al.*, 2000; Huber *et al.*, 2003). Thus, the CO dehydrogenase/acetyl-CoA synthetase pathway may reflect an ancient chemistry that existed before life evolved.

7.2. Methanogenesis

Methanogens growing with H_2/CO_2, methylamines, or methanol use CO dehydrogenase/acetyl-CoA synthetase to synthesize acetyl-CoA that can then be used for biosynthesis (Figure 4) (Ferry, 1999; Deppenmeier, 2002a; 2002b). In methanogens, the methyl group is formed on tetrahydromethanopterin (H_4MTP) rather than THF. Like THF, tetrahydromethanopterin has a central pterin ring that is able to carry one-carbon moieties between the formyl and methyl oxidation states. The first step in the process is the formation of formyl-methanofuran (MFR) from carbon dioxide and electrons derived from a membrane-bound hydrogenase complex called Ech. Under autotrophic conditions, Ech hydrogenase uses the energy of a transmembrane ion gradient to catalyze the energetically unfavorable reduction of ferredoxin at low H_2 concentrations (Meuer *et al.*, 2002). The reduced ferredoxin serves as the electron donor for carbon dioxide reduction to formyl-MFR. The formyl group is then transferred to H_4MTP and reduced in a stepwise manner to 5-methyl-H_4MTP. The electron donor for the reductive reactions is reduced Factor 420 (F_{420}), which accepts electrons from hydrogenases.

The synthesis of acetyl-CoA from 5-methyl-H_4MTP, CO, and CoA probably occurs as outlined above for acetogens (Ragsdale, 1991). The methyl group on H_4MTP is transferred to a corrinoid protein by a methyltransferase (reaction 12, Figure 4). The methyl-corriniod is then transferred to acetyl-CoA synthetase. Mutants of *Methanococcus maripaludis* that lost the ability to grow autotrophically were associated with a mutation in CO dehydrogenase/acetyl-CoA synthetase (Ladapo and Whitman, 1990). Revertants that regained CO dehydrogenase activity also regained the ability to grow autotrophically.

A membrane-bound methyltransferase (reaction 7, Figure 4) catalyzes the exergonic transfer of the methyl moiety from 5-methyl-H_4MTP to coenzyme M (CoM) to form methyl-CoM. This reaction results in the formation of a sodium gradient that can be used to make ATP. The methyl-CoM is reduced to methane with coenzyme B (CoB) serving as the electron donor. This reaction results in the formation of a mixed heterodisulfide called CoM-S-S-CoB. The reduction of the heterodisulfide is important in the bioenergetics of methanogens and will be discussed later in this chapter.

The formation of methane from methyl-CoM also involves the formation of a methyl-nickel intermediate (Ermler *et al.*, 1997). Methyl-CoM reductase contains two molecules of an unusual nickel-porphinoid prosthetic group called F_{430}. Each F_{430} is positioned at the bottom of identical channels located about 50 angstroms apart. Thus, the enzyme has two separate active sites. A nucleophilic attack of Ni(I) on CH_3-S-CoM results in the formation of a $[F_{430}]$Ni(III)-CH_3 intermediate. The Ni(III) oxidizes HS-CoM to form a thiyl radical. Protonolysis releases the methyl group as methane and the thiyl radical reacts with CoB to form the heterodisulfide.

The degradation of acetate to carbon dioxide and methane also involves CO dehydrogenase/acetyl-CoA synthetase. For acetate degradation, CO dehydrogenase/acetyl-CoA synthetase disassembles acetyl-CoA into 5-methyl-tetrahydrosarcinapterin, CO_2, and CoA. Acetate is activated to acetyl-CoA either by an acetate:CoA ligase reaction or by the acetate kinase/phosphotransacetylase system. The CO dehydrogenase/acetyl-CoA synthetase from *Methanoseata thermophila* has three metal clusters, which have electron paramagnetic resonance properties that are indistinguishable from those of the clostridial enzyme (Ferry, 1999).

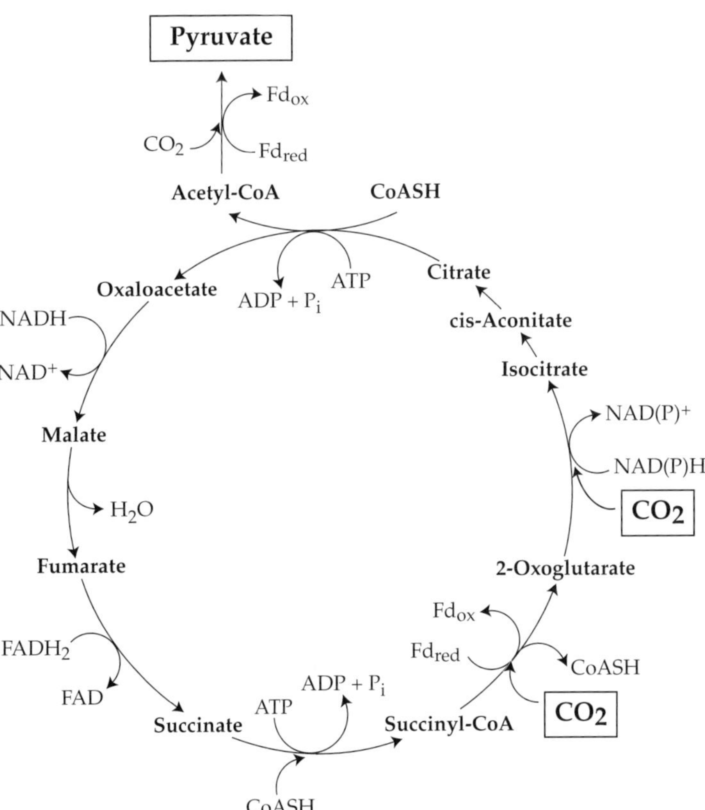

Figure 5. Reductive citric acid cycle for CO_2 fixation.

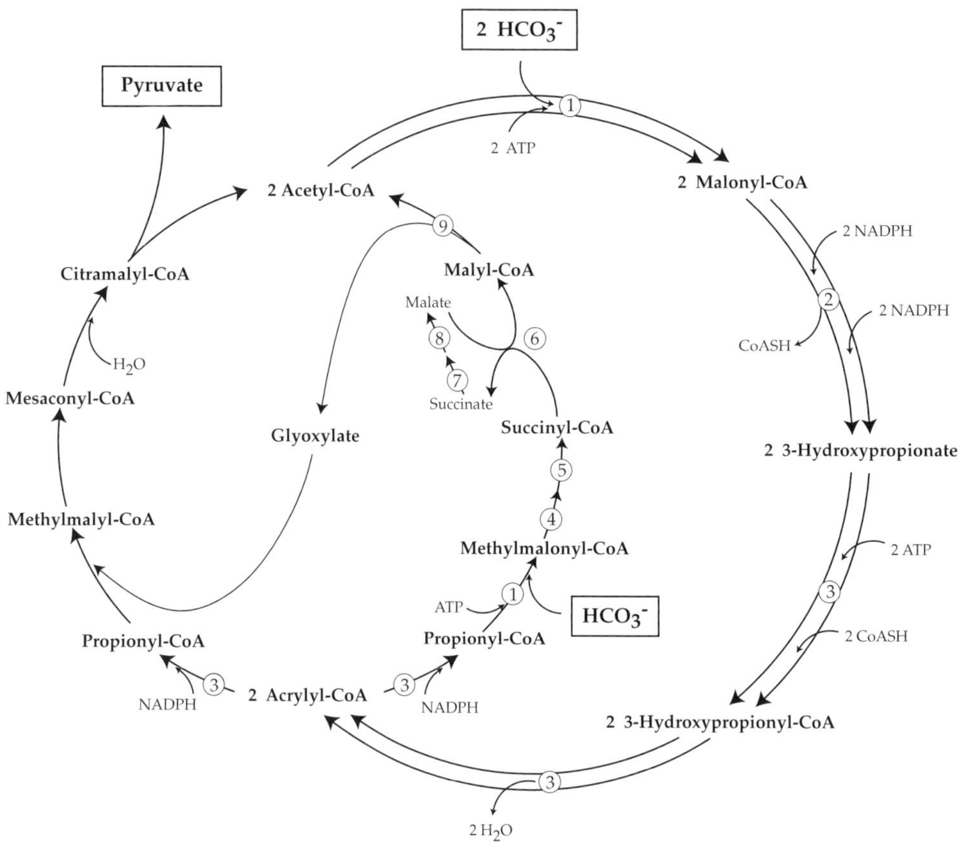

Figure 6. The bicyclic pathway for CO_2 fixation by the 3-hydroxypropionate pathway (adapted from Herter *et al*., 2002 and Hügler *et al*., 2003b). Enzymes: (1), acetyl-CoA/propionyl-CoA carboxylase; (2), malonyl-CoA reductase (a bifunctional enzyme); (3), propionyl-CoA synthease (a trifunctional enzyme); (4), methylmalonyl-CoA epimerase;, (5), methylmalonyl-CoA mutase; (6), succinyl-CoA:malate CoA transferase; (7), succinate dehydrogenase; (8), fumarase; and (9), L-malyl-CoA lyase.

7.3. Reductive Citric Acid Cycle

The reductive citric acid cycle was first discovered in green sulfur bacteria, namely *Chlorobium* species (Evans *et al.*, 1966; Eisen *et al.*, 2002). The pathway is also found in *Desulfobacter hydrogenophilus, Hydrogenobacter thermophilus, Thermoproteus neutrophilus, Pyrobaculum islandicum,* and *Aquifex pyrophilus* (Schauder *et al.*, 1987; Hügler *et al.*, 2003a). The pathway results in the formation of acetyl-CoA from two carbon dioxide molecules (Figure 5). The reductive citric acid cycle involves enzymes found in the oxidative citric acid cycle except for three steps (Eisen *et al.*, 2002). Two of the steps unique to the reductive citric acid cycle involve ferredoxin-dependent reactions including the synthesis of pyruvate and 2-oxoglutarate by a NifJ homolog and a complex of CT0163/CT0162, respectively (Eisen *et al.*, 2002). The other unique enzyme is ATP-citrate lyase that forms oxaloacetate and acetyl-CoA from citrate. In contrast to other prokaryotic citrate lyases, the enzyme involved in the reductive citric acid cycle is both ATP- and CoA-dependent (Schauder *et al.*, 1987; Wahlund and Tabita, 1997).

Desulfobacter postgatei oxidizes acetate to carbon dioxide by the citric acid cycle and uses ATP-citrate lyase. This allows net ATP synthesis by substrate-level phosphorylation (Brandis-Heep *et al.*, 1983; Möller *et al.*, 1987). The reduction of sulfite (HSO_3^-/H_2S: E°' of –116 mV) by succinate (fumarate/succinate: E°' of +33 mV) or malate (oxaloacetate/malate: E°' of –172 mV) is either endogonic or not sufficiently exergonic to drive ATP synthesis. The reduction of sulfide by NADPH (coenzyme of isocitrate dehydrogenase) or ferredoxin (coenzyme of 2-oxoglutarate:ferredoxin oxidoreductase) is exergonic; however, the free energy changes suggest that at most 2 ATP would be formed (Möller-Zinkhan and Thauer, 1988). If two ATP are needed for the activation of sulfate, insufficient energy would be generated by electron transport to support the growth of the organism. Thus, a mechanism for net ATP synthesis by substrate-level phosphorylation is needed. Interestingly, *Desulfuromonas acetoxidans* also oxidizes acetate to carbon dioxide coupled to the reduction of elemental sulfur to sulfide. This organism uses the citric acid cycle with a citrate synthase rather than with ATP-citrate lyase (Gebhardt *et al.*, 1985). Since sulfur does not have to be activated with ATP before it can be reduced to sulfide, sufficient energy for growth can be obtained from electron transport. The hyperthermophilic archaea *Thermoproteus tenax* and *Pyrobaculum islandicum* also contain a complete citric acid cycle (Selig and Schönheit, 1994).

7.4. 3-Hydroxypropionate Cycle

The 3-hydroxypropionate cycle was first discovered in *Cloroflexus aurantiacus* (Holo, 1989; Strauss and Fuchs, 1993). This pathway also appears to be present in autotrophic members of the phylum Crenarchaeota including *Acidanus brierleyi, Acidianus ambivalens, Sulfolobus metallicus, Sulfolobus* strain VE6, and *Metallosphaera sedula* (Menendez *et al.*, 1999; Hügler *et al.*, 2003a). The inner cycle results in the net fixation of two carbon dioxide molecules into glyoxlyate (Figure 6). The two carboxylation reactions, acetyl-CoA to malonyl-CoA and propionyl-CoA to methylmalonyl-CoA, are catalyzed by a single, large, biotin-containing enzyme (Hügler *et al.*, 2003b). Malonyl-CoA reductase is a bifunctional enzyme that reduces malonyl-CoA to 3-hydroxypropionate (Hügler *et al.*, 2002). Malonyl-CoA reductase contains aldehyde and alcohol reductase activities that convert malonyl-CoA first to malonate semialdehyde and then to 3-hydroxypropionate with NADPH and $NADP^+$, respectively, as cofactors. Propionyl-CoA synthase converts 3-hydroxypropionate to propionyl-CoA and has CoA ligase, enoyl-CoA hydratase, and enoyl-CoA reductase domains (Alber and Fuchs, 2002). Consistent with a ligase mechanism for the synthesis of an acyl-CoA (propionyl-CoA) from an acid anion (3-hydroxypropionate), ATP is cleaved to AMP and pyrophosphate. NADPH is the electron donor for the reductase reaction. Methylmalonyl-CoA epimerase converts methylmalonyl-CoA to succinyl-CoA. Succinyl-CoA:L-malate CoA transferase forms malyl-CoA and succinate from malate and succinyl-CoA (Herter *et al.*, 2001). Succinate is converted to malate by citric acid cycle enzymes. Malyl-CoA lyase then cleaves malyl-CoA to acetyl-CoA, reforming the starting compound, and glyoxylate (Herter *et al.*, 2001).

The question of how glyoxylate is used for biosynthesis has recently been solved by the realization that the 3-hydroxypropionate cycle is really a bicyclic pathway (Figure 6) (Herter *et al.*, 2002). ^{13}C-Labeling studies indicate that glyoxylate condenses with propionyl-CoA to form methylmalyl-CoA. Methylmalyl-CoA is converted to mesaconyl-CoA and citramalyl-CoA; the latter compound is then cleaved to acetyl-CoA and pryuvate. Pyruvate is the compound that becomes available for biosynthetic purposes.

8. Problems Anaerobes Face in the Degradation of Carbon Compounds

Carbon-carbon bond cleavage requires the formation of an intermediate that has a β–carbonyl located to another oxygen-bearing carbon atom (Dagley, 1989). The β–carbonyl allows enolization to occur that stabilizes the carbanion intermediate formed after abstraction of a proton. For many organic compounds that contain oxygen such as sugars, fatty acids, amino acids, and organic acids, this intermediate can be formed anaerobically by a series of oxidation-reduction and/ or hydration reactions. One of the exceptions to the above mechanism is the decarboxylation of α–keto acids. Enzymes that catalyze these reactions have thiamine pyrophosphate as a coenzyme. Thiamine pyrophosphate has a $=N^+$ moiety that serves as the electron-attracting group (White, 2000). Hydrocarbons lack an activating group needed for carbon-carbon cleavage. Aerobes use oxygenases to add oxygen to hydrocarbons so that they can be degraded. This is not possible for anaerobes. The aromatic ring is also another moiety that presents challenges for anaerobes. Aerobes will destabilize the resonance of the aromatic ring by insertion of oxygen. Again, anaerobes must use other approaches. Lastly, anaerobes use interesting approaches to catalyze dehydrations. The degradation of 2, 4, or 5-hydroxyacyl-CoA intermediates involves the removal of a hydrogen atom from the unactivated β or γ positions. 1,2-Diols lack any activating group. This section will discuss the approaches that anaerobes use to degrade hydrocarbons and aromatic compounds and catalyze difficult dehydrations.

8.1. Anaerobic Hydrocarbon Metabolism

The anaerobic decomposition of hydrocarbons is of great environmental importance, since most terrestrial freshwater and marine ecosystems contaminated with petroleum mixtures become anaerobic. It is now known that numerous hydrocarbon classes, including monoaromatic compounds, straight-chained and cyclic alkanes, and polycyclic aromatic hydrocarbons can be biodegraded under redox conditions ranging from nitrate-reducing to methanogenic. Novel biochemical strategies utilized by anaerobic microorganisms to overcome the energy required to cleave high energy C-H bonds of these hydrocarbon classes are outlined below and are also the subject of several recent reviews (Heider *et al.*, 1999; Widdel and Rabus, 2001; Boll *et al.*, 2002).

Although the earliest investigations into the anaerobic metabolism of toluene suggested methyl group or aromatic ring hydroxylation as an initial mechanism of activation, the detection of benzylsuccinic and benzylfumaric acids during anaerobic incubations with this substrate (Evans *et al.*, 1992; Beller *et al.*, 1992; Chee-Sanford *et al.*, 1996) led to the discovery of a novel mechanism of hydrocarbon activation. More detailed studies with the denitrifying organisms *Thauera aromatica* strain K172 (Biegert *et al.*, 1996) and *Azoarcus* sp. strain T (Beller and Spormann, 1997a) showed that toluene is anaerobically activated by an enzymatic mechanism involving the addition of the double bond of fumarate to the methyl group of the toluene to form benzylsuccinic acid (Figure 7A). This oxygen-independent reaction did not require CoA nor ATP (Beller and Spormann, 1998). Mass spectral

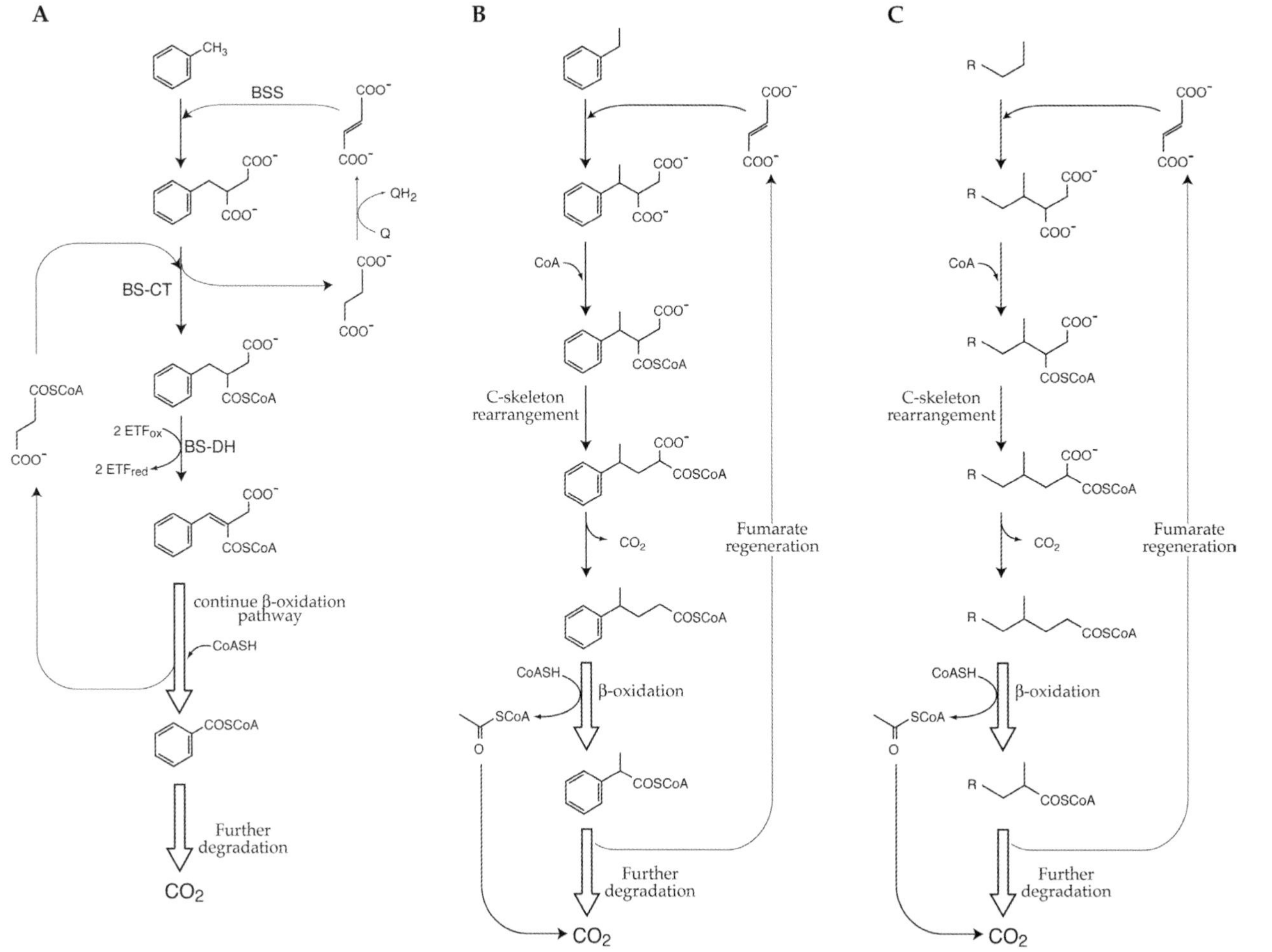

Figure 7. Schematics of anaerobic hydrocarbon metabolism via activation with fumarate for (A) toluene (adapted from Leutwein and Heider, 2002); (B) ethylbenzene under sulfate-reducing conditions (based on Kniemeyer *et al*., 2003); and (C) alkanes (modified from Wilkes *et al*., 2003). Benzylsuccinate synthase (BSS), succinyl-CoA:benzylsuccinate CoA transferase (BS-CT), and benzylsuccinyl-CoA dehydrogenase (BS-DH) indicated in (A) have been definitively characterized (summarized in Boll *et al*., 2002). ETF, electron-transferring flavoprotein; Q, quinone functioning in the regeneration of fumarate from succinate.

studies with strain T and a sulfate-reducer, PRTOL1, further showed that a H atom abstracted from the methyl group of toluene was retained in the succinyl portion of benzylsuccinate (Beller and Spormann, 1997a; Beller and Spormann, 1997b). The enzyme responsible for this novel reaction, benzylsuccinate synthase (BSS) was found to be highly stereospecific, producing almost exclusively (*R*)-(+) benzylsuccinic acid from toluene (Beller and Spormann, 1998). It was postulated that BSS was a novel glycyl radical enzyme by independent biochemical and genetic characterization in three different nitrate-reducing organisms [*Azoarcus* sp. strain T, (Beller and Spormann, 1998); *T. aromatica* strain K172, (Leuthner *et al*., 1998); *T. aromatica* strain T1, (Coschigano *et al*., 1998)]. Purification of BSS has shown this enzyme to be a heterohexamer ($\alpha_2\beta_2\gamma_2$) consisting of three subunits. For strain K172, the genes encoding BSS (*bssCAB*), along with a putative radical-generating enzyme (*bssD*) and an ATP-dependent chaperone (*bssE*) are encoded on a single toluene-induced operon (Leuthner *et al*., 1998; Hermuth *et al*., 2002). Similar gene organization was also observed for *Azoarcus* sp. strain T (Achong *et al*., 2001), whereas analogous genes from strain T1 are encoded on two operons (*tutE* and *tutFDGH*) (Coschigano, 2000). Genetic sequencing of the largest subunit (α) of BSS has shown high homology with the known glycyl-radical enzymes, pyruvate-formate lyase and anaerobic ribonucleotide reductase and has thus been proposed to be a new member of the glycyl radical-containing enzyme family (Coschigano *et al*., 1998; Leuthner *et al*., 1998). Specifically, cysteine and glycine residues characterizing the active portions of the latter enzymes are also conserved in the BSS enzymes of three nitrate-reducing species and a toluene-degrading iron-reducing bacterium (Coschigano *et al*., 1998; Leuthner *et al*., 1998; Kane *et al*., 2002). Studies using electron paramagnetic resonance spectroscopy and electrospray MS have pinpointed that the BSS radical is produced at the conserved glycine residue (Leuthner and Heider, 1998; Krieger *et al*., 2001), whereas the conserved cysteine residue functions at the active site. As with other glycyl-radical enzymes, *S*-adenosylmethionine is thought to function as a cosubstrate for radical generation (Leuthner *et al*., 1998). Collectively, these observations indicate a novel means of both hydrocarbon activation under anaerobic conditions and C-C bond formation.

Aside from BSS, other enzymes in the pathway leading from toluene to benozyl-CoA have been described in studies with *T. aromatica* strain K172 (Figure 7A). Benzylsuccinate is subsequently thioesterified to its CoA derivative by a succinyl-CoA:benzylsuccinate CoA-transferase (Leutwein and Heider, 1999; Leutwein and Heider, 2001) and further transformed to the CoA thioester of (*E*)-phenylitaconic acid by (*R*)-benzylsuccinyl-CoA dehydrogenase (Leutwein and Heider, 2002). This enzyme is similar to other known acyl-CoA dehydrogenases in that it is a homotetramer containing one FAD per subunit but is unique in its high substrate- and enantiomer-specificity for (*R*)-benzylsuccinyl-CoA. (*E*)-Phenylitaconyl-CoA then undergoes modified β-oxidation yielding benzoyl-CoA, which is presumably further metabolized by relatively well-characterized ring reduction and cleavage mechanisms (Harwood *et al*., 1999). Nine genes encoding the enzymes responsible for β-oxidation of benzylsuccinate are reported to be under the control of another toluene-induced operon in *T. aromatica* strain K172 (Leuthner and Heider, 2000). Fumarate regeneration presumably occurs via the action of succinate dehydrogenase (Figure 7A) (Boll *et al*., 2002).

Aside from denitrifying conditions, the fumarate addition mechanism of anaerobic toluene decay has also been reported to occur with anaerobes incubated under iron-reducing (Kane *et al*., 2002), sulfate-reducing (Beller *et al*., 1996; Rabus and Heider, 1998), methanogenic (Beller and Edwards, 2000), and anoxic phototrophic conditions (Zengler *et al*., 1999). Although toluene has been the paradigm for anaerobic hydrocarbon decay, other alkylbenzenes have been shown to be activated by fumarate addition to corresponding methylbenzylsuccinates. While it was reported that *Azoarcus* sp. strain T only co-metabolizes *o*-xylene to 2-methylbenzylsuccinic acid (Beller and Spormann, 1997a), this strain can mineralize *m*-xylene via 3-methylbenzylsuccinic acid as a transient intermediate (Krieger *et al*., 1999). Genetic studies have shown that this denitrifier uses BSS to activate both toluene and *m*-xylene via fumarate addition (Achong *et al*., 2001). The appropriate methylbenzylsuccinate isomer was also transiently detected in sulfidogenic incubations containing aquifer-derived sediments amended with either *o*-, *m*-, or *p*-xylene (Elshahed *et al*., 2001a).

While most reports of anaerobic alkylbenzene metabolism to date have demonstrated the universality of fumarate addition reactions regardless of the anaerobic terminal electron-accepting process, ethylbenzene activation occurs differently under nitrate- or sulfate-reducing conditions. Ethylbenzene metabolism by two denitrifying isolates proceeds via a mechanism involving the oxidation of the α-carbon to 1-phenylethanol and subsequent dehydrogenation of the latter to acetophenone (Rabus and Widdel, 1995; Ball *et al*., 1996). It was postulated that acetophenone is carboxylated to benzoylacetate, which is subsequently cleaved to acetate and benzoate (or their CoA derivatives). Two different ethylbenzene dehydrogenases, which stereospecifically hydroxylate ethylbenzene to (*S*)-(-)-1-phenylethanol, have recently been purified from two *Azoarcus* sp., strains EB1 and EbN1 (Johnson *et al*., 2001; Kniemeyer and Heider, 2001b). While both proteins are novel molybdenum-iron-sulfur enzymes existing as heterotrimers, the dehydrogenase from strain EbN1was found to be periplasmic and contained heme *b* as a cofactor (Kniemeyer and Heider, 2001b), whereas that from strain EB1 was membrane-associated with molybdopterin as a cofactor (Johnson *et al*., 2001). The subsequent enzyme, an NAD^+-dependent (*S*)-1-phenylethanol dehydrogenase from strain EbN1, has also been characterized (Kniemeyer and Heider, 2001a).

In contrast, Elshahed *et al*. (2001a) positively detected (1-phenylethyl)succinate in sulfidogenic aquifer-derived cultures incubated with ethylbenzene suggesting that fumarate addition to the α (benzyl) carbon occurred under highly-reduced conditions. Kniemeyer *et al*. (2003) confirmed this finding during studies with a pure ethylbenzene-degrading sulfate-reducer, strain EbS7. Although this initial reaction is akin to anaerobic toluene decay (Figure 7B), the addition of fumarate to the benzyl carbon of the ethyl group gives rise to a tertiary structure which precludes straightforward β-oxidation reactions as for toluene. Indeed, Kniemeyer *et al*. (2003) also positively detected 4-phenylpentanoic acid as a metabolite, suggesting that further decomposition of (1-phenylethyl)succinate proceeds via a C-skeleton rearrangement process similar to that recently postulated for anaerobic alkane metabolism [Figure 7B; described below; (Wilkes *et al*., 2002)]. The difference in activation reactions for ethylbenzene under nitrate- and sulfate-reducing conditions may lie in the fact that dehydrogenase reactions require a more positive redox potential (> 0 V) than is conducive in a sulfate-reducing environment (< 0 V). In contrast, activation of ethylbenzene involving a radical mechanism such as fumarate addition would be feasible at lower redox potentials (Kniemeyer *et al*., 2003).

Recent investigations have demonstrated that fumarate addition is also a mechanism whereby alkanes can be activated for anaerobic decomposition. Initial studies with a highly-purified sulfate-reducing enrichment (Kropp *et al*., 2000) and a nitrate-reducing isolate (strain HxN1) (Rabus *et al*., 2001) demonstrated the formation of a fumarate addition metabolite from dodecane and hexane, respectively. Based on gas chromatographic-mass spectral (GC-MS) comparisons with authentic standards, fumarate addition was shown to occur at a subterminal position on the alkane (Kropp *et al.*, 2000; Rabus *et al.*, 2001), presumably at the C-2 position (Figure 7C). Such alkylsuccinic acids contain two chiral centers, and GC detection as two peaks suggested the production of diastereomers. This observation may be attributed to a somewhat relaxed stereospecificity at the carbon addition site, assuming the enzyme responsible for fumarate addition to alkanes is similar to BSS (Widdel and Rabus, 2001). Electron paramagnetic resonance analysis suggested the involvement of a radical mechanism in the initial reaction of *n*-hexane metabolism by strain HxN1 (Rabus *et al.*, 2001). Further metabolic studies with the *n*-hexane-utilizing denitrifier demonstrated that (1-methylpentyl)succinate is converted to 4-methyloctanoate. Mass spectral studies using stable isotopes showed that this latter metabolite is not merely produced by decarboxylation, but possibly via a carbon skeleton rearrangement reaction (Figure 7C) analogous to that known for succinyl-CoA conversion to propionyl-CoA via malonyl-CoA (Wilkes *et al.*, 2002). 4-Methyloctanoate is thought to further decay by a β-oxidation pathway, allowing for the regeneration of fumarate (Figure 7C) (Wilkes *et al.*, 2002). Analogous metabolites observed during the sulfidogenic degradation of ethylcyclopentane suggested that cyclic alkanes can also be activated by radical fumarate addition reactions (Rios-Hernandez *et al.*, 2003). Wilkes *et al.* (2003) further observed that when strain HxN1 was incubated with crude oil, a series of C_4 to C_8 *n*-alkanes as well as cyclic alkanes were activated to their appropriate alkylsuccinates and subsequent methyl-branched fatty acids.

To date, all reports of anaerobic alkane activation have pointed to fumarate addition, with one notable exception. The first alkane-degrading anaerobe isolated, the sulfate-reducing strain Hxd3 (Aeckersberg *et al.*, 1991), uses carboxylation as an initial mechanism of alkane activation (So *et al.*, 2003). ^{13}C-Bicarbonate was found to be incorporated into the C-3 position of hexadecane, followed by the elimination of the two adjacent terminal C atoms yielding a fatty acid one C shorter than the parent alkane. The enzymes carrying out such transformations and the nature of the C_2 compound released have yet to be described. The apparent trend of subterminal fumarate addition or carboxylation reaction for alkanes (and for alkylated aromatics such as ethylbenzene) may be explained by lower bond dissociation energies associated with methylene C-H bonds relative to methyl C-H bonds (Rabus and Widdel, 1995; Widdel and Rabus, 2001; Kniemeyer *et al.*, 2003).

The anaerobic activation of polycyclic aromatic hydrocarbons (PAHs) can also proceed via fumarate addition or carboxylation. The identification of naphthyl-2-methyl-succinic acid in the supernatants of a sulfate-reducing enrichment culture incubated with 2-methylnaphthalene demonstrates that fumarate addition can occur for alkylated PAHs (Annweiler *et al.*, 2000). Naphthyl-2-methyl-succinate synthase activity was also demonstrated for this reaction. The fumarate addition metabolite was found to further decompose to naphthyl-2-methylene-succinic and 2-naphthoic acids. In the absence of such methylation, carboxylation has been reported as an initial activation mechanism, also forming 2-naphthoic acid from naphthalene (Zhang and Young, 1997; Meckenstock *et al.*, 2000) and phenanthrenecarboxylic acid from phenanthrene (Zhang and Young, 1997). 2-Naphthoic acid has been found to undergo ring reduction reactions to form tetrahydro-, octahydro-, and decahydronaphthoic acids, seemingly analogous to the ring reduction occurring during anaerobic benzoate decay (Annweiler *et al.*, 2000; Meckenstock *et al.*, 2000; Zhang *et al.*, 2000). The detection of identical ring fission products produced during naphthalene, 2-methylnaphthalene, or 1, 2, 3, 4-tetrahydronaphthalene degradation demonstrated that ring cleavage proceeded through cyclohexane structures and not via monoaromatic structures during sulfidogenic PAH decay (Annweiler *et al.*, 2002).

Thus, to date, fumarate addition and carboxylation reactions have emerged as dominant mechanisms used by anaerobic microorganisms to initiate the biodegradation of alkylbenzenes, alkanes, and PAHs. However, studies examining the anaerobic metabolism of benzene have thus far revealed no experimental data to support such seemingly universal hydrocarbon activation mechanisms. The benzene molecule presents a unique challenge for anaerobes in that it is characterized by high resonance energy and lacks pi electron withdrawing groups that help destabilize the aromatic nucleus. A recent review by (Coates *et al.*, 2002) discusses several pathways by which anaerobic benzene decay may theoretically proceed, with the majority of studies supporting initial hydroxylation to phenol followed by conversion to benzoate. The lack of an anaerobic benzene-degrading isolate has hampered detailed benzene activation studies, although this may change with the recent unprecedented isolation of two *Dechloromonas* strains capable of anaerobic benzene oxidation (Coates *et al.*, 2002).

8.2. Anaerobic Degradation of Aromatic Compounds

The benzene ring is the second most abundant chemical structure on the planet and many compounds that contain this structural feature enter anaerobic environments. Initial anaerobic transformations of aromatic compounds often involve the oxidation or reduction of substituent groups, carbon-carbon cleavage of substituents from the ring, decarboxylations, or the removal of O-methyl, sulfur, nitrogen, or halogen groups from the ring (Heider and Fuchs, 1997a; 1997b; Schink and Müller, 2000; Gibson and Harwood, 2002). Most of the time, these transformations lead to the conversion of diverse aromatic compounds to benzoate or its CoA thioester (Gallert *et al.*, 1991; Gibson *et al.*, 1994; 1997; Breese and Fuchs, 1998; Hirsch *et al.*, 1998). Exceptions include the conversion of methoxylated aromatic compounds such as lignin monomers to polyphenolic compounds that are then transformed to phloroglucinol prior to ring cleavage (Kaiser and Hanselmann, 1982; Krumholz *et al.*, 1987). Some dihydroxyphenols and benzoates are metabolized to resorcinol prior to ring cleavage (Schink and Müller, 2000). The polyphenolic compounds that have two or three hydroxyl groups in *meta*-positions relative to each other have little aromaticity and ring reduction does not require the input of energy (Haddock and Ferry, 1989; Kluge *et al.*, 1990; Brune and Schink, 1992). Recently, hydroxyhydroquinone has been shown to be a central intermediate in the degradation of resorcinol and 3,5-dihydroxybenzoate in denitrifying bacteria and its oxidation to hydroxybenzoquinone should make it prone to ring fission (Schink and Müller, 2000).

The reduction of benzoyl-CoA represents a considerable energy barrier for anaerobic microorganisms. The presence of the thioester group on the CoA moiety substantially lowers the mid-point potential of the first electron transfer, from -3.1 V for benzene to -1.9 V for benzoyl-CoA (Heider and Fuchs, 1997a; Boll and Fuchs, 1998; Boll *et al.*, 2002). However, the reduction of benzoyl-CoA still represents a major energy barrier since most cellular electron donors would have mid-point potentials near -0.4 V. In *T. aromatica*, electrons for benzoyl-CoA reduction are supplied by a low-potential ferredoxin that accepts electrons from 2-oxoglutarate: ferredoxin oxidoreductase in the citric acid cycle (Boll and Fuchs, 1998; Ebenau-Jehle *et al.*, 2003). Benzoyl reductase in *T. aromatica* catalyzes the reduction of benzyol-CoA to cyclohexa-1, 5-diene-1-carboxyl-CoA coupled to the hydrolysis of ATP to ADP and P_i (Koch *et al.*, 1993; Boll and Fuchs, 1995; Boll *et al.*, 1997; 2000). *Rhodopseudomonas palustris* contains *badDEFG* genes whose deduced amino acid sequences have high degree of identity to the BcrCBAD subunits of the benzoyl-CoA reductase of *T. aromatica* (Boll and Fuchs, 1995; Egland *et al.*, 1997; Breese *et al.*, 1998). The amino acid sequence of benzoyl-CoA reductase has similarity to only one other protein, 2-hydroxyglutaryl-CoA dehydratase, which catalyzes the reversed dehydration of (*R*)-2-hydroxyglutaryl-CoA to glutaconyl-CoA (Egland *et al.*, 1997; Harwood *et al.*, 1999). Activation of 2-hydroxyglutaryl-CoA dehydratase requires the ATP-driven electron transfer from a protein called HgdC.

The ATP-driven electron transfer by benzoyl-CoA reductase and HgdC are similar to that catalyzed by the Fe-protein in nitrogenase (see above). However, no sequence homology exists between nitrogenase and benzoyl-CoA reductase. The ATP-binding fold of benzoyl-CoA reductase is predicted to be similar to the ASKHA super family (acetate and sugar kinases, heat shock protein cognate, actin) where an enzyme-phosphate intermediate is involved in the reaction (Unciuleac and Boll, 2001). An enzyme-phosphate intermediate has been detected and its formation and turnover depends on the redox state of the enzyme and the presence of substrate, consistent with the role of an enzyme-phosphate intermediate in the catalytic cycle of the enzyme (Unciuleac and Boll, 2001). ATP hydrolysis initiates the switch of a 4Fe-4S cluster from a low spin (S = 1/2) to a high spin (S = 7/2) state along with magnetic coupling of two 4Fe-4S clusters (Boll *et al.*, 2001; 2002). Electron paramagnetic resonance studies detected two radical signals that have *g* values greater than 2.01, suggesting a thiyl or disulfide radicals rather than carbon radicals (Boll *et al.*, 2001). A disulfide radical is attractive since its redox potential would be very negative ($E^{o\prime}$ of –1.6 V), sufficient for electron transfer to the aromatic ring of benzoyl-CoA.

8.3. Unusual Dehydration Reactions

The most common type of dehydration is the β–elimination of water. The hydrogen that is removed is located adjacent (alpha) to a carboxyl, oxo, or thiol ester group and the hydroxyl leaves from the β–position (Buckel, 1996). Examples of such reactions are the dehydration of citrate to *cis*-aconitate and its hydration to (2R,3S)-isocitrate in the citric acid cycle, the metabolism of 2-phosphoglycerate to phosphoenolpyruvate in the Embden-Meyerhoff-Parnas pathway, and the β-oxidation of fatty acids. However, many anaerobes catalyze dehydrations where the hydrogen atom that is removed is not adjacent to an activated carbon. As we will see, the dehydration of substrates with hydroxyl groups in the 2, 3, 4, 5, or *n* positions transiently transfer *n*-3 electrons to enable the dehydration (Buckel, 2001). Two of these dehydrations require radical formation.

Acidaminococcus fermentans, Clostridium sporosphaeroides, Clostridium symbiosum, Fusobacterium nucleatum, and *Peptostreptococcus asaccharolyticus* degrade glutamate to acetate and butyrate by the 2-hydroxyglutarate pathway (Figure 8A) (Buckel, 1996; 2001). The 2-hydroxyglutaryl-CoA dehydratase (HgdAB) is a heterodimer that contains an 4Fe-4S cluster, riboflavin-5'-monophosphate, 0.3 riboflavin, and 0.1 to 0.2 molybdenum per heterodimer (Buckel, 1980). An extremely oxygen-sensitive protein, HgdC, is required for activation of 2-hydroxyglutaryl-CoA dehydratase. HgdC uses ATP to provide low potential electrons for the activation of 2-hydroxyglutaryl-CoA dehydratase. The proposed mechanism involves a one-electron reduction of 2-hydroxyglutaryl-CoA to form a ketyl radical. The hydroxyl group is eliminated forming an enoxy radical, which is deprotonated to form the corresponding ketyl radical of glutaconyl-CoA. A final

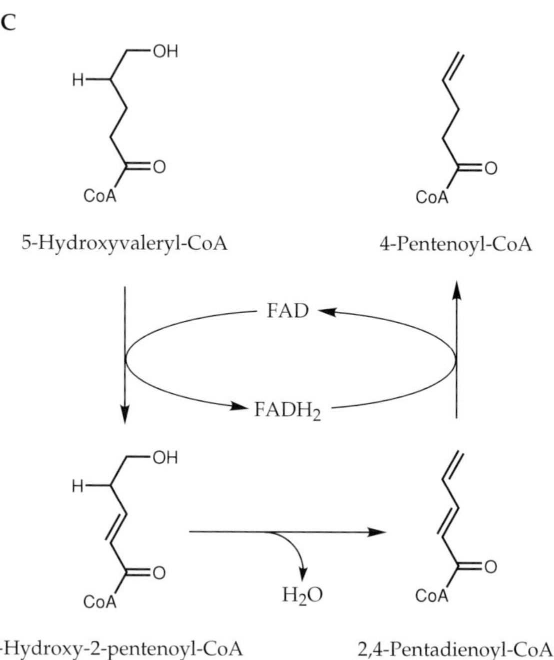

Figure 8. Free radical reactions occurring in the dehydration of (R)-2-hydroxyglutarate (A), 4-hydroxybutyrate (B), and 5-hydroxybutyrate (C) (adapted from Buckel, 1996 and 2001).

oxidation forms glutaconyl-CoA and the electron is recycled back to the enzyme. The one-electron cycle can last for many turnovers of the enzyme, which explains why only catalytic amounts of ATP are required for the reactions. Similar mechanisms are probably operative in the degradation of lactate to propionate by the acrylate pathway and phenyllactate to phenylpropionate (Buckel, 2001).

Escherichia coli decarboxylates glutamate to 4-aminobutyrate during acid stress. The latter compound is degraded to acetate and butyrate (equation 10) by *Clostridium aminobutyricum.*

$$2\ \text{4-aminobutyrate} + 2\ H_2O \rightarrow 2\ \text{acetate}^- + \text{butyrate}^- + 2\ NH_4^+ + H^+ \qquad (10)$$

After removal of the amino group and the addition of CoA, 4-hydroxybutyryl-CoA is formed (Figure 8B). The unactivated β–hydrogen has to be removed. In this case, 4-hydroxybutyryl-CoA is oxidized by FAD present on the dehydratase to form an enoxyl radical. This makes the β–hydrogen acidic enough for deprotonation resulting in a ketyl radical, from which the hydroxyl group is removed to form a dienoxy radical (Buckel, 1996; 2001). Reduction of the dienoxy radical and protonation results in the formation of crotonyl-CoA.

Clostridium viride ferments 5-aminovalerate to acetate, propionate, and valerate (equation 11).

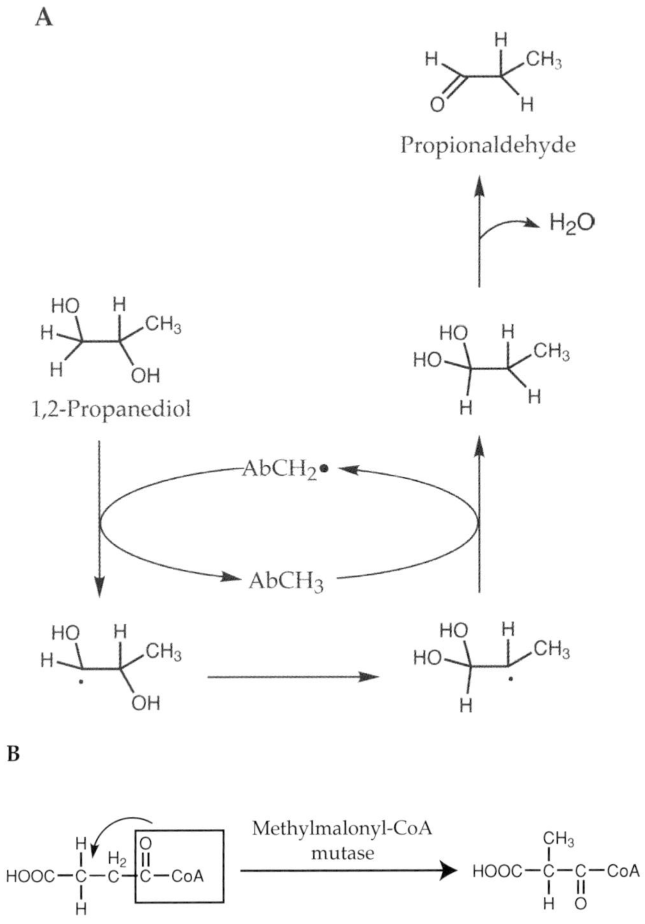

A

Propionaldehyde

H_2O

1,2-Propanediol

$AbCH_2\bullet$

$AbCH_3$

B

Methylmalonyl-CoA mutase

Figure 9. Cofactor B_{12} catalyzed reactions involving dehydration and carbon-carbon bond cleavage (adapted from Buckel, 1996 and Halpern, 1985). $AbCH_3$, 5'-deoxyadenosine.

$$2\ \text{5-aminovalerate} + 2\ H_2O \rightarrow \text{acetate}^- + \text{propionate}^- + \text{valerate}^- + 2\ NH_4^+ + H^+ \quad (11)$$

An aminotransferase and a CoA transferase convert 5-aminovalerate to 5-hydroxyvaleryl-CoA (Figure 8C) (Buckel, 1996; Buckel, 2001). A bifunctional enzyme, 5-hydroxyvaleryl-CoA dehydrogenase/dehydratase, uses FAD to introduce a double bond to form 5-hydroxyl-2-pentenoyl-CoA. The double bond allows the activation of the γ-hydrogen so it can be deprotonated. This is followed by dehydration to form 2,4-pentadienoyl-CoA, which is further reduced by $FADH_2$ to 4-pentenoyl-CoA.

8.4. Cobalamin-Mediated Reactions

Ethylene glycol, 1,2-propanediol, and glycerol are anaerobically metabolized to acetaldehyde, propionaldehyde, and 3-hydroxypropionaldehyde, respectively. These substrates lack any activating group, so how does dehydration occur? As with the dehydrations discussed above, a radical mechanism is used; however, the radical is formed by cleavage of the carbon-cobalt bond of cofactor B_{12}, 5'-deoxyadenosylcobalamin (Halpern, 1985; Buckel, 1996; Roth *et al.*, 1996). A common feature of cofactor B_{12}-mediated reactions is the 1,2-interchange of a hydrogen atom and another substituent [OH, NH_2, C(=O)-CoA, $C(=CH_2)COOH$, or $CH(NH_2)COOH$] on the adjacent carbon atom of the substrate (Halpern, 1985). For example, in the dehydration of 1,2-propanediol (Buckel, 1996), the homolysis of the carbon-cobalt bond forms a 5'-deoxyadenosine radical that abstracts a hydrogen atom at carbon-1 (Figure 9A). The hydroxyl group on carbon-2 migrates to carbon-1, forming a *gem*-1,1-diol. The hydrogen on 5'deoxyadenosine is donated to carbon-2, regenerating the 5'-deoxyadenosine radical. The *gem*-1,1-diol loses H_2O to form propionaldehyde. The reaction sequence may include the formation of ketyl and enoxy radicals as discussed above (Buckel, 2001).

Cofactor B_{12} is also involved in a number of carbon-carbon bond cleavage/rearrangement reactions catalyzed by various mutases (Halpern, 1985; Roth *et al.*, 1996). An important cofactor B_{12}-mediated reaction in many anaerobes is the conversion of succinyl-CoA to methylmalonyl-CoA by methylmalonyl-CoA mutase (Figure 9B). This reaction is found in anaerobes that utilize the randomizing or methylmalonyl-CoA pathway to make propionate. Other cofactor B_{12}-mediated reactions occur in the anaerobic metabolism of amino acids, including the formation of methylaspartate from glutamate, and 2,5-diaminohexanoate and 3,6-diaminohexanoate from α- and β-lysine, respectively (McInerney, 1988; Buckel, 2001).

9. Fermentations and Substrate-Level Phosphorylation

The metabolism of sugar, amino acids, and nucleotide bases results in the formation of phosphoacyl anhydrides (acetyl phosphate, 1,3-bisphosphoglycerate and carbamyl phosphate), acyl anilide (N^{10}-formyl-tetrahydrofolate), and/or the phosphoenol ester, phosphoenol pyruvate (Thauer *et al.*, 1977). All of these compounds have large negative free energies of hydrolysis and thus can be used to synthesize ATP by phosphoryl transfer. These phosphorylated "high energy" intermediates are often synthesized by oxidation-reduction reactions. The exceptions are the formation of acetyl-phosphate by lyase reactions involved in xylulose-5-phosphate metabolism and by pyruvate-formate lyase. The latter reaction forms acetyl-CoA that can be converted to acetyl-phosphate by phosphotransacetylase. Carbamyl phosphate and N^{10}-formyl-tetrahydrofolate are also synthesized by lyase reactions. Carbamyl phosphate is synthesized during arginine metabolism and N^{10}-formyl-tetrahydrofolate is made during purine degradation. Two pathways for sugar, amino acid, and nucleotide metabolism are discussed in detail by Gottschalk (1985) and White (2000) and space prohibits a detailed discussion here. However, some salient points concerning the factors that affect the energy gain during fermentations will be discussed.

The mechanisms by which fermentative bacteria form "high energy" intermediates and reoxidize reduced cofactors dictate a net ATP gain. As an example, streptococci ferment glucose to two lactate molecules using the Embden-Myerhof-Parnas pathway. Two ATP are used to activate glucose to fructose-1,6-bisphosphate. The pathway forms two glyceraldehyde-3-phosphates that are metabolized to two pyruvates. This portion of the pathway forms four ATP by substrate-level phosphorylation and 2 NADH. The latter are reoxidized by reducing 2 pyruvates to 2 lactates. As a result, streptococci have a net gain of 2 ATP per glucose fermented. *Leuconostoc* species use the hexose monophosphate pathway where glucose is oxidized to xylulose-5-phosphate and CO_2 using 1 ATP for activation and generating 2 NADPH. Xylulose-5-phosphate is cleaved to acetyl-phosphate and glyceraldehyde-3-phosphate by phosphoketolase. The latter compound is metabolized in an identical manner as in the Embden-Myerhof-Parnas pathway forming pyruvate, NADH, and 2 ATP. The hexose monophosphate pathway forms 3 reduced cofactors (1 NADH and 2 NADPH) that are reoxidized by reducing pyruvate to lactate and acetyl-phosphate to ethanol (consuming 2 NADPH). Thus, while *Leuconostoc* species generate the "high-energy" intermediate, acetyl-phosphate, they use this compound as an electron acceptor rather than for ATP synthesis. *Leuconostoc* species thus only gain 1 ATP per glucose fermented.

Propionate and butyrate are more reduced than the starting substrate, glucose. Thus, the formation of these products will spare some of the pyruvate from being used as an electron acceptor. In propionate and butyrate fermentations (Table 3), some of the pyruvate is metabolized to acetate and CO_2. Each additional acetate formed results in the synthesis of an additional ATP. The formation of propionate involves the reduction of fumarate to succinate and ATP can be synthesized from the proton motive force generated by the electron transport chain involved in fumarate reduction. The net ATP gain per glucose for butyrate and propionate producers is 3.3 and 4.0, respectively. Bifidobacteria form acetyl-phosphate by action of phosphoketolases that do not lead to the formation of reduced cofactors. The ability to generate a "high energy" intermediate without concomitant formation of reduced cofactors results in a net gain of 2.5 ATP per glucose.

As discussed above, pyruvate metabolism is central to the energetics of fermentative bacteria. There are two enzymes that anaerobes use to cleave pyruvate, both by radical mechanisms. Pyruvate:ferredoxin (Fd) oxidoreductase (equation 12) uses thiamine pyrophosphate as a cofactor to decarboxylate pyruvate by a radical mechanism (Furdui and Ragsdale, 2002). The rate of electron transfer from the hydroylethyl-thiamine pyrophosphate radical to the iron-sulfur clusters is enhanced by the presence of CoA.

$$\text{Pyruvate} + \text{CoA} + \text{Fd}_{ox} \rightarrow \text{acetyl-CoA} + \text{CO}_2 + \text{Fd}_{red} \quad \Delta G^{o\prime} = -19.2 \text{ kJ/mol} \qquad (12)$$

This reaction is favorable enough to allow the reoxidation of reduced ferredoxin by hydrogen production. Pyruvate-formate lyase cleaves pyruvate to acetyl-CoA and formate without the formation of any reduced cofactors (equation 13).

$$\text{Pyruvate} + \text{CoA} \rightarrow \text{acetyl-CoA} + \text{formate} \quad \Delta G^{o\prime} = -16.3 \text{ kJ/mol} \qquad (13)$$

Thus, acetyl-CoA can be converted to acetyl-phosphate that is then used for ATP synthesis. Pyruvate-formate lyase is a member of the glycyl radical-forming enzymes (Wagner *et al.*, 1992; Eklund and Fontecave, 1999). The glycyl radical is used to form a sulfur radical at the active site that participates in the carbon-carbon bond cleavage.

Archaea use variations of the Embden-Myerhof-Parnas and Entner-Doudoroff pathways to degrade carbohydrates (Danson, 1988; Adams, 1994). *Pyrococcus furiosus* uses a modified Embden-Myerhof-Parnas pathway to degrade glucose where glyceraldehyde-3-phosphate is oxidized to 3-phosphoglycerate by a glyceraldehyde-3-phosphate:ferredoxin oxidoreductase and is not directly coupled to a phosphorylation step. The conversion of glucose to two pyruvate molecules does not result in net ATP synthesis. Pyruvate is converted to acetyl-CoA by pyruvate:ferredoxin oxidoreductase and acetyl-CoA synthetases I and II that form ATP and acetate. The formation of hydrogen by a membrane-bound hydrogenase is coupled to the formation of a proton motive force that leads to additional ATP synthesis (Sapra *et al.*, 2003). The use of ferredoxin in place of NAD as the electron acceptor for glyceraldehyde-3-phosphate enables energy to be conserved by hydrogen production.

10. Generation of a Chemiosmotic Potential

Mitchell's original formulation of the chemiosmotic theory proposed that proton translocation occurred by a redox loop where two electrons are transferred from the positive side of the membrane to the negative side of the membrane where they reduce quinone to quinol in conjunction with the uptake of two protons. The quinol diffuses back to the positive side of the membrane where it is oxidized to quinone with the release of two protons on the outside of the membrane. While the mitochondrial respiratory chain in aerobic eukaryotes does not contain such redox loops, many bacteria do. Anaerobes use

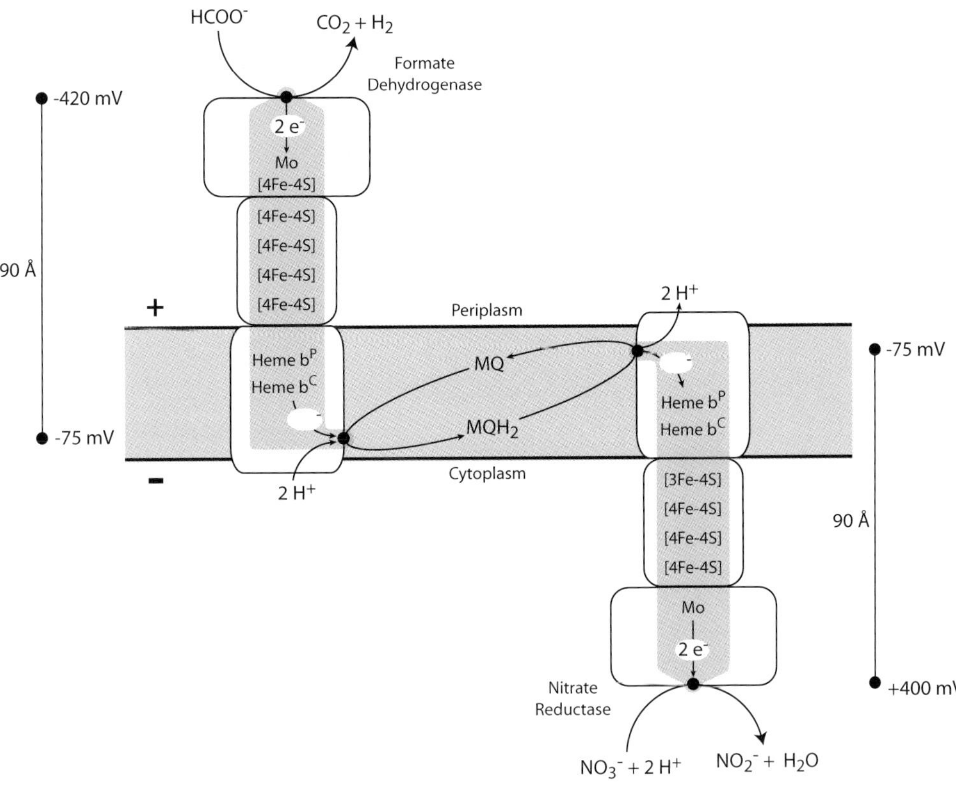

Figure 10. Generation of a proton motive force through the redox loop composed of formate dehydrogenase and nitrate reductase in *Escherichia coli* (adapted from Richardson and Sawers, 2002; Jormakka *et al.*, 2002).

menaquinone instead of ubiquinone to link primary dehydrogenases with terminal reductases since the redox potential of menaquinone (E°' of –74 mV) allows energetically favorable electron transfer to acceptors such as fumarate (E°' of +33 mV for the fumarate/succinate couple). Several examples of redox loops are discussed below. The separation of proton-consuming and proton-producing reactions on opposite sides of the membrane connected by electron transfer across the membrane also results in the formation of a chemiosmotic potential. Such a mechanism appears to be important in the energetics of sulfate-reducing bacteria. Redox-driven ion pumps are another way to form a chemiosmotic potential. Two unusual, primary sodium pumps, biotin-dependent decarboxylases and methyl transferases, are discussed. Lastly, substrate and product gradients can be used to generate a chemiosmotic potential. Examples of such processes are oxalate metabolism and lactate efflux.

10.1. Redox Loops

10.1.1. Nitrate Reduction

The crystal structure of the formate dehydrogenase-nitrate reductase system of *Escherichia coli* shows how such redox loops function (Figure 10) (Jormakka *et al.*, 2002; Richardson and Sawers, 2002). Formate dehydrogenase (Fdh-N for nitrate inducible) is a seleno-molybdenum enzyme. Molybdenum is coordinated with selenocysteine and two molybdenum guanidine dinucleotide cofactors on the periplasmic α subunit and is the site of formate oxidation. The electrons pass from the active site on the α subunit through an electron wire composed of five iron-sulfur clusters, one on the α subunit and four on the β subunit, to two *b*-type hemes located in the integral membrane γ subunit. The electron wire is approximately 90 angstroms long. Electrons are then transferred to menaquinone to form menaquinol on the cytoplasmic side of the membrane. The menaquinol diffuses across the membrane where it is oxidized to menaquinone by nitrate reductase with the release of protons into the periplasm. The similarities in cofactor composition between formate dehydrogenase and nitrate reductase suggest that a similar electron wire transfers electrons released by menaquinol oxidation on the periplamic side of the enzyme to reduce nitrate in the cytoplasm.

10.1.2. Fumarate Reduction

Fumarate is an intermediate in the degradation of a number of different amino acids, organic acids, and glucose (Kröger, 1978). Some bacteria such as *Wolinella succinogenes* rely on fumarate reduction as the sole means of energy conservation. Fumarate respiration is also important for certain anaerobic protozoa and helminthes. In *W. succinogenes*, fumarate reduction with hydrogen as the electron donor leads to the generation of a proton motive force and ultimately ATP synthesis (Kröger, 1978). Fumarate reductase from *W. succinogenes* has three subunits. Subunit A has a covalently-bound FAD

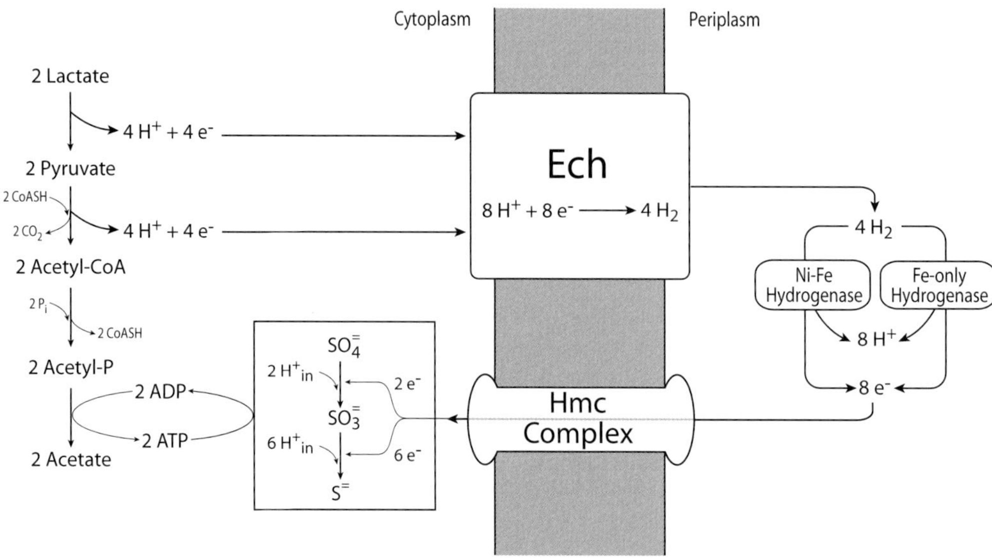

Figure 11. Scheme for energy conservation by hydrogen recycling in *Desulfovibrio vulgaris* (adapted from Voordouw, 2002).

prosthetic group and is the site of fumarate reduction (Lancaster and Kröger, 2000). Subunit B contains three iron-sulfur clusters and is involved in electron transfer. Subunit C is an integral membrane protein with two *b*-type hemes and is the site of menaquinol oxidation. The orientation of the hydrogenase on the periplasmic side and fumarate reductase on the cytoplasmic side of the membrane results in the proton production in the periplasm and proton consumption in the cytoplasm. The location of the menaquinone binding sites on these two enzyme complexes also suggests the presence of a redox loop (Lancaster and Kröger, 2000).

10.1.3. Heterodisulfide Reductase

The heterodisulfide, CoM-S-S-CoB, is the terminal electron acceptor for the redox reactions involved in methane production (Deppenmeier, 2002a; Deppenmeier, 2002b). With hydrogen as the substrate, a membrane-bound Ni-Fe hydrogenase, called F_{420}-nonreducing, oxidizes hydrogen and releases protons to the outside of the cell. The electrons are transferred through an electron wire consisting of iron-sulfur clusters and heme *b*. The electrons and protons from the cytoplasm reduce methanophenazine, a membrane-diffusible redox carrier. Reduced methanophenazine then diffuses to the heterodisulfide reductase system where it is oxidized releasing two protons to the outside of the cell. The heterodisulfide reductase has an integral membrane subunit containing two heme *b* prosthetic groups and a membrane-bound subunit facing the cytoplasm that contains two iron-sulfur clusters and the active site for disulfide reduction. The reduction of CoM-S-S-CoB to HS-CoM and HS-CoB consumes two protons in the cytoplasm. The overall reaction leads to the translocation of four protons to the outside of the cell. The orientation of the components and the location of methanophenazine binding sites are consistent with the F_{420}-nonreducing hydrogenase and heterodisulfide reductase forming a redox loop.

10.2. Separation of Proton-Producing and Consuming Reactions: Sulfate Reduction

Sulfate-reducing bacteria use diverse electron donors to reduce sulfate to sulfide. We will investigate the mechanism by which a proton motive force is created during the oxidation of lactate coupled to sulfate reduction (equation 14).

$$2\ \text{Lactate}^- + SO_4^= \rightarrow 2\ \text{Acetate}^- + 2\ HCO_3^- + HS^- + H^+ \quad (14)$$

The redox potential for the sulfate/sulfite couple ($SO_4^=/HSO_3^-$) is very low (E°' of –516 mV) which makes reduction of sulfate by common physiological donors such as NADH (NAD^+/NADH; E°' of –320 mV) difficult without energy input. For this reason, sulfate is activated to adensosine-5'-phosphosulfate (APS) by ATP sulfurylase (equation 15) before it is reduced to sulfite by adensosine-5'-phosphosulfate reductase (equation 16). Pyrophosphate is then hydrolyzed to phosphate (equation 17).

$$\text{ATP} + SO_4^= + 2\ H^+ \rightarrow \text{APS} + PP_i \quad (15)$$

$$\text{APS} + 2\ e^- + 2\ H^+ \rightarrow \text{AMP} + SO_3^= \quad (16)$$

$$PP_i + H_2O \rightarrow 2\ P_i \quad (17)$$

This sequence of reactions converts ATP to AMP and phosphate, wherein two "high-energy" bonds are consumed. The sulfate reducer obtains 2 ATP by substrate-level phosphorylation during the conversion

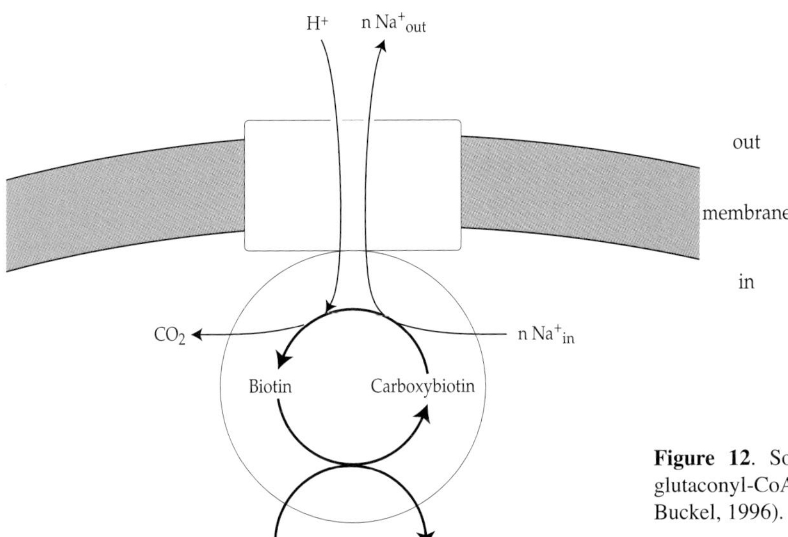

Figure 12. Sodium efflux coupled to the decarboxylation of glutaconyl-CoA in *Acidaminococcus fermentans* (Härtel and Buckel, 1996).

of two lactate molecules to two acetate molecules. This is sufficient for sulfate activation but not growth. Molar growth yields indicate that a net of one ATP is made during sulfate reduction coupled to lactate oxidation (Magee *et al.*, 1978). Thus, ATP must be formed by the electron transport chain involved in sulfite reduction.

A hydrogen recycling model has been proposed for energy production coupled to the reduction of sulfate (Figure 11) (Odum and Peck Jr., 1981). This model proposes that the reducing equivalents generated in the cytoplasm during lactate oxidation to acetate are released as hydrogen. Periplasmic hydrogenases use the hydrogen as it escapes from the cell and electrons flow back to the cytoplasm to reduce sulfate to sulfide. The consumption of 8 protons in the cytoplasm during sulfate reduction along with the production of protons in the periplasm during hydrogen oxidation (8 protons) would form a proton motive force. The question is whether the architecture of the electron transport chain supports such a model.

Sulfate reducers can have three different kinds of hydrogenases, one with only 4Fe-4S clusters designated the Fe hydrogenase, one with nickel and iron (NiFe hydrogenase) and one with nickel, selenium and iron (NiFeSe hydrogenase) (Voordouw, 1995). The Fe hydrogenase and the NiFe hydrogenase are located in periplasm so that the oxidation of hydrogen would produce protons on the outside of the membrane. Deletion of the Fe hydrogenase leads to hydrogen accumulation with lactate as the donor and impairs growth with hydrogen as the donor (Pohorelic *et al.*, 2002; Voordouw, 2002). The periplasm also contains large amounts of a tetraheme *c*-type cytochrome called cytochrome c_3 that may act to shuttle electrons from the hydrogenases to membrane-bound carriers. Cytochrome c_3 mutants have impaired growth with pyruvate but not with lactate as the electron donor (Rapp-Giles *et al.*, 2000). The membrane complex called Hmc that contains *c*-type hemes and iron-sulfur clusters is probably involved in transfer of electrons though the membrane to APS reductase and bisulfite reductase (Voordouw, 1995; Dolla *et al.*, 2000). Deletion of the *hmc* operon leads to hydrogen accumulation with lactate as the donor (Voordouw, 2002) and impairs growth with hydrogen as the electron donor (Dolla *et al.*, 2000). Thus, the arrangement of electron transport proteins involved in sulfate reduction is consistent with the hydrogen recycling model. A recent study even suggests that sulfate reducers may cycle carbon monoxide as well (Voordouw, 2002).

10.3. Sodium-Pumps

10.3.1. Biotin-Dependent Decarboxylases

A number of facultative and anaerobic bacteria use the energy of decarboxylation to generate a sodium gradient (Buckel and Semmler, 1982.; 1983; Dimroth and Schink, 1998). Oxaloacetate decarboxylase in *Klebsiella pneumoniae* was the first system shown to function as a sodium pump. Oxaloacetate decarboxylase has one soluble (α) and two membrane-bound subunits (β and γ) that form a complex in the membrane. A biotin group bound to the α subunit accepts the carboxyl group from oxaloacetate; the carboxybiotin moves to the beta subunit where the decarboxylation occurs. During the decarboxylation, one to two sodium ions are translocated across the membrane into the periplasm and one proton moves from the periplasm into the cytoplasm (Figure 12) (Dimroth and Thomer, 1993). Biotin-dependent decarboxylases that act as primary sodium pumps are involved in the metabolism of succinate by *Propionigenium modestum*, glutamate metabolism by *Acidaminococcus fermentans* and strain

WoG13, and malonate metabolism by *Malonomonas rubra* and *Sporomusa malonica* (Buckel and Semmler, 1982.; Buckel and Semmler, 1983; Dehning and Schink, 1989; Beatrix *et al.*, 1990; Matthies and Schink, 1992; Dimroth and Schink, 1998). Hydrogen production from glutamate by *A. fermentans* is sodium dependent indicating the involvement of a sodium- and biotin-dependent decarboxylase (Härtel and Buckel, 1996). At high sodium concentrations, the proton consumed during the decarboxylation of glutaconyl-CoA comes from the cytoplasm rather than from the periplasm. This leads to the formation of a pH gradient that is used to drive the unfavorable transfer of electrons from NADH to hydrogen by membrane-bound NADH dehydrogenase and hydrogenase.

10.3.2. Membrane-Bound Methyltransferases

The methyl-H_4MTP:CoM methyltransferase acts as a primary sodium pump where 2 sodium ions are translocated per methyl group (Deppenmeier, 2002a). The protein generates a sodium ion gradient during methanogenesis from CO_2 or acetate. During growth on methylated one-carbon compounds, the sodium gradient is used to drive the unfavorable methyl transfer from CH_3-S-CoM to H_4MTP.

10.4. Energy from Substrate or Product Gradients

Gradients of substrate or products can be used to drive the net movement of negative charges inwards or the net movement of positive charges outwards from the cell (Konings *et al.*, 1994). Both cases will lead to the development of a membrane potential. Examples of such systems are the exchange of formate for oxalate and the efflux of end-products such as lactate and succinate.

10.4.1. Oxalate Metabolism

Oxalobacter formigenes is a rumen bacterium that metabolizes oxalate to formate and carbon dioxide (Anantharam *et al.*, 1989). Growth yields are about 1 g (dry wt) of cells per mole of oxalate degraded. The pathway for oxalate metabolism does not involve substrate-level phosphorylation. Oxalate is activated to oxalyl-CoA by transfer of the CoA group from formyl-CoA. Oxalyl-CoA is decarboxylated to formyl-CoA and carbon dioxide by a cytoplasmic decarboxylate. This reaction also consumes one proton. The bacterium possesses an active, membrane-bound oxalate$^{=}$: formate^{-} antiporter that catalyzes the rapid electrogenic exchange of these two compounds across the cell membrane (Anantharam *et al.*, 1989). The exchange of two negative charges for one negative charge generates a membrane potential. The consumption of a proton during decarboxylation leads to a more alkaline cytoplasm and the development of a pH gradient. Other examples of precursor/product antiporters forming chemiosmotic potentials are discussed by Konings *et al.* (1994; 1995).

10.4.2. End-product Efflux

The analysis of microorganisms as open systems that exchange materials with the environment discussed above leads to the realization that mass transfer of end products could be energy yielding, e. g., lead to the generation of a proton motive force (McInerney and Beaty, 1988). The excretion of lactate by lactic acid bacteria is an example of such a process (Michels *et al.*, 1979; Konings *et al.*, 1994). In the early phases of growth, lactate concentrations inside of *Streptococcus cremoris* approach 200 mM corresponding to a chemical potential of about 80 mV (ten Brink and Konings, 1982). This chemical gradient could be used to drive the electrogenic efflux of protons from the cell (two protons per lactate). The efflux of lactate from whole cells and membrane vesicles is carrier-mediated and leads to the generation of a proton motive force (Otto *et al.*, 1980; 1982). Maintaining a lower external lactate concentration by a second bacterium that used lactate increased the molar growth yield of *S. cremoris* (Otto *et al.*, 1980).

Succinate efflux by membrane vesicles of the rumen bacterium, *Selenomonas ruminantium,* was linked to sodium excretion and resulted in the development of a membrane potential (Michel and Macy, 1990). The molar growth yields of *S. ruminantium* were higher under conditions where succinate production was maximal, suggesting that the process may generate additional energy for biosynthesis. Succinate is an important intermediate in the rumen and is rapidly decarboxylated to propionate. The rapid decarboxylation of succinate would keep the extracellular concentration of succinate low; this would allow succinate excreters to generate additional energy by electrogenic efflux of succinate. The succinate-decarboxylating bacteria would benefit by obtaining additional substrate for growth.

10.5. Operating with Small Energy Budgets

Syntrophic benzoate metabolism represents a paradox. The free energy change is low, about –33 kJ/mol (Jackson and McInerney, 2002), much less than that needed to synthesize an ATP. How do these associations gain energy for growth from reactions that yield very little energy (Table 1)? The activation of benzoate to benzoyl-CoA by a benzoate-CoA ligase results in the formation of AMP and pyrophosphate from ATP (equation 18).

$$\text{Benzoic acid} + \text{CoA} + \text{ATP} \rightarrow \text{benzoyl-CoA} + \text{AMP} + \text{PP}_i \qquad (18)$$

The metabolism of benzoate to acetate forms 3 ATP molecules by substrate-level phosphorylation from acetyl-phosphate. Schöcke and Schink (1998) found that *S. gentianae* contains a membrane-bound, proton-translocating pyrophosphatase that can synthesize about 1 ATP for every 3 pyrophosphate molecules hydrolyzed. Thus, about a third of the energy lost in reaction 17 can be converted into ATP by this reaction. The decarboxylation of glutaconyl-CoA is linked to sodium-ion translocation (Schöcke and Schink, 1999), which may result in the synthesis of one-third of an ATP by the generation of a sodium gradient (Buckel and Semmler, 1982.; Buckel and Semmler, 1983). Thus, about 3 2/3 ATP equivalents could be formed per benzoate consumed (degraded).

There are several steps where significant investment of ATP or chemiosmotic energy may be needed. The activation of benzoate to benzoyl-CoA by a benzoate-CoA ligase results in the formation of AMP and pyrophosphate from ATP (reaction 17), equivalent to the loss of two ATP molecules. The redox potential of pimelyl-CoA and glutaryl-CoA ($E^{o\prime}$ of -120 mV) (Schink, 1997) is much more positive than the hydrogen couple ($E^{o\prime}$ of -414 mV). Hydrogen production from these high-potential donors would be endogonic at the hydrogen levels (about 10 Pa) found in growing cocultures of *S. gentianae* (Schöcke and Schink, 1997). About two-thirds of an ATP would be needed for the oxidation of each of these acyl-CoA intermediates (Schöcke and Schink, 1999). If the reduction of benzoyl-CoA to a diene intermediate is similar to that found in *T. aromatica*, where two additional ATP molecules are required for the initial reduction of the ring (Boll *et al.*, 1997), more than 4 ATP equivalents would be needed to metabolize benzoate syntrophically. The only way that net ATP synthesis can occur is if benzoyl-CoA reduction does not require energy input. The addition of four or more electrons onto the aromatic ring is exergonic (Schöcke and Schink, 1999; Elshahed and McInerney, 2001a). *S. aciditrophicus* transiently produces large amounts of cyclohexane carboxylate during syntrophic benzoate metabolism (Elshahed *et al.*, 2001b) consistent with the idea that ring reduction involves a reaction that would not be energy requiring. Cyclohexane carboxylate is subsequently degraded to acetate and CO_2 by *S. aciditrophicus* if a hydrogen user is present. The production of large amounts of cyclohexane carboxylate suggests that the initial reactions involved in syntrophic benzoate metabolism differ from those found in denitrifying and photosynthetic bacteria. Syntrophic associations use a combination of substrate-level phosphorylation and ion translocation reactions in order to conserve energy from reactions that operate near thermodynamic equilibrium (Jackson and McInerney, 2002).

11. Physiological Basis of Anaerobiosis

Oxygen is an interesting molecule in that its thermodynamic potential as an oxidant is very large ($E^{o\prime}$ of +818 mV for O_2/H_2O). However, its orbital structure restricts its reactivity with other molecules. Oxygen has two unpaired electrons of identical spin in its outer bonding orbitals. Molecules that react with oxygen would have to transfer two electrons of the same spin but opposite to that of electrons already present in the outer shell (Farr and Kogoma, 1991; Imlay, 2002). Kinetically, this makes oxygen unreactive with most biomolecules except radicals. The spin restriction can be overcome by interaction with transition metals such as iron or copper that catalyze one-electron reductions. Aerobically, respiratory flavoproteins and the cytochrome bc_1 complex react with oxygen to form superoxide (O_2^-) and hydrogen peroxide (H_2O_2) (Farr and Kogoma, 1991; Imlay, 2002). These enzymes have semiquinones as part of their normal catalytic cycle and can catalyze the reduction of oxygen to superoxide or hydrogen peroxide. Superoxide reacts with the solvent-exposed, iron-sulfur clusters present in many enzymes resulting in the destruction of the cluster by loss of a ferrous atom. Superoxide and hydrogen peroxide can then react with free iron in a Fenton-type reaction to generate hydroxyl radicals. Hydroxyl radicals are very reactive and are responsible for much of the damage that occurs from oxygen exposure. This section will discuss the effect of oxygen on anaerobes and the kinds of defense mechanisms that anaerobes have to deal with reactive oxygen species.

Aerobes have evolved a number of mechanisms to protect themselves from the reactive oxygen species discussed above. These include non-enzymatic detoxification by glutathione as well as enzymatic detoxification by superoxide dismutase (equation 19), catalase (equation 20) and a number of peroxidases (equation 21) (Farr and Kogoma, 1991; Imlay, 2002; Pomposiello and Demple, 2002).

$$O_2^- + O_2^- + 2\,H^+ \rightarrow H_2O_2 \quad (19)$$

$$2\,H_2O_2 \rightarrow 2\,H_2O + O_2 \quad (20)$$

$$H_2O_2 + RH_2 \rightarrow 2\,H_2O + R \quad (21)$$

Traditionally, anaerobes have been defined as organisms that are unable to survive and grow in the presence of oxygen. However, it is now clear that some anaerobes survive for long periods of time in the presence of oxygen and even use oxygen as an electron acceptor (see Chapter 8). Acetogens have been isolated from well-drained, oxic soils (Wagner *et al.*, 1996) and are active in termite guts where steep oxygen gradients exist (Tholen and Brune, 1999). Several acetogens can grow aerobically (Karnholz *et al.*, 2002) and reduce oxygen coupled to the oxidation of hydrogen (Boga and Brune, 2003). Sulfate-reducing bacteria are often found at the oxic-anoxic interfaces of stratified systems (Cypionka,

2000). *Desulfovibrio desulfuricans* CSN uses oxygen faster than a number of other electron acceptors including oxyanions of sulfur (Krekeler and Cypionka, 1995). *Desulfovibrio termitidis* is abundant in the gut of termites where its rapid use of oxygen may allow the formation of anaerobic conditions (Cypionka, 2000). The use of oxygen by some *Desulfovibrio* species results in proton translocation across the cell membrane. While some sulfate-reducing bacteria have the ability to use oxygen, aerobic growth of sulfate-reducing bacteria is often poor resulting in no more than one doubling of cells. However, oxygen-dependent growth of *Desulfovibrio vulgaris* has been reported (Johnson *et al.*, 1997). Even in the deep ocean, episodic exposure to oxygen from down-welling cold waters occurs (Lloyd, 1999). Thus, while we have thought that anaerobes would not normally encounter oxygen in their natural habitats, it is clear that some do. Thus, they must have evolved mechanisms to protect themselves against reactive oxygen species.

11.1. What Makes Anaerobes so Sensitive to Oxygen?

The basis of obligate anaerobiosis was thought to reside in whether the organism had a complete set of detoxification enzymes. Initial studies indicated that strict anaerobes lacked both superoxide dismutase and catalase while aerotolerant anaerobes contained low levels of superoxide dismutase but not catalase (McCord *et al.*, 1971). However, further studies indicated that some strict anaerobes contain both enzymes (Morris, 1975). While the presence of detoxification enzyme systems is a critical factor for oxygen tolerance, the dependency of anaerobes on low-potential biochemistry in order to generate energy also makes them vulnerable to oxygen. Exposure to air disrupted the energy metabolism of *Bacteroides thetaiotaomicron* because of the loss of two central enzymes, fumarase and pyruvate:ferredoxin oxidoreductase (Pan and Imlay, 2001). Fumarase is needed to form succinate from pyruvate. Succinate formation allows the cell to reoxidize its reduced NADH as well as make ATP. Pyruvate:ferredoxin oxidoreductase is needed for acetate and ATP production. Fumarase belongs to the 4Fe-4S dehydratase family and contains an iron-sulfur cluster sensitive to superoxide. Pyruvate: ferredoxin oxidoreductase is also an iron-sulfur protein and is inactivated by oxygen but not superoxide.

There are a number of highly oxygen sensitive enzyme systems that anaerobes use for metabolism (Imlay, 2002). The iron-sulfur dehydratases such as fumarase discussed above is one example of a highly oxygen sensitive enzyme. The sensitivity of iron-sulfur dehydratases probably limits the oxygen tolerance of aerobes as well. Ferrodoxin-dependent dehydrogenases are also highly oxygen sensitive. These enzymes are involved in the degradation of a number of important compounds such as pyruvate, 2-oxoketoglutarate, indolepyruvate, and branched-chain 2-oxoacids. The oxidoreductases involved in these reactions contain multiple iron-sulfur clusters, some of which are solvent-exposed. This may be why they are rapidly inactivated by oxygen since 2-oxoacid:ferredoxin oxidoreductases of aerobes and archaea, which are not sensitive to oxygen, have only one iron-sulfur cluster that is not solvent-exposed. The ferredoxin-dependent oxidoreductases that contain tungsten and pterin cofactors are also very oxygen sensitive. The flavin-dependent dehydratases discussed above are highly oxygen sensitive. These enzymes are involved in difficult dehydration reactions such as the metabolism of lactyl-CoA, phenyllactyl-CoA, 2-hydroxyglutaryl-CoA, and 4-hydroxyglutaryl-CoA. The dehydratase is not particularly oxygen sensitive, but the activating protein has a 4Fe-4S cluster that bridges the dimer interface. This cluster is solvent-exposed and readily degrades to a 2Fe-2S cluster upon oxygen exposure. All of the above enzymes have one common feature; they contain a solvent-exposed 4Fe-4S cluster that rapidly degrades when exposed to oxygen. The last type of highly oxygen sensitive enzyme includes those that involve glycyl radicals (Imlay, 2002). These include pyruvate-formate lyase, anaerobic ribonucleotide reductase, benzylsuccinate synthase, and 2-oxoacid-formate lyase (for threonine degradation).

11.2. Protective Enzymes

Many anaerobes have an enzyme system that protects them from reactive oxygen species without the regeneration of oxygen. First, superoxide reductase (Sor) (rubredoxin oxidoreductase) reduces superoxide to hydrogen peroxide (equation 22) (Jenny Jr. *et al.*, 1999; Das *et al.*, 2001; Lumppio *et al.*, 2001; Silva *et al.*, 2001; Fournier *et al.*, 2003).

$$e^- + O_2^- + 2\,H^+ \rightarrow H_2O_2 \qquad (22)$$

Superoxide reductase is a homodimeric protein and each subunit contains two mononuclear nonheme iron centers. The first center has a distorted rubredoxin-like (Fe-4Cys) arrangement that may be involved in electron transfer. The second center has iron coordinated with histidine and cysteine and may be the site of superoxide reduction. Deletion of the *sor* gene made *Desulfovibrio vulgaris* more air senstitive (Voordouw and Voordouw, 1998) whereas the *sor* gene from this organism restored the growth of superoxide dismutase-defective mutants of *E. coli* (Silva *et al.*, 2001).

The second protein, rubrerythrin (Rbr), is a nonheme iron protein that has a rubredoxin-like iron center and a non-sulfur, di-iron site (Das *et al.*, 2001; Lumppio *et al.*, 2001; Fournier *et al.*, 2003). Rbr has peroxidase activity catalyzing the reduction of hydrogen peroxide to water (equation 23):

$$NADH + H^+ + H_2O_2 \rightarrow NAD^+ + 2\ H_2O \quad (23)$$

Expression of two different plasmid-borne Rbr from *D. vulgaris* increased the viability of a catalase-deficient strain of *E. coli* after exposure to hydrogen peroxide (Lumppio *et al.*, 2001). A *rbr* mutant of *Porphyromonas gingivalis* was more oxygen and hydrogen peroxide sensitive (Sztukowska *et al.*, 2002). The presence of *rbr* in 26 different bacterial and archaeal genomes of obligate anaerobes or microaerophiles indicates that Rbr may be a common protective mechanism against reactive oxygen species (Sztukowska *et al.*, 2002).

Genetic analysis showed that two genes orthologous to the *ydaD and ycdF* genes of *Bacillus subtilis* were required for efficient resistance to oxidative stress in *Clostridium perfringens* (Briolat and Reysset, 2002). These two genes encode for a putative NADPH dehydrogenase. High concentrations of NADPH may be needed to recycle glutathione to its reduced form and suggests the importance of maintaining intracellular redox balance in response to oxidative stress.

Bacteroides fragilis is an aerotolerant, opportunistic pathogen that has a complex oxidative stress response system (Rocha and Smith, 1996). Genetic analysis has implicated roles for catalase and alkyl hydroperoxide reductase in response to peroxide exposure (Rocha and Smith, 1998; 1999). The redox-sensitive, transcriptional activator, OxyR, controls the peroxide-inducible genes *ahpCF, dps*, and *katB* (Rocha *et al.*, 2000). Thus, anaerobes have evolved efficient and diverse protective mechanisms against reactive oxygen species. However, many anaerobes rely on a number of very oxygen-sensitive enzyme systems for metabolism and energy production. The balance between the activity of the protective systems versus the rate of inactivation of critical enzyme systems governs the degree to which anaerobes can tolerate oxygen.

12. Conclusion

The Gibbs free energy change of most anaerobic processes is small compared to aerobic metabolism or denitrification. Because of this, there is a strong effect of concentration (particularly pH) on the Gibbs free energy of the catabolic reaction of some anaerobic processes. This results in the formation of tightly coupled associations of microorganisms that are thermodynamically-based. Thus, in many anaerobic ecosystems, a consortium of microorganisms rather than a single species is the catalytic unit responsible for the degradation of carbon substrates. The amount of ATP that an anaerobe obtains from its substrate depends on the mechanisms by which it is able to reoxidize its reduced cofactors. Some fermentative anaerobes produce products more reduced than the original substrate (propionate or butyrate) and have higher ATP yields from carbohydrates compared to organisms such as the lactic acid-producing bacteria that produce a product, lactate, that has the same oxidation value as the carbohydrate substrate. Many anaerobes reoxidize their reduced cofactors by hydrogen or formate production. When these bacteria grow in association with hydrogen-utilizing bacteria such as methanogens or sulfate-reducing bacteria, hydrogen and formate levels are kept low and all of the reduced cofactors generated in the metabolism of the substrate can be reoxidized by hydrogen or formate synthesis. This maximizes ATP gain by maximizing acetate production. Some anaerobes such as syntrophic bacteria are only able to reoxidize their reduced cofactors by hydrogen or formate production. These bacteria are thus dependent on the activity of hydrogen/formate-using bacteria for their metabolism. The above discussion illustrates the importance of hydrogen and formate in regulating the flow of carbon and electrons in anaerobic ecosystems.

In addition to fermentative metabolism where ATP production occurs mainly by substrate-level phosphorylation, anaerobes can use a diverse number of inorganic or organic compounds as electron acceptors. Energy production by these respiratory processes results in the formation of a chemiosmotic potential that can be used to synthesize ATP or do other kinds of work such as drive transport systems or flagellar movement. Many anaerobic respiratory chains rely on redox loops to create the chemiosmotic potential. Furthermore, redox-driven pumps and the separation of proton-consuming and proton-producing reactions across the membrane are used to generate a chemiosmotic potential. Interestingly, some anaerobes are able to generate a chemiosmotic potential by the electrogenic efflux of compounds across the membrane.

Many anaerobic redox reactions involve a variety of metals such as iron, nickel, tungsten, and molybdenum. These metals perform critical functions in the metabolism of inorganic compounds such as H_2 and N_2, or in CO_2 fixation, methanogenesis, and acetogenesis. The lack of oxygen does present a challenge for the metabolism of certain compounds such as hydrocarbons, aromatic compounds, and compounds with unactivated hydroxyl groups. Anaerobes have evolved novel approaches to activate and metabolize these compounds such as the addition of fumarate or a carboxyl group to initiate hydrocarbon metabolism, the use of ATP to provide low potential electrons needed for aromatic ring reduction, and the oxidation-reduction reactions or cofactor B_{12}-interactions to provide free radicals needed for dehydration reactions.

The dependence of anaerobes on low potential and free radical biochemistry for metabolism and energy production makes these organisms very sensitive to oxygen. Enzymes involved in these types of reactions are very oxygen-labile. The degree to which an organism depends on such oxygen-sensitive systems versus the degree to which reactive oxygen compounds are detoxified will determine oxygen tolerance. Recent work shows that anaerobes have evolved highly effective systems to detoxify reactive oxygen compounds and thus are able to survive and metabolize in aerobic environments.

References

Achong, G.R., Rodriguez, A.M., and Spormann, A.M. 2001. Benzylsuccinate synthase of *Azoarcus* sp. strain T: cloning, sequencing, transcriptional organization, and its role in anaerobic toluene and *m*-xylene mineralization. J. Bacteriol. 183: 6763-6770.

Adams, M.W. 1994. Biochemical diversity among sulfur-dependent, hyperthermophilic microorganisms. FEMS Microbiol. Rev. 15: 261-277.

Aeckersberg, F., Bak, F., and Widdel, F. 1991. Anaerobic oxidation of saturated hydrocarbons to CO_2 by a new type of sulfate-reducing bacterium. Arch. Microbiol. 156: 5-14.

Ahring, B.K., and Westermann, P. 1988. Product inhibition of butyrate metabolism by acetate and hydrogen in a thermophilic coculture. Appl. Environ. Microbiol. 54: 2393-2397.

Alber, B.E., and Fuchs, G. 2002. Propionyl-coenzyme A synthase from *Chloroflexus aurantiacus*, a key enzyme of the 3-hydroxypropionate cycle for autotrophic CO_2 fixation. J. Biol. Chem. 277: 12137-12143.

Allison, M.J., Bryant, M.P., and Doetsch, R.N. 1962a. Studies on the metabolic function of branched-chain volatile fatty acids in ruminococci. I. Incorporation of isovalerate into leucine. J. Bacteriol. 83: 523-532.

Allison, M.J., Bryant, M.P., Katz, I., and Keeney, M. 1962b. Metabolic function of branched-chain volatile fatty acids, growth factors for ruminococci. II. Biosynthesis of higher branched-chain fatty acids and aldehydes. J. Bacteriol. 83: 1084-1093.

Anantharam, V., Allison, M.J., and Maloney, P.C. 1989. Oxalate:formate exchange. The basis for energy coupling in *Oxalobacter*. J. Biol. Chem. 264: 7244-7250.

Andereesen, J.R. 1994. Glycine metabolism in anaerobes. Antonie von Leeuwenhoek. 66: 223-227.

Annweiler, E., Michaelis, W., and Meckenstock, R.U. 2002. Identical ring cleavage products during anaerobic degradation of naphthalene, 2-methylnaphthalene, and tetralin indicate a new metabolic pathway. Appl. Environ. Microbiol. 68: 852-858.

Annweiler, E., Materna, A., Safinowski, M., Kappler, A., Richnow, H.H., Michaelis, W., and Meckenstock, R.U. 2000. Anaerobic degradation of 2-methylnaphthalene by a sulfate-reducing enrichment culture. Appl. Environ. Microbiol. 66: 5329-5333.

Ashida, H., Saito, Y., Kojima, C., Kobayashi, K., Ogasawara, N., and Yokota, A. 2003. A functional link between RuBisCO-like protein of *Bacillus* and photosynthetic RuBisCO. Science. 302: 286-290.

Balch, W.E., and Wolfe, R.S. 1976. New approach to the cultivation of methanogenic bacteria: 2-mercaptoethanesulfonic acid (HS-CoM)-dependent growth of *Methanobacterium ruminantium* in a pressurized atmosphere. Appl. Environ. Microbiol. 32: 781-791.

Ball, H.A., Johnson, H.A., Reinhard, M., and Spormann, A.M. 1996. Initial reactions in anaerobic ethylbenzene oxidation by a denitrifying bacterium, strain EB1. J. Bacteriol. 178: 5755-5761.

Banat, I.M., and Nedwell, D.B. 1983. Mechanism of turnover of C_2-C_4 fatty acids in high-sulfate and low-sulfate anaerobic sediments. FEMS Microbiol. Lett. 17: 107-110.

Barton, L.L., Plunkett, R.M., and Thomson, B.M. 2003. Reduction of metals and nonessential elements. In: Biochemistry and Physiology of Anaerobic Bacteria. L.G. Ljungdahl, M.W. Adams, L.L. Baron, J.G. Ferry, and M.K. Johnson, eds. Springer, New York. p. 220-234.

Bauchop, T., and Mountfort, D.O. 1981. Cellulose fermentation by a rumen anaerobic fungus in both the absence and the presence of rumen methanogens. Appl. Environ. Microbiol. 42: 1103-1110.

Beatrix, B., Bendrat, K., Rospert, S., and Buckel, W. 1990. The biotin-dependent sodium ion pump glutaconyl-CoA decarboxylase from *Fusobacterium nucleatum* (subsp. *nucleatum*). Arch. Microbiol. 154: 362-369.

Beguin, P., and Aubert, J.P. 1994. Biological degradation of cellulose. FEMS Microbiol. Rev. 13: 25-58.

Beller, H.R., and Spormann, A.M. 1997a. Anaerobic activation of toluene and *o*-xylene by addition to fumarate in denitrifying strain T. J. Bacteriol. 179: 670-676.

Beller, H.R., and Spormann, A.M. 1997b. Benzylsuccinate formation as a means of anaerobic toluene activation by sulfate-reducing strain PRTOL1. Appl. Environ. Microbiol. 63: 3729-3741.

Beller, H.R., and Spormann, A.M. 1998. Analysis of the novel benzylsuccinate synthase reaction for anaerobic toluene activation based on structural studies of the product. J. Bacteriol. 180: 5454-5457.

Beller, H.R., and Edwards, E.A. 2000. Anaerobic toluene activation by benzylsuccinate synthase in a highly enriched methanogenic culture. Appl. Environ. Microbiol. 66: 5503-5505.

Beller, H.R., Reinhard, M., and Grbic-Galic, D. 1992. Metabolic by-products of anaerobic toluene degradation by sulfate-reducing enrichment cultures. Appl. Environ. Microbiol. 58: 3192-3195.

Beller, H.R., Spormann, A.M., Sharma, P.K., Cole, J.R., and Reinhard, M. 1996. Isolation and characterization of a novel toluene-degrading, sulfate-reducing bacterium. Appl. Environ. Microbiol. 62: 1188-1196.

Biegert, T., Fuchs, G., and Heider, J. 1996. Evidence that anaerobic oxidation of toluene in the denitrifying bacterium *Thauera aromatica* is initiated by formation of benzylsuccinate from toluene and fumarate. Eur. J. Biochem. 238: 661-668.

Boga, H.I., and Brune, A. 2003. Hydrogen-dependent oxygen reduction by homoacetogenic bacteria isolated from termite guts. Appl. Environ. Microbiol. 69: 779-786.

Boll, M., and Fuchs, G. 1995. Benzoyl-Coenzyme A reductase (dearomatizing), a key enzyme of anaerobic aromatic metabolism. ATP dependence of the reaction, purification and some properties of the enzyme from *Thauera aromatica* strain K172. Eur. J. Biochem. 234: 921-933.

Boll, M., and Fuchs, G. 1998. Identification and characterization of the natural electron donor ferredoxin and of FAD as a possible prosthetic group of benzoyl-CoA reductase (dearomatizing), a key enzyme of anaerobic aromatic metabolism. Eur. J. Biochem. 251: 946-954.

Boll, M., Albracht, S.S., and Fuchs, G. 1997. Benzoyl-CoA reductase (dearomatizing), a key enzyme of anaerobic aromatic metabolism. A study of adenosinetriphosphate activity, ATP stoichiometry of the reaction and EPR properties of the enzyme. Eur. J. Biochem. 244: 840-851.

Boll, M., Laempe, D., Eisenreich, W., Bacher, A., Mittelbergert, T., Heinze, J., and Fuchs, G. 2000. Nonaromatic products from anoxic conversion of benzoyl-CoA with benzoyl-CoA reductase and cyclohexa-1, 5-diene-1-carboxyl-CoA hydratase. J. Biol. Chem. 275: 21889-21895.

Boll, M., Fuchs, G., and Lowe, D.J. 2001. Single turnover EPR studies of benzoyl-CoA reductase. Biochemistry. 40: 7612-7620.

Boll, M., Fuchs, G., and Heider, J. 2002. Anaerobic oxidation of aromatic compounds and hydrocarbons. Curr. Opin. Chem. Biol. 6: 604-611.

Bond, J.H., Engel, R.R., and Levitt, M.D. 1971. Factors influencing pulmonary methane excretion in man. J. Exp. Med. 133: 572-578.

Boone, D.R., Johnson, R.L., and Liu, Y. 1989. Diffusion of interspecies electron carriers H_2 and formate in methanogenic ecosystems and its implications in the measurement of K_m for H_2 or formate uptake. Appl. Environ. Microbiol. 55: 1735-1741.

Boone, D.R., Whitman, W.B., and Rouviere, P. 1993. Diversity and taxonomy of methanogens. In: Methanogenesis: Ecology, Physiology, Biochemistry, Genetics. J.G. Ferry, ed. Chapman and Hall, New York. p. 35-80.

Boyington, J.C., Gladyshev, V.N., Khangulov, S.V., Stadtman, T.C., and Sun, P.D. 1997. Crystal structure of formate dehydrogenase H: Catalysis involving Mo, molybdopterin, selenocysteine, and an Fe_4S_4 cluster. Science. 275: 1305-1308.

Brandis-Heep, A., Gebhardt, N.A., Thauer, R.K., Widdel, F., and Pfennig, N. 1983. Anaerobic acetate oxidation to CO_2 by *Desulfovibrio postgatei* 1. Demonstration of all enzymes required for the operation of the citric acid cycle. Arch. Microbiol. 136: 222-229.

Breese, K., and Fuchs, G. 1998. 4-Hydroxybenzoyl-CoA reductase (dehydroxylating) from the denitrifying bacterium *Thauera aromatica*--prosthetic groups, electron donor, and genes of a member of the molybdenum-flavin-iron-sulfur proteins. Eur. J. Biochem. 251: 916-923.

Breese, K., Boll, M., Alt-Morbe, J., Schagger, H., and Fuchs, G. 1998. Genes coding for the benzoyl-CoA pathway of anaerobic aromatic metabolism in the bacterium *Thauera aromatica*. Eur. J. Biochem. 256: 148-154.

Breznak, J.A. 1994. Acetogenesis from carbon dioxide in termite guts. In: Acetogenesis. H.L. Drake, ed. Chapman & Hall, New York. p. 303-330.

Briolat, V., and Reysset, G. 2002. Identification of the *Clostridium perfringens* genes involved in the adaptive response to oxidative stress. J. Bacteriol. 184: 2333-2343.

Brune, A., and Schink, B. 1992. Phloroglucinol pathway in the strictly anaerobic *Pelobacter acidigallici*: fermentation of trihydroxybenzenes to acetate via triacetic acid. Arch. Microbiol. 157: 417-424.

Bryant, M.P. 1974. Nutritional features and ecology of predominant anaerobic bacteria of the intestinal tract. Amer. J. Clin. Nutrit. 27: 1313-1319.

Bryant, M.P., and Robinson, I.M. 1962. Some nutritional characteristics of predominant culturable ruminal bacteria. J. Bacteriol. 84: 605-614.

Buckel, W. 1980. The reversible dehydration of (*R*)-2-hydroxyglutarate to (*E*)-glutaconate. Eur. J. Biochem. 106: 439-447.

Buckel, W. 1996. Unusual dehydrations in anaerobic bacteria: considering ketyls (radical anions) as reactive intermediates in enzymatic reactions. FEBS Lett. 389: 20-24.

Buckel, W. 2001. Unusual enzymes involved in five pathways of glutamate fermentation. Arch. Microbiol. 57: 263-273.

Buckel, W., and Semmler, R. 1982. A biotin-dependent sodium pump: glutaconyl-CoA decarboxylase from *Acidaminococcus fermentans*. FEBS Lett. 148: 35-38.

Buckel, W., and Semmler, R. 1983. Purification, characterisation, and reconstitution of glutaconyl-CoA decarboxylase, a biotin-dependent sodium pump from anaerobic bacteria. Eur. J. Biochem. 136: 427-434.

Caccavo, J., F., Blakemore, R.P., and Lovley, D.R. 1992. A hydrogen-oxidizing, Fe(III)-reducing microorganism from the Great Bay Esturary, New Hampshire. Appl. Environ. Microbiol. 58: 3211-3216.

Caccavo, J., F., Lonergan, D.J., Lovley, D.R., Dovis, M., Stolz, J.F., and McInerney, M.J. 1994. *Geobacter sulfurreducens* sp. nov., a hydrogen- and acetate-oxidizing dissimilatory metal-reducing microorganism. Appl. Environ. Microbiol. 60: 3752-3759.

Caldwell, D.R., White, D.C., Bryant, M.P., and Doetsch, R.N. 1965. Specificity of the heme requirement for growth of *Bacteroides ruminocola*. J. Bacteriol. 90: 1645-1654.

Chee-Sanford, J.C., Frost, J.W., Fries, M.R., Zhou, J., and Tiedje, J.M. 1996. Evidence for acetyl coenzyme A and cinnamoyl coenzyme A in the anaerobic toluene mineralization pathway in *Azoarcus tolulyticus* Tol-4. Appl. Environ. Microbiol. 62: 964-973.

Coates, J.D., Chakraborty, R., and McInerney, M.J. 2002. Anaerobic benzene biodegradation- a new era. Research Microbiol. 153: 621-628.

Cody, G.D., Boctor, N.Z., Filley, T.R., Hazen, R.M., Scott, J.H., Sharma, A., and Yoder, H.S., Jr. 2000. Primordial carbonylated iron-sulfur compounds and the synthesis of pyruvate. Science. 289: 1337-1340.

Conrad, R., and Wetter, B. 1990. Influence of temperature on energetics of hydrogen metabolism in homoacetogenic, methanogenic, and other anaerobic bacteria. Arch. Microbiol. 155: 94-98.

Conrad, R., Goodwin, S., and Zeikus, J.G. 1987. Hydrogen metabolism in a mildly acidic lake sediment (Knaack Lake). FEMS Microbiol. Ecol. 45: 243-249.

Cord-Ruwisch, R., Ollivier, B., and Garcia, J.-L. 1986. Fructose degradation by *Desulfovibrio* sp. in pure culture and in coculture with *Methanospirillum hungatei*. Curr. Microbiol. 13: 285-289.

Cord-Ruwisch, R., Seitz, H.-J., and Conrad, R. 1988. The capacity of hydrogenotrophic anaerobic bacteria to compete for traces of hydrogen depends on the redox potential of the terminal electron acceptor. Arch. Microbiol. 149: 350-357.

Cord-Ruwisch, R., Mercz, T.I., Hoh, C.-Y., and Strong, G.E. 1997. Dissolved hydrogen concentration as an on-line control parameter for the automated operation and optimization of anaerobic digesters. Biotech. Bioeng. 56: 626-634.

Coschigano, P.W. 2000. Transcriptional analysis of the tutE tutFDGH gene cluster from *Thauera aromatica* strain T1. Appl. Environ. Microbiol. 66: 1147-1151.

Coschigano, P.W., Wehrman, T.S., and Young, L.Y. 1998. Identification and analysis of genes involved in anaerobic toluene metabolism by strain T1: putative role of a glycine free radical. Appl. Environ. Microbiol. 64: 1650-1656.

Costilow, R.N. 1977. Selenium requirement for the growth of *Clostridium sporogenes* with glycine as the oxidant in Stickland reaction systems. J. Bacteriol. 131: 366-368.

Cypionka, H. 2000. Oxygen respiration by *Desulfovibrio* species. Ann. Rev. Microbiol. 54: 827-848.

Dagley, S. 1989. Chemical unity and diversity in bacterial catabolism. In: Bacteria in Nature. J.S. Poindexter, and E.R. Leadbetter, eds. Plenum Press, New York. p. 259-291.

Danson, M.J. 1988. Archaebacteria: The comparative enzymology of their central metabolic pathways. Adv. Micro. Physiol. 29: 166-231.

Das, A., Coulter, E.D., Kurtz Jr., D.M., and Ljungdahl, L.G. 2001. Five-gene cluster in *Clostridium thermoaceticum* consisting of two different operons encoding rubredoxin oxidoreductase-rubredoxin and rubrerythrin-type A flavoprotein-high-molecular-weight rubredoxin. J. Bacteriol. 183: 1560-1567.

de Bok, F.A.M., Luijten, M.L.G.C., and Stams, A.J.M. 2002. Biochemical evidence for formate transfer in syntrophic propionate-oxidizing coculture of *Syntrophobacter fumaroxidans* and *Methanospirillum hungatei*. Appl. Environ. Microbiol. 68: 4247-4252.

Dehning, I., and Schink, B. 1989. *Malonomonas rubra* ge. nov. sp. now., a microaerotolerant anaerobic bacterium growing by decarboxylation of malonate. Arch. Microbiol. 151: 427-433.

Dehority, B.A. 1971. Carbon dioxide requirement of various species of rumen bacteria. J. Bacteriol. 105: 70-76.

Deppenmeier, U. 2002a. Redox-driven proton translocation in methanogenic Archaea. CMLS Cell. Mol. Life. Sci. 59: 1513-1533.

Deppenmeier, U. 2002b. The unique biochemistry of methanogenesis. Prog. Nucl. Acid. Res. 71: 223-283.

Diekert, G., and Ritter, M. 1982. Nickel requirement of *Acetobacterium woodii*. J. Bacteriol. 151: 1043-1045.

Dilworth, G.L. 1983. Occurence of molybdenum in nicotinic acid hydroxylase from *Clostridium barkeri*. Arch. Biochem. Biophys. 221: 565-569.

Dimroth, P., and Thomer, A. 1993. On the mechanism of sodium ion translocation by oxaloacetate decarboxylase of *Klebsiella pneumoniae*. Biochemistry. 32: 1734-1739.

Dimroth, P., and Schink, B. 1998. Energy conservation in the decarboxylation of dicarboxylic acids. Arch. Microbiol. 170: 69-77.

Dobbek, H., Svetlitchnyi, V., Gremer, L., Huber, R., and Meyer, O. 2001. Crystal structure of a carbon monoxide dehydrogenase reveals a [Ni-4Fe-5S] cluster. Science. 293: 1281-1285.

Dolfing, J. 1988. Acetogenesis. In: Biology of Anaerobic Microorganisms. A.J.B. Zehnder, ed. Wiley-Liss, New York. p. 417-468.

Dolla, A., Pohorelic, B.K., Voordouw, J.K., and Voordouw, G. 2000. Deletion of the *hmc* operon of *Desulfovibrio vulgaris* subsp. Hildenborough hampers hydrogen metabolism and low-redox-potenial niche establishment. Arch. Microbiol. 174: 143-151.

Dong, X., and Stams, A.J.M. 1995. Evidence for H_2 and formate formation during syntrophic butyrate and propionate degradation. Anaerobe. 1: 35-39.

Dong, X., Plugge, C.M., and Stams, A.J.M. 1994a. Anaerobic degradation of propionate by a mesophilic acetogenic bacterium in coculture and triculture with different methanogens. Appl. Environ. Microbiol. 60: 2834-2838.

Dong, X., Cheng, G., and Stams, A.J.M. 1994b. Butyrate oxidation by *Syntrophospora bryantii* in co-culture with different methanogens and in pure culture with pentenoate as electron acceptor. Appl. Microbiol. Biotechnol. 42: 647-652.

Doré, J., and Bryant, M.P. 1989. Lipid growth requirement and influence of lipid supplement on fatty acid and aldehyde composition of *Syntrophococcus sucromutans*. Appl. Environ. Microbiol. 55: 927-933.

Doukov, T.I., Iverson, T.M., Seravalli, J., Ragsdale, S.W., and Drennan, C.L. 2002. A Ni-Fe-Cu center in a bifunctional carbon monoxide dehydrogenase/acetyl-CoA synthase. Science. 298: 567-572.

Dworkin, M. 1999. The Prokaryotes: An Evolving Electronic Resource for the Microbiological Community. 3rd edn. Springer-Verlag, New York.

Dwyer, D.F., Weeg-Aerssens, E., Shelton, D.R., and Tiedje, J.M. 1988. Bioenergetic conditions of butyrate metabolism by a syntrophic, anaerobic bacterium in coculture with hydrogen-oxidizing methanogenic and sulfidogenic bacteria. Appl. Environ. Microbiol. 54: 1354-1359.

Ebenau-Jehle, C., Boll, M., and Fuchs, G. 2003. 2-Oxoglutarate: NADP(+) oxidoreductase in *Azoarcus evansii*: properties and function in electron transfer reactions in aromatic ring reduction. J. Bacteriol. 185: 6119-6129.

Egland, P.G., Pelletier, D.A., Dispensa, M., Gibson, J., and Harwood, C.S. 1997. A cluster of bacterial genes for anaerobic benzene ring biodegradation. Proc. Natl. Acad. Sci. USA. 94: 6484-6489.

Einsle, O., Tezcan, F.A., Andrade, S.L., Schmid, B., Yoshida, M., Howard, J.B., and Rees, D.C. 2002. Nitrogenase MoFe-protein at 1.16 A resolution: a central ligand in the FeMo-cofactor. Science. 297: 1696-1700.

Eisen, J.A., Nelson, K.E., Paulsen, I.T., Heidelberg, J.F., Wu, M., Dodson, R.J., Deboy, R., Gwinn, M.L., Nelson, W.C., Haft, D.H., Hickey, E.K., Peterson, J.D., Durkin, A.S., Kolonay, J.L., Yang, F., Holt, I., Umayam, L.A., Mason, T., Brenner, M., Shea, T.P., Parksey, D., Nierman, W.C., Feldblyum, T.V., Hansen, C.L., Craven, M.B., Radune, D., Vamathevan, J., Khouri, H., White, O., Gruber, T.M., Ketchum, K.A., Venter, J.C., Tettelin, H., Bryant, D.A., and Fraser, C.M. 2002. The complete genome sequence of *Chlorobium tepidum* TLS, a photosynthetic, anaerobic, green-sulfur bacterium. Proc. Natl. Acad. Sci. USA. 99: 9509-9514.

Eklund, H., and Fontecave, M. 1999. Glycyl radical enzymes: a conservative structural basis for radicals. Structure Fold. Des. 7: R257-262.

Elshahed, M.S., and McInerney, M.J. 2001a. Benzoate fermentation by the anaerobic bacterium Syntrophus aciditrophicus in the absence of hydrogen-using microorganisms. Appl Environ Microbiol. 67: 5520-5525.

Elshahed, M.S., and McInerney, M.J. 2001b. Is interspecies hydrogen transfer needed for toluene degradation under sulfate-reducing conditions? FEMS Microbiol. Ecol. 35: 163-169.

Elshahed, M.S., Gieg, L.M., McInerney, M.J., and Suflita, J.M. 2001a. Signature metabolites attesting to the in situ attenuation of alkylbenzenes in anaerobic environments. Environ. Sci. Technol. 35: 682-689.

Elshahed, M.S., Bhupathiraju, V.K., Wofford, N.Q., Nanny, M.A., and McInerney, M.J. 2001b. Metabolism of benzoate, cyclohex-1-ene carboxylate, and cyclohexane carboxylate by *Syntrophus aciditrophicus* strain SB in syntrophic association with H_2-using microorganisms. Appl. Environ. Microbiol. 67: 1728-1738.

Ermler, U., Grabarse, W., Shima, S., Goubeaud, M., and Thauer, R.K. 1997. Crystal structure of methyl-coenzyme M reductase: the key enzyme of biological methane formation. Science. 278: 1457-1462.

Evans, M.C.W., Buchanan, B.B., and Arnon, D.I. 1966. A new ferredoxin-dependent carbon reduction cycle in photosynthetic bacteria. Proc. Natl. Acad. Sci. USA. 55: 928-934.

Evans, P.J., Ling, W., Goldschmidt, B., Ritter, E.R., and Young, L.Y. 1992. Metabolites formed during anaerobic transformation of toluene and *o*-xylene and their proposed relationship to the initial steps of toluene mineralization. Appl. Environ. Microbiol. 58: 496-501.

Farr, S.B., and Kogoma, T. 1991. Oxidative stress responses in *Escherichia coli* and *Salmonella typhimurium*. Microbiol. Rev. 55: 561-585.

Felix, C.R., and Ljungdahl, L.G. 1993. The cellulosome: The exocellular organelle of *Clostridium*. Ann. Rev. Microbiol. 47: 791-819.

Ferry, J.G. 1999. Enzymology of one-carbon metabolism in methanogenic pathways. FEMS Microbiol. Rev. 23: 13-38.

Fournier, M., Zhang, Y., Wildschut, J.D., Dolla, A., Voordouw, J.K., Schriemer, D.C., and Voordouw, G. 2003. Function of oxygen resistance proteins in the anaerobic, sulfate-reducing bacterium *Desulfovibrio vulgaris* Hildenborough. J. Bacteriol. 185: 71-79.

Frazer, A.C. 1994. O-Demethylation and other transformations of aromatic compounds by acetogenic bacteria. In: Acetogenesis. H.L. Drake, ed. Chapman & Hall, New York. p. 445-483.

Friedrich, M., and Schink, B. 1993. Hydrogen formation from glycolate driven by reversed transport in membrane vesicles of a syntrophic glycolate-oxidizing bacterium. Eur. J. Biochem. 217: 233-240.

Friedrich, M., and Schink, B. 1995. Electron transport phosphorylation driven by glyoxylate respiration with hydrogen as electron donor in membrane vesicles of a glyoxylate-fermenting bacterium. Arch. Microbiol. 163: 268-275.

Fuller, R.C. 1978. Photosynthetic carbon metabolism in the green and purple bacteria. In: The Photosynthetic Bacteria. R.K. Clayton, R.K., and W.R. Sistrom, eds. Plenum Press, New York. p. 691-705.

Furdui, C., and Ragsdale, S.W. 2002. The role of coenzyme A in the pyruvate:ferredoxin oxidoreductase reaction mechanism: rate enhancement of electron transfer from a radical intermediate to an iron-sulfur cluster. Biochemistry. 41: 9921-9937.

Gallert, C., Knoll, G., and Winter, J. 1991. Anaerobic carboxylation of phenol to benzoate: use of deuterated phenols revealed carboxylation exclusively in the C4-position. Appl. Microbiol. Biotechnol. 36: 124-129.

Gebhardt, N.A., Thauer, R.K., Linder, D., Kaulfers, P.-M., and Pfennig, N. 1985. Mechanism of acetate oxidation to CO_2 with elemental sulfur in *Desulfuromonas acetoxidans*. Arch. Microbiol. 141: 392-398.

Georgiadis, M.M., Komiya, H., Chakrabarti, P., Woo, D., Kornuc, J.J., and Rees, D.C. 1992. Crystallographic structure of the nitrogenase iron protein from *Azotobacter vinelandii*. Science. 257: 1653-1659.

Gibson, J., and Harwood, C.S. 2002. Metabolic diversity in aromatic compound utilization by anaerobic microbes. Ann. Rev. Microbiol. 56: 345-369.

Gibson, J., Dispensa, M., and Harwood, C.S. 1997. 4-Hydroxybenzoyl coenzyme A reductase (dehydroxylating) is required for anaerobic degradation of 4-hydroxybenzoate by *Rhodopseudomonas palustris* and shares features with molybdenum-containing hydroxylases. J. Bacteriol. 179: 634-642.

Gibson, J., Dispensa, M., Fogg, G.C., Evans, D.T., and Harwood, C.S. 1994. 4-Hydroxybenzoate-coenzyme A ligase from *Rhodopseudomonas palustris*: purification, gene sequence, and role in anaerobic degradation. J. Bacteriol. 176: 634-641.

Gomez-Alarcon, R.A., O'Dowd, C., Leedle, J.A.Z., and Bryant, M.P. 1982. 1,4-Naphthoquinone and other nutrient requirements of *Succinovibrio dextrinosolvens*. Appl. Environ. Microbiol. 44: 346-350.

Gottschalk, G. 1985. Bacterial Metabolism, 2nd edn. Springer-Verlag, New York.

Gunsalus, R.P. 1992. Control of electron flow in *Escherichia coli*: coordinated transcription of respiratory pathway genes. J. Bacteriol. 174: 7069-7074.

Gustafson, W.G., Feinberg, A., and McFarland, J.T. 1986. Energetics of β-oxidation. Reduction potentials of general fatty acyl-CoA dehydrogenase, electron transfer flavoprotein, and fatty acyl-CoA substrates. J. Biol. Chem. 261: 7733-7741.

Guwy, A.J., Hawkes, F.R., Hawkes, D.L., and Rozzi, A.G. 1997. Hydrogen production in a high rate fluidised bed anaerobic digester. Water Res. 31: 1291-1298.

Haddock, J.D., and Ferry, J.G. 1989. Purification and properties of phloroglucinol reductase from *Eubacterium oxidoreducens* G-41. J. Biol. Chem. 264: 4423-4427.

Halpern, J. 1985. Mechanisms of coenzyme B_{12}-dependent rearrangements. Science. 227: 869-875.

Härtel, U., and Buckel, W. 1996. Sodium ion-dependent hydrogen production in *Acidaminococcus fermentans*. Arch. Microbiol. 166: 350-356.

Harwood, C.S., Burchhardt, G., Herrmann, H., and Fuchs, G. 1999. Anaerobic metabolism of aromatic compounds via benzoyl-CoA pathway. FEMS Microbiol. Rev. 22: 439-458.

Heider, J., and Fuchs, G. 1997a. Anaerobic metabolism of aromatic compounds. Eur. J. Biochem. 243: 577-596.

Heider, J., and Fuchs, G. 1997b. Microbial anaerobic aromatic metabolism. Anaerobe. 3: 1-22.

Heider, J., Spormann, A.M., Beller, H.R., and Widdel, F. 1999. Anaerobic bacterial metabolism of hydrocarbons. FEMS Microbiol. Rev. 22: 459-473.

Heijnen, J.J. 1994. Thermodynamics of microbial growth and its implications for process design. Trends Biotechnol. 12: 483-492.

Heijnen, J.J. 1999. Bioenergetics of microbial growth. In: Encyclopedia of Bioprocess Technology: Fermentation,

Biocatalysis, and Bioseparation. M.C. Flickinger, and S.W. Derew, eds. John Wiley & Sons, Inc., New York. p. 267-291.

Herbeck, J.L., and Bryant, M.P. 1974. Nutritional features of the intestinal anaerobe *Ruminococcus bromii*. Appl. Microbiol. 28: 1018-1022.

Hermuth, K., Leuthner, B., and Heider, J. 2002. Operon structure and expression of the genes for benzylsuccinate synthase in *Thauera aromatica* strain K172. Arch. Microbiol. 177: 132-138.

Herter, S., Farfsing, J., Gad'on, N., Rieder, C., Eisenreich, W., Bacher, A., and Fuchs, G. 2001. Autotrophic CO_2 fixation by *Chloroflexus aurantiacus*: study of glyoxylate formation and assimilation via the 3-hydroxypropionate cycle. J. Bacteriol. 183: 4305-4316.

Herter, S., Fuchs, G., Bacher, A., and Eisenreich, W. 2002. A bicyclic autotrophic CO_2 fixation pathway in *Chloroflexus aurantiacus*. J. Biol. Chem. 277: 20277-20283.

Hirsch, W., Schägger, H., and Fuchs, G. 1998. Phenylglyxoylate: NAD+ oxidoreductase (CoA benzoylating), a new enzyme of anaerobic phenylalanine metabolism in the denitrifying bacterium *Azoarcus evansii*. Eur. J. Biochem. 251: 907-915.

Holo, H. 1989. *Chloroflexus aurantiacus* secretes 3-hydroxypropionate, a possible intermediate in the assimilation of CO_2 and acetate. Arch. Microbiol. 151: 252-256.

Houwen, F., Plokker, P., Stams, A.J.M., and Zehnder, A.J.B. 1990. Enzymatic evidence for the involvement of the methylmalonyl-CoA pathway in propionate oxidation by *Syntrophobacter wolinii*. Arch. Microbiol. 155: 52-55.

Huber, C., and Wächtershäuser, G. 1997. Activated acetic acid by carbon fixation on (Fe,Ni)S under primordial conditions. Science. 276: 245-247.

Huber, C., Eisenreich, W., Hecht, S., and Wächtershäuser, G. 2003. A possible primordial peptide cycle. Science. 301: 938-940.

Hügler, M., Menendez, C., Schägger, H., and Fuchs, G. 2002. Malonyl-coenzyme A reductase from *Chloroflexus aurantiacus*, a key enzyme of the 3-hydroxypropionate cycle for autotrophic CO_2 fixation. J. Bacteriol. 184: 2404-2410.

Hügler, M., Huber, H., Stetter, K.O., and Fuchs, G. 2003a. Autotrophic CO_2 fixation pathways in archaea (Crenarchaeota). Arch. Microbiol. 179: 160-173.

Hügler, M., Krieger, R.S., Jahn, M., and Fuchs, G. 2003b. Characterization of acetyl-CoA/propionyl-CoA carboxylase in *Metallosphaera sedula*. Eur. J. Biochem. 270: 736-744.

Hungate, R.E., and Stack, R.J. 1982. Phenylpropanoic acid: growth factor for *Ruminococcus albus*. Appl. Environ. Microbiol. 44: 79-83.

Imlay, J.A. 2002. How oxygen damages microbes: oxygen tolerance and obligate anaerobiosis. Adv. Microb. Physiol. 46: 111-153.

Jackson, B.E., and McInerney, M.J. 2002. Anaerobic metabolism can proceed close to thermodynamic limits. Nature. 415: 454-456.

Jenny Jr., F.E., Verhagen, M.F.J.M., Cui, X., and Adams, M.W. 1999. Anaerobic microbes: oxygen detoxification without superoxide dismutase. Science. 286: 306-309.

Johnson, H.A., Pelletier, D.A., and Spormann, A.M. 2001. Isolation and characterization of anaerobic ethylbenzene dehydrogenase, a novel Mo-Fe-S enzyme. J. Bacteriol. 183: 4536-4542.

Johnson, M.S., Zhulin, I.B., Gapuzan, M.-E.R., and Taylor, B.L. 1997. Oxygen-dependent growth of the obligate anaerobe *Desulfovibrio vulgaris* Hildenborough. J. Bacteriol. 179: 5598-5601.

Jones, J.B., and Stadtman, T.C. 1977. *Methanococcus vannielii*: culture and effects of selenium and tungsten on growth. J. Bacteriol. 130: 1404-1405.

Jormakka, M., Törnroth, S., Byrne, B., and Iwata, S. 2002. Molecular basis of proton motive force generation: structure of formate dehydrogenase-N. Science. 295: 1863-1868.

Jungermann, K., Thauer, R.K., Leimenstoll, G., and Decker, K. 1973. Function of reduced pyridine nucleotide-ferredoxin oxidoreductases in saccharolytic clostridia. Biochem. Biophys. Acta. 304: 268-280.

Kaiser, J.-L., and Hanselmann, K.W. 1982. Fermentative metabolism of substituted monoaromatic compounds by a bacterial community from anaerobic sediments. Arch. Microbiol. 133: 185-194.

Kane, S.R., Beller, H.R., Legler, T.C., and Anderson, R.T. 2002. Biochemical and genetic evidence of benzylsuccinate synthase in toluene-degrading, ferric iron-reducing *Geobacter metallireducens*. Biodegradation. 13: 149-154.

Karnholz, A., Küsel, K., Göβner, A., Schramm, A., and Drake, H.L. 2002. Tolerance and metabolic response of acetogenic bacteria toward oxygen. Appl. Environ. Microbiol. 68: 1005-1009.

Kaspar, H.F., and Wuhrmann, K. 1978. Kinetic parameters and relative turnover of some important catabolic reactions in digestor sludge. Appl. Environ. Microbiol. 36: 1-7.

Kisker, C., Schindelin, H., and Rees, D.C. 1997. Molybdenum-cofactor-containing enzymes: structure and mechanism. Annu. Rev. Biochem. 66: 233-267.

Kluge, C., Tschech, A., and Fuchs, G. 1990. Anaerobic metabolism of resorcyclic acids (*m*-dihydroxybenzoic acids) and resorcinol (1,3-benzendiol) in a fermenting and in a denitrifying bacterium. Arch. Microbiol. 155: 68-74.

Kniemeyer, O., and Heider, J. 2001a. (*S*)-1-phenylethanol dehydrogenase of *Azoarcus* sp. strain EbN1, an enzyme of anaerobic ethylbenzene catabolism. Arch. Microbiol. 176: 129-135.

Kniemeyer, O., and Heider, J. 2001b. Ethylbenzene dehydrogenase, a novel hydrocarbon-oxidizing molybdenum/iron-sulfur/heme enzyme. J. Biol. Chem. 276: 21381-21386.

Kniemeyer, O., Leutwein, C., Schulz, H., Horth, P., Haehnel, W., Schiltz, E., Schagger, H., and Heider, J. 2003. Anaerobic degradation of ethylbenzene by a new type of marine sulfate-reducing bacterium. Appl. Environ. Microbiol. 69: 760-768.

Koch, J., Eisenreich, W., Bacher, A., and Fuchs, G. 1993. Products of enzymatic reduction of benzoyl-CoA, a key reaction in anaerobic aromatic metabolism. Eur. J. Biochem. 211: 649-661.

König, H., Nusser, E., and Stetter, K.O. 1985. Glycogen in *Methanolobus* and *Methanococcus*. FEMS Microbiol. Lett. 28: 265-269.

Konings, W.N., Poolman, B., and van Veen, H.W. 1994. Solute transport and energy transduction in bacteria. Antonie von Leeuwenhoek. 65: 369-380.

Konings, W.N., Lolkema, J.S., and Poolman, B. 1995. The generation of metabolic energy by solute transport. Arch. Microbiol. 164: 235-242.

Kotsyurbenko, O.R., Nozhevnikova, A.N., Soloviova, T.I., and Zavarzin, G.A. 1996. Methanogenesis at low temperatures by microflora of tundra wetland soil. Antonie van Leeuwenhoek. 69: 75-86.

Krekeler, D., and Cypionka, H. 1995. The preferred electron acceptor of *Desulfovibrio desulfuricans* CSN. FEMS Microbiol. Ecol. 17: 271-278.

Krieger, C.J., Beller, H.R., Reinhard, M., and Spormann, A.M. 1999. Initial reactions in anaerobic oxidation of *m*-xylene by the denitrifying bacterium *Azoarcus* sp. strain T. J. Bacteriol. 181: 6403-6410.

Krieger, C.J., Roseboom, W., Albracht, S.P., and Spormann, A.M. 2001. A stable organic free radical in anaerobic benzylsuccinate synthase of *Azoarcus* sp. strain T. J. Biol. Chem. 276: 12924-12927.

Kröger, A. 1978. Fumarate as terminal acceptor of phosphorylative electron transport. Biochem. Biophys. Acta. 505: 129-145.

Kropp, K.G., Davidova, I.A., and Suflita, J.M. 2000. Anaerobic oxidation of *n*-dodecane by an addition reaction in a sulfate-reducing bacterial enrichment culture. Appl. Environ. Microbiol. 66: 5393-5398.

Krumholz, L.R., Crawford, R.L., Hemling, M.E., and Bryant, M.P. 1987. Metabolism of gallate and phloroglucinol in *Eubacterium oxidoreducens* via 3-hydroxy-5-oxohexanoate. J. Bacteriol. 169: 1886-1890.

Kuhner, C.A., Smith, S.S., Noll, K.M., Tanner, R.S., and Wolfe, R.S. 1991. 7-Mercaptoheptanoylthreonine phosphate substitutes for heat-stable factor (mobile factor) for growth of *Methanomicrobium mobilie*. Appl. Environ. Microbiol. 57: 2891-2895.

Ladapo, J., and Whitman, W.B. 1990. Method for isolation of auxotrophs in the methanogenic archaebacteria: role of the acetyl-CoA pathway of autotrophic CO_2 fixation in *Methanococcus maripaludis*. Proc. Natl. Acad. Sci. USA. 87: 5598-5602.

Lajoie, S.F., Bank, S., Miller, T.L., and Wolin, M.J. 1988. Acetate production from hydrogen and [^{13}C] carbon dioxide by microflora of human feces. Appl. Environ. Microbiol. 54: 2723-2727.

Lancaster, C.R.D., and Kröger, A. 2000. Succinate: quinone oxidoreductases: new insights from X-ray crystal structures. Biochem. Biophys. Acta. 1459: 422-431.

Länge, S., Scholtz, R., and Fuchs, G. 1989. Oxidative and reductive acetyl-CoA/carbon monoxide dehydrogenase pathway in *Desulfobacterium autotrophicum* 1. Characterization and metabolic function of cellular tetrahydropterin. Arch. Microbiol. 151: 77-83.

Leschine, S.B. 1995. Cellulose degradation in anaerobic environments. Ann. Rev. Microbiol. 49: 399-426.

Leuthner, B., and Heider, J. 1998. A two-component system involved in regulation of anaerobic toluene metabolism in *Thauera aromatica*. FEMS Microbiol. Lett. 166: 35-41.

Leuthner, B., and Heider, J. 2000. Anaerobic toluene catabolism of *Thauera aromatica*: the bbs operon codes for enzymes of beta oxidation of the intermediate benzylsuccinate. J. Bacteriol. 182: 272-277.

Leuthner, B., Leutwein, C., Schulz, H., Horth, P., Haehnel, W., Schiltz, E., Schagger, H., and Heider, J. 1998. Biochemical and genetic characterization of benzylsuccinate synthase from *Thauera aromatica*: a new glycyl radical enzyme catalysing the first step in anaerobic toluene metabolism. Mol. Microbiol. 28: 615-628.

Leutwein, C., and Heider, J. 1999. Anaerobic toluene-catabolic pathway in denitrifying *Thauera aromatica*: activation and beta-oxidation of the first intermediate, (*R*)-(+)-benzylsuccinate. Microbiology. 145: 3265-3271.

Leutwein, C., and Heider, J. 2001. Succinyl-CoA:(*R*)-benzylsuccinate CoA-transferase: an enzyme of the anaerobic toluene catabolic pathway in denitrifying bacteria. J. Bacteriol. 183: 4288-4295.

Leutwein, C., and Heider, J. 2002. (*R*)-Benzylsuccinyl-CoA dehydrogenase of *Thauera aromatica*, an enzyme of the anaerobic toluene catabolic pathway. Arch. Microbiol. 178: 517-524.

Ljungdahl, L.G. 1986. The autotrophic pathway of acetate synthesis in acetogenic bacteria. Ann. Rev. Microbiol. 40: 415-450.

Ljungdahl, L.G., and Eriksson, K.-E. 1985. Ecology of microbial cellulose degradation. Adv. Micro. Ecol. 8: 237-299.

Lloyd, D. 1999. How to avoid oxygen. Science. 286: 249.

Lovley, D.R. 1985. Minimum threshold for hydrogen metabolism in methanogenic bacteria. Appl. Environ. Microbiol. 49: 1530-1531.

Lovley, D.R. 1991. Dissimilatory Fe(III) and Mn(IV) reduction. Microbiol. Rev. 55: 259-287.

Lovley, D.R., and Goodwin, S. 1988. Hydrogen concentration as an indicator of the predominant terminal electron-accepting reactions in aquatic sediments. Geochim. Cosmochim. Acta. 52: 2993-3003.

Lovley, D.R., and Phillips, E.J.P. 1989. Requirement for a microbial consortium to completely oxidize glucose in Fe(III)-reducing sediments. Appl. Environ. Microbiol. 55: 3234-3236.

Lumppio, H.L., Shenvi, N.V., Summers, A.O., Voordouw, G., and Kurtz Jr., D.M. 2001. Rubrerythrin and rubredoxin oxidoreductase in *Desulfovibrio vulgaris*: a novel oxidative stress protection system. J. Bacteriol. 183: 101-108.

Mackie, R.I., and Bryant, M.P. 1994. Acetogenesis and the rumen: syntrophic relationships. In: Acetogenesis. H.L. Drake, ed. Chapman & Hall, New York. p. 331-364.

Mackie, R.I., White, B.A., and Bryant, M.P. 1991. Lipid metabolism in anaerobic ecosystems. Crit. Rev. Microbiol. 17: 449-479.

Magee, E.L., Ensley, B.D., and Barton, L.L. 1978. An assessment of growth yields and energy coupling in *Desulfovibrio*. Arch. Microbiol. 117: 21-26.

Matthies, C., and Schink, B. 1992. Energy conservation in fermentative glutarate degradation by the bacterial strain WoG13. FEMS Microbiol. Lett. 100: 221-226.

McCarty, P.L. 1971. Energetics and kinetics of anaerobic treatment. In: Anaerobic Biological Treatment Processes. R.F. Gould, ed. American Chemical Society, Washington, DC. p. 91-107.

McCord, J.M., Keele Jr., B.B., and Fridovich, I. 1971. An enzyme-based theory of obligate anaerobiosis: the physiological function of superoxide dismutase. Proc. Natl. Acad. Sci. USA. 68: 1024-1027.

McInerney, M.J. 1986. Transient and persistent associations among prokaryotes. In: Bacteria in Nature. J.S. Poindexter, and E.R. Leadbetter, eds. Plenum Publishing Co., New York. p. 293-338.

McInerney, M.J. 1988. Anaerobic hydrolysis and fermentation of fats and proteins. In: Biology of Anaerobic Microorganisms. A.J.B. Zehnder, ed. John Wiley & Sons, New York. p. 373-415.

McInerney, M.J. 1999. Anaerobic metabolism and its regulation. In: Biotechnology, Second Completely Revised Edition. Vol. 11 a. Environmental processes-Wastewater and waste treatment (J. Winter, vol ed.). (ed.), H.-J. Rehm and G. Reed, eds. Wiley-VCH, Weinheim, Germany. p. 455-478.

McInerney, M.J., and Bryant, M.P. 1981. Basic principles of bioconversions in anaerobic digestion and methanogenesis. In: Biomass Conversion Processes for Energy and Fuels. S.S. Sofer, and O.R. Zaborsky, eds. Plenum Publishing Corporation, New York. p. 277-296.

McInerney, M.J., and Beaty, P.S. 1988. Anaerobic community structure from a nonequilibrium thermodynamic perspective. Can. J. Microbiol. 34: 487-493.

Meckenstock, R.U., Annweiler, E., Michaelis, W., Richnow, H.H., and Schink, B. 2000. Anaerobic naphthalene degradation by a sulfate-reducing enrichment culture. Appl. Environ. Microbiol. 66: 2743-2747.

Menendez, C., Bauer, Z., Huber, H., Gad'on, N., Stetter, K.O., and Fuchs, G. 1999. Presence of acetyl coenzyme A (CoA) carboxylase and propionyl-CoA carboxylase in autotrophic Crenarchaeota and indication for operation of a 3-hydroxypropionate cycle in autotrophic carbon fixation. J. Bacteriol. 181: 1088-1098.

Meuer, J., Kuettner, H.C., Zhang, J.K., Hedderich, R., and Metcalf, W.W. 2002. Genetic analysis of the archaeon *Methanosarcina barkeri* Fusaro reveals a central role for Ech hydrogenase and ferredoxin in methanogenesis and carbon fixation. Proc. Natl. Acad. Sci. USA. 99: 5632-5637.

Michel, T.A., and Macy, J.M. 1990. Generation of a membrane potential by sodium-dependent succinate efflux in *Selenomonas ruminantium*. J. Bacteriol. 172: 1430-1435.

Michels, P.A.M., Michels, J.P.J., Boonstra, J., and Konings, W.N. 1979. Generation of an electrochemical proton gradient in bacteria by the excretion of metabolic end products. FEMS Microbiol. Lett. 5: 357-364.

Miller, T.L., and Wolin, M.J. 1973. Formation of hydrogen and formate by *Ruminococcus albus*. J. Bacteriol. 116: 836-846.

Möller, D., Schauder, R., Fuchs, G., and Thauer, R.K. 1987. Acetate oxidation to CO_2 via a citric acid cycle involving an ATP-citrate lyase: a mechanism for the synthesis of ATP via substrate-level phosphorylation in *Desulfovibrio postgatei* growing on acetate and sulfate. Arch. Microbiol. 148: 202-207.

Möller-Zinkhan, D., and Thauer, R. 1988. Membrane-bound NADPH dehydrogenase- and ferredoxin:NADP oxidoreductase activity involved in electron transport during acetate oxidation to CO_2 in *Desulfobacter postgatei*. Arch. Microbiol. 150: 145-154.

Morris, J.G. 1975. The physiology of obligate anaerobiosis. Adv. Micro. Physiol. 16: 169-246.

Mosey, F.E., and Fernandes, X.A. 1989. Patterns of hydrogen in biogas from the anaerobic digestion of milk-sugars. Wat. Sci. Tech. 21: 187-196.

Müller, V., and Gottschalk, G. 1994. The sodium ion cycle in acetogenic and methanogenic bacteria: Generation and utilization of a primary electrochemical gradient. In: Acetogenesis. H.L. Drake, ed. Chapman & Hall, New York. p. 127-156.

Murray, P.A., and Zinder, S.H. 1987. Polysaccharide reserve material in the acetotrophic methanogen, *Methanosarcina thermophila* strain TM-1: accumulation and mobilization. Arch. Microbiol. 147: 109-116.

Nachaiyasit, S., and Stuckey, D.C. 1997. Effect of low temperatures on the performance of an anaerobic baffled reactor (ABR). J. Chem. Tech. Biotechnol. 69: 276-284.

Nealson, K.H., and Saffarini, D. 1994. Iron and manganese in anaerobic respiration: environmental significance. Ann. Rev. Microbiol. 48: 311-343.

Odum, J.M., and Peck Jr., H.D. 1981. Hydrogen cycling as a general mechanism for energy coupling in the sulfate-reducing bacteria, *Desulfovibrio* sp. FEMS Microbiol. Lett. 12: 47-50.

Ormen-Johnson, W.H. 1992. Nitrogenase structure: where to now? Science. 257: 1639-1640.

Otto, R., Sonnenberg, A.S.M., Veldkamp, H., and Konings, W.N. 1980. Generation of an electrochemical proton gradient in *Streptococcus cremoris* by lactate efflux. Proc. Natl.Acad. Sci. USA. 77: 5502-5506.

Otto, R., Lageveen, R.G., Veldkamp, H., and Konings, W.N. 1982. Lactate efflux-induced electrical potential inmembrane vesicles of *Streptococcus cremoris*. J. Bacteriol. 149: 733-738.

Oude Elferink, S.J.W.H., Visser, J., Hulshoff Pol, L.W., and Stams, A.J.M. 1994. Sulfate reduction in methanogenic bioreactors. FEMS Microbiol. Rev. 15: 119-136.

Pan, N., and Imlay, J.A. 2001. How does oxygen inhibit central metabolism in the obligate anaerobe *Bacteroides thetaiotaomicron*? Mol. Microbiol. 39: 1562-1571.

Paulsen, J., Kröger, A., and Thauer, R.K. 1986. ATP-driven succinate oxidation in the catabolism of *Desulfuromonas acetoxidans*. Arch. Microbiol. 144: 78-83.

Peters, J.W., Fisher, K., and Dean, D.R. 1995. Nitrogenase structure and function: a biochemical-genetic perspective. Ann. Rev. Microbiol. 49: 335-366.

Pezacka, E., and Wood, H.G. 1984. The synthesis of acetyl-CoA by *Clostridium thermoaceticum* form carbon dioxide, hydrogen, coenzyme A and methyltetrahydrofolate. Arch. Microbiol. 137: 63-69.

Pfeiffer, T., Schuster, S., and Bonhoeffer, S. 2001. Cooperation and competition in the evolution of ATP-producing pathways. Science. 292: 504-507.

Phelps, T.J., and Zeikus, J.G. 1984. Influence of pH on terminal carbon metabolism in anoxic sediments from a mildly acidic lake. Appl. Environ. Microbiol. 48: 1088-1095.

Pittman, K.A., Lakshmanan, S., and Bryant, M.P. 1967. Oligopeptide uptake by *Bacteroides ruminicola*. J. Bacteriol. 93: 1499-1508.

Pohorelic, B.K., Voordouw, J.K., Lojou, E., Dolla, A., Harder, J., and Voordouw, G. 2002. Effects of deletion of genes encoding Fe-only hydrogenase of *Desulfovibrio vulgaris* Hildenborough on hydrogen and lactate metabolism. J. Bacteriol. 184: 679-686.

Pomposiello, P.J., and Demple, B. 2002. Adjustment of microbial physiology to free radical stress. Adv. Micro. Physiol. 46: 319-341.

Prins, R.A., and Lankhorst, A. 1977. Synthesis of acetate from CO_2 in the cecum of some rodents. FEMS Microbiol. Lett. 1: 255-258.

Rabus, R., and Widdel, F. 1995. Anaerobic degradation of ethylbenzene and other aromatic hydrocarbons by new denitrifying bacteria. Arch. Microbiol. 163: 96-103.

Rabus, R., and Heider, J. 1998. Initial reactions of anaerobic metabolism of alkylbenzenes in denitrifying and sulfate-reducing bacteria. Arch. Microbiol. 170: 377-384.

Rabus, R., Wilkes, H., Behrends, A., Armstroff, A., Fischer, T., Pierik, A.J., and Widdel, F. 2001. Anaerobic initial reaction of *n*-alkanes in a denitrifying bacterium: evidence for (1-methylpentyl)succinate as initial product and for involvement of an organic radical in *n*-hexane metabolism. J. Bacteriol. 183: 1707-1715.

Ragsdale, S.W. 1991. Enzymology of the acetyl-CoA pathway of CO_2 fixation. Crit. Rev. Biochem. Mol. Biol. 26: 261-300.

Ragsdale, S.W., Clark, J.E., Ljungdahl, L.G., Lundie, L.L., and Drake, H.L. 1983. Properties of purified carbon monoxide dehydrogenase from *Clostridium thermoaceticum*, a nickel, iron-sulfur protein. J. Biol. Chem. 258: 2364-2369.

Rapp-Giles, B.J., Casalot, L., English, R.A., Ringbauer, J.A., Dolla, A., and Wall, J.D. 2000. Cytochrome c_3 mutants of *Desulfovibrio desulfuricans*. Appl. Environ. Microbiol. 66: 671-677.

Rebac, S., van Lier, J.B., Janssen, M.G.J., Dekkers, F., Swinkels, K.T.M., and Lettinga, G. 1997. High-rate anaerobic treatment of malting waste water in a pilot-scale EGSB system under psychrophilic conditions. J. Chem. Tech. Biotechnol. 68: 135-146.

Richardson, D., and Sawers, G. 2002. PMF through the redox loop. Science. 295: 1842-1843.

Rios-Hernandez, L.A., Gieg, L.M., and Suflita, J.M. 2003. Biodegradation of an alicyclic hydrocarbon by a sulfate-reducing enrichment from a gas condensate-contaminated aquifer. Appl. Environ. Microbiol. 69: 434-443.

Robinson, J.A., and Tiedje, J.M. 1982. Kinetics of hydrogen consumption by rumen fluid, anaerobic digestor sludge, and sediment. Appl. Environ. Microbiol. 44: 1374-1384.

Robinson, J.A., and Tiedje, J.M. 1984. Competition between sulfate-reducing and methanogenic bacteria for H_2 under resting and growing conditions. Arch. Microbiol. 137: 26-32.

Rocha, E.R., and Smith, C.J. 1996. Oxidative stress response in an anaerobe, *Bacteroides fragilis*: a role for catalase in protection against hydrogen peroxide. J. Bacteriol. 178: 6895-6903.

Rocha, E.R., and Smith, C.J. 1998. Characterization of a peroxide-resistant mutant of the anaerobic bacterium *Bacteroides fragilis*. J. Bacteriol. 180: 5906-5912.

Rocha, E.R., and Smith, C.J. 1999. Role of the alkyl hydroperoxide reductase (*ahpCF*) gene in oxidative stress defense of the obligate anaerobe *Bacteroides fragilis*. J. Bacteriol. 181: 5701-5710.

Rocha, E.R., Owens Jr., G., and Smith, C.J. 2000. The redox-sensitive transcriptional activator OxyR regulates the peroxide response regulon in the obligate anaerobe *Bacteroides fragilis*. J. Bacteriol. 182: 5059-5069.

Rosner, B.M., and Schink, B. 1995. Purification and characterization of acetylene hydratase of *Pelobacter acetylenicus*, a tungsten iron-sulfur protein. J. Bacteriol. 177: 5767-5772.

Roth, J.R., Lawrence, J.G., and Bobik, T.A. 1996. Cobalamin (coenzyme B_{12}): Synthesis and biological significance. Ann. Rev. Microbiol. 50: 137-181.

Sapra, R., Bagramyan, K., and Adams, M.W. 2003. A simple energy-conserving system: proton reduction coupled to proton translocation. Proc. Natl. Acad. Sci. USA. 100: 7545-7550.

Schauder, R., Widdel, F., and Fuchs, G. 1987. Carbon assimilation pathways in sulfate-reducing bacteria II. Enzymes of a reductive citric acid cycle in the autotrophic *Desulfovibrio hydrogenophilus*. Arch. Microbiol. 148: 218-225.

Schink, B. 1997. Energetics of syntrophic cooperation in methanogenic degradation. Microbiol. Mol. Biol. Rev. 61: 262-280.

Schink, B., and Müller, J. 2000. Anaerobic degradation of phenolic compounds. Naturwissenschaften. 87: 12-23.

Schöcke, L., and Schink, B. 1997. Energetics of methanogenic benzoate degradation by *Syntrophus gentianae*. Microbiology. 143: 2345-2351.

Schöcke, L., and Schink, B. 1998. Membrane-bound proton-translocating pyrophosphatase of *Syntrophus gentianae*, a syntrophically benzoate-degrading fermenting bacterium. Eur. J. Biochem. 256: 589-594.

Schöcke, L., and Schink, B. 1999. Energetics and biochemistry of fermentative benzoate degradation by *Syntrophus gentianae*. Arch. Microbiol. 171: 331-337.

Scholten, J.C.M., and Conrad, R. 2000. Energetics of syntrophic propionate oxidation in defined batch and chemostat cocultures. Appl. Environ. Microbiol. 66: 2934-2942.

Schönheit, P., Moll, J., and Thauer, R.K. 1979. Nickel, cobalt and molybdenum requirement for growth of *Methanobacterium thermoautotrophicum*. Arch. Microbiol. 123: 105-107.

Schwartz, E., and Friedrich, B. 2003. H_2-metabolizing procaryotes. In: The Prokaryotes: An Evolving Electronic Resource for the Microbial Community. release 3.14, July 31, 2003. Dworkin, M., ed. Springer-Verlag, New York. http://link.springer-ny.com/link/service/books/10125/.

Schwarz, W.H. 2001. The cellulosome and cellulose degradation by anaerobic bacteria. Appl. Microbiol. Biotechnol. 56: 634-649.

Self, W.T. 2003. Selenium-dependent enzymes from *Clostridia*. In: Biochemistry and Physiology of Anaerobic Bacteria. Ljungdahl, L.G., Adams, M.W., Baron, L.L., Ferry, J.G., and Johnson, M.K., eds. Springer, New York. p. 157-170.

Selig, M., and Schönheit, P. 1994. Oxidation of organic compounds to CO_2 with sulfur or thiosulfate as electron acceptor in the anaerobic hyperthermophilic archaea *Thermoproteus tenax* and *Pyrobaculum islandicum* proceeds via the citric acid cycle. Arch. Microbiol. 162: 286-294.

Shea, T.G., Prinorius, W.A., Cole, R.D., and Pearson, E.A. 1968. Kinetics of hydrogen assimilation in the methane fermentation. Water Res. 2: 833-848.

Silva, G., LeGall, J., Xavier, A.V., Teixeira, M., and Rodrigues-Pousada, C. 2001. Molecular characterization of *Desulfovibrio gigas* neelaredoxin, a protein involved in oxygen detoxification in anaerobes. J. Bacteriol. 183: 4413-4420.

Smith, D.P., and McCarty, P.L. 1989. Energetic and rate effects on methanogenesis of ethanol and propionate in perturbed CSTRs. Biotechnol. Bioengineer. 34: 39-54.

Smith, R.L. 2002. Determining the terminal electron-accepting reaction in the saturated subsurface. In: Manual of Environmental Microbiology. American Society for Microbiology, Washington, D.C. p. 743-752.

So, C.M., Phelps, C.D., and Young, L.Y. 2003. Anaerobic transformation of alkanes to fatty acids by a sulfate-reducing bacterium, strain Hxd3. Appl. Environ. Microbiol. 69: 3892-3900.

Spormann, A., and Thauer, R.K. 1988. Anaerobic acetate oxidation to CO_2 by *Desulfotomaculum acetoxidans*. Arch. Microbiol. 150: 374-380.

Stams, A.J.M. 1994. Metabolic interactions between anaerobic bacteria in methanogenic environments. Antonie van Leeuwenhoek. 66: 271-294.

Stiefel, E.I. 1996. Molybdenum bolsters the bioorganic brigade. Science. 272: 1599-1600.

Strauss, G., and Fuchs, G. 1993. Enzymes of a novel autotrophic CO_2 fixation pathway in the phototrophic bacterium *Chloroflexus aurantiacus*, the 3-hydroxypropionate cycle. Eur. J. Biochem. 215: 633-643.

Strayer, R.F., and Tiedje, J.M. 1978. Kinetic parameters of the conversion of methane precursors to methane in a hypereutrophic lake sediment. Appl. Environ. Microbiol. 36: 330-340.

Strong, G.E., and Cord-Ruwisch, R. 1995. An in situ dissolved-hydrogen probe for monitoring anaerobic digesters under overload conditions. Biotech. Bioeng. 45: 63-68.

Sztukowska, M., Bugno, M., Potempa, J., Travis, J., and Kurtz Jr., D.M. 2002. Role of rubrerythrin in the oxidative stress response of *Porphoyromonas gingivalis*. Mol. Microbiol. 44: 479-488.

Tanaka, H., and Stadtman, T.C. 1979. Selenium-dependent clostridial glycine reductase. Purification and characterization of the two membrane-bound protein components. J. Biol. Chem. 254: 447-452.

Tanner, R.S., and Wolfe, R.S. 1988. Nutritional requirements of *Methanomicrobium mobile*. Appl. Environ. Microbiol. 54: 625-628.

ten Brink, B., and Konings, W.N. 1982. Electrochemical proton gradient and lactate concentration gradient in *Streptococcus cremoris* cell grown in batch culture. J. Bacteriol. 152: 682-686.

Tewes, F.J., and Thauer, R.K. 1980. Regulation of ATP synthesis in glucose fermenting bacteria involved in interspecies hydrogen transfer. In: Anaerobes and Anaerobic Infections. S. Gottschalk, N. Pfennig, and H. Werner, eds. Gustav Fischer Verlag, Stuttgart, Germany. p. 97-104.

Thamdrup, B., Finster, K., Würgler, J., and Bak, F. 1993. Bacterial disproportionation of elemental sulfur coupled to chemical reduction of iron or maganese. Appl. Environ. Microbiol. 59: 101-108.

Thauer, R.K. 2001. Enzymology. Nickel to the fore. Science. 293: 1264-1265.

Thauer, R.K., and Morris, J.G. 1984. Metabolism of chemotrophic anaerobes: old views and new aspects. Symp. Soc. Gen. Microbiol. 36: 123-168.

Thauer, R.K., Jungermann, K., and Decker, K. 1977. Energy conservation in chemotrophic anaerobic bacteria. Bacteriol. Rev. 41: 100-180.

Thauer, R.K., Jungermann, K., Rupprecht, E., and Decker, K. 1969. Hydrogen formation from NADH in cell-free extracts of *Clostridium kluyveri*. FEBS Lett. 4: 108-112.

Thiele, J.H., and Zeikus, J.G. 1988. Control of interspecies electron flow during anaerobic digestion: significance of formate transfer versus hydrogen transfer during syntrophic methanogenesis in flocs. Appl. Environ. Microbiol. 54: 20-29.

Tholen, A., and Brune, A. 1999. Localization and in situ activities of homoacetogenic bacteria in highly compartmentalized hindgut of soil-feeding higher termites (*Cubitermes* spp.). Appl. Environ. Microbiol. 65: 4497-4505.

Unciuleac, M., and Boll, M. 2001. Mechanism of ATP-driven electron transfer catalyzed by the benzene ring-reducing enzyme benzoyl-CoA reductase. Proc. Natl. Acad. Sci. USA. 98: 13619-13624.

Unden, G., Becker, S., Bongaerts, J., Schirawski, J., and Six, S. 1994. Oxygen regulated expression in facultatively anaerobic bacteria. Antonie von Leeuwenhoek. 66: 3-23.

van Lier, J.B., Rebac, S., and Lettinga, G. 1997. High-rate anaerobic wastewater treatment under psychrophilic and thermophilic conditions. Water Sci. Technol. 35: 199-206.

Varel, V.H., and Bryant, M.P. 1974. Nutritional features of *Bacteroides fragilis* subsp. *fragilis*. Appl. Microbiol. 18: 251-257.

Voordouw, G. 1995. The genus *Desulfovibrio*: The centennial. Appl. Environ. Microbiol. 61: 2813-2819.

Voordouw, G. 2002. Carbon monoxide cycling by *Desulfovibrio vulgaris* Hildenborough. J. Bacteriol. 184: 5903-5911.

Voordouw, J.K., and Voordouw, G. 1998. Deletion of the *rbo* gene increases the oxygen sensitivity of the sulfate-reducing bacterium *Desulfovibrio vulgaris* Hildenborough. Appl. Environ. Microbiol. 64: 2882-2887.

Wagner, A.F., Frey, M., Neugebauer, F.A., Schafer, W., and Knappe, J. 1992. The free radical in pyruvate formate-lyase is located on glycine-734. Proc. Natl. Acad. Sci. USA. 89: 996-1000.

Wagner, C., GriessHammer, A., and Drake, H.L. 1996. Acetogenic capacities and the anaerobic turnover of carbon in a Kansas prairie soil. Appl. Environ. Microbiol. 62: 494-500.

Wagner, R., and Andreesen, J.R. 1987. Accumulation and incorporation of ^{185}W into proteins of *Clostrdium acidiurici* and *Clostridium cylindrosporum*. Arch. Microbiol. 147: 295-299.

Wahlund, T.M., and Tabita, F.R. 1997. The reductive tricarboxylic acid cycle of carbon dioxide assimilation: initial studies and purification of ATP-citrate lyase from the green sulfur bacterium *Chlorobium tepidum*. J. Bacteriol. 179: 4859-4867.

Wallrabenstein, C., and Schink, B. 1994. Evidence of reversed electron transport in syntrophic butyrate or benzoate oxidation by *Syntrophomonas wolfei* and *Syntrophus buswellii*. Arch. Microbiol. 162: 136-142.

Ward, D.M., and Winfrey, M.R. 1985. Interactions between methanogenic and sulfate-reducing bacteria in sediments. Adv. Aquat. Microbiol. 3: 141-179.

Warren, R.A.J. 1996. Microbial hydrolysis of polysaccharides. Ann. Rev. Microbiol. 50: 183-212.

White, D.C. 2000. The Physiology and Biochemistry of Prokaryotes. Oxford University Press, New York.

Whitman, W.B., Ankwanda, E., and Wolfe, R.S. 1982. Nutrition and carbon metabolism of *Methanococcus voltae*. J. Bacteriol. 149: 852-863.

Widdel, F., and Rabus, R. 2001. Anaerobic biodegradation of saturated and aromatic hydrocarbons. Curr. Opin. Biotechnol. 12: 259-276.

Wilkes, H., Rabus, R., Fischer, T., Armstroff, A., Behrends, A., and Widdel, F. 2002. Anaerobic degradation of *n*-hexane in a denitrifying bacterium: further degradation of the initial intermediate (1-methylpentyl)succinate via C-skeleton rearrangement. Arch. Microbiol. 177: 235-243.

Wilkes, H., Kuhner, S., Bolm, C., Fischer, T., Classen, A., Widdel, F., and Rabus, R. 2003. Formation of *n*-alkane- and cycloalkane-derived organic acids during anaerobic growth of a denitrifying bacterium with crude oil. Org. Geochem. 34: 1313-1323.

Wofford, N.Q., Beaty, P.S., and McInerney, M.J. 1986. Preparation of cell-free extracts and the enzymes involved in fatty acid metabolism in *Syntrophomonas wolfei*. J. Bacteriol. 167: 179-185.

Wolin, M.J. 1982. Hydrogen transfer in microbial communities. In: Microbial Interactions and Communities. A.T. Bull, and J.H. Slater, eds. Academic Press, London. p. 323-356.

Wolin, M.J., and Miller, T.L. 1989. Carbohydrate fermentation. In: Human Intestinal Microflora in Health and Disease. D.F. Hentges, ed. Academic Press, London. p. 147-165.

Wolin, M.J., and Miller, T.L. 1994. Acetogenesis from CO_2 in the human colonic ecosystem. In: Acetogenesis. H.L. Drake, ed. Chapman & Hall, New York. p. 365-385.

Wu, L.-F., and Mandrand, M.A. 1993. Microbial hydrogenases: primary structure, classification, signatures and phylogeny. FEMS Microbiol. Lett. 104: 243-270.

Yamamoto, I., Saiki, T., Liu, S.-M., and Ljungdahl, L.G. 1983. Purification and properties of NADP-dependent formate dehydrogenase from *Clostridium thermoaceticum*, a tungsten-selenium-iron protein. J. Biol. Chem. 258: 1826-1832.

Yandulov, D.V., and Schrock, R.R. 2003. Catalytic reduction of dinitrogen to ammonia at a single molybdenum center. Science. 301: 76-78.

Zehnder, A.J.B., and Stumm, W. 1988. Geochemistry and biogeochemistry of anaerobic habitats. In: Biology of Anaerobic Microorganisms. A.J.B. Zehnder, ed. John Wiley & Sons, Inc., New York. p. 1-38.

Zengler, K., Heider, J., Rossello-Mora, R., and Widdel, F. 1999. Phototrophic utilization of toluene under anoxic conditions by a new strain of *Blastochloris sulfoviridis*. Arch Microbiol. 172: 204-212.

Zhang, X., and Young, L.Y. 1997. Carboxylation as an initial reaction in the anaerobic metabolism of naphthalene and phenanthrene by sulfidogenic consortia. Appl. Environ. Microbiol. 63: 4759-4764.

Zhang, X., Sullivan, E.R., and Young, L.Y. 2000. Evidence for aromatic ring reduction in the biodegradation pathway of carboxylated naphthalene by a sulfate reducing consortium. Biodegradation. 11: 117-124.

Zindel, U., Freudenberg, W., Rieth, M., Andreesen, J.R., Schnell, J., and Widdel, F. 1988. *Eubacterium acidaminophilum* sp. nov., a versatile amino acid-degrading anaerobe producing or utilizing H_2 or formate. Arch. Microbiol. 150: 254-266.

Zinder, S.H. 1986. Patterns of carbon flow from glucose to methane in a thermophilic anaerobic bioreactor. FEMS Microbiol. Ecol. 38: 243-250.

Zinder, S.H. 1993. Physiological ecology of methanogens. In: Methanogenesis: Ecology, Physiology, Biochemistry, Genetics. J.G. Ferry, ed. Chapman and Hall, New York. p. 128-206.

Zwolinski, M.D., Harris, R.F., and Hickey, W.J. 2000. Microbial consortia involved in the anaerobic degradation of hydrocarbons. Biodegradation. 11: 141-158.

Chapter 3

Redox (Oxygen)-dependent Gene Regulation in Facultative Anaerobes

R. Gary Sawers* and Michiko M. Nakano

Abstract

Our goal in this chapter is to review recent and current themes in the area of redox-dependent gene regulation. Over the last few years there has been an explosion in research looking at mechanisms of oxygen and redox sensing, transmission of these signals to the genome and their consequences on metabolism. These cover Gram-positives and -negatives, facultative anaerobes as well as obligate aerobes and phototrophs. We focus on recent aspects of redox sensing and gene regulation only and highlight new areas of research in the field. This research field is flourishing and revealing some very surprising and exciting new science.

1. Introduction

It is now almost 30 years since an *Escherichia coli* mutant was discovered that grew normally with oxygen as respiratory terminal electron but failed to grow anaerobically with glycerol as the carbon source and either fumarate or nitrate as the exogenous electron acceptor (Lambden and Guest, 1976). The mutant was designated fumarate and nitrate reduction defective and the mutant gene encoded the FNR protein. Remarkably, in those pre-bioinformatics days, there was a characterised protein that exhibited similarity to FNR and that was the cAMP receptor protein, CRP, (Shaw *et al.*, 1983). Based on this, the identification of a helix-turn-helix motif, which is characteristic of a DNA-binding protein, and the pleiotropy of the *fnr* mutation, the authors proposed that FNR was a transcriptional regulator. Through the seminal work over the last two decades in the laboratories of Guest, Green and more recently Kiley, FNR remains *en vogue* and is the paradigm for redox-responsive transcriptional regulators. Indeed, the advent of genome sequencing has now increased the number of close CRP/FNR homologues in the database to over 350, such that they have acquired the 'superfamily' status (Green *et al.*, 2001; Körner *et al.*, 2003).

Work on FNR revealed rather quickly that, at least for *E. coli*, it could not be the sole regulator of redox-regulated genes and this spawned a number of searches to identify new redox regulators and previously undiscovered mechanisms of redox-sensing. It became clear comparatively quickly that a facultative anaerobic bacterium like *E. coli* has a large variety of oxygen- or redox- sensitive proteins, some of which function as signal transducers, while others function as DNA-binding proteins. This variety gives a microorganism the remarkable capacity to adapt to a huge range of oxygen levels and to adjust its physiology rapidly to the demands set by the environment. More recent work has revealed that it is likely that all microorganisms will prove to have some means of sensing their 'redox environment', even obligate aerobes.

Our aim in this review is not to provide an exhaustive literature survey but to highlight very recent new aspects of redox regulation in bacteria. Wherever possible throughout this chapter, the authors have cited reviews that provide more detailed literature, often with different points of view, in a number of key areas of redox regulation. We apologise in advance to those scientists whose work we did not have the opportunity to cite due to constraints of space. For the same reason, this review will not cover oxidative stress, which is also redox regulation, because of the extensive nature of the literature in this area. There are a number of excellent recent reviews on this topic, which the authors recommend (Pomposiello and Demple, 2001; Imlay, 2002; Paget and Buttner, 2003). Related topics are also covered in Chapters 4 and 7 in this book.

2. The FNR/CRP Superfamily of Transcriptional Regulators

Until recently, the FNR/CRP family of transcriptional regulators comprised comparatively few members (Green *et al.*, 2001). The advent of genome sequencing, however, has resulted in the identification of many new members, which has expanded this superfamily dramatically. Indeed, a recent phylogenetic analysis of

*For correspondence email: gary.sawers@bbsrc.ac.uk

Table 1. Groups of FNR proteins with Fe-S clusters

Group	Features of cluster	Location of cluster in protein	Examples of bacteria[a]
FNR_{Ec}	$[4Fe\text{-}4S]^{2+}$	N-terminal	*Escherichia coli Pseudomonas* spp. *Azotobacter vinelandii Shewanella* spp.
FNR_{Bac}	$[4Fe\text{-}4S]^{2+}$	C-terminal	*Bacillus* spp.
FnrN	$[4Fe\text{-}4S]^{2+}$ altered Cys spacing	N-terminal	*Rhodobacter* spp. *Rhodopseudomonas palustris* *Rhizobium* spp. *Agrobacterium tumefaciens* *Brucella* spp.
YeiL	$[4Fe\text{-}4S]^{2+}$ aconitase type	Central	*Escherichia coli* *Desulfitobacterium hafniense* *Bacillus anthracis* *Streptococcus mutans*
FlpA	$[4Fe\text{-}4S]^{2+}$ Cys + His ligation	N-terminal	*Lactococcus lactis*

[a] Only some examples are given for each class. Data is taken from Körner *et al.* (2003).

almost 370 members has revealed that they are found in microorganisms of diverse origin (Körner *et al.*, 2003). Current evidence based on bioinformatic analyses suggests that FNR/CRP regulators might also be present in the archaea; however, the amino acid similarities with either FNR or CRP are low. So far there are over 20 branches of the phylogenetic tree, which has been used to classify the various members. Interestingly, the modular nature of this family of transcription factors has enabled them to adopt a variety of sensory functions. The huge diversity of the signals sensed by these regulators is large, and although initially surprising, upon consideration this should not be, because the original members of this superfamily, CRP and FNR, bind very different substrates. The signal molecules processed range from cAMP (CRP), CO (CooA), 2-oxoglutarate (NtcA) and aromatic compounds (HbaR). However, identification of the physiological signal sensed by the majority of family members is outstanding. Notably, one organism can have several superfamily members, each with a different regulatory function. For example, *Bradyrhizobium japonicum* has 14 members, and *Rhodopseudomonas palustris* has 13 members (Körner *et al.*, 2003).

2.1. Role in Redox Regulation: Responding to Anoxia

The phylogenetic analysis discussed above (Körner *et al.*, 2003) revealed that there are five groups, each containing at least one member characterised as being an oxygen- or redox-responsive transcriptional regulator (Table 1). A characteristic of all of these groups is that the members contain an iron-sulphur cluster. The paradigm is FNR, which is a dimeric protein with one $[4Fe\text{-}4S]^{2+}$ cluster per subunit; only the dimeric form of FNR is able to bind DNA specifically and activate transcription (Green *et al.*, 2001; Kiley and Beinert, 2003). The $[4Fe\text{-}4S]^{2+}$ clusters are directly involved in sensing dioxygen (Jordan *et al.*, 1997; Khoroshilova *et al.*, 1997). In the presence of oxygen the cluster disassembles to a $[2Fe\text{-}2S]^{2+}$ cluster, which no longer can bind DNA and *in vitro* can no longer dimerise. It is thought that FNR switches between a $[4Fe\text{-}4S]^{2+}$-containing, transcriptionally active species and a $[2Fe\text{-}2S]^{2+}$-containing transcriptionally inactive species, depending upon the cellular oxygen levels. The switch from a cubane $[4Fe\text{-}4S]^{2+}$ to a planar $[2Fe\text{-}2S]^{2+}$ species presumably results in a major conformational change in the protein, preventing dimer formation (Green *et al.*, 2001).

2.2. Members of the FNR Group

Over 35 members of the FNR group of transcription factors have been identified (Körner *et al.*, 2003) and include ANR from *Pseudomonas* species (Pessi and Haas, 2000; see also Chapter 4 for a discussion of ANR). All members of this group show a high degree of amino acid similarity. Structural characteristics of this group include a N-terminal $[4Fe\text{-}4S]^{2+}$ cluster ensconced in an eight-stranded β-roll, a long α-helix, which provides the dimerisation interface (Moore and Kiley, 2001) and a C-terminal helix-turn-helix domain. The DNA recognition sequence for this group is $TTGATN_4ATCAA$ and usually is located centered at either -41.5 bp relative to the transcription initiation site in Class II activators, at ~ -61.5 bp in Class I activators or overlapping the RNA polymerase recognition sequences when the protein acts as a transcriptional repressor (Green *et al.*, 2001). Note, however, the repression of the *E. coli ndh* and *yfiD* (Marshall *et al.*, 2001) promoters by FNR involves unusual mechanisms. Major advances in our understanding of FNR dimerisation (Moore and Kiley,

2001) and its interaction with RNA polymerase have been made recently (Blake *et al*., 2002; Scott and Green, 2002).

Of all the members of this class, detailed knowledge of mechanism is known for only a few and the details are covered in recent excellent reviews (Green *et al*., 2001; Körner *et al*., 2003). One FNR-type regulator worthy of mention is CydR from *Azotobacter vinelandii*, which controls expression of cytochrome *bd* oxidase synthesis and polyhydroxyalkanoate metabolism (Wu *et al*., 2000, 2001). *A. vinelandii* is an obligate aerobe, yet it performs nitrogen fixation through the action of an extremely oxygen-labile nitrogenase. The paradox in this system is how nitrogenase is protected from the deleterious effects of dioxygen. This is thought to occur through the phenomenon of respiratory protection, whereby the highly active cytochrome *bd* oxidase is expressed at high levels as the oxygen levels in the immediate environment increase and it prevents oxygen entering the cytoplasm and inactivating the nitrogen-fixing reaction. CydR has an extremely oxygen-sensitive iron-sulphur cluster thereby conferring upon it increased oxygen responsiveness *in vivo*. As the oxygen concentration decreases, CydR binds to the promoter region of the *cydAB* operon and reduces expression. At high oxygen concentration, the $[4Fe\text{-}4S]^{2+}$ cluster in CydR is rapidly converted to the $[2Fe\text{-}2S]^{2+}$ species, releasing CydR from the promoter.

An interesting new development in the field of FNR-dependent regulation involves the control of the NifL-NifA regulatory cascade in *Klebsiella pneumoniae* (Schmitz *et al*., 2002). NifA is a transcriptional regulator that activates *nif* gene expression. The activity of NifA is controlled by the NifL protein, which responds to the cellular dioxygen and nitrogen status. In the presence of dioxygen and/or fixed nitrogen, NifL complexes NifA and inhibits its activity. NifL is sequestered to the cytoplasmic membrane when oxygen is limiting and FNR is required for this to occur (Klopprogge *et al*., 2002). It is likely that FNR acts indirectly by controlling expression of membrane-associated dehydrogenases, e.g. formate dehydrogenase or NADH dehydrogenase (Grabbe and Schmitz, 2003), which in turn maintain NifL in a reduced, membrane-associated form. Whether FNR additionally controls expression of a NifL-specific receptor is yet to be determined.

2.3. The FnrN, $FixK_2$, and YeiL Groups

Like the FNR family, the members of this group of transcription factors have the conserved cysteinyl residues, which would be required for the assembly of an oxygen-sensing $[4Fe\text{-}4S]^{2+}$ cluster. So far, only two members of these sub-families have been shown to possess an iron-sulphur cluster; FnrP from *Paracoccus denitrificans* (Hutchings *et al*., 2002), and YeiL from *E. coli* (Anjum *et al*., 2000). FnrP controls synthesis of cytochrome cbb_3 oxidase, respiratory nitrate reductase and cytochrome *c* peroxidase (Otten *et al*., 2001) in response to oxygen limitation and it is related to FnrL from *Rhodobacter* (Oh *et al*., 2000). As with FnrN in the rhizobia, this group of regulators controls expression of respiratory genes important in controlling the shift from heterotrophy to phototrophy in genera such as *Rhodopseudomonas* and dinitrogen metabolism in the genera *Rhizobium* and *Paracoccus*. Another member that has recently been identified in *Bradyrhizobium japonicum* is $FixK_2$, which, in contrast to other FixK homologues, has more similarity to the FnrN group and appears to have an iron-sulphur cluster (Durmowicz and Meier, 1998). FixK's normally do not have the N-terminal domain characteristic of the FNR family, but instead are found as components in complex regulatory cascades that control various aspects of nitrogen fixation in the rhizobia.

The final member of this group is the YeiL protein of *E. coli*, which is one of three FNR/CRP proteins in this organism. The physiological function of YeiL is not clear but expression of its gene is controlled in a complex manner and high levels of the YeiL protein are found in stationary phase cells (Anjum *et al*., 2000). It has been demonstrated that a $[4Fe\text{-}4S]^{2+}$ cluster can be reconstituted into YeiL *in vitro*; however, this cluster may have more properties akin to that found in aconitase than FNR. Furthermore, the putative DNA-binding helix of YeiL is different to that of FNR suggesting the recognition sequence might be uncharacteristic of other FNR family members.

2.4. FLP: Dithiol-Disulphide Exchange Versus $[4Fe\text{-}4S]^{2+}$

Until 1993 no FNR homologue had been identified in a Gram-positive bacterium. This changed with the discovery of an FNR-like protein (FLP) in *Lactobacillus casei* (Irvine and Guest, 1993). Although FLP has the DNA-binding domain characteristic of the FNR family it has a rather uncharacteristic two cysteinyl residues rather than the five found in FNR. It is now clear that FLP undergoes reversible dithiol-disulphide exchange and consequently is involved in sensing oxidative stress rather than anaerobiosis *per se* (Gostick *et al*., 1998). Through genome sequencing this group has expanded considerably and perhaps unsurprisingly is found primarily in homo- and heterofermentative lactic acid bacteria but also in some clostridia and the sulphate-reducing bacterium *Desulfitobacterium hafniense* (Körner *et al*, 2003). The potential medical importance of FLP is suggested by the existence of five homologues encoded in the small genome of *Clostridium perfringens* (Shimizu *et al*., 2002).

There is one example, FlpA (Table 1), which at least *in vitro*, can be reconstituted to contain a $[4Fe\text{-}4S]^{2+}$

cluster (Scott *et al.*, 2000). FlpA has one N-terminal Cys and one centrally located Cys residue, so the suggestion has been made that His residues provide the other ligands for the cluster. The protein is dimeric and contains Zn when isolated from the heterologous host *E. coli* and it is this form of the protein that binds DNA. The Fe-S form of the protein is inactive.

2.5. FNR and *Bacillus subtilis*

An FNR homologue was first identified in *B. subtilis* in 1995 (Cruz-Ramos *et al.*, 1995) and it was shown to be necessary to allow the organism to respire with nitrate. Unlike *E. coli* FNR, which has a N-terminal $[4Fe\text{-}4S]^{2+}$ cluster, the cysteinyl residues proposed to coordinate the $[4Fe\text{-}4S]^{2+}$ cluster are located in the C-terminus of the protein (for a review see Nakano and Zuber, 2002). Phylogenetic analysis of FNR from *B. subtilis* places it close to PfrA from the *Listeria*. PfrA is an important pathogenicity determinant but it lacks a $[4Fe\text{-}4S]^{2+}$ cluster (Körner *et al.*, 2003).

A further difference between *E. coli* and *B. subtilis fnr* is that the synthesis of FNR in *B. subtilis* is induced by anaerobiosis and this is controlled by the ResDE two-component regulatory system (see below). Moreover, FNR, as well as activating expression of the *narGHJI* operon also controls the expression of another regulator called ArfM, which is an activator of a number of other genes involved in heme biosynthesis, as well as those required for lactate and acetoin fermentation (Cruz Ramos *et al.*, 2000; Marino *et al.*, 2001).

2.6. Functional Genomics and FNR Homologues

Probably one of the first proteomic studies performed to examine the consequences of the shift from aerobic to anaerobic growth in *E. coli* was carried out over 20 years ago (Smith and Neidhardt, 1983a,b). A total of 170 polypeptides were examined and the steady-state levels of approximately 20% of these were found to change in response to a shift between aerobic and anaerobic growth. Another study extended these findings by looking at the effect of an *fnr* mutation on the aerobic and anaerobic protein profile of *E. coli* (Sawers *et al.*, 1988). In that particular study, the levels of 125 polypeptides were identified as changing in response to anaerobiosis and of these the levels of a total of 45 polypeptides were altered in an *fnr* mutant. Intriguingly, this study indicated that FNR exerted repressor function and suggested that it also functioned in aerobically grown cells.

More recent transcription profiling experiments have revealed that as much as a third of all genes expressed in aerobic cells are subject to change in concentration upon shift to anaerobiosis and of these as many as 1000 may be regulated directly or indirectly by FNR (Salmon *et al.*, 2003). Surprisingly, a large number of genes shown to be FNR-dependent through conventional studies were not identified in these experiments. The reasons for this are not clear, but it suggests that some optimisation of conditions is required. Although the results of these studies require substantiation through other experimental approaches, it seems clear that FNR has a very large influence on the physiology in the *E. coli* cell.

The Gram-negative facultative anaerobe *Shewanella oneidensis* has remarkable respiratory capacity, and is able to use a large number of terminal electron acceptors, including metal oxides (Nealson and Saffarini, 1994). The *etrA* (electron transport regulator) gene product is highly similar to FNR from *E. coli* and originally was believed to be required for anaerobic growth and respiration (Saffarini and Nealson, 1993). More recent knock-out studies have demonstrated, however, that EtrA is not essential for either anaerobic growth or reduction of most electron acceptors (Maier and Myers, 2001). Transcription profiling of an *etrA* mutant of *S. oneidensis* identified 69 genes that are either directly or indirectly controlled by EtrA (Beliaev *et al.*, 2002). A large number of genes controlled by EtrA are indeed involved directly or indirectly in some electron transport processes, for example the Ni/Fe hydrogenase and fumarate reductase. Other EtrA-regulated genes include 11 involved in central metabolism, 9 encode transcriptional regulators, and 12 are involved in transport processes. These findings for *S. oneidensis* suggest the existence of other redox-responsive transcriptional regulators that are involved in controlling respiratory gene expression (see below).

3. Transcriptional Regulation Via ArcA-ArcB

3.1. The Arc Modulon

The Arc (anaerobic redox control) two-component system was originally discovered in *E. coli* (reviewed in Iuchi and Lin, 1995) as a redox-responsive regulatory system that does not sense oxygen *per se*, rather the respiratory and metabolic consequences of oxygen deprivation (Figure 1). The redox status is monitored by the membrane-associated ArcB histidine kinase. This protein is a member of a class of complex, tripartite sensor kinases with a PAS domain and primary and secondary transmitter domains, and a complex phosphorelay system coordinates kinase (and minor phosphatase) activity and phosphoryl group delivery to ArcA (Kwon *et al.*, 2000). Although it was originally thought to be involved more or less exclusively in controlling anaerobic gene expression, it has transpired that the Arc system has a more pervasive role in controlling gene expression in response to the cellular redox status (reviewed in Sawers, 1999). In addition to plasmid maintenance and stationary phase survival it also controls chromosomal replication initiation by binding near *oriC* and inhibiting open complex formation, which is a prerequisite for the replication process to initiate (Lee *et al.*, 2001).

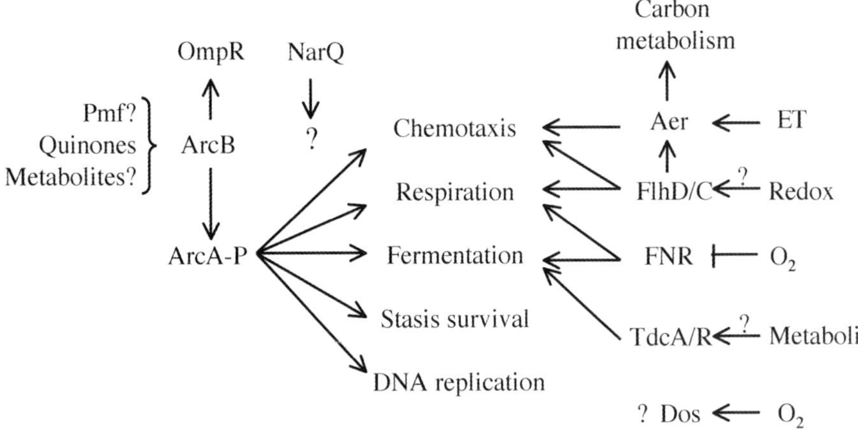

Figure 1. Interplay between redox-regulatory pathways in *E. coli* and their involvement in controlling cellular processes. Arrows indicate activation of gene expression, while lines terminated with a bar indicate repression. Question marks signify that the role of the regulator is unclear or the signal to which it responds has not been defined.

Recent transcriptional profiling of two-component regulatory system knock-outs (Oshima *et al.*, 2002) has identified the Arc system as being important for chemotaxis, osmotic adaptation, purine metabolism and peptide secretion. It also impacts significantly on the RpoS regulon (Sevcík *et al.*, 2001; Oshima *et al.*, 2002). Further study has concentrated exclusively on transcriptome analysis of anaerobically grown wild type *E. coli* versus an isogenic *arcA* deletion mutant (Liu and De Wulf, 2004). Interestingly, these authors combined transcription profiling, with the development of an ArcA-P-binding site weight matrix to identify new binding sites and correlated this data with real-time PCR analyses of a selected set of operons to obtain a stringent data set for putative ArcA-P-regulated genes. The final outcome of this study was the identification of 372 open reading frames that are regulated directly or indirectly by ArcA-P. This figure represents 9% of the open reading frames in the *E. coli* genome. Combining data derived from the location of putative ArcA-P-binding sites with the transcription profiling analysis, the authors confidently predicted the identification of 51 newly identified operons that are likely to be regulated directly by the Arc system. Of these, 13 are activated and 38 repressed (Liu and De Wulf, 2004). This study identified Arc-dependent control of flagellum biosynthesis, which is consistent with the findings of Oshima *et al.* (2002). However, the fact that ArcA-P controls different aspects of the PTS system, as well as carnitine, betaine and lysine transport, suggests a role in modulating major cellular transport processes. Metal ion utilisation and fatty acid biosynthesis are also identified as being controlled by Arc based on these studies. Clearly, there is also a hugely important role in control of central metabolic processes. Despite the fact that these findings require further experimental confirmation, they give a broad insight into the general importance of the Arc system in controlling gene expression.

One further intriguing observation made in the transcription profiling study conducted on two-component systems revealed that a knock-out in *arcB* affects roughly 2.5 fold more genes than a knock-out in *arcA* (Oshima *et al.*, 2002). The implication from these findings is that ArcB cross-regulates with other two-component regulators. One example of this is that under anaerobic condition both ArcB and EnvZ transfer phosphoryl groups to OmpR to control porin synthesis. A phenotypic analysis of *arcA* and *arcB* mutants confirmed this finding and demonstrated that mutations in these genes result in very pleiotropic phenotypes (Zhou *et al.*, 2003); especially membrane-associated functions are affected, which is in accord with the proposed role of the Arc system in controlling major aspects of respiratory and fermentative metabolism.

What is clear from a number of studies is that there is significant overlap between various transcriptional regulators with regard to their influence in controlling target-specific operons. Hence, many of the operons identified as being regulated by ArcA-P are also controlled by FNR (Oshima *et al.*, 2002; Salmon *et al.*, 2003; Liu and De Wulf, 2004). This functional overlap is important because it means that more than one redox signal can be sensed by the organism and transduced to the level of gene expression. This gives the bacterium much greater flexibility in controlling gene expression in response to fluctuating environmental oxygen concentrations.

3.2. Redox-Sensing and Microaerobiosis

The role of the PAS domain in ArcB is still unclear (although see below) and it is still not absolutely clear the precise number of signals sensed by ArcB. One signal that is clearly sensed by ArcB, however, is the oxidation state of the quinone pool, which reflects the rate of electron transport through the respiratory chain and consequently can be considered a monitor of the availability of exogenous electron acceptors (Georgellis *et al.*, 2001). A correlation has been shown between increased oxidation of the quinone pool and decreased autophosphorylation activity of ArcB kinase during aerobiosis. The implication from this finding is that as oxygen levels decrease the quinone pool becomes more reduced and ArcB autophosphorylation is no longer inhibited.

In the early 1990's Iuchi (1993) reported that certain metabolites characteristic of fermentation stimulated ArcB autophosphorylation *in vitro*. A more recent study demonstrated that this effect was due to accelerated autophosphorylation of ArcB by D-lactate, pyruvate or acetate, accompanied by enhanced trans-phosphorylation of ArcA (Georgellis *et al.*, 1999). The effect of these metabolites on autophosphorylation required the presence of the ArcB receiver module. This indicates that there is intramolecular domain-domain communication within this complex kinase. Nevertheless, a clear demonstration that these metabolites have an effect *in vivo* must be provided.

Recent studies using steady-state glucose-limited chemostat cultures have provided strong evidence that the Arc system has an important role in sensing microaerobiosis (Alexeeva *et al.*, 2000; 2002; 2003). This work extends an earlier investigation that examined the effects of different oxygen concentrations on expression of a selection of aerobically and anaerobically regulated promoters (Tseng *et al.*, 1996). FNR-regulated promoters were activated essentially under anaerobic conditions, while certain ArcA-regulated promoters were activated when the cultures became microaerobic.

Transcription profiling studies have revealed that a considerable proportion (36%) of Arc-controlled genes are activated by ArcA-P (Oshima *et al.*, 2002; Liu and De Wulf, 2004). Two of the first operons that were shown to be activated by ArcA-P were the *cydAB* and *focApfl* operons (Cotter and Gunsalus, 1992; Sawers and Suppmann, 1992). Using controlled levels of oxygen availability in glucose-limited chemostats, Alexeeva *et al.* (2000) demonstrated for the first time that pyruvate formate-lyase (PFL) synthesis actually is induced under microaerobic conditions, and not, as previously surmised exclusively anaerobically. This correlated with the synthesis of cytochrome *bd* oxidase. In an *arcA* mutant, this induced synthesis was abolished, which led the authors to suggest that the synthesis of the highly active cytochrome *bd* oxidase is coordinated with that of PFL to afford respiratory protection from the deleterious effects that dioxygen has on the radical-bearing species of the PFL enzyme. This is analogous to the respiratory protection afforded by cytochrome *d* oxidase to nitrogenase in *A. vinelandii* (Poole and Hill, 1997). This work of Alexeeva *et al.* (2000) also highlighted the fact that the $NADH/NAD^+$ ratio varied 10-fold over the aerobic to anaerobic extremes of growth. Such a dramatic change in the $NADH/NAD^+$ ratio could potentially be a further signal sensed by the Arc system, as originally suggested by Iuchi (1993).

More recent work has focussed on developing an *in vivo* method of monitoring the phosphorylation status of Arc, which is not proving to be an easy task (Alexeeva, 2002). It will be important in future studies to use steady-state culture methodologies to study this system and this will be particularly relevant for transcription profiling experiments. Nevertheless, it seems clear that the Arc system has an important role in redox regulation in the microaerobic range, which, together with anaerobiosis, is likely to be the most frequently encountered environment of bacteria like *E. coli* (Tseng *et al.*, 1996; Alexeeva *et al.*, 2000; 2003).

3.3. Distribution of the Arc System

All early studies on the Arc two-component system have focussed on *E. coli*. This obviously raises questions as to whether the Arc system is restricted to enterobacteria or whether it is more widespread amongst facultative anaerobes. Genome sequence analysis identified several close homologues of ArcB and ArcA in the chromosomes of *Haemophilus influenzae*, *Vibrio cholerae, Salmonella enterica,* and *Yersinia pestis*. The *H. influenzae* ArcB lacks a redox-sensing PAS domain, yet in the heterologous host *E. coli* it can complement an *arcB* mutation with no apparent loss of redox-sensing capacity (Georgellis *et al.*, 2001). Indeed *arcA* and *arcB* mutants of *E. coli* were complemented by the corresponding allele from *H. influenzae*. This intriguing finding raises important questions concerning the role of the PAS domain in *E. coli* ArcB.

An *arcA* mutant of *S. enterica* serovar Enteritidis has increased sensitivity to reactive nitrogen and oxygen intermediates (Lu *et al.*, 2002). This might be a further indication of overlapping control of gene expression by more than one redox-responsive transcriptional regulator, since this phenotype has not been observed for an *arcA* mutant of *E. coli*. Alternatively, it could indicate that *S. enterica* and *E. coli* differ markedly in the redox functions controlled by the Arc system.

A surprising role for ArcA in controlling ferric oxide reduction has been suggested recently based on a proteomic study of the metal-reducing facultative anaerobe *Shewanella oneidensis* MR-1 (Vanrobaeys *et al.*, 2003). ArcA was one of several abundant proteins whose synthesis was up-regulated after growth on iron oxide. So far, no ArcB homologue has been annotated in the *S. oneidensis* genome sequence, although several candidate histidine kinases with weak similarity are present. It will be intriguing to determine the associated response regulator of ArcA in *S. oneidensis* and to determine how its gene is regulated in response to Fe^{3+}.

3.4. Arc and Modulation of Virulence Factors

Two reports have been published recently that point to an important role of the Arc system in controlling virulence in *Haemophilus influenzae* and *Vibrio cholerae* (De Souza-Hart *et al.*, 2003; Sengupta *et al.*, 2003).

In *H. influenzae* an *arcA* deletion mutant is severely impaired in resisting the bactericidal activities present in human serum. Closer examination of the factors involved suggested that resistance to the complement system was one of the phenotypes lost in the *arcA* mutant (De Souza-Hart *et al.*, 2003). The authors also demonstrated by conventional 1-D and 2-D gel electrophoretic analyses that ArcA affects the synthesis of a number of cytoplasmic and membrane proteins in *H. influenzae*.

In *V. cholerae* production of cholera toxin (CT) and a toxin-coregulated pilus (TCP) is controlled directly by the transcriptional activator ToxT. Expression of *toxT* is induced 4-fold anaerobically (Sengupta *et al.*, 2003) and in an *arcA* deletion mutant this induction was abolished. In an infant mouse cholera model, *arcA* mutants were significantly attenuated for virulence, indicating that ArcA can be classified as a virulence factor in *V. cholerae*. The extent to which ArcA is involved in controlling other virulence factors in this organism must await more detailed experimentation. Clearly, however, these recent findings broaden the scope of influence of this important redox-responsive transcription factor.

4. Other Mechanisms of Redox Control of Gene Expression

4.1. Aer and the Link to the Flagellum Regulator FlhC/FlhD

Aer is a flavoprotein identified in *E. coli* that is thought to detect perturbations in aerobic and anaerobic electron transport systems. In some respects, this resembles one of the proposed sensory functions of ArcB (Georgellis *et al.*, 2001). Aer has an N-terminal flavin-containing domain and a C-terminal domain that exhibits similarity to the serine chemotaxis protein, Tsr (Bibikov *et al.*, 1997; Rebbapragada *et al.*, 1997). Thus, Aer couples the intracellular energy status with the methyl-accepting chemotaxis proteins CheW, CheA, and CheY and consequently controls aerotaxis (Rebbapragada *et al.*, 1997). Until recently Aer was thought to be principally involved in signal communication with the chemotaxis machinery, but a new study revealed some remarkable insights into anaerobic, redox-dependent gene regulation (Prüß *et al.*, 2003). Two transcriptional regulators FlhC and FlhD combine to form a heterotetrameric transcription factor that controls expression of 14 operons involved in synthesis and control of the flagellar machinery (Prüß, 2000). A combination of transcription profiling, real-time-PCR, enzyme assays and gene fusion analyses revealed several important new findings: First, FlhD can exert positive control of gene expression in the absence of FlhC and the genes and operons controlled by FlhD are not part of the flagellar regulon (Prüß *et al.*, 2003); second, the FlhC/D complex controls the expression of a large number of genes and operons whose products have various functions, which include components of anaerobic respiratory pathways in *E. coli*; third, FlhC/FlhD induces expression of Aer in anaerobic cultures and Aer, in turn, activates expression of 5 operons that encode anaerobic dehydrogenases and reductases, as well as the *edd* gene that encodes 6-phosphogluconate dehydratase, an enzyme in the Entner-Douderoff pathway. The 5 operons encode dimethyl sulphoxide reductase, formate dehydrogenase-N, fumarate reductase, glycerol-3-phosphate dehydrogenase. Based solely on transcription profiling studies there are other genes and operons involved in anaerobic respiration, which are regulated by Aer. Of course, since FlhCD controls Aer synthesis, by inference, the complex is also a regulator of these anaerobic respiratory pathways (Figure 1). Clearly, much work still needs to be done to determine whether these regulators function directly or indirectly. For example, to which transcriptional activator(s) does Aer transduce its signal, or is Aer itself a transcriptional activator? This discovery might provide a common link between the control of anaerobic respiratory activity and motility. Furthermore, and as discussed above, another obvious link between flagellar biosynthesis and redox-dependent gene expression is provided by the Arc system.

4.2. Dos

The hemoprotein Dos (direct oxygen sensor) from *E. coli* was identified through homology searches (Delgado-Nixon *et al.*, 2000). The N-terminus of the protein has a PAS domain and binds heme, while the C-terminal domain has similarity to a phosphodiesterase. It is still not clear what the targets of the Dos protein in *E. coli* are, or what its role in redox regulation might be. Dos is highly similar over its entire length to a protein called AxPDEA1 from *Acetobacter xylinum,* which controls cellulose synthesis (Chang *et al.*, 2001). Both Dos and AxPDEA1 bind dioxygen through the heme moiety and the binding of O_2 is believed to control the phosphodiesterase activity of the proteins (Liebl *et al.*, 2003).

4.3. The Global Regulator CRP and Anaerobic Gene Regulation

The heptacistronic *tdc* operon of *E. coli* encodes most of the components of a metabolic pathway that allows the bacterium to ferment L-threonine or L-serine as energy sources (Datta *et al.*, 1987; Sumantran *et al.*, 1990; Heßlinger *et al.*, 1998). Expression of the operon is anaerobically inducible and a number of transcription factors are required to mediate this control, including CRP (Wu *et al.*, 1992; Sawers, 2001). In a *crp* mutant no anaerobic induction of *tdc* expression occurs. It is not thought that CRP *per se* senses anaerobiosis, but rather one of two other activators of the *tdc* operon, termed TdcA and TdcR. TdcA is a LysR-like DNA-binding protein and is encoded by the first gene in the *tdc* operon, while the *tdcR* gene is divergently transcribed from it (Hagewood *et al.*, 1994). CRP functions at the *tdc* promoter as a Class II transcription factor binding just upstream of the -35 RNA polymerase recognition sequence. It is presumed that TdcR and/or TdcA, upon 'sensing' the change in the cellular redox status, interact with the upstream DNA and contact CRP to activate transcription (Figure 1; Sawers, 2001). Essentially this is similar to the dual control mechanism of regulation of a number of FNR-dependent promoters, with the difference that the FNR protein detects the redox status directly and the upstream regulator is responsive to another signal, for example nitrate in the case of the NarXL-regulated *nar* operon of *E. coli* (Stewart, 2003). We currently do not know what signal is sensed either by TdcA or TdcR, but it is conceivable that it could be a small metabolite, which accumulates anaerobically.

The significance of CRP-dependent control of anaerobic gene expression has recently been exemplified in *S. oneidensis*, where, as mentioned above the FNR homologue EtrA only controls a subset of genes involved in anaerobic respiration. Deletion of the *crp* gene in *S. oneidensis* resulted in loss of the ability to use FeIII, MnIV, nitrate, fumarate or DMSO as terminal electron acceptors (Saffarini *et al.*, 2003). Remarkably, addition of cAMP to aerobic cultures induced FeIII and fumarate reductase activities, suggesting that cAMP-CRP in *S. oneidensis* might be a signal of altered redox state in the organism. Clearly, further work will be required to determine whether CRP is functioning in *S. oneidensis* in a different manner compared to other bacteria or whether, as is the case of the *E. coli tdc* operon, CRP is acting in concert with a redox-responsive transcription factor.

5. The Complexity of Redox Regulation in Purple Non-Sulphur Bacteria

The purple non-sulphur bacteria show remarkable metabolic versatility. They can generate energy from light, organic as well as inorganic compounds and they can respire using a wide variety of electron acceptors. They are also able to fix nitrogen and carbon.

Anaerobic regulation of photosynthetic gene expression in purple bacteria has been studied primarily in *Rhodobacter sphaeroides* and *Rhodobacter capsulatus* and most of what has been discovered concerning regulatory mechanisms applies to either organism. Expression of photosynthetic genes is tightly regulated by oxygen and, to a lesser extent, by light intensity. The redox state of the cells during phototrophic anaerobic respiration is controlled by interactive regulatory mechanisms (Tichi and Tabita, 2001). Global analysis of gene expression in *R. sphaeroides* in response to oxygen and light intensity has been studied using transcriptome experiments (Roh *et al.*, 2004) and the detailed regulatory mechanism of photosynthesis gene expression has emerged from recent studies as described below. Detailed reviews on the topic are available (Bauer *et al.*, 2003; Gregor and Klug, 1999, 2002; Oh and Kaplan, 2001; Pemberton *et al.*, 1998; Zeilstra-Ryalls *et al.*, 1998).

5.1. The RegA-RegB/PrrA-PrrB Global Regulatory Systems

A two-component signal transduction system has been found both in *R. sphaeroides* and *R. capsulatus*, which is required for photosynthetic gene expression in response to oxygen limitation (Figure 2). A response regulator, RegA, and its orthologue, PrrA, were isolated from *R. capsulatus* (Sganga and Bauer, 1992) and *R. sphaeroides* (Eraso and Kaplan, 1994; Phillips-Jones and Hunter, 1994), respectively. A null mutation of the gene had an adverse effect on photosynthetic gene expression as well as on growth under phototrophic conditions, although the *prrA* mutation conferred a more severe effect than the *regA* mutation. The target photosynthetic genes include *puf*, *puc*, and *puhA*. The cognate sensor kinases, RegB (Mosley *et al.*, 1994) and PrrB (Eraso and Kaplan, 1995) are membrane proteins with six membrane-spanning helices.

5.2. Transcriptional Activation by RegA

Wild-type RegA exhibits weak or no DNA-binding activity and the first evidence that it is a DNA-binding protein came from studies using a mutant, RegA* (S97D), which is constitutively active and supports the expression of photosynthesis genes independently of its cognate kinase RegB. DNase I footprinting analyses using purified RegA* showed that it binds to the regulatory regions of the *puf* and *puc* operons, indicating that RegA interacts with the promoter region to activate transcription (Bird *et al.*, 1999; Du *et al.*, 1998). Phosphorylated RegA also exerts autoregulation by negatively controlling the transcription of *regB* and the *senC-regA-hvrA* operon through direct interaction with a region between *regB* and *senC* (Du *et al.*, 1999).
Surprisingly, introducing the amino acid substitution D63K in RegA, which is the site of phosphorylation, results in a null phenotype; however, the mutant protein was still able to bind DNA, suggesting that phosphorylation is not required for DNA binding but may be needed for a subsequent step of transcriptional initiation (Hemschemeier *et al.*, 2000). In contrast, another report showed that phosphorylation increased binding of wild-type RegA by approximately 16-fold (Bird *et al.*, 1999). Clearly, this apparent paradox needs to be resolved.

Recent NMR structural analysis of the C-terminal effector domain of PrrA has revealed that it forms a three-

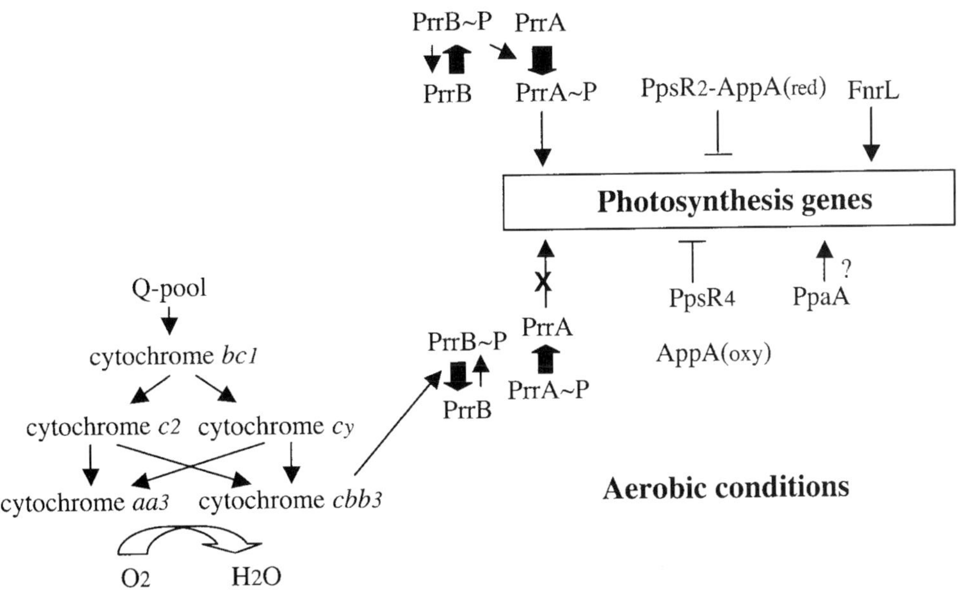

Figure 2. Regulatory pathways in the control of photosynthetic gene expression in *R. sphaeroides* in response to oxygen. Arrows indicate activation of gene expression, while lines terminated with a bar indicate repression. An inhibitory signal generated by cbb_3 oxidase controls the activity of PrrB. Switches to kinase- or phosphatase-dominant mode of PrrB are shown by thickness of the arrows. Question mark signifies that the role of PpaA is unclear. Note that AerR, the PpaA homolog in *R. capsulatus*, negatively regulates photosynthetic gene expression under aerobic conditions (see text).

helix bundle containing a helix-turn-helix motif, the fold of which is similar to that of the FIS protein (Laguri *et al.*, 2003). Two GCGNC inverted repeats with variable spacing between the half sites are considered to comprise the consensus recognition sequence for PrrA/RegA and it has been proposed that residues R171, R172, Q175, and R176 in PrrA make contact with the core consensus motif (GCG) (Laguri *et al.*, 2003).

5.3. Activation Mechanism of RegB/Prr

Mutants lacking the cbb_3 cytochrome *c* oxidase exhibit elevated photosynthetic gene expression under aerobic conditions (O'Gara and Kaplan, 1997) and epistatic analysis showed that the signal(s) originating from the cbb_3 terminal oxidase negatively affect gene expression through the PrrAB system (O'Gara *et al.*, 1998). *R. sphaeroides*, but not *R. capsulatus*, has another terminal oxidase, the aa_3 cytochrome *c* oxidase, which is responsible for the bulk of the total cytochrome *c* oxidase activity under highly aerobic conditions. Inactivation of the aa_3 oxidase has no effect on photosynthetic gene expression, indicating that the inhibitory role is unique to the cbb_3 oxidase (Oh and Kaplan, 2000). Electrons are transferred from the quinone pool to cbb_3 oxidase through the bc_1 complex, and to cytochromes c_2 and c_y (Figure 2). Strains lacking these cytochromes also exhibit aerobic derepression of photosynthesis genes, which indicates that it is electron flow to the cbb_3 oxidase that is being sensed (Oh and Kaplan, 2000). The default state of PrrB is the kinase-dominant mode and it is switched to the phosphatase-dominant mode upon receiving the inhibitory signal from the cbb_3 oxidase under aerobic conditions (Oh *et al.*, 2001).

5.4. A Redox-Active Cysteine

A recent study showed that a truncated soluble RegB kinase retains redox-sensing activity and this is mediated by a redox-reactive cysteine (Swem *et al.*, 2003). RegB, under oxidizing conditions, forms an inactive tetramer from active dimers via intermolecular disulfide bond formation. Mutational analysis has shown that the cysteine is involved in redox-sensing both *in vivo* and *in vitro* (Swem *et al.*, 2003). Disulfide bond formation is metal-dependent, which could explain why a mutation in the putative copper chaperone gene *prrC* in *R. sphaeroides* (Eraso and Kaplan, 1995) and *senC* in *R. capsulatus* (Buggy and Bauer, 1995) renders RegB constitutively active (Eraso and Kaplan, 2000). It is also consistent with the result that mutations in the putative copper transport proteins RdxI, RdxH and RdxS lead to derepression of the PrrAB pathway (Roh and Kaplan, 2000). Alternatively, the mutational effects of *prrC* and *rdx* may be due to the copper requirement of the cbb_3 cytochrome oxidase (Roh and Kaplan, 2000).

5.5. The RegAB/PrrA Regulon beyond Photosynthesis

RegAB/PrrAB function as global regulatory systems that activate expression of genes involved in CO_2 fixation (*cbb*$_I$ and *cbb*$_{II}$ operons) (Dubbs *et al.*, 2000), nitrogen fixation (*nifA2*) (Elsen *et al.*, 2000; Joshi and Tabita, 1996), denitrification (*nirK*) (Laratta *et al.*, 2002), the negative aerotactic response (Romagnoli *et al.*, 2002), electron transfer (*petABC, cycA, cycY*) (Comolli *et al.*, 2002; Eraso and Kaplan, 1994; Swem *et al.*, 2001), and aerobic respiration (*cydABC, ccoNOPQ*) (Swem *et al.*, 2003). It is noteworthy that expression of both the *cbb*$_I$ and *cbb*$_{II}$ operons, which encode enzymes in duplicated Calvin-Benson-Bassham CO_2 fixation-reductive pentose phosphate cycles, is dependent on the Reg/Prr system during photoautotrophic growth in a 1.5% CO_2/98.5% H_2 atmosphere (Qian and Tabita, 1996), whereas, during chemoautotrophic growth with 5% CO_2/45% H_2/50% air, expression of *cbb*$_{II}$ is greatly reduced in the PrrA mutant, but *cbb*$_I$ expression is either unaffected or enhanced (Gibson *et al.*, 2002). Besides its positive role, phosphorylated RegA negatively regulates expression of hydrogenase (*hupSLC*) (Elsen *et al.*, 2000) and DMSO reductase in cells grown under phototrophic conditions in the absence of DMSO (Kappler *et al.*, 2002).

RegAB/PrrAB often functions along with other transcriptional regulators to regulate certain genes. The *puc* operon, for example, is activated by RegA under anaerobic conditions and aerobically is repressed by CrtJ (see below). Expression of *hup*, which is required for hydrogen oxidation, is activated by the response regulator HupR and the histone-like IHF (Dischert *et al.*, 1999) protein, while it is repressed by RegA. In addition to RegA, *nifA2* activation requires NtrC (Elsen *et al.*, 2000) and the *cbb* operons are activated by CbbR (Dubbs *et al.*, 2000; Gibson *et al.*, 2002; Vichivanives *et al.*, 2000). The expression of *nirK* is dependent on the NnrR transcriptional activator (Tosques *et al.*, 1996), a member of the FNR/CRP family. Therefore, it appears that the RegAB/PrrAB system serves as a global redox regulator that can operate in conjunction with other regulators, which are specific for each set of genes. This is a common theme that has also been discussed in the context of *E. coli* above.

5.6. Reg/Prr Homologues

Reg/Prr orthologues, which are functionally interchangeable, were found in other bacteria. *Rhodovulum sulfidophilum* and *Roseobacter denitrificans* can photosynthesise even under aerobic conditions. Furthermore, the *reg* genes from these aerobic photosynthetic bacteria can complement *R. capsulatus reg* mutants and the complemented strains still retain regulation in response to oxygen (Masuda *et al.*, 1999).

The symbiotic bacteria *Bradyrhizobium japonicum* and *Sinorhizobium meliloti* have RegA homologues, called RegR and ActR. *In vitro* and *in vivo* studies showed that RegA, RegR, and ActR are functionally interchangeable (Emmerich *et al.*, 2000).

The response regulator RoxR from *Pseudomonas aeruginosa* activates transcription of *cioAB*, encoding a cyanide-insensitive oxidase. PrrA rescues the defect conferred by a *roxR* mutation and *vice versa*, indicating that RoxR is a functional homolog of PrrA (Comolli and Donohue, 2002).

5.7. The Roles of CrtJ /PpsR and AppA

Redox-sensitive transcriptional repressors for photosynthetic genes, PpsR (Penfold and Pemberton, 1994), and its homologue, CtrJ (Ponnampalam *et al.*, 1995), were found in *R. sphaeroides* and *R. capsulatus*, respectively. CrtJ represses transcription in aerobically grown cells by binding to the consensus sequence TGTN_{12}ACA in the promoters of *bch, crt*, and *puc*. At these promoters, two CrtJ dimers interact to form tetramers (Elsen *et al.*, 1998; Ponnampalam and Bauer, 1997; Ponnampalam *et al.*, 1998). PpsR/CrtJ is composed of a C-terminal DNA binding region, a central region with two PAS domains that are likely to be involved in protein oligomerization, and an N-terminal region of unknown function (Gomelsky *et al.*, 2000). CrtJ binds to its target DNA more efficiently when it is preincubated with oxygen-saturated buffer (Ponnampalam and Bauer, 1997), suggesting that CrtJ is capable of sensing redox changes. PpsR/CrtJ does this by forming an intra-molecular disulfide bond between conserved Cys249 and Cys420 residues *in vivo* and *in vitro* when exposed to oxygen and this is essential for it to bind to its target promoter (Masuda and Bauer, 2002; Masuda *et al.*, 2002).

AppA has been identified as another important component involved in photosynthetic gene regulation (Gomelsky and Kaplan, 1995). However, AppA is present in *R. sphaeroides* but not in *R. capsulatus*. Genetic analysis suggested that AppA antagonizes PpsR repressor activity by interacting with it (Gomelsky and Kaplan, 1997). AppA has a novel FAD-binding domain at its N-terminus and FAD binds noncovalently with an apparent 1:1 stoichiometry (Gomelsky and Kaplan, 1998; Gomelsky and Klug, 2002).

Studies of AppA have provided important clues to the mechanism by which photosynthetic genes are regulated both by oxygen and light (Braatsch *et al.*, 2002). Under low oxygen tension, the expression of photosynthetic genes was increased independently of light both in *R. sphaeroides* and *R. capsulatus*. Under semi-aerobic conditions, the expression was strongly repressed by blue light in *R. sphaeroides* but not in *R. capsulatus*. The difference in the response of the two organisms to blue light is attributed to AppA. The blue light-mediated repression seen in *R. sphaeroides* is dependent on the FAD cofactor of AppA, the possible photoreceptor for blue light. NMR analysis of AppA revealed that Tyr21 of AppA forms π-π stacking interactions with the isoalloxazine ring of FAD and photochemical excitation of the flavin leads to a stable local conformational change in AppA (Kraft *et al.*, 2003).

Elegant *in vitro* experiments by Masuda and Bauer demonstrated how AppA functions as an antirepressor of PpsR (Masuda and Bauer, 2002). In aerobically growing cells, oxidised PpsR binds to its target promoters and represses transcription. As oxygen becomes limiting, AppA becomes reduced, which then allows it to reduce the disulfide bond in PpsR. Reduced AppA also forms a stable AppA-$PpsR_2$ antirepressor-repressor complex, which releases the transcriptional repression exerted by PpsR. Under anaerobic conditions, blue light excitation of AppA results in a conformation that is incapable of forming the AppA-$PpsR_2$ complex leading to repression of the target genes. Therefore, AppA functions as a transcriptional regulator that senses two different stimuli, blue light and a redox signal.

5.8. TspO

The tryptophan-rich TspO protein is an outermembrane protein that shows a significant sequence homology to the peripheral-type benzodiazepine receptor from mammalian cells. It has a moderate negative effect on transcription of several photosynthetic genes in *R. sphaeroides* (Yeliseev and Kaplan, 1995). TspO likely modulates the PpsR/AppA system by acting upstream of PpsR/AppA activity (Zeng and Kaplan, 2001).

5.9. FnrL

FnrL probably activates transcription of the *puc* operon by binding to the FNR-consensus sequence present in the promoter; however, FnrL appears to have an additional indirect role in *puc* operon expression (Zeilstra-Ryalls and Kaplan, 1998). FnrL does not function in transduction of the signal derived by the *cbb*$_3$ oxidase, but functions together with other regulators in photosynthetic gene expression. Expression of *hemA, hemZ,* and *hemN* is activated by FnrL and the Prr pathway, while *bchEJG* operon expression seems to be coordinately regulated by FnrL, PrrA, and PpsR/AppA (Oh *et al.*, 2000).

5.10. PpaA/AerR

Another newly discovered regulatory protein of photosystem gene expression in *R. sphaeroides* is PpaA (Gomelsky *et al.*, 2003), termed AerR in *R. capsulatus* (Dong *et al.*, 2002). Disruption of *aerR* results in an increase in aerobic photosynthetic gene expression, indicating that AerR, like CrtJ, functions as a repressor under aerobic conditions (Dong *et al.*, 2002). AerR binds to the *crtI*, *puc*, and *puf* promoters but not to the *bchC* promoter. AerR and CrtJ bind cooperatively to the *puc* promoter region (Dong *et al.*, 2002).

In *R. sphaeroides*, PpaA activates photopigment production and *puc* operon expression under aerobic conditions. Sequence analysis of PpaA homologues showed a putative corrinoid-binding domain, suggesting that a corrinoid cofactor affects PpaA activity (Gomelsky *et al.*, 2003).

6. Redox Regulation in Gram-Positives

6.1. *Bacillus subtilis*

B. subtilis grows anaerobically either by respiration or by fermentation. Unlike other facultative anaerobes such as *E. coli*, the bacterium is restricted to using only oxygen or nitrate as respiratory electron acceptor. Fermentative growth is also not robust and requires supplementation with pyruvate in addition to glucose to support optimal growth (Nakano *et al.*, 1997). Much of our current understanding of anaerobiosis in *B. subtilis* has been reviewed recently (Nakano and Zuber, 1998, 2002).

6.1.1. Global Analysis of Gene Regulation Upon a Shift to Anaerobiosis

DNA microarray work has revealed that in *B. subtilis* expression of more than 100 genes is induced by oxygen limitation, while expression of nearly 30 genes is repressed (Ye *et al.*, 2000). Genes induced by anaerobiosis include those encoding regulatory proteins (*fnr, resDE, yclJK,* and *arfM*; see below), those involved in carbon metabolism, electron transport and cytochrome metabolism, respiratory nitrate/nitrite reduction, and antibiotic biosynthesis. Genes required for iron transport/utilization are among those repressed during anaerobic growth (Ye *et al.*, 2000). Proteomic analyses (Clements *et al.*, 2002; Marino *et al.*, 2000) confirmed that the concentration of products for some of the genes identified by microarray analysis is also affected.

The primary regulatory pathway for anaerobic gene expression in *B. subtilis* when oxygen becomes limited is first activated by the ResDE signal transduction system (Sun *et al.*, 1996). ResD is a response regulator that activates transcription of *fnr* (Nakano *et al.*, 1996). FNR then auto-activates transcription of the *narK-fnr* operon from the operon promoter, and positively controls *narGHJI* (respiratory nitrate reductase operon) (Glaser *et al.*, 1995). FNR also induces expression of *arfM*, which is needed for anaerobic induction of *lctEP* (lactate fermentation), *alsSD* (acetoin formation) (Marino *et al.*, 2001), and heme biosynthetic genes (Homuth *et al.*, 1999). Little is known about ArfM

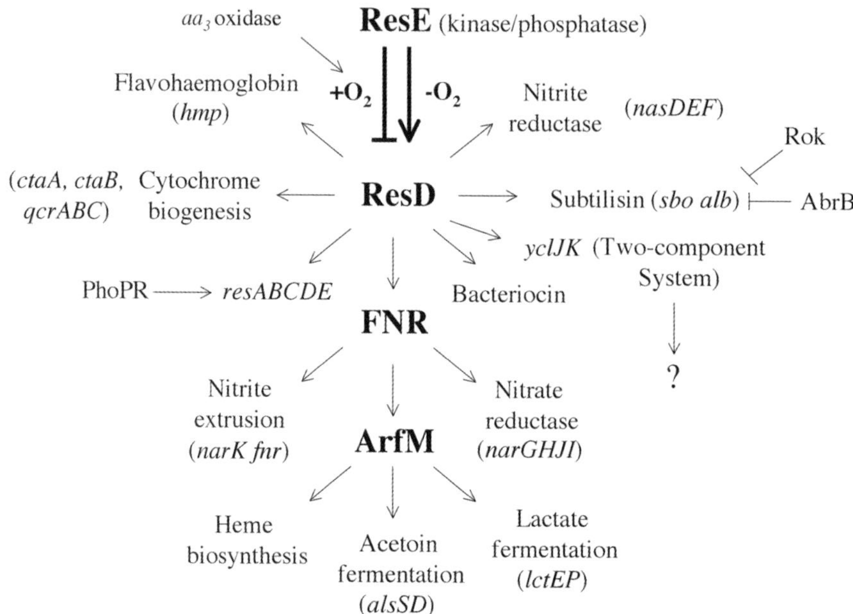

Figure 3. The regulatory cascade controlling redox-dependent gene expression in *B. subtilis*. Arrows indicate activation of gene expression, while lines terminated with a bar indicate repression. PhoPR is required for expression of the *res* operon only under phosphate depleted conditions.

other than it is an 18 kD protein that has no similarity to known transcriptional regulators. Important features of the regulatory cascade controlling redox regulation in *B. subtilis* are shown in Figure 3.

6.1.2. The ResDE Regulon

The ResDE signal transduction system plays a key role in anaerobic gene regulation in *B. subtilis*. Interestingly, it also has a role in controlling aerobic respiration (Sun *et al.*, 1996) and production of bacteriocin (Nakano *et al.*, 2000a). However, all the ResDE-dependent genes examined so far are induced by oxygen limitation. Target genes include *ctaA* (Sun *et al.*, 1996) and *ctaB* (Liu and Taber, 1998), both of which are required for cytochrome caa_3 oxidase synthesis, *qcrABC* (menaquinol:cytochrome *c* biogenesis) (Sun *et al.*, 1996), *sbo-alb* (Nakano *et al.*, 2000a) (subtilosin biosynthesis), *resABCDE* (Sun *et al.*, 1996), *fnr* (Nakano *et al.*, 1996), *nasDEF* (nitrite reductase) (Nakano *et al.*, 1995), *hmp* (flavohemoglobin) (LaCelle *et al.*, 1996), and *yclJK* (two-component regulatory proteins) (Härtig *et al.*, 2004). Expression of the *yclJK* operon is activated by ResD upon oxygen limitation, but target genes of YclJK and their role in anaerobiosis are currently unknown (Härtig *et al.*, 2004).

6.1.3. Activation of ResE Autophosphorylation

ResE functions as both a kinase and a phosphatase (Nakano *et al.*, 1999). Analyses of a ResE mutant (T378R) that lacks phosphatase activity but retains kinase activity indicated that kinase and phosphatase activities of ResE are reciprocally regulated by oxygen (Nakano and Zhu, 2001). Under aerobic conditions phosphatase activity is dominant, and when oxygen becomes limiting, ResE is converted to a kinase-dominant mode.

What is the actual signal perceived by ResE? Aerobic expression of the ResDE regulon is partially derepressed by a mutation in a terminal oxidase gene (Nakano and Zuber, 1998), which resembles the derepressed, aerobic expression of *Rhodobacter* photosynthetic genes in the cbb_3 oxidase mutants (see above). Increased aerobic expression is caused by a mutation in the *qox* operon that codes for a major terminal cytochrome aa_3-type quinol oxidase. Mutations in other terminal oxidase genes do not have this phenotype. Because the cytochrome aa_3 oxidase oxidizes menaquinone, it was proposed that oxidation/reduction of menaquinone might be the signal affecting ResE activity (Nakano and Zuber, 1998). There are analogies here with the redox-sensing mechanism proposed for *E. coli* ArcB as discussed above (Georgellis *et al.*, 2001).

It was noted that some ResDE-dependent genes were not fully activated in an anaerobically grown *B. subtilis narG* mutant unless nitrite was added (LaCelle *et al.*, 1996; Nakano *et al.*, 1998). This result was explained by the stimulatory effect of NO, which is generated from nitrite, on the ResDE regulon (Nakano, 2002). Expression of the ResDE-controlled genes is highly induced by an NO donor or NO and this induction is dependent on ResE (except for the *hmp* gene, which is stimulated by NO by both ResE-dependent and ResE-independent mechanisms). Further studies are needed to determine whether NO is a ligand of ResE or whether NO together with oxygen limitation, for example through reduced menaquinone, generate a signal for ResE.

The ResE histidine kinase is composed of two transmembrane helices and a long extracytoplasmic loop. The second transmembrane region is followed by a HAMP (histidine kinase, adenylyl cyclase, MCP, and phosphatase) linker (Appleman and Stewart, 2003), and a PAS domain, which in other proteins has been shown to function in sensing light, oxygen, and the intracellular redox state (Taylor and Zhulin, 1999). A recent study showed that a truncated cytoplasmic form of ResE, which lacks the transmembrane helices and the extracytoplasmic region, is capable of sensing oxygen limitation (Baruah *et al.*, 2003). Deletion of the PAS domain abolished ResE activity, suggesting that the PAS domain is required for sensing, for protein integrity, or for both. The cytoplasmic form of ResE has less activity than the full-length ResE, suggesting that either membrane anchoring is important for efficient signal detection and/or transduction, or an additional signal-sensing domain resides in the extracytoplasmic region.

6.1.4. Transcriptional Activation by ResD

Direct binding of ResD to regulatory DNA sequences has been demonstrated for *ctaA* (Zhang and Hulett, 2000), *resA* (Zhang and Hulett, 2000), *nasD* (Nakano *et al.*, 2000b), *hmp* (Nakano *et al.*, 2000b), fnr (Nakano *et al.*, 2000b), and *yclJ* (Härtig *et al.*, 2004) by gel shift analysis and DNase I footprinting experiments. Hydroxyl radical footprinting analysis showed that five ResD monomers tandemly bind to the promoter region of *nasD* and *hmp*. Each monomer binds to the same face of the DNA helix at intervals of 10 bp except for one monomer that binds on the opposite face of the helix at the most promoter-proximal site of the *hmp* regulatory region (Geng *et al.*, 2003).

In vitro transcription using *B. subtilis* RNA polymerase has shown that ResD alone is sufficient to activate the transcription of *ctaA* (Paul *et al.*, 2001), *fnr* (Geng *et al.*, 2003), *nasD* (Geng *et al.*, 2003), and *hmp* (Geng *et al.*, 2003). Phosphorylation of ResD greatly stimulates transcription, although DNA binding is only moderately stimulated by phosphorylation. ResD increases the DNA-binding affinity of RNA polymerase at the *hmp* promoter, suggesting that ResD recruits RNA polymerase to the promoter (Nakano *et al.*, 2003). A single amino acid substitution in the C-terminal domain (CTD) of RNA polymerase α subunit that affects *fnr* and *nasD* transcription but not *hmp* was isolated (Nakano, unpublished result), which suggests that ResD interacts with αCTD to activate transcription at certain promoters.

The ResD regulon is also controlled by other transcriptional regulators. For example, the *sbo-alb* operon is under multiple control (Nakano *et al.*, 2000a); positive control by ResDE and negative control by AbrB, the transition state regulator (Strauch, 1993) and by Rok (Albano and Dubnau, unpublished results) that negatively regulates genetic competence (Hoa, 2002). However, *in vitro* studies have not been carried out to determine how the multiple regulators bind to the regulatory region (cooperatively or non-cooperatively) and activate/repress transcription. Expression of the *res* operon is dependent on another two-component system, PhoPR, under phosphate-limiting conditions, but not under oxygen limitation (Birkey *et al.*, 1998).

6.1.5. Phosphorylation-Independent Activation of ResD

A D57A phosphorylation-defective mutant of ResD still retains some transcription activation ability *in vitro*. When the gene encoding the D57A mutant ResD protein is expressed from an inducible promoter, it is also able to activate *nasD* and *hmp* expression *in vivo* in response to oxygen limitation (Geng *et al.*, 2003). This result indicated that ResD itself, in addition to its activation through a phosphorylation-mediated conformational change, senses oxygen levels via an unknown mechanism. Mutational analyses showed that a single cysteine present in the N-terminus of ResD is indispensable for phosphorylation-independent activation (Nakano, unpublished result). Because C69S or C69A mutations did not result in aerobic derepression of *nasD* or *hmp*, the possibility is unlikely that intermolecular disulfide bond formation via Cys69 is responsible for aerobic repression. Other possibilities, such as a modification of the cysteinyl residue need to be tested.

6.2. Redox regulation and Anaerobic Stress Survival in *Streptomyces*

Like many antibiotic-producing soil bacteria *Streptomyces* are exposed to a constantly changing environment and this also holds for levels of oxygen. Despite the fact that *Streptomyces* spp. are obligate aerobes, like *Azotobacter vinelandii*, they must cope with low oxygen tensions or even anaerobiosis. The genome sequence of *Streptomyces coelicolor* reveals a number of genes and operons encoding anaerobic enzymes (Bentley *et al.*, 2002). There are three operons encoding respiratory nitrate reductase that exhibit a high degree of similarity to those in *E. coli*. It is unclear when *S. coelicolor* utilses nitrate as an electron acceptor, although transcription studies have determined that expression of one of the *narGHJI* operons is induced by anaerobiosis. We, and others, have also determined that *S. coelicolor* is able to survive long periods of anaerobiosis (van Keulen *et al.*, 2003; G. Sawers, J. Alderson and J. White, unpublished); the mechanism underlying how *S. coelicolor* senses redox changes is still unclear. One recent study, however, has identified a protein termed Rex, which is a redox-sensing repressor protein that controls expression of the *cydABCD* operon encoding a cytochrome *d* oxidase (Brekasis and Paget, 2003). Expression of *cydABCD* is induced by oxygen limitation and Rex functions as a repressor when the NAD^+:NADH ratio is high. However, when oxygen becomes limiting that ratio changes and NADH inhibits DNA-binding, and thus repressor activity of Rex. NAD^+ competes with NADH for Rex binding and so Rex actually senses the redox poise of the cell. Interestingly, although superficially *narGHJI* expression is regulated in a similar manner to *cydABCD*, Rex does not affect regulation (G. Sawers, J. Alderson and J. White, unpublished). Clearly, more than one redox-sensing system is present in *Streptomyces* species and it will be a challenge to identify their roles in streptomycete biology.

7. Perspectives

What is clear from the foregoing discussion is that the more we delve into the mechanisms underlying redox regulation of gene expression, the more we uncover complex interplay between multiple factors. This ranges from multiple transcription factors transducing different metabolic signals converging to control a single gene, as in the case of the *cydAB* operon of *E. coli* (Govantes *et al.*, 2000), the initiation of regulatory cascades coordinating multiple transcription factors to activate or repress large numbers of target operons, as in the control of symbiotic gene expression in *Bradyrhizobium japonicum* (Sciotti *et al.*, 2003), and the discovery of hitherto unexpected functions of sensors or transcription factors, for example the recent discovery of redox-sensing function in the nitrate/nitrite sensor NarQ (Stewart *et al.*, 2003) or the redox control exerted by the master flagellar regulator FlhD/FlhC (Prüß *et al.*, 2003). Another regulatory mechanism involving cross-regulation between various two-component systems that was suggested a number of years ago (Wanner, 1992) has recently gained substantial experimental support (Oshima *et al.*, 2002; Soupene, 2003; Zhou *et al.*, 2003). It is clear that control of not only the activity but also the amount of transcription factors in a cell, as well as their sub-cellular localisation, are crucial to the strict and specific control of gene expression. This is clearly not exclusive to redox-dependent gene expression but holds for all transcriptional control. Nevertheless, if we wish in the future to understand the mechanism underlying transcriptional control due cognizance of these various phenomena will be crucial.

Many of the advances alluded to have been made through the development and use of new high-throughput technologies such as transcriptome and proteome profiling, with real-time PCR in transcript analysis, large-scale rapid knock-out procedures and phenotypic microarray analysis. All of these techniques cannot be relied on exclusively to provide clear and unequivocal information on the physiological function of proteins, but as many of these techniques as possible, along with classical methodologies, must be employed to provide as solid a foundation for premises as possible.

As the biology of more microorganisms that were previously recalcitrant to study become more tractable through the use of the techniques mentioned above, so our understanding of the mechanisms underlying redox regulation of gene expression will reveal how important this area is in our understanding of bacterial physiology.

Acknowledgements

We thank David Dubnau for communicating unpublished results. The work in the lab of R.G.S. is supported by the BBSRC and the work in the lab of M.M.N. is supported by a grant from National Science Foundation (MCB0110513).

References

Alexeeva, S., de Kort, B., Sawers, G., Hellingwerf, K.J., and Teixeira de Mattos, M.J. 2001. Effects of limited aeration and of the ArcAB system on intermediary pyruvate catabolism in *Escherichia coli*. J. Bacteriol. 182: 4934-4940.

Alexeeva, S. 2000. Molecular physiology of responses to oxygen in *Escherichia coli*: the role of the ArcAB system. PhD thesis, University of Amsterdam.

Alexeeva, S., Hellingwerf, K.J., and Teixeira de Mattos, M.J. 2002. Quantitative assessment of oxygen availability: perceived aerobiosis and its effect on flux distribution in the respiratory chain of *Escherichia coli*. J. Bacteriol. 184: 1402-1406.

Alexeeva, S., Hellingwerf, K.J., and Teixeira de Mattos, M.J. 2003. Requirement of ArcA for redox regulation in *Escherichia coli* under microaerobic but not anaerobic or aerobic conditions. J. Bacteriol. 185: 204-209.

Anjum, M.F., Green, J., and Guest, J.R. 2000. YeiL, the third member of the CRP-FNR family in *Escherichia coli*. Microbiology. 146: 3157-3170.

Appleman, J.A., and Stewart, V. 2003. Mutational analysis of a conserved signal-transducing element: the HAMP linker of the *Escherichia coli* nitrate sensor NarX. J. Bacteriol. 185: 89-97.

Baruah, A., Lindsey, B., Zhu, Y., and Nakano, M.M. 2004. Mutational analysis of the signal-sensing domain of ResE histidine kinase from *Bacillus subtilis*. J. Bacteriol. 186: 1694-1704.

Bauer, C., Elsen, S., Swem, L.R., Swem, D.L., and Masuda, S. 2003. Redox and light regulation of gene expression in photosynthetic prokaryotes. Philos. Trans. R. Soc. Lond. B. Biol. Sci. 358: 147-153; discussion 153-154.

Beliaev, A.S., Thompson, D.K., Fields, M.W., Wu, L., Lies, D.P., Nealson, K.H., and Zhou, J. 2002. Microarray transcription profiling of a *Shewanella oneidensis etrA* mutant. J. Bacteriol. 184: 4612-4616.

Bentley, S.D., *et al.* 2002. Complete genome sequence of the model actinomycete *Streptomyces coelicolor* A3(2). Nature. 417: 141-147.

Bibikov, S.I., Biran, R., Rudd, K.E., and Parkinson, J.S. 1997. A signal transducer for aerotaxis in *Escherichia coli*. J. Bacteriol. 179: 4075-4079.

Bird, T.H., Du, S., and Bauer, C.E. 1999. Autophosphorylation, phosphotransfer, and DNA-binding properties of the RegB/RegA two-component regulatory system in *Rhodobacter capsulatus*. J. Biol. Chem. 274: 16343-16348.

Birkey, S.M., Liu, W., Zhang, X., Duggan, M.F., and Hulett, F.M. 1998. Pho signal transduction network reveals direct transcriptional regulation of one two-component system by another two-component regulator: *Bacillus subtilis* PhoP directly regulates production of ResD. Mol. Microbiol. 30: 943-953.

Blake, T., Barnard, A., Busby, S.J.W., and Green, J. 2002. Transcription activation by FNR: evidence for a functional activating region 2. J. Bacteriol. 184: 5855-5861.

Braatsch, S., Gomelsky, M., Kuphal, S., and Klug, G. 2002. A single flavoprotein, AppA, integrates both redox and light signals in *Rhodobacter sphaeroides*. Mol. Microbiol. 45: 827-836.

Brekasis, D., and Paget, M.S.B. 2003. A novel sensor of $NADH/NAD^+$ redox poise in *Streptomyces coelicolor* A3(2). EMBO J. 22: 4856-4865.

Buggy, J., and Bauer, C.E. 1995. Cloning and characterization of *senC*, a gene involved in both aerobic respiration and photosynthesis gene expression in *Rhodobacter capsulatus*. J. Bacteriol. 177: 6958-6965.

Chang, A.L., Tuckerman, J.R., Gonzalez, G., Mayer, R., Weinhouse, H., Volman, G., Amikam, D., Benziman, M., and Gilles-Gonzalez, M.-A. 2001. Phosphodiesterase A1, a regulator of cellulose synthesis in *Acetobacter xylinum*, is a heme-based sensor. Biochemistry. 40: 3420-3426.

Clements, L.D., Streips, U.N., and Miller, B.S. 2002. Differential proteomic analysis of *Bacillus subtilis* nitrate respiration and fermentation in defined medium. Proteomics. 2: 1724-1734.

Comolli, J.C., Carl, A.J., Hall, C., and Donohue, T. 2002. Transcriptional activation of the *Rhodobacter sphaeroides* cytochrome c(2) gene P2 promoter by the response regulator PrrA. J. Bacteriol. 184: 390-399.

Comolli, J.C., and Donohue, T.J. 2002. *Pseudomonas aeruginosa* RoxR, a response regulator related to *Rhodobacter sphaeroides* PrrA, activates expression of the cyanide-insensitive terminal oxidase. Mol. Microbiol. 45: 755-768.

Cotter, P.A., and Gunsalus, R.P. 1992. Contribution of the *fnr* and *arcA* gene products in coordinate regulation of cytochrome *o* and cytochrome *d* oxidase (*cyoABCD* and *cydAB*) in *Escherichia coli*. FEMS Microbiol. Lett. 70: 31-36.

Cruz Ramos, H., Boursier, L., Moszer, I., Kunst, I., Danchin, A., and Glaser, P. 1995. Anaerobic transcription activation in *Bacillus subtilis*: identification of distinct FNR-dependent and -independent regulatory mechanisms. EMBO J. 23: 5984-5994.

Cruz Ramos, H., Hoffmann, T., Marino, M., Nedjari, H., Presecanp-Siedel, E., Dreesen, O., Glaser, P., and Jahn, D. 2000. Fermentative metabolism of *Bacillus subtilis*: physiology and regulation of gene expression. J. Bacteriol. 182: 3072-3080.

Datta, P., Goss, T.J., Omnass, J.R., and Patil, R.V. 1987. Covalent structure of biodegradative threonine dehydratase of *Escherichia coli*: homology with other dehydratases. Proc. Natl. Acad. Sci. USA. 84: 393-397.

Delgado-Nixon, V.M., Gonzalez, G., and Gilles-Gonzalez, M.-A. 2000. Dos, a heme-binding PAS protein from *Escherichia coli*, is a direct oxygen sensor. Biochemistry. 39: 2685-2691.

De Sousa-Hart, J.A., Blackstock, W., Di Modugno, V., Holland, I.B., and Kok, M. 2003. Two-component systems in *Haemophilus influenzae*: a regulatory role for ArcA in serum resistance. Infect. Immun. 71: 163-172.

Dischert, W., Vignais, P.M., and Colbeau, A. 1999. The synthesis of *Rhodobacter capsulatus* HupSL hydrogenase is regulated by the two-component HupT/HupR system. Mol. Microbiol. 34: 995-1006.

Dong, C., Elsen, S., Swem, L.R., and Bauer, C.E. 2002. AerR, a second aerobic repressor of photosynthesis gene expression in *Rhodobacter capsulatus*. J. Bacteriol. 184: 2805-2814.

Du, S., Bird, T.H., and Bauer, C.E. 1998. DNA binding characteristics of RegA. A constitutively active anaerobic activator of photosynthesis gene expression in *Rhodobacter capsulatus*. J. Biol. Chem. 273: 18509-18513.

Du, S., Kouadio, J.L., and Bauer, C.E. 1999. Regulated expression of a highly conserved regulatory gene cluster is necessary for controlling photosynthesis gene expression in response to anaerobiosis in *Rhodobacter capsulatus*. J. Bacteriol. 181: 4334-4341.

Dubbs, J.M., Bird, T.H., Bauer, C.E., and Tabita, F.R. 2000. Interaction of CbbR and RegA* transcription regulators with the *Rhodobacter sphaeroides cbbI* promoter-operator region. J. Biol. Chem. 275: 19224-19230.

Durmowicz, M.C., and Maier, R.J. 1998. The FixK2 protein is involved in regulation of symbiotic hydrogenase expression in *Bradyrhizobium japonicum*. J. Bacteriol. 180: 3253-3256.

Elsen, S., Ponnampalam, S.N., and Bauer, C.E. 1998. CrtJ bound to distant binding sites interacts cooperatively to aerobically repress photopigment biosynthesis and light harvesting II gene expression in *Rhodobacter capsulatus*. J. Biol. Chem. 273: 30762-30769.

Elsen, S., Dischert, W., Colbeau, A., and Bauer, C.E. 2000. Expression of uptake hydrogenase and molybdenum nitrogenase in *Rhodobacter capsulatus* is coregulated by the RegB-RegA two-component regulatory system. J. Bacteriol. 182: 2831-2837.

Emmerich, R., Hennecke, H., and Fischer, H.M. 2000. Evidence for a functional similarity between the two-component regulatory systems RegSR, ActSR, and RegBA (PrrBA) in alpha-Proteobacteria. Arch. Microbiol. 174: 307-313.

Eraso, J.M., and Kaplan, S. 1994. *prrA*, a putative response regulator involved in oxygen regulation of photosynthesis gene expression in *Rhodobacter sphaeroides*. J. Bacteriol. 176: 32-43.

Eraso, J.M., and Kaplan, S. 1995. Oxygen-insensitive synthesis of the photsynthetic membranes of *Rhodobacter sphaeroides*: a mutant histidine kinase. J. Bacteriol. 177: 2695-2706.

Eraso, J.M., and Kaplan, S. 2000. From redox flow to gene regulation: Role of the PrrC protein of *Rhodobacter sphaeroides* 2.4.1. Biochemistry. 39: 2052-2062.

Geng, H., Nakano, S., and Nakano, M.M. 2004. Transcriptional activation by *Bacillus subtilis* ResD: Tandem binding to target elements and phosphorylation-dependent and -independent transcriptional activation. J. Bacteriol. 186: 2028-2037.

Georgellis, D., Kwon, O., and Lin, E.C.C. 2001. Quinones as the redox signal for the Arc two-component system of bacteria. Science. 292: 2314-2316.

Georgellis, D., Kwon, O., Lin, E.C.C., Wong, S.M., and Akerley, B.J. 2001. Redox signal transduction by the ArcB sensor kinase of *Haemophilus influenzae* lacking the PAS domain. J. Bacteriol. 183: 7206-7212.

Gibson, J.L., Dubbs, J.M., and Tabita, F.R. 2002. Differential expression of the CO_2 fixation operons of *Rhodobacter sphaeroides* by the Prr/Reg two-component system during chemoautotrophic growth. J. Bacteriol. 184: 6654-6664.

Glaser, P., Danchin, A., Kunst, F., Zuber, P., and Nakano, M.M. 1995. Identification and isolation of a gene required for nitrate assimilation and anaerobic growth of *Bacillus subtilis*. J. Bacteriol. 177: 1112-1115.

Gomelsky, M., and Kaplan, S. 1995. *appA*, a novel gene encoding a trans-acting factor involved in the regulation of photosynthesis gene expression in *Rhodobacter sphaeroides* 2.4.1. J. Bacteriol. 177: 4609-4618.

Gomelsky, M., and Kaplan, S. 1997. Molecular genetic analysis suggesting interactions between AppA and PpsR in regulation of photosynthesis gene expression in *Rhodobacter sphaeroides* 2.4.1. J. Bacteriol. 179: 128-134.

Gomelsky, M., and Kaplan, S. 1998. AppA, a redox regulator of photosystem formation in *Rhodobacter sphaeroides* 2.4.1, is a flavoprotein. Identification of a novel FAD binding domain. J. Biol. Chem. 273: 35319-35325.

Gomelsky, M., Horne, I.M., Lee, H.J., Pemberton, J.M., McEwan, A.G., and Kaplan, S. 2000. Domain structure, oligomeric state, and mutational analysis of PpsR, the *Rhodobacter sphaeroides* repressor of photosystem gene expression. J. Bacteriol. 182: 2253-2261.

Gomelsky, M., and Klug, G. 2002. BLUF: a novel FAD-binding domain involved in sensory transduction in microorganisms. Trends Biochem. Sci. 27: 497-500.

Gomelsky, L., Sram, J., Moskvin, O.V., Horne, I.M., Dodd, H.N., Pemberton, J.M., McEwan, A.G., Kaplan, S., and Gomelsky, M. 2003. Identification and *in vivo* characterization of PpaA, a regulator of photosystem formation in *Rhodobacter sphaeroides*. Microbiology. 149: 377-388.

Gostick, D.O., Green, J., Irvine, A.S., Gasson, M.J., and Guest, J.R. 1998. A novel regulatory switch mediated by the FNR-like protein of *Lactobacillus casei*. Microbiology. 144: 705-717.

Govantes, F., Orjalo, A.V., and Gunsalus, R.P. 2000. Interplay between three global regulatory proteins mediates oxygen regulation of the *Escherichia coli* cytochrome *d* oxidase (*cydAB*) operon. Mol. Microbiol. 38: 1061-1073.

Grabbe, R., and Schmitz, R.A. 2003. Oxygen control of nif gene expression in *Klebsiella pneumoniae* depends on NifL reduction at the cytoplasmic membrane by electrons derived from the reduced quinone pool. Eur. J. Biochem. 270: 1555-1566.

Green, J., Scott, C., and Guest, J.R. 2001. Functional versatility in the CRP-FNR superfamily of transcription factors: FNR and FLP. Adv. Microbiol. Physiol. 44: 1-34.

Gregor, J., and Klug, G. 1999. Regulation of bacterial photosynthesis genes by oxygen and light. FEMS Microbiol. Lett. 179: 1-9.

Gregor, J., and Klug, G. 2002. Oxygen-regulated expression of genes for pigment binding proteins in *Rhodobacter capsulatus*. J. Mol. Microbiol. Biotechnol. 4: 249-253.

Hagewood, B.T., Ganduri, Y.L., and Datta, P. 1994. Functional analysis of the *tdcABC* promoter of *Escherichia coli*: roles of TdcA and TdcR. J. Bacteriol. 176: 6214-6220.

Härtig, E., Geng, H., Hubacek, A., Münch, R., Jahn, D., Ye, R.W., and Nakano, M.M. 2004. *Bacillus subtilis* ResD induces regulatory *yclJK* genes upon oxygen limitation. J. Bacteriol. Submitted.

Hemschemeier, S.K., Ebel, U., Jager, A., Balzer, A., Kirndorfer, M., and Klug, G. 2000. *In vivo* and *in vitro* analysis of RegA response regulator mutants of *Rhodobacter capsulatus*. J. Mol. Microbiol. Biotechnol. 2: 291-300.

Heßlinger, C., Fairhurst, S.A., and Sawers, G. 1998. Novel keto acid formate-lyase and propionate kinase enzymes are components of an anaerobic pathway in *Escherichia coli* that degrades L-threonine to propionate. Mol. Microbiol. 27: 477-492.

Hoa, T.T., Tortosa, P., Albano, M., and Dubnau, D. 2002. Rok (YkuW) regulates genetic competence in *Bacillus subtilis* by directly reprressing *comK*. Mol. Microbiol. 43:15-26.

Homuth, G., Rompf, A., Schumann, W., and Jahn, D. 1999. Transcriptional control of *Bacillus subtilis hemN* and *hemZ*. J. Bacteriol. 181: 5922-5929.

Hutchings, M.I., Crack, J.C., Shearer, N., Thompson, B.J., Thomson, A.J., and Spiro, S. 2002. Transcription factor FnrP from *Paracoccus denitrificans* contains an iron-sulfur cluster and is activated by anoxia: identification of essential cysteine residues. J. Bacteriol. 184: 503-508.

Imlay, J.A. 2002. How oxygen damages microbes: oxygen tolerance and obligate anaerobiosis. Adv. Microb. Physiol. 46: 111-153.

Irvine, A.S., and Guest, J.R. 1993. *Lactobacillus casei* contains a member of the CRP-FNR family. Nucleic Acids Res. 21: 753.

Iuchi, S. 1993. Phosphorylation/ dephosphorylation of the receiver module at the conserved aspartate residue controls transphosphorylation activity of histidine kinase in sensor protein ArcB of *Escherichia coli*. J. Biol. Chem. 268: 23972-23980.

Iuchi, S., and Lin, E.C.C. 1995. Signal transduction in the Arc system for control of operons encoding aerobic respiratory enzymes. In: Two-Component Signal Transduction. J.A. Hoch, and T.J. Silhavy, eds. ASM Press, Washington, D.C. p. 223-231.

Jordan, P.A., Thomson, A.J., Ralph, E.T., Guest, J.R., and Green, J. 1997. FNR is a direct oxygen sensor having a biphasic response curve. FEBS Lett. 416: 349-352.

Joshi, H.M., and Tabita, F.R. 1996. A global two component signal transduction system that integrates the control of photosynthesis, carbon dioxide assimilation, and nitrogen fixation. Proc. Natl. Acad. Sci. USA. 93: 14515-14520.

Kappler, U., Huston, W.M., and McEwan, A.G. 2002. Control of dimethylsulfoxide reductase expression in *Rhodobacter capsulatus*: the role of carbon metabolites and the response regulators DorR and RegA. Microbiology. 148: 605-614.

Khoroshilova, N., Popescu, C., Münck, E., Beinert, H., and Kiley, P.J. 1997. Iron-sulphur cluster disassembley in the FNR protein of *Escherichia coli* by O_2: [4Fe-4S] to [2Fe-

2S] conversion with loss of biological activity. Proc. Natl. Acad. Sci. USA. 94: 6087-6092.

Kiley, P.J., and Beinert, H. 2003. The role of Fe-S proteins in sensing and regulation in bacteria. Curr. Opin. Microbiol. 6: 181-185.

Klopprogge, K., Grabbe, R., Hoppert, M., Schmitz, R.A. 2002. Membrane association of *Klebsiella pneumoniae* NifL is affected by molecular oxygen and combined nitrogen. Arch Microbiol. 177: 223-234.

Körner, H., Sofia, H.J., and Zumft, W.G. 2003. Phylogeny of the bacterial superfamily of Crp-Fnr transcription regulators: exploiting the metabolic spectrum by controlling alternative gene programs. FEMS Microbiol. Rev. 27: 559-592.

Kraft, B.J., Masuda, S., Kikuchi, J., Dragnea, V., Tollin, G., Zaleski, J.M., and Bauer, C.E. 2003. Spectroscopic and mutational analysis of the blue-light photoreceptor AppA: a novel photocycle involving flavin stacking with an aromatic amino acid. Biochemistry. 42: 6726-6734.

Kwon, O., Georgellis, D., and Lin, E.C.C. 2000. Phosphorelay as the sole physiological route of signal transmission by the Arc two-component system of *Escheichia coli*. J. Bacteriol. 182: 3858-3862.

LaCelle, M., Kumano, M., Kurita, K., Yamane, K., Zuber, P., and Nakano, M.M. 1996. Oxygen-controlled regulation of flavohemoglobin gene in *Bacillus subtilis*. J. Bacteriol. 178: 3803-3808.

Laguri, C., Phillips-Jones, M.K., and Williamson, M.P. 2003. Solution structure and DNA binding of the effector domain from the global regulator PrrA (RegA) from *Rhodobacter sphaeroides*: insights into DNA binding specificity. Nucleic Acids Res. 31: 6778-6787.

Lambden, P.R., and Guest, J.R. 1976. Mutants of *Escherichia coli* K-12 unable to use fumarate as an anaerobic electron acceptor. J. Gen. Microbiol. 97: 145-160.

Laratta, W.P., Choi, P.S., Tosques, I.E., and Shapleigh, J.P. 2002. Involvement of the PrrB/PrrA two-component system in nitrite respiration in *Rhodobacter sphaeroides* 2.4.3: evidence for transcriptional regulation. J. Bacteriol. 184: 3521-3529.

Lee, Y.S., Han, J.S., Jeon, Y., and Hwang, D.S. 2001. The Arc two-component signal transduction system inhibits *in vitro Escherichia coli* chromosomal initiation. J. Biol. Chem. 276: 9917-9923.

Liebl, U., Bouzhir-Sima, L., Kiger, L., Marden, M.C., Lambry, J.C., Negrerie, M., and Vos, M.H. 2003. Ligand binding dynamics to the heme domain of the oxygen sensor Dos from *Escherichia coli*. Biochemistry. 42: 6527-6535.

Liu, X., and Taber, H.W. 1998. Catabolite regulation of the *Bacillus subtilis ctaBCDEF* gene cluster. J. Bacteriol. 180: 6154-6163.

Liu, X., and De Wulf, P. 2004. Probing the ArcA-P modulon of *Escherichia coli* by whole-genome transcriptional analysis and sequence-recognition profiling. J. Biol. Chem. 279: 12588-12597.

Lu, S., Killoran, P.B., Fang, F.C., and Riley, L.W. 2002. The global regulator ArcA controls resistance to reactive nitrogen and oxygen intermediates in *Salmonella enterica* serovar enteritidis. Infect. Immun. 70: 451-461.

Maier, T.M., and Myers, C.R. 2001. Isolation and characterisation of a *Shewanella putrefaciens* MR-1 electron transport regulator *etrA* mutant: reassessment of the role of EtrA. J. Bacteriol. 183: 4918-4926.

Marino, M., Hoffmann, T., Schmid, R., Möbitz, H., and Jahn, D. 2000. Changes in protein synthesis during the adaptation of *Bacillus subtilis* to anaerobic growth conditions. Microbiology. 146: 97-105.

Marino, M., Ramos, H.C., Hoffmann, T., Glaser, P., and Jahn, D. 2001. Modulation of anaerobic energy metabolism of *Bacillus subtilis* by *arfM* (*ywiD*). J. Bacteriol. 183: 6815-6821.

Marshall, F.A., Messenger, S.L., Wyborn, N.R., Guest, J.R., Wing, H., Busby, S.J.W., and Green, J. 2001. A novel promoter architecture for microaerobic activation of the anaerobic transcription factor FNR. Mol. Microbiol. 39: 747-753.

Masuda, S., Matsumoto, Y., Nagashima, K.V., Shimada, K., Inoue, K., Bauer, C.E., and Matsuura, K. 1999. Structural and functional analyses of photosynthetic regulatory genes *regA* and *regB* from *Rhodovulum sulfidophilum, Roseobacter denitrificans*, and *Rhodobacter capsulatus*. J. Bacteriol. 181: 4205-4215.

Masuda, S., and Bauer, C.E. 2002. AppA is a blue light photoreceptor that antirepresses photosynthesis gene expression in *Rhodobacter sphaeroides*. Cell. 110: 613-623.

Masuda, S., Dong, C., Swem, D., Setterdahl, A.T., Knaff, D.B., and Bauer, C.E. 2002. Repression of photosynthesis gene expression by formation of a disulfide bond in CrtJ. Proc. Natl. Acad. Sci. USA. 99: 7078-7083.

Moore, L.J., and Kiley, P.J. 2001. Characterization of the dimerization domain in the FNR transcription factor. J. Biol. Chem. 276: 45744-45750.

Mosley, C.S., Suzuki, J.Y., and Bauer, C.E. 1994. Identification and molecular genetic characterization of a sensor kinase responsible for coordinately regulating light harvesting and reaction center gene expression in response to anaerobiosis. J. Bacteriol. 176: 7566-7573.

Nakano, M.M., Yang, F., Hardin, P., and Zuber, P. 1995. Nitrogen regulation of *nasA* and the *nasB* operon, which encode genes required for nitrate assimilation in *Bacillus subtilis*. J. Bacteriol. 177: 573-579.

Nakano, M.M., Zuber, P., Glaser, P., Danchin, A., and Hulett, F.M. 1996. Two-component regulatory proteins ResD-ResE are required for transcriptional activation of *fnr* upon oxygen limitation in *Bacillus subtilis*. J. Bacteriol. 178: 3796-3802.

Nakano, M.M., Dailly, Y.P., Zuber, P., and Clark, D.P. 1997. Characterization of anaerobic fermentative growth in *Bacillus subtilis*: Identification of fermentation end products and genes required for the growth. J. Bacteriol. 179: 6749-6755.

Nakano, M.M., Hoffmann, T., Zhu, Y., and Jahn, D. 1998. Nitrogen and oxygen regulation of *Bacillus subtilis nasDEF* encoding NADH-dependent nitrite reductase by TnrA and ResDE. J. Bacteriol. 180: 5344-5350.

Nakano, M.M., and Zuber, P. 1998. Anaerobic growth of a "strict aerobe" (*Bacillus subtilis*). Annu. Rev. Microbiol. 52: 165-190.

Nakano, M.M., Zhu, Y., Haga, K., Yoshikawa, H., Sonenshein, A.L., and Zuber, P. 1999. A mutation in the 3-phosphoglycerate kinase gene allows anaerobic growth of *Bacillus subtilis* in the absence of ResE kinase. J. Bacteriol. 181: 7087-7097.

Nakano, M.M., Zheng, G., and Zuber, P. 2000a. Dual control of *sbo-alb* operon expression by the Spo0 and ResDE systems of signal transduction under anaerobic conditions in *Bacillus subtilis*. J. Bacteriol. 182: 3274-3277.

Nakano, M.M., Zhu, Y., LaCelle, M., Zhang, X., and Hulett, F.M. 2000b. Interaction of ResD with regulatory regions of anaerobically induced genes in *Bacillus subtilis*. Mol. Microbiol. 37: 1198-1207.

Nakano, M.M., and Zhu, Y. 2001. Involvement of the ResE phosphatase activity in down-regulation of ResD-controlled genes in *Bacillus subtilis* during aerobic growth. J. Bacteriol. 183: 1938-1944.

Nakano, M.M. 2002. Induction of ResDE-dependent gene expression in *Bacillus subtilis* in response to nitric oxide and nitrosative stress. J. Bacteriol. 184: 1783-1787.

Nakano, M.M., and Zuber, P. 2002. Anaerobiosis. In: *Bacillus subtilis* and Its Closest Relatives: From Genes to Cells. A.L. Sonenshein, J.A. Hoch, and R. Losick, eds. ASM Press, Washington, D.C. p. 393-404.

Nakano, S., Nakano, M. M., Zhang, Y., Leelakriangsak, M., and Zuber, P. 2003. A regulatory protein that interferes with activator-stimulated transcription in bacteria. Proc. Natl. Acad. Sci. USA. 100: 4233-4238.

Nealson, K.H., and Saffarini, D.A. 1994. Iron and manganese in anaerobic respiration: environmental significance, physiology, and regulation. Annu. Rev. Microbiol. 48: 311-343.

O'Gara, J.P., and Kaplan, S. 1997. Evidence for the role of redox carriers in photosynthesis gene expresion and carotenoid biosynthesis in *Rhodobacter sphaerobides* 2.4.1. J. Bacteriol. 179: 1951-1961.

O'Gara, J.P., Eraso, J.M., and Kaplan, S. 1998. A redox-responsive pathway for aerobic regulation of photosynthesis gene expression in *Rhodobacter sphaeroides* 2.4.1. J. Bacteriol. 180: 4044-4050.

Oh, J.I., Eraso, J.M., and Kaplan, S. 2000. Interacting regulatory circuits involved in orderly control of photosynthesis gene expression in *Rhodobacter sphaeroides* 2.4.1. J. Bacteriol. 182: 3081-3087.

Oh, J.I., and Kaplan, S. 2000. Redox signaling: globalization of gene expression. EMBO J. 19: 4237-4247.

Oh, J.I., and Kaplan, S. 2001. Generalized approach to the regulation and integration of gene expression. Mol. Microbiol. 39: 1116-1123.

Oh, J.I., Ko, I.J., and Kaplan, S. 2001. The default state of the membrane-localized histidine kinase PrrB of *Rhodobacter sphaeroides* 2.4.1 is in the kinase-positive mode. J. Bacteriol. 183: 6807-6814.

Oshima, T., Aiba, H., Masuda, Y., Kanaya, S., Sugiura, M.,Wanner, B.L., Mori, H., and Mizuno, T. 2002. Transcriptome analysis of all two-component regulatory system mutants of *Escherichia coli* K-12. Mol. Microbiol. 46: 281-291.

Otten, M.F., Stork, D.M., Reijnders, W.N.M., Westerhoff, H.V., and van Spanning, R.J.M. 2001. Regulation of expression of terminal oxidases in *Paracoccus denitrificans*. Eur. J. Biochem. 268: 2486-2497.

Paget M.S., and Buttner M.J. 2003. Thiol-based regulatory switches. Annu. Rev. Genet. 37: 91-121.

Paul, S., Zhang, X., and Hulett, F.M. 2001. Two ResD-controlled promoters regulate *ctaA* expression in *Bacillus subtilis*. J. Bacteriol. 183: 3237-3246.

Pemberton, J.M., Horne, I.M., and McEwan, A.G. 1998. Regulation of photosynthetic gene expression in purple bacteria. Microbiology. 144: 267-278.

Penfold, R.J., and Pemberton, J.M. 1994. Sequencing, chromosomal inactivation, and functional expression in *Escherichia coli* of *ppsR*, a gene which represses carotenoid and bacteriochlorophyll synthesis in *Rhodobacter sphaeroides*. J. Bacteriol. 176: 2869-2876.

Pessi, G., and Haas, D. 2000. Transcriptional control of the hydrogen cyanide biosynthetic genes *hcnABC* by the anaerobic regulator ANR and the quorum-sensing regulators LasR and RhlR in *Pseudomonas aeruginosa*. J. Bacteriol. 182: 6940-6949.

Phillips-Jones, M.K., and Hunter, C.N. 1994. Cloning and nucleotide sequence of *regA*, a putative response regulator gene of *Rhodobacter sphaeroides*. FEMS Microbiol. Lett. 116: 269-275.

Pomposiello P.J., and Demple B. 2001. Redox-operated genetic switches: the SoxR and OxyR transcription factors. Trends Biotechnol. 19: 109-114.

Ponnampalam, S.N., Buggy, J.J., and Bauer, C.E. 1995. Characterization of an aerobic repressor that coordinately regulates bacteriochlorophyll, carotenoid, and light harvesting-II expression in *Rhodobacter capsulatus*. J. Bacteriol. 177: 2990-2997.

Ponnampalam, S.N., and Bauer, C.E. 1997. DNA binding characteristics of CrtJ. A redox-responding repressor of bacteriochlorophyll, carotenoid, and light harvesting-II gene expression in *Rhodobacter capsulatus*. J. Biol. Chem. 272: 18391-18396.

Ponnampalam, S.N., Elsen, S., and Bauer, C.E. 1998. Aerobic repression of the *Rhodobacter capsulatus bchC* promoter involves cooperative interactions between CrtJ bound to neighboring palindromes. J. Biol. Chem. 273: 30757-30761.

Poole, R.K., and Hill, S. 1997. Respiratory protection of nitrogenase activity in *Azotobacter vinelandii*-roles of the terminal oxidases. Bioscience Rep. 17: 303-317.

Prüß, B.M. 2000. FlhD, a transcriptional regulator in bacteria. Recent Res. Dev. Microbiol. 4: 31-42.

Prüß, B.M., Campbell, J.W., Van Dyk, T.K., Zhu, C., Kogan, Y., and Matsumura, P. 2003. FlhD/FlhC is a regulator of anaerobic respiration and the Entner-Douderoff pathway through induction of the methyl-accepting chemotaxis protein Aer. J. Bacteriol. 185: 534-543.

Qian, Y., and Tabita, F.R. 1996. A global signal transduction system regulates aerobic and anaerobic CO_2 fixation in *Rhodobacter sphaeroides*. J. Bacteriol. 178: 12-18.

Rebbapragada, A., Johnson, M.S., Harding, G.P., Zuccarellli, A.J., Fletcher, H.M., Zhulin, I.B., and Taylor, B.L. 1997. The Aer protein and the serine chemoreceptor Tsr independently sense intracellular energy levels and transduce oxygen, redox, and energy levels for *Escherichia coli*. Proc. Natl. Acad. Sci. USA. 94: 10541-10546.

Roh, J.H., and Kaplan, S. 2000. Genetic and phenotypic analyses of the *rdx* locus of *Rhodobacter sphaeroides* 2.4.1. J. Bacteriol. 182: 3475-3781.

Roh, J.H., Smith, W.E., and Kaplan, S. 2004. Effects of oxygen and light intensity on transcriptome expression in *Rhodobacter sphaeroides* 2.4.1: redox active gene expression profile. J. Biol. Chem. 279: 9146-9155.

Romagnoli, S., Packer, H.L., and Armitage, J.P. 2002. Tactic responses to oxygen in the phototrophic bacterium *Rhodobacter sphaeroides* WS8N. J. Bacteriol. 184: 5590-5598.

Saffarini, D.A., and Nealson, K.H. 1993. Sequence and genetic characterisation of *etrA*, and *fnr* analog that regulates anaerobic respiration in *Shewanella putrefaciens* MR-1. J. Bacteriol. 175: 7938-7944.

Saffarini, D.A., Schulz, R., and Beliaev, A. 2003. Involvement of cyclic AMP (cAMP) and cAMP receptor protein in anaerobic respiration of *Shewanella oneidensis*. J. Bacteriol. 185: 3668-3671.

Salmon, K., Hung, S.-P., Mekjian, K., Baldi, P., Hatfield, G.W., and Gunsalus, R.P. 2003. Global gene expression profiling in *Escherichia coli* K 12. the effects of oxygen availability and FNR. J. Biol. Chem. 278: 29837-29855.

Sawers, R.G., Zehelein, E., and Böck, A. 1988. Two-dimensional gel electrophoretic analysis of *Escherichia coli* proteins: influence of various anerobic growth conditions and the *fnr* gene product on cellular protein composition. Arch. Microbiol. 149: 240-244.

Sawers, G., and Suppmann, B. 1992. Anaerobic induction of pyruvate formate-lyase gene expression is mediated by the ArcA and FNR proteins. J. Bacteriol. 174: 3474-3478.

Sawers, G. 1999. The aerobic/anaerobic interface. Curr. Opin. Microbiol. 2: 181-187.

Sawers, G. 2001. A novel mechanism controls anaerobic and catabolite regulation of the *Escherichia coli tdc* operon. Mol. Microbiol. 39: 1285-1298.

Schmitz, R.A., Klopprogge, K., and Grabbe, R. 2002. Regulation of nitrogen fixation in *Klebsiella pneumoniae* and *Azotobacter vinelandii*: NifL, transducing two environmental signals to the nif transcriptional activator NifA. J. Mol. Microbiol. Biotechnol. 4: 235-242.

Sciotti, M.-A., Chanfon, A., Hennecke, H., and Fischer, H.-M. 2003. Disparate oxygen responsiveness of two regulatory cascades that control expression of symbiotic genes in *Bradyrhizobium japonicum*. J. Bacteriol. 185: 5639-5642.

Scott, C., Guest, J.R., and Green, J. 2000. Characterization of the *Lactococcus lactis* transcription factor FlpA and demonstration of an *in vitro* switch. Mol. Microbiol. 35: 1383-1393.

Scott, C., and Green, J. 2002. Miscoordination of the iron-sulfur clusters of the anaerobic transcription factor, FNR, allows simple repression but not activation. J. Biol. Chem. 277: 1749-1754.

Sengupta, N., Paul, K., and Chowdhury, R. 2003. The global regulator ArcA modulates expression of virulence factors in *Vibrio cholerae*. Infect. Immun. 71: 5583-5589.

Sevcík, M., Sebková, A., Volf, J., and Rychlík, I. 2001. Transcription of *arcA* and *rpoS* during growth of *Salmonella typhimurium* under aerobic and microaerobic conditions. Microbiology. 147: 701-708.

Sganga, M.W., and Bauer, C.E. 1992. Regulatory factors controlling photosynthetic reaction center and light-harvesting gene expression in *Rhodobacter capsulatus*. Cell. 68: 945-954.

Shaw, D.J., Rice, D.W., and Guest, J.R. 1983. Homology between CAP and Fnr, a regulator of anaerobic respiration in *Escherichia coli*. J. Mol. Biol. 166: 241-247.

Shimizu, T., Ohtani, K., Hirakawa, H., Ohshima, K., Yamashita, A., Shiba, T., Ogasawara, N., Hattori, M., Kuhara, S., and Hayashi, H. 2002. Complete genome sequence of *Clostridium perfringens*, an anaerobic flesh-eater. Proc. Natl. Acad. Sci. USA. 99: 996-1001.

Smith, M.W., and Neidhardt, F.C. 1983a. Proteins induced by anaerobiosis in *Escherichia coli*. J. Bacteriol. 154: 336-343.

Smith, M.W., and Neidhardt, F.C. 1983b. Proteins induced by aerobiosis in *Escherichia coli*. J. Bacteriol. 154: 344-350.

Soupene, E., van Heeswijk, W.C., Plumbridge, J., Stewart, V., Bertenthal, D., Lee, H., Prasad, G., Paliy, O., Charernnoppakul, P., and Kustu, S. 2003. Physiological studies of *Escherichia coli* strain MG1655: growth defects and apparent cross-regulation of gene expression. J. Bacteriol. 185: 5611-5626.

Stewart, V. 2003. Nitrate- and nitrite-responsive sensors NarX and NarQ of proteobacteria. Biochem. Soc. Trans. 31: 1-10.

Stewart, V., Chen, L.-L., and Wu, H.-C. 2003. Response to culture aeration mediated by the nitrate and nitrite sensor NarQ of *Escherichia coli* K-12. Mol. Microbiol. 50: 1391-1399.

Strauch, M.A. 1993. AbrB, a transition state regulator. In: *Bacillus subtilis* and Other Gram-Positive Bacteria: Physiology, Biochemistry, and Molecular Biology. A.L. Sonenshein, J.A. Hoch, R. and Losick, eds. American Society for Microbiology, Washington, D.C. p. 757-764.

Sumantran, V.N., Schweizer, H.P., and Datta, P. 1990. A novel membrane-associated threonine permease encoded by the *tdcC* gene of *Escherichia coli*. J. Bacteriol. 172: 4288-4294.

Sun, G., Sharkova, E., Chesnut, R., Birkey, S., Duggan, M.F., Sorokin, A., Pujic, P., Ehrlich, S.D., and Hulett, F.M. 1996. Regulators of aerobic and anaerobic respiration in *Bacillus subtilis*. J. Bacteriol. 178: 1374-1385.

Swem, L.R., Elsen, S., Bird, T.H., Swem, D.L., Koch, H.G., Myllykallio, H., Daldal, F., and Bauer, C.E. 2001. The RegB/RegA two-component regulatory system controls synthesis of photosynthesis and respiratory electron transfer components in *Rhodobacter capsulatus*. J. Mol. Biol. 309: 121-138.

Swem, L.R., Kraft, B.J., Swem, D.L., Setterdahl, A.T., Masuda, S., Knaff, D.B., Zaleski, J.M., and Bauer, C.E. 2003. Signal transduction by the global regulator RegB is mediated by a redox-active cysteine. EMBO J. 22: 4699-4708.

Taylor, B.L., and Zhulin, I.B. 1999. PAS domains: Internal sensors of oxygen, redox potential, and light. Microbiol. Mol. Biol. Rev. 63: 479-506.

Tichi, M.A., and Tabita, F.R. 2001. Interactive control of *Rhodobacter capsulatus* redox-balancing systems during phototrophic metabolism. J. Bacteriol. 183: 6344-6354.

Tosques, I.E., Shi, J., and Shapleigh, J.P. 1996. Cloning and characterization of *nnrR*, whose product is required for the expression of proteins involved in nitric oxide metabolism in *Rhodobacter sphaeroides* 2.4.3. J. Bacteriol. 178: 4958-4964.

Tseng, C.P., Albrecht, J., and Gunsalus, R.P. 1996. Effect of microaerophilic cell growth conditions on expression of the aerobic (*cyoABCDE* and *cydAB*) and anaerobic (*narGHJI, frdABCD,* and *dmsABC*) respiratory pathway genes in *Escherichia coli*. J. Bacteriol. 178: 1094-1098.

Van Keulen, G., Jonkers, H.M., Claessen, D., Dijkhuizen, L., and Wösten, H.A.B. 2003. Differentiation and anaerobiosis in standing liquid cultures of *Streptomyces coelicolor*. J. Bacteriol. 185: 1455-1458.

Vanrobaeys, F., Devreese, B., Lecocq, E., Rychlewski, L., De Smet, L., and van Beeumen, J. 2003. Proteomics of the dissimilatory iron-reducing bacterium *Shewanella oneidensis* MR-1, using a matrix-assisted laser desorption/ ionization-tnadem-time of flight mass spectrometer. Proteomics. 3: 2249-2257.

Vichivanives, P., Bird, T.H., Bauer, C.E., and Tabita, F.R. 2000. Multiple regulators and their interactions *in vivo* and in vitro with the cbb regulons of *Rhodobacter capsulatus*. J. Mol. Biol. 300: 1079-1099.

Wanner, B.L. 1992. Is cross-regulation by phosphorylation of two-component response regulator proteins important in bacteria? J. Bacteriol. 174: 2053-2058.

Wu, Y., Patil, R.V., and Datta, P. 1992. Catabolite gene activator protein and integration host factor act in concert to regulate *tdc* operon expression in *Escherichia coli*. J. Bacteriol. 174: 6918-6927.

Wu, G., Cruz-Ramos, H., Hill, S., Green, J., Sawers, G., and Poole R.K. 2000. Regulation of cytochrome *bd* expression in the obligate aerobe *Azotobacter vinelandii* by CydR (Fnr): extreme sensitivity to oxygen, reactive oxygen species and nitric oxide. J. Biol. Chem. 275: 4679-4686.

Wu, G., Moir,A.J.G., Hill, S., Sawers, G., and Poole R.K. 2001. Biosynthesis of poly-β-hydroxbutyrate is controlled by CydR (FNR) in the obligate aerobe *Azotobacter vinelandii*. FEMS. Lett. 194: 215-220.

Ye, R.W., Tao, W., Bedzyk, L., Young, T., Chen, M., and Li, L. 2000. Global gene expression profiles of *Bacillus subtilis* grown under anaerobic conditions. J. Bacteriol. 182: 4458-4465.

Yeliseev, A.A., and Kaplan, S. 1995. A sensory transducer homologous to the mammalian peripheral-type benzodiazepine receptor regulates photosynthetic membrane complex formation in *Rhodobacter sphaeroides* 2.4.1. J. Biol. Chem. 270: 21167-21175.

Zeilstra-Ryalls, J., Gomelsky, M., Eraso, J.M., Yeliseev, A., O'Gara, J., and Kaplan, S. 1998. Control of photosystem formation in *Rhodobacter sphaeroides*. J. Bacteriol. 180: 2801-2809.

Zeilstra-Ryalls, J.H., and Kaplan, S. 1998. Role of the *fnrL* gene in photosystem gene expression and photosynthetic growth of *Rhodobacter sphaeroides* 2.4.1. J. Bacteriol. 180: 1496-1503.

Zeng, X., and Kaplan, S. 2001. TspO as a modulator of the repressor/antirepressor (PpsR/AppA) regulatory system in *Rhodobacter sphaeroides* 2.4.1. J. Bacteriol. 183: 6355-6364.

Zhang, X., and Hulett, F.M. 2000. ResD signal transduction regulator of aerobic respiration in *Bacillus subtilis*; *cta* promoter regulation. Mol. Microbiol. 37: 1208-1219.

Zhou, L., Lei, X.-H., Bochner, B.R., and Wanner, B.L. 2003 Phenotype microarray analysis of *Escherichia coli* K-12 mutants with deletions of all two-component systems. J. Bacteriol. 185: 4956-4972.

Chapter 4

Anaerobic Metabolism by *Pseudomonas aeruginosa* in Cystic Fibrosis Airway Biofilms: Role of Nitric Oxide, Quorum Sensing, and Alginate Production

Daniel J. Hassett*, Sergei V. Lymar, John J. Rowe, Michael J. Schurr, Luciano Passador, Andrew B. Herr, Geoffrey L. Winsor, Fiona S. L. Brinkman, Sang Sun Yoon, Gee W. Lau, and Sung Hei Hwang

Abstract

The goal of this review is to highlight new developments that could lead to better treatment strategies for complications associated with chronic *Pseudomonas aeruginosa* infections in cystic fibrosis (CF) airway disease. Recent data suggest that *P. aeruginosa* grows anaerobically in "biofilms" enmeshed in the thick airway mucus of CF patients. The most energy-efficient form of anaerobic growth by *P. aeruginosa* is via respiration using nitrate (NO_3^-) or nitrite (NO_2^-) as a terminal electron acceptor. Both of these nitrogen oxides are amply present in CF airway surface liquid. In this review, we discuss how the anaerobic biofilm mode of growth actually benefits *P. aeruginosa* in the context of CF airway disease. We will also describe in detail the anaerobic respiratory pathway and how enzymatic production and disposal of a gaseous by-product of anaerobic growth, nitric oxide (NO), is tightly regulated and critical for anaerobic survival during respiration. We next describe how the process of intercellular communication known as quorum sensing is necessary for anaerobic survival of *P. aeruginosa*. Finally, when *P. aeruginosa* shifts to a mucoid, alginate-overproducing form, the anaerobic biofilm mode of growth and anaerobic metabolism is further promoted. Thus, it is our belief that new therapeutic strategies aimed at targeting anaerobic gene products and determining the effectiveness of various anti-*P. aeruginosa* antibiotics under anaerobic conditions need to be developed to help combat these infections.

1.CF Airway Infection by *P. aeruginosa*

The function of lung epithelial cells in normal individuals is to elicit transport of Cl^-, Na^+ and water from the basolateral to the apical surface. One of the major roles of the **c**ystic **f**ibrosis **t**ransmembane **r**egulator (CFTR) is to transport Cl^- across the apical surface of secretory cells. Thus, of the primary defects associated with CF is little or no Cl^- transport across the apical surface. Three potential scenarios can be envisioned; either (i) no CFTR produced, (ii) mutated CFTR, (iii) or mutated and truncated CFTR. Invariably, abnormal CFTR processing, membrane positioning, or defects in function allow for Na^+, Cl^- and water reabsorbtion in an unopposed and accelerated manner. Goblet cells also contribute to CF airway disease in that they almost continuously secrete mucus to the apical surface even though the mucus already on the epithelial surface has not been cleared. Unlike the free beating and highly efficient clearing function of airway cilia in normal cells, the cilia on CF epithelia are matted down by the thick mucus and either do not beat or beat erratically, thus exacerbating the disease. Thus, the thickened, highly inspissated, and poorly cleared mucus becomes a haven for opportunistic pathogens in CF that include *P. aeruginosa, Staphylococcus aureus, Hemophilus influenzae, Burkholderia cepacia,* and other opportunists, as well as some true pathogens (e.g., *Mycobacterium tuberculosis*). However, in most CF centers in the U.S., with progression of CF airway disease, *P. aeruginosa* typically predominates and its mode of growth *in vitro* is dramatically different than that *in vivo*. The *in vivo* mode of growth in CF airway disease has been termed a "biofilm" (Singh *et al.*, 2000). Below, we will discuss what have been gaining dogmatic momentum as the two major forms of biofilms, the second of which we believe best represents that which is formed in CF airway disease.

*For correspondence email: Daniel.Hassett@UC.Edu.

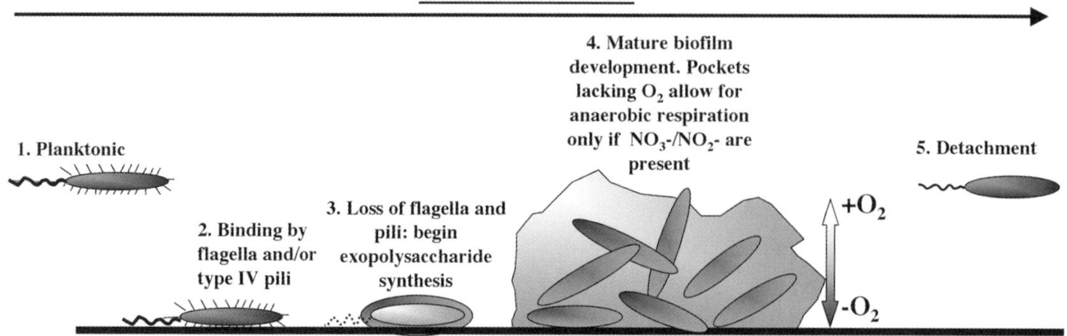

Figure 1. Model of the steps in the formation of a non-CF *P. aeruginosa* biofilm. 1. Free-swimming (planktonic) bacteria. 2. Attachment of bacteria by flagella/type IV pili. 3. Loss of flagellum and initiation of exopolysaccharide (EPS) production. 4. Cell division and accumulation of EPS, leading to the development of an oxygen gradient but little or no growth in anaerobic zones due to the absence of NO_3^-/NO_2^- or arginine. 5. Mechanical purturbation or dispersion signals allows for detached cells to synthesize a new flagellum and migrates to new site for colonization. This model figure was initially published by Hassett *et al.* (2002).

2. Biofilm Development: Two Contrasting Models

2.1. Model #1: Surface Attachment

The "father" of biofilm research, Dr. Bill Costerton of Montana State University, has defined biofilms as a highly organized yet diverse growth of microorganisms on virtually any substratum. In one model of biofilm development, bacteria seek to attach to surfaces or cells. O'Toole and Kolter (1998) first described that the surface appendages, flagella and type IV pili, are required for optimal biofilm formation on plastic and glass substrata (Figure 1). Other gene products including polyphosphate kinase, the catabolite repressor control protein (Crc), LasR, RhlR, GacA, RpoS, Crc, and PvrR have been reported to be important in various biofilm culture systems. Most recently, Finelli et al. (Finelli *et al.*, 2003), using IBET (**in**-**b**iofilm **e**xpression **t**echnology), revealed several more gene products involved in biofilm development including an alcohol dehydrogenase, a putative porin, and a homologue of the *Streptomyces griseus* developmental regulator AdpA that were not produced during planktonic growth.

2.2. Model #2: *P. aeruginosa* Enmeshed in Thick, Anaerobic CF Airway Mucus

The surface attachment model of biofilm formation depicted in Figure 1 is vastly different that the biofilms observed in the thick CF airway mucus. More evidence related to this is discussed below in the next paragraph. First, *P. aeruginosa* is almost never observed on the epithelial cell surface during the course of CF airway disease (Figures 2 and 3). This fairly bold statement is based upon solid published data (Worlitzsch *et al.*, 2002) and upon microscopic examination of thin sections of airways removed from multiple CF patients that received lung transplants (R.C. Boucher and D.J. Hassett, unpublished observations). In fact, at the annual Cystic Fibrosis Foundation meeting at Williamsburg, VA in June of 2003, many former naysayers finally agreed that this is a very important, unique and poorly understood feature of the disease. Thus, the fundamental difference between biofilm model #1 and the CF airway mucus biofilm is that the latter organisms are deeply embedded in the thick, highly inspissated airway mucus. In support of this theory, R. Boucher, G. Doring and colleagues (Worlitzsch *et al.*, 2002) have shown that there are "steep hypoxic gradients" generated within the CF airway mucus. A hypoxic and/or strict anaerobic model of how such conditions occur within the CF mucus is depicted in Figure 2. This figure was redrawn from the work of Worlitzsch *et al.*, (2002) and has also been used in another review (Hassett *et al.*, 2002). The surface of normal airway epithelial cells has a very thin, hydrated mucus layer directly on top of what is termed a periciliary liquid layer (PCL, Figure 2A). The viscosity of the PCL is very low and this feature facilitates highly efficient mucociliary clearance (see arrow). It is for this reason that normal individuals, with some exceptions such as the elderly and the immunocompromised with pneumonia, and those infected by "true" lung pathogens such as *M. tuberculosis* or *Francisella tularensis*, do not get airway infections. A typical rate of oxygen consumption (QO_2; left of top figure) produces no significant O_2 gradients within the "thin" airway surface liquid (ASL) [denoted by white (O_2 rich) or black (hypoxic/anaerobic) bars to the right of each panel]. In contrast to the normal epithelium seen in part A, parts B-F depict CF airway epithelia. Part B indicates that water is excessively reabsorbed (see vertical H_2O arrow). This, in turn, decreases the volume of the PCL, resulting in the mucus adhering tightly to the epithelium. At the same

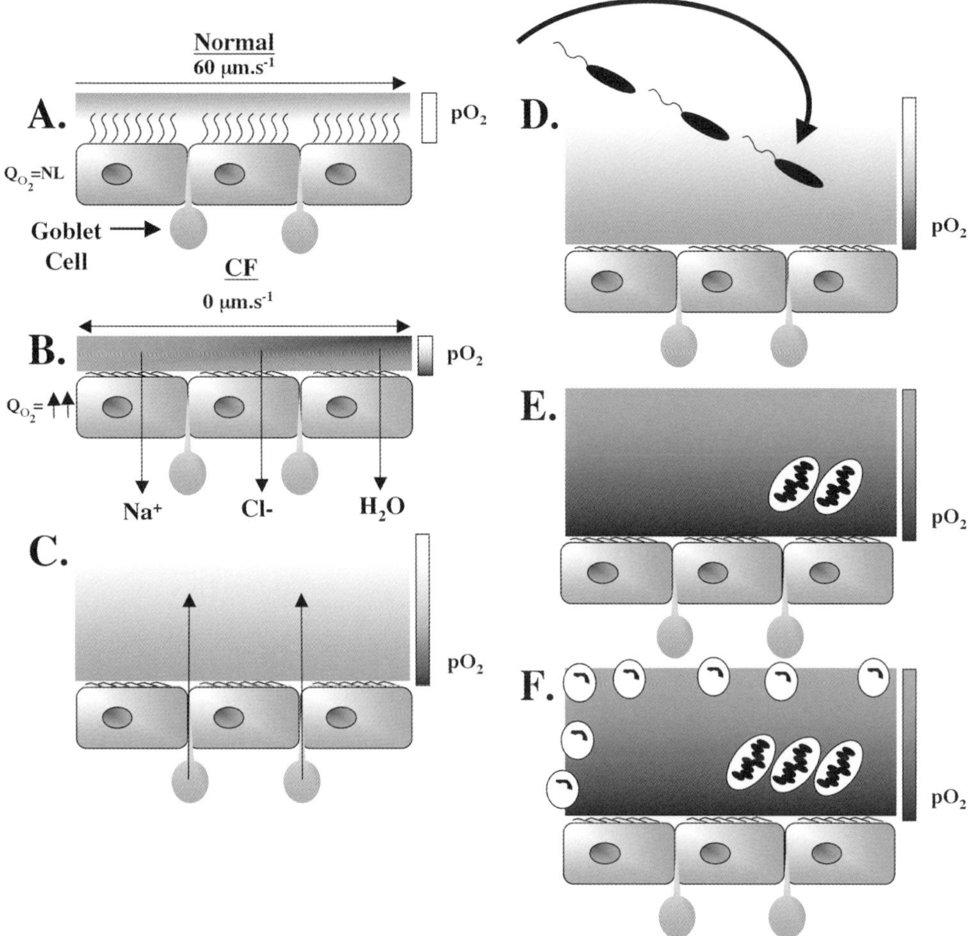

Figure 2. Model of *P. aeruginosa* "biofilms" in anaerobic CF mucus relative to normal mucus. This figure was redrawn in PowerPoint from Worlitzsch *et al.* (2002). See text for a description of each caption A-F.

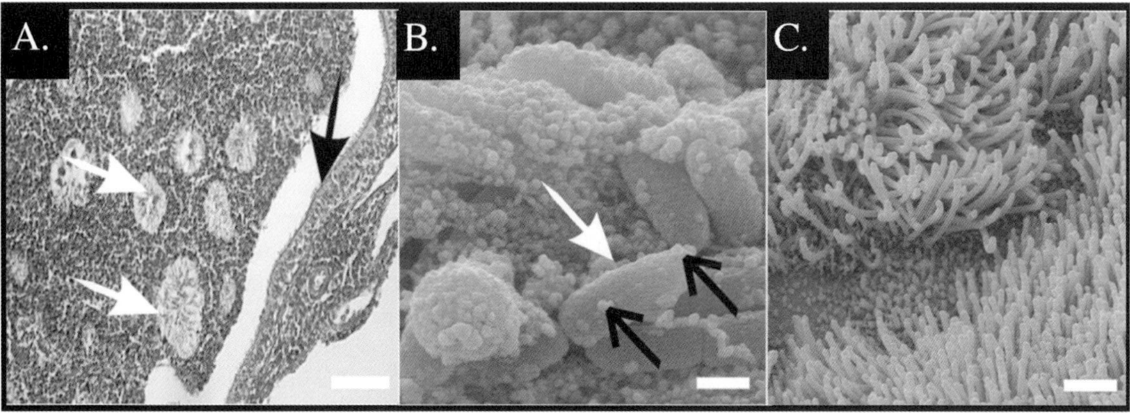

Figure 3. *P. aeruginosa* in the CF airways is not associated with the epithelium but embedded in the thick airway mucus. Note, parts A-C are from the work of Worlitzsch *et al.* (2002) as is the captions describing each figure. **A.** Thin section of an obstructed CF bronchus, stained with hematoxalin/eosin. Note the absence of *P. aeruginosa* on epithelial surface (black arrow) and presence of *P. aeruginosa* macrocolonies within intraluminal material (white arrows). The gap is an artifact due to fixation. **B.** Scanning electron micrograph of mucus-coated spheroid derived from CF respiratory epithelium. *P. aeruginosa* (white arrow) were enmeshed in mucus (black arrows) following a 2 hour incubation. **C.** Spheroid with adherent mucus removed by prewash, then incubated with *P. aeruginosa* for 2 hours. Note the absence of bacteria on ciliated epithelial cell surfaces. Bars: **B**, 0.6 μm; **C**, 2.5 μm.

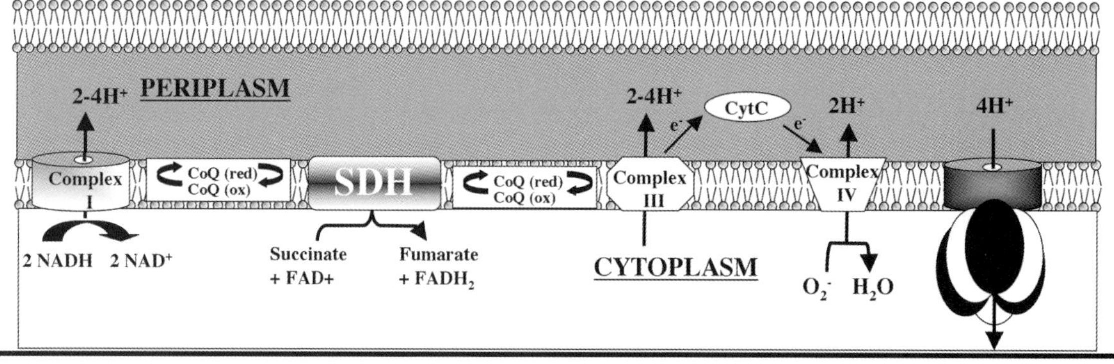

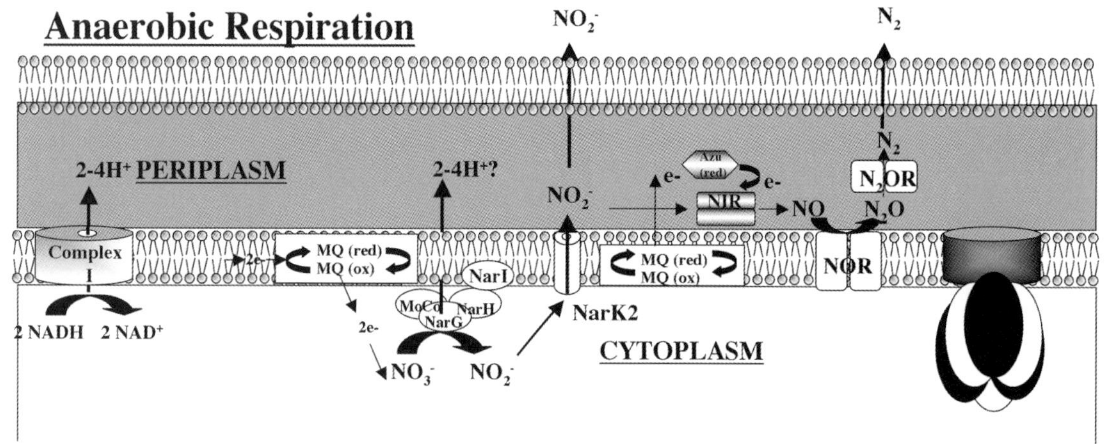

Figure 4. Schematic diagram of the machinery involved in aerobic versus anaerobic respiration in *P. aeruginosa*. **A.** Aerobic respiration, complex I (NADH dehydrogenase), SDH (succinate dehydrogenase), CoQ (coenzyme Q), complex III (cytochrome *c*:coenzyme Q oxidoreductase), CytC (cytochrome *c*), complex IV (cytochrome oxidase). **B.** Anaerobic respiration, MQ (menaquinone), NIR (NO_2^- reductase), NOR (NO reductase), N_2OR (nitrous oxide reductase), Azu (azurin), MoCo (molybdopterin cofactor).

time, the cilia, that facilitate clearance of bacteria and other opportunists in hydrated mucus, become matted down onto the epithelial cell surface. Finally, transport of the mucus slows or ceases (see bi-directional vector). Stutts and Boucher (1986) showed that CF epithelia have a 2- to 3-fold increased rate of O_2 consumption relative to normal epithelial cells and this is directly correlated with accelerated CF ion transport and increased Na^+/K^+-ATPase activity. Part C indicates that persistent mucus hypersecretion (denoted as mucus secretory gland/goblet cell "teardrops" below epithelium) occurs over time, an event which triggers increased luminal mucus thickness and associated masses and plugs. The elevated CF epithelial QO_2 generates steep hypoxic gradients (white-to-black transition in bar on right) in the thickened mucus masses. In part D, *P. aeruginosa* are seeded onto the mucus surface by inhalation and the bacteria can penetrate the mucus actively (presumably using moving surface appendages including flagella and type IV pili) and/or passively (due to mucus turbulence) into hypoxic zones within the mucus. Part E shows a schematic that is meant to indicate that *P. aeruginosa* can adapt to such confines by undergoing a rapid transition from aerobic to anaerobic metabolism, the latter process involving utilization of the alternative electron acceptors, NO_3^- or NO_2^-, that are present in CF airway mucus (see discussion below in section 6). We have recently shown that NO_3^- stabilizes the mucoid form of *P. aeruginosa* during anaerobic respiration and prevents mucoid-to-nonmucoid conversion under aerobic conditions (Hassett, 1996; Wyckoff *et al.*, 2002). Finally, Part F demonstrates that the developing macrocolonies can resist secondary defenses, including stimulated human neutrophils that migrate aggressively to the CF airway lumen through tight junctures, macrophages, and many first-tier antibiotics (ceftazidime, tobramycin, ciprofloxacin, etc.), setting the stage for chronic infection. It is our view that the presence of increased macrocolony density and, to a lesser extent, neutrophils, renders the now mucopurulent mass anaerobic (black bar).

In the work of Worlitzsch *et al.* (2002), the authors also demonstrated that the organisms are rarely, if ever, associated with the epithelium. Note in Figure 3A that there are pockets of organisms in oblong spheroid plugs enmeshed in the thick mucus directly produced from tissue of an obstructed CF bronchus, consistent with biofilm model #2. In fact, the distances from the epithelium for >95% of all bacteria were between 5 and 17 μm. In contrast, there were only ~5% of the bacteria between 2 to 5 μm from the cell surface. The nature

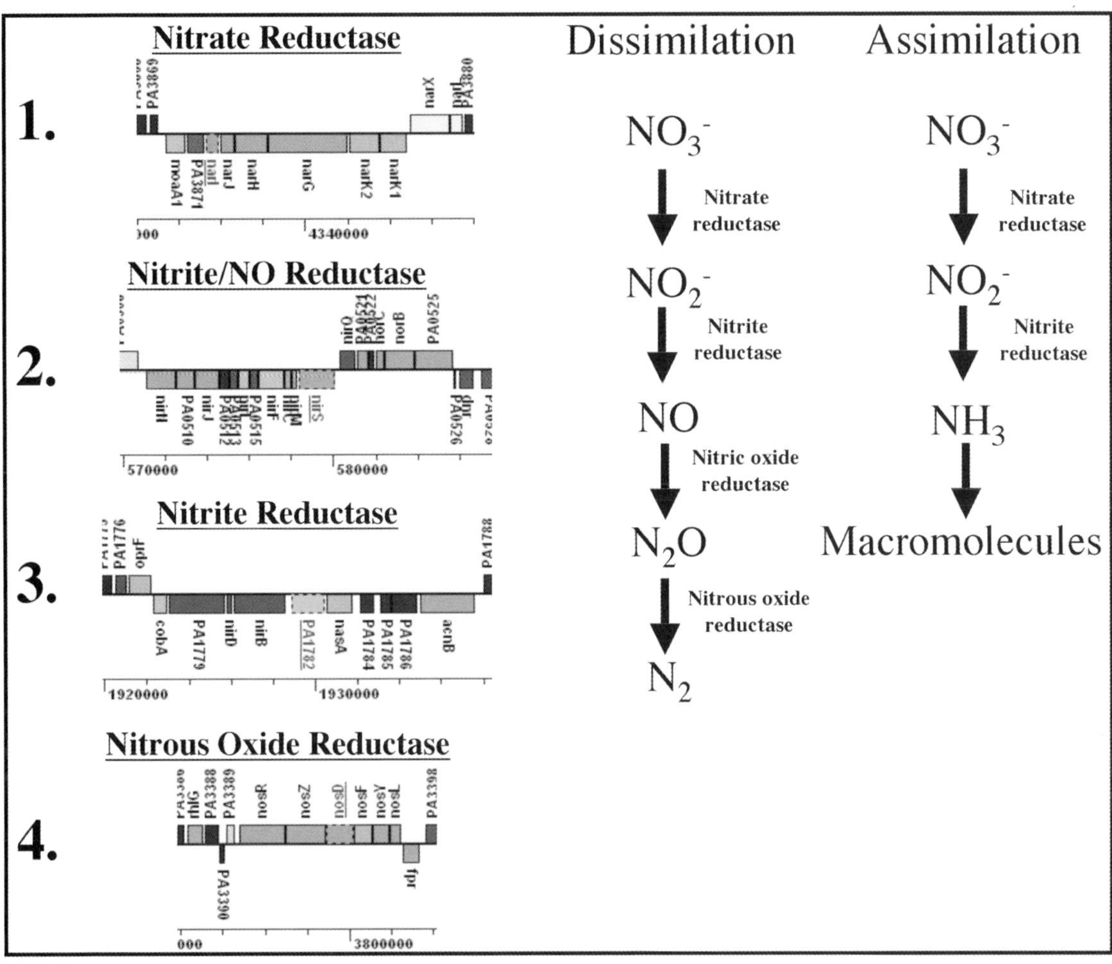

Figure 5. Localization of many of the genes involved in anaerobic respiration in operons in *P. aeruginosa*. These include those involved in: 1, NO_3^- reduction; 2, NO_2^- and nitric oxide (NO) reduction; 3, NO_2^- reduction and 4, nitrous oxide reduction. To the right, note the different steps in the dissimilatory vs. assimilatory NO_3^- reduction pathways.

of *P. aeruginosa* interaction with mucus is depicted in Figure 3B where the organism binds to and becomes enmeshed in the thick mucus. Finally, bacteria were unable to bind to ciliated epithelial cells that had been washed free of mucus (Figure 3C). These data clearly implicate the mucus as an important initiating binding factor for *P. aeruginosa* biofilm formation in CF airway disease. While enmeshed in the thick CF mucus layer, there is likely a tendency for the mucus to become anaerobic, especially during chronic infection. Thus, a description of the machinery involved in anaerobic respiration by *P. aeruginosa* and regulation of this process for optimal survival and growth is warranted and presented in the following subsection.

3. Anaerobic Growth of *P. aeruginosa*

P. aeruginosa, although once erroneously thought to be an obligate aerobe, is now defined as an "obligate respirer." The organism cannot ferment in the classic sense, (gain energy from organic sources and independent of respiratory chains), although it can undergo a substrate level phosphorylation using arginine (Gamper *et al.*, 1991) (discussed below). The most facile means to generate anaerobic energy is through anaerobic NO_3^- respiration. Respiratory NO_3^- reduction by *P. aeruginosa* occurs only under anaerobic conditions and involves the sequential 8-electron reduction of NO_3- → NO_2- → nitric oxide ($NO^\cdot$) → nitrous oxide (N_2O) → nitrogen gas (N_2). Organisms such as *P. aeruginosa* which are able to reduce NO_3^- to N_2 are called denitrifiers. Each reduction step may be viewed as a terminal step in the respiratory transfer of two electrons from the oxidation of NADH to the reduction of each nitrogen oxide in the pathway.

During aerobic respiration (Figure 4A), there are 3 steps that are termed coupling steps involved in proton extrusion from the cytoplasmic to the periplasmic space. The first of which is at the NADH dehydrogenase complex (complex I), the second at the cytochrome *c*: ubiquinone oxidoreductase (complex III), and the last at cytochrome oxidase (complex IV). In contrast, it is postulated that there are only one to two coupling steps during anaerobic respiration (Figure 4B). Like aerobic respiration, the first is the NADH dehydrogenase complex. The second is believed to be at each of the terminal nitrogen oxide reduction steps. In addition,

Table 1. Identification of *P. aeruginosa* genes with promoter regions harboring consensus ANR binding domains (TTGA-N_6-TCAA)

Anr box consensus	Box location	Gene name/ PA#	Putative function of gene downstream of Anr box	Link to Anaerobic growth	Notes
TTGACGGCCGTCAA	25630 to 25643	*qor*/ PA0023	quinone oxidoreductase		*hemF* and *qor* are divergently transcribed*
TTGACGGCCGTCAA	25630 to 25643	*hemF*/ PA0024	coproporphyrinogen III oxidase, aerobic	Known	*hemF* and *qor* are divergently transcribed
TTGACCCCCATCAA	161697 to 161710	PA0141	putative regulator		PA0141 and PA0142 are divergently transcribed
TTGACCCCCATCAA	161697 to 161710	PA0142	hypothetical protein		PA0141 and PA0142 are divergently transcribed
TTGACCGGAATCAA	580224 to 580237	*nirS*/ PA0519	nitrite reductase precursor	Known	*nirS* and *nirQ* are divergently transcribed
TTGACCGGAATCAA	580224 to 580237	*nirQ*/ PA0520	regulatory protein NirQ	Known	*nirS* and *nirQ* are divergently transcribed
TTGATTGCCATCAA	581944 to 581957	*norC*/ PA0523	nitric-oxide reductase subunit C	Known	
TTGACCGCGATCAA	585953 to 585966	PA0526	hypothetical protein		Upstream is *dnr*.
TTGAGGTGGTTCAA	796814 to 796827	PA0729	probable bacteriophage integrase		
TTGATTTCCATCAA	921474 to 921487	*plcH*/ PA0845	hemolytic phospholipase C precursor		
TTGAACAGGCTCAA	1175533 to 1175546	PA1088	putative methyltransferase		within last 27 bp of *flgL* and upstream of PA1088
TTGACCTGTCTCAA	1554939 to 1554952	PA1429	probable cation-transporting P-type ATPase		
TTGATACAAATCAA	1684013 to 1684026	*hemN*/ PA1546	oxygen-independent coproporphyrinogen III oxidase	Known	
TTGATCCCGATCAA	1696414 to 1696427	PA1557	probable cytochrome oxidase subunit		PA1557 and PA1558 are divergently transcribed
TTGATCCCGATCAA	1696414 to 1696427	PA1558	hypothetical protein		PA1557 and PA1558 are divergently transcribed
TTGATACGAATCAA	1803437 to 1803450	PA1656	hypothetical protein		
TTGACGCCGATCAA	1824875 to 1824888	PA1672	hypothetical protein		PA1672 and PA1673 are divergently transcribed
TTGACGCCGATCAA	1824875 to 1824888	PA1673	hypothetical protein		PA1672 and PA1673 are divergently transcribed
TTGATGCGTATCAA	1888310 to 1888323	PA1745	hypothetical protein		PA1745 and PA1746 are divergently transcribed
TTGATGCGTATCAA	1888310 to 1888323	PA1746	hypothetical protein		PA1745 and PA1746 are divergently transcribed
TTGATATGCATCAA	1939478 to 1939491	PA1789	hypothetical protein		PA1789 and PA1790 are divergently transcribed
TTGATATGCATCAA	1939478 to 1939491	PA1790	hypothetical protein		PA1789 and PA1790 are divergently transcribed
TTGACCTGAATCAA	2911985 to 2911998	*alkB1*/ PA2574	alkane-1-monooxygenase		*alkB1* and PA2575 are divergently transcribed
TTGACCTGAATCAA	2911985 to 2911998	PA2575	hypothetical protein		*alkB1* and PA2575 are divergently transcribed

TTGACCCGCATCAA	3219521 to 3219534	PA2867	probable chemotaxis transducer		
TTGACCTGGATCAA	3705554 to 3705567	PA3305	hypothetical protein		
TTGACTTTCATCAA	3794684 to 3794697	PA3390	hypothetical protein		PA3390 and *nosR* are divergently transcribed
TTGACTTTCATCAA	3794684 to 3794697	*nosR*/ PA3391	regulatory protein NosR	Known	PA3390 and *nosR* are divergently transcribed
TTGATCATTCTCAA	4092220 to 4092233	*frr*/ PA3653	ribosome recycling factor		overlaps 6 bp start of gene
TTGATAGGAATCAA	4344897 to 4344910	*narK1*/ PA3877	nitrite extrusion protein 1	Known	*narK1* and *narX* are divergently transcribed
TTGATAGGAATCAA	4344897 to 4344910	*narX*/ PA3878	two-component sensor NarX	Known	*narK1* and *narX* are divergently transcribed
TTGATTCCGGTCAA	4347547 to 4347560	PA3880	conserved hypothetical protein		PA3880 is downstream of *narXL*
TTGACCGGGATCAA	4384000 to 4384013	PA3913	probable protease		
TTGAGCTGTGTCAA	4544522 to 4544535	*oprG*/ PA4067	Outer membrane protein OprG precursor		
TTGATGTGCATCAA	4881244 to 4881257	PA4352	conserved hypothetical protein		
TTGATCCACGTCAA	5136235 to 5136248	*ccpR*/ PA4587	cytochrome c551 peroxidase precursor	Known	
TTGAACAACTTCAA	5442257 to 5442270	*accB*/ PA4847	biotin carboxyl carrier protein (BCCP)		within the last 23 bp of *aroQ1* gene, upstream of *accB* gene
TTGATGTCATTCAA	5579616 to 5579629	PA4969	conserved hypothetical protein		
TTGATGCCGATCAA	5820627 to 5820640	*arcD*/ PA5170	arginine/ornithine antiporter	Known	
TTGATGTCGATCAA	6166052 to 6166065	PA5475	hypothetical protein		

* It is noted when a putative ANR box is identified between two divergently transcribed genes, since in such cases it is often unclear whether just one

either a quinone loop or quinone cycle-mediated proton extrusion could further enhance the generation of the proton motive force (ΔP). In contrast to aerobic respiration where coenzyme Q (CoQ) is used as the inner membrane quinone pool, menaquinone (MQ) is used during anaerobic respiration. In *P. aeruginosa*, each of the nitrogen oxides in the sequential reduction of NO_3^- to N_2 can also act as terminal electron acceptors for anaerobic respiration. However, the amount of free energy, as reflected by the redox potential (E'_o) of the various couples, limits the amount of ΔP, and ultimately ATP that is generated. In addition, the location of the specific nitrogen oxide reductase also dictates the degree of ΔP generated. If such enzymes are located on the cytoplasmic side of the inner membrane, an additional two protons are generated during the reduction process, thus increasing the ΔP. If the reductase is located on the periplasmic side, as is the case for nitrite reductase (NIR) and nitrous oxide reductase (N_2OR), protons are consumed and the ΔP is diminished. The amount of free energy potentially available is reflected in the difference between the NAD/NADH couple (E'_0 = -0.32 V) and the nitrogen oxide couple (NO_3^- /NO_2^- = +0.42 V; NO_2^-/NO = +0.36V; NO/N_2O = +1.18 V and N_2O/N_2 = +1.38 V). In general, growth rates are compromised by the inherent toxicity of both NO_2^- and NO; thus growth yield studies have not met expected theoretical calculations.

The best studied genes and their respective gene products are involved in the first step, the reduction of NO_3^- to NO_2^- and are termed the *nar* (**n**itr**a**te **r**eduction) genes. Those involved in reduction of NO_2^- to NO are termed *nir* (**n**itr**i**te **r**eductase). Those involved in removal of NO are termed *nor* (**n**itric **o**xide **r**eductase), and finally those involved in the reduction of N_2O are termed *nos* (**n**itr**o**u**s** oxide reductase). The *nar*, *nir*, *nor*, and *nos* genes encoding proteins required for anaerobic growth and survival during the denitrification process are localized in operons dispersed throughout the genome (Figure 5).

3.1. Dissimilatory NO_3^- Reduction

Much of what we already know of anaerobic respiratory systems comes from studies in one of the respiratory NO_3^- reduction steps in *E. coli*. There have been limited studies in *P. aeruginosa* and, therefore, the content of this section is primarily by extrapolation of *E. coli* homologs found in the *P. aeruginosa* genome. The dissimilatory NAR is embedded in the cytoplasmic side of the inner membrane (Figure 4). In the case of *E. coli*, the genes coding for the enzyme complex *narGHJI* are contained within the *nar* operon. All of the gene products encoded from the *nar* operon are membrane-bound and, together, are involved in the overall reduction of NO_3^- to NO_2^-. NarG is the NAR enzyme which contains the catalytic molybdenum center. Electrons are passed from the quinone pool and/or perhaps directly from formate reduction to NarI (cytochrome b_{556}). NarI is oxidized by NarH, which contains 4 Fe-S centers. In *E. coli*, the initial oxidation of NADH or formate represents the primary entry of electrons to the anaerobic electron transport chain. The *nar* operon of *P. aeruginosa* differs from *E. coli* in terms of gene content. The operon contains, in order, the *narK1K2GHJI* genes. The function of the *narK1* and *narK2* gene products is proposed to be the exporter of NO_2^- and import of NO_3^-, respectively, and are discussed in the following section.

3.2. Nitrogen Oxide Transport

The transport of NO_3^- across the cytoplasmic membrane in *P. aeruginosa* is thought to be mediated via a NO_3^- transporter. Virtually nothing is known about the NO_3^- transporters in *P. aeruginosa* except by way of extrapolation from gene sequence. The primary reason for this is that there is no long-lived isotope of nitrogen or oxygen to allow such studies to be performed reliably. Although some very indirect approaches have been used to examine NO_3^- transport, in most cases, the studies were performed using whole cells in a wild-type metabolic background. Thus, the transport process *per se* was not separated from subsequent metabolism. Initially, it was proposed that transport occurred via a NO_3^-/NO_2^- antiport system because there was a clear external accumulation of NO_2^- associated with NO_3^- uptake in whole cells of *E. coli*. A phenotype representing a *nar* mutation was identified early on and designated *narK*. A definitive study using $^{13}NO_3^-$ in vesicle and proteoliposome systems (Rowe *et al.*, 1994) demonstrated that NarK was responsible for the export of NO_2^- but not the uptake of NO_3^-. Information can only be inferred from the genes of the *P. aeruginosa* PAO1 genome (www.pseudomonas.com) and what is already known in *E. coli*. Rowe and colleagues (Rowe *et al.*, 1994) have shown that NO_3^- uptake can be stimulated by the protonophores carbonyl cyanide *m*-chlorophenylhydrazone and 2,4-dinitrophenol, indicating that NO_3^- uptake may be affected by proton motive force (PMF). Oxygen regulation of NO_3^- uptake may, in part, be through redox-sensitive thiol groups since N-ethylmaleimide at high concentrations decreased the rate of NO_3^- transport. Cells grown with tungstate (deficient in NAR activity) and azide-treated cells transported NO_3^- at significantly lower rates than untreated cells, indicating that physiological rates of NO_3^- transport are dependent on NO_3^- reduction. Furthermore, tungstate-grown cells transported NO_3^- only in the presence of NO_2^- (Rowe *et al.*, 1994), lending support to the NO_3^-/NO_2^- antiport model for transport. Oxygen regulation of NO_3^- transport was relieved (to 10% that of typical anaerobic rates) by the cytochrome oxidase inhibitors, carbon monoxide and cyanide.

3.3. The Arginine Deiminase Pathway for Anaerobic Growth

Another means by which *P. aeruginosa* can generate ATP is via the arginine deiminase (ADI) pathway. Three enzymatic reactions occur in the ADI pathway that catabolizes L-arginine to L-ornithine. These include ADI, catabolic ornithine carbamoyltransferase, and carbamate kinase (Galimand *et al.*, 1991; Gamper *et al.*, 1991). Thus, in the absence of the preferred terminal electron acceptor (O_2) or an alternative one (NO_3^-, NO_2^-, N_2O), the ADI pathway allows for anaerobic growth. However, this particular mode of growth is abysmally slow. The typical aerobic generation time in rich medium for *P. aeruginosa* is ~30 minutes while anaerobic growth with 100 mM NO_3^- requires ~50 minutes. In contrast, the time required for a single doubling while using the anaerobic ADI pathway is ~40 hours (Vander Wauven *et al.*, 1984). The structural genes (*arcABC*) for the three enzymes involved in *P. aeruginosa* ADI pathway are preceded by a gene (*arcD*) encoding an arginine-ornithine antiporter (Verhoogt *et al.*, 1992). The induction of transcription of the operon is mediated not by arginine, but by oxygen limitation (Mercenier *et al.*, 1980). Mutants lacking any of the *arcDABC* genes are unable to grow anaerobically via the ADI pathway (Vander Wauven *et al.*, 1984).

4. Regulatory Hierarchy in Anaerobic Respiration and Arginine Substrate Level Phosphorylation: ANR, DNR and ArgR

Not surprisingly, the anaerobic respiratory pathway is tightly regulated. The major "player" in activation of genes involved in anaerobic respiration is a homolog of the *E. coli* FNR protein called the **a**naerobic **n**itrate **r**egulator, ANR (Gamper *et al.*, 1991; Ye *et al.*, 1995). ANR, like FNR, is a dimeric $[4Fe\text{-}4S]^{2+}$ cluster protein that controls both denitrification genes (Ye *et al.*, 1995) and the *arcABCD* genes that are involved in arginine substrate level phosphorylation (Gamper *et al.*, 1991). A consensus FNR/ANR signature sequence of [TTGA-N_6-TCAA] in the promoter region of ANR-controlled genes was identified by Galimand et al. (Galimand *et al.*, 1991) in studies designed to better understand how genes involved in the arginine deiminase pathway are controlled. We performed a comprehensive search of the *P. aeruginosa* PAO1 genome using a customized DNA search tool (www.pseudomonas.com) to identify potential ANR-controlled genes and the results are listed in Table 1. Many of the genes with ANR boxes are predictable. These include *narX*, *nirS*, *nirQ*, *norC*, *norL*, and *oprG*. Other genes, including PA0729 (probable bacteriophage integrase), PA3913 (probable protease) and *plcH* (hemolytic phospholipase C) were unexpected and their potential role in anaerobic metabolism is unknown. Putative operons of hypothetical genes such as a four-gene cluster starting with PA1656 were found to contain a putative ANR box upstream, as well as genes encoding putative efflux/transport membrane components such as PA3305-PA3304, clusters of genes which may be involved in cell wall metabolism (PA1088-PA1091), and genes such as PA4352 that contain a universal stress protein family sequence motif within the deduced protein. These genes should be further investigated for their potential role in the *Pseudomonas* anaerobic lifestyle. Interestingly, an ANR box was located upstream of PA2867, a probable chemotaxis transducer. This may indicate that *P. aeruginosa* might be coerced to move toward a more anaerobic microniche, such as the conditions within CF airway biofilms.

The *nar* operon is also transcriptionally regulated through the action of a two-component regulatory signal transduction system mediated by the products of *narX* and *narL,* whose expression is regulated by ANR. NarX, located within the cytoplasmic membrane, functions to sense the external presence of nitrate and subsequently phosphorylates NarL which acts as a transcriptional activator. Post-translational regulation by oxygen has been well documented and exposure to oxygen results in the immediate cessation of anaerobic nitrogen oxide reduction in *P. aeruginosa* (Hernandez and Rowe, 1987, 1988; Hernandez *et al.*, 1991) and *E. coli.*

A second-tier regulator DNR is directly controlled by ANR is DNR (Arai *et al.*, 1995; 1997). DNR, like ANR, is a CRP/FNR-like protein but lacks an Fe-S cluster. DNR is known to participate in the regulation of the *nos* loci, encoding the N_2OR enzyme (Arai *et al.*, 2003), the *hemA* gene (Krieger *et al.*, 2002), as well as *hemF* and *hemN* (Rompf *et al.*, 1998). No structural information is currently available for either ANR or DNR.

During anaerobic growth via arginine substrate level phosphorylation, ANR is responsible for transcriptional activation of *argR* encoding a transcriptional activator (Lu *et al.*, 1999). ArgR is responsible for activation of the *arcDABC* operon. From electrophoretic mobility shift assays and DNase I footprinting analyses, a model was proposed to explain how ANR interacts with ArgR to activate *arcDABC* (Figure 6). Two ArgR proteins bind to regions I and II that are upstream of two consensus ANR binding sites. Together, ArgR and ANR participate in the transcriptional control of the *arcDABC* operon.

5. Why is NO_3^- and NO_2^- Present in CF Airway Mucus and Sputa?: Relationships with Production and Metabolism of Nitric Oxide (NO)

What is the source of NO_3^- and NO_2^- in the CF airways? They must be produced via oxidation of NO, which is the major source of inorganic nitrogen. The oxidation of NO requires molecular oxygen and several mechanisms can contribute to NO_2^- and NO_3^- production. First, the very rapid addition of superoxide (O_2^-) to NO (Goldstein and Czapski, 1995; Huie and Padmaja, 1993; Kobayashi

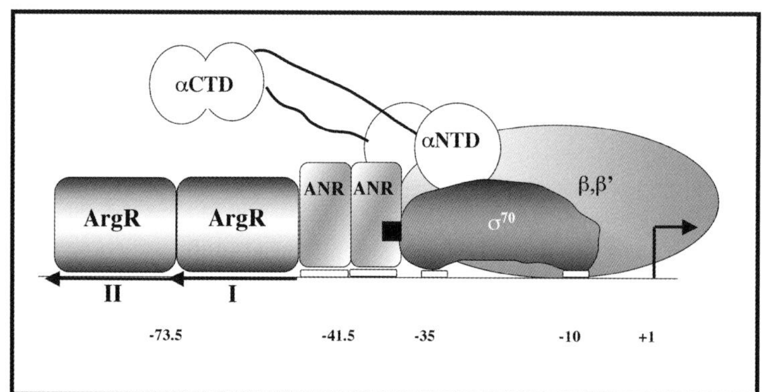

Figure 6. Redrawn model representing the predicted interaction between ArgR and ANR at the *P. aeruginos arc* promoter after the work of Lu *et al.* (1999). The figure legend provided here is that of the Lu *et al.* (1999). Given the overlapping specificity of FNR and ANR and the functional conservation of the two regulators, this model is based on the structural data elaborated for FNR and RNA polymerase of *E. coli*. The σ^{70}, β, and β' subunits of the RNA polymerase were labeled accordingly. CTD and NTD represent the carboxy-terminal domain and the amino-terminal domain of the subunits, respectively. The possible contact between ANR and the σ^{70} subunit is represented by a solid square.

et al., 1995; Lymar *et al.*, 2003a), which can be formed by NO-synthases present in airway epithelia, leads to production of peroxynitrite ($ONOO^-$).

$$O_2^- + NO \leftrightarrow ONOO^-$$

$ONOO^-$ can also be produced by the addition of oxygen to the nearest redox neighbor of NO, nitroxyl (HNO/NO^-) (Liochev and Fridovich, 2003; Miranda *et al.*, 2003; Shafirovich and Lymar, 2002, 2003)

$$O_2 + HNO/NO^- \leftrightarrow ONOOH/ONOO^-$$

The nitroxyl species are probable intermediates generated *en route* to NO from arginine via human nitric oxide synthases, during denitrification in *P. aeruginosa*, and as the intermediates of NO metabolism. $ONOO^-$ is unstable and decomposes both spontaneously and catalytically (Lymar and Poskrebyshev, 2003), with CO_2 being the most important physiological catalyst (Lymar and Hurst, 1995; 1996), through a complex set of free radical reactions (Coddington *et al.*, 1999; Gerasimov and Lymar, 1999; Hodges and Ingold, 1999; Lymar and Hurst, 1998; Merényi and Lind, 1998; Merényi *et al.*, 1999). In most biological environments, these reactions will ultimately lead to both NO_3^- and NO_2^- with a peculiar ~2/1 overall stoichiometry with respect to nitrogen (Coddington *et al.*, 1999; Hodges and Ingold, 1999; Lymar and Hurst, 1998; Lymar *et al.*, 2003b)

$$ONOO^- \leftrightarrow 0.7NO_3^- + 0.3NO_2^-$$

Only in the environments that are very rich in thiols will this stoichiometry be slightly shifted toward NO_2^- at the expense of NO_3^- (Briviba *et al.*, 1998; Lymar and Hurst, 1996).

The second process that can lead to NO_3^- involves oxidation of NO by oxyhemoglobin and its analogues [for a recent review, see Ford and Lorkovic (2002)].

$$Hb(O_2) + NO \leftrightarrow metHb + NO_3^-$$

However, a significant fraction of NO produced in cells is expected to be converted to NO_2^- through oxidation by metalloproteins (Ford and Lorkovic, 2002; Radi, 1996; Wink and Mitchel, 1998), e.g.,

$$Cyt^{III} + NO + H_2O \leftrightarrow Cyt^{II} + NO_2^- + 2H^+$$

Outside the cells or in the NO-rich cellular environments that can occur in and around stimulated macrophages at the sites of inflammation, the well-known and mechanistically complex autoxidation of NO

$$4NO + O_2 + 2H_2O \leftrightarrow 4NO_2^- + 4H^+$$

which is kinetically second-order in [NO] acquires biological significance.

Once produced, NO_2^- and NO_3^- can be metabolized by *P. aeruginosa*. In addition, NO_2^- exhibits rather complex chemistry in the acidic aqueous environments typical of the inflammation sites. This chemistry is associated with the instability of nitrous acid with respect to its disproportionation.

$$HNO_2 + HNO_2 \leftrightarrow NO + NO_2 + H_2O$$
$$NO_2 + NO_2 + H_2O \leftrightarrow NO_3^- + HNO_2 + H^+$$
$$\text{Net: } 3HNO_2 \leftrightarrow 2NO + NO_3^- + H^+ + H_2O$$

The first reaction is reversible, but the second irreversible reaction leads to the overall decomposition of HNO_2 and attendant accumulation of NO. The salient feature of this mechanism is that the rate of NO formation constantly decreases as NO accumulates, but the accumulation never ceases. This situation is known as the persistent radical effect and is associated with the shift of competition for NO_2 in favor of the reverse reaction at higher [NO]. Obviously, the NO accumulation rate is also dependent upon pH, NO_2- concentration, and the presence of oxygen (see NO autoxidation reaction above).

In Figure 7A, we show the NO accumulation kinetics over the first 24 hours in pure buffer at various pH levels computed by numerical simulations using the literature rate data for the individual reactions (Goldstein and Czapski, 1996; Park and Lee, 1988). Under anaerobic conditions, the accumulation of NO rapidly decelerates but continues, as expected. In contrast, due to the autoxidation reaction under aerobic conditions, the NO concentration rapidly attains a constant limiting value. The calculations also reveal that both the initial rate of NO accumulation and its amount at any given time are exponentially dependent upon pH. Both aerobically and anaerobically, the initial rate increases 100-fold upon lowering the pH by one unit; at pH 6.5, the rate is ~2.5 μM/hour, indicating that acidic NO_2^- is a potent NO source. Indeed, NO will be generated and consumed at

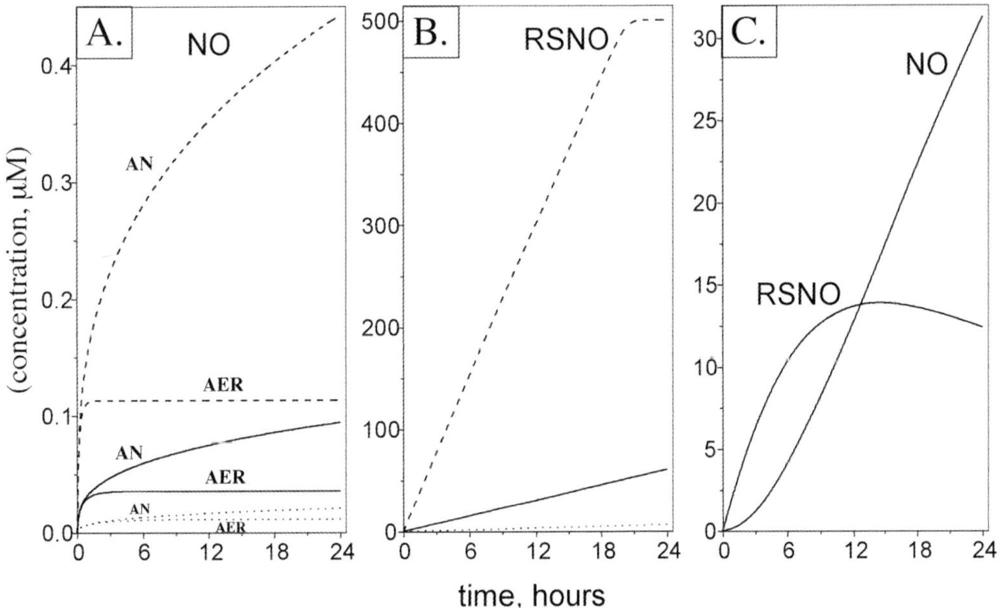

Figure 7. Simulated kinetic traces for various products in 10 mM nitrite buffered solution. Dashed lines – at pH 6.0; solid lines – at pH 6.5; dotted lines – at pH 7.0. **A.**, accumulation of NO in pure buffer under anaerobic (AN) and aerobic (AER, 0.26 mM O_2, corresponding to air-saturation) conditions. **B.**, anaerobic accumulation of RSNO in the presence of added 0.5 mM RSH. **C.**, anaerobic accumulation of both NO and RSNO in the presence of added 0.5 mM RSH and with the RSNO decomposition (6 hours half-life).

this rate, when there is a rapid sink for NO. Similarly, the NO amount increases 20-fold anaerobically and 10-fold aerobically upon lowering the pH from 7 to 6.

This picture changes significantly and becomes more complicated in biological milieus that contain antioxidants. Of particular importance are the thiols (RSH), such as cysteine and glutathione, whose cytosolic levels can be in the millimolar range (Wahllander *et al.*, 1979); cysteine residues in proteins may also play a role. Thiols rapidly react with the NO_2 radical, producing thiyl radical (Ford *et al.*, 2002)

$$NO_2 + RSH \leftrightarrow HNO_2 + RS$$

Although the rates of NO addition to the thiyl radicals yielding nitrosothiols

$$NO + RSH \leftrightarrow RSNO$$

have not been measured, this radical recombination reaction is strongly energetically favorable (Bartberger *et al.*, 2001), uncomplicated, and should be nearly diffusion controlled. With this assumption and using typical rates for thiol oxidation by NO_2 at various pH levels (Ford *et al.*, 2002), the kinetics of RSNO accumulation under anaerobic conditions can be modeled and the results are presented in Figure 7B.

It is evident that RSNO is predicted to accumulate at a constant rate and to much higher levels than NO. As with NO accumulation, there is strong pH dependency; the rates increase exactly 10 times per pH unit decrease. The rate will remain constant until there is complete consumption of the thiol, as seen from the kinetics at pH 6, upon which the normal NO accumulation shown in Figure 7A will set in. This indicates that acidic NO_2^- can lead to a very deep and relatively rapid depletion of the cellular antioxidants, making the cells vulnerable to oxidative damage. Thus, under aerobic conditions, the bacterial flavohemoglobin (or nitric oxide dioxygenase), participates in removal of NO to NO_3^- (Gardner *et al.*, 1998), while anaerobically the nitric oxide reductase (NOR) removes anaerobically generated NO (Yoon *et al.*, 2002). In this respect, we should recall that the thiyl radical itself is a potent oxidant, whose injurious effectiveness increases at low antioxidant content and in the presence of oxygen (Zhao *et al.*, 1994), when the formation of organic peroxides (ROOH) and O_2^- is expected. The latter should be of particular concern in the presence of NO, because it will lead to $ONOO^-$ (see the O_2^- + NO reaction above), which exhibits an extreme toxicity toward bacteria (Hurst and Lymar, 1997; 1999). Unfortunately, sensible kinetic modeling of thiol oxidations by acidic NO_2^- under aerobic conditions is presently not feasible due to the lack of necessary rate data.

Formation of RSNO constitutes the thiol nitrosation and is not innocuous; when occurring at the cysteine residues in an enzyme, this reaction can seriously compromise its metabolic function. For instance, modification of only Cys_{349} in the sarcoplasmic reticulum Ca-ATPase is sufficient for modulating its activity (Viner *et al.*, 1999). At the same time, extensive nitrosation of the nonprotein thiols, especially cysteine and glutathione, is in a sense equivalent to the accumulation of NO itself because the nitrosothiols are regarded as the NO "pool" in biological systems. Many nitrosothiols decompose to release free NO under certain *in vitro* conditions and their biological effects are similar to those of NO, which has lead to the suggestions that biologically occurring nitrosothiols are capable of *in vivo* trafficking and releasing NO [see, e.g., (Bartberger *et al.*, 2001; Williams, 1999) and references therein]. Our simulations in Figure 7B clearly show that even weakly acidic NO_2^- can very significantly contribute to filling up this NO "pool". To illustrate these points, we have included into our simulations the nitrosothiol decomposition reaction

$$RSNO \leftrightarrow 0.5RSSR + NO$$

with the assumed half-life of 6 hours and plotted the results for pH 6.5 in Figure 7C. It is evident that the RSH-dependent NO accumulation is about 300 times greater than in the RSH-free buffer under otherwise identical conditions. Thus, both extra- and intrabacterial thiols have the capacity of increasing toxicity of acidic NO_2^-, if it is mediated by NO.

6. Quorum Sensing in *P. aeruginosa*

In *P. aeruginosa* the expression of a large number of genes appears to be regulated by a cell density-dependent system known as quorum sensing. Quorum sensing allows bacterial populations to respond to increases in cell density by coordinating the expression of various gene products (Whitehead *et al.*, 2001). In its most basic form, quorum sensing requires only two components; a transcriptional activator protein (R-protein) and its cognate acylated homoserine lactone (AHL) signal molecule. The current paradigm proposes that the R-protein remains in an inactive state until it binds its cognate AHL. Following AHL binding the AHL/R-protein complex can function as a transcriptional regulator by binding at a specific element (*las* box) upstream of the target gene and interacting with RNA polymerase to affect gene expression. The connection between gene expression and cell density is based on the observation that the AHL molecule(s) produced by the bacterium can move across the bacterial cell wall (Pearson *et al.*, 1999). As a result, the concentration of AHL able to enter the bacteria reflects the number of bacterial cells within a population and a group of cells can measure their population density simply by monitoring the concentration of AHL molecules present. Hence, during periods of low cell density, the concentration of AHL molecules is also low, resulting in a limitation on the numbers of AHL/R-protein complexes that can form. As the density of the population increases, there is a concomitant increase in both internal and external AHL levels. Upon reaching a threshold concentration of AHL, sufficient AHL/R-protein complexes are formed such that the bacterial population senses it has achieved a "quorum" and the appropriate response can be triggered.

While the paradigm outlined above is relatively simple, *P. aeruginosa* presents a much more complicated scenario, perhaps as an indication of its metabolic and pathogenic versatility. *P. aeruginosa* contains at least two linked quorum sensing circuits (Whitehead *et al.*, 2001). The first circuit was initially shown to positively regulate the expression of the *lasA* and *lasB* elastase genes and thus was termed the *las* system (Gambello and Iglewski, 1991; Toder *et al.*, 1991). This circuit consists of the LasR transcriptional activator protein encoded by *lasR* and its cognate AHL *N*-3-oxo-dodecanoyl-homoserine lactone (3O-C_{12}-HSL) whose synthesis is directed by the *lasI*-encoded AI synthase (Passador *et al.*, 1993; Pearson *et al.*, 1994). The expression of *lasI* is extremely sensitive to the presence of LasR and 3O-C_{12}-HSL and this finding has led to the suggestion of the existence of a hierarchy of expression by quorum sensing-regulated genes (Seed *et al.*, 1995). This control provides an autoregulatory loop in which the expression of LasR/3O-C_{12}-HSL-dependent target genes is tightly regulated by the concentration of 3O-C_{12}-HSL present and thus by cell density.

Subsequent studies of the *las* system indicate that it actually regulates the expression of a large number of genes many of which encode virulence factors (Parsek and Greenberg, 2000; Smith and Iglewski, 2003). A key target is the *rhlR* gene, which along with the *rhlI* gene, encode homologs of LasR and LasI and form the components of the second quorum sensing circuit termed the *rhl* system (Whitehead *et al.*, 2001). The *rhlRI* genes were initially identified via their being required for the expression of *rhlAB* which encode products involved in the production of the biosurfactant/ciliotoxin known as rhamnolipid (Ochsner *et al.*, 1994; Ochsner and Reiser, 1995). Analogous to the *las* system, RhlI is the synthase responsible for the synthesis of *N*-butanoyl-L-homoserine lactone (C_4-HSL), the second major autoinducer of *P. aeruginosa* and the cognate signal for RhlR. The *rhl* system also regulates a variety of genes not the least of which is *rhlI* (Parsek and Greenberg, 2000). Hence the *rhl* system also contains an autoregulatory loop linking expression of the target genes to cell density via the presence of C_4-HSL.

In addition to regulation by the *las* system, the two systems are also linked by their regulation of at least several common target genes. Among this subset of genes are those for elastase (*lasB*), alkaline protease (*aprA*), and some components of the protein secretion apparatus (*xcpR* and *xcpP*) (Whitehead *et al.*, 2001). While the reason for this sharing of target genes is unclear, it may reflect a situation in which different environmental signals are used to preferentially stimulate a specific circuit. In that regard it must be noted that the quorum sensing circuits are also regulated by various other global regulatory factors including Vfr, the two component regulatory system GacA/LemA and the stress sigma factor RpoS (Whitehead *et al.*, 2001).

It is clear that the presence of two distinct but interconnected quorum sensing circuits provides *P. aeruginosa* with a complex network of gene regulation. This likely allows the organism to respond to a wide variety of environmental signals in addition to and perhaps concomitant with, cell density.

The importance of QS as a regulatory system is underscored by a number of studies that indicate that quorum sensing may regulate a significant percentage of all *P. aeruginosa* genes. Initial studies using transposon mutagenesis (Whiteley *et al.*, 1999) suggested that as much as 4% of the genes within the *P. aeruginosa* genome might be quorum sensing regulated. More recent studies using DNA microarray analyses (Schuster *et al.*, 2003; Wagner *et al.*, 2003) indicate that the quorum sensing regulon may actually consist of almost twice the initial estimates.

The actual number of genes that are regulated by quorum sensing mechanisms in *P. aeruginosa* may actually surpass even the most recent reports since the *las* and *rhl* systems may not be the sole quorum sensing circuits present in *P. aeruginosa*. Protein homology searches of the information encoded by the *P. aeruginosa* genome indicate that other R-protein homologs may exist. One of these, QscR, has already been described, but to date no cognate AI synthase homolog has been identified (Chugani *et al.*, 2001). As an added twist, it appears that QscR may function as a negative regulator in contrast to the positive effect of both LasR and RhlR (Chugani *et al.*, 2001). In addition, the identification of non-AHL signal molecules (Holden *et al.*, 1999; Pesci *et al.*, 1999) may provide a link between quorum sensing and other regulatory pathways.

6.1. *P. aeruginosa* QS and Anaerobiosis

While quorum sensing regulation of many genes has been described, there is only limited data with respect to genes that are expressed when *P. aeruginosa* is growing anaerobically, a condition that occurs within the thick, inspissated mucus lining the CF airways (Yoon *et al.*, 2002). The ability of the organism to adapt and grow within anoxic environments (e.g. CF mucus or biofilms) may partly account for its persistence in various infections (Costerton *et al.*, 1987; Worlitzsch *et al.*, 2002; Yoon *et al.*, 2002).

One of the first links between quorum sensing and anaerobiosis for *P. aeruginosa* was the finding that both growth phase and reduced oxygen levels were important for the synthesis of hydrogen cyanide (HCN) (Castric, 1983) which requires the products encoded by the genes of the *hcn* locus. Subsequent data indicated that LasR and RhlR could function as regulators of *hcn* expression (Pessi and Haas, 2000; Whiteley *et al.*, 1999). While *rhlR* mutants exhibited decreased HCN production, *lasR* mutants appeared to be completely unable to produce HCN (Pessi and Haas, 2000). This regulation was linked to the presence of specific elements within the promoter region of the *hcn* locus (Pessi and Haas, 2000). However the LasR and RhlR proteins are not the only factors necessary. The same study showed that the anaerobic expression of *hcn* also required the anaerobic regulatory factor ANR. As discussed earlier, it acts as a transcriptional activator of genes that function in anaerobic respiration with nitrogen oxides as the electron acceptors and also acts as an activator for the anaerobically inducible *arcDABC* operon involved in the arginine deiminase pathway (Arai *et al.*, 1997; Zimmermann *et al.*, 1991). Interestingly, whereas ANR can be sufficient for expression of other anaerobically-expressed genes it appears to function as an auxiliary factor in the case of *hcn,* whereas the quorum sensing regulators appear to play the major role (Pessi and Haas, 2000).

The biofilms formed during *P. aeruginosa* infection consist of many microenvironments including areas which are oxygen-limited (Costerton *et al.*, 1995; Worlitzsch *et al.*, 2002; Yoon *et al.*, 2002). Early studies indicated that quorum sensing plays an integral part in biofilm formation (Davies *et al.*, 1998) resulting in *P. aeruginosa* becoming an intensively studied model for biofilm development (Costerton, 2001; Hoiby *et al.*, 2001). However, most of these studies were carried out under aerobic growth conditions. To date there has been only one definitive study linking anaerobiosis, quorum sensing and biofilm development. In that study, the ability of *P. aeruginosa* and various mutants to form biofilms under anaerobic conditions was examined (Yoon *et al.*, 2002). The authors demonstrated that *P. aeruginosa* actually forms more robust biofilms anaerobically than those formed aerobically, suggesting that *P. aeruginosa* may actually prefer the anaerobic biofilm mode of growth. Interestingly, *lasR, rhlR* and *lasRrhlR* mutants were able to form very robust biofilms anaerobically. However, these biofilms were found to mainly consist of dead cells, indicating that these mutants can survive for brief periods of time but die more quickly than wild-type cells (data not shown). The mechanism of killing of anaerobic biofilm *rhlR* mutants was via a 35-fold increase in net NAR and NIR activity, resulting in an enormous overproduction of toxic NO. The paltry 2-fold induction of potentially protective NOR activity in the *rhlR* mutant was not enough to save the organism from NO-mediated death. This suggests that the *rhl* quorum sensing circuit could be a viable target for therapeutic intervention in anaerobic CF airway biofilms. However, Yoon *et al.* (Yoon *et al.*, 2002) have also shown that a mutant lacking NOR, which would predictably be dead upon exposure to anaerobic respiratory conditions, actually survives. Thus, current research in the corresponding authors laboratory is designed to address two important questions; (i) why is there such a dramatic activation of NAR and NIR activity

Table 2. Anaerobic upregulation of genes involved in bacteriophage formation in *P. aeruginosa*.

Fold Change	sig log ratio - Median	sig - T-Test P Value	Description
29.4	4.88	0.006	PA0618 /DEF=probable bacteriophage protein
26.5	4.73	0	PA0619 /DEF=probable bacteriophage protein
36.7	5.2	0.01	PA0620 /DEF=probable bacteriophage protein
21.4	4.42	0.013	PA0621 /DEF=conserved hypothetical protein
36.7	5.2	0.001	PA0622 /DEF=probable bacteriophage protein
25.5	4.67	0.01	PA0623 /DEF=probable bacteriophage protein
34.8	5.12	0.008	PA0624 /DEF=hypothetical protein
28.4	4.83	0.005	PA0625 /DEF=hypothetical protein
25.6	4.68	0	PA0626 /DEF=hypothetical protein
36.8	5.2	0.002	PA0627 /DEF=conserved hypothetical protein
39.1	5.29	0.001	PA0628 /DEF=conserved hypothetical protein
47.8	5.58	0.004	PA0629 /DEF=conserved hypothetical protein
39.9	5.32	0	PA0630 /DEF=hypothetical protein
46.2	5.53	0.006	PA0631 /DEF=hypothetical protein
68.1	6.09	0	PA0632 /DEF=hypothetical protein
28.2	4.82	0.004	PA0633 /DEF=hypothetical protein
38.9	5.28	0.016	PA0634 /DEF=hypothetical protein
58.9	5.88	0	PA0635 /DEF=hypothetical protein
50.5	5.66	0	PA0636 /DEF=hypothetical protein
35.2	5.14	0	PA0637 /DEF=conserved hypothetical protein
60.1	5.91	0	PA0638 /DEF=probable bacteriophage protein
53.4	5.74	0	PA0639 /DEF=conserved hypothetical protein
82.1	6.36	0.002	PA0640 /DEF=probable bacteriophage protein
50.2	5.65	0.001	PA0641 /DEF=probable bacteriophage protein
38.3	5.26	0.002	PA0642 /DEF=hypothetical protein
30.7	4.94	0.002	PA0643 /DEF=hypothetical protein
46.2	5.53	0.001	PA0644 /DEF=hypothetical protein
22.5	4.49	0.01	PA0645 /DEF=hypothetical protein
27.1	4.76	0.035	PA0646 /DEF=hypothetical protein
11.6	3.54	0.005	PA0647 /DEF=hypothetical protein
13.2	3.73	0.023	PA0648 /DEF=hypothetical protein
62.7	5.97	0.008	PA0720 /DEF=helix destabilizing protein of bacteriophage Pf1
58	5.86	0.005	PA0723 /GENE=*coaB* /DEF=coat protein B of bacteriophage Pf1
154.3	7.27	0.013	PA0724 /DEF=probable coat protein A of bacteriophage Pf1
-2.4	-1.29	0.083	PA3143 /DEF=hypothetical protein
-2.8	-1.5	0.002	PA3144 /DEF=hypothetical protein

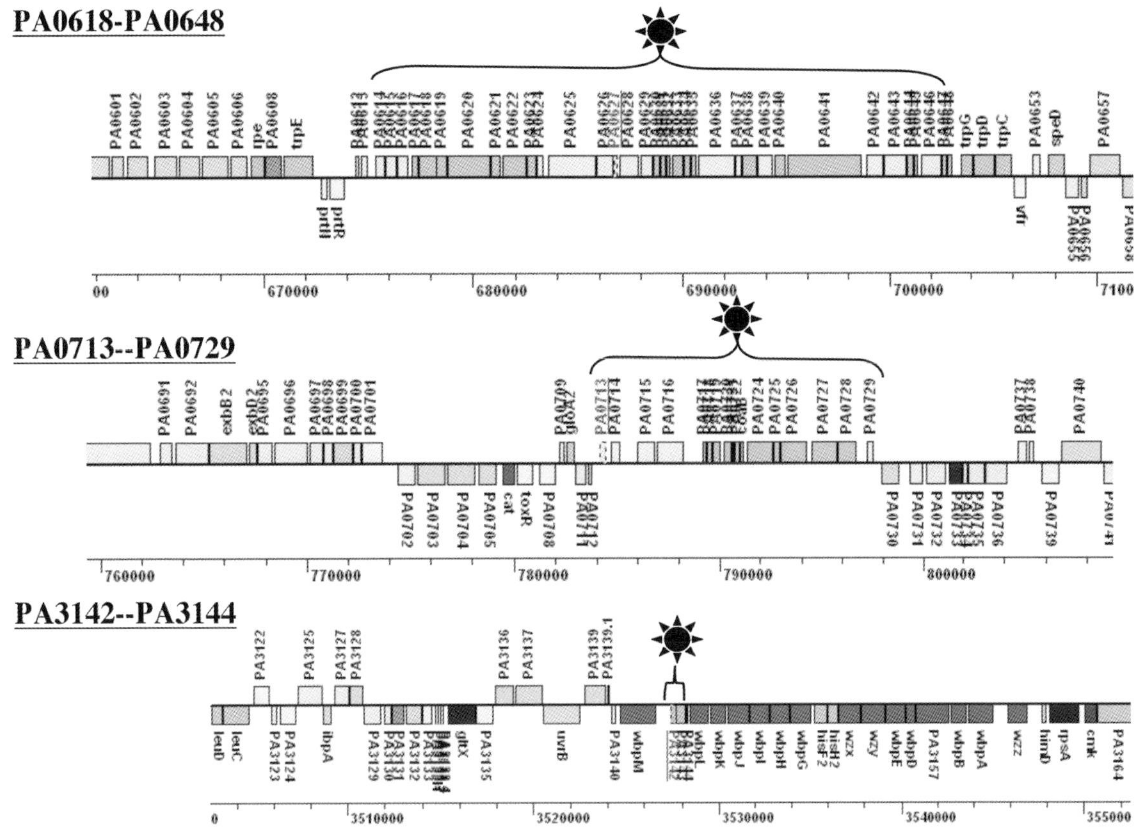

Figure 8. Location of bacteriophage genes shown in Table 2 to be upregulated during anaerobic NO_3^-/NO_2^- respiration. The bacteriophage related operons were downloaded from www.pseudomonas.com and the genes that were most upregulated are in brackets designated by a star.

in the anaerobic *rhlR* mutant leading to NO-mediated cell death?, and (ii) how can a mutant lacking NOR during anaerobic conditions survive?

Finally, a more recent study using microarrays appears to substantiate the quorum sensing regulation of many genes involved in anaerobic growth (Schuster *et al.*, 2003; Wagner *et al.*, 2003). Interestingly, one study that addresses anaerobiosis indicated that approximately 18% of the transcripts (110 transcripts of the total 616 identified) were not detectable under anaerobic growth conditions. Most recently, we have performed some aerobic vs. anaerobic microarray experiments using *P. aeruginosa* that had been grown in medium containing NO_3^- or NO_2^-. Many of the genes found to be upregulated in NO_3^- and especially NO_2^- grown organisms are bacteriophage-related genes (Table 2 and Figure 8). Interestingly, some of the same bacteriophage genes were dramatically upregulated in biofilms (Whiteley *et al.*, 2001). Thus, one feature of the complex and unknown physiology of biofilm bacteria may involve significant expression of anaerobic metabolism and bacteriophage related genes. The findings reported in the microarray studies are most exciting in that they suggest that many more genes important for anaerobic growth await discovery that may also be linked to QS.

7. Alginate Production by *P. aeruginosa* Correlates Directly with Anaerobic Metabolism

One unique feature of *P. aeruginosa* that is undeniably unique to CF is their phenotypic conversion to a mucoid, alginate-overproducing form (Figure 9). Alginate is a linear co-polymer of acetylated β-D-mannuronate and its C-5 epimer, α-L-guluronate. Arguably, the most significant virulence determinant affecting the overall clinical demise of CF patients is alginate. Alginate production is directly correlated with decreased pulmonary function (as respiratory quotient) and an increase in lung tissue damage. Chronically infected CF patients can harbor ~85-90% mucoid *P. aeruginosa* (Govan and Deretic, 1996). Once mucoid organisms are observed in CF patient sputa, they are almost never eradicated, and the mucoid phenotype is maintained. The rare instance where mucoid organisms disappear in CF is when a powerful competitor such as *B. cepacia* infects the lungs. The mechanism behind the successful takeover of *P. aeruginosa* by *B. cepacia* in the CF airways is unknown. When mucoid organisms from CF patients are grown with vigorous aeration *in vitro*, the mucoid phenotype is maintained. Similarly, when such organisms are grown under strict anaerobic conditions, mucoidy is also maintained. However, one instance where mucoid bacteria revert to their

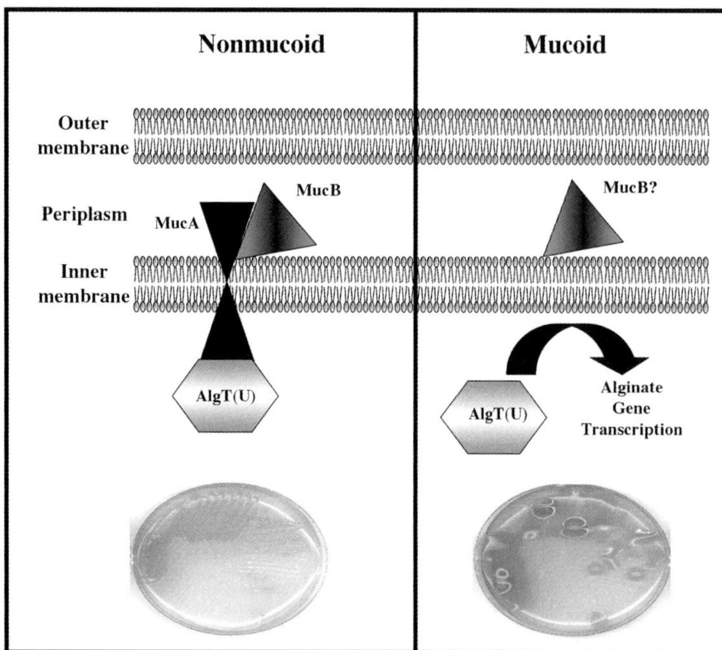

Figure 9. Mutations in the *mucA* locus leads to mucoid conversion in *P. aeruginosa*. Deretic and colleagues (Martin *et al.*, 1993b) showed that conversion to mucoidy in 84% of CF clinical isolates is mediated by mutations in the *mucA*, encoding an anti-sigma factor. In the presence of wild-type MucA, the cytoplasmic domain binds the ECF sigma factor AlgT(U) (or σ^{22}), thereby preventing mucoid conversion. However, during chronic CF lung disease, mucoid subpopulations emerge because of *mucA* mutants, leaving AlgT(U) free to transcribe alginate regulatory and biosynthetic genes. Although MucA has been shown to interact with the periplasmic domain of MucB (Xie *et al.*, 1996), the role of this interaction in the context of alginate production is unclear.

nonmucoid counterparts *in vitro* is when the bacteria are grown under static conditions in the presence of oxygen. The only obvious difference between each of the three aforementioned growth conditions is that static growth in the presence of oxygen allows the bacteria to sense an oxygen gradient. Obviously, under anaerobic conditions, there is no oxygen, and the organisms cannot sense an oxygen gradient in a rapidly agitating Erlenmeyer flask. The static scenario where organisms sense an oxygen gradient is reasonable in that significantly more ATP can be generated with increased oxygen levels. Because mucoid organisms are non-flagellated, mutations in the *algT(U)* gene, encoding an alternative sigma factor essential for alginate production, allow for flagellar gene transcription and motility toward the highly oxygenated meniscus of statically grown organisms (Garrett *et al.*, 1999). Thus, the maintenance of mucoidy in CF during the course of disease may be a result of the fact that mucoid bacteria cannot sense an oxygen gradient when enmeshed in the thick, anaerobic CF mucus. How might an oxygen gradient be created in CF to enhance mucoid reversion? One very simple treatment comes to mind that has been used successfully to kill *P. aeruginosa* biofilms *in vitro*, hydrogen peroxide (H_2O_2) (Hassett *et al.*, 1999). Direct application of H_2O_2 to the upper airway mucus would be immediately degraded to oxygen and water by the catalases of *P. aeruginosa* and that of human neutrophils present within the airway mucus. Thus, application of H_2O_2 would not only serve to hydrate the airway mucus, but also aerate it. Although H_2O_2 is mutagenic at high concentrations, it is relatively harmless at low concentrations. We are currently testing whether application of H_2O_2 to anaerobic biofilms of mucoid bacteria cause mucoid reversion. This reversion process is important in that nonmucoid bacteria are far more susceptible to current antibiotic regimens and phagocytosis by human neutrophils.

8. Role of *P. aeruginosa mucA* Mutations in Anaerobic Metabolism and Susceptibility to NO

Mucoid conversion in the vast majority of CF isolates is via mutations in *mucA*, encoding an intrinsic inner membrane anti-sigma factor (Martin *et al.*, 1993a). Once MucA is absent or incapable of binding its cognate sigma factor (Figure 9), AlgT(U), AlgT(U) is free to direct transcription of genes under its control. Recently, Deretic and colleagues have found 10 new members of the AlgT(U)-dependent "sigmulon" including two lipoprotein genes (*lptA* and *lptB*), *osmC* (osmolarity induced protein), *slyB* (putative porin), *asm* genes (A6, B2, C1, D2, D3) and *oprF* (outer membrane protein OprF) (Firoved *et al.*, 2002). Bacterial lipoproteins have been reported to be proinflammatory agents and both LptA and LptB increase production of interleukin-8 (IL-8), a cytokine whose levels are elevated in CF airway disease (Tabary *et al.*, 2001). Interestingly, one role of IL-8 is to recruit neutrophils to the site of infection, the toxic products from which can further exacerbate inflammation and tissue destruction.

How is mucoidy related to anaerobiosis? First, there is a direct correlation between pulmonary insufficiency and mucoid subpopulations. Second, mucoid reversion cannot occur during strict anaerobic conditions (Hassett, 1996) or in M9 minimal medium during aerobic growth (D. Hassett and D. Wozniak, unpublished observations). Interestingly, we recently discovered that mucoid organisms lacking MucA are exquisitely susceptible

Table 3. Effect of AlgR on transcription of genes involved in anaerobic metabolism. *Table 1 PAO1* vs. *DD*

Description	**Fold Change**	**Significance Mann Whitney**
PA2664 *fhp* flavohemoprotein	*-24.3*	*0.05*
PA0523 *norC* nitric-oxide reductase subunit C	*-18.1*	*0.05*
PA3875 *narG* respiratory nitrate reductase alpha chain	*-10.7*	*0.05*
PA0525 probable dinitrification protein NorD	*-10.7*	*0.05*
PA0524 *norB* nitric oxide reductase subunit B	*-10.3*	*0.05*
PA5170 *arcD* arginine/ornithine antiporter	*-9.3*	*0.05*
PA5171 *arcA* arginine deiminase	*-8.6*	*0.05*
PA5172 *arcB* ornithine carbamoyltransferase, catabolic	*-7.1*	*0.05*
PA1546 *hemN* oxygen-independent coproporphyrinogen III oxidase	*-6.7*	*0.05*
PA0520 *nirQ* regulatory protein	*-6.6*	*0.05*
PA1557 *probable cytochrome oxidase subunit (cbb3-type)*	*-4.1*	*0.05*
PA2194 *hcnB* hydrogen cyanide synthase	*-2.8*	*0.05*
PA3879 *narL* two-component response regulator	*-2.7*	*0.05*
PA0527 *dnr* transcriptional regulator	*-2.6*	*0.05*
PA3391 *nosR* regulatory protein	*-2.1*	*0.05*
PA4922 *azu* azurin precursor	*-1.8*	*0.05*

Table 4. Effect of AlgR on transcription of genes involved in anaerobic metabolism in a *mucB* mutant background. *PAO6857 (mucB::Tc^r)* vs. *PAR6857 (mucB::Tc^r, ΔalgR)*

PA#/Gene/Gene Product	**Fold Change**	**Significance Mann Whitney**
PA3879 *narL* two-component response regulator	-4.3	0.05
PA5170 *arcD* arginine/ornithine antiporter	-3.3	0.05
PA0524 *norB* nitric-oxide reductase subunit B	13.7	0.05
PA0523 *norC* nitric-oxide reductase subunit C	8.7	0.05
PA3391 *nosR* regulatory protein	2.8	0.05
PA0520 *nirQ* regulatory protein	2.1	0.05

to NO_2^- at pH 6.5, the pH of the CF ASL (Yoon et al., submitted). The molecular basis behind this phenomenon is a rapid activation of *nir* gene transcription by NO_2^-, overproduction of toxic NO, and little or no activation of the protective *norCB* genes. Provision of the *mucA* gene *in trans* to the susceptible strain rescues these organisms from NO-mediated death. Our laboratory is aggressively focusing on research designed to unravel the role of MucA in this process.

In nonmucoid *P. aeruginosa* strains, the cytoplasmic domain of MucA sequesters AlgT(U), preventing its association with RNA polymerase and transcription of mucoidy-related genes. However, this cytoplasmic interaction between MucA and AlgT(U) is regulated by the periplasmic region of MucA. MucB, a periplasmic protein whose structural gene is co-transcribed with *mucA*, has been shown to be critical for regulating the MucA-AlgT(U) interaction by an unknown mechanism (Rowen and Deretic, 2000). In the absence of MucB or upon truncation of the periplasmic domain of MucA (which abrogates MucB binding), MucA levels decrease significantly and AlgT(U) is released from its interaction with the MucA cytoplasmic domain, allowing transcription of mucoidy-related genes by the holo-RNA polymerase. In a related anti-sigma factor system in *E. coli*, the MucA homolog RseA is cleaved sequentially by periplasmic and inner membrane proteases, respectively, upon loss of interaction with RseB (a MucB homolog) (Ades *et al.*, 1999; 2003; Alba *et al.*, 2002). The intramembrane RseA cleavage event is thought to result in formation of a substrate for the ClpAP/ClpXP proteases, which could unfold and cleave RseA, resulting in release of σ^E and transcription of mucoidy genes.

By analogy to the RseA/RseB system, we suspect that MucB functions to stabilize the fold of the periplasmic domain of MucA and/or physically mask a proteolysis site within MucA. An intriguing possibility is that the periplasmic domain of MucA could be intrinsically unstable in the absence of MucB, leading to partial unfolding and sensitivity to periplasmic proteases. Such an intrinsically unstable periplasmic domain would be particularly sensitive to thermal denaturation, which would make sense given that AlgT(U) has been implicated in extreme heat-shock response as well as mucoid conversion of *P. aeruginosa* (Schurr and Deretic, 1997). A complete understanding of the nature of the interaction between MucA and MucB, as well as the mechanism by which MucB in the periplasmic space regulates the cytoplasmic sequestration of AlgT(U) by MucA, will require careful biophysical experiments. In particular, determination of high-resolution atomic structures of MucA and MucB and their complex, combined with careful biophysical studies of their binding affinities, assembly states, and complex stoichiometries, will lead to a better understanding of this complex system.

9. AlgR Regulation of Anaerobic Gene Expression: Another Connection Between Mucoidy and the Anaerobic Biofilm Mode of Growth

As discussed above, there is good evidence that *P. aeruginosa* may be in a microaerophilic or anaerobic microenvironment trapped within the airway mucus of CF patients (Worlitzsch *et al.*, 2002). There are also two studies demonstrating that *P. aeruginosa* is able to produce alginate under these conditions and maintain mucoidy (Hassett, 1996). Recently, we have discovered that AlgR, one of the critical response regulators responsible for activation of alginate production, may also be involved in regulating anaerobic metabolism genes in *P. aeruginosa*. We have examined the transcriptional profiles of the nonmucoid strain PAO1, a mucoid derivative of this strain PAO6857 (*mucB*::Tc^r), and *algR* deletion mutants of these strains using the Affymetrix technology. These analyses revealed that AlgR may regulate at least 16 different anaerobic metabolism genes (Tables 3 and 4). RT-PCR results confirmed the results obtained from the Affymetrix analyses from independent samples (data not shown). One might predict that the *algR* mutant would grow faster under anaerobic conditions and we did observe a slight but not significant increase in the growth rate of the *algR* mutant under anaerobic conditions. The Affymetrix data indicates that there are several genes regulated by AlgR that are involved in nitrogen metabolism but attempts to confirm these results by examining the growth of nonmucoid *P. aeruginosa* using different nitrogen sources and anaerobic conditions compared to their *algR* mutants have been inconclusive to date. This could be explained by the fact that our Affymetrix data indicates that AlgR is repressing the function of anaerobic metabolism genes under aerobic conditions but under anaerobic conditions, AlgR-dependent negative control is relieved and, therefore, has no regulatory role.

Interestingly, examination of the genes regulated by AlgR shows that AlgR acts mostly as a repressor in nonmucoid *P. aeruginosa* but as an activator in mucoid *P. aeruginosa*. This is consistent with regulation of the *algD* promoter where it is activated in mucoid cells but transcriptionally silent in nonmucoid cells. We observed the same trends with some of the anaerobic metabolism genes. For instance, in strain PAO1, *norBC*, *nirQ*, and *nosR* are repressed but these genes are activated by AlgR in mucoid cells. We have observed that PAR6857 (*mucB*::Tc^r Δ*algR*) has an extended lag phase when grown aerobically using NO_3^- as the nitrogen source, indicating that AlgR is necessary but not essential for growth when NO_3^- is the only nitrogen source. A isogenic *mucB*::Tc^r mutant grows as the wild-type strain PAO1 under the same conditions, indicating that AlgR plays a role in activation of denitrification gene transcription.

Some of the genes that AlgR regulates (*arcDABC*, *hemN*, *hcnB*, PA1557) contain either ANR binding sites within their promoters and/or potential AlgR binding sites. Also, the denitrification genes of *P. aeruginosa* are regulated by DNR and ANR. Since we obtained *narG*, *narL*, *nirQ*, *norBC*, and *nosR* on our Affymetrix list we examined the promoter regions of *anr* and *dnr* for AlgR binding sites. We found that *dnr* contains three putative consensus AlgR binding sites but no AlgR binding sites were observed within the *anr* promoter region. S1 nuclease protection assays were performed on the *dnr* promoter to map the transcriptional start sites and to examine the role of AlgR on *dnr* transcription. This analysis showed AlgR indirectly affects transcription of *dnr* approximately two-fold when *P. aeruginosa* is grown aerobically regardless of the presence of NO_3^-. Taken together, these data indicate that AlgR may regulate denitrification genes indirectly through DNR or other transcriptional regulators such as NarL or NosR.

10. Where do We Go From Here?

Because the "anaerobic model" of CF lung disease is by no means dogma in the eyes of most of the CF research community, our group is planning some very simple experiments to address definitively whether anaerobic respiration is occurring during the chronic stage of CF lung disease. First, Storey and colleagues have for many years been examining transcript levels of *P. aeruginosa* directly in sputum from CF patients (Erickson *et al.*, 2002; Storey *et al.*, 1992; 1998). Transcript levels of *nar*, *nir* and *nor* genes, as well as those of the *arcDABC* operon would indicate that anaerobic gene transcription is operable during infection. Quantitative RT-PCR can also be performed that would correlate transcript levels *in vivo* versus those expressed *in vitro* after aerobic and anaerobic growth. Furthermore, because humans do not have NAR, NIR or NOR enzymes, these activities can be measured directly from cell lysates of sputum samples or transplant pus containing high titers of *P. aeruginosa* and activity measurements correlated with colony forming units. Finally, if for some reason such enzymes are inactivated during harvest and processing, serum from CF patients can be used in quantitative Western blot analyses to assess whether the proteins are produced *in vivo* during infection. If anaerobic-specific proteins are detected by these various methods, it is our belief that new strategies need to be developed to paralyze the anaerobic respiratory machinery in *P. aeruginosa* as a supplement with current drug regimens to combat anaerobic *P. aeruginosa* biofilms CF lung disease.

Acknowledgments

This work was supported in part by grants AI-40541 and AI-55487 from the National Institutes of Health and a Cystic Fibrosis Foundation New Technology Grant to D.J.H.

References

Ades, S.E., Connolly, L.E., Alba, B.M., and Gross, C.A. 1999. The *Escherichia coli* sigma E-dependent extracytoplasmic stress response is controlled by the regulated proteolysis of an anti-sigma factor. Genes Dev. 13: 2449-2461.

Ades, S.E., Grigorova, I.L., and Gross, C.A. 2003. Regulation of the alternative sigma factor sigma E during initiation, adaptation, and shutoff of the extracytoplasmic heat shock response in *Escherichia coli*. J. Bacteriol. 185: 2512-2519.

Alba, B.M., Leeds, J.A., Onufryk, C., Lu, C.Z., and Gross, C.A. 2002. DegS and YaeL participate sequentially in the cleavage of RseA to activate the sigma E-dependent extracytoplasmic stress response. Genes Dev. 16: 2156-2168.

Arai, H., Igarashi, Y., and Kodama, T. 1995. Expression of the *nir* and *nor* genes for denitrification of *Pseudomonas aeruginosa* requires a novel CRP/FNR-related transcriptional regulator, DNR, in addition to ANR. FEBS Lett. 371: 73-76.

Arai, H., Kodama, T., and Igarashi, Y. 1997. Cascade regulation of the two CRP/FNR-related transcriptional regulators ANR and DNR and the denitrification enzymes in *Pseudomonas aeruginosa*. Mol. Microbiol. 25: 1141-1148.

Arai, H., Mizutani, M., and Igarashi, Y. 2003. Transcriptional regulation of the nos genes for nitrous oxide reductase in *Pseudomonas aeruginosa*. Microbiology. 149: 29-36.

Bartberger, M.D., Mannion, J.D., Powell, S.C., Stamler, J.S., Houk, K.N., and Toone, E.J. 2001. S-N Dissociation energies of S-nitrosothiols: on the origins of nitrosothiol decomposition rates. J. Am. Chem. Soc. 123: 8868-8869.

Briviba, K., Kissner, R., Koppenol, W.H., and Sies, H. 1998. Kinetic-study of the reaction of glutathione-peroxidase with peroxynitrite. Chem. Res. Toxicol. 11: 1398-1401.

Castric, P.A. 1983. Hydrogen cyanide production by *Pseudomonas aeruginosa* at reduced oxygen levels. Can. J. Microbiol. 29: 1344-1349.

Chugani, S.A., Whiteley, M., Lee, K.M., D'Argenio, D., Manoil, C., and Greenberg, E.P. 2001. QscR, a modulator of quorum-sensing signal synthesis and virulence in *Pseudomonas aeruginosa*. Proc. Natl. Acad. Sci. USA. 98: 2752-2757.

Coddington, J.W., Hurst, J.K., and Lymar, S.V. 1999. Hydroxyl radical formation during peroxynitrous acid decomposition. J. Am. Chem. Soc. 121: 2438-2443.

Costerton, J.W., Cheng, K.-J., Geesey, G.G., Ladd, T.I., Nickel, J.C., Dasgupta, M., and Marrie, T.J. 1987. Bacterial biofilms in nature and disease. Annu. Rev. Microbiol. 41: 435-464.

Costerton, J.W., Lewandowski, Z., Caldwell, D.E., Korber, D.R., and Lappin-Scott, H.M. 1995. Microbial biofilms. Annu. Rev. Microbiol. 49: 711-745.

Costerton, J.W. 2001. Cystic fibrosis pathogenesis and the role of biofilms in persistent infection. Trends Microbiol. 9: 50-52.

Davies, D.G., Parsek, M.R., Pearson, J.P., Iglewski, B.H., Costerton, J.W., and Greenberg, E.P. 1998. The involvement of cell-to-cell signals in the development of a bacterial biofilm. Science. 280: 295-298.

Erickson, D.L., Endersby, R., Kirkham, A., Stuber, K., Vollman, D.D., Rabin, H.R., Mitchell, I., and Storey, D.G. 2002 *Pseudomonas aeruginosa* quorum-sensing systems may control virulence factor expression in the lungs of patients with cystic fibrosis. Infect. Immun. 70: 1783-1790.

Finelli, A., Gallant, C.V., Jarvi, K., and Burrows, L.L. 2003. Use of in-biofilm expression technology to identify genes involved in *Pseudomonas aeruginosa* biofilm development. J. Bacteriol. 185: 2700-2710.

Firoved, A.M., Boucher, J.C., and Deretic, V. 2002. Global genomic analysis of AlgU sigmaE-dependent promoters sigmulon in *Pseudomonas aeruginosa* and implications for inflammatory processes in cystic fibrosis. J. Bacteriol. 184: 1057-1064.

Ford, E., Hughes, M.N., and Wardman, P. 2002. Kinetics of the reactions of nitrogen dioxide with glutathione, cysteine, and uric acid at physiological pH. Free Radic. Biol. Med. 32: 1314-1323.

Ford, P.C., and Lorkovic, I.M. 2002. Mechanistic aspects of the reactions of nitric oxide with transition-metal complexes. Chem. Rev. 102: 993-1017.

Galimand, M., Gamper, M., Zimmermann, A., and Hass, D. 1991. Positive FNR-like control of anaerobic arginine degradation and nitrate respiration in *Pseudomonas aeruginosa*. J. Bacteriol. 173: 1598-1606.

Gambello, M.J., and Iglewski, B.H. 1991. Cloning and characterization of the *Pseudomonas aeruginosa lasR* gene, a transciptional activator of elastase expression. J. Bacteriol. 173: 3000-3009.

Gamper, M., Zimmermann, A., and Haas, D. 1991. Anaerobic regulation of transcription initiation in the *arcDABC* operon of *Pseudomonas aeruginosa*. J. Bacteriol. 173: 4742-4750.

Gardner, P.R., Gardner, A.M., Martin, L.A., and Salzman, A.L. 1998. Nitric oxide dioxygenase: an enzymic function for flavohemoglobin. Proc. Natl. Acad. Sci. USA. 95: 10378-10383.

Garrett, E.S., Perlegas, D., and Wozniak, D.J. 1999. Negative control of flagellum synthesis in *Pseudomonas aeruginosa* is modulated by the alternative sigma factor AlgT (AlgU). J. Bacteriol. 181: 7401-7404.

Gerasimov, O.V., and Lymar, S.V. 1999. The yield of hydroxyl radical from the decomposition of peroxynitrous acid. Inorg. Chem. 38: 4317-4321.

Goldstein, S., and Czapski, G. 1995. The reaction of NO. with O2.- and HO2.: a pulse radiolysis study. Free Radic. Biol. Med. 19: 505-510.

Goldstein, S., and Czapski, G. 1996. Mechanism of the nitrosation of thiols and amines by oxygenated NO solutions: the nature of the nitrosating intermediates. J. Am. Chem. Soc. 118: 3419-3425.

Govan, J.R.W., and Deretic, V. 1996. Microbial pathogenesis in cystic fibrosis: mucoid *Pseudomonas aeruginosa* and *Burkholderia cepacia*. Microbiol. Rev. 60: 539-574.

Hassett, D.J. 1996. Anaerobic production of alginate by *Pseudomonas aeruginosa*: alginate restricts diffusion of oxygen. J. Bacteriol. 178: 7322-7325.

Hassett, D.J., Ma, J.-F., Elkins, J.G., McDermott, T.R., Ochsner, U.A., West, S.E.H., Huang, C.-T., Fredericks, J., Burnett, S., Stewart, P.S., McPheters, G., Passador, L., and Iglewski, B.H. 1999. Quorum sensing in *Pseudomonas aeruginosa* controls expression of catalase and superoxide dismutase genes and mediates biofilm susceptibility to hydrogen peroxide. Mol. Microbiol. 34: 1082-1093.

Hassett, D.J., Cuppoletti, J., Trapnell, B., Lymar, S.V., Rowe, J.J., Sun Yoon, S., Hilliard, G.M., Parvatiyar, K., Kamani, M.C., Wozniak, D.J., Hwang, S.H., McDermott, T.R., and Ochsner, U.A. 2002. Anaerobic metabolism and quorum sensing by *Pseudomonas aeruginosa* biofilms in chronically infected cystic fibrosis airways: rethinking antibiotic treatment strategies and drug targets. Adv. Drug Deliv. Rev. 54: 1425-1443.

Hernandez, D., and Rowe, J.J. 1987. Oxygen regulation of nitrate uptake in denitrifying *Pseudomonas aeruginosa*. Appl. Environ. Microbiol. 53: 745-750.

Hernandez, D., and Rowe, J.J. 1988. Oxygen inhibition of nitrate uptake is a general regulatory mechanism in nitrate respiration. J. Biol. Chem. 263: 7937-7939.

Hernandez, D., Dias, F.M., and Rowe, J.J. 1991. Nitrate transport and its regulation by O_2 in *Pseudomonas aeruginosa*. Arch. Biochem. Biophys. 286: 159-163.

Hodges, G.R., and Ingold, K.U. 1999. Cage-escape of geminate radical pairs can produce peroxynitrate from peroxynitrite under a wide variety of experimental conditions. J. Am. Chem. Soc. 121: 10695-10701.

Hoiby, N., Krogh Johansen, H., Moser, C., Song, Z., Ciofu, O., and Kharazmi, A. 2001. *Pseudomonas aeruginosa* and the *in vitro* and *in vivo* biofilm mode of growth. Microbes Infect. 3: 23-35.

Holden, M.T., Ram Chhabra, S., de Nys, R., Stead, P., Bainton, N.J., Hill, P.J., Manefield, M., Kumar, N., Labatte, M., England, D., Rice, S., Givskov, M., Salmond, G.P., Stewart, G.S., Bycroft, B.W., Kjelleberg, S., and Williams, P. 1999. Quorum-sensing cross talk: isolation and chemical characterization of cyclic dipeptides from *Pseudomonas aeruginosa* and other gram-negative bacteria. Mol. Microbiol. 33: 1254-1266.

Huie, R.E., and Padmaja, S. 1993. The reaction of no with superoxide. Free Radic. Res. Commun. 18: 195-199.

Hurst, J.K., and Lymar, S.V. 1997. Toxicity of peroxynitrite and related reactive nitrogen species toward *Escherichia coli*. Chem. Res. Toxicol. 10: 802-810.

Hurst, J.K., and Lymar, S.V. 1999. Cellularly generated inorganic oxidants as natural microbicidal agents. Acc. Chem. Res. 32: 520-528.

Kobayashi, K., Miki, M., and Tagawa, S. 1995. Pulse-radiolysis study of the reaction of nitric oxide with superoxide. J. Chem. Soc., Dalton Trans. 17: 2885-2889.

Krieger, R., Rompf, A., Schobert, M., and Jahn, D. 2002. The *Pseudomonas aeruginosa hemA* promoter is regulated by Anr, Dnr, NarL and Integration Host Factor. Mol. Genet. Genomics. 267: 409-417.

Liochev, S.I., and Fridovich, I. 2003. The mode of decomposition of Angeli's salt $Na_2N_2O_3$ and the effects thereon of oxygen, nitrite, superoxide dismutase, and glutathione. Free Radic. Biol. Med. 34: 1399-1404.

Lu, C.D., Winteler, H., Abdelal, A., and Haas, D. 1999. The ArgR regulatory protein, a helper to the anaerobic regulator ANR during transcriptional activation of the *arcD* promoter in *Pseudomonas aeruginosa*. J. Bacteriol. 181: 2459-2464.

Lymar, S.V., and Hurst, J.K. 1995. Rapid reaction between peroxonitrite ion and carbon dioxide: Implications for biological activity. J. Am. Chem. Soc. 117: 8867-8868.

Lymar, S.V., and Hurst, J.K. 1996. Carbon dioxide: Physiological catalyst for peroxynitrite- mediated cellular damage or cellular protectant? Chem. Res. Toxicol. 9: 845.

Lymar, S.V., and Hurst, J.K. 1998. CO_2-catalyzed one-electron oxidations by peroxynitrite: properties of the reactive intermediate. Inorg. Chem. 37: 294-301.

Lymar, S.V., Khairutdinov, R.F., and Hurst, J.K. 2003a. Hydroxyl radical formation by O-O bond homolysis in peroxynitrous acid. Inorg. Chem. 42: 5259-5266.

Lymar, S.V., Khairutdinov, R.F., and Hurst, J.K. 2003b. Hydroxyl radical formation by O-O bond homolysis in peroxynitrous acid. Inorg. Chem. 42: 5259-5266.

Lymar, S.V., and Poskrebyshev, G.A. 2003. Rate of $ON-OO^-$ bond homolysis and the Gibbs energy of formation of peroxynitrite. J. Phys. Chem. A 107: 7991-7996.

Martin, D.W., Holloway, B.W., and Deretic, V. 1993a. Characterization of a locus determining the mucoid status of *Pseudomonas aeruginosa*: AlgU shows sequence similarities with a *Bacillus* sigma factor. J. Bacteriol. 175: 1153-1164.

Martin, D.W., Schurr, M.J., Mudd, M.H., and Deretic, V. 1993b. Mechanism of conversion to mucoidy in *Pseudomonas aeruginosa* infecting cystic fibrosis patients. Proc. Natl. Acad. Sci. USA. 90: 8377-8381.

Mercenier, A., Simon, J.-P., Vander Wauven, C., Haas, D., and Stalon, V. 1980. Regulation of enzyme synthesis in the arginine deiminase pathway of *Pseudomonas aeruginosa*. J. Bacteriol. 144: 159-163.

Merényi, G., and Lind, J. 1998. Free radical formation in the peroxynitrous acid ONOOH peroxynitrite ONOO- system. Chem. Res. Toxicol. 11: 243-246.

Merényi, G., Lind, J., Goldstein, S., and Czapski, G. 1999. Mechanism and thermochemistry of peroxynitrite decomposition in water. J. Phys. Chem. A 103: 5685-5691.

Miranda, K.M., Paolocci, N., Katori, T., Thomas, D.D., Ford, E., Bartberger, M.D., Espey, M.G., Kass, D.A., Feelisch, M., Fukuto, J.M., and Wink, D.A. 2003. A biochemical rationale for the discrete behavior of nitroxyl and nitric oxide in the cardiovascular system. Proc. Natl. Acad. Sci. USA. 100: 9196-9201.

O'Toole, G.A., and Kolter, R. 1998. Flagellar and twitching motility are necessary for *Pseudomonas aeruginosa* biofilm development. Mol. Microbiol. 30: 295-304.

Ochsner, U.A., A.K., K., Fiechter, A., and Reiser, J. 1994. Isolation and characterization of a regulatory gene affecting rhamnolipid biosurfactant synthesis in *Pseudomonas aeruginosa*. J. Bacteriol. 176: 2044-2054.

Ochsner, U.A., and Reiser, J. 1995. Autoinducer-mediated regulation of rhamnolipid biosurfactant synthesis in *Pseudomonas aeruginosa*. Proc. Natl. Acad. Sci. USA. 92: 6424-6428.

Park, J.-Y., and Lee, Y.-N. 1988. Solubility and decomposition kinetics of nitrous acid in aqueous solution. J. Phys. Chem. A 92: 6294-6302.

Parsek, M.R., and Greenberg, E.P. 2000. Acyl-homoserine lactone quorum sensing in gram-negative bacteria: a signaling mechanism involved in associations with higher organisms. Proc. Natl. Acad. Sci. USA. 97: 8789-8793.

Passador, L., Cook, J.M., Gambello, M.J., Rust, L., and Iglewski, B.H. 1993. Expression of *Pseudomonas aeruginosa* virulence genes requires cell-to-cell communication. Science. 260: 1127-1130.

Pearson, J.P., Gray, K.M., Passador, L., Tucker, K.D., Eberhard, A., Igkewski, B.H., and Greenberg, E.P. 1994. Structure of the autoinducer required for expression of *Pseudomonas aeruginosa* virulence genes. Proc. Natl. Acad. Sci. USA. 91: 197-201.

Pearson, J.P., Van Delden, C., and Iglewski, B.H. 1999. Active efflux and diffusion are involved in transport of *Pseudomonas aeruginosa* cell-to-cell signals. J. Bacteriol. 181: 1203-1210.

Pesci, E.C., Milbank, J.B., Pearson, J.P., McKnight, S., Kende, A.S., Greenberg, E.P., and Iglewski, B.H. 1999. Quinolone signaling in the cell-to-cell communication system of *Pseudomonas aeruginosa*. Proc. Natl. Acad. Sci. USA. 96: 11229-11234.

Pessi, G., and Haas, D. 2000. Transcriptional control of the hydrogen cyanide biosynthetic genes *hcnABC* by the anaerobic regulator ANR and the quorum-sensing regulators LasR and RhlR in *Pseudomonas aeruginosa*. J. Bacteriol. 182: 6940-6949.

Radi, R. 1996. Reactions of nitric oxide with metalloproteins. Chem. Res. Toxicol. 9: 828-835.

Rompf, A., Hungerer, C., Hoffmann, T., Lindenmeyer, M., Romling, U., Gross, U., Doss, M.O., Arai, H., Igarashi, Y., and Jahn, D. 1998. Regulation of *Pseudomonas aeruginosa hemF* and *hemN* by the dual action of the redox response regulators Anr and Dnr. Mol. Microbiol. 29: 985-997.

Rowe, J.J., Ubbink-Kok, T., Molenaar, D., Konings, W.N., and Driessen, A.J. 1994. NarK is a nitrite-extrusion system involved in anaerobic nitrate respiration by *Escherichia coli*. Mol. Microbiol. 12: 579-586.

Rowen, D.W., and Deretic, V. 2000. Membrane-to-cytosol redistribution of ECF sigma factor AlgU and conversion to mucoidy in *Pseudomonas aeruginosa* isolates from cystic fibrosis patients. Mol. Microbiol. 36: 314-327.

Schurr, M.J., and Deretic, V. 1997. Microbial pathogenesis in cystic fibrosis: co-ordinate regulation of heat-shock response and conversion to mucoidy in *Pseudomonas aeruginosa*. Mol. Microbiol. 24: 411-420.

Schuster, M., Lostroh, C.P., Ogi, T., and Greenberg, E.P. 2003. Identification, timing, and signal specificity of *Pseudomonas aeruginosa* quorum-controlled genes: a transcriptome analysis. J. Bacteriol. 185: 2066-2079.

Seed, P.C., Passador, L., and Iglewski, B.H. 1995. Activation of the *Pseudomonas aeruginosa lasI* gene by LasR and the *Pseudomonas* autoinducer PAI: an autoinduction regulatory hierarchy. J. Bacteriol. 177: 654-659.

Shafirovich, V., and Lymar, S.V. 2002. Nitroxyl and its anion in aqueous solutions: Spin states, protic equilibria, and reactivities toward oxygen and nitric oxide. Proc. Natl. Acad. Sci. USA. 99: 7340-7345.

Shafirovich, V., and Lymar, S.V. 2003. Spin-forbidden deprotonation of aqueous nitroxyl HNO. J. Am. Chem. Soc. 125: 6547-6552.

Singh, P.K., Schaefer, A.L., Parsek, M.R., Moninger, T.O., Welsh, M.J., and Greenberg, E.P. 2000. Quorum-sensing signals indicate that cystic fibrosis lungs are infected with bacterial biofilms. Nature. 407: 762-764.

Smith, R.S., and Iglewski, B.H. 2003. *P. aeruginosa* quorum-sensing systems and virulence. Curr. Opin. Microbiol. 6: 56-60.

Storey, D.G., Ujack, E.E., and Rabin, H.R. 1992. Population transcript accumulation of *Pseudomonas aeruginosa* exotoxin A and elastase in sputa from patients with cystic fibrosis. Infect. Immun. 60: 4687-4694.

Storey, D.G., Ujack, E.E., Rabin, H.R., and Mitchell, I. 1998. *Pseudomonas aeruginosa lasR* transcription correlates with the transcription of *lasA*, *lasB*, and *toxA* in chronic lung infections associated with cystic fibrosis. Infect. Immun. 66: 2521-2528.

Tabary, O., Escotte, S., Couetil, J.P., Hubert, D., Dusser, D., Puchelle, E., and Jacquot, J. 2001. Relationship between IkappaBalpha deficiency, NFkappaB activity and interleukin-8 production in CF human airway epithelial cells. Pflugers Arch. 443: S40-44.

Toder, D.S., Gambello, M.J., and Iglewski, B.H. 1991. *Pseudomonas aeruginosa* LasA: a second elastase under the transcriptional control of *lasR*. Mol. Microbiol. 5: 2003-2010.

Vander Wauven, C., Pierard, A., Kley-Raymann, M., and Haas, D. 1984. *Pseudomonas aeruginosa* mutants affected in anaerobic growth on arginine: evidence for a four-gene cluster encoding the arginine deiminase pathway. J. Bacteriol. 160: 928-934.

Verhoogt, H.J., Smit, H., Abee, T., Gamper, M., Driessen, A.J., Haas, D., and Konings, W.N. 1992. *arcD*, the first gene of the arc operon for anaerobic arginine catabolism in *Pseudomonas aeruginosa*, encodes an arginine-ornithine exchanger. J. Bacteriol. 174: 1568-1573.

Viner, R.I., Williams, T.D., and Schoneich, C. 1999. Peroxynitrite modification of protein thiols: Oxidation, nitrosylation, and S-glutathiolation of functionally important cysteine residues in the sarcoplasmic reticulum Ca-ATPase. Biochemistry. 38: 12408-12415.

Wagner, V.E., Bushnell, D., Passador, L., Brooks, A.I., and Iglewski, B.H. 2003. Microarray analysis of *Pseudomonas aeruginosa* quorum-sensing regulons: effects of growth phase and environment. J. Bacteriol. 185: 2080-2095.

Wahllander, A., Soboll, S., Sies, H., Linke, I., and Muller, M. 1979. Hepatic mitochondrial and cytosolic glutathione content and the subcellular-distribution of GSH-S-transferases. FEBS Lett. 97: 138-140.

Whitehead, N.A., Barnard, A.M., Slater, H., Simpson, N.J., and Salmond, G.P. 2001. Quorum-sensing in Gram-negative bacteria. FEMS Microbiol. Rev. 25: 365-404.

Whiteley, M., Lee, K.M., and Greenberg, E.P. 1999. Identification of genes controlled by quorum sensing in *Pseudomonas aeruginosa*. Proc. Natl. Acad. Sci. USA. 96: 13904-13909.

Whiteley, M., Bangera, M.G., Bumgarner, R.E., Parsek, M.R., Teitzel, G.M., Lory, S., and Greenberg, E.P. 2001. Gene expression in *Pseudomonas aeruginosa* biofilms. Nature. 413: 860-864.

Williams, D.L.H. 1999. The Chemistry of S-Nitrosothiols. Acc. Chem. Res. 32: 869-876.

Wink, D.A., and Mitchel, J.B. 1998. Chemical biology of nitric oxide: insights into regulatory, cytotoxic, and cytoprotective mechanisms of nitric oxide. Free Radic. Biol. Med. 25: 434-456.

Worlitzsch, D., Tarran, R., Ulrich, M., Schwab, U., Cekici, A., Meyer, K.C., Birrer, P., Bellon, G., Berger, J., Wei, T., Botzenhart, K., Yankaskas, J.R., Randell, S., Boucher, R.C., and Doring, G. 2002. Reduced oxygen concentrations in airway mucus contribute to the early and late pathogenesis of *Pseudomonas aeruginosa* cystic fibrosis airway infection. J. Clin. Invest. 109: 317-325.

Wyckoff, T.J., Thomas, B., Hassett, D.J., and Wozniak, D.J. 2002. Static growth of mucoid *Pseudomonas aeruginosa* selects for non-mucoid variants that have acquired flagellum-dependent motility. Microbiology. 148: 3423-3430.

Xie, Z.-D., Hershberger, C.D., Shankar, S., Ye, R.W., and Chakrabarty, A.M. 1996. Sigma factor-anti sigma factor interaction in alginate synthesis: inhibition of AlgT by MucA. J. Bacteriol. 178: 4990-4996.

Ye, R.W., Haas, D., Ka, J.-O., Krishnalillai, V., Zimerman, A., Baird, C., and Tiedje, J.M. 1995. Anaerobic activation of the entire denitrification pathway in *Pseudomonas aeruginosa* requires Anr, an analog of Fnr. J. Bacteriol. 177: 3606-3609.

Yoon, S.S., Hennigan, R.F., Hilliard, G.M., Ochsner, U.A., Parvatiyar, K., Kamani, M.C., Allen, H.L., DeKievit, T.R., Gardner, P.R., Schwab, U., Rowe, J.J., Iglewski, B.H., McDermott, T.R., Mason, R.P., Wozniak, D.J., Hancock, R.E., Parsek, M.R., Noah, T.L., Boucher, R.C., and Hassett, D.J. 2002. *Pseudomonas aeruginosa* anaerobic respiration in biofilms: Relationships to cystic fibrosis pathogenesis. Dev. Cell. 3: 593-603.

Zhao, R., Lind, J., Merényi, G., and Eriksen, T.E. 1994. Kinetics of one-electron oxidation of thiols and hydrogen abstraction by thiyl radicals from alpha-amino C-H Bonds. J. Am. Chem. Soc. 116: 12010-12015.

Zimmermann, A., Reimmann, C., Galimand, M., and Haas, D. 1991. Anaerobic growth and cyanide synthesis of *Pseudomonas aeruginosa* depend on *anr*, a regulatory gene homologous with fnr of *Escherichia coli*. Mol. Microbiol. 5: 1483-1490.

Chapter 5

Oral Microbial Communities: Genetic Analysis of Oral Biofilms

Howard K. Kuramitsu*

Abstract

Dental plaque represents one of the most complex bacterial biofilms that exist in nature. These bacterial communities represent an ideal system to investigate the interactions between different members of a heterogeneous population. The utilization of both molecular genetic as well as various microscopic approaches has suggested that biofilm formation represents a genetically regulated developmental program. Genes involved in attachment to inert surfaces, extracellular polysaccharide synthesis, quorum sensing, detachment, as well as those involved in microcolony interactions have all been demonstrated to play a role in biofilm development in bacteria. Some of these genes have also been shown to be important in biofilm development by both gram-positive facultative anaerobes involved in supragingival plaque formation and gram-negative obligate anaerobic bacteria present in subgingival plaque. In addition, evidence for cell-cell communication by means of gene transfer as well as signaling molecules has also been demonstrated. This information may be useful in designing new strategies to regulate dental plaque formation and subsequently dental caries and periodontitis.

1. Introduction-Dental Plaque

Microbiologists have examined the physiology of bacteria both as model cellular systems as well as for its own intrinsic scientific value. This endeavor has basically focused on mimicking the natural environments of these organisms as much as possible in the laboratory. However, it has now become abundantly clear that the standard procedure of growing up bacteria in liquid broth cultures does not duplicate the normal physiological state of many bacteria found in our environment. Instead, many of these organisms grow in heterogeneous bacterial communities adherent to solid surfaces in structures which are now termed biofilms (Costerton *et al.*, 1995). Therefore, an examination of the biology of biofilms may be more relevant for our understanding of "real time" bacteria than examination of broth-grown organisms. This information is also very relevant to infectious diseases and industrial processes where biofilms are particular problems.

The earliest recorded observations of biofilms were carried out by van Leeuwenhoek using scrapings from his own teeth observed under one of his newly constructed microscopes. These biofilms associated with human teeth later were termed dental plaque and are responsible for two of the most common infectious diseases afflicting mankind: dental caries and periodontal diseases (gingivitis and various forms of periodontitis). Though neither of these conditions is considered to be life threatening (however; see 6. below), they do cause billions of dollars of damage worldwide and extensive suffering. Despite the fact that improved oral hygiene together with the introduction of flouridation has significantly reduced the incidence of dental caries in the developed world, there remains a large population of individuals who still suffer from this disease (Oral Health US, 2002). In addition, the increased retention of teeth by adult populations due to the reduction in dental caries also will likely increase the incidence of periodontal diseases as well as root surface caries. Thus, information leading to the control of dental plaque formation would have significant impact on the quality of life for many individuals. Although a multitude of organisms are found in human saliva, the constant swallowing of this bacterial milieu prevents the formation of climax communities in this important biological fluid. Stable growing communities in the oral cavity are found attached to soft tissues or to teeth (dental plaque) (Figure 1). As documented below, since dental plaque is a common form of human biofilms, a better understanding of the molecular basis for biofilm formation should improve our efforts to further reduce the incidence of oral diseases.

Since dental caries primarily affects the tooth surfaces above the gingival margin while gingivitis and periodontitis affect the soft and hard tissues below this region, it is not surprising that both diseases have different etiologies. Dental plaque located on the smooth, approximal (between teeth), or chewing surfaces of

*For correspondence email: Kuramits@buffalo.edu

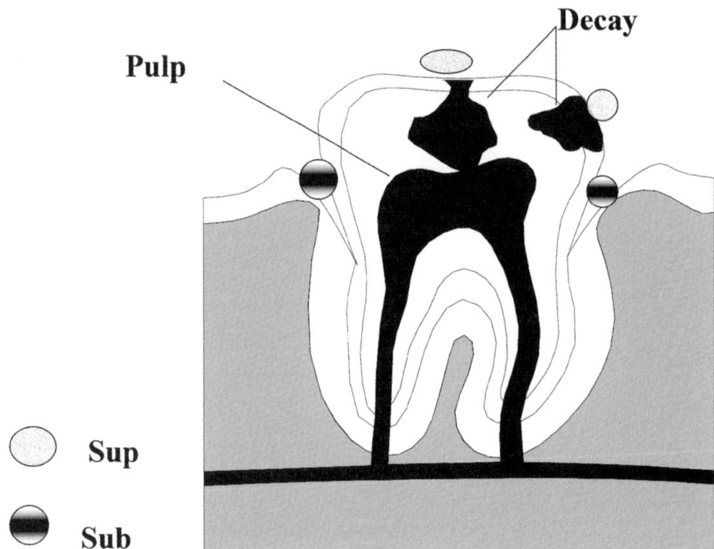

Figure 1. Dental plaque formation. Sup, supragingival plaque located on chewing surfaces, between teeth and on smooth surfaces. Dark regions beneath plaque indicate dental decay and invasion reaching the dental pulp region. Sub, subgingival plaque located between the gingival epithelium and root surfaces of teeth which can lead to gingival tissue destruction and ultimately bone resorbtion.

teeth is referred to as supragingival plaque (Figure 1). Likewise, plaque observed at or below the gum line is termed subgingival plaque and is associated with periodontal diseases. The composition of dental plaque associated with these two surfaces of teeth is quite distinct (Table 1). Supragingival plaque is composed primarily of Gram-positive facultative anaerobic organisms, predominantly belonging to the genus *Streptococcus*. Among these organisms, *S. mutans* (and to a lesser extent another member of the mutans streptococcal group *S. sobrinus*) have been strongly implicated in the etiology of dental caries (Loesche, 1986). However, other studies (van Houte *et al*., 1996) have suggested that nonmutans streptococci may also play a role in disease initiation. Evidence is beginning to emerge that the nonmutans streptococci are not "innocent bystanders" in the disease process initiated by supragingival plaque but may also affect the caries potential of these biofilms (Kleinberg, 2002).

Evidence has been obtained that the early colonizing streptococci are transmitted from mother to child. Several approaches have documented such a mode of transmission for *S. mutans* (Caufield *et al*., 1993). Since these organisms require teeth for optimal colonization of the oral cavity, it is not surprising that a "window of opportunity" for colonization by these organisms occurs around the age of eighteen months when significant tooth eruption has occurred. Interestingly, evidence has also been presented that for another group of oral streptococci, *S. sanguinis* (previously named *S. sanguis*), an earlier window of colonization occurs in infants at around nine months-the period when teeth begin to erupt (Caufield *et al*., 2000). Furthermore, there appears to be an antagonistic relationship between *S. mutans* and *S. sanguinis*, ie., children with high levels of one of these organisms appear to be poorly colonized by the other. However, the molecular basis for such antagonism still remains unclear.

Table 1. Bacterial species isolated from dental plaque[a]

Supragingival	Subgingival
Streptococcus sanguinis	*Fusobacterium nucleatum*
Streptococcus gordonii	*Treponema denticola*
Streptococcus mitior	*Porphyromonas gingivalis*
Streptococcus milleri	*Actinobacillus actinomycetemcomitans*
Streptococcus sobrinus	*Tannerella forsythensis*
Streptococcus oralis	*Eikenella corrodens*
Streptococcus mutans	*Campylobacter rectus*
Actinomyces naeslundii	Unknown spirochetes
Veillonella spp.	*Prevotella intermedia*

[a] The composition of plaque can vary and individual organisms may be found in both types of plaque samples. For a more comprehensive listing of bacterial species identified in dental plaque please see Becker *et al*., 2002; Socransky *et al*., 1998.

The composition of subgingival plaque is not only distinct from that of supragingival plaque but also changes depending upon the health status of the gingival margin. During the earlier stages of subgingival plaque formation, gram-positive bacteria (similar to those found in supragingival plaque) appear to be the predominant organisms. This type of plaque is associated with the milder forms of periodontal diseases known as gingivitis and is quite widespread. However, if not controlled, the plaque can be converted into one containing primarily gram-negative anaerobic organisms (Table 1) which is associated with more severe forms of periodontitis (Darveau *et al.,* 2000). Little information is currently available to explain how a predominantly gram-positive facultative anaerobic biofilm is converted to one containing primarily gram-negative obligate anaerobic bacteria. This may occur as the result of inflammation induced by the early colonizing bacteria leading to increased crevicular flow and bleeding which in turn might stimulate the growth of more anaerobic gram-negative organisms.

Several studies have implicated *Porphyromonas gingivalis, Tannerella forsythensis* (previously named *Bacteroides forsythus*), and the oral spirochete *Treponema denticola* as the primary organisms associated with periodontitis (Socransky *et al.,* 1998). In addition, other gram-negative anaerobic bacteria have also been implicated in these polymicrobial diseases (Kumar *et al.,* 2003). Likewise, there is an association between particular anaerobic organisms and specific forms of periodontitis (ie., *Actinobacillus actinomycetemcomitans* in localized aggressive periodontitis). Since low levels of some of these periodontopathogens can be detected at healthy sites in the gingival margin as part of the endogenous oral flora it has been suggested that these organisms are normally present at low levels and proliferate under conditions of "ecological catastrophes" (Marsh, 2003). There is also some evidence that these organisms may be transmitted by close contact (care-giver to child transmission) (Tanner *et al.,* 2002). The predominance of obligate anaerobic organisms in mature subgingival plaque reflects the relative anaerobiosis of biofilms in the deep pockets between the gingival epithelium and the teeth. Therefore, it is clear that anaerobic bacteria, both facultative and obligate, are involved in the two major oral diseases, dental caries and periodontitis.

2. Biofilm Formation

2.1. Structure of Biofilms

Biofilms which exist in nature can be classified into two major groups: homogeneous and heterogeneous biofilms. The former structures are observed on surgically implanted devices and in some diseased tissue (Donlan and Costerton, 2002). Complex biofilms are more common in nature and exhibit varying degrees of heterogeneity. Dental plaque, which can be composed of hundreds of bacterial species (Kolenbrander, 2000), may represent one of the most complex types of naturally occurring biofilms. Although much of the recent emphasis has been directed toward examining homogeneous biofilms *in vitro,* attempts to investigate more heterogeneous biofilms have also been initiated (Kinniment *et al.,* 1996).

The utilization of both scanning electron microscopy (SEM) and confocal laser scanning microscopy (CLSM) has revealed that homogeneous biofilms do not consist merely of monolayers of bacteria piled on top of each other on inert surfaces. Instead, these studies have revealed that biofilms can consist of three-dimensional structures involving "mushroom-like" columns separated by aqueous channels (Costerton *et al.,* 1995). The later may serve as a means of allowing for nutrient uptake and waste removal from biofilms. The presence of such structures depends, in part, on the relative flow-rate of the media which bathe the biofilms (Purevdorj *et al.,* 2002). However, it is not yet clear if all organisms are capable of producing similar appearing biofilms *in vitro* or *in vivo*. The columnar nature of biofilms appears to be dependent upon the synthesis of exopolysaccharides by some of these biofilm-forming organisms (Bomchil *et al.,* 2003).

For several decades the important role of extracellular polysaccharides (EPS) in biofilm formation has been recognized for the supragingival plaque associated with dental caries (Loesche, 1986). The synthesis of water-insoluble glucan polymers from dietary sucrose by *S. mutans* is recognized as an important factor in the accumulation of these organisms on tooth surfaces. These polysaccharides can also entrap other bacteria into the plaque matrix. Glucans, as well as fructans (polysaccharides also synthesized from sucrose by plaque bacteria), make up a significant portion of the supragingival plaque matrix, likely contribute to the structure of these biofilms, and may also serve as nutrients for plaque organisms.

In terms of subgingival plaque, a role for exopolysaccharides in the structure of these biofilms has not yet been demonstrated. However, there is no reason to assume that such a role may not also be important in these structures as suggested for other gram-negative bacterial biofilms (Sutherland, 2001).

2.2. Models for Biofilm Formation

An examination of homogeneous biofilms using a variety of *in vitro* model systems (static microtiter plate assays, flow-cells, chemostat-rotary systems) has suggested that biofilm formation may represent a form of microbial development (O'Toole *et al.,* 2000). According to such a model, biofilm formation may

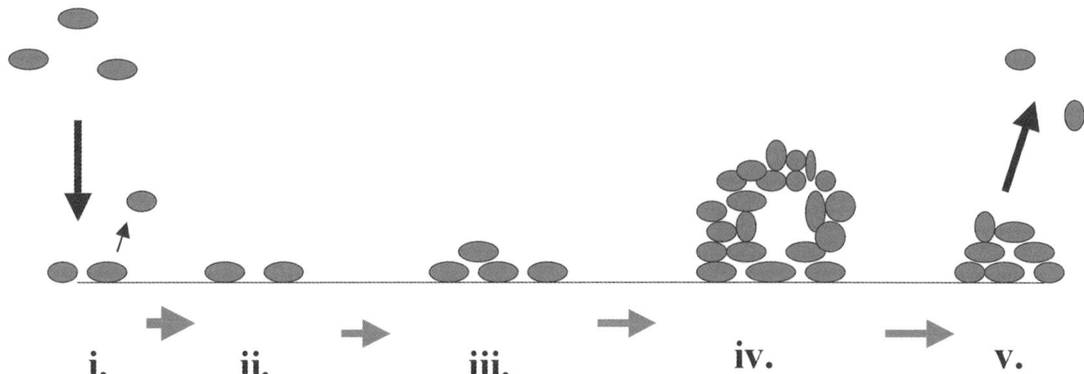

Figure 2. Simplified model for biofilm formation. Five discrete stages of biofilm development are depicted: i., initial reversible attachment of bacteria; ii., irreversible attachment; iii. microcolony interactions; iv. biofilm maturation including the formation of channels; v. detachment of biofilm bacteria.

require a series of genetically programmed sequential stages (analogous to sporulation in *Bacillus subtilis*). For example, this process may involve at least five discrete stages (Figure 2): i) initial reversible attachment of bacteria to inert surfaces; ii) irreversible attachment and microcolony formation; iii) spreading and interaction with other microcolonies; iv) maturation of the biofilms into three-dimensional structures; v) detachment of some biofilm cells and their return to the planktonic state. As described below, the recent utilization of both proteomic and genomic approaches has indeed revealed differential gene expression during biofilm formation.

2.3. Environmental Effects on Biofilm Formation

As with any cellular programmed developmental process, the ultimate question which needs to be answered is what initially triggers such a series of events. Just as each dividing *B. subtilis* cell undergoes either normal replication or initiates sporulation, one can envision a scenario where a bacterium which attaches to an inert surface either detaches or becomes irreversibly bound to initiate the formation of a biofilm. What then is the signal(s) that determines which pathway is followed? As with sporulation, is there an initial trigger analogous to the *spo0A* gene product that needs to be phosphorylated in order to initiate biofilm formation? At present, there is very little data available relevant to this fundamental question.

One possible explanation for how biofilm formation is triggered is suggested by the observation that for many, but not all, bacteria, biofilm formation occurs more efficiently in nutrient-limited media (Korber *et al.,* 1995). This suggests that biofilm formation may occur, in part, in response to nutrient stress and that the formation of compact biofilms helps to protect bacteria under these conditions. In this regard, it is interesting to note that some bacteria aggregate or clump under stress conditions (Baev *et al.,* 1999). In addition, interference with quorum sensing has also been observed to influence biofilm formation for some bacteria (Davies *et al.,* 1998), suggesting the importance of cell density in this process. One of the key environmental factors affecting biofilm formation by anaerobic bacteria may be the relative levels of oxygen. Direct measurements of oxygen concentrations in model biofilms indicate that oxygen is present in the interior of *in vitro* biofilms (Marquis, 1995). Furthermore, the utilization of the oxygen-dependent green fluorescent protein reporter system in biofilm forming bacteria demonstrated that fluorescence can be detected at the lowest depths of the *in vitro* biofilms (Whiteley *et al.,* 2001). The presence of channels within biofilms may allow for penetration of oxygen into the interior of biofilms. Nevertheless, direct measurements of oxygen levels in periodontal pockets (Tanaka *et al.,* 1998) have also revealed that these are around 10% of the dissolved oxygen concentrations detected in aqueous solutions.

Several investigations have suggested that the survival of obligate anaerobes in either planktonic or biofilm stages is dependent upon the presence of more oxygen tolerant organisms (Bradshaw *et al.,* 1998). In regard to subgingival plaque, it has been demonstrated that the survival of the periodontitis-associated obligate anaerobes *P. gingivalis* and *Prevotella nigrescens* in chemostat cultures is dependent upon the presence of the more oxygen tolerant, yet obligate anaerobic, *Fusobacterium nucleatum.* The presence of the later organism enhances the survival of the two black-pigmented organisms in both the planktonic and biofilm growth stages. More recent results have suggested that *F. nucleatum* can both utilize oxygen and produce carbon dioxide which can each contribute to the survivability of *P. gingivalis* (Diaz *et al.*, 2002). Since *F. nucleatum* has been demonstrated to aggregate with many oral bacteria including *P. gingivalis* and *P. nigrescens*, it is suggested that these aggregates produce microenvironments with relatively low levels of oxygen and high concentrations of carbon dioxide. Therefore, it is not surprising that there is a strong positive correlation between the presence of *P. gingivalis* and *F. nucleatum* in subgingival plaque samples (Socransky *et al.*, 1998). The highly proteolytic *P. gingivalis* may also provide nutrient peptides for metabolism by weakly proteolytic *F. nucleatum* strains. Thus, subgingival plaque appears to represent a biofilm-containing microenvironment consisting of mutually compatible organisms displaying symbiotic interactions.

2.4. Role of Extracellular Polymers in Biofilm Formation

As indicated previously, it has been well-established that the synthesis of insoluble glucans from dietary sucrose by the mutans streptococci is an important virulence property related to dental caries etiology (Loesche, 1986). Both *in vitro* as well as *in vivo* analyses utilizing *gtf* mutants defective in glucan synthesis have confirmed the important role of these polymers in *S. mutans* –induced biofilm formation (Kuramitsu, 1993). In addition, the expression of glucan-binding proteins on the cell surface of these organisms may also play important roles in plaque morphology (Banas *et al.*, 2001). These later proteins may help to bind the streptococci to each other via glucan molecules and influence the architecture of the resulting biofilm. Several approaches (Hudson and Curtis, 1990; Yoshida and Kuramitsu, 2002b) have also demonstrated that the *gtfB* and *gtfC* genes (coding for enzymes involved in insoluble glucan synthesis) are upregulated during biofilm formation which is consistent with a significant role for glucans in *S. mutans* biofilm formation.

Evidence has also been obtained implicating a role for glucan synthesis in the attachment of the early colonizing oral bacterium *S. gordonii* to teeth (Vickerman *et al.*, 1991) and host tissue (Vacca-Smith *et al.*, 1994). Although not considered a primary pathogen, these organisms can produce carious lesions in experimental animals (Tanzer *et al.*, 2001) and may also contribute to dental caries or periodontitis by interacting with other more virulent organisms (McNab *et al.*, 2003). In contrast to the mutans streptococci, strains of *S. gordonii* evidently express only a single *gtf* gene, *gtfG*, which is regulated in part by the *rgg* (regulator gene for glucosyltransferases) gene (Vickerman and Minnick, 2002).

It has been proposed that biofilm formation by gram-negative bacteria is dependent upon EPS synthesis (Costerton *et al.*, 1995). In fact, mutants that are defective in such polymer biosynthesis have been isolated and are attenuated in biofilm formation (Yildiz and Schoolnik, 1999). In that study, it was demonstrated that mutagenesis of a rugose *V. cholerae* El Tor strain which forms biofilms on glass and PVC surfaces resulted in the attenuation of biofilm formation when genes involved in EPS formation were inactivated. This is consistent with more recent observations that a mutant of *V. cholerae* which overproduces EPS produces aberrant biofilms (Bomchil *et al.*, 2003). Thus, at least for one gram-negative organism, genetic evidence for a role for EPS synthesis in biofilm formation has been confirmed. In addition, in several bacteria (Sutherland, 2001) the genes involved in EPS synthesis are upregulated during biofilm formation which is consistent with an important role for these polymers in biofilm development. However, no data has yet been published showing an essential role for EPS biosynthesis in biofilm formation by gram-negative anaerobic oral bacteria. Interestingly, strains of *P. gingivalis,* which exhibit enhanced EPS synthesis such as strain W50, are attenuated in biofilm formation *in vitro* (Chen and Kuramitsu, unpublished results). However, these strains are also poorly fimbriated, so it is not yet clear which of these phenotypes is principally responsible for their reduced ability to form biofilms.

Another polymer which has recently been implicated in biofilm formation is exogenous DNA (Whitchurch *et al.*, 2002). Since DNA has been detected on the cell surface of one plaque constituent, *A. actinomycetemcomitans* (Ohta *et al.*, 1993) and these polymers might be released from dying cells, these molecules could be present as a normal component of dental plaque. However, it is not clear if this is also a factor in dental plaque formation.

2.5. Molecular Genetics of Biofilm Formation

Obviously, genes that are involved in the expression of adhesin molecules on the surface of a bacterium will influence biofilm formation. This has been experimentally confirmed in a number of bacterial systems. For example, the major fimbrae of *P. gingivalis* have been implicated in the attachment of these organisms to teeth, host cells, and other bacteria (Genco *et al.*, 1994). Mutants that are defective in the expression of these adhesins are unable to form biofilms *in vitro* (Chen and Kuramitsu, unpublished results). Similarly and as mentioned above, strains of *P. gingivalis*, such as W50, which express low levels of fimbrae on their surfaces display attenuated biofilm formation (Chen and Kuramitsu, unpublished results). Likewise, the SpaP protein has been implicated in the attachment of *S. mutans* to the pellicle of teeth (Crowley *et al.*, 1999). Mutants defective in SpaP expression do not colonize saliva-coated hydroxyapatite beads (model system for teeth) as well as the parental strain and form smaller biofilms (Lee *et al.*, 1989).

Recent results have further suggested that the motility of the oral spirochete *T. denticola* is also crucial for biofilm formation *in vitro* (Vesey and Kuramitsu, 2002). A *flgE* mutant of *T. denticola* 35405 which lacks the periplasmic flagella and is nonmotile cannot form biofilms on fibronectin or *P. gingivalis*-coated microtiter plates. Motility may be required for the spirochete to overcome the close-range repulsive forces between the organism and various surfaces for attachment and/or in subsequent interactions with other microcolonies.

Beginning with the pioneering work of the Gibbons laboratory (Slots and Gibbons, 1978), the role of interbacterial aggregation in biofilm formation has been strongly implicated (Kolenbrander, 1988). Such interactions allow the accumulation into dental plaque of organisms that do not independently colonize teeth efficiently. These interactions appear to be mediated by both cell surface protein-protein and protein-carbohydrate interactions. Since biofilm maturation appears to be dependent upon the interactions between microcolonies (see Figure 2), it is likely that both intra- and interbacterial aggregation play important roles in both homogenous and heterologous biofilm formation, respectively.

Recently, it has been demonstrated that interspecies interactions may be an important factor in oral biofilm formation (Palmer *et al.*, 2001). This study demonstrated that two early colonizers of the tooth surface, *A. naeslundii* and *S. oralis* formed biofilms on saliva-conditioned glass surfaces while either organism alone could not. This appears to be the result of a mutual nutritional interaction which may allow growth of both organisms in the presence of saliva as the sole nutrient. More recently, *in vitro* biofilm formation by mixtures of oral periodontopathic bacteria has revealed synergistic effects. *T. denticola* cannot form detectable biofilms on polyvinyl chloride (PVC) surfaces while *P. gingivalis* forms biofilms which can be readily removed by vigorous washing. However, mixtures of the two organisms form tenacious biofilms on PVC surfaces that are resistant to washing (Ikegami and Kuramitsu, unpublished results). A similar synergistic effect was observed when *P. gingivalis* and *T. forsythensis* were grown together (A. Sharma, personal communication). The molecular basis for such interactions remains to be determined but may be related to the observations that *T. denticola, P. gingivalis, and T. fosythensis* are consistently observed together in subgingival plaque associated with periodontitis (Socransky *et al.*, 1998).

The sequential differential expression of genes in sessile (biofilm) cells relative to planktonic cells would support the hypothesis that biofilm formation represents a programmed developmental process. Several different approaches have been recently undertaken to identify such genes that might constitute a "biofilmon" (genes differentially expressed during biofilm formation). Initially, in order to identify genes that are required for biofilm formation, relatively simple static microtiter plate biofilm assays were utilized in conjunction with random mutagenesis of a biofilm-forming organism such as *Escherichia coli* (Pratt and Kolter, 1998) or *Pseudomonas aeruginosa* (O'Toole and Kolter, 1998). Utilizing a mini Tn10 transposon mutagenesis strategy in the former study, a number of genes required for this process were identified. Included in this group were genes involved in swarming, motility, and chemotaxis. Flagellar genes involved in motility and chemotaxis may be required for initial attachment of the bacteria to inert surfaces, while type I pili appear to be involved in subsequent swarming over the surfaces. A similar mutagenesis approach has also been carried out with other microorganisms including *Vibrio cholerae* (Watnick and Kolter, 1999) as well as the oral bacteria *Streptococcus gordonii* (Loo *et al.*, 2000) and *S. mutans* (Yoshida and Kuramitsu, 2002a). A comparison of the genes required for sucrose-independent biofilm formation by the latter two organisms has revealed that genes involved in quorum-sensing (the *com* regulon) as well as cell-wall synthesis are required for this process. In addition, evidence for a role for a manganese transport system in *S. gordonii* biofilm formation has recently been reported (Loo *et al.*, 2003). Biofilm formation by *S. mutans* also appears to be dependent upon the *sgp* gene coding for an essential GTPase activity. The *tcbR* gene of *S. mutans,* encoding a protein of a two-component signal transduction system with an unknown function, also appears to be involved in this developmental process (Bhagwat *et al.*, 2001). Furthermore, two different approaches (Wen and Burne, 2002; Yoshida and Kuramitsu, 2002a) have identified the *brpA* (biofilm regulatory protein A) gene as playing an essential role in *S. mutans* biofilm formation. An examination of several of the streptococcal mutants attenuated in biofilm formation suggests that some of these genes are involved in the maturation stages of biofilm formation (steps iii to iv in Figure 2). Thus, it is clear that biofilm formation is influenced by a variety of genes in a given organism.

Polyphosphate kinase activity appears to be ubiquitous in plants, animals, as well as in microorganisms (Kornberg *et al.*, 1999). Interestingly, the *ppk* gene also is essential for normal biofilm formation by *P. aeuruginosa* and virulence in animal model systems (Rashid *et al.*, 2000). Since *ppk* mutants of the organisms express pleiotropic effects, it is not clear how polyphosphate accumulation affects biofilm formation in this organism. More recent studies (Chen and Kuramitsu, 2002) have indicated that the *ppk* gene is also required for biofilm formation by the oral periodontopathic bacterium *P. gingivalis*. This suggests the possibility that the *ppk* gene may play an important role in biofilm formation for several Gram-negative pathogens and that the development of specific PPK inhibitors could be useful in controlling biofilm-dependent bacterial infections by these organisms.

More recently, a novel IVET (in vivo expression technology) system to detect genes of *P. aeruginosa* which are expressed in biofilms, but not during planktonic growth, has identified several novel genes involved in biofilm formation which were not previously recognized with microarrays or proteomic analyses (Finelli *et al.*, 2003). It may be possible to utilize similar systems in oral bacteria implanted into the oral cavities of experimental animals or human subjects.

The recent development of microarrays for bacteria whose genomes have now been sequenced has allowed for an examination of differential gene expression during biofilm formation. The initial application of this approach with *P. aeruginosa* has identified a number of genes that are either up- or down-regulated during this process (Whiteley *et al.*, 2001). Using a flow-cell system, it was demonstrated that 73 of the 5500 genes examined were differentially regulated during biofilm formation. Approximately half of these regulated genes were induced while the remainder was repressed during biofilm development. Other than the genes for fimbrae, which were repressed during this process, no obvious relationship between biofilm formation and the functions of these genes could be discerned. Interestingly, 34 of the regulated genes corresponded to hypothetical proteins of unknown function.

Recently, microarrays for *P. gingivalis* have also been utilized to screen for differentially regulated genes during biofilm formation (Chen *et al.*, 2003). Using a static anaerobic biofilm system, it was observed that the changes in gene expression were not as great as those identified in flow cell systems. This could be due to the incorporation of planktonic cells into static biofilm cultures which could mask the genetic changes which normally occur during this process. Moreover, the results from microarray analysis of biofilm formation are subject to the limitations of this technique (Vasil, 2003) as well as variability in the requirements for this process depending upon whether flow-cells or static model systems are utilized (DeKiviet *et al.*, 2001). Nevertheless, a number of *P. gingivalis* genes were identified which appear to be upregulated during biofilm formation. The most prominent of these appeared to be the gene coding for a homologue of a universal stress protein (USP) (Kvint *et al.*, 2003). This was of interest relative to biofilm formation since the expression of this gene also appeared to be dependent upon the *ppk* gene of *P. gingivalis* (Chen and Kuramitsu, unpublished results). However, mutagenesis of the *usp* gene revealed that the mutation had no noticeable effect on quantitative biofilm formation *in vitro*. This indicates that not all differentially expressed genes identified during biofilm formation are required for this process. It is possible that the upregulation of the *usp* gene during biofilm formation may reflect the "stressed" nature of cells undergoing this developmental process. Because of the wide-spread nature of *usp* homologues in a variety of bacteria, it will be of interest to determine if these genes are also upregulated during biofilm formation by other bacteria. Prominent among the genes that are downregulated during this process are those coding for the minor and major fimbrae of *P. gingivalis* 381 (Dickinson *et al.*, 1988; Arai *et al.*, 2000). Since these proteins are likely involved in the earlier stages of biofilm formation (attachment to surfaces, interactions between the organisms) it is likely that these changes reflect later stages of biofilm formation. The development of microarrays made with the genes of other oral bacteria such as *S. mutans* (R. Quivey, personal communication) will allow for similar approaches with these organisms as well.

Since the transcription of a gene may not always correlate with its cognate protein levels, proteomic approaches have also been utilized to investigate biofilm formation. Using 2-D gels to resolve proteins from *P. aeruginosa*, Sauer and colleagues (2002) identified protein changes in five defined stages of biofilm development. In contrast to the earlier study on transcriptional regulation during biofilm formation (Whiteley *et al.*, 2001), these studies suggested that expression of over 50% of the proteins of the organism were significantly altered during this developmental process. Interestingly, this approach detected differences in protein expression between each of the five proposed stages of biofilm development. These results also demonstrated that the LasIR-dependent quorum-sensing system was important in the later stages (maturation) of biofilm formation following irreversible attachment. Again, these differences in the interpretation of microarray and proteome analyses must be tempered by the limitations of the techniques involved as well as influences of the distinct environmental conditions utilized in these experiments.

Proteomic analysis of *P. aeruginosa* biofilm formation at different stages has further indicated that the cells present on inert surfaces at stage I (Figure 2) display protein profiles very similar to those of planktonic cells (Sauer *et al.*, 2002). Therefore, this may suggest that the initial crucial gene expression required for the activation of the biofilmon occurs following surface attachment.

In a two-stage biofilm study involving *S. mutans*, Svanstater *et al.* (2001) have examined the differences in protein profiles of cells growing as biofilms relative to planktonic cells. Interestingly, the 2-D gel patterns identified a number of enzymes involved in glycolysis as differentially expressed during biofilm formation *in vitro*. However, the relationship of these changes to biofilm development remains unclear and it would be interesting to determine if such changes also occur *in vivo*.

2.6. Maturation of Biofilms

Although various models of biofilm formation (see Figure 2) propose that microcolonies initially attached to an inert surface, spread, and eventually form "mature" biofilms, very little information is available concerning how this takes place. Is the spreading merely the result of bacterial reproduction or are active processes akin to "swarming" over the surfaces involved? Since some bacteria which are nonmotile and do not spread on agar surfaces form biofilms *in vitro*, it would appear that classical forms of swarming are not required for biofilm formation. However, the twitching motility involved in the swarming of *P. aeruginosa* appears to play a significant role in static biofilm formation but not in some flow-cell systems (Heydorn *et al.*, 2002). Nevertheless, some organisms such as *V. cholerae* display swarming phenotypes and this process appears to play a role in biofilm formation, at least *in vitro* (Watnick and Kolter, 1999).

For some organisms (Costerton *et al.*, 1995), the three dimensional structure of mature biofilms involves columnar structures and even mushroom-like forms with "caps" and "stalks". What controls the formation of these communal "high rises"? What environmental as well as genetic factors regulate the formation of such structures? Recent data has suggested that EPS formation is essential for maintaining the three dimensional structure of biofilms (Bomchil *et al.*, 2003). These studies suggested that a *V. cholerae* gene, *mbaA*, is important for maintaining the structure of biofilms by perhaps regulating the formation of EPS. In addition, CLSM has shown the presence of channels between the columnar structures in some biofilms (Costerton *et al.*, 1995). One recent study has suggested how such channels may remain open and not be filled by growing microcolonies (Davey *et al.*, 2003). In *P. aeruginosa,* the *rhlA* quorum-sensing gene regulates the formation of rhamnolipids. It was observed that *rhlA* mutants formed defective biofilms that lacked the normal channels associated with *P. aeruginosa* biofilms. These results suggested that the rhamnolipid surfactants coated the channel-lining cells and prevented further colonization to fill the channels. Furthermore, rhamnolipid biosynthesis occurs relatively late during biofilm maturation, which coincides with the activation of the *rhl* quorum-sensing system during this process. These recent findings have not yet been explored relative to the structure of subgingival dental plaque.

2.7. Detachment of Sessile Cells in Biofilms

The last step proposed in biofilm development involves the detachment of sessile cells which can then revert to the planktonic mode of growth (see Figure 2). This process has not yet been extensively studied but recent information has suggested several mechanisms for such a process in dental plaque. Since interspecies interactions play an important role in the formation of biofilms, the release of cell surface molecules involved in such interactions could affect the stability of biofilms. Lee and Boran (2003) have identified an enzyme, SrtA sortase, of *S. mutans* which is responsible for the maintenance of the adhesin P1 on the cell surface of these plaque constituents. Mutants defective in the *srtA* gene are incorporated into biofilms both *in vitro* and *in vivo* at significantly lower levels than the parental strain NG8. It is possible that other *S. mutans* cell surface molecules involved in biofilm formation may also require SrtA for proper localization. Therefore, the decrease in the activity of this enzyme at later stages of biofilm development could play a role in the release of some organisms from a biofilm during the detachment phase. It would be of interest to examine the regulation of expression of *srtA* and correlate this to the development of biofilms.

In addition, recent results with *A. actinomycetemcomitans* (Kaplan *et al.*, 2003) have suggested that the release of cells from biofilms is dependent upon EPS synthesis. An interesting observation from these studies was that cells released from the biofilm originated from the interior of the structures and were not simply sloughed off from the surface. Thus, the EPS may play a role in embedding the bacteria into stable biofilm structures. Whether or not this is the case for other organisms still remains to be determined. These results, taken together, further suggest that the release of biofilm cells into the surrounding medium may also depend upon differential gene expression late in this developmental process.

3. Communication between Bacteria in Dental Plaque

The close proximity between bacteria in biofilms allows for high probabilities for cell-cell communication. Since many bacteria including both aerobic and anaerobic organisms are capable of forming such structures, it is likely that microorganisms have evolved to maximize survivability in biofilms. Thus, cell-cell signaling within biofilms to optimize growth and survival would be predicted. In this regard, much attention has been focused on cell signaling mediated by diffusible signaling molecules that are produced as a function of cell density (quorum-sensing). Such signaling appears to be mediated by homoserine lactones (Fuqua *et al.*, 1994) in some Gram-negative bacteria, while small peptides play a similar role for Gram-positive bacteria (Pestova *et al.*, 1996). In addition, furanone-like molecules may function as signal mediators in both Gram-positive and Gram-negative bacteria (Bassler, 2002). In fact, these latter molecules have been proposed to act as mediators of interactions between bacterial species. The production of signaling molecules appears to be essential for the formation of biofilms by some bacteria. However, such a general model for biofilm formation by all bacteria has recently been questioned since this developmental program appears to be a complex process influenced by both direct and indirect effectors (Kjelleberg and Molin, 2002). Nevertheless, the *lasR-lasI* quorum-sensing system involved in the formation of oxododecanoyl homoserine lactone has been suggested to play a major role in *P. aeruginosa* biofilm formation (Davies *et al.*, 1998). The second overlapping quorum- sensing system in these organisms involving *rhlI-rhlR* is also operative in biofilms and appears to affect the three-dimensional structure of the biofilms (Davey *et al.*, 2003). Interestingly, acylhomoserine lactones have been detected within *in vitro* biofilms as well as *in vivo* (Stickler *et al.*, 1998). However, as recently documented (Kjelleberg and Molin, 2002), the role of quorum-sensing in biofilm formation for a given organism can vary depending upon the experimental conditions utilized.

Up to now, no dental plaque bacterium has been demonstrated to secrete a homoserine lactone quorum-sensing signal. Furthermore, an examination of the available genome data bases for the gram-negative anaerobic plaque organisms *P. gingivalis, T. denticola,* and *A. actinomycetemcomitans* has failed to identify homologues of the *luxI* gene involved in homoserine lactone synthesis. However, this does not rule out the possibility of the presence of low sequence identity *luxI* homologues in these organisms. By contrast, a recent survey (Frias *et al.*, 2001) has indicated the ability of a variety of periodontal plaque organisms to secrete furanone-like autoinducer-2 (AI-2) homologues. More recently, AI-2 production in *S. mutans* has been demonstrated (Merritt *et al.*, 2003). These results are not surprising if the primary function of AI-2 is to serve as a signal of "foreignness" between bacteria. However, more recent results have suggested that the primary function of AI-2 may be to serve as a signal for metabolic stress within the producing organism (Beeston and Surrette, 2002). Regardless of the primary role of this signaling system in bacteria, it is clear that it is present in a wide variety of bacteria and thus has been conserved as an important feature of cell metabolism.

The results from the utilization of *luxS* mutants in several oral bacteria are consistent with the recently proposed hypothesis that the primary function of the *luxS* system may be to serve as a monitor of self, rather than nonself, activity. Mutations in the *luxS* gene of *S. mutans* do not affect sucrose-independent biofilm formation *in vitro* (Yoshida and Kuramitsu, 2002) but significantly altered biofilm formation in the presence of sucrose (Merritt *et al.*, 2003; Yoshida and Kuramitsu, 2002b). Furthermore, the mutant biofilm displays much larger aggregates than the parental strain on solid surfaces and is much more readily dislodged by sodium dodecylsulfate treatment. This latter result might have some practical relevance since many toothpastes contain small amounts of detergents. These effects apparently result from the induction of the *gtfB* and *gtfC* genes by the *luxS* mutation (Yoshida and Kuramitsu, unpublished results), leading to enhanced autoaggregation of the *S. mutans* planktonic cells in the presence of sucrose which attenuates biofilm formation. Culture supernatant fluids from some, but not all, oral streptococci can complement the *luxS* mutation in *S. mutans* GS-5 (Yoshida and Kuramitsu, unpublished results), suggesting the possibility of AI-2 mediated interspecies communication in dental plaque. In addition, a *luxS* mutant of *P. gingivalis* 381 is unaffected for biofilm formation (Chen and Kuramitsu, unpublished results). However, this mutant has been demonstrated to upregulate the expression of the RgpA gingipain protease and downregulate *hemR* transcription (Chung *et al.*, 2001) both potential virulence factors in these organisms. Furthermore, culture fluids of *A. actinomycetemcomitans* (presumably containing AI-2) can complement this mutational defect (Fong *et al.*, 2001).

A. actinomycetemcomitans luxS mutants have also been constructed and were shown to be defective in leukotoxin production (Fong *et al.*, 2001). Moreover, this defect could be complemented by culture supernatants of the parental strain. Relative to interspecies signaling, it should be pointed out that no direct effects of heterologous organisms on *luxS*-mediated properties have yet to be reported for wild-type recipient oral bacteria. However, conditioned media from *A. actinomycetemcomitans* (presumably containing AI-2) could induce increased levels of leukotoxin in early log-phase cultures of the organism (Fong *et al.*, 2001). The effects of AI-2 depletion on *A. actinomycetemcomitans* biofilm formation have yet to be reported.

Despite the fact that the mutation of the *luxS* gene in several oral bacteria does not appear to affect biofilm formation, an interesting recent study has revealed that this signaling system can markedly affect the formation of a heterogeneous biofilm *in vitro* (McNab *et al.*, 2003). In this system, *P. gingivalis* interacts with *S. gordonii* to form mixed biofilms as may occur *in vivo* (Kolenbrander, 2000). However, mixed biofilm formation by *luxS* mutants of both organisms, but not of either alone, is markedly inhibited in a flow-cell system using saliva-coated glass slides. These results suggest that at least for these two organisms, the AI-2 signaling molecule is required for mixed biofilm formation. Such interactions may be involved in the conversion of a primarily Gram-positive plaque into a Gram-negative biofilm associated with periodontal diseases. It will be interesting to determine if similar signaling is also involved in other interactions that could play a role in dental plaque formation. In this regard, one organism that may play a key role in the formation of dental plaque is *F. nucleatum*. This gram-negative anaerobic organism is found in many plaque samples and has been demonstrated to interact with a variety of bacteria (Kolenbrander, 1988). Thus, the presence of *F. nucleatum* in developing dental plaque may allow the association of bacteria which individually may not be able to colonize the tooth surface. However, since a gene inactivation system for these organisms has not yet been developed, it has not been possible to determine the genetic basis for the ubiquitous interaction of this organism with a variety of different bacteria, both gram-positive and gram-negative.

Beginning with investigations of *Enterococcus faecalis* it is now apparent that many gram-positive bacteria utilize small peptides for cell signaling (Dunny and Winans, 1999). For *Streptococcus pneumoniae,* a quorum sensing peptide which is secreted by the cells has been shown to play a crucial role in the development of genetic competence (Pestova *et al.,* 1996). This peptide, CSP (competence stimulating peptide) serves as a signal to activate a two-component signal transduction system (mediated by the *comD* and *E* genes) leading to subsequent activation of a number of genes involved in genetic transformation. Similar signaling systems have been identified in the oral bacteria *S. gordonii* (Lundsford and Robie, 1997) and *S. mutans* (Li *et al.,* 2001). Random, as well as targeted, mutagenesis of both organisms has revealed that mutants defective in the *com* regulon are attenuated in normal biofilm formation *in vitro* (Loo *et al.,* 2000; Yoshida and Kuramitsu, 2002a; Li *et al.,* 2002b). However, the mechanism by which CSP regulates biofilm formation at the molecular level still remains to be determined. It is also of interest that sucrose-mediated biofilm formation by *S. mutans* GS5 is not affected by the *comC* mutation (Yoshida and Kuramitsu, 2002). Apparently, insoluble adhesive glucan synthesis by these organisms can readily compensate for the attenuation of the CSP-mediated requirements for biofilm formation. However, the structures of these biofilms have not yet been compared.

Interestingly, unlike the *luxS* system, the *com* system appears to be regulated by highly specific CSP factors. A comparative examination of the sequences of the *comC* genes in the available streptococcal genome databases revealed no apparent homology between the CSPs. This has been confirmed by recent demonstrations that there is no complementation of the *comC* mutation in *S. mutans* GS5 by culture supernatant fluids from other oral streptococci or the synthetic CSP of *S. gordonii* (Wang and Kuramitsu, 2003). It is also of interest that the CSP-mediated regulatory system also appears to affect bacteriocin production by *S. mutans* GS-5 (Wang and Kuramitsu, 2003) as well as the acid tolerance of some strains of these organisms (Li *et al.,* 2002b). Thus the *com* regulon in streptococci appears to function as a signaling system to mediate density- dependent stress related responses including biofilm formation.

Although recent results have suggested that activation of the streptococcal *com* regulons by heterologous CSPs is not common, evidence for the attenuation of *com*-dependent properties in the presence of heterologous oral streptococci has now been obtained (Wang and Kuramitsu, 2003). These results have shown that a variety of nonmutans streptococci are able to inhibit the genetic transformation and bacteriocin production of some strains of *S. mutans*. The molecular basis for such effects has not yet been delineated but these results suggest the possibility that such interactions may also occur within dental plaque.

Another recent example of intercellular communication between oral streptococci was suggested for *S. pyogenes* and *S. salivarius* (Upton *et al.*, 2001), organisms which could coexist together on mucosal surfaces. Strains of each organism contain genes for homologous lantiobiotic molecules which can affect gene expression of the other via two-component signaling systems and ultimately affect growth. It was suggested that such a mechanism may play a role in limiting the growth of the virulent *S. pyogenes* by the commensal *S. salivarius* strains.

4. Gene Transfer Between Dental Plaque Bacteria

Another possible form of interspecies communication between biofilm bacteria is that of genetic exchange. Such interactions within dental plaque may be important in terms of understanding the evolution of oral bacteria as well as for controlling the emergence of drug-resistant bacteria in the human host. For example, one earlier investigation (Curtis *et al.,* 1999) has revealed that the periodontopathogen *P. gingivalis* contains a "pathogenicity island" on its chromosome which may code for genes involved in the virulence of these organisms. The G+C ratio of these genes suggests that they may have originated in another bacterial species and were transferred into *P. gingivalis*. However, the source of these genes and the mechanism of their putative transfer still remain to be determined. Gene transfer between some oral bacteria can occur either as a consequence of natural transformation or conjugation (Leblanc *et al.,* 1978; Kuramitsu and Trappa, 1984). In addition, the demonstration that some strains of *Actinomyces naeslundii* can be infected with bacteriophage extracted from dental plaque (Yeung, 1999) suggests the possibility that transduction may occur between selected bacteria present in dental plaque.

Gene transfer between oral streptococci has been demonstrated by both conjugation (Leblanc *et al.*, 1978) as well as by transformation (Kuramitsu and Trappa, 1984). In addition, the transformation of organisms growing in preformed biofilms *in vitro* has also been observed (Robertson *et al.*, 2001; Li *et al.*, 2001). Moreover, direct transfer of a broad host range shuttle plasmid from the oral spirochete *T. denticola* into *S. gordonii* has recently been demonstrated in mixed biofilms *in vitro* (Wang *et al.*, 2002). It is likely that such a transfer mediated by transformation between different oral streptococci in dental plaque also occurs although this has not yet been directly documented. This process, in addition to transfer by conjugation, may be responsible in part for the spread of antibiotic resistance genes between bacteria in the oral cavity and subsequently to organisms that transiently inhabit the oral cavity, but colonize other regions of the body (especially the throat and lungs). Whether or not genetic exchange occurs between bacteria other than streptococci within dental plaque still remains to be determined. However, the demonstration that a similar, if not identical, plasmid is present in oral isolates of *T. denticola* and *T. socranskii* (Chan *et al.*, 1996) makes this a likely possibility. Whether chromosomal markers can be transferred between heterogeneous plaque organisms is much more difficult to demonstrate since stable maintenance of the transferred DNA requires homologous recombination with the recipient's chromosome.

5. Role of Dental Plaque in Influencing Systemic Diseases

It has long been recognized that dental procedures are a contributing factor to the development of endocarditis (Herzberg *et al.*, 1997). These manipulations release oral bacteria into the circulation and microorganisms can then migrate and attach to damaged heart valves. Some of these organisms may reside in saliva but others could be released from dental plaque as a result of mechanical manipulations.

Recent interest has also been directed at the possible contribution of periodontitis to systemic diseases (Beck *et al.*, 1996). Several epidemiological studies have shown an increased risk of developing atherosclerosis, diabetes, stroke, and induction of preterm births for individuals suffering from periodontitis (Wu *et al.*, 2000; Champagne *et al.*, 2000). However, not all of the epidemiological data support such relationships (Hujoel *et al.*, 2002; Davenport *et al.*, 2002). Therefore, although still somewhat controversial, these suggestive studies, in conjunction with more recent animal model studies (Li *et al.*, 2002a), make a case for a role for periodontitis in atherosclerosis. According to such a model, the extensive bleeding associated with severe periodontitis could lead to a transient bacteremia by which subgingival plaque anaerobic bacteria and their products may migrate via the vasculature to cardiac tissue and contribute to the development of atheromas (Kuramitsu *et al.*, 2001). Alternatively, the bacteremia could induce a host inflammatory response (induction of C-reactive protein, fibrinogen, inflammatory cytokines) which could also exacerbate coronary lesion formation (Slade *et al.*, 2003). Clearly, additional approaches will be necessary to confirm a role for periodontitis and anaerobic subgingival plaque bacteria in systemic diseases.

6. Strategies to Inhibit Plaque Formation Based upon Biofilm Physiology

At present, the primary means of reducing the incidence of dental caries and periodontitis is to physically remove dental plaque that is associated with teeth. Thus, brushing and flossing together with routine professional plaque control and the utilization of sealants and fluoridation are our primary means of defending against these diseases. The fact that the most recent dental health survey in the USA (Oral Health U.S., 2002) has revealed that there has been almost a 60% reduction in dental caries in the 6-18 year old population in the past three decades speaks to the relative effectiveness of such strategies. However, since a large fraction of children still experience high caries rates, especially in medically underserved communities, it would be judicious to entertain other plaque control approaches. In this regard, one possibility is to interfere with plaque formation in addition to attempting to eliminate the microorganisms involved. For example, the demonstration that the *ppk* gene is important for biofilm formation by several distinct gram-negative bacteria (Rashid *et al.*, 2000; Chen and Kuramitsu, 2002) suggests that the development of PPK inhibitors may provide a novel means of reducing subgingival plaque formation. Likewise, the recent identification of antagonists of *P. aerugionosa* biofilm formation by synthetic quorum-sensing inhibitors (Smith *et al.*, 2003) raises the possibility that a similar strategy may be employed to identify inhibitors which may also impact subgingival plaque formation. Furthermore, since the AI-2 product of the *luxS* gene appears to play a role in biofilm formation by *S. mutans* (Yoshida *et al.*, 2002; Merritt *et al.*, 2003) as well as *P. gingivalis-S. sanguinis* mixed biofilms (McNab *et al.*, 2003), it may be possible to identify antagonists of AI-2 activity which might be utilized to reduce plaque formation by organisms involved in dental caries or periodontitis. Whether or not such approaches are feasible will await further investigation.

Several other strategies to regulate plaque formation by eliminating virulent oral bacteria have also been proposed based upon more traditional approaches. For example, investigations leading to the development of an anti-caries vaccine have been progressing (Michalek *et al.*, 2001). Although much has been learned about mucosal immunity in these studies, testing of such vaccines in humans still awaits evaluation. In addition, in order to avoid potential complications from active immunization against dental caries, alternate passive immunization approaches have been suggested but have not yet been evaluated for their protective effects in humans (Koga *et al.*, 2002). Attempts to develop an anti-periodontitis vaccine have been initiated (Evans *et al.*, 1992; Gibson and Genco, 2001) but comparable studies with an anticaries vaccine have not yet progressed.

Recently, a novel probiotic approach has been proposed to control dental caries (Hillman, 2002).

This replacement therapy approach would involve the implantation into a susceptible subject of a *S. mutans* strain that has been genetically engineered to be unable to produce lactic acid and is also capable of producing a mutacin which would antagonize the growth of many other strains of this organism. Although demonstrated to provide protection against dental caries when tested in rats, an evaluation of this approach in humans has not yet been carried out.

7. Future Approaches for Oral Biofilm Research

The rapid increase in the literature devoted to biofilm research indicates the recognition that these microbial communities have practical implications for medicine as well as industrial applications. Although current efforts are focused on identification of biofilmon genes of individual organisms, the recent demonstration of relevant interactions between heterogeneous organisms in mixed biofilms suggests that this will be a fruitful area of research in the near future. In addition, the interactions between biofilms and host tissue will be of great interest in terms of potential signaling between prokaryotic and eukaryotic cells. Such effects may be especially significant in inflammatory diseases such as periodontitis. The complexity of such studies will require the development of novel techniques to investigate gene changes in heterogeneous biofilms. Likewise, it is becoming increasingly recognized that biofilms themselves are composed of mixtures of different microenvironments. For anaerobic organisms within biofilms, their relative location compared to aerobic organisms may play important roles in altering the physiology of each group of organisms. Therefore, techniques to investigate gene changes in individual or small groups of organisms in biofilms will need to be developed. The development of ultrasensitive reporter systems in conjunction with high-resolution confocal microscopy may be available in the future for this purpose. Finally, direct access to dental biofilms (Palmer *et al.*, 2003) and their facile manipulation suggests that the oral cavity may represent an ideal environment for testing the predictions gained from the study of *in vitro* biofilms in the human host.

In addition, since oral biofilm research has exclusively focused on dental plaque bacteria, it will be important to expand such efforts to biofilms formed on soft tissue (mucosal surfaces). These biofilms may also influence the oral immune responses as well as reveal unsuspected interactions between bacteria and oral fungi. Therefore, it is likely that, despite much progress, we currently have only a glimpse of the complex interactions between oral biofilms and host tissue.

Acknowledgments

This monograph is dedicated to the memory of Ronald J. Doyle who was a pioneer in examining the molecular basis for plaque formation by *S. mutans*. The author also wishes to thank numerous colleagues, both inter- and intra-institutionally based, for stimulating some of the ideas presented in this review. Data cited from the author's laboratory was supported by National Institutes of Health grants DE03258, DE08293, and DE09821.

References

Arai, M., Hamada, N., and Umemoto, T. 2000. Purification and characterization of a novel secondary fimbrial protein from *Porphyromonas gingivalis* strain 381. FEMS Microbiol. Lett. 193: 75-81.

Baev, D., England, R., and Kuramitsu, H.K. 1999. Stress-induced membrane association of the *Streptococcus mutans* GTP-binding protein, an essential G protein, and investigation of its physiological role by utilization of an antisense RNA strategy. Infect. Immun. 67: 4510-4516.

Banas, J.A., Hazlett, K.R., and Mazurkiewicz, J.E. 2001. An in vitro model for studying the contributions of the *Streptococcus mutans* glucan-binding protein A to biofilm structure. Meths. Enzymol. 337: 425-433.

Bassler, B.L. 2002. Small talk. Cell-to-cell communication in bacteria. Cell. 109: 421-424.

Beck, J., Garcia, R., Heiss, G., Vokonas, P.S., and Offenbacher, S. 1996. Periodontal disease and cardiovascular disease. J. Periodont. 67 (Suppl.): 1123-1137.

Beeston, A.L., and Surette, M.G. 2002. *pfs*-Dependent regulation of autoinducer 2 production in *Salmonella enterica* serovar typhimurium. J. Bacteriol. 184: 3450-3456.

Bomchil, N., Watnick, P., and Kolter, R. 2003. Identification and characterization of a *Vibrio cholera* gene, *mbaA*, involved in the maintenance of biofilm architecture. J. Bacteriol. 185: 1384-1390.

Bradshaw, D.J., Marsh, P.D., Watson, G.K., and Allison, C. 1998. Role of *Fusobacterium nucleatum* and coaggregates in anaerobe survival in planktonic and biofilm oral communities during aeration. Infect. Immun. 66: 4729-4732.

Caufield, P.W., Cutter, G.R., and Dasanayake, A.P. 1993. Initial acquisition of mutans streptococci by infants: evidence for a discrete window of infectivity. J. Dent. Res. 72: 37-45.

Caufield, P.W., Dasanayake, A.P., Li, Y. Pan, Y., Hsu, J., and Hardin J.M. 2000. Natural history of *Streptococcus sanguinis* in the oral cavity of infants: evidence for a discrete window of infectivity. Infect. Immun. 68: 4018-4023.

Champagne, C.M., Madianos, P.N., Lieff, S., Murtha, A.P., Beck, J.D., and Offenbacher, S. 2000. Periodontal medicine: emerging concepts in pregnancy outcomes. J. Int. Acad. Periodontol. 2: 9-13.

Chan, E.C., Klitorinos, A., Gharbia, S., Caudry, S.D., Rahal, M.D., and Siboo, R. 1996. Characterization of a 4.2-kb plasmid isolated from periodontopathic spirochetes. Oral Microbiol. Immunol. 11: 365-368.

Chen, W., and Kuramitsu, H.K. 2002. Role of polyphosphate kinase in biofilm formation by *Porphyromonas gingivalis*. Infect. Immun. 70: 4708-4715.

Chen, W., Walling, J., and Kuramitsu, H.K. 2003. Biofilm-associated genes of *P. gingivalis* identified by microarray analysis. J. Dent. Res. 82 (Spec. Issue A): Abst. 0365.

Chung, W.O., Park, Y., Lamont, R.J., McNab, R., Barbieri, B., and Demuth, D.R. 2001. Signaling system in *Porphyromonas gingivalis* based upon a LuxS protein. J. Bacteriol. 183: 3903-3909.

Costerton, J.W., Lewandowski, Z., Caldwell, D.E., Kerber, K.R., and Lappin-Scott, H.M. 1995. Microbial biofilms. Annu. Rev. Microbiol. 49: 711-745.

Crowley, P.J., Brady, L.J., Michalek, S.M., and Bleiweis, A.S. 1999. Virulence of a *spaP* mutant of *Streptococcus mutans* in a gnotobiotic rat model. Infect. Immun. 67: 1201-1206.

Curtis, M.A., Hanley, S.A., and Aduse-Opoku, J. 1999. The *rag* locus of *Porphyromonas gingivalis:* a novel pathogenicity island. J. Periodont. Res. 34: 400-405.

Darveau, R.P., Tanner, A., and Page, R.C. 1997. The microbial challenge in periodontitis. Periodontal. 2000. 14: 12-32.

Davenport, E.S., Williams, C.E., Sterne, J.A., Murad, S., Sivatapathasundram, V., and Curtis, M.A. 2002. Maternal periodontal disease and preterm birth weight: case-control study. J. Dent. Res. 81: 313-318.

Davey, M.E., Caiazza, N.C., and O'Toole, G.A. 2003. Rhamnolipid surfactant production affects biofilm architecture in *Pseudomonas aeruginosa* PA01. J. Bacteriol. 185: 1027-1036.

Davies, D.G., Parsek, M.R., Pearson, J.P., Iglewski, B.H., Costerton, J.W., and Greenberg, E.P. 1998. The involvement of cell-to-cell signals in the development of a bacterial biofilm. Science 280: 295-298.

DeKierit, T.R., Gillis, R., Marx, S., Brown, C., and Iglewski, B.H. 2001. Quorum-sensing genes in *Pseudomonas aeruginosa* biofilms: their role and expression patterns. Appl. Environ. Microbiol. 67: 1865-1873.

Diaz, P.I., Zilm, P.S., and Rogers, A.H. 2002. *Fusobacterium nucleatum* supports the growth of *Porphyromonas gingivalis* in oxygenated and carbon-dioxide-depleted environments. Microbiol. 148: 467-472.

Dickinson, D.P., Kubiniec, M.K., Yoshimura, F., and Genco, R.J. 1988. Molecular cloning and sequencing of the gene encoding the fimbrial subunit protein of *Bacteroides gingivalis*. J. Bacteriol. 170: 1658-1665.

Donlan, R.M., and Costeron, J.W. 2002. Biofilms: survival mechanisms of clinically relevant microorganisms. Clin. Microbiol. Rev. 15: 167-193.

Dunny, G.M., and Winans, S.C. 1999. Cell-cell signaling in bacteria. Amer. Soc. Microbiol. Press, Washington, D.C.

Evans, R.T., Klauseen, B., Sojar, H.T., Bedi, G.S., Sfintescu, C., Ramamurthy, N.S., Golub, L.M., and Genco, R.J. 1992. Immunization with *Porphyromonas* (*Bacteroides*) *gingivalis* fimbrae protects against periodontal destruction. Infect. Immun. 60: 2926-2935.

Finelli, A., Gallant, C.V., Jarvi, K., and Burrows, L.L. 2003. Use of in-biofilm expression technology to identify genes involved in *Pseudomonas aeruginosa* biofilm development. J. Bacteriol. 185: 2700-2710.

Fong, K.P., Chung, W.O., Lamont, R.J., and Demuth, D.R. 2001. Intra- and interspecies regulation of gene expression by *Actinobacillus actinomycetemcomitans* LuxS. Infect. Immun. 69: 7625-7634.

Frias, J., Olle, E., and Alsina, M. 2001. Periodontal pathogens produce quorum sensing signal molecules. Infect. Immun. 69: 3431-3434.

Fuqua, W.C., Winans, S.C., and Greenberg, E. P. 1994. Quorum sensing in bacteria: the *luxR-luxI* family of cell density-responsive transcriptional regulators. J. Bacteriol. 176: 269-275.

Genco, R.J., Sojar, H., Lee J-Y., Sharma, A., Bedi, G., Cho, M-I., and Dyer, D.W. 1994. *Porphyromonas gingivalis* fimbrae: structure, function, and insertional inactivation mutants. In: Molecular Pathogenesis of Periodontal Diseases. R. Genco, S. Hamada, T. Lehner, J. McGhee, and S. Mergenhagen, eds. Amer. Soc. Microbiol. Press, Washington, D.C. p. 13-23.

Gibson, F.C. 3rd, and Genco, C.A. 2001. Prevention of *Porphyromonas gingivalis*-induced bone loss following immunization with gingipains R1. Infect. Immun. 69: 7959-7963.

Herzberg, M.C., Meyer, M.W., Kilic, A., and Tao, L. 1997. Host-pathogen interactions in bacterial endocarditis: streptococcal virulence in the host. Adv. Dent. Res. 11: 69-74.

Heydorn, A., Ersboll, B., Kato, J., Hentzer, M., Parsek, M.R. Tolker-Nielsen, T., Givskov, M., and Molin, S. 2002. Statistical analysis of *Pseudomonas aeruginosa* biofilm development: impact of mutations in genes involved in twitching motility, cell-to-cell signaling, and stationary-phase sigma factor expression. Appl. Environ. Microbiol. 68: 2008-2017.

Hillman, J.D. 2002. Genetically modified *Streptococcus mutans* for the prevention of dental caries. Antoine Van Leewenhoek. 82: 361-366.

Hudson, M.C., and Curtiss, R. 3rd. 1990. Regulation of expression of *Streptococcus mutans* genes important to virulence. Infect. Immun. 58: 464-470.

Hujoel, P.P., Drangsholt, M., Spiekmann, C., and DeRouen, T.A. 2002. Periodontitis-systemic disease associations in the presence of smoking-causal or coincidental? Periodontol 2000. 30: 51-60.

Kaplan, J.B., Meyenhofer, M.F., and Fine, D.H. 2003. Biofilm growth and detachment of *Actinobacillus actinomycetemcomitans*. J. Bacteriol. 185: 1399-1404.

Kinniment, S.L., Wimpenny, J.W., Adams, D., and Marsh. P.D. 1996. Development of a steady-state oral microbial biofilm community using the constant-depth film fermenter. Microbiology. 142: 631-638.

Kjelleberg, S., and Molin, S. 2002. Is there a role for quorum sensing signals in bacterial biofilms? Curr. Opin. Microbiol. 5: 254-258.

Kleinberg, I. 2002. A mixed-bacterial ecological approach to understanding the role of oral bacteria in dental caries causation: an alternate to *Streptococcus mutans* and the specific-plaque hypothesis. Crit. Rev. Oral Biol. Med. 13: 108-125.

Koga, T., Oho, T., Shimazaki, Y., and Nakano, Y. 2002. Immunization against dental caries. Vaccine. 20: 2027-2044.

Kolenbrander, P.E. 1988. Intergeneric coaggregation among human oral baceria and ecology of dental plaque. Annu. Rev. Microbiol. 42: 627-656.

Kolenbrander, P.E. 2000. Oral microbial communities: biofilms, interactions, and genetic systems. Annu. Rev. Microbiol. 54: 413-437.

Korber, D.R., Lawrence, J.R., Lappin-Scott, H.M., and Costerton, J.W. 1995. Growth of microorganisms on surfaces. In: Microbial Biofilms. H.M. Lappin-Scott, and J.W. Costerton, eds. Cambridge University Press, Cambridge. p. 15-45.

Kornberg, A., Rao, N., and Ault-Riche, D. 1999. Inorganic polyphosphate: a molecule of many functions. Annu. Rev. Biochem. 68: 89-125.

Kumar, P.S., Giffen, A.L., Barton, J.A., Paster, B.J., Moeschberger, M.L., and Leys, E.J. 2003. New bacterial species associated with chronic periodontitis. J. Dent. Res. 82: 338-344.

Kuramitsu, H.K. 1993. Virulence factors of mutans streptococci: role of molecular genetics. Crit. Rev. Oral Biol. Med. 4: 159-176.

Kuramitsu, H.K., and Trappa, V. 1984. Genetic exchange between oral streptococci during mixed growth. J. Gen. Microbiol. 130: 2497-2500.

Kuramitsu, H.K., Qi, M., Kang, I-C., and Chen, W. 2001. Role for periodontal bacteria in cardiovascular diseases. Ann. Periodontol. 6: 41-47.

Kvint, K., Nachin, I., Diez, A., and Nystrom, T. 2003. The bacterial universal stress protein: function and regulation. Curr. Opin. Microbiol. 6: 140-145.

Leblanc, D.J., Hawley, R.J., Lee, L.N., and St. Martin, E.J. 1978. "Conjugal" transfer of plasmid DNA among oral streptococci. Proc. Natl. Acad. Sci. USA. 75: 3484-3487.

Lee, S.F., and Boran, T.L. 2003. Roles of sortase in surface expression of the major protein adhesin P1, saliva-induced aggregation and adherence and cariogenicity of *Streptococcus mutans*. Infect. Immun. 71: 676-681.

Lee, S.F., Progulske-Fox, A., Erdos, G.W., Piacentini, D.A., Ayakawa, G.Y., Crowley, P.J., and Bleisweis, A.S. 1989. Construction and characterization of isogenic mutants of *Streptococcus mutans* deficient in major surface protein antigen P1 (I/II). Infect. Immun. 57: 3306-3313.

Li, L., Messas, E., Batista, E.L. Jr., Levine, R.A., and Amar, S. 2002a. *Porphyromonas gingivalis* infection accelerates the progression of atherosclerosis in a heterozygous apolipoprotein E-deficient murine model. Circulation. 105: 861-867.

Li, Y.H., Lau, P.C., Lee, J.H., Ellen, R.P., and Cvitkovitch, D.G. 2001. Natural genetic transformation of *Streptococcus mutans* growing in biofilms. J. Bacteriol. 183: 897-908.

Li, Y.H., Lau, P.C., Tang, N., Svensater, G., Ellen, R.P., and Cvitkovitch, D.G. 2002b. Novel two-component regulatory system involved in biofilm formation and acid resistance in *Streptococcus mutans*. J. Bacteriol. 184: 6333-6342.

Loesche, W.J. 1986. Role of *Streptococcus mutans* in human dental decay. Microbiol. Rev. 50: 353-380.

Loo, C.Y., Corliss, D.A., and Ganeshkumar, N. 2000. *Streptococcus gordonii* biofilm formation: identification of genes that code for biofilm phenotypes. J. Bacteriol. 182: 1374-1382.

Loo, C.Y., Mitrakul, K., Voss, I.B., Hughes, C.V., and Ganeshkumar, N. 2003. Involvement of the *adc* operon and manganese homeostasis in *Streptococcus gordonii* biofilm formation. J. Bacteriol. 185: 2887-2900.

Lundsford, R.D., and Robic, A.G. 1997. *comYA*, a gene similar to *comGA* of *Bacillus subtilis*, is essential for competence-factor-dependent DNA transformation in *Streptococcus gordonii*. J. Bacteriol. 179: 3122-3126.

Marquis, R.E. 1995. Oxygen metabolism, oxidative stress and acid-base physiology of dental plaque biofilms. J. Indust. Microbiol. 15: 198-207.

Marsh, P.D., 2003. Are dental diseases examples of ecological catastrophes? Microbiology. 149: 279-294.

McNab, R., Ford, S.K., El-Sabaney, A., Barbieri, B., Cook, G.S. and Lamont, R.J. 2003. LuxS-based signaling in *Streptococcus gordonii*: autoinducer-2 controls carbohydrate metabolism and biofilm formation with *Porphyromonas gingivalis*. J. Bacteriol. 185: 274-284.

Merritt, J.E., Qi, F., Goodman, S.D., Anderson, M.H., and Shi, W. 2003. Mutation of *luxS* affects biofilm formation in *Streptococcus mutans*. Infect. Immun. 71: 1972-1979.

Michalek, S.M., Katz, J., and Childers, N.K. 2001. A vaccine against dental caries: an overview. BioDrugs. 15: 501-508.

Ohta, H., Hara, H., Fukui, K., Kurhihara, H., Murayama, Y., and Kato, K. 1993. Association of *Actinobacillus actinomycetemcomitans* leukotoxin with nucleic acids on the bacterial cell surface. Infect. Immun. 61: 4878-4884.

Oral Health, U.S. 2002. Dental, Oral and Carniofacial Data Resource Center of the National Institute of Dental and Craniofacial Research, National Institutes of Health, Bethesda, Maryland.

O'Toole, G.A., and Kolter, R. 1998. Flagellar and twitching motility are necessary for *Pseudomonas aeruginosa* biofilm development. Mol. Microbiol. 30: 295-304.

O'Toole, G.A., Kaplan, H.B., and Kolter, R. 2000. Biofilm formation as microbial development. Annu. Rev. Microbiol. 54: 49-79.

Palmer, R.J. Jr., Gordon, S.M., Cisar, J.O. and Kolebrander, P.E. 2003. Coaggregation-mediated interactions of streptococci and actinomyces detected in initial human dental plaque. J. Bacteriol. 185: 3400-3409.

Palmer, R.J., Jr., Kazmerzak, K., Hansen, M.C., and Kolenbrander P.E. 2001. Mutualism versus independence: strategies of mixed-species oral biofilms in vitro using saliva as the sole nutrient source. Infect. Immun. 69: 5794-5804.

Pestova, E.V., Havarstein, L.S., and Morrison, D.A. 1996. Regulation of competence for genetic transformation in *Streptococcus pneumoniae* by an auto-induced peptide pheromone two-component regulatory system. Mol. Microbiol. 21: 853-862.

Pratt, L.A., and Kolter, R. 1998. Genetic analysis of *Escherichia coli* biofilm formation: roles of flagella, motility, chemotaxis, and type I pili. Mol. Microbiol. 30: 285-293.

Purevdorj, B., Costerton, J.W., and Stoodley, P. 2002. Influence of hydrodynamics and cell signaling on the structure and behavior of *Pseudomonas aeruginosa* biofilms. Appl. Environ. Microbiol. 68: 4457-4464.

Rashid, M.H., Rumbaugh, K. Passador, L., Davies, D.G., Hamood, A.N., Iglewski, B.H., and Kornberg, A. 2000. Polyphosphate kinase is essential for biofilm formation, quorum sensing, and virulence of *Pseudomonas aeruginosa*. Proc. Natl. Acad. Sci. USA. 97: 9636-9641.

Roberts, A.P., Cheah, G., Ready, D, Pratten, J., Wilson, M., and Mullany, P. 2001. Transfer of Tn*916*-like elements in microcosm dental plaque. Antimicrobiol. Agents Chemother. 45: 2943-2946.

Sauer, K., Camper, A.K., Ehrlich, G.D., Costerton, J.W., and Davies, D.G. 2002. *Pseudomonas aeruginosa* displays multiple phenotypes during development as a biofilm. J. Bacteriol. 184: 1140-1154.

Slade, G.D., Ghezzi, G.M., Heiss, G., Beck, J.D., Richie, E., and Offenbacher, S. 2003. Relationship between periodontal disease and C-reactive protein among adults in the atherosclerosis risk in communities study. Arch. Intern. Med. 163: 1172-1179.

Slots, J., and Gibbons, R.J. 1978. Attachment of *Bacteroides melaninogenicus* subsp. asaccharolyticus to oral surfaces and its possible role in colonization of the mouth and of periodontal pockets. Infect. Immun. 19: 254-264.

Smith, K.H., Bu, Y., and Suga, H. 2003. Induction and inhibition of *Pseudomonas aeruginosa* quorum sensing by synthetic autoinducer analogs. Chem. Biol. 10: 81-89.

Socransky, S.S., Haffajee, A.D., Cugini, M.A., Smith, C., and Kent, R.L. Jr. 1998. Microbial complexes in subgingival plaque. J. Clin. Periodontol. 25: 134-144.

Stichler, D.J., Morris, N.S., McLean, R.J.C., and Fuqua, C. 1998. Biofilms on indwelling urethral catheters produce quorum-sensing signal molecules *in situ* and *in vitro*. Appl. Environ. Microbiol. 64: 3486-3490.

Sutherland, I.W. 2001. Biofilm exopolysaccharides: a strong and sticky framework. Microbiology. 141: 3-9.

Svensater, G., Welin, J., Wilkins, J.C. Beighton, D., and Hamilton, I.R. 2001. Protein expression by planktonic and biofilm cells of *Streptococcus mutans*. FEMS Microbiol. Lett. 205: 139-146.

Tanaka, M., Hanoi, T., Takaya, K., and Shizukuishi, S. 1998. Association of oxygen tension in human periodontal pockets with subgingival inflammation. J. Periodontol. 69: 1127-1130.

Tanner, A.C., Milgrom, P.M., Kent, R. Jr., Mokeem, S.A., Page, R.C., Liao, S.I., Riedy, C.A., and Bruss, J.B. 2002. Similarity of the oral microbiota of pre-school children with that of their caregivers in a population based study. Oral Microbiol. Immunol. 17: 379-387.

Tanzer, J.M., Baranowski, L.K., Rogers, J.D., Hasse, E.M., and Scannapieco, F.A 2001. Oral colonization and cariogenicity of *Streptococcus gordonii* in specific pathogen-free TAN: SPZOM (OM) BR rats consuming starch or sucrose diets. Arch. Oral Biol. 46: 323-333.

Upton, M., Tagg, J.R., Wescmbe, P., and Jenkinson, H.F. 2001. Intra-and interspecies signaling between *Streptococcus salivarius* and *Streptococcus pyogenes* mediated by SalA and SalA1 lantibiotic peptides. J. Bacteriol. 183: 3931-3938.

Vacca-Smith, A.M., Jones, C.A., Levine, M.J., and Stinson, M.W. 1994. Glucosyltransferase mediates adhesion of *Streptococcus gordonii* to human endothelial cells *in vitro*. Infect. Immun. 62: 2187-2194.

Van Houte, J., Lopman, J., and Kent, R. 1994. The final pH of bacteria comprising the predominant flora of sound and carious human root and enamel surfaces. J. Dent. Res. 75: 1008-1014.

Vasil, M.L. 2003. DNA microarrays in analysis of quorum sensing: strengths and limitations. J. Bacteriol. 185: 2061-2065.

Vesey, P.M., and Kuramitsu, H.K. 2002. Interaction of *Treponema denticola* with *Porphyromonas gingivalis, Fusobacterium nucleatum* and fibronectin in an *in vitro* biofilm model. J. Dent. Res. 81 (Spec. Issue A): 364.

Vickerman, M.M.,and Minick, P.E. 2002. Genetic analysis of the *rgg-gtfG* junctional region and its role in *Streptococcus gordonii* glucosyltransferases activity. Infect. Immun. 70: 1703-1714.

Vickerman, M.N., Clewell, D.B., and Jones, G.W. 1991. Sucrose-promoted accumulation of growing glucosyltransferases variants of *Streptococcus gordonii* on hydroxyapatite surfaces. Infect. Immun. 59: 3523-3530.

Wang, B-Y., Chi, B., and Kuramitsu, H.K. 2002. Genetic exchange between *Treponema denticola* and *Streptococcus gordonii* in biofilms. Oral Microbiol. Immunol. 17: 108-112.

Watnick, P.I., and Kolter, R. 1999. Steps in the development of *Vibrio cholera* El Tor biofilms. Mol. Microbiol. 34: 586-595.

Whitchurch, C.B., Toker-Nielson, T., Rogas, P.C., and Mattick, J.S. 2002. Extracellular DNA required for bacterial biofilm formation. Science. 295: 1487.

Whiteley, M., Bangera, M.G., Bumgarner, R.E., Parsek, M.R., Teitzel, G.M., Lory, S., and Greenberg, E.P. 2001. Gene expression in *Pseudomonas aeruginosa* biofilms. Nature. 413: 860-864.

Wu, T., Trevisan, M., Genco, R.J., Dorn, J.P., Falkner, K.L., and Sempos, C.T. 2000. Periodontal disease and risk of cerebrovascular disease. The First National Health and Nutrition Examination Survey and its follow-up study. Arch. Intern. Med. 160: 2749-2755.

Yeung, M.K. 1999. Molecular and genetic analyses of *Actinomyces spp*. Crit. Rev. Oral Biol. Med. 10: 120-138.

Yildiz, F.H., and Schoolnik, G.K. 1999. *Vibrio cholerae* 01 El Tor: identification of a gene cluster required for the rugose colony type, exopolysaccharide production, chlorine resistance, and biofilm formation. Proc. Natl. Acad. Sci. USA. 96: 4028-4033.

Yoshida, A., and Kuramitsu, H.K. 2002a. Multiple *Streptococcus mutans* genes are involved in biofilm formation. Appl. Environ. Microbiol. 68: 6283-6291.

Yoshida, A., and Kuramitsu, H.K. 2002b. *Streptococcus mutans* biofilm formation: utilization of a *gtfB* promoter-green fluorescent protein (P*gtfB*:*gfp*) construct to monitor development. Microbiology. 148: 3385-3394.

Yoshida, A., Merritt, J., and Kuramitsu, H.K. 2002. Role of the *Streptococcus mutans luxS* gene in biofilm formation. Abst. Amer. Soc. Microbiol. p. 161.

Chapter 6

The Gut Microflora

Rodrigo Bibiloni, Jens Walter and Gerald W. Tannock*

Abstract

The gut of monogastric animals is colonised in the distal regions (ileum and colon) by a complex bacterial community in which anaerobic bacteria predominate numerically. Analysis of the composition of this community (generally referred to as the gut microflora) by the use of nucleic acid-based methods has revealed that many of the bacterial inhabitants have not yet been cultivated in the laboratory. Despite this handicap, information about the overall impact of the bacterial community on the host has been obtained by comparing the characteristics of germfree and conventional animals. In studying host-microflora relationships, however, it is essential to work with bacterial species that establish and persist within (colonise) the gut ecosystem rather than species that are merely transient. Differentiating between autochthonous and allochthonous species is therefore critical in investigations aimed at revealing the bacterial traits that are essential for life in the gut. Members of the genus *Lactobacillus* are ideal model bacteria with which to carry out such investigations because they predominate in the proximal regions of the gut of mice, poultry and pigs. It is estimated that hundreds of autochthonous bacterial species reside in the gut, yet the antigenic load associated with their cells does not stimulate a marked inflammatory response in the gut mucosa. In contrast, the presence of gut pathogens results in stimulation of the innate and adaptive immune systems and the eventual destruction of the pathogenic cells. Investigations to resolve the question as to how the mucosal immune system differentiates gut microflora from pathogenic species suggest that Toll-like receptors, oral tolerance mechanisms, and the production of secretory IgA molecules that coat the cells of members of the gut microflora, are involved.

1. An Introduction to the Gut Microflora

The animal body, including that of humans, is home to a vast collection of microbial species, mostly bacteria, that inhabits regions that are accessible to the microbes by one or more body orifices. This collection, known as the normal microflora, is acquired soon after birth and persists throughout life. Because of the variation in physical and chemical properties of the various body sites, different microbial communities exist in the oral cavity, upper respiratory tract, gastrointestinal tract (gut), vagina, and on the skin. It is possible, therefore, to recognize microbial communities characteristic of each site (oral microflora, gut microflora and so on). The largest numbers of bacteria reside in the distal gut (ileum and colon) of monogastric animals, but some animal species have relatively large numbers of lactic acid-producing bacteria in the proximal gut (forestomach of rodents, crop of chickens, pars oesophagea of pigs). This special foregut association is due to the adherence of lactobacilli to the surface of the non-secretory epthelium lining these sites, enabling the bacteria to form biofilms that provide a bacterial inoculum of the digesta. Although a complete catalogue of the inhabitants of the gut ecosystem is not yet available, hundreds of bacterial types, predominantly obligately anaerobic species, are estimated to be capable of residing in the distal regions (Tannock, 1995).

An ecosystem containing hundreds of bacterial species for study should be paradise for a bacteriologist. Yet analysis of this complex bacterial community is fraught with difficulties. Until the 1990s, analysis of the composition of the gut microflora relied on the use of traditional bacteriological methods of culture, microscopy and identification (O'Sullivan, 1999). Selective bacteriological culture media were essential for accurate analysis of the microflora because they enabled enumeration of specific bacterial populations to be made (Summanen *et al.*, 1993). Unfortunately, few culture media used in the analysis of the microflora are absolutely selective and misinterpretations by the novice are easily made. Not all of the species comprising a population may be able to proliferate with equal ease on the selective medium. This introduces bias to the results.

Even in the 1970s, researchers had observed that the total microscopic count of bacterial cells in human faecal smears was always higher than the total viable count (CFU, colony-forming units) obtained by culture on a non-selective agar medium. It was claimed, however, that

*For correspondence email: gerald.tannock@stonebow.otago.ac.nz

good bacteriological methods would permit the culture of 88% of the total microscopic count. But this comparison was obtained by using total microscopic 'clump' counts (aggregates of bacterial cells) rather than by counting individual bacterial cells in smears (Moore and Holdeman, 1974). While valid from the point of view that 'colony forming units' on agar plates have not necessarily arisen from a single bacterial cell, the comparisons gave a false sense of confidence with regard to analytical results at that time. Total bacteria microscopic counts, utilising the 4', 6-diamidino-2-phenylindole (DAPI) stain and computer imaging, have revealed average total bacterial cell counts in human faeces approaching 1×10^{11} per gram (wet weight) (Tannock *et al.*, 2000). State-of-the-art bacteriological methodologies still only permit about 40% of this bacterial community to be cultivated on non-selective agar medium in the laboratory (Tannock *et al.*, 2000). Thus a large proportion of the bacterial cells seen in microscope smears have never been investigated. Although some of these cells may be non-viable, it is likely that many are viable but non-cultivable due to their fastidious requirements for anaerobiosis or, more likely, due to the complex nutritional interactions that can occur between the inhabitants of bacterial communities (Suau *et al.*, 2000). These nutritional complexities may be difficult, if not impossible, to achieve in laboratory culture media.

2. Compositional Complexity of the Gut Microflora

Carl Woese had revealed that small ribosomal subunit RNA (16S rRNA in the case of bacteria) contained regions of nucleotide base sequence that were highly conserved across the bacterial world and that these were interspersed with variable to hypervariable regions (V regions). These V regions contained the signatures of phylogenetic groups and even species (Woese, 1987). With this knowledge in hand, new methods for the analysis of bacterial communities became available. Bacterial DNA or RNA could be extracted (in theory nucleic acid from all of the bacterial types in the sample will be represented in the extracts) and PCR amplification (reverse transcription-PCR in the case of RNA extracts) of the 16S rRNA gene in part or complete, could be accomplished. Clone libraries of the 16S rRNA genes could be made, and the clones sequenced, thus producing a kind of catalogue of the bacterial constituents of the ecosystem. As reported by Suau *et al.* (1999), the majority of cloned sequences (76%) from human faeces did not correspond to known organisms and were therefore derived from hitherto unknown species comprising the human gut microflora. From 16S rDNA sequence information, it was possible to derive DNA probes that specifically targeted variable regions of the 16S rRNA gene. Now it was possible to enumerate the various phylogenetic groups of bacteria inhabiting the human gut regardless of whether they could be cultured or not (Franks *et al.*, 1998; Sghir *et al.*, 2000). But making a library of hundreds of clones for every sample that needed to be investigated was logistically impossible. Even microscope counts using DNA probes is a serious undertaking and really requires an automated system for unbiased results to be obtained. A screening method to compare the bacterial composition of samples was needed, and PCR combined with denaturing gradient gel electrophoresis (DGGE) filled the bill (Zoetendal *et al.*, 1998). DNA or RNA can be extracted directly from intestinal or faecal samples. Then hypervariable 16S rDNA sequences can be amplified using PCR primers that anneal with conserved sequences that span the selected V regions. One of the PCR primers has a GC-rich 5′ end (GC clamp) to prevent complete denaturation of the DNA fragments during gradient gel electrophoresis. To separate the component 16S fragments in the PCR product, a polyacrylamide gel is used. The double-stranded 16S fragments migrate through a polyacrylamide gel containing a gradient of urea and formamide until they are partially denatured by the chemical conditions. The fragments do not completely denature because of the GC clamp, and migration is radically slowed when partial denaturation occurs. Because of the variation in the 16S sequences of different bacterial species, chemical stability is also different; therefore different 16S 'species' can be differentiated by this electrophoretic method. This method generates a profile of the numerically predominant members of the microflora (Muyzer and Smalla, 1998). Individual fragments of DNA can be cut from the gels, further amplified and cloned, then sequenced. The sequence can be compared to those in gene databanks in order to obtain identification of the bacterium from which the 16S sequence originated. Depending on the length of the sequence, identification to at least the level of bacterial phylogenetic group can be made (Zoetendal *et al.*, 1998). In a further development of this methodology, PCR primers specific for bacterial groups can be derived. These primers generate a profile of the species comprising a specific bacterial genus, for example, within the bacterial community (Walter *et al.*, 2001; Requena *et al.*, 2002).

Studies using PCR/DGGE have revealed that each human has a bacterial community of unique composition in the faeces that provides a further 'fingerprint' in addition to that of the fingers and the genome (Zoetendal *et al.*, 1998). This uniqueness has been confirmed by the use of another nucleic acid-based analytical method: fluorescent *in situ* hybridisation (FISH) (Stebbings *et al.*, 2002). The unique profiles are, moreover, conserved over time.

Ecosystems in general are recognised to have a high degree of stability (Alexander, 1971). This is because, when the environment changes, homeostatic reactions come into play to restore the relationships that pre-existed among the populations forming the community. Homeostasis makes modification of the composition of the gut microflora a difficult prospect because of this tendency of ecosystems to resist alteration. This has been demonstrated in detailed studies that have shown that consumption of "probiotic" bacteria results in their transient detection in the faeces. Once consumption ceases, the probiotic bacteria are no longer detected (Spanhaak *et al.* 1998; Dunne *et al.* 1999; Tannock *et al.*, 2000; Satokari *et al.*, 2001).

Three phylogenetic groups of bacteria predominate in the faecal microflora of humans (*Bacteroides-Prevotella* group, *Clostridium coccoides* group, *Clostridium leptum* group). Some of these phylogenetic groups contain a variety of genera and species, so molecular analytical approaches are amassing a huge catalogue of bacterial species that can be detected in samples from humans (Suau *et al.*, 1999; Sghir *et al.*, 2000). This information, however, may not be very helpful in understanding the functioning of the gut ecosystem. Knowledge of functional aspects of the colonic ecosystem has changed little during the past 30 years: food and secretions go in and bacterial cells and their metabolites are the products. Fifty per cent of faecal mass is made up of bacterial cells. What impact this mass of bacteria, regardless of taxonomy, has on the human host is what is probably important.

3. Determining the Impact of the Gut Microflora on the Animal Host

The function of the microflora in relation to its impact on the host has long been of interest in microbiology. Louis Pasteur thought that gut microbes were indispensable to life and proposed the derivation of germfree animals so that host-microflora relationships could be studied scientifically. Comparisons of the characteristics of germfree and conventional animals have clearly demonstrated that the total microflora has considerable influences on host biochemistry, physiology, immunology and low-level resistance to gut infections (Gordon and Pesti, 1971). Gnotobiotic research such as this can be done at a very sophisticated level nowadays because of the convergence of genome sequencing of animals and bacteria, and the consequent manufacture of DNA microarrays that provide sequences representative of the entire genome of the animal or bacterium. It is now possible, therefore, to measure the transcription of mammalian genes resulting from exposure of the animal to specific bacteria and vice versa. The potential of this approach has been demonstrated by the work of Lora Hooper in Jeffrey Gordon's laboratory, in which the impact of colonisation of ex-germfree mice by *Bacteroides thetaiotaomicron* was recorded (Hooper *et al.*, 2001). But if we are honest, major responses of the previously germfree animal to abrupt exposure to large numbers of bacteria is hardly surprising. It is much different in the case of the conventional animal where a complex bacterial community exists and the animal host has been exposed to bacteria of many types from soon after birth.

The composition of the human infant differs from that of the adult. Particularly noticeable are the higher numbers of facultative anaerobes (enterococci, enterobacteria) and bifidobacteria in relation to the total microflora when compared to that of adults (Tannock, 1994). A general decline in the numbers of enterococci and enterobacteria in the faeces occurs as the intestinal microflora of infants matures and as short-chain fatty acids increase in quantity and diversity in the intestine (Tannock and Cook, 2002). Bifidobacteria are predominant members of the infant intestinal microflora, forming between 60 and 91% of the total bacterial community in breast-fed babies and 28 to 75% (average 50%) in formula-fed infants (Harmsen *et al.*, 2000). Clearly, the molecular impact of this group of bacteria that predominate in the intestine during the formative years may be worthy of further attention.

The influence of the gut microflora on the immune system is of special interest because of the observed increase in allergies in children in affluent countries over recent decades (Burr *et al.*, 1989; Burney *et al.*, 1990). Pediatricians, some time ago, developed the hygiene hypothesis to try to explain this increased prevalence. Strachan advanced the hypothesis in 1989, but Gerrard had published earlier work that supports it (Gerrard, 1976). In brief, the hypothesis states that in affluent countries, the pattern of microbial exposure in early life has changed. Families are smaller, epidemics of childhood infectious diseases are rare, infections with helminths are rare, living standards are higher, and antibiotics are widely used to treat infants. This altered exposure to microbes, it is proposed, has predisposed the immune system of children to react to environmental antigens. Correlation between the occurrence of an atopic disease, asthma, and the frequency of treatments with antibiotics during the first year of life has been reported. In New Zealand, epidemiological data show that children are four times more likely to have asthma if they have been treated with three to four courses of antibiotics during the first year of life (Wickens *et al.*, 1999).

4. Determining the Impact of Bacteria Requires Knowledge of Autochthony and Allochthony

Lactobacillus species can be cultured from human faeces at population levels of up to 10^9 CFU per gram (Mitsuoka, 1992; Kimura *et al.*, 1997; Tannock *et al.*, 2000). Examination of *Lactobacillus* populations over extended periods of time has revealed marked variation in the complexity and stability of these populations among human subjects. Studies using genetic fingerprinting methods such as ribotyping and pulsed-field gel electrophoresis of DNA digests prepared from lactobacilli isolated on selective agar media, for example, permitted the evaluation of these populations and demonstrated, in some subjects, a succession of strains over time (Kimura *et al.*, 1997; Tannock *et al.*, 2000). These studies indicated that individual humans harboured a unique population of *Lactobacillus* strains. Moreover, the subjects could be divided into two groups: one group harboured relatively small populations of lactobacilli ($<10^6$ CFU per gram of faeces) that fluctuated in size and composition. Lactobacilli could not be cultured from the faeces of some of these subjects at intervals during the study period. These individuals also often harboured strains that could be detected consistently, but unpredictably (intermittently). Other *Lactobacillus* strains were detected in only a single sample. In the second group of subjects, lactobacilli were harboured in larger numbers ($>10^6$ CFU per gram of faeces) and one strain predominated thoughout the period of investigation (up to 15 months).

Research carried out at Rockefeller University in the 1960s demonstrated that certain bacterial populations inhabiting the gut of rodents were acquired soon after birth and persisted throughout the life of the animal (Dubos *et al.*, 1965). All members of the rodent colony harboured these bacteria in similar numbers. Other bacterial populations were present in only some colonies of animals of the same species. These observations led Rene Dubos and colleagues to postulate that microbial populations detected in the digestive tract could be divided into three categories.

- "Autochthonous microbiota" composed of persistent populations that were represented in every individual.
- "Normal microbiota" composed of populations that were frequently (but not always) detected in the ecosystem, but of variable population size.
- "True pathogens" that were accidentally acquired and produced disease.

The terminology was not adopted at the time, but the concept of "autochthony" is a valid one in the light of recent ecological observations of the intestinal ecosystem. Applying Dubos's terminology in a slightly different way, the persisting strains of lactobacilli discussed above could be described as "autochthonous" to the intestinal tract of the particular individual, meaning the strain was found where it was formed. Strains that were detected consistently but unpredictably could be termed "opportunistic" strains because they represent autochthonous, but numerically subdominant populations whose numbers vary over time so that their detection is intermittent among the numerically dominant strains. They could be said to be opportunistic because their population density may be determined by changes in the intestinal milieu produced by endogenous or exogenous factors. For example, the amount of non-digestible oligosaccharides and polysaccharides in the diet might influence the availability of certain substrates to bacteria inhabiting the intestinal ecosystem. When appropriate substrates became available, the subpopulations of opportunistic strains would increase and become detectable by the molecular typing and detection strategy. An opportunistic strain would not be detected when numerically subdominant because the molecular typing techniques and PCR-DGGE only monitor the dominant members of the population. "Foreign" strains transiently present in the gut could be called "allochthonous" to the intestinal tract (found in a place other than where they were formed). They could be strains whose habitat was located in a more proximal region of the digestive tract (e.g. the oral cavity) or organisms that had been ingested with food. Allochthonous strains ingested on a regular basis would be difficult to distinguish from opportunistic or autochthonous strains without extensive knowledge of the microbial ecology of the members of the genus *Lactobacillus*.

There are sixteen *Lactobacillus* species that are putative inhabitants of the human gut (Vaughan *et al.*, 2002), some of which were only recently detected by

Table 1. *Lactobacillus* species commonly detected in human faeces, saliva, and in food

Species	Faeces	Saliva	Food
L. acidophilus	+	+	
L. crispatus	+	+	
L. gasseri	+	+	
L. johnsonii	+		+
L. salivarius	+	+	
L. ruminis	+		
L. casei	+	+	+
L. paracasei	+	+	+
L. rhamnosus	+	+	+
L. plantarum	+	+	+
L. reuteri	+		+
L. fermentum	+	+	+
L. brevis	+	+	+
L. delbrueckii	+		+
L. sakei	+		+
L. curvatus	+		+

molecular techniques using PCR primers specific for lactic acid bacteria (Table 1). Studies conducted between 1960 and 1980 indicated that *Lactobacillus acidophilus*, *Lactobacillus fermentum*, *Lactobacillus salivarius* and an anaerobic bacterium, previously named *Catenabacterium catenaforme*, were the dominant autochthonous *Lactobacillus* species of humans (Lerche and Reuter, 1961; Reuter 1965b; Mitsuoka 1969; Mitsuoka *et al.*, 1975). Other species were commonly detected but were transient (Table 1). Based on current taxonomic criteria, most of the *L. acidophilus* isolates are now classified as *Lactobacillus gasseri* and *Lactobacillus crispatus* and most of the *L. fermentum* strains belong to the species *Lactobacillus reuteri* (Mitsuoka, 1992; Reuter, 2001). Isolates identified as *C. catenaforme* were later identified as non-motile variants of *Lactobacillus ruminis* (Reuter, 2001). The dominant and persistent species identified by Tannock *et al.* (2000) in four subjects over a period of fifteen months belonged to the species *L. ruminis* and *L. salivarius*. *L. acidophilus, L. crispatus,* and *L. gasseri* were regularly detected in the faeces, but strains of these species did not persist over time. Investigation of monthly faecal samples of humans using PCR-DGGE in combination with specific primers for lactic acid bacteria revealed that fluctuations were observed at the species level (Walter *et al.*, 2001; Heilig *et al.*, 2002). This was in contrast to PCR-DGGE profiles generated with universal bacterial PCR primers, which indicated that the composition of the microflora with regard to the numerically dominant bacterial species was very stable (Zoetendal *et al.*, 1998; Tannock *et al.*, 2000). PCR-DGGE with group-specific primers confirmed that *L. ruminis* was the dominant species in the faeces of some human subjects over several months and *L. salivarius, L. acidophilus, L. crispatus,* and *L. gasseri* could be regularly detected (Walter *et al.*, 2001; Heilig *et al.*, 2002). *L. gasseri* has been detected in association with human colonic biopsies (Zoetendal *et al.*, 2002). *L. reuteri*, once considered to be part of the autochthonous *Lactobacillus* microflora of humans (Mitsouka, 1992; Reuter, 2001), has been rarely detected in human faecal samples in recent studies either by culture or by nucleic acid-based methods of analysis (Ahrne *et al.*, 1998; Tannock *et al.*, 2000; Walter *et al.*, 2001; Heilig *et al.*, 2002; Dal Bello *et al.*, 2003).

All of the reports described above indicated that species such as *Lactobacillus paracasei, Lactobacillus rhamnosus, Lactobacillus delbrueckii, Lactobacillus brevis, Lactobacillus johnsonii, Lactobacillus plantarum* and *L. fermentum* were transient organisms and therefore allochthonous to the intestinal tract. Strains of these species are regularly present in fermented foods and are used as starter organisms in food production. Studies by Reuter (1965a), Jacobsen *et al.* (1999) and Bunte *et al.* (2000) showed that food-associated lactobacilli survive passage through the intestine and can be cultured in numbers comparable to that of autochthonous lactobacilli. Application of PCR-DGGE in combination with group-specific primers revealed that other food-associated lactic acid bacteria, such as *Lactobacillus sakei, Lactobacillus curvatus, Leuconostoc mesenteroides, Leuconostoc argentinum, Pediococcus pentosaceus*, and *Pediococcus acidilactici*, were regularly present in human faecal samples (Walter *et al.*, 2001; Heilig *et al.*, 2002). Most of these bacteria can be detected at about 10^6 CFU per gram of faeces (Walter *et al.*, 2001), especially if incubation conditions are used that are optimal for the growth of lactic acid bacteria associated with fermented food (30°C; atmosphere containing 2% oxygen) (Dal Bello *et al.*, 2003).

These observations raise fundamental questions concerning the ecological role of lactobacilli in the human intestinal tract and the use of probiotics (dietary supplements that contain viable bacteria that are putatively beneficial to health when consumed). It has been proposed that *Lactobacillus* strains used as probiotics should be of human origin (Dunne *et al.*, 1999). However, the observations discussed above demonstrate that it is hard to define whether lactobacilli detected in faeces are indeed inhabitants of the gut ecosystem. In turn, this questions the necessity for such a prerequisite. Strains used for probiotics are isolated from human faeces but belong to species which have never been shown to persist in the human intestine and are instead food-associated bacteria (*L. plantarum, Lactobacillus paracasei, Lactobacillus casei, Lactobacillus rhamnosus, L. johnsonii* and *L. fermentum*). Nevertheless, these species have been reported to confer health benefits when ingested as probiotics (Ouwehand *et al.*, 2002). Strains of bacteria incorporated in probiotic products can usually only be detected in the faeces of human subjects while they continue to consume the product. They behave like lactobacilli of food origin in that they transit the human gut alive but cannot colonise (persist) in the ecosystem. In this context, several questions remain to be answered: are lactobacilli of food origin suitable candidates for probiotics? Has food prepared by fermentations using lactic acid bacteria probiotic potential? What happens if an autochthonous *Lactobacillus* strain is ingested (strains of the species *L. ruminis* or *L. salivarius* which have been shown to persist in the human intestinal tract have never been included in probiotic trials)? Is a strain autochthonous for one person able to become autochthonous for another individual?

Due to the difficulties in obtaining samples of intestinal contents, most attention in gut microflora research has focused on the bacteria present in faeces. Considering that the human gut is several meters in length and constituted of several organs of significantly different physiology, it is impossible to predict the bacterial composition and the metabolic state of the bacteria in the different parts of the gut by investigating faecal samples. While Moore *et al.* (1978) demonstrated that the composition of the faecal microflora reflected that of the distal colon, the microflora is more metabolically active in the proximal colon compared to the distal colon (Cummings and MacFarlane, 1991). Some bacteria may be numerically predominant in the proximal large bowel or in the ileum, or associated with the mucosal surface, but may be outnumbered by other species in faeces and therefore difficult to detect. Lactobacilli can be detected in all parts of the human digestive tract including the mouth, stomach and small intestine (Mitsuoka, 1992; Reuter, 2001). They are present in human saliva in variable numbers but sometimes attain a population level exceeding 10^6 CFU per ml (Ahola *et al.*, 2002). The average output of saliva is 1000-1500 ml per day which, when swallowed, potentially introduces large numbers of oral lactobacilli into the gastrointestinal tract. The species composition present in the oral cavity and in faecal samples of humans coincides to some extent (Mikelsaar *et al.*, 1998; Ahrné *et al.*, 1999), but this topic has not been studied thoroughly. The species that predominate in the oral cavity, such as *L. acidophilus* (and closely related species like *L. gasseri* and *L. crispatus*), *L. plantarum, L. salivarius, L. brevis, L. rhamnosus* and *L. paracasei* (Ahrné *et al.*, 1998; Marsh and Martin, 1999) are frequently isolated from human faeces (Ahrné *et al.*, 1998; Tannock *et al.*, 2000; Walter *et al.*, 2001; Dal Bello *et al.*, 2003). Strains of these species can be detected consistently, but unpredictably, and in numbers less than 10^6 CFU per gram of faeces (Tannock *et al.* 2000), suggesting that they enter the gut on a relatively regular basis. The small numbers of lactobacilli present in the stomach and proximal small intestine of humans are presumably transient bacteria because the conditions in this part of the gut are too hostile for colonisation to occur. It has been reported that lactic acid bacteria are more numerous in the caecum and proximal colon than in the faeces of humans (Reuter, 1965b; Marteau *et al.* 2001). Perhaps the lactobacilli detected in faeces represent remnants of much larger populations inhabiting the proximal colon? Clearly there is a requirement for more sophisticated and more invasive sampling methods to study the microflora in more proximal parts of the intestinal tract. Bacteria resident in the ileum might have an influence on the immunology and physiology of the host because Peyer's patches, which play a major role in sampling the antigenic milieu, are numerous in the distal ileum (Van Kruiningen *et al.*, 2002).

The bacteria residing in the gut of mammals have co-evolved with their host and have developed a high degree of adaptation and specialization. These bacteria must possess traits that enable them to establish and maintain themselves in a highly competitive environment. The challenge for the bacteria is to satisfy their own growth requirements as well as coping with the hostile conditions generated by the competing members of the microflora and by the defense mechanisms of the host.

Lactobacilli constitute a tiny proportion of the human faecal microflora (<1% of the total bacteria). Molecular characterization of the microbial composition of human faeces as well as of the ileal and colonic mucosa by direct analysis of 16S rDNA sequences, has revealed that clones derived from lactobacilli were absent from these libraries (Suau *et al.*, 1999; Hayashi *et al.*, 2002; Wang *et al.*, 2003). The situation is different in other animal species such as pigs, mice and chickens, where lactobacilli can be detected by direct cloning and can be cultured in relatively large numbers from throughout the intestinal tract (Mitsuoka, 1992; Gong *et al.*, 2002; Leser *et al.*, 2002; Salzman *et al.*, 2002). Unlike the human stomach, which is lined with a glandular mucosa, the stomach of pigs, mice, rats and the crop of birds are lined, at least partly, with a non-glandular, squamous stratified epithelium (Tannock, 1992). These regions are densely colonised by lactobacilli adhering directly to the epithelium forming a layer of bacterial cells. Lactobacilli shed from this layer inoculate continuously the digesta, and lactobacilli are therefore detected in large numbers throughout the gut (Tannock, 1997a). The presence of a squamous stratified epithelium in some region of the gut seems to be a prerequisite for lactobacilli to form a significant proportion of the gut microflora.

The mechanism by which *Lactobacillus* strains adhere to these epithelia has not yet been determined, but preliminary *in vitro* investigations have shown that both carbohydrate and protein molecules are involved (Tannock, 1997a). *Lactobacillus* strains that adhere to epithelial cells show specificity for an animal host. Strains originating from the rodent forestomach do not adhere to crop epithelial cells, while isolates from poultry do not adhere to epithelial cells from the rodent forestomach or the pars oesophagea of pigs (Tannock, 1997a). However, some exceptions occur. For example, *Lactobacillus* strains isolated from chicken adhered to pig squamous epithelial cells (Tannock *et al.*, 1982) and a strain of *L. reuteri*, isolated from calf faeces, adhered strongly to the squamous epithelium of the mouse stomach (Sherman and Savage, 1986).

When attached and replicating on a host surface, a microbe can persist in a flowing habitat whereas non-adherent microbes are carried away by the flow of secretions (Tannock, 1995). Therefore, adherence to intestinal epithelia as described for rodents, pigs and birds has been considered to play a significant role in persistence in the gut ecosystem after being ingested as a probiotic (Vaughan *et al.*, 1999). A stratified squamous epithelium is present in the mouth and the vagina of humans (Marieb, 1989) and lactobacilli are commonly isolated from these habitats in relatively large numbers (Tannock, 1995; Ahola *et al.* 2002). It would be interesting to learn whether adherence to stratified epithelia is as significant a mechanism for oral and vaginal lactobacilli to persist in these ecosystems as it is for lactobacilli colonising the proximal parts of the murine, avian and porcine gut. A stratified epithelium is not present in the intestinal tract of humans, but it has been shown that some lactobacilli have the ability to bind to intestinal mucus and polymers associated with the surface of enterocytes and this trait might be involved in intestinal colonisation (Roos and Jonsson, 2002). Lactobacilli adhere to human intestinal cell lines (Caco-2 and HT-29) (Haller *et al.*, 2001) or epithelial cells (enterocytes) harvested from the intestinal tract (Tannock, 1997b). On the other hand, colonisation of mucus associated with tissue surfaces by members of the normal microflora is very limited in humans and the numbers of bacteria obtained from washed tissue surfaces are considerably lower than those observed in studies of rodents (Tannock, 1995). Evidence for significant *in vivo* association of lactobacilli with the columnar epithelium of the human gut is inconclusive. Furthermore, *in vitro* tests have shown that adherence is highly dependent on certain physiological conditions (e.g. pH), and correlation was not observed between the *in vivo* findings and adherence of a particular bacterial strain to enterocytes or cell lines. Thus it appears impossible to predict the potential of a strain to persist in the human gut by this method (Wold, 1999; Morelli, 2000).

Comparative genomic analysis between *Lactobacillus* species that occupy the gut is expected to identify operons that are necessary for survival and activity in this environment. In addition, comparative genomics of intestinal lactobacilli with other lactic acid bacteria from distinctive environments will hopefully identify the genomic regions that are involved in the adaptation of lactobacilli to many specialized and varied environments including the gut (McAuliffe and Klaenhammer, 2002). For example, comparative genomics have revealed the presence of an operon for the utilisation of fructo-oligosaccharides (FOS) in *L. acidophilus* (Barrangou *et al.*, 2003). It would be interesting to compare the genome sequences of *Lactobacillus* strains that are allochthonous to the gut with those belonging to *L. ruminis* which have been shown to persist in the human intestinal tract in relatively large numbers (Tannock *et al.*, 2000).

The complete genome sequence of the type strain of *Bacteroides thetaiotaomicron*, which was isolated from human faeces, revealed that the proteome of the bacterium includes an elaborate apparatus for acquiring and hydrolyzing dietary and host-derived polysaccharides and an associated environment-sensing system consisting of a large repertoire of environmentally regulated expression systems (Xu *et al.*, 2003). Genes conferring the ability to adhere to enterocytes or mucus were not identified. This strain has the ability to induce the fucosylation of glycoconjugates produced by enterocytes in the bowel of mice. (Bry *et al.*, 1996). It is proposed that the *Bacteroides* cells, which utilise L-fucose as an energy source, control the availability of this growth substrate in the gut environment (Hooper *et al.*, 1999). Regulation of host gene expression is mediated through a soluble factor produced by the bacteria (Bry *et al.*, 1996; Hooper *et al.* 1999). As in the case of *Bacteroides thetaiotaomicron,* the genome of *Bifidobacterium longum* contains a large number of genes encoding proteins specialized in the catabolism of a variety of oligosaccharides, glycoproteins and glycoconjugates. Many of the genes for oligosaccharide metabolism occur in self-regulated modules. Complete pathways for the synthesis of amino acids, nucleotides, and some key vitamins were identified. Since biofilm formation in the human intestine has not been reported to date, it appears that *Bacteroides thetaiotaomicron* and *Bifidobacterium longum* base their ecological competitiveness on the utilization of complex nutrients using well-regulated pathways to save energy and assure high proliferation rates in the lumen of the gut. Lactobacilli have complex nutritional requirements such as amino acids, peptides, nucleic acid derivatives, vitamins, salts, fatty acid esters, and fermentable carbohydrates for growth (Kandler and Weiss, 1986). It appears that lactobacilli are numerous in habitats where these nutrients are readily available. In the proximal parts of the gut of pigs, rodents and birds and in the mouths of humans, substrates for the colonizing lactobacilli are abundant and are probably supplied, for the most part, by the host's food. Easily accessible nutrients are probably in very low supply in the colon due to their absorption in the small intestine, hence the minor contribution of lactobacilli to the total faecal microflora. It would be interesting to know which nutrients are utilized by autochthonous lactobacilli for cell maintenance in distal parts of the gut. Sugars and amino acids are present in the large intestine due to the metabolism of macromolecules by other members of the microflora (MacFarlane *et al.*, 2000), but competition for these substrates must be intense. Although lactobacilli have always been considered to colonise the large bowel of various animals including humans, the availability of growth factors available for their growth in this habitat has rarely been addressed.

In the last decade, promoter-trapping technologies have been developed to overcome the limitation of *in vitro* models to study the traits that enhance ecological performance in complex ecosystems. *In vivo* expression technology (IVET) was developed by Mahan and co-workers (1993) to study gene expression by *Salmonella typhimurium* during infection of mice. IVET has been used to identify *in vivo* induced (*ivi*) genes for a number of other pathogens (Rainey and Preston, 2000) and mutations within a subset of these *ivi* genes resulted in a decrease in virulence. IVET has been used to identify *Lactobacillus reuteri* strain 100-23 genes specifically induced in the murine gut (Walter *et al.*, 2003). A plasmid-based system was constructed containing *'ermGT* (confers lincomycin resistance) as the primary reporter gene for selection of promoters active in the gastrointestinal tract of mice treated with lincomycin. A second reporter gene *'bglM* (beta-glucanase) allowed the differentiation between constitutive and *in vivo* inducible promoters. Application of the IVET system using *L. reuteri* and reconstituted lactobacillus-free mice detected three genes that were induced specifically during colonisation. Sequences showing homologies to xylose isomerase (*xylA*) and methionine sulfoxide reductase (*msrB*) were detected. The third locus showed homology to a protein of unknown function. Xylose is a typical plant-derived sugar commonly found in straw and bran and is introduced into the gut via the ingested food. Additionally, xylose is a product of the breakdown of xylans and pectins by the gut microflora (MacFarlane *et al.*, 2000). *L. reuteri* may also utilise isoprimeverose, which is the major building block of xyloglucans, as described for *L. pentosus* (Chaillou *et al.*, 1998). The selective expression of xylose isomerase suggests that *L. reuteri* 100-23 meets its energy requirements in the gut at least partly by the fermentation of xylose or isoprimeverose.

Methionine sulfoxide reductase (MsrB) is a repair enzyme protecting bacteria against oxidative damage caused by reactive nitrogen and oxygen intermediates. Nitric oxide (NO) is produced by epithelial cells of the ileum and colon and possibly acts as an oxidative barrier, maintaining intestinal homeostasis, reducing bacterial translocation and providing a means of defense against pathogenic microorganisms (Hoffman *et al.*, 1997; Roberts *et al.*, 2001). NO production by human bronchial epithelial cells reduced *Pseudomonas aeruginosa* adherence to a cultured cell line (Darling and Evans, 2003). It would be interesting to investigate the role of methionine-sulfoxide reductase in the adherence of *L. reuteri* 100-23 to the non-secretory epithelium of the murine forestomach.

Methods are now available to study selective gene expression in ecosystems and are a powerful way of determining the interaction of bacterial cells with their environment and to identify traits that are necessary for the persistence of the bacteria. The bacterial transcriptome is a dynamic entity that reflects the cell's immediate, ongoing and genome-wide response to its environment. Techniques such as DNA microarray expression profiling, selective-capture-of-transcribed-sequences (SCOTS) and direct quantitative transcript analysis using RT-PCR have already been applied to study the gene expression of pathogens under certain conditions, in host cells and during infection (Graham and Clark-Curtiss, 1999; Goerke *et al.*, 2000; Conway and Schoolnik, 2003). SCOTS has already been used in conjunction with genomic array hybridization to characterize gene expression of *Helicobacter pylori* in association with the gastric mucosa (Graham *et al.*, 2002). Additionally, the expression of several specific *H. pylori* genes were monitored by real time RT-PCR in stomach biopsies during the course of a 6 month infection in mice (Rokbi *et al.*, 2001). The application of these techniques to bacteria inhabiting the gut ecosystem requires rigorous and rapid methods for mRNA purification. mRNA-based methods and IVET have the potential to identify genes of lactobacilli and other intestinal bacteria that have not previously been functionally characterized but might play a role in bacterial persistence in the gut ecosystem.

Perhaps the greatest enigma concerning the gut microflora is the mechanism by which large numbers of bacterial cells can persist in intimate association with the bodies of animals without inducing a marked inflammatory response on the part of the host. Antigens from gut bacteria are detected by the host's immune system because antibodies that react with them are present in the sera of healthy humans (Gillespie *et al.*, 1950; Evans *et al.*, 1966; Hoiby and Hertz, 1979; Cohen and Norins, 1996; Kimura *et al.*, 1997). IgM antibodies reactive with the cells of lactobacilli and bifidobacteria have been detected in the serum of the majority of humans tested (Kimura *et al.*, 1997). Some autochthonous members of the gut microflora do not elicit as great an immune response from the host as do allochthonous bacteria (Berg and Savage, 1972). The immunological environment of the host may provide a selective force for the evolution of a special relationship. Immunological selection of genetic variants less antigenically foreign to the host may favor the establishment of certain bacterial strains as members of the gut microflora.

Using PCR-DGGE, it has been shown that the host genotype affects the bacterial community in the human gut because twins harbor more similar bacterial communities in their intestine than do unrelated individuals (Zoetendal *et al.*, 2001). Additionally, *Bacteroides fragilis* has been demonstrated to be capable of modulating its antigenicity by producing at least 8 distinct capsular polysaccharides, and to be able to regulate their expression by the reversible inversion of DNA segments containing the promoters directing expression of the corresponding polysaccharide biosynthesis genes (phase variation) (Krinos *et al.*, 2001). Seven capsular polysaccharide synthesis loci were also identified in the genome of *Bacteroides thetaiotaomicron* (Xu *et al.*, 2003). Phase variation with respect to capsular polysaccharide may help the *Bacteroides* cells maintain their residence in the gut in the face of constant immune pressure from the host. Berg (1983) observed that *Lactobacillus* strains isolated from the murine gut (autochthonous) elicited a relatively weak immune response when administered parenterally to mice. Strains from humans (allochthonous to the mouse) produced a stronger immune response in the animals.

L. acidophilus cells express an S-layer protein which crystallizes into a regular monolayer on the bacterial cell surface. Interestingly, the strain harbours two S-protein-encoding genes, and the expression of the genes can be interchanged by gene inversion (Boot *et al.*, 1996a). Other studies have revealed the presence of two S-layer proteins in *L. crispatus* (Boot *et al.*, 1996b). These lactobacilli might therefore change their antigenicity by altering their dominant surface proteins. Studies aimed at recognizing the molecular mechanisms of persistence in the gut ecosystem would increase knowledge about homeostasis, the degree to which the host develops immunological tolerance to members of the microflora, and to the degree to which bacteria adapt to the conditions provided by the host's immune system.

5. How Does the Immune System Tell Microbial Friend from Foe?

Several questions that are of interest to microbiologists and immunologists alike are posed by consideration of the host-microflora relationship. How does the immune system tolerate the presence of the huge antigenic burden associated with the gut microflora? How does the immune system analyse an immense and diverse collection of bacterial cells and differentiate between nonpathogens and pathogens? Is there any genetic predisposition of the host to tolerate certain kinds of bacteria more than others? The description of the Toll-like receptors (TLR) as a means of sensing the outside world and the role of dendritic cells (DC) in the initiation and orchestration of both the innate and adaptive immune responses, has paved the way to a much better understanding of how the immune system responds appropriately to a wide variety of antigens (Medzhitov, 2001).

The containment of the gut microflora and the elimination of pathogens without disrupting systemic homeostasis require the presence of a meticulously organized system composed of cellular components, molecular mediators and receptors that continuously sense the microbial environment. Moreover, the system must be able to discriminate between potentially dangerous and harmless antigens and generate appropriate responses. Surveillance of the gut ecosystem occurs at different levels: in the intestinal lumen or in the surrounding tissue, even systemically, and by different mechanisms, including elements of innate and adaptive immunity. The application of methods such as DNA microarrays, laser capture micro-dissection and proteomics, in combination with the use of gnotobiotic animals, can help reveal the mechanisms involved and help derive therapies that can be used when mucosal homeostasis fails.

Due to the continuous antigenic stimulus derived from bacterial cells inhabiting the intestinal tract, it is not surprising that a large proportion of the immune cells in the body is associated with the gut. These immune cells are referred to as the gut-associated lymphoid tissue (GALT). The antigens present in the gut contents are physically separated from the systemic milieu by a highly selective barrier composed of enterocytes. There is a low rate of translocation of bacteria across the gut epithelium but these bacterial cells are normally intercepted by the innate immune system and killed by phagocytosis. Providing that the epithelium remains intact, an initial barrier to the passage of potentially harmful microbes is present. The epithelium is also a suitable site to carry out a surveillance/sampling process. Though many epithelial cell lines have been derived to study microbial-cell interactions, some care must be used in interpreting the results of *in vitro* studies. There is evidence that there is cell diversity within epithelial tissue and, consequently, epithelial function results from a combination of individual performances of each cell type. Any speculations based on the use of pure cell lines in culture may be of limited validity (Gordon *et al.*, 1997).

Three mechanisms by which the GALT makes contact with luminal antigens have been described. First, located above lymphoid follicles and interspersed between enterocytes, M cells deliver a wide variety of substances and microbes (pathogenic and nonpathogenic alike) to the underlying antigen-presenting cells (APC) (Neutra, 1998), such as DC, which in turn process and present (in association with co-stimulatory molecules) antigenic peptides to T cells by a major histocompatibility complex (MHC). Second, although the *in vivo* situation is still unclear, enterocytes may behave as APC by taking up antigens and expressing the processed fragments on MHC class II molecules (Mayer, 1998). Though they cannot express co-stimulatory molecules required for T-cell activation, they have the capacity to produce chemokines and cytokines that could launch an innate mucosal immune response (Kagnoff, 1997). The production of interleukin-8 (IL-8), which is a potent polymorphonuclear chemo-attractant, by enterocytes is an example of this. Third, through a pathway described in mice, luminal contents can be sampled by a subset of sub-epithelial DC, which are able to extend projections through tight junctions between enterocytes into the lumen to sample the contents without disrupting the epithelial barrier function. Through ligated-loop experiments in mice, in which 3-cm-long stretches of the small intestine were ligated at both extremities and injected with bacteria, Rescigno and co-workers (Rescigno *et al.*, 2001) showed that the number of CD11c+ DC recruited to the lumen doubled when pathogenic *Salmonella typhimurium* was administered in addition to non pathogenic *Escherichia coli* DH5α. The number then decreased over time. By contrast DC in intestinal loops injected with *E. coli* alone remained constant. These results suggested that, whereas DC that interacted with the pathogen left the epithelium, those that encountered avirulent bacteria remained in place.

Enterocytes and DC sense the presence of microbial cells and their products by recognition of unique

Table 2. Ligands of TLR family members

Member	Ligand	Reference
TLR -1	Unknown[1]	(Takeuchi *et al.*, 2002; Zarember and Godowski, 2002)
TLR -2	LPS (*Leptospira, Porphyromonas*), peptidoglycan, diacylated mycoplasmal lipopeptide	(Ozinsky *et al.*, 2000; Takeuchi *et al.*, 2001)
TLR -3	dsRNA, polyinosine-polycytidylic acid	(Alexopoulou *et al.*, 2001)
TLR -4	LPS, lipoteichoic acids	(Poltorak *et al.*, 1998)
TLR -5	flagellin	(Hayashi *et al.* 2001)
TLR -6	peptidoglycan, zymozan (yeast)	(Ozinsky *et al.*, 2000)
TLR -7	unknown, (imidazoquinoline compounds)	(Hemmi *et al.*, 2002)
TLR -8	unknown, (imidazoquinoline compounds)	(Jurk *et al.*, 2002)
TLR -9	unmethylated CpG-DNA	(Bauer *et al.*, 2001)
TLR -10	unknown	(Chuang and Ulevitch, 2001)

TLR-1 was associated with TLR-2 in response to triacylated lipopeptides but not diacylated lipopeptides.

molecular patterns. These "signature molecules" are often referred to as pathogen-associated molecular patterns (PAMP). However, as pointed out by Didierlaurent and co-workers (2002), it would be more appropriate to use the term microbe-associated molecular patterns (MAMP) since non-pathogenic microbes share these patterns with pathogens. These structures are unique to microbes, are evolutionarily conserved, and are not present in the host cells. This makes them ideal targets for recognition by the innate immune system. MAMP are recognized by a heterogeneous group of pattern-recognition receptors (PRR) that are expressed either on the surface or in the cytosol of the "gatekeeper" eukaryotic cells, or are secreted. Several recognition domains have been described including C-type and mannose binding lectins, cysteine-rich domains and leucine-rich repeats (LRR), the latter comprising the extracellular domain of the TLR family. TLR (10 members described so far) can recognize diverse molecules such as lipoproteins and lipopolysaccharides (LPS), flagellin, viral dsRNA and bacterial DNA, and even synthetic compounds (Hemmi, 2002) (Table 2). The binding of MAMP to the corresponding PRR triggers signal transduction pathways that, although only partially understood, result in the activation of transcription factors essential for the expression of several cytokine, co-stimulation molecule, and chemokine genes. The production of effector molecules such as nitric oxide and antimicrobial peptides that destroy pathogens can also be induced. Despite differences in the activation cascades initiated by individual TLR, nuclear factor-kappa B (NF-κB) has a pivotal role in the initiation of a variety of immune responses. NF-κB is generally kept inactive in the cytoplasm through its association with IκB (inhibitor of NF-κB). A shared pathway involves the degradation of IκB which results in the release of NF-κB. It migrates to the nucleus and induces transcription of a variety of genes including those encoding the synthesis and production of pro-inflammatory cytokines and chemokines, such as those for tumour necrosis factor–α (TNF-α), IL-8 and interleukin-12 (IL-12) (Barnes and Karin, 1997).

The recognition of MAMP associated with overt pathogens, such as viral dsRNA which is produced by most viruses during their replication, or antigens derived from helminths, is clearly critical to the defense of the host. But how can TLR distinguish between LPS from *Salmonella* species that could cause infection and LPS from nonpathogenic *E. coli*?

Transgenic mice and gene knock-out mice have been useful in deciphering TLR specificities. Experiments with TLR6- and TLR1-deficient mice have revealed that TLR can recognize subtle structural differences in lipopeptides (Takeuchi *et al.*, 2001). In addition, TLR-2 recognizes atypical LPS from *Leptospira interrogans* or *Porphyromonas gingivalis* (Hirschfeld *et al.*, 2001; Werts *et al.*, 2001), both of which have different biochemical and physical properties. The combined effect of different TLR, or the effect of additional molecules (CD14 or MD-2), can add an extra degree of specificity to the discrimination of MAMP. It seems, therefore, that TLR might be endowed with the capacity to differentiate between MAMP from pathogenic and non-pathogenic microbes and thus elicit appropriate responses in each case. It is important to note that the use of transgenic animal methods could alter other components of the immune system such as lymph-node and splenic development and/or systemic responses (Mowat, 2003). Observations based on these methodologies have to be carefully evaluated.

A genetic variation in TLR-4 and a variation in the expression pattern of receptors among individuals have been described (Raby *et al.*, 2002). This may help to explain the uniqueness of the bacterial community that each human can harbor in their gut. Several groups have studied the genetic polymorphisms in TLR-4 by sequencing the complete coding region of TLR-4 in individuals of diverse ethnic origin. This research has identified amino-acid sequence variations, some resulting in a reduction in cell-surface expression of the receptor, and subsequent disruption of LPS-mediated signaling (Arbour *et al.*, 2000; Smirnova *et al.*, 2001).

Flow cytometric analysis and confocal laser microscopy can be used to monitor the expression and localization of different TLR in target cells. Using monoclonal antibodies against TLR, the surface localization of TLR-1, TLR-2 and TLR-4 has been shown (Yoshimura *et al.*, 1999; Akashi *et al.*, 2000; Ochoa *et al.*, 2003), whereas TLR-5 is expressed exclusively on the basolateral side of enterocytes (Gewirtz *et al.*, 2001). In contrast, TLR-9 is located intracellularly (Hemmi *et al.*, 2000), suggesting a different recognition mechanism than those operating for the other TLR. Flow cytometry results rely on the specificity of the labeled antibodies directed against distinctive epitopes in the different TLR. Unfortunately, there is a commercial scarcity of such antibodies.

Recent work has suggested that signaling through other DC surface receptors may interfere with TLR-mediated activation resulting in inhibition of an inflammatory response. Glycolipid (lipo-arabinomannan) associated with mycobacterial cells binds to a C-type lectin, specific to DC and induces the production of IL-10 thereby compromising DC function (Geijtenbeek *et al.*, 2003). It is possible that components of the gut microflora may be similarly immunosuppressive.

Oligonucleotide microarrays are a valuable tool in exploring cellular responses to the presence of microbial cells. DC or enterocytes can be exposed to the microbial cells, then eukaryotic mRNA is extracted, converted to labeled cDNA, and hybridized to an appropriate microarray containing immobilized oligonucelotides that represent the genes whose expression is to be measured (cytokines, antigen presentation, T-cell regulation, and signaling). Hybridized targets are measured and quantified. The advantage of DNA microarray technology is that the transcription of a large number of genes can be monitored simultaneously. The effect of cellular components such as cell wall, LPS or nucleic acids can be analyzed. Applying this methodology to measure gene expression profiles of DC in response to several kinds of microbes, Huang and co-workers (2001) reported that these cells generated common responses to all of the microbes that were tested, but also specific responses to some of them that might lead to different physiological consequences. In this elegant approach, human DC were exposed to different microbes and microbial components (*E. coli* and LPS, *Candida albicans* and cell wall mannan, influenza A virus and dsRNA) for between 1 and 36 hours. The impact of exposure to these cells and substances was monitored for approximately 6800 genes. Altered expression of 1330 genes after exposure to at least one of the pathogens or components was recorded. The authors showed that specific genes were regulated by each pathogen. The intersection of the responses induced by the three pathogens revealed a common set of 166 regulated genes whose responses were elicited independently of pathogen characteristics.

Laser capture microdissection (LCM) can be used to isolate specific groups of cells from a heterogeneous tissue specimen. The LCM system is basically an inverted microscope fitted with a low-energy infrared laser. Samples (tissue fragments) are mounted on glass slides and covered with a thermoplastic film. The laser transiently (short periods avoid any biological damage to the cells) melts the film in a precise location, binding it to the individual cells or clusters. The film with adherent cells is removed, leaving the unselected tissue in contact with the glass slide. These cells can then be investigated using molecular methods (2D PAGE, microarray, sequencing). Temporal studies (taking samples at successive points) can identify not only the transcriptional changes but also the sequence in which they take place. This provides evidence about which genes control a specific response directly, and which are indirectly affected. The major disadvantage of the approach is that minute amounts of material are extracted, which limits the extent of subsequent analysis.

Gene expression studies have shown that mRNA and expressed protein concentrations do not necessarily correlate. Thus proteomic analysis has emerged as a useful tool to study the production of proteins, their functionality, and protein-protein interactions. The scope is extremely broad: proteomics might be used to measure the abundance of cytokines produced as a result of a bacterial stimulus, the presence and localization of signaling components in the TLR transduction cascade, ligand specificities of receptors and their interactions with adaptor molecules.

In spite of the evident advantages of these sophisticated methodologies, classical molecular methods are still important in investigating the differential recognition of non-pathogens and pathogens by the immune system. Northern hybridization has been used to show that, while most enteric pathogens activate

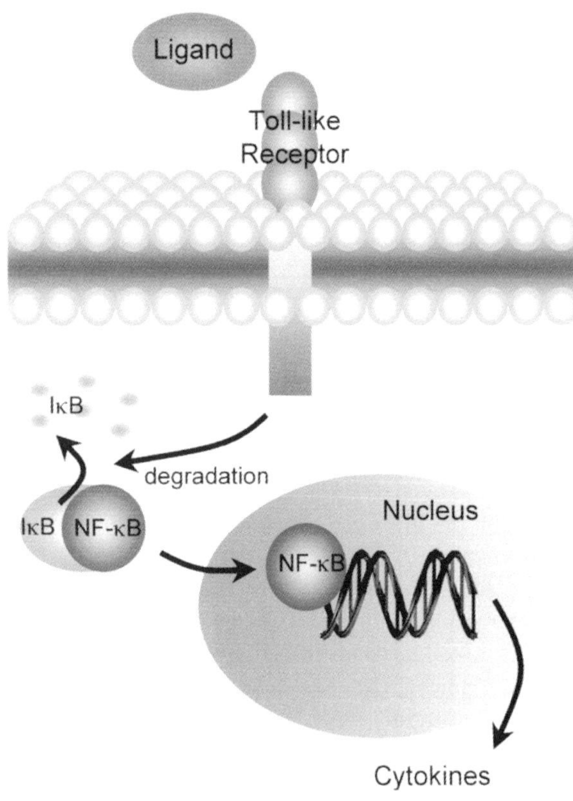

Figure 1. Diagram representing the release of NF-κB resulting in the induction of an immune response.

the NF-κB pathway resulting in the secretion of pro-inflammatory chemokines, nonpathogens may be able to selectively attenuate this pathway by inhibiting the polyubiquitination step required for IκB degradation, thereby retaining NF-κB in the cytoplasm (see Figure 1) (Neish *et al.*, 2000). The production of an inflammatory response is consequently avoided. Such studies suggest that the modulation of pro-inflammatory responses could occur at the NF-κB level. The use of nonpathogenic bacteria that utilize this mechanism of control could possibly be employed in the treatment of inflammatory diseases.

It is possible that APC receptors might be able to recognize soluble compounds produced by bacteria as intercellular signaling messengers. The bacteria secrete diffusible chemical signals that allow bacterial cells to take a census of their population size and, as a result, express particular target genes. This inter-bacterial communication phenomenon, sometimes referred to as quorum-sensing, may operate in some intestinal pathogens in order to activate genes essential for infection (Speradino *et al.*, 2003). It has also been described in the production of virulence factors by staphylococci, the production of antibiotics, and in the formation of biofilms (Winans, 2002). Although originally described for microbe-microbe communication, it has been reported that cell-cell signaling crosses the prokaryote-eukaryote boundary. Smith and co-workers (2001) showed that human epithelial cells, *in vitro*, sense and respond to a quorum-sensing molecule (homoserine lactone; AHL) produced by *Pseudomonas aeruginosa*, which as a result secrete IL-8. The legume *Medicargo truncatula*, a close relative of alfalfa, has been shown to distinguish between AHL from both symbiotic and pathogenic bacteria and respond to each in different ways (Mathesius *et al.*, 2003). The power of DNA microarray analysis could again be exploited in this particular case to identify the eukaryotic genes activated by the bacterial signalling molecules.

TLR can recognize "danger signals" produced by the host in response to cell damage or cellular stress. These endogenous signals can be heat shock proteins, nucleotides or extra-cellular matrix breakdown products. Heat shock protein-60 (HSP60) binds to TLR-4 and elicits similar immune responses to that of LPS (Ohashi *et al.*, 2000). The response includes the up-regulation of adhesion molecule gene expression and the release of pro-inflammatory mediators such as IL-12 and TNF-α. Surface expression of autologous HSP60 may be triggered in response to cellular stress. HSP60 can also be released by cell necrosis during inflammation (Soltys and Gupta, 1997). Bacterial HSP60/65 also triggers pro-inflammatory responses suggesting that both damaged host cells, and the pathogens that cause the damage, can stimulate the same recognition system to initiate an immune response (Friedland *et al.*, 1993).

Once DC have detected microbial cells or their products, the expression of co-stimulatory molecules such as CD40, CD80 and CD86, and adhesion molecules such as CD44, may be up-regulated. The antigen is processed and the information is transmitted to T cells in secondary lymphoid organs leading to the activation and differentiation of naïve T helper cells into either T_H1 or T_H2 effector cells. This initiates the adaptive immune response. DC in an immature form capture the antigenic material, then migrate towards T cell areas (Peyer's patches or mesenteric lymph nodes) undergoing a maturation process along the way, and interact with T cells in an activated or mature form triggering the production of cytokines. The cytokines produced locally have a key role in determining the type of T_H response that will result: while IL-12 generates a T_H1 response, interleukin-4 (IL-4) generates T_H2 responses (Seder and Paul, 1994; Trincheri, 1995). However, cell-to-cell signaling might also influence the T_H1/T_H2 pattern. T_H1 cells produce interferon-γ (INF-γ) and induce B cells to produce antibodies of the IgG_2 isotype. T_H2 cells produce IL-4, interleukin-5 (IL-5) and interleukin-13 (IL-13) and induce the production of IgE antibodies. Despite conflicting data on this point, it seems that different subpopulations of DC induce distinct T_H responses: lymphoid DC induce T_H1 differentiation whereas myeloid DC induce T_H2 differentiation[1] (Rissoan *et al.*, 1999; Moser and Murphy, 2000). In addition, cytokines secreted by activated T cells can also modulate DC function. Thus T_H1-inducing DC, when exposed to interleukin-10 (IL-10) or transforming growth factor-β (TGF-β), induce T_H2-like responses. INF-γ can stimulate DC to have some T_H1-inducing capacity. Consequently, the basis of a differential response to nonpathogens and pathogens might be related to the conditions, and probably the location, of antigen presentation. In other words, how and where the antigens are taken up and presented. In the context of appropriate co-stimulation, the presentation of antigens to T cells by APC results in T cell priming and the development of a response. On the other hand, antigen presentation in the absence of co-stimulation results in anergy (lack of immune response). Cells that express appropriate co-stimulatory molecules and therefore efficiently prime T cells are activated DC, macrophages and B cells whereas those that express low or no levels of co-stimulatory molecules include immature DC, naïve B cells and enterocytes. Therefore, antigen presentation by enterocytes or immature DC results in tolerance by T cells due to the lack of appropriate mediating signals. Regarding the location of presentation, antigens that enter the lamina propria can be transferred by local DC and presented to T cells in Peyer's patches or in the mesenteric lymph nodes. Alternatively, antigens can be presented to the neighboring T cells by MHC II+ enterocytes, or those that pass into the bloodstream could interact with T cells in the peripheral lymphoid tissues.

The study of intestinal DC is not a simple task. The relatively low number of DC in the tissues and the difficulty of obtaining intestinal tissues other than colonic mucosa, such as Peyer's patches and mesenteric lymph nodes, has impeded ex vivo studies. Methods for the purification of cells from solid tissue samples, such as biopsies, are limited and may affect DC phenotype and function. DC preparations can be prepared from progenitors: bone marrow stem cells stimulated with appropriate cytokines (GM-CSF and eventually IL-4) yield reasonable numbers of DC for study. As for secondary lymphoid tissues, a combination of gradient centrifugation with magnetic bead separation may yield acceptable quantities of DC without modifying their physiological activity. DC can then be maintained in cell culture medium with an appropriate cocktail of cytokines in order to promote maturation or differentiation. Nevertheless, the availability of different subsets of DC is still limited and therefore many of the questions regarding their function remain unanswered.

[1]Myeloid and lymphoid dendritic cells are the two main subsets that differ in the precursors from which they arise, and are distinguished on the basis of their differential CD8α expression.

Up to this point, it is clear that even though there may be immune discrimination between nonpathogens and pathogens through TLR, most of nonpathogen-derived patterns are perceived by the innate immune system, and their recognition by the PRR triggers an immune response. In other words, recognition of nonpathogen antigens does not mean non-response to them, but the induction of both local and systemic states of immunological tolerance. This mechanism is normally referred to as oral tolerance and it is generally accepted that T cells are the major target. Oral tolerance can be achieved by the functional or physical elimination of T cells through deletion and anergy of T cells that respond to antigens in gut contents (Garside and Mowat, 2001). Clonal deletion of antigen-specific T cells via apoptosis has been considered to operate when they are exposed to a large dose of antigen (which could be the case with the gut microflora). At low doses, the induction of regulatory cells (T_H3 and T_R1) would suppress the recruitment of naïve T cells through the production of inhibitory cytokines such as IL-10 and transforming growth factor-β (TGF-β) (Weiner, 2001). IL-10 may suppress T_H1 activity via down-regulation of the expression of co-stimulatory molecules and IL-12 production by APC. TGF-β and the IL-12/INF-γ pathway may play opposing roles (Smith *et al.*, 2000). Anergy of T cells is believed to be induced when cells are stimulated without co-stimulatory molecules. Anergized T cells do not proliferate or produce cytokines when re-stimulated with the antigens. A breakdown of tolerance to components of the gut microflora is considered to be an important factor in the aetiopathogenesis of inflammatory bowel diseases.

The production of secretory IgA (sIgA) molecules that react with members of the gut microflora deserves special attention. The production of this class of immunoglobulins (Ig) is induced in response to antigenic changes within the microflora. MacPherson *et al.* (2000) found that in contrast to sIgA responses against pathogen-derived epitopes and exotoxins which require co-stimulation by antigen-specific T cells, anti-microflora sIgA induction in mice is through a pathway that is independent of T-cell help and involves B lymphocytes from the lamina propria. sIgA is apparently produced to prevent the passage of members of the gut microflora into the intestinal mucosa by blocking the adhesion of the bacteria to the mucosal surface. It is also speculated that this mechanism allows the host to respond to shifts in the composition of the microflora without eliciting an inflammatory response. Studies with gnotobiotic mice have suggested that maternal IgA antibodies in milk, by coating bacteria in the neonatal intestine, shield or block the neonatal immune system from stimulation and delay the active development of sIgA responses to these bacterial antigens. The underlying events leading to the production of anti-microflora sIgA should focus on the activation of B cells, and on the resulting education of the immune system.

The expression of self-peptides presented by APC does not trigger an immune response because T cells specific for these peptides are almost all eliminated in the thymus (negative selection). Therefore negative selection provides a means of recognizing self-antigens and avoiding an immune response towards them. This educational process starts before birth of the host and continues throughout life, but it is crucial during the early months and years of life when the immune development of the mucosa is incomplete. The stimuli received immediately after birth (microflora, environmental antigens, sub-clinical infections) drive the production of cytokines that influence the maturation of neonatal T cells (Rook and Stanford, 1998).

The immunological events described so far are associated with the gut-associated lymphoid tissue (GALT), which is the largest and probably the most intensively studied component of the mucosal immune system. Similar events regarding the induction of an immune response may take place in the bronchus-associated lymphoid tissue (BALT). Gut and bronchus mucosal surfaces are continuously exposed to environmental antigens but the lungs are maintained sterile whereas the gut is not. It has been shown that the immunoglobulin composition in respiratory secretions differs depending on the region (Tlaskalova-Hogenova *et al.*, 2002) and that DC from the respiratory tract differ from the gut DC by pronounced IL-10 production (Akbari *et al.*, 2001). In addition, pulmonary DC have been shown to induce T_R1 regulatory $CD4^+$ cells, whereas DC from the gut induce T_H3 regulatory cells (Weiner, 2001). This is further evidence of the central role of DC in orchestrating the mucosal tolerogenic immune response. The comparison of DC activity in these two systems might help to identify mechanisms by which the gut immune system distinguishes microbial friend from foe.

Acknowledgement

We thank Dr Margaret Baird for reviewing the manuscript during preparation.

References

Ahola, A.J., Yli-Knuuttila, H., Suomalainen, T., Poussa, T., Ahlström, A., Meurman, J.H., and Korpela, R. 2002. Short-term consumption of probiotic-containing cheese and its effect on dental caries risk factors. Arch. Oral. Biol. 47: 799-804.

Ahrné S., Nobaek, S., Jeppsson, B., Adlerberth, I., Wold, A.E., and Molin, G. 1998. The normal *Lactobacillus* flora of healthy human rectal and oral mucosa. J. Appl. Microbiol. 85: 88-94.

Akashi, S., Shimazu, R., Ogata, H., Nagai, Y., Takeda, K., Kimoto, M., and Miyake, K. 2000. Cutting edge: cell surface expression and lipopolysaccharide signaling via the toll-like receptor 4-MD-2 complex on mouse peritoneal macrophages. J. Immunol. 164: 3471-3475.

Akbari, O., DeKruyff, R.H., and Umetsu, D.T. 2001. Pulmonary dendritic cells producing IL-10 mediate tolerance induced by respiratory exposure to antigen. Nat. Immunol. 2: 725-731.

Alexander, M. 1971. Microbial Ecology. John Wiley and Sons, New York.

Alexopoulou, L., Holt, A.C., Medzhitov, R., and Flavell, R.A. 2001. Recognition of double-stranded RNA and activation of NF-kappaB by Toll-like receptor 3. Nature. 413: 732-738.

Arbour, N.C., Lorenz, E., Schutte, B.C., Zabner, J., Kline, J.N., Jones, M., Frees, K, Watt, J.L., and Schwartz, D.A. 2000. TLR4 mutations are associated with endotoxin hyporesponsiveness in humans. Nature Genet. 25: 187-191.

Barnes, P.J., and Karin, M. 1997. Nuclear factor-kappaB: a pivotal transcription factor in chronic inflammatory diseases. N. Engl. J. Med. 336: 1066-1071.

Barrangou, R., Altermann, E., Hutkins, R., Cano, R., and Klaenhammer, T.R. 2003. Functional and comparative gnomic analyses of an operon involved in fructooligosaccharide utilization by *Lactobacillus acidophilus*. Proc. Natl. Acad. Sci. USA. 100: 8957-8962.

Bauer, S., Kirschning, C.J., Hacker, H., Redecke, V., Hausmann, S., Akira, S., Wagner, H., and Lipford, G.B. 2001. Human TLR9 confers responsiveness to bacterial DNA via species-specific CpG motif recognition. Proc. Natl. Acad. Sci. USA. 98: 9237-9242.

Berg, R.D., and Savage, D.C. 1972. Immunological responses to microorganisms indigenous to the gastrointestinal tract. Am. J. Clin. Nutr. 25: 1364-1371.

Berg, R.D. 1983. Host immune response to antigens of the indigenous intestinal flora. In: Human Intestinal Microflora in Health and Disease. D.J. Hentges, ed. Academic Press, New York, New York. p. 101-126.

Boot, H.J., Kolen, C.P., and Pouwels, P.H. 1996a. Interchange of the active and silent s-layer protein of *Lactobacillus acidophilus* by inversion of the chromosomal slp segment. Mol. Microbiol. 21:799-809.

Boot, H.J., Kolen, C.P., Pot, B., Kersters, K., and Pouwels, P.H. 1996b. The presence of two s-layer-protein-encoding genes is conserved among species related to *Lactobacillus acidophilus*. Microbiology. 142: 2375-2384.

Bry, L., Falk, P.G., and Gordon, J.I. 1996. A model of host-microbial interactions in an open mammalian ecosystem. Science. 273: 1380-1383.

Bunte, C., Hertel C., and Hammes, W.P. 2000. Monitoring and survival of *Lactobacillus paracasei* LTH 2579 in food and the human intestinal tract. Syst. Appl. Microbiol. 23: 260-266.

Burney, P.G., Chinn, S., and Rona, R.J. 1990. Has the prevalence of asthma increased in children? Evidence from the national study of health and growth 1973-1986. Brit. Med. J. 300: 1306-1310.

Burr, M.L., Butland, B.K., King, S., and Vaughan, W.E. 1989. Changes in asthma prevalence: two surveys 15 years apart. Arch. Dis. Child. 61: 1452-1456.

Chaillou, S., Lokman, B.C., Leer, R.J., Posthuma, C., Postma, P.W., and Pouwels, P.H. 1998. Cloning, sequence analysis, and characterization of the genes involved in isoprimeverose metabolism in *Lactobacillus pentosus*. J. Bacteriol. 180: 2312-2320.

Chuang, T., and Ulevitch, R.J. 2001. Identification of hTLR10: a novel human Toll-like receptor preferentially expressed in immune cells. Biochim. Biophys. Acta. 1518: 157-161.

Cohen, I.R., and Norins, L.C. 1966. Natural antibodies to gram-negative bacteria: immunoglobulins G, A, and M. Science. 152:1257-1259.

Conway, T., and Schoolnik, G.K. 2003. Microarray expression profiling, capturing a genome-wide portrait of the transcriptome. Mol. Microbiol. 47: 879-889.

Cummings, J. H., and MacFarlane, G.T. 1991. The control and consequences of bacterial fermentation in the human colon. J. Appl. Bacteriol. 70: 443-459.

Dal Bello, F., Walter, J., Hammes, W.P., and Hertel, C. 2003. Increased complexity of the species composition of lactic acid bacteria in human feces revealed by alternative incubation condition. Microb. Ecol. 45: 455-463.

Darling, K.E., and Evans, T.J. 2003. Effects of nitric oxide on *Pseudomonas aeruginosa* infection of epithelial cells from a human respiratory cell line derived from a patient with cystic fibrosis. Infect. Immun. 71: 2341-2349.

Didierlaurent, A., Sirard, J.C., Kraehenbuhl, J.P., and Neutra, M.R. 2002. How gut senses its content. Cell Microbiol. 4: 61-71.

Dubos, R., Schaedler, R.W., Costello, R., and Hoet, P. 1965. Indigenous, normal, and autochthonous flora of the gastrointestinal tract. J. Exp. Med. 122: 67-76.

Dunne, C., Murphy, L., Flynn, S., O'Mahony, L., O'Halloran, S., Feeney, M., Morrissey, D., Thornton, G., Fitzgerald, F., Daly, C., Kiely, B., Quigley, E.M., O'Sullivan, G.C., Shanahan, F., and Collins, J.K. 1999. Probiotics: from myth to reality. Demonstration of functionality in animal models of disease and in human clinical trials. Anton. Van Leeuwen. 76: 279-292.

Evans, R.T., Spaeth, S., and Mergenhagen, S.E. 1966. Bactericidal antibody in mammalian serum to obligately anaerobic gram-negative bacteria. J. Immunol. 97: 112-119.

Franks, A.H., Harmsen, H.J., Raangs, G.C., Jansen, G. J., Schut, F., and Welling, G.W. 1998. Variations of bacterial populations in human feces measured by fluorescent *in situ* hybridization with group-specific 16S rRNA-targeted oligonucleotide probes. Appl. Environ. Microbiol. 64: 3336-3345.

Friedland, J.S., Shattock, R., Remick, D.G., and Griffin, G.E. 1993. Mycobacterial 65-kD heat shock protein induces release of proinflammatory cytokines from human monocytic cells. Clin. Exp. Immunol. 91: 58-62.

Garside, P., and Mowat, A.M. 2001. Oral tolerance. Semin. Immunol. 13: 177-185.

Geijtenbeek T.B., Van Vliet S.J., Koppel E.A., Sanchez-Hernandez M., Vandenbroucke-Grauls C.M., Appelmelk B., Van Kooyk, Y. 2003. Mycobacteria target DC-SIGN to suppress dendritic cell function. J. Exp. Med. 197: 7-17.

Gerrard, J.W., Geddes, C.A., Reggin, P.L., Gerrard, C.D., and Horne, S. 1976. Serum IgE levels in white and metis communities in Saskatchewan. Ann. Allergy 37: 91-100.

Gewirtz, A.T., Simon, P.O. Jr, Schmitt, C.K., Taylor, L.J., Hagedorn, C.H., O'Brien, A.D., Neish, A.S., and Madara, J.L. 2001. *Salmonella typhimurium* translocates flagellin across intestinal epithelia, inducing a proinflammatory response. J Clin Invest. 107: 99-109.

Gillespie, H.B., Steber, M.S., Scott, E.N., and Christ, Y.S. 1950. Serological relationships existing between bacterial parasites and their host. I. Antibodies in human blood serum for native intestinal bacteria. J. Immunol. 65: 105-113.

Goerke, C., Campana, S., Bayer, M.G., Doring, G., Botzenhart, K., Wolz, C. 2000. Direct quantitative transcript analysis of the *agr* regulon of *Staphylococcus aureus* during human infection in comparison to the expression profile *in vitro*. Infect. Immun. 68: 1304-1311.

Gong, J., Matsutera, E., Kanda, H., Yamaguchi N., Tani, K., and Nasu, M. 2002. Diversity and phylogenetic analysis of bacteria in the mucosa of chicken ceca and comparison with bacteria in the cecal lumen. FEMS Microb. Lett. 208: 1-7.

Gordon, H.A., and Pesti, L. 1971. The gnotobiotic animal as a tool in the study of host microbial relationships. Bacteriol. Rev. 35: 390-429.

Gordon, J.I., Hooper, L.V., McNevin, M.S., Wong, M., and Bry, L. 1997. Epithelial cell growth and differentiation. III. Promoting diversity in the intestine: conversations between the microflora, epithelium, and diffuse GALT. Am. J. Physiol. 273: G565-G570.

Graham, J.E., and Clark-Curtis, J. E. 1999. Identification of *Mycobacterium tuberculosis* RNAs synthesized in response to phagocytosis by human macrophages by selective capture of transcribed sequences (SCOTS). Proc. Natl. Acad. Sci. USA. 96: 11554-11559.

Graham, J.E., Peek, R.M., Drishna, U., and Cover, T.L. 2002. Global analysis of *Helicobacteri pylori* gene expression in human gastric mucosa. Gastroenterology. 123: 1637-1648.

Haller, D., Colbus, H., Gänzle, M.G., Scherenbacher, P., Bode, C., and Hammes, W. P. 2001. Metabolic and functional properties of lactic acid bacteria in the gastrointestinal ecosystem: a comparative *in vitro* study between bacteria of intestinal and fermented food origin. System. Appl. Microbiol. 24: 218-226.

Harmsen, H.J.M., Wildeboer, A.C.M., Raangs, G.C., Wagendorp, A.A., Klijn, N., Bindels, J.G., and Welling, G.W. 2000. Analysis of intestinal flora development in breast-fed and formula-fed infants by using molecular identification and detection methods. J. Pediatr. Gastroenterol. Nutr. 30: 61-67.

Hayashi, F., Smith, K.D., Ozinsky, A., Hawn, T.R., Yi, E.C., Goodlett, D.R., Eng, J.K., Akira, S., Underhill, D.M., and Aderem, A. 2001. The innate immune response to bacterial flagellin is mediated by Toll-like receptor 5. Nature. 410: 1099-1103.

Hayashi, H., Sakamoto, M., and Benno, Y. 2002. Phylogenetic analysis of the human gut microbiota using 16S rDNA clone libraries and strictly anaerobic culture methods. Microbiol. Immunol. 46: 535-548.

Heilig, H.G.J., Zoetendal, E.G., Vaughan, E.E., Marteau, P., Akkermans, A.D.L., and de Vos, W. M. 2002. Molecular diversity of *Lactobacillus* spp. and other lactic acid bacteria in the human intestine as determined by specific amplification of 16S ribosomal DNA. Appl. Environ. Microbiol. 68: 114-123.

Hemmi, H., Takeuchi, O., Kawai, T., Kaisho, T., Sato, S., Sanjo, H., Matsumoto, M., Hoshino, K., Wagner, H., Takeda, K., and Akira, S. 2000. A Toll-like receptor recognizes bacterial DNA. Nature. 408: 740-745.

Hemmi, H., Kaisho, T., Takeuchi, O., Sato, S., Sanjo, H., Hoshino, K., Horiuchi, T., Tomizawa, H., Takeda, K., and Akira, S. 2002. Small anti-viral compounds activate immune cells via the TLR7 MyD88-dependent signaling pathway. Nature Immunol. 3: 196-200.

Hirschfeld, M., Weis, J.J., Toshchakov, V., Salkowski, C.A., Cody, M.J., Ward, D.C., Qureshi, N., Michalek, S.M., and Vogel, S.N. 2001. Signaling by toll-like receptor 2 and 4 agonists results in differential gene expression in murine macrophages. Infect. Immunol. 69: 1477-1482.

Hoffman, R.A., Zhang, G., Nüssler, N.C., Gleixner, S.L., Ford, H.R., Simmons R.L., and Watkins S.C. 1997. Constitutive expression of inducible nitric oxide synthase in the mouse ileal mucosa. Am. J. Physiol. 272: G383-G392.

Hoiby, N., and Hertz, J.B. 1979. Precipitating antibodies against *Escherichia coli*, *Bacteroides fragilis* ss. *thetaiotaomicron* and *Pseudomonas aeruginosa* in serum from normal persons and cystic fibrosis patients, determined by means of crossed electrophoresis. Acta Paediatrica. Scand. 68: 495-500.

Hooper, L.V., Wong, M.H., Thelin, A., Hansson, L., Falk, P.G., and Gordon, J.I. 2001. Molecular analysis of commensal host-microbial relationships in the intestine. Science. 291: 881-884.

Hooper, L.V., Xu, J., Falk, P.G., Midtvedt, T., and Gordon, J.I. 1999. A molecular sensor that allows a gut commensal to control its nutrient foundation in a competitive ecosystem. Proc. Natl. Acad. Sci. USA. 96: 9833-9838.

Huang, Q., Liu, D., Majewski, P., Schulte, L.C., Korn, J.M., Young, R.A., Lander, E.S., and Hacohen, N. 2001. The plasticity of dendritic cell responses to pathogens and their components. Science. 294: 870-875.

Jacobsen, C.N., Rosenfeldt Nielsen, V., Hayford A.E., Moller, P.L., Michaelsen, K.F., Paerregaard, A., Sandström, B., Tvede, M., and Jakobsen, M. 1999. Screening of probiotic activities of forty-seven strains of *Lactobacillus* spp. by *in vitro* techniques and evaluation of the colonisation ability of five selected strains in humans. Appl. Environ. Microbiol. 65: 4949-4956.

Jurk, M., Heil, F., Vollmer, J., Schetter, C., Krieg, A.M., Wagner, H., Lipford, G., and Bauer, S. 2002. Human TLR7 or TLR8 independently confer responsiveness to the antiviral compound R-848. Nature Immunol. 3: 499.

Kagnoff, M.F. 1997. Epithelial cells as sensors for microbial infection. J. Clin. Invest. 100: 6-10.

Kandler, O., and Weiss, N. 1986. Regular, nonsporing gram-positive rods. In: Bergey's Manual of Systematic Bacteriology, vol. 2. P.H.A. Sneath, N.S. Mair, M.E. Sharpe, and J.G. Holt, eds. Williams and Wilkins, Baltimore, Maryland. p. 1208-1234.

Kimura, K., McCartney, A.L., McConnell, M.A., and Tannock, G.W. 1997. Analysis of fecal populations of bifidobacteria and lactobacilli and investigation of the immunological responses of their human hosts to the predominant strains. Appl. Environ. Microbiol. 63: 3394-3398.

Krinos, C.M., Coyne, M.J., Weinacht, K.G., Tzianabos, A.O., Kasper, D.L., and Comstock, L.E. 2001. Extensive surface diversity of a commensal microorgansim by multiple DNA inversions. Nature. 414:555-558.

Lerche, M., and Reuter, G. 1961. Isolierung und differenzierung anaerober *Lactobacillaceae* aus dem darm erwachsener Menschen. (Beitrag zum *Lactobacillus bifidus*-Problem). Zentralblatt für Bakeriologie, Parasitenkunde, Infektionskrankheiten und Hygiene, Abteilune 1. Originale. 180: 324-356.

Leser, T.D., Amenuvor, A.E., Jensen, T.K., Lindecrona, R.H., Boye, M., and Moller, K. 2002. Culture-independent analysis of gut bacteria: the pig gastrointestinal tract microbiota revisited. Appl. Environ. Microbiol. 68: 673-690.

MacFarlane, S., Hopkins, M.J., and MacFarlane, G.T. 2000. Bacterial growth and metabolism on surfaces in the large intestine. Microb. Ecol. Health Dis. 2: S64-72.

MacPherson, A.J., Gatto, D., Sainsbury, E., Harriman, G.R., Hengartner, H., and Zinkernagel, R.M. 2000. A primitive T cell-independent mechanism of intestinal mucosal IgA responses to commensal bacteria. Science. 288: 2222-2226.

Mahan, M.J., Slauch, J.M., and Mekalanos. J.J. 1993. Selection of bacterial virulence genes that are specifically induced in host tissues. Science. 259:686-688.

Marieb, E.N. 1989. Human anatomy and physiology. Benjamin/Cummings Publishing Company, Inc., Redwood City.

Marsh, P., and Martin, M.V. 1999. Oral Microbiology. Fourth edition. Butterworth-Heinemann, Oxford.

Marteau, P., Pochart, P., Dore, J., Bera-Maillet, C., Bernalier, A., and Corthier, G. 2001. Comparative study of bacterial groups within the human cecal and fecal microbiota. Appl. Environ. Microbiol. 67: 4939-4942.

Mathesius, U., Mulders, S., Gao, M., Teplitski, M., Caetano-Anolles, G., Rolfe, B.G., and Bauer, W.D. 2003. Extensive and specific responses of a eukaryote to bacterial quorum-sensing signals. Proc. Natl. Acad. Sci. USA. 100: 1444-1449.

Mayer, L. 1998. Antigen presentation in the intestine: new rules and regulations. Am. J. Physiol. Gastrointest. Liver Physiol. 274: G7-G9.

McAuliffe, O.E., and Klaenhammer, T.R. 2002. Genomic perspectives on probiotics and the gastrointestinal microflora. In: Probiotics and Prebiotics: Where are we going? G.W. Tannock, ed. Caister Academic Press, Wymondham, UK. p. 263-309.

Medzhitov, R. 2001. Toll-like receptors and innate immunity. Nat. Rev. Immunol. 1: 135-145.

Mikelsaar, M., Mändar, R.M., and Sepp, E. 1998. Lactic acid microflora in the human microbial ecosystem and its development. In: Lactic Acid Bacteria: Microbiology and Functional Aspects. S. Salminen, and A. von Wright, eds. Marcel Dekker Inc., New York, N.Y. p. 279-342.

Mitsuoka, T. 1969. Vergleichende Untersuchungen über die Laktobazillen aus den Faeces von Menschen, Schweinen, und Hühnern. Zentralblatt für Bakeriologie, Parasitenkunde, Infektionskrankheiten und Hygiene, Abteilune 1. Originale. 210: 32-51.

Mitsuoka, T. 1992. The human gastrointestinal tract. In: The Lactic Acid Bacteria. Vol. 1 The Lactic Acid Bacteria in Health and Disease. B.J.B. Wood ed. Elsevier Applied Science, London, p. 69-114.

Mitsuoka, T., Hayakawa, K., and Eida, T. 1975. Die Faekalflora bei Menschen. III. Mitteilung: Die Zusammensetzung der Laktobazillenflora der verschiedenen Altergruppen.

Zentralblatt für Bakeriologie, Parasitenkunde, Infektionskrankheiten und Hygiene, Abteilung 1. Originale. A232: 499-511.

Moore, W.E., Cato, E.P., and Holdeman, L.V. 1978. Some current concepts in intestinal bacteriology. Am. J. Clin. Nutr. 31: S33-42.

Moore, W.E., and Holdeman, L.V. 1974. Special problems associated with the isolation and identification of intestinal bacteria in fecal flora studies. Am. J. Clin. Nutr. 27: 1450-1455.

Morelli, L. 2000. *In vitro* selection of probiotic lactobacilli: A critical appraisal. Curr. Issues Intest. Microbiol. 1: 59-67.

Moser, M., and Murphy, K.M. 2000. Dendritic cell regulation of TH1-TH2 development. Nature Immunol. 1: 199-205.

Mowat, A. M. 2003. Anatomical basis of tolerance and immunity to intestinal antigens. Nature Rev. Immunol. 3: 331-341.

Muyzer, G., and Smalla, K. 1998. Application of denaturing gradient gel electrophoresis (DGGE) and temperature gradient gel electrophoresis (TGGE) in microbial ecology. Anton. Van Leeuwen. 73: 127-141.

Neish, A.S., Gewirtz, A.T., Zeng, H., Young, A.N., Hobert, M.E., Karmali, V., Rao, A.S., and Madara, J.L. 2000. Prokaryotic regulation of epithelial responses by inhibition of IkappaB-alpha ubiquitination. Science. 289: 1560-1563.

Neutra, M.R. 1998. Current concepts in mucosal immunity. V. Role of M cells in transepithelial transport of antigens and pathogens to the mucosal immune system. Am. J. Physiol. 274: G785-G791.

Ochoa, M.T., Legaspi, A.J., Hatziris, Z., Godowski, P.J., Modlin, R.L., and Sieling, P.A. 2003. Distribution of Toll-like receptor 1 and Toll-like receptor 2 in human lymphoid tissue. Immunol. 108: 10-15.

Ohashi, K., Burkart, V., Flohe, S., and Kolb, H. 2000. Cutting edge: heat shock protein 60 is a putative endogenous ligand of the toll-like receptor-4 complex. J. Immunol. 164: 558-561.

O'Sullivan, D.J. 1999. Methods of analysis of the intestinal microflora. In: Probiotics: A Critical Review. G.W. Tannock, ed. Horizon Scientific Press., Wymondham, p. 23-44.

Ouwehand, A.C., Salminen, S., and Isolauri, E. 2002. Probiotics: an overview of beneficial effects. Anton. van Leeuwen. 82: 279-289.

Ozinsky, A., Underhill, D.M., Fontenot, J.D., Hajjar, A.M., Smith, K.D., Wilson, C.B., Schroeder, L., and Aderem, A. 2000. The repertoire for pattern recognition of pathogens by the innate immune system is defined by cooperation between toll-like receptors. Proc. Natl. Acad. Sci. USA. 97: 13766-13771.

Poltorak, A., He, X., Smirnova, I., Liu, M.Y., Van Huffel, C., Du, X., Birdwell, D., Alejos, E., Silva, M., Galanos, C., Freudenberg, M., Ricciardi-Castagnoli, P., Layton, B., and Beutler, B. 1998. Defective LPS signaling in C3H/HeJ and C57BL/10ScCr mice: mutations in Tlr4 gene. Science. 282: 2085-1088.

Raby, B.A., Klimecki, W.T., Laprise, C., Renaud, Y., Faith, J., Lemire, M., Greenwood, C., Weiland, K.M., Lange, C., Palmer, L.J., Lazarus, R., Vercelli, D., Kwiatkowski, D.J., Rainey, P.B., and Preston, G.M. 2000. *In vivo* expression technology strategies: valuable tools for biotechnology. Curr. Opin. Biotechnol. 11: 440-444.

Requena, T., Burton, J., Matsuki, T., Munro, K., Simon, M.A., Tanaka, R., Watanabe, K., and Tannock, G.W. 2002. Identification, detection, and enumeration of human *Bifidobacterium* species by PCR targeting the transaldolase gene. Appl. Environ. Microbiol. 68: 2420-2427.

Rescigno, M., Urbano, M., Valzasina, B., Francolini, M., Rotta, G., Bonasio, R., Granucci, F., Kraehenbuhl, J.P., and Ricciardi-Castagnoli, P. 2001. Dendritic cells express tight junction proteins and penetrate gut epithelial monolayers to sample bacteria. Nature Immunol. 2: 361-367.

Reuter, G. 1965a. Das Vorkommen von Laktobazillen in Lebensmitteln und ihr Verhalten im menschlichen Intestinaltrakt. Zentralblatt für Bakeriologie, Parasitenkunde, Infektionskrankheiten und Hygiene, Abteilung 1. Originale. 197: 468-87.

Reuter, G. 1965b. Untersuchungen über die Zusammensetzung und die Beeinflussbarkeit der menschlichen Magen- und Darmflora unter besonderer Berücksichtigung der Laktobazillen. Ernährungsforschung 10: 429-435.

Reuter, G. 2001. The *Lactobacillus* and *Bifidobacterium* microflora of the human intestine: composition and succession. Curr. Issues Intest. Microbiol. 2: 43-53.

Rissoan, M.C., Soumelis, V., Kadowaki, N., Grouard, G., Briere, F., de Waal Malefyt, R., and Liu, Y.J. 1999. Reciprocal control of T helper cell and dendritic cell differentiation. Science. 283: 1183-1186.

Roberts, P.J., Riley, G.P., Morgan, K., Miller, R., Hunter, J.O., and Middleton, S.J. 2001. The physiological expression of inducible nitric oxide synthase (iNOS) in the human colon. J. Clin. Pathol. 54: 293-297.

Rokbi, B., Seguin, D., Guy, B., Mazarin, V., Vidor, E., Mion, F., Cadoz, M., and Quentin-Millet, M.J. 2001. Assessment of *Helicobacter pylori* gene expression within mouse and human gastric mucosae by real-time reverse transcriptase PCR. Infect. Immun. 69:4759-4766.

Rook, G.A., and Stanford, J.L. 1998. Give us this day our daily germs. Immunol. Today. 19: 113-116.

Roos, S., and Jonsson, H. 2002. A high-molecular-mass cell-surface protein from *Lactobacillus reuteri* 1063 adheres to mucus components. Microbiology. 148:433-442.

Salzman, N.H., de Jong, H.,Paterson, Y., Harmsen, H. J., Welling, G.W., and Bos, N.A. 2002. Analysis of 16S libraries of mouse gastrointestinal microflora reveals a large new group of mouse intestinal bacteria. Microbiology. 148: 3651-3660.

Satokari, R.M., Vaughan, E.E., Akkermans, A.D.L., Sareela, M., and De Vos, W.M. 2001. Bifidobacterial diversity in human feces detected by genus-specific PCR and denaturing gradient gel electrophoresis. Appl. Environ. Microbiol. 67: 504-513.

Seder, R.A., and Paul, W.E. 1994. Acquisition of lymphokine-producing phenotype by CD4+ T cells. Annu. Rev. Immunol. 12: 635-673

Sghir, A., Gramet, G., Suau, A., Rochet, V., Pochart, P., and Dore, J. 2000. Quantification of bacterial groups within human fecal flora by oligonucleotide probe hybridization. Appl. Environ. Microbiol. 66: 2263-2266.

Sherman, L.A., and Savage, D.C. 1986. Lipoteichoic acids in *Lactobacillus* strains that colonise the mouse gastric epithelium. Appl. Environ. Microbiol. 52: 302-304.

Silverman, E.K., Martinez, F.D., Hudson, T.J., and Weiss, S.T. 2002. Polymorphisms in toll-like receptor 4 are not associated with asthma or atopy-related phenotypes. Am. J. Respir. Crit. Care. Med. 166: 1449-1456.

Smirnova, I., Hamblin, M.T., McBride, C., Beutler, B., and Di Rienzo, A. 2001. Excess of rare amino acid polymorphisms in the Toll-like receptor 4 in humans. Genetics. 158: 1657-1664.

Smith, K.M., Eaton, A.D., Finlayson, L.M., and Garside, P. 2000. Oral tolerance. Am. J. Respir. Crit. Care. Med. 162: S175-S178.

Smith, R.S., Fedyk, E.R., Springer, T.A., Mukaida, N., Iglewski, B.H., Phipps, R.P. 2001. IL-8 production in human lung fibroblasts and epithelial cells activated by the *Pseudomonas* autoinducer N-3-oxododecanoyl homoserine lactone is transcriptionally regulated by NF-kappa B and activator protein-2. J. Immunol. 167: 366-374.

Soltys, B.J., and Gupta, R.S. 1997. Cell surface localization of the 60 kDa heat shock chaperonin protein (hsp60) in mammalian cells. Cell Biol. Int. 21: 315-320.

Spanhaak, S., Havenaar, R., and Schaafsma, G. 1998. The effect of consumption of milk fermented by *Lactobacillus casei* strain Shirota on the intestinal microflora and immune parameters in humans. Eur. J. Clin. Nutr. 52: 899-907.

Sperandio, V., Torres, A.G., Jarvis, B., Nataro, J.P., and Kaper, J.B. 2003. Bacteria-host communication: the language of hormones. Proc. Natl. Acad. Sci. USA. 100: 8951-8956.

Stebbings, S., Munro, K., Simon, M., Tannock, G., Highton, J., Harmsen, H., Welling, G., Seksik, P., Dore, J., Gramet, G., and Tilsala-Timisjarvi, A. 2002. Comparison of the faecal microflora of patients with ankylosing spondylitis and controls using molecular methods of analysis. Rheumatol. 41: 1395-1401

Strachan, D.P. 1989. Hay fever, hygiene, and household size. Brit. Med. J. 18: 1259-1260.

Suau, A., Bonnet, R., Sutren, M., Godon, J.-J., Gibson, G.R., Collins, M.D., and Doré, J. 1999. Direct analysis of genes encoding 16S rRNA from complex communities reveals many novel molecular species within the human gut. Appl. Environ. Microbiol. 65: 4799-4807.

Summanen, P., Baron, E.J., Citron, D.M., Strong, C., Wexler, H.M., and Finegold, S.M. 1993. Wadsworth Anaerobic Bacteriology Manual, 5th edn. Star Publishing Company, Belmont.

Takeuchi, O., Kawai, T., Muhlradt, P.F., Morr, M., Radolf, J.D., Zychlinsky, A., Takeda, K., and Akira, S. 2001. Discrimination of bacterial lipoproteins by Toll-like receptor 6. Int. Immunol. 13: 933-940.

Takeuchi, O., Sato, S., Horiuchi, T., Hoshino, K., Takeda, K., Dong, Z., Modlin, R.L., and Akira, S. 2002. Cutting edge: role of Toll-like receptor 1 in mediating immune response to microbial lipoproteins. J. Immunol. 169: 10-14.

Tannock, G.W. 1992. Lactic microflora of pigs, mice and rats. In: The Lactic Acid Bacteria. Vol 1. The Lactic Acid Bacteria in Health and Disease. B.J.B. Wood, ed. Elsevier Applied Science, London. p. 21-48.

Tannock, G.W. 1994. The acquisition of the normal microflora in the gastrointestinal tract. In: Human Health: The Contribution of Microorganisms. S.A.W. Gibson. ed. Springer-Verlag, London. p. 1-16.

Tannock, G.W. 1995. Normal Microflora. An Introduction to Microbes Inhabiting the Human Body. Chapman and Hall, London.

Tannock, G.W. 1997a. Normal microbiota of the gastrointestinal tract of rodents. In: Gastrointestinal Microbiology. vol II. R.I. Mackie, B.A. White, and R.E. Isaacson, eds. Chapman and Hall, London. p. 187-215.

Tannock, G. W. 1997b. Influences of the normal microbiota on the animal host. In: Gastrointestinal Microbiology. vol II. R.I. Mackie, B.A. White, and R.E. Isaacson, eds. Chapman and Hall, London, United Kingdom. p. 466-497.

Tannock, G.W., and Cook, G. 2002. Enterococci as members of the intestinal microflora of humans. In: The Enterococci: Pathogenesis, Molecular Biology, and Antibiotic Resistance. M.S. Gilmore *et al.* eds. ASM Press, Washington, DC. p. 101-132.

Tannock, G.W., Munro, K., Harmsen, H.J.M., Welling, G.W., Smart, J., and Gopal, P.K. 2000. Analyses of the fecal microflora of human subjects consuming a probiotic product containing *Lactobacillus rhamnosus* DR20. Appl. Environ. Microbiol. 66: 2578-2588.

Tannock, G.W., Szylit, O., Duval, Y., and Raibaud, P. 1982. Colonisation of tissue surfaces in the gastrointestinal tract of gnotobiotic animals by lactobacillus strains. Can. J. Microbiol. 28: 1196-1198.

Trinchieri, G. 1995. Interleukin-12: a proinflammatory cytokine with immunoregulatory functions that bridge innate resistance and antigen-specific adaptive immunity. Annu. Rev. Immunol. 13: 251-276.

Van Kruiningen, H.J., West, A.B., Freda, B.J., and Holmes, K.A. 2002. Distribution of Peyer's patches in the distal ileum. Inflamm. Bowel. Dis. 8: 180-185.

Vaughan, E.E., Mollet, B., and deVos, V.M. 1999. Functionality of probiotics and intestinal lactobacilli: light in the intestinal tract tunnel. Curr. Opin. Biotechnol. 58: 505-510.

Vaughan, E.E., de Vries, M.C., Zoetendal, E.G., Ben-Amor, K., Akkermans, A.D.L., and de Vos, W.M. 2002. The intestinal LAB. Anton. van Leeuwen. 82: 341-352.

Walter, J., Heng, N.C.K., Hammes, W.P., Loach, D.M., Tannock, G.W., and Hertel, C. 2003. Identification of *Lactobacillus reuteri* genes specifically induced in the mouse gastrointestinal tract. Appl. Environ. Microbiol. 69: 2044-2051.

Walter, J., Hertel, C., Tannock, G.W., Lis, C.M., Munro, K., and Hammes, W.P. 2001. Detection of *Lactobacillus, Pediococcus, Leuconostoc,* and *Weissella* species in human feces by using group-specific PCR primers and denaturing gradient gel electrophoresis. Appl. Environ. Microbiol. 67: 2578-2585.

Wang, X., Heazlewood, S.P., Krause, D.O., and Florin, T.H.J. 2003. Molecular characterization of the microbial species that colonize human ileal and colonic mucosa by using 16S rDNA sequence analysis. J. Appl. Micobiol. 95: 508-520.

Weiner, H.L. 2001. The mucosal milieu creates tolerogenic dendritic cells and T(R)1 and T(H)3 regulatory cells. Nature Immunol. 2: 671-672.

Werts, C., Werts, C., Tapping, R.I., Mathison, J.C., Chuang, T.H., Kravchenko, V., Saint Girons, I., Haake, D.A., Godowski, P.J., Hayashi, F., Ozinsky, A., Underhill, D.M., Kirschning, C.J., Wagner, H., Aderem, A., Tobias, P.S., and Ulevitch, R.J. 2001. Leptospiral lipopolysaccharide activates cells through a TLR2-dependent mechanism. Nature Immunol. 2: 346-352.

Wickens, K., Pearce, N., Crane, J., and Beasley, R. 1999. Antibiotic use in early chidhood and the development of asthma. Clin. Exp. Allergy 29: 766-771.

Winans, S.C. 2002. Bacterial esperanto. Nat. Struct. Biol. 9: 83-84

Woese, C.R. 1987. Bacterial evolution. Microbiol. Rev. 51: 221-271.

Wold, A.E. 1999. Role of bacterial adherence in the establishment of the normal intestinal microflora. In: Nestle Nutrition Workshop Series, vol. 42, L.A. Hanson, and R.H. Yolken, eds. Nestec Ltd., Vevey/Lippincott-Raven Publishers, Philadelphia. p. 47-61.

Xu, J., Bjursell, M.K., Himrod, J., Deng, S., Carmichael, L.K., Chiang, H.C., Hooper, L.V., and Gordon, J.I. 2003. A genomic view of the human-*Bacteroides thetaiotaomicron* symbiosis. Science. 299:2074-2076.

Yoshimura, A., Lien, E., Ingalls, R.R., Tuomanen, E., Dziarski, R., and Golenbock, D. 1999. Cutting edge: recognition of Gram-positive bacterial cell wall components by the innate immune system occurs via Toll-like receptor 2. J. Immunol. 163: 1-5.

Zarember, K.A., and Godowski, P.J. 2002. Tissue expression of human Toll-like receptors and differential regulation of Toll-like receptor mRNAs in leukocytes in response to microbes, their products, and cytokines. J. Immunol. 168: 554-561.

Zoetendal, E.G., Akkermans A.D., and de Vos, W.M. 1998. Temperature gradient gel electrophorisis analysis of 16S rRNA from human fecal samples reveals stable and host-specific communities of active bacteria. Appl. Environ. Microbiol. 64: 3854-3859.

Zoetendal, E.G., Akkermans, A.D.L., Akkermans-van Vliet, W.M., de Viosser, J.A.G.M., and de Vos, W.M. 2001. The host genotype affects the bacterial community in the human gastrointestinal tract. Microbial Ecol. Health Dis. 13: 129-134.

Zoetendal, E.G., von Wright, A., Vilpponen-Salmela, T., Ben-Amor, K., Akkermans, A.D., and de Vos, W.M. 2002. Mucosa-associated bacteria in the human gastrointestinal tract are uniformly distributed along the colon and differ from the community recovered from feces. Appl. Environ. Microbiol. 68: 3401-3407.

Chapter 7

The Bacillus Holds Its Breath – Latency and the Hypoxic Response of *Mycobacterium tuberculosis*

David R. Sherman* and David M. Roberts

Abstract

Tuberculosis (TB) has plagued mankind for millennia, and with more than two million deaths annually it remains in the upper-most echelon of infectious killers. Central to the pathogenic success of *Mycobacterium tuberculosis* (MTB) is its ability to persist within humans for long periods without causing any overt disease symptoms. A positive tuberculin skin test and/or chest X-ray indicative of MTB infection in the absence of disease symptoms define the clinical syndrome known as latent tuberculosis. Current models state that oxygen tension within the host is critical to the balance between latent TB and active disease, but several aspects of these models including the role of hypoxia are still untested. In this chapter we summarize and critically review the data accumulated over many decades associating TB with oxygen tension. We highlight both the strong evidence linking oxygen levels with MTB growth and disease as well as the substantial gaps in our understanding of TB latency.

1. The Organism and the Disease

With more than 8 million new cases of active disease and nearly 2 million deaths annually (Corbett *et al.*, 2003), tuberculosis (TB) is a global health emergency of staggering proportions. While the great brunt of these deaths fall in the developing world, several factors, including synergy with the AIDS epidemic (Corbett *et al.*, 2003), immigration from TB-endemic countries (Raviglione, 2003), and the recent surge of multi-drug-resistant (MDR) cases (Espinal, 2003) have insured that TB remains a top public health priority in the developed world as well. Among infectious diseases, only AIDS and malaria have a comparable impact. And unlike AIDS, which requires specific high-risk behaviors for its transmission, or malaria, which is geographically restricted by the range of its mosquito vector, TB transmission occurs worldwide and requires only sharing air with an infected individual. With another person dying in roughly the time it takes to read this sentence, the need for improved understanding of TB and the factors that promote its virulence has never been more urgent.

The causative agent of TB is *Mycobacterium tuberculosis* (MTB), a weakly Gram-positive, strongly acid fast bacillus. MTB grows quite slowly, with a maximum reported doubling time of ~18 hrs. Like all mycobacteria, MTB is covered by a dense lipid-rich outer layer comprised primarily of alpha-branched beta-hydroxy poly-substituted fatty acids called mycolic acids. The mycolic acids of mycobacteria are the largest known, 70 – 90 carbons in length, and comprise nearly one-third of the dry weight of the bacterium (Barry *et al.*, 1998). As befits an organism with such a sheath, a large portion of the genome codes for enzymes involved in lipid biosynthesis or metabolism. There are about 250 such genes in the MTB genome, roughly 5-fold more than in *E. coli* (Cole *et al.*, 1998). TB spreads when an acutely infected individual coughs, sneezes or speaks, releasing infectious bacilli in the form of droplet nuclei. The number of bacilli per droplet is generally fewer than 10, but could be as many as 400 (Garay, 1996). It has been argued that only particles with three or fewer bacilli are small enough to reach the lung alveoli and start an infection (Riley *et al.*, 1962). The infectious dose for humans is probably quite low, in the range of one (McKinney *et al.*, 1998) to fifty (Dannenberg and Rook, 1994) droplets. The highly attenuated *Mycobacterium bovis* BCG is widely used as a TB vaccine, with over 100 million doses administered annually, but its efficacy in preventing adult pulmonary TB is questionable (Mostowy *et al.*, 2003).

Once deposited in the lung, the first human cells to encounter MTB are usually alveolar macrophages, and this exposure can lead to several potential outcomes (Parrish *et al.*, 1998; Cosma *et al.*, 2003). About 70% of people exposed to TB (close contacts of active cases) never show immunological or radiological evidence of infection (Parrish *et al.*, 1998). The basis for this innate protection is entirely unknown. In other persons, bacilli multiply within macrophages and survive long enough to

*For correspondence. Email: dsherman@u.washington.edu

elicit an adaptive immune response, usually measured as a T lymphocyte-mediated reaction to a dermal injection of processed proteins from MTB culture supernatants called Purified Protein Derivative (PPD) (Gedde-Dahl, 1952). People who convert to PPD+ in the absence of BCG vaccination are said to be infected with TB, even in the absence of any clinical symptoms. Of these, about 40% develop disease soon after infection while an unknown proportion harbor the bacteria for long periods without overt signs of disease (Gedde-Dahl, 1952; Parrish *et al.*, 1998). It is unclear how many people clear their infection but remain PPD+. However, in a real sense latency shapes the current TB pandemic. The WHO estimates that 1.86 billion people (32% of the world's population) are infected with MTB, with 16.2 million cases of active disease. The remaining 99% may carry latent infections, as they are asymptomatic but PPD+ (Dye *et al.*, 1999). A person with latent tuberculosis has roughly a 10% lifetime chance of developing active disease (Enarson and Murray, 1996), though co-infection with HIV increases the risk to 8 – 10% per year (Zumla *et al.*, 1999; Corbett *et al.*, 2003).

Thus, unlike most bacterial pathogens that are overtly toxic to their hosts, the striking success of MTB as a pathogen is closely associated with its ability to persist for extended periods without causing disease. The ability of MTB to persist both *in vitro* and *in vivo* is truly extraordinary. In 1920, Corper and Cohn inoculated various clinical isolates of MTB into broth cultures, sealed the bottles with paraffin, and placed them at 37°C. Twelve years later they broke the seals, and were able to subculture MTB from 20 of 47 bottles (Corper and Cohn, 1933). Importantly, none of the positive cultures suffered any attenuation of virulence, as measured by inoculation into guinea pigs. Perhaps even more surprising is the case of the Danish man diagnosed with active TB in 1994. Molecular fingerprinting revealed that the infecting strain was unique among modern Danish isolates. However, the man's father had been treated for TB in 1961, and a sample of that strain had been freeze-dried and stored. By restriction fragment–length polymorphism, strains of father and son were a perfect match, providing molecular evidence for TB reactivation following a latency period of 33 years (Lillebaek *et al.*, 2002). Clearly, questions of how MTB persists and for how long are important not only to researchers but also to those who shape policies for TB treatment and control.

2. The TB Latency Model

Technical advances, scientific curiosity and the huge public health need have all fueled a recent surge of interest in TB latency, and a model has emerged in which hypoxic conditions within the host play a prominent role. Very briefly, this model holds that MTB deposited in the lung grows unabated for days or weeks within alveolar macrophages until an appropriate adaptive immune response can be mounted. Then activated macrophages and other host components surround the infected cells in an organized display known as a granuloma, creating conditions that are no longer permissive for MTB replication. Several features of the granuloma may help inhibit MTB, including enhanced levels of nitric oxide, acidic pH, nutrient starvation and oxygen limitation. However, the bacilli are not always eradicated. Instead, they may adapt to a metabolically altered, possibly dormant state that can persist for many years. These metabolically altered bacteria within the granuloma are viewed as the seeds of reactivation of TB, waiting until immunosuppression or some other factor restores conditions permissive for fulminant disease. This model is now the starting point for many discussions of TB latency (Dannenberg, 1993; Parrish *et al.*, 1998; Manabe and Bishai, 2000; McKinney *et al.*, 2000; Flynn and Chan, 2001; Wayne and Sohaskey, 2001; Cosma *et al.*, 2003). However, despite this widespread acceptance, many prominent features must be viewed with caution due to a lack of relevant data.

First, the granuloma is not necessarily the site from which TB reactivates (Cosma *et al.*, 2003). Older human studies showed that in the absence of active disease the great majority of granulomas are sterile (Opie and Aronson, 1927; Feldman and Baggenstoss, 1938). In addition, bacilli (Opie and Aronson, 1927; Feldman, 1939) or mycobacterial DNA (Hernandez-Pando *et al.*, 2000) have been detected in superficially normal lung tissue, raising the possibility that reactivation disease could occur in areas that were not previously granulomatous.

A second controversy surrounds the metabolic state of bacilli during a latent infection (Parrish *et al.*, 1998; Cosma *et al.*, 2003). The argument that latent TB involves physiologically dormant bacilli is oft-encountered (Dannenberg, 1993; McKinney *et al.*, 2000; Glickman and Jacobs, 2001; Wayne and Sohaskey, 2001), stemming from the observations that latency in humans involves few bacilli, minimal pathology, and no disease symptoms. In support are studies where diseased tissues that stained positive for acid fast bacteria either did not yield positive cultures or the bacteria grew very slowly (Parrish *et al.*, 1998; Manabe and Bishai, 2000; McKinney *et al.*, 2000). Also, in many animal models, chronic TB achieves a state in which bacterial numbers do not change for extended periods. In experiments with the mouse model, Rees and Hart found no evidence of bacterial turnover during the chronic period (Rees and Hart, 1961). On the other hand, Orme has dismissed the very concept of the latent bacillus, saying "I'll let you know if I ever meet one" (Orme, 2001). Arguing for Orme: an antibiotic (isoniazid) that targets actively growing MTB dramatically reduces the rate of reactivation tuberculosis in PPD+ individuals (Comstock *et al.*, 1979). Also, recent animal studies demonstrate the presence of bacterial metabolism during latency. MTB mRNA was detected in lungs of mice in a drug-induced latency model (Hu *et al.*, 2000; Pai *et al.*, 2000). Similarly, many mycobacterial genes involved in a variety of active cellular processes are highly expressed in seventeen month-old granulomas of chronically infected frogs, even as the viable bacterial counts remain constant (Chan *et al.*, 2002). Of course, bacterial gene expression can occur in the absence of replication, and during latency the bacteria need not all be in the same metabolic state. At present, it seems prudent to conclude that arguments over the existence of latent bacilli during clinical latency are partly semantic (how latent is latent?), and the question of whether bacteria replicate during latency remains unanswered.

A third issue with the current model of TB latency is the role played by oxygen tension. Oxygen limitation within the granuloma is frequently invoked as central to producing and maintaining a latent state (Dannenberg, 1993; Wayne and Sohaskey, 2001). However, as described below that view is based on a long history of clever experiments and interesting observations with human disease and animal models that still are not decisive, as well as an extrapolation from *in vitro* models that are not yet validated. In the rest of this chapter, we will critically review the evidence linking hypoxia and TB latency. We highlight older studies that shed light on current issues, as well as recent work to decipher the molecular details of the MTB hypoxic response that may at last provide a means to resolve this question.

3. TB and Hypoxia – the Early Years

Throughout the millennia, well before it was even clear that disease could have an infectious origin, physicians associated TB preferentially with the most oxygenated regions of the body. The ancient Greeks, Hindus and Romans all recognized that TB (variously called phthisis, consumption, sosha, rajayakshma, etc.) is primarily a disease of the respiratory tract (Haas and Haas, 1996). Experiments to address this issue were greatly accelerated by the discovery of the tubercle bacillus by Koch in 1882. Following intravenous inoculation, Corper and Lurie determined the localization of tubercle bacilli in the organs of dogs, guinea pigs, rabbits, and monkeys (Corper and Lurie, 1926). The degree of tuberculous involvement in the spleen, liver, and kidney differed markedly in each species, but the lungs of all animals revealed the most overall involvement and the largest tubercles. The authors concluded that "the lung occupies such a position of peculiar susceptibility to tuberculosis after intravenous inoculation...that in this organ there exists a responsible factor differing from that present in the other organs." (Corper *et al.*, 1927). In their search for this "responsible factor", Corper, Lurie and Uyei looked to the fundamental differences between the physiology of the organs and their tuberculous lesions. They concluded that one mitigating factor, "would seem to be the differences in gases available for the respiration and growth of the tubercle bacilli" (Corper *et al.*, 1927).

As the role of oxygen *in vivo* was being explored, it was also becoming clear that the growth of MTB *in vitro* was directly dependent on the level of oxygenation (reviewed in Canetti, 1955). In 1924, Webb and colleagues demonstrated that growth of MTB in broth cultures correlated directly with the amount of available oxygen. They concluded that oxygen "appears to be the only gas which is essential to the normal life and growth of the tubercle bacillus" (Webb *et al.*, 1924). A subsequent study by Corper, Lurie and Uyei found that cultures of MTB grew better under gaseous tensions that mimic those found in alveolar air as opposed to those found in venous blood (Corper *et al.*, 1927). In a more detailed study, Kempner maintained MTB *in vitro* at 14 different oxygen tensions and determined that the rate of MTB respiration correlated directly with the oxygen tension (Kempner, 1939). At an oxygen tension of ~1%, MTB respiration was inhibited by 70% in comparison to controls grown at normal atmospheric oxygen levels. Levels of oxygen higher than 20% resulted in an increase in respiration. Thus, it appeared that one way to impede growth of MTB was to limit its oxygen supply.

But does artificially restricting available oxygen also hinder growth of MTB *in vivo?* Using guinea pigs, animals exquisitely susceptible to TB, and continuous airflow chambers, Rich and Follis sought to determine the effects of different oxygen tensions on MTB growth and development (Rich and Follis, 1942). Guinea pigs were infected with MTB and then placed in atmospheres containing 9-10%, 12-14%, or 20% oxygen corresponding to low, alveolar, and atmospheric oxygen tensions, respectively. At an oxygen level of 9-10%, "the development of the experimental infection was markedly inhibited in comparison to the controls kept in open cages in the same room." This inhibitory effect was present at 12-14% oxygen, but to a much lesser degree than at 9-10%, and the disease progression in the animals kept at 20% was the same as that observed in the open-air control animals. The authors concluded that reduced oxygen tension can curb the growth of MTB and ameliorate the development of TB disease (Rich and Follis, 1942). Human TB probably responds the same way. A very recent report compared TB transmission rates in high altitude Peruvian villages to those at sea level (Olender *et al.*, 2003). Even after adjusting for differences in age, education, BCG vaccination and exposure to active TB cases, villagers living ~11,000 ft. above sea level were nearly five-fold less likely to be PPD+ than those in sea level areas (prevalence rates of ~6% and ~29% respectively). Of course, like the earlier study with guinea pigs, this work does not demonstrate that hypoxia is directly responsible for the effect observed, and confounding factors (differences in humidity, UV light, etc.) are easy to imagine.

To assess whether this phenomenon could be exploited therapeutically, Adams and Vorwald conducted further animal studies (Adams and Vorwald, 1934). Dogs were infected with MTB either prior to or 6 weeks after blocking the lung bronchi by cauterization. This treatment induced collapse in one lung while the other remained inflated. Dogs infected after lung collapse displayed very little tuberculous disease in the collapsed lung whereas the inflated lung was greatly involved. In dogs infected prior to lung collapse, there was little disease progression and even some healing in the collapsed lung. The inflated lung of these dogs displayed normal disease progression with no signs of healing. The authors concluded that lung collapse could have therapeutic benefit, and that oxygen levels affect TB disease progression *in vivo*. Of course, other changes brought on by lung collapse such as reduced flow of blood or lymph or possible changes in immuno-surveillance were not considered in this study.

As is often the case, the need to relieve human suffering drove clinicians ahead of the available data. Well before the practice could be rationalized experimentally, depriving the bacillus of oxygen lay at the heart of some of the earliest TB therapies. The first successful TB treatment to see widespread use was bed rest, part of the rationale for which was to reduce the amount of oxygen consumed (Boyd *et al.*, 1996). More radical approaches involved collapse of the infected lung, an idea that pre-dated the experiments of Adams and Vorwald by at least 150 years. Lung collapse as TB therapy was suggested by Edmond Bourru in 1770 and again by James Carson in 1821, but this approach was not really practical until surgical methods improved in the late 19th and early 20th centuries (Davis, 1996). The most popular methods of collapse were pneumothorax (placing air, gas, oil or solid material such as Lucite spheres the size of ping pong balls between the lung and the chest wall) and thoracoplasty (removal of a portion of the ribs). Lung collapse was quite widespread before the advent of chemotherapy for TB. There were probably more than 100,000 pneumothorax procedures performed in the early 20th century, and thoracoplasty may have been just as popular (Davis, 1996). Though fraught with serious complications, lung collapse was considered efficacious in 50 – 80% of cases when optimal conditions were met (Pinner, 1947; Davis, 1996).

Collectively, these studies demonstrated that good oxygen tension is necessary for robust growth of MTB both *in vivo* and *in vitro*, and that artificially limiting the available oxygen could have therapeutic benefit. Some researchers went further to speculate that even in the absence of any therapeutic intervention, local differences in oxygen tension within the lung might affect TB outcome, and that hypoxia could play a role in TB latency. This idea was fueled by longstanding observations of TB pathology. Initially MTB seeds all regions of the lung in an approximately random pattern (McKinney *et al.*, 1998). However, before he succumbed to his own TB in 1826, inventor of the stethoscope Rene Laënnec (Farrell, 1998) reported that TB disease was more frequent and more severe in the upper portion or apex of the lung (Haas and Haas, 1996). In an upright animal, the lung apices have the highest oxygen tension. Laënnec's observations were confirmed and extended by many subsequent investigators. In 1882, after autopsies on 152 persons who died of TB, Ewart concluded that cavitary disease was 25 times more likely to strike the upper lobes (reviewed in (McKinney *et al.*, 1998). In 1931, Sweany, Cook and Kegerreis studied the chest X-rays of about 200 TB patients and confirmed these observations (Sweany *et al.*, 1931). Medlar used clever animal studies to show that the localization of severe TB disease is largely determined by posture (Medlar and Sasano, 1936; Medlar, 1940). In both rabbits and cows, the majority of TB lesions were in the dorsal region of the lungs. The dorsal regions in these animals are analogous to the apical lung regions in humans, in that they are furthest from the ground and most oxygenated. In a fascinating twist that would not receive animal care committee approval today, Medlar infected rabbits with TB and kept them in an upright position for 11 hours a day (Medlar and Sasano, 1936). Remarkably, the TB lesions in these animals were predominantly in the apical areas of the lungs.

According to William Dock, it was Johannes Orth in 1887 who first argued in print that the lung apex is most susceptible to TB specifically because oxygen tension is highest there (Dock, 1954). Several authors since then have reached the same conclusion (Dock, 1946; Dock, 1954; Riley, 1960). However, robust oxygenation of the upper lung lobes is linked to poor perfusion of this region with blood and lymph. The unique TB susceptibility of the lung apex may stem from some other consequence of poor perfusion, for example a difference in the amount or quality of subsequent immune surveillance (McKinney *et al.*, 1998).

None of the experiments or observations described above can distinguish from among these possibilities. However, extraordinary experiments reported in the early 1950s sought to address the issue of perfusion by tweaking lung oxygenation in creative ways (Hanlon *et al.*, 1950; Scott *et al.*, 1950; Olson *et al.*, 1951). Monkeys infected with MTB received surgeries to increase blood flow and oxygenation in one lung or one lobe as opposed to the other. In one such surgery, a systemic artery was spliced into a pulmonary artery (which normally carries deoxygenated blood) so that one lung was perfused with more blood containing more oxygen. The opposite lung was left untouched as a control (Hanlon *et al.*, 1950). If TB susceptibility of the lung apex stems from poor perfusion, the higher blood pressure in affected areas should result in signs of protection. If however the lung apex is sensitive because of higher oxygen levels there, affected regions might see worse disease. In fact, TB developed much more quickly and severely in the lung of the animals receiving the more oxygenated blood. The untouched lung was comparable to the lungs of control animals that received no surgery. In a continuation of these studies Scott and colleagues closed a pulmonary artery to one lung in infected monkeys, again resulting in more oxygenated blood in the affected area (Scott *et al.*, 1950). Whether the animals were infected before or after the operation, TB disease was much more severe in the lung or lobes in which the pulmonary artery was sealed (Olson *et al.*, 1951). These results indicate that oxygenation is more important than perfusion in determining the outcome of a TB infection and show that even without collapsing a lung TB is limited by the amount of oxygen available *in vivo*.

Not only is severe TB predominantly associated with the richly oxygenated upper lobes, but this region is far more commonly associated with reactivation disease (McKinney *et al.*, 1998). Given the sensitivity of MTB to low oxygen *in vitro*, might oxygen be important in activation from latency? With the advent of surgical resection to treat TB, researchers had an unprecedented view on this process in infected human tissue. Vandiviere and colleagues surgically resected lung tissue from TB patients and classified each lesion based on its macroscopic physiology (Vandiviere *et al.*, 1956). "Open" lesions were described as having a connection to bronchial airways while "closed" lesions were described as having no airway connection. Interestingly, when cultured, 94% of the open lesions were positive for MTB growth after a normal incubation period (8 weeks). In contrast, none of the closed lesions were positive for MTB growth in 8 weeks. However 41% were positive for MTB growth after 3 to 10 months of incubation. Vandiviere proposed that the difference in oxygen availability within the lesions led to the difference in the ability to recover organisms in culture. His work suggested that hypoxic microenvironments within the host lead directly to dormant bacteria and latent disease.

4. Wayne's World of Non-Replicating Persistence

Understanding latency has long been viewed as crucial to TB control, but it is surpassingly difficult to study a phenomenon defined by the lack of any symptoms. With advances in microbiological and biochemical techniques, researchers began to pay more attention to factors that might promote or maintain TB latency. Dubos studied the role of oxygen by growing MTB in sealed containers with germinating oats that rapidly consume the oxygen supply (Dubos, 1953). After three weeks, the bacilli failed to grow when transferred to fresh aerated medium and failed to produce disease when they were injected into experimental animals. These data seem to contradict earlier observations on the long-term stability of MTB in the absence of oxygen. For example, Corper and Cohn showed that nearly half of liquid cultures of MTB sealed with paraffin retained viability and virulence after twelve years of incubation (Corper and Cohn, 1933). The method used by Dubos to remove oxygen was very rapid compared to the cultures of Corper and Cohn, which presumably settled slowly and depleted the available oxygen over several days. Wayne and Lin addressed this issue directly, comparing the anaerobic survival of log-phase MTB with that of "resting bacilli" in late stationary phase (Wayne and Lin, 1982). When actively growing aerated cultures were suddenly switched to an anaerobic jar, the MTB died rapidly, with a half-life ~10 hours. In contrast, bacilli maintained for 19 days without agitation gradually depleted the available oxygen and stopped growing as they slowly settled through a column of media. These sedimented bacilli were much more tolerant of the anaerobic jar, with a half-life ~116 hours. They concluded that long-term anaerobic survival of MTB required a period of adaptation, and suggested that this process might be important for persistence in caseous or blocked lesions in man (Wayne and Lin, 1982).

Wayne and colleagues embarked on a series of studies to characterize slowly settled cultures of MTB, making several observations of note. Slowly settled bacilli become more susceptible to autolysis in the presence of glycerol (Wayne and Diaz, 1967), and display altered carbon metabolism, with induction of the enzymes isocitrate lyase (Icl) and glycine dehydrogenase (Wayne and Lin, 1982). Icl is the first enzyme of the glyoxalate shunt, which bypasses the oxygen-dependent TCA cycle. In MTB, the glyoxalate produced can be reductively aminated by glycine dehydrogenase to yield glycine and NAD (Goldman and Wagner, 1962), which may be required to fuel the adaptation to anaerobiosis. Another study showed that when adapted bacilli are re-aerated, they undergo at least three rounds of synchronized replication (Wayne, 1977), suggesting that the adaptation to anaerobic tolerance is an ordered process. Wayne coined the phrase "nonreplicating persistence" to distinguish these bacilli from the bacteria that result in TB latency *in vivo* (Wayne and Sohaskey, 2001).

Despite this progress, limitations of using the slow-settling model to study MTB persistence soon became apparent. The bacilli in these settled cultures are heterogeneous, with some sedimented at the bottom of the tubes whiles others are still settling. Oxygen is consumed by the bacteria, but other changes doubtless occur as well. Reproducible conditions and hence reproducible results were difficult to obtain. To address this shortfall, Wayne and Hayes revamped the settling method and produced what is now known as the Wayne model of mycobacterial persistence (Wayne and Hayes, 1996). Cultures are slowly stirred to keep the bacilli suspended uniformly in the media and are incubated in sealed tubes or flasks with a defined culture to headspace ratio so that each is subjected to the same initial oxygen concentration. Employing this revamped model, Wayne and Hayes defined two distinct phases as the bacilli adapt to non-replicating persistence. The first stage of non-replicating persistence, NRP1, begins as the oxygen concentration reaches 1%. Replication ceases but the turbidity of the cultures continues to increase as existing bacterial cells elongate. This stage is also marked by an increase in glycine dehydrogenase similar to that observed in the settling model, a steady low level of intracellular ATP, and detectable RNA synthesis. At an oxygen concentration of about 0.06%, cells enter NRP2. Cellular elongation and culture turbidity no longer increase. Also, glycine dehydrogenase levels and ATP concentration both decline. Throughout this stage, bacterial colony forming units drop slowly, but RNA synthesis continues at a modest rate.

Attracted by the convenience of the system as well as the magnitude of the TB latency problem, new researchers have helped further characterize the Wayne model bacilli (Dick *et al.*, 1998; Hu *et al.*, 1998; 1999; Lim *et al.*, 1999; Michele *et al.*, 1999; Florczyk *et al.*, 2001). As might be expected, protein synthesis is severely reduced in non-replicating bacilli (Hu *et al.*, 1998). However, against this background of declining protein synthesis, expression of certain genes is enhanced. The alternative sigma factors σ^B, σ^E and σ^F are all reported to be up-regulated by hypoxia (DeMaio *et al.*, 1996; Hu and Coates, 1999; Manganelli *et al.*, 1999). As alternative sigma factors promote broad changes in transcription (Gomez *et al.*, 1997), expression of these genes may help drive the adaptation to persistence. In addition, as oxygen is depleted, the ability of MTB to reduce nitrate is dramatically increased (Wayne and Hayes, 1998), and a gene (*acg* or Rv2032c) with a nitroreductase signature is powerfully induced (Purkayastha *et al.*, 2002). These changes may indicate a switch to the use of nitrate as the terminal electron acceptor for energy production (Wayne and Hayes, 1998). The *acg* gene product may be involved in energy production at low oxygen tension, or it may be needed for detoxification of nitro compounds encountered in this environment (Purkayastha *et al.*, 2002). Interestingly, a gene with homology to bacterial flavohemoglobins that may protect against nitrosative stress is also induced (Hu *et al.*, 1999).

Nonreplicating persistent bacilli also increase expression of a protein later identified as α–crystallin (Acr) (Wayne and Lin, 1982). The crystallins were originally found in mammalian eye tissue, where their abundant expression functions in the long-term maintenance of lens protein stability and lens clarity (Graw, 1997). In MTB, increased expression of Acr (also called HspX and Rv2031c) in response to reduced oxygen tension has now been reported by several groups using a variety of experimental approaches (Yuan and Barry, 1996; Cunningham and Spreadbury, 1998; Imboden and Schoolnik, 1998; Yuan *et al.*, 1998; Desjardin *et al.*, 2001; Florczyk *et al.*, 2001; Sherman *et al.*, 2001; Shi *et al.*, 2003). Purified MTB α-crystallin (Acr) showed a robust ability to suppress thermal denaturation and aggregation of other proteins *in vitro*, similar to the activity of eukaryotic α-crystallins (Yuan and Barry, 1996). In addition, Acr expression has been associated with thickening of the MTB cell wall under microaerophilic conditions (Cunningham and Spreadbury, 1998). Still, despite speculation that this protein must be important for long-term survival of MTB (Wayne and Sohaskey, 2001), the precise role of Acr *in vivo* remains undefined.

Though improvements have engendered much more widespread use, the Wayne model is still dogged by issues of reproducibility. In response, a few researchers interested in hypoxia and TB latency have abandoned the self-generated oxygen gradients of Wayne. Instead, in a throwback to the early experiments of Kempner (1939) and Rich and Follis (Rich and Follis, 1942), bacilli are exposed to some defined hypoxic atmosphere that remains constant throughout the experiment. In one such report, a cDNA library of genes expressed at reduced oxygen tension was screened and several transcripts were identified (Imboden and Schoolnik, 1998). However, the oxygen tension employed (5%) mimics that of venous blood and is probably too high to induce a nonreplicating state. Another study demonstrated that reduced oxygen alone is sufficient to induce the *acr* gene, and that induction was maximal at oxygen concentrations below 1% (Yuan *et al.*, 1998). A third study used metabolic labeling and 2D gel analysis of proteins from MTB maintained at 1%, 5% or 20% oxygen to identify a group of proteins differentially expressed at low oxygen tension (Rosenkrands *et al.*, 2002).

In one series of defined hypoxic culture experiments, early log-phase bacilli were cultured under 0.2% oxygen in nitrogen, with RNA isolated at various times and analyzed by whole genome microarray (Sherman *et al.*, 2001). Expression of ~100 genes was altered significantly in the first two hours. Predicted functions for many of the 60 repressed genes indicate that low O_2 tension is associated with broad adaptation to reduced metabolic activity. Repressed loci encode components of protein synthesis, DNA synthesis/cell division, lipid or amino acid synthesis, production of polyketides and other complex hydrophobic molecules and aerobic metabolism. Fourteen (~23 %) of the repressed genes are of unknown function. In comparison, more than two-thirds (~68%) of the 47 induced genes are of unknown function, suggesting that the adaptation to hypoxia is not yet well characterized. Several induced genes have postulated functions that could help promote survival in a latent state: *acr* (as described above, stabilizing partially denatured proteins); *narX*, *nark2*, and *fdxA* (nitrate accumulation and alternative electron transport); *nrdZ* (dNTP synthesis under microaerophilic conditions). Also induced were six MTB orthologues of the universal stress protein (Usp) family. In *E. coli*, Usp proteins are induced by starvation and stationary phase, where they confer resistance to DNA damage (Diez *et al.*, 2000; Gustavsson *et al.*, 2002). Usp proteins may be important for the long-term survival of MTB as well (O'Toole and Williams, 2003; Voskuil *et al.*, 2003). Little overlap exists between the genes induced by hypoxia and those induced in the starvation model of non-replicative persistence (Betts *et al.*, 2002). However, some of the genes identified by microarray are also induced in the Wayne model and using other approaches that generate hypoxic conditions (Yuan *et al.*, 1998; Hutter and Dick, 1999; Lim *et al.*, 1999; Boon *et al.*, 2001; Florczyk *et al.*, 2001; Rosenkrands *et al.*, 2002; Florczyk *et al.*, 2003).

5. DosR

Clearly MTB responds to oxygen deprivation with a coordinated genetic program. How is the MTB hypoxic response controlled? Among the loci first induced by reduced oxygen tension is a gene, Rv3133c, encoding a putative transcription factor called DevR or DosR (Dasgupta *et al.*, 2000; Sherman *et al.*, 2001; Boon and Dick, 2002). To assess the role of this protein, we performed targeted replacement of the *dosR* locus followed by expression analysis of wild-type (Sherman *et al.*, 2001) and *dosR* mutant bacilli (Park *et al.*, 2003). In two separate studies, promoters of representative genes induced by hypoxia were cloned upstream of reporter genes. In each case, reporter gene expression depended on an intact *dosR* gene (Sherman *et al.*, 2001; Florczyk *et al.*, 2003). In fact, transcriptome analysis by microarray indicated that nearly all the genes powerfully up-regulated by hypoxia in wild-type MTB require DosR for their induction. Of the 27 genes most powerfully induced by hypoxia, 26 require the presence of DosR (Park *et al.*, 2003). Genetic complementation demonstrated that these effects are due to mutation of DosR and not some unrelated MTB gene.

In silico analysis predicted that DosR is a transcription factor of the two-component response regulator family (Dasgupta *et al.*, 2000; Sherman *et al.*, 2001). Such proteins are activated by phosphorylation at a conserved aspartate, after which they bind to DNA upstream of the genes they control (Hoch and Silhavy, 1995; West and Stock, 2001). We and others sought to test these predictions. We identified a 20-mer degenerate palindromic motif associated with the genes that respond to hypoxia: 5′ TT(C/G)GGGACT(T/A)(A/T)AGTCCC(G/C)AA (Park *et al.*, 2003). McDonough and colleagues identified essentially the same sequence using a different computer algorithm (Florczyk *et al.*, 2003). While a perfect match to this sequence is not found anywhere in the MTB genome, a variant is located upstream of nearly all MTB genes rapidly induced by hypoxia. We also showed that DosR binds to both copies of this motif upstream of the hypoxic response gene *acr*. Mutations within the sites abolish both DosR binding as well as hypoxic induction of a downstream reporter gene. Additional mutation experiments confirmed sequence-based predictions that the DosR C-terminus is responsible for DNA binding and that the conserved aspartate at position 54 is essential for function. Altogether these results demonstrate that DosR is a transcription factor of the two-component response regulator class, and that it is the primary mediator of a hypoxic signal within *M. tuberculosis* (Park *et al.*, 2003).

6. DosR, Hypoxia and Nitric Oxide

Significant overlap has been demonstrated between the MTB responses to hypoxia and nitric oxide (NO) (Ohno *et al.*, 2003; Voskuil *et al.*, 2003). NO is one of few host products known to be directly toxic to mycobacteria at clinically relevant concentrations (Nathan, 2002), but at sub-lethal concentrations NO reversibly inhibits aerobic respiration and has a distinct role as a signaling molecule in both prokaryotes (Zumft, 2002) and eukaryotes (Martin *et al.*, 2000). Recent work shows that, like hypoxia, low levels of NO produce reversible growth arrest of MTB *in vitro* and that NO and hypoxia are additive in their inhibitory effects (Voskuil *et al.*, 2003). Further, concentrations of NO that impede MTB growth were shown to induce the DosR regulon in a DosR-dependent fashion (Ohno *et al.*, 2003; Voskuil *et al.*, 2003). A model was proposed in which host-generated NO and hypoxia act together to limit aerobic respiration and impair growth of MTB, which then responds and adapts in a DosR-dependent manner (Voskuil *et al.*, 2003). Of course, this overlap in the MTB responses to NO and oxygen limitation complicates attempts to define an independent role for hypoxia as a trigger for TB latency *in vivo*.

7. Completing the Circle – Relating Hypoxic MTB *In Vitro* with Conditions *In Vivo*

Despite significant progress in understanding the effects of hypoxia on MTB *in vitro*, it remains an open question how well these experiments model TB latency *in vivo*. Several studies shed light on that question. Perhaps the most compelling association between anaerobically-adapted MTB *in vitro* and bacilli during latency *in vivo* is the altered susceptibility to drugs that is evident in both cases. Numerous compounds, including rifampin, isoniazid, ofloxacin, ampicillin and mitomycin C show much better activity against log-phase MTB than against nonreplicating persistent bacilli (Wayne and Sramek, 1994; Herbert *et al.*, 1996; Wayne and Hayes, 1996; Stover *et al.*, 2000; Peh *et al.*, 2001). Similarly, TB patients who receive appropriate chemotherapy are quickly rendered non-infectious, but fully efficacious treatment requires 6 – 12 months. Early studies showed that some bacilli survived chemotherapy and could be viewed microscopically in resected tissue (Medlar *et al.*, 1952; Wayne, 1960). McCune and colleagues demonstrated that infected mice treated with anti-TB drugs could be superficially sterilized with no viable bacilli detected, yet after immunosuppressive treatment a high proportion of the animals developed TB (McCune *et al.*, 1956; 1966a; 1966b). As is true *in vitro*, the bacilli cultured from these animals retained their sensitivity to the drugs, indicating they were drug tolerant as opposed to drug resistant. McDermott has argued for drug sensitive bacilli that somehow persist in the face of intensive chemotherapy as a primary reason that TB is so difficult to treat (McDermott, 1958). Updating this idea, Mitchison proposed that MTB is heterogeneous *in vivo*, with a few dormant bacilli that are highly resistant to chemotherapeutic eradication in virtually every population (Mitchison, 1979; 1992).

Of course, the link between the bacilli of latent infections and those that survive prolonged chemotherapy *in vivo* is purely hypothetical. Still, there is a striking parallel between drug tolerant organisms that persist *in vivo* and the drug tolerance of nonreplicating persistent bacilli *in vitro* (Hu *et al.*, 2000; Wayne and Sohaskey, 2001). In both cases the bacteria are not eliminated by drugs to which they remain genetically susceptible, and they fail to replicate, though modest RNA synthesis has been detected (Wayne and Hayes, 1996; Hu *et al.*, 2000; Pai *et al.*, 2000). At the level of gene expression, a sigma factor induced by hypoxia is also induced by exposure to various TB drugs (Michele *et al.*, 1999). Nonetheless, this correlation would be more compelling if there were a drug effective against nonreplicating persistent bacilli *in vitro* that also shortened the course of therapy or was effective in an animal model of TB latency. To date, there are very few compounds with known activity against nonreplicating bacilli (Wayne and Sramek, 1994; Wayne and Hayes, 1996; Murugasu-Oei and Dick, 2000; Stover *et al.*, 2000; Peh *et al.*, 2001), and only metronidazole is more effective against these bacteria than against log-phase cells (Wayne and Sramek, 1994). Tested twice in animal models of TB, metronidazole has failed to show any *in vivo* activity (Dhillon *et al.*, 1998; Brooks *et al.*, 1999). Conversion of metronidazole to an active form requires microaerophilic conditions (Wayne and Sramek, 1994). Apparently lesions in mice are insufficiently hypoxic for this drug to be effective.

As mentioned above, activities of isocitrate lyase and glycine dehydrogenase are induced in the Wayne model and glycine/alanine dehydrogenase is induced in the defined hypoxia models (Rosenkrands *et al.*, 2002)(data not shown), indicating an induction of the glyoxylate shunt pathway (Wayne and Lin, 1982). A subsequent study showed that the isocitrate lyase gene *icl* is required specifically for persistence in the mouse, suggesting that the glyoxylate shunt may indeed be activated at this stage of infection (McKinney *et al.*, 2000). However, *icl* is not induced directly by hypoxia (Sherman *et al.*, 2001), and the mutant was not attenuated under hypoxic conditions *in vitro* (McKinney *et al.*, 2000). In addition, *icl* is expressed in human TB granulomas, but this material was derived from a patient with active rather than latent disease (Fenhalls *et al.*, 2002). With regard to glycine dehydrogenase, another study showed that the *Mycobacterium marinum* gene *ald* encoding glycine/alanine dehydrogenase is also induced in granulomas (Chan *et al.*, 2002; Davis *et al.*, 2002). However, while data are not yet available regarding the MTB granuloma, the *M. marinum* granuloma is clearly a dynamic environment for both host and pathogen (Bouley *et al.*, 2001, Chan *et al.*, 2002; Davis *et al.*, 2002) and so cannot be considered to represent mycobacterial latency.

Acr is induced in nearly every *in vitro* model of hypoxia induced latency (Yuan and Barry, 1996; Cunningham and Spreadbury, 1998; Imboden and Schoolnik, 1998; Yuan *et al.*, 1998; Desjardin *et al.*, 2001; Florczyk *et al.*, 2001; Sherman *et al.*, 2001; Shi *et al.*, 2003), and there is compelling evidence that *acr* is also expressed *in vivo*. More than three-quarters of TB patients express antibodies directed against Acr (Lee *et al.*, 1992; Verbon *et al.*, 1992), and abundant *acr* transcript has been detected in infected mice (Shi *et al.*, 2003). However, *acr* mRNA appears at least one week prior to cessation of bacterial growth. Acr is also powerfully induced when MTB is growing within macrophages (Yuan *et al.*, 1998) or in interferon-γ-deficient mice (Shi *et al.*, 2003) in which there is no latent phase. Similarly, other genes induced by hypoxia *in vitro* are also expressed in either animal models (Weber *et al.*, 2000; Shi *et al.*, 2003) or in man (Fenhalls *et al.*, 2002), but always in the context of active bacterial growth or fulminant disease.

8. The DosR Response *In Vitro* and *In Vivo*

Identifying the primary MTB transcription factor to mediate a hypoxic or NO signal should help clarify the role of these signals *in vitro* and *in vivo*. To assess the significance of the DosR response to long-term survival in reduced oxygen, Boon and Dick tested a *dosR* mutant of BCG [the avirulent derivative of *M. bovis* used worldwide as a TB vaccine (Mostowy *et al.*, 2003)] in a series of Wayne model experiments (Boon and Dick,

2002). In their hands, depletion of oxygen from the cultures took 15 days as determined by decolorization of a methelyne blue indicator dye. The number of wild-type bacilli increased for 10 days and then remained constant for the remainder of the experiment. The *dosR* mutant BCG behaved similarly for 5 days, but by day 10 the colony forming units (CFU) recovered were 10-fold lower than for wild-type. This trend continued until by day 40 the CFU of the mutant were 1500-fold lower than wild-type. Our own experiments with mutant MTB unable to mount a DosR response yielded a much more modest 10-fold difference in survival at nearly the same time point (data not shown). There is no obvious explanation for this discrepancy, though it may stem from differences in strain (BCG vs. MTB) or the nature of the mutations (in the BCG mutant, the very small portion of the *dosR*-coding region that remains could potentially exert a negative effect). In addition, these experiments assessed the ability to grow on solid media as opposed to survival per se. Bacilli cultured for long periods in low oxygen may require a period of re-adaptation to ambient air before forming colonies on plates. We have found that the *dosR*-mutant may be more sensitive than wild-type to this adaptation period (data not shown). Thus the impaired ability of *dosR*-mutant bacilli to form colonies following hypoxic incubation may be due in part to a greater need for re-adaptation rather than a difference in survival.

Several observations argue that the DosR response is not simply an *in vitro* phenomenon. Along with the data summarized above suggesting that MTB encounters hypoxic conditions within the host, nitric oxide is generated both by MTB-infected mice (MacMicking *et al.*, 1997; Scanga *et al.*, 2000) and in the tuberculous human lung (Choi *et al.*, 2002). Interfering with host NO production with either specific inhibitors (Chan *et al.*, 1995; Ehlers *et al.*, 1999) or by gene disruption (MacMicking *et al.*, 1997) greatly exacerbates TB infection. It should be no surprise then that genes of the DosR regulon are induced during MTB infections. In three different studies, mice were infected with MTB by aerosol and expression of various genes was monitored by real-time PCR (Shi *et al.*, 2003; Voskuil *et al.*, 2003). Each of the nine DosR response genes examined was powerfully expressed. Also, in humans the DosR response genes *acr* (Lee *et al.*, 1992) and *narX* (Fenhalls *et al.*, 2002) are expressed by MTB during infection, as measured by antibody response or *in situ* hybridization. Still, since these instances of DosR regulon gene expression occur in the context of active TB disease, it is clear that the DosR response by itself is not sufficient to induce latency. The relevance of the DosR response to TB *in vivo* must still be examined.

DosR-mutant bacilli should help address this question. Since they fare poorly under hypoxic stress *in vitro*, *dosR* mutants should provide a measure of the importance of hypoxia in various *in vivo* models. Results thus far are equivocal. Parish, Stoker and colleagues used homologous recombination to generate a *dosR* mutant strain of MTB, which they call *devR*Δ. Surprisingly, following intravenous (iv) infection of severe combined immunodeficient (SCID) mice, the *devR*Δ bacilli displayed hypervirulence – the mutant killed SCID mice in ~30 days while wild-type MTB killed in ~40 days (Parish *et al.*, 2003). The same trend was evident on iv infection of immunocompetent but susceptible DBA/2 mice. At 15 days post-infection, the numbers of mutant bacteria in lungs, liver and spleen were elevated ~10-fold. Since initial depositions were not reported, it is unclear whether the mutant grew better or if it seeded the organs better following iv infection. This difference in bacterial burden narrowed by 30 days post-infection and had virtually disappeared by 60 days in all organs but the liver. In contrast, we introduced our *dosR* mutant MTB into C57BL/6 mice by aerosol and monitored bacterial burden in lung and spleen, histopathology and time to morbidity. In another surprise, we found no differences between mutant and wild-type in any parameter at any time point (data not shown).

The difference in these data may stem from subtle differences in the mutants or the routes of infection. However, these results may not be as discrepant as they at first appear. The mice used thus far to test the DosR response present a spectrum of TB susceptibility, from exquisitely sensitive to highly resistant. Lacking both T cells and B cells, SCID mice are unable to mount any adaptive immune response. Though more resistant than SCID mice, DBA/2 mice are still extremely susceptible to MTB, while C57BL/6 mice are considered highly resistant (Medina and North, 1998; Turner *et al.*, 2001; Cardona *et al.*, 2003; Chackerian and Behar, 2003). If the DosR response prepares MTB for non-replicative persistence as has been suggested (Sherman *et al.*, 2001; Boon and Dick, 2002; Parish *et al.*, 2003; Park *et al.*, 2003; Voskuil *et al.*, 2003), disruption of this response in a severely immunocompromised animal might lead to unchecked growth and hypervirulence. Similarly, the DBA/2 mouse may be so vulnerable to MTB infection that the *dosR* mutant can outgrow wild-type, even as the resistant C57BL/6 mouse contains both strains equally well. To resolve these issues, different laboratories must test the DosR mutants in more than one animal model each.

Since the *dosR* mutants are not impeded for growth under the *in vivo* conditions tested to date, these experiments suggest that hypoxia does not limit MTB growth in these systems. Perhaps DosR controls an initial response of bacilli to *in vivo* conditions independent of other signals/responses necessary for latency. Alternatively, conditions in SCID, DBA/2 and C57BL/6 mice may be hypoxic enough to induce the DosR response but not severe enough to induce latency.

When infected with MTB, none of these mice produce caseating granulomas, the lung environment where oxygen is likely most limiting (Canetti, 1955). It will be important to test the DosR mutants in animal models of TB that feature caseation, such as the guinea pig (Turner *et al.*, 2003), the rabbit (Dannenberg, 1993) or the fish (Talaat *et al.*, 1998). Like the other hypotheses based on *in vitro* models, the link between hypoxia, NO, DosR and latency cannot be substantiated without a better understanding of latent TB in humans and an animal model that clearly reproduces major components of the human condition.

9. Conclusions and Future Prospects

Persuasive evidence accumulated over several decades indicates that oxygen tension is a key force that helps determine the growth of MTB and the severity of disease. This effect can be measured *in vitro*, with both respiration (Kempner, 1939) and replication rates (Canetti, 1955) directly dependent on the level of available oxygen. The evidence is less direct *in vivo*, but no less compelling. In both animals (Medlar and Sasano, 1936; Medlar, 1940) and people (Sweany *et al.*, 1931), TB is most devastating in the most oxygenated regions of the lungs. In guinea pigs, rabbits and dogs, artificially reducing oxygen tension led to less severe TB disease (Adams and Vorwald, 1934; Rich and Follis, 1942) while in monkeys locally increasing available oxygen exacerbated disease (Hanlon *et al.*, 1950; Scott *et al.*, 1950; Olson *et al.*, 1951). The qualified success of lung collapse as a TB therapy (Davis, 1996) and the remarkable work of Vandivere with open and closed human lesions (Vandiviere *et al.*, 1956) argue that oxygen has the same effect on human TB.

Based on these observations, Larry Wayne and others have developed *in vitro* models of TB latency in which oxygen deprivation plays a key role (Dick *et al.*, 1998; Hu *et al.*, 1998; Yuan *et al.*, 1998; Michele *et al.*, 1999; Florczyk *et al.*, 2001; Sherman *et al.*, 2001; Wayne, 2001). In recent years experiments with these systems have substantially deepened our knowledge of how MTB reacts to hypoxia. As oxygen tension drops, MTB responds with a coordinated genetic program to adapt to this stress. DosR is the first MTB transcription factor to recognize hypoxic stress and alter gene expression (Sherman *et al.*, 2001; Park *et al.*, 2003). As researchers move beyond the initial hypoxic response, other transcription factors and other effector genes will surely be implicated in the adaptation to nonreplicating persistence *in vitro*.

Some light has also been shed on hypoxia and TB latency *in vivo*, though this link remains uncertain. MTB hypoxic response genes are powerfully induced *in vivo*, however their expression alone is not enough to initiate latency. *In vivo* environments are probably all somewhat hypoxic relative to the aerated roller bottle or flask that MTB usually inhabits *in vitro*. Increased hypoxia within caseating granulomas or elsewhere may be one signal that promotes latency, but the experiments to test that hypothesis have not yet been reported. To date, the only MTB genes implicated in persistence *in vivo*, *icl* (McKinney *et al.*, 2000) and *relA* (Dahl *et al.*, 2003), both respond to changes in carbon utilization and not to hypoxia.

To advance our understanding of TB latency, we must address the controversies and assumptions of the current model. Where are the bacilli during a latent infection and what is their metabolic state? What aspects of host immunology or physiology alter the balance between latency and reactivation? Is oxygen tension a key determinant of latency? Answering these questions will require much more work with tissues from chronically MTB-infected animals and humans. Also, as both targeted and random mutagenesis are applied more frequently to MTB, we can expect that many more genes important for hypoxic survival *in vitro* will be identified, and each of these will be tested *in vivo*. Of course, one way to test the link between hypoxia and TB latency would be to identify drugs that kill nonreplicating persistent bacilli in a hypoxic *in vitro* model, and then assess the activity of those agents *in vivo*. Such compounds might do much more than help solve the riddle of TB latency. They may shorten the course of therapy, provide effective short-course chemotherapy to combat latency and finally help stem the tide against this pernicious and elusive bacterial pathogen. Screening for compounds with activity against nonreplicating persistent bacilli *in vitro* should be a major research and public health priority.

Acknowledgements

We apologize for any errors or omissions, which are solely our own. We thank Eric Rubin, Lalita Ramakrishnan, Tige Rustad, Mark Hickey, Kristi Guinn, Maria Isabel Harrell, Jennifer Chang and Nada Harik for critical reading of the manuscript, and members of the Sherman, Cangelosi and Ramakrishnan labs for stimulating discussions. DRS is supported by NIH (HL64550 and HL68533) and the Sequella Global Tuberculosis Foundation. DMR is supported by an NIH training grant (AI07509) awarded to the University of Washington Pathobiology Department.

References

Adams, W.E., and Vorwald, A.J. 1934. The treatment of pulmonary tuberculosis by bronchial occlusion. J. Thoracic Surg. 3: 633-666.

Barry, C.E., 3rd, Lee, R.E., Mdluli, K., Sampson, A.E., Schroeder, B.G., Slayden, R.A., and Yuan, Y. 1998. Mycolic acids: structure, biosynthesis and physiological functions. Prog. Lipid Res. 37: 143-179.

Betts, J.C., Lukey, P.T., Robb, L.C., McAdam, R.A., and Duncan, K. 2002. Evaluation of a nutrient starvation model of *Mycobacterium tuberculosis* persistence by gene and protein expression profiling. Mol. Microbiol. 43: 717-731.

Boon, C., and Dick, T. 2002. *Mycobacterium bovis* response regulator essential for hypoxic dormancy. J. Bacteriol. 184: 6760-6767.

Boon, C., Li, R., Qi, R., and Dick, T. 2001. Proteins of *Mycobacterium bovis* BCG induced in the Wayne dormancy model. J. Bacteriol. 183: 2672-2676.

Bouley, D.M., Ghori, N., Mercer, K.L., Falkow, S., and Ramakrishnan, L. 2001. Dynamic nature of host-pathogen interactions in *Mycobacterium marinum* granulomas. Infect. Immun. 69: 7820-7831.

Boyd, A.D., Crawford, B.K., and Glassman, L. 1996. Surgical therapy of tuberculosis. In: *Tuberculosis*. W.N. Rom, and S.M. Garay, eds. Little, Brown and Co., Boston, Massachusetts. p. 513-523.

Brooks, J.V., Furney, S.K., and Orme, I.M. 1999. Metronidazole therapy in mice infected with tuberculosis. Antimicrob. Agents Chemother. 43: 1285-1288.

Canetti, G. 1955. Growth of the tubercle bacilus in the tuberculosis lesion. In: The Tubercle Bacillus in the Pulmonary Lesion of Man. Springer Publishing Co., New York. p. 111-126.

Cardona, P.J., Gordillo, S., Diaz, J., Tapia, G., Amat, I., Pallares, A., Vilaplana, C., Ariza, A., and Ausina, V. 2003. Widespread bronchogenic dissemination makes DBA/2 mice more susceptible than C57BL/6 mice to experimental aerosol infection with *Mycobacterium tuberculosis*. Infect. Immun. 71: 5845-5854.

Chackerian, A.A., and Behar, S.M. 2003. Susceptibility to *Mycobacterium tuberculosis*: lessons from inbred strains of mice. Tuberculosis (Edinb). 83: 279-285.

Chan, J., Tanaka, K., Carroll, D., Flynn, J., and Bloom, B.R. 1995. Effects of nitric oxide synthase inhibitors on murine infection with *Mycobacterium tuberculosis*. Infect. Immun. 63: 736-740.

Chan, K., Knaak, T., Satkamp, L., Humbert, O., Falkow, S., and Ramakrishnan, L. 2002. Complex pattern of *Mycobacterium marinum* gene expression during long-term granulomatous infection. Proc. Natl. Acad. Sci. USA. 99: 3920-3925.

Choi, H.S., Rai, P.R., Chu, H.W., Cool, C., and Chan, E.D. 2002. Analysis of nitric oxide synthase and nitrotyrosine expression in human pulmonary tuberculosis. Am. J. Respir. Crit. Care Med. 166: 178-186.

Cole, S.T., Brosch, R., Parkhill, J., Garnier, T., Churcher, C., Harris, D., *et al.* 1998. Deciphering the biology of *Mycobacterium tuberculosis* from the complete genome sequence. Nature. 393: 537-544.

Comstock, G.W., Baum, C., and Snider, D.E. 1979. Isoniazid prophylaxis among alaskan eskimos: a final report of the Bethel isoniazid studies. Am. Rev. Respir. Dis. 119: 827-830.

Corbett, E.L., Watt, C.J., Walker, N., Maher, D., Williams, B.G., Raviglione, M.C., and Dye, C. 2003. The growing burden of tuberculosis: global trends and interactions with the HIV epidemic. Arch. Intern. Med. 163: 1009-1021.

Corper, H.J., and Lurie, M.B. 1926. The variability of localization of tuberculosis in the organs of different animals. I. Quantitative relations in the rabbit, guinea pig, dog and monkey. Am. Rev. Tuberc. 14: 662-679.

Corper, H.J., and Cohn, M.L. 1933. The viability and virulence of old cultures of tubercle bacilli. Am. Rev. Tuberc. 28: 856-874.

Corper, H.J., Lurie, M.B., and Uyei, N. 1927. The variability of localization of tuberculosis in the organs of different animals. III. The importance of the growth of the tubercle bacilli as determined by gaseous tension. Am. Rev. Tuberc. 15: 65-87.

Cosma, C.L., Sherman, D.R., and Ramakrishnan, L. 2003. The secret lives of pathogenic mycobacteria. Annu. Rev. Microbiol. 57: 641-676.

Cunningham, A.F., and Spreadbury, C.L. 1998. Mycobacterial stationary phase induced by low oxygen tension: cell wall thickening and localization of the 16-kilodalton alpha-crystallin homolog. J. Bacteriol. 180: 801-808.

Dahl, J.L., Kraus, C.N., Boshoff, H.I.M., Doan, B., Foley, K., Avarbock, D., Kaplan, G., Mizrahi, V., Rubin, H., and Barry C.E., 3rd. 2003. The role of Rel_{Mtb}-mediated adaptation to stationary phase in long-term persistence of *Mycobacterium tuberculosis* in mice. Proc. Natl. Acad. Sci. USA. 100: 10026-10031.

Dannenberg, A.M., and Rook, G.A.W. 1994. Pathogenesis of pulmonary tuberculosis: an interplay of tissue-damaging and macrophage-activating immune responses - dual mechanisms that control bacillary multiplication. In: Tuberculosis: Pathogenesis, Protection and Control. B.R. Bloom, ed. ASM Press, Washington, D.C. p. 459-483.

Dannenberg, A.M., Jr. 1993. Immunopathogenesis of pulmonary tuberculosis. Hosp. Pract. 28: 51-58.

Dasgupta, N., Kapur, V., Singh, K.K., Das, T.K., Sachdeva, S., Jyothisri, K., and Tyagi, J.S. 2000. Characterization of a two-component system, *devR-devS*, of *Mycobacterium tuberculosis*. Tuber. Lung Dis. 80: 141-159.

Davis, A.L. 1996. History of the sanatorium movement. In: Tuberculosis. W.N. Rom, and S.M. Garay, eds. Little, Brown and Co., Boston, Massachusetts. p. 35-54.

Davis, J.M., Clay, H., Lewis, J.L., Ghori, N., Herbomel, P., and Ramakrishnan, L. 2002. Real-time visualization of mycobacterium-macrophage interactions leading to initiation of granuloma formation in zebrafish embryos. Immunity. 17: 693-702.

DeMaio, J., Zhang, Y., Ko, C., Young, D.B., and Bishai, W.R. 1996. A stationary-phase stress-response sigma factor from *Mycobacterium tuberculosis*. Proc. Natl. Acad. Sci. USA. 93: 2790-2794.

Desjardin, L.E., Hayes, L.G., Sohaskey, C.D., Wayne, L.G., and Eisenach, K.D. 2001. Microaerophilic induction of the alpha-crystallin chaperone protein homologue (hspX) mRNA of *Mycobacterium tuberculosis*. J. Bacteriol. 183: 5311-5316.

Dhillon, J., Allen, B.W., Hu, Y.M., Coates, A.R., and Mitchison, D.A. 1998. Metronidazole has no antibacterial effect in Cornell model murine tuberculosis. Int. J. Tuberc. Lung Dis. 2: 736-742.

Dick, T., Lee, B.H. and Murugasu-Oei, B. 1998. Oxygen depletion induced dormancy in *Mycobacterium smegmatis*. FEMS Microbiol. Lett. 163: 159-164.

Diez, A., Gustavsson, N., and Nystrom, T. 2000. The universal stress protein A of *Escherichia coli* is required for resistance to DNA damaging agents and is regulated by a RecA/FtsK-dependent regulatory pathway. Mol. Microbiol. 36: 1494-1503.

Dock, W. 1946. Apical localization of phthisis: Its significance in treament by prolonged rest in bed. Am. Rev. Tuberc. 53: 297-305.

Dock, W. 1954. Effect of posture on alveolar gas tension in tuberculosis. AMA Arch. Int. Med. 94: 700-708.

Dubos, R.J. 1953. Effect of the composition of the gaseous and aqueous environments on the survival of tubercle bacilli *in vitro*. J. Exp. Med. 97: 357-366.

Dye, C., Scheele, S., Dolin, P., Pathania, V., and Raviglione, M.C. 1999. Consensus statement. Global burden of tuberculosis: estimated incidence, prevalence, and mortality by country. WHO Global Surveillance and Monitoring Project. JAMA. 282: 677-686.

Ehlers, S., Kutsch, S., Benini, J., Cooper, A., Hahn, C., Gerdes, J., Orme, I., Martin, C., and Rietschel, E.T. 1999. NOS2-derived nitric oxide regulates the size, quantity and quality of granuloma formation in *Mycobacterium avium*-infected mice without affecting bacterial loads. Immunology. 98: 313-323.

Enarson, D.A., and Murray, J.F. 1996. Global epidemiology of tuberculosis. In: Tuberculosis. W.N. Rom, and S. Garay, eds. Little, Brown and Co., Boston. p. 55-75.

Espinal, M.A. 2003. The global situation of MDR-TB. Tuberculosis (Edinb). 83: 44-51.

Farrell, J. 1998. Invisible Enemies: Stories of Infectious Disease. Farrar Straus & Giroux, New York, New York.

Feldman, W.H. 1939. The occurrence of virulent tubercule bacilli in presumably non-tuberculous lung tissue. Am. J. Pathol. 15: 501-515.

Feldman, W.H., and Baggenstoss, A.H. 1938. The residual infectivity of the primary complex of tuberculosis. Am. J. Pathol. 14: 473-490.

Fenhalls, G., Stevens, L., Moses, L., Bezuidenhout, J., Betts, J.C., Helden Pv, P., *et al.* 2002. *In situ* detection of *Mycobacterium tuberculosis* transcripts in human lung granulomas reveals differential gene expression in necrotic lesions. Infect. Immun. 70: 6330-6338.

Florczyk, M.A., McCue, L.A., Stack, R.F., Hauer, C.R., and McDonough, K.A. 2001. Identification and characterization of mycobacterial proteins differentially expressed under standing and shaking culture conditions, including Rv2623 from a novel class of putative ATP-binding proteins. Infect. Immun. 69: 5777-5785.

Florczyk, M.A., McCue, L.A., Purkayastha, A., Currenti, E., Wolin, M.J., and McDonough, K.A. 2003. A family of acr-coregulated *Mycobacterium tuberculosis* genes shares a common DNA motif and requires Rv3133c (*dosR* or *devR*) for expression. Infect. Immun. 71: 5332-5343.

Flynn, J.L., and Chan, J. 2001. Tuberculosis: latency and reactivation. Infect. Immun. 69: 4195-4201.

Garay, S.M. 1996. Pulmonary Tuberculosis. In: Tuberculosis. W.N. Rom, and S.M. Garay, eds. Little, Brown and Co., Boston. p. 373-412.

Gedde-Dahl, T. 1952. Tuberculous infection in the light of tuberculin matriculation. Am. J. Hyg. 56: 139-214.

Glickman, M.S., and Jacobs, W.R., Jr. 2001. Microbial pathogenesis of *Mycobacterium tuberculosis*: dawn of a discipline. Cell. 104: 477-485.

Goldman, D., and Wagner, M. 1962. Enzymes systems in the mycobacteria XIII. Glycine dehydrogenase and the glyoxylic acid cycle. Biochim. Biophys. Acta. 65: 297-306.

Gomez, J.E., Chen, J.M., and Bishai, W.R. 1997. Sigma factors of *Mycobacterium tuberculosis*. Tuber. Lung Dis. 78: 175-183.

Graw, J. 1997. The crystallins: genes, proteins and diseases. Biol. Chem. 378: 1331-1348.

Gustavsson, N., Diez, A., and Nystrom, T. 2002. The universal stress protein paralogues of *Escherichia coli* are co-ordinately regulated and co-operate in the defence against DNA damage. Mol. Microbiol. 43: 107-117.

Haas, F., and Haas, S.S. 1996. The origin of *Mycobacterium tuberculosis* and the notion of its contagiousness. In: Tuberculosis. W.N. Rom, and S.M. Garay, eds. Little, Brown and Co., Boston. p. 3-19.

Hanlon, C.R., Scott, J., H.W., and Olson, B.J. 1950. Experimental tuberculosis. I. Effects of anastomosis between systemic and pulmonary arteries on tuberculosis in Monkeys. Surgery. 28: 209-224.

Herbert, D., Paramasivan, C.N., Venkatesan, P., Kubendiran, G., Prabhakar, R., and Mitchison, D.A. 1996. Bactericidal action of ofloxacin, sulbactam-ampicillin, rifampin, and isoniazid on logarithmic- and stationary-phase cultures of *Mycobacterium tuberculosis*. Antimicrob. Agents Chemother. 40: 2296-2299.

Hernandez-Pando, R., Jeyanathan, M., Mengistu, G., Aguilar, D., Orozco, H., Harboe, M., Rook, G.A., and Bjune, G. 2000. Persistence of DNA from *Mycobacterium tuberculosis* in superficially normal lung tissue during latent infection. Lancet. 356: 2133-2138.

Hoch, J.A., and Silhavy, T.J. 1995. Two-Component Signal Transduction. ASM Press, Washington D.C.

Hu, Y., and Coates, A.R.M. 1999. Transcription of two sigma 70 homologue genes, *sigA* and *sigB*, in stationary-phase *Mycobacterium tuberculosis*. J. Bacteriol. 181: 469-476.

Hu, Y., Mangan, J.A., Dhillon, J., Sole, K.M., Mitchison, D.A., Butcher, P.D., and Coates, A.R. 2000. Detection of mRNA transcripts and active transcription in persistent *Mycobacterium tuberculosis* induced by exposure to rifampin or pyrazinamide. J. Bacteriol. 182: 6358-6365.

Hu, Y.M., Butcher, P.D., Sole, K., Mitchison, D.A., and Coates, A.R. 1998. Protein synthesis is shutdown in dormant *Mycobacterium tuberculosis* and is reversed by oxygen or heat shock. FEMS Microbiol. Lett. 158: 139-145.

Hu, Y.M., Butcher, P.D., Mangan, J.A., Rajandream, M.A., and Coates, A.R.M. 1999. Regulation of *hmp* gene transcription in *Mycobacterium tuberculosis*: Effects of oxygen limitation and nitrosative and oxidative stress. J. Bacteriol. 181: 3486-3493.

Hutter, B., and Dick, T. 1999. Up-regulation of *narX*, encoding a putative 'fused nitrate reductase' in anaerobic dormant *Mycobacterium bovis* BCG. FEMS Microbiol. Lett. 178: 63-69.

Imboden, P., and Schoolnik, G.K. 1998. Construction and characterization of a partial *Mycobacterium tuberculosis* cDNA library of genes expressed at reduced oxygen tension. Gene. 213: 107-117.

Kempner, W. 1939. Oxygen tension and the tubercle bacillus. *Am. Rev. Tubercul.* 40: 157-168.

Lee, B.Y., Hefta, S.A., and Brennan, P.J. 1992. Characterization of the major membrane protein of virulent *Mycobacterium tuberculosis*. Infect. Immun. 60: 2066-2074.

Lillebaek, T., Dirksen, A., Baess, I., Strunge, B., Thomsen, V.O., and Andersen, A.B. 2002. Molecular evidence of endogenous reactivation of *Mycobacterium tuberculosis* after 33 years of latent infection. J. Infect. Dis. 185: 401-404.

Lim, A., Eleuterio, M., Hutter, B., Murugasu-Oei, B., and Dick, T. 1999. Oxygen depletion-induced dormancy in *Mycobacterium bovis* BCG. J. Bacteriol. 181: 2252-2256.

MacMicking, J.D., North, R.J., LaCourse, R., Mudgett, J.S., Shah, S.K., and Nathan, C.F. 1997. Identification of nitric oxide synthase as a protective locus against tuberculosis. Proc. Natl. Acad. Sci. USA. 94: 5243-5248.

Manabe, Y.C., and Bishai, W.R. 2000. Latent *Mycobacterium tuberculosis*-persistence, patience, and winning by waiting. Nat. Med. 6: 1327-1329.

Manganelli, R., Dubnau, E., Tyagi, S., Kramer, F.R., and Smith, I. 1999. Differential expression of 10 sigma factor genes in *Mycobacterium tuberculosis*. Mol. Microbiol. 31: 715-724.

Martin, E., Davis, K., Bian, K., Lee, Y.C., and Murad, F. 2000. Cellular signaling with nitric oxide and cyclic guanosine monophosphate. Semin. Perinatol. 24: 2-6.

McCune, R.M., Tompsett, R., and McDermott, W. 1956. The fate of *Mycobacterium tuberculosis* in mouse tissues as determined by the microbial enumeration technique. J. Exp. Med. 104: 763-803.

McCune, R.M., Feldmann, F.M., and McDermott, W. 1966a. Microbial persistence. II. Characteristics of the sterile state of tubercle bacilli. J. Exp. Med. 123: 469-486.

McCune, R.M., Feldmann, F.M., Lambert, H.P., and McDermott, W. 1966b. Microbial persistence. I. The capacity of tubercle bacilli to survive sterilization in mouse tissues. J. Exp. Med. 123: 445-468.

McDermott, W. 1958. Microbial persistence. Yale J. Biol. Med. 30: 257.

McKinney, J.D., Jacobs, W.R., Jr., and Bloom, B.R. 1998. Persisting problems in tuberculosis. In: Emerging Infections. R.M. Krause, A.S. Fauci, and J. Gallin, eds. Academic Press, San Diego, California. p. 51-146.

McKinney, J.D., Honer zu Bentrup, K., Munoz-Elias, E.J., Miczak, A., Chen, B., Chan, W.T., Swenson, D., Sacchettini, J.C., Jacobs, W.R. Jr., and Russell, D.G. 2000. Persistence of *Mycobacterium tuberculosis* in macrophages and mice requires the glyoxylate shunt enzyme isocitrate lyase. Nature. 406: 735-738.

Medina, E., and North, R.J. 1998. Resistance ranking of some common inbred mouse strains to *Mycobacterium tuberculosis* and relationship to major histocompatibility complex haplotype and *Nramp1* genotype. Immunology. 93: 270-274.

Medlar, E.M. 1940. Pulmonary tuberculosis in cattle. The location and type of lesions in naturally aquired tuberculosis. Am. Rev. Tuberc. 41: 283-306.

Medlar, E.M., and Sasano, K.T. 1936. A study of the pathology of experimental pulmonary tuberculosis in the rabbit. Am. Rev. Tuberc. 34: 456-476.

Medlar, E.M., Berstein, S., and Steward, D.M. 1952. A bacteriologic study of resected tuberculous lesions. Am. Rev. Tuberc. 66: 36-43.

Michele, T.M., Ko, C., and Bishai, W.R. 1999. Exposure to antibiotics induces expression of the *Mycobacterium tuberculosis sigF* gene: implications for chemotherapy against mycobacterial persistors. Antimicrob. Agents Chemother. 43: 218-225.

Mitchison, D.A. 1979. Basic mechanisms of chemotherapy. Chest. 76: 771-781.

Mitchison, D.A. 1992. The Garrod Lecture: understanding the chemotherapy of tuberculosis -- current problems. J. Antimicrob. Chemother. 29: 477-493.

Mostowy, S., Tsolaki, A.G., Small, P.M., and Behr, M.A. 2003. The *in vitro* evolution of BCG vaccines. Vaccine. 21: 4270-4274.

Murugasu-Oei, B., and Dick, T. 2000. Bactericidal activity of nitrofurans against growing and dormant *Mycobacterium bovis* BCG. J. Antimicrob. Chemother. 46: 917-919.

Nathan, C. 2002. Inducible nitric oxide synthase in the tuberculous human lung. Am. J. Respir. Crit. Care Med. 166: 130-131.

Ohno, H., Zhu, G., Mohan, V.P., Chu, D., Kohno, S., Jacobs, W.R., Jr., and Chan, J. 2003. The effects of reactive nitrogen intermediates on gene expression in *Mycobacterium tuberculosis*. Cell Microbiol. 5: 637-648.

Olender, S., Saito, M., Apgar, J., Gillenwater, K., Bautista, C.T., Lescano, A.G., Moro, P., Caviedes, L., Hsieh, E.J., and Gilman R.H. 2003. Low prevalence and increased household clustering of *Mycobacterium tuberculosis* infection in high altitude villages in Peru. Am. J. Trop. Med. Hyg. 68: 721-727.

Olson, B.J., Scott, J., H.W., Hanlon, C.R., and Mattern, C.F.T. 1951. Experimental tuberculosis. III. Further observations on the effects of alteration of the pulmonary arterial circulation on tuberculosis in monkeys. Am. Rev. Tuberc. 65: 48-63.

Opie, E.L., and Aronson, J.D. 1927. Tubercle bacilli in latent tuberculous lesions and in lung tissue without tuberculous lesions. Arch. Pathol. Lab. Med. 4: 1-21.

Orme, M. 2001. The latent tuberculosis bacillus (I'll let you know if I ever meet one). Int. J. Tuberc. Lung Dis. 5: 589-593.

O'Toole, R., and Williams, H.D. 2003. Universal stress proteins and *Mycobacterium tuberculosis*. Res. Microbiol. 154: 387-392.

Pai, S.R., Actor, J.K., Sepulveda, E., Hunter, R.L., and Jagannath, C. 2000. Identification of viable and non-viable *Mycobacterium tuberculosis* in mouse organs by directed RT-PCR for antigen 85B mRNA. Microb. Pathog. 28: 335-342.

Parish, T., Smith, D.A., Kendall, S., Casali, N., Bancroft, G.J., and Stoker, N.G. 2003. Deletion of two-component regulatory systems increases the virulence of *Mycobacterium tuberculosis*. Infect. Immun. 71: 1134-1140.

Park, H.D., Guinn, K.M., Harrell, M.I., Liao, R., Voskuil, M.I., Tompa, M., *et al.* 2003. Rv3133c/dosR is a transcription factor that mediates the hypoxic response of *Mycobacterium tuberculosis*. Mol. Microbiol. 48: 833-843.

Parrish, N.M., Dick, J.D., and Bishai, W.R. 1998. Mechanisms of latency in *Mycobacterium tuberculosis*. Trends Microbiol. 6: 107-112.

Peh, H.L., Toh, A., Murugasu-Oei, B., and Dick, T. 2001. *In vitro* activities of mitomycin C against growing and hypoxic dormant tubercle bacilli. Antimicrob. Agents Chemother. 45: 2403-2404.

Pinner, M. 1947. Evaluation of results of treatment. In: Pulmonary Tuberculosis in the Adult. Charles C. Thomas, Springfield. p. 478-486.

Purkayastha, A., McCue, L.A., and McDonough, K.A. 2002. Identification of a *Mycobacterium tuberculosis* putative classical nitroreductase gene whose expression is coregulated with that of the *acr* gene within macrophages, in standing versus shaking cultures, and under low oxygen conditions. Infect. Immun. 70: 1518-1529.

Raviglione, M.C. 2003. The TB epidemic from 1992 to 2002. Tuberculosis (Edinb). 83: 4-14.

Rees, R.J.M., and Hart, P.D. 1961. Analysis of the host-parasite equilibrium in chronic murine tuberculosis by total and viable bacillary counts. Brit. J. Exp. Pathol. 42: 83-88.

Rich, A.R., and Follis, R.H., Jr. 1942. The effect of low oxygen tension upon the development of experimental tuberculosis. Bull. Johns Hopkins Hosp. 71: 345-363.

Riley, R.L. 1960. Apical localization of pulmonary tuberculosis. Bull. Johns Hopkins Hosp. 106: 232-239.

Riley, R.L., Mills, C.C., O'Grady, F., Sultan, L.U., Wittstadt, F., and Shivpuri, D.N. 1962. Infectiousness of air from a tuberculosis ward. Ultraviolet irradiation of infected air: comparative infectiousness of different patients. Am. Rev. Respir. Dis. 85: 511-525.

Rosenkrands, I., Slayden, R.A., Crawford, J., Aagaard, C., Barry, C.E., 3rd, and Andersen, P. 2002. Hypoxic response of *Mycobacterium tuberculosis* studied by metabolic labeling and proteome analysis of cellular and extracellular proteins. J. Bacteriol. 184: 3485-3491.

Scanga, C.A., Mohan, V.P., Yu, K., Joseph, H., Tanaka, K., Chan, J., and Flynn, J.L. 2000. Depletion of CD4(+) T cells causes reactivation of murine persistent tuberculosis despite continued expression of interferon gamma and nitric oxide synthase 2. J. Exp. Med. 192: 347-358.

Scott, J., H.W., Hanlon, C.R., and Olson, B.J. 1950. Experimental tuberculosis. II. Effects of ligation of pulmonary arteries on tuberculosis in monkeys. J. Thoracic Surg. 20: 761-773.

Sherman, D.R., Voskuil, M., Schnappinger, D., Liao, R., Harrell, M.I., and Schoolnik, G.K. 2001. Regulation of the *Mycobacterium tuberculosis* hypoxic response gene encoding alpha -crystallin. Proc. Natl. Acad. Sci. USA. 98: 7534-7539.

Shi, L., Jung, Y.J., Tyagi, S., Gennaro, M.L., and North, R.J. 2003. Expression of Th1-mediated immunity in mouse lungs induces a *Mycobacterium tuberculosis* transcription pattern characteristic of nonreplicating persistence. Proc. Natl. Acad. Sci. USA. 100: 241-246.

Stover, C.K., Warrener, P., VanDevanter, D.R., Sherman, D.R., Arain, T.M., Langhorne, M.H., Anderson, S.W., Towell, J.A., Yuan, Y., McMurray, D.N., Kreiswirth, B.N., Barry, C.E., and Baker, W.R. 2000. A small-molecule nitroimidazopyran drug candidate for the treatment of tuberculosis. Nature. 405: 962-966.

Sweany, H.C., Cook, C.E., and Kegerreis, R. 1931. A study on the position of primary cavities in pulmonary tuberculosis. Am. Rev. Tuberc. 24: 558-582.

Talaat, A.M., Reimschuessel, R., Wasserman, S.S., and Trucksis, M. 1998. Goldfish, Carassius auratus, a novel animal model for the study of *Mycobacterium marinum* pathogenesis. Infect. Immun. 66: 2938-2942.

Turner, J., Gonzalez-Juarrero, M., Saunders, B.M., Brooks, J.V., Marietta, P., Ellis, D.L., Frank, A.A., Cooper, A.M., and Orme, I.M. 2001. Immunological basis for reactivation of tuberculosis in mice. Infect. Immun. 69: 3264-3270.

Turner, O.C., Basaraba, R.J., and Orme, I.M. 2003. Immunopathogenesis of pulmonary granulomas in the guinea pig after infection with *Mycobacterium tuberculosis*. Infect. Immun. 71: 864-871.

Vandiviere, H.M., Loring, W.E., Melvin, I., and Willis, S. 1956. The treated pulmonary lesion and its tubercle bacillus. II. The death and resurrection. Am. J. Med. Sci. 232: 30-37.

Verbon, A., Hartskeerl, R.A., Schuitema, A., Kolk, A.H., Young, D.B., and Lathigra, R. 1992. The 14,000-molecular-weight antigen of *Mycobacterium tuberculosis* is related to the alpha-crystallin family of low-molecular-weight heat shock proteins. J. Bacteriol. 174: 1352-1359.

Voskuil, M.I., Schnappinger, D., Harrell, M.I., Visconti, K.C., Dolganov, G., Sherman, D.R., and Schoolnik, G.K. 2003. Inhibition of respiration by nitric oxide induces a *Mycobacterium tuberculosis* dormancy program. J. Exp. Med. 198: 705-713.

Wayne, L.G. 1960. The bacteriology of resected tuberculosis pulmonary lesions. II. Observations on bacilli which are stainable but which cannot be cultured. Am. Rev. Respir. Dis. 82: 370-377.

Wayne, L.G. 1977. Synchronized replication of *Mycobacterium tuberculosis*. Infect. Immun. 17: 528-530.

Wayne, L.G. 2001. *In vitro* model of hypoxically induced nonreplicating persistence of *Mycobacterium tuberculosis*. In: Methods in Molecular Medicine. *Mycobacterium tuberculosis* Protocols. T. Parish, and N.G. Stoker, eds. Humana Press Inc., Totowa, New Jersey. p. 247-269.

Wayne, L.G., and Diaz, G.A. 1967. Autolysis and secondary growth of *Mycobacterium tuberculosis* in submerged culture. J. Bacteriol. 93: 1374-1381.

Wayne, L.G., and Lin, K.Y. 1982. Glyoxylate metabolism and adaptation of *Mycobacterium tuberculosis* to survival under anaerobic conditions. Infect. Immun. 37: 1042-1049.

Wayne, L.G., and Sramek, H.A. 1994. Metronidazole is bactericidal to dormant cells of *Mycobacterium tuberculosis*. Antimicrob. Agents Chemother. 38: 2054-2058.

Wayne, L.G., and Hayes, L.G. 1996. An *in vitro* model for sequential study of shiftdown of *Mycobacterium tuberculosis* through two stages of nonreplicating persistence. Infect. Immun. 64: 2062-2069.

Wayne, L.G., and Hayes, L.G. 1998. Nitrate reduction as a marker for hypoxic shiftdown of *Mycobacterium tuberculosis*. Tuber. Lung Dis. 79: 127-132.

Wayne, L.G., and Sohaskey, C.D. 2001. Nonreplicating persistence of *Mycobacterium tuberculosis*. Annu. Rev. Microbiol. 55: 139-163.

Webb, G.B., Boissevain, C.H., and Ryder, C.T. 1924. Gas requirements of the tubercle bacillus. Am. Rev. Tuberc. 9: 534-537.

Weber, I., Fritz, C., Ruttkowski, S., Kreft, A., and Bange, F.C. 2000. Anaerobic nitrate reductase (*narGHJI*) activity of *Mycobacterium bovis* BCG in vitro and its contribution to virulence in immunodeficient mice. Mol. Microbiol. 35: 1017-1025.

West, A.H., and Stock, A.M. 2001. Histidine kinases and response regulator proteins in two-component signaling systems. Trends Biochem. Sci. 26: 369-376.

Yuan, Y., and Barry, C.E.3rd. 1996. A common mechanism for the biosynthesis of methoxy and cyclopropyl mycolic acids in *Mycobacterium tuberculosis*. Proc. Natl. Acad. Sci. USA. 93: 12828-12833.

Yuan, Y., Crane, D.D., Simpson, R.M., Zhu, Y., Hickey, M.J., Sherman, D.R., and Barry III, C.E. 1998. The 16-kDa a-crystallin (Acr) protein of *Mycobacterium tuberculosis* is required for growth in macrophages. Proc. Natl. Acad. Sci. USA. 95: 9578-9583.

Zumft, W.G. 2002. Nitric oxide signaling and NO dependent transcriptional control in bacterial denitrification by members of the FNR-CRP regulator family. J. Mol. Microbiol. Biotechnol. 4: 277-286.

Zumla, A., Mwaba, P., Squire, S.B., and Grange, J.M. 1999. The tuberculosis pandemic - which way now? J. Infect. 38: 74-79.

Chapter 8

Molecular Basis for Aerotolerance of the Obligately Anaerobic *Bacteroides* Spp.

Anthony D. Baughn and Michael H. Malamy*

Abstract

Obligate anaerobes cannot grow in the presence of atmospheric concentrations of oxygen. It is likely that this deficit is due, in part, to the presence of oxygen-labile targets within the cell. Despite the inability to grow under aerobic conditions, many obligate anaerobes can survive transient exposure to O_2 and reactive oxygen species (ROS). In the presence of O_2 and ROS, obligately anaerobic *Bacteroides* species elicit a coordinated response that is essential for survival during periods of oxidative stress. This oxidative stress response (OSR) includes expression of ROS quenching enzymes, such as superoxide dismutase and multiple peroxidases, as well as the DNA protective protein Dps. Similar to other eubacteria, expression of *Bacteroides* OSR proteins is subject to both OxyR-dependent and OxyR-independent regulation. In addition to the ability to quench ROS, *B. fragilis* can consume O_2 via a cytochrome *bd* oxidase respiratory chain. In the presence of nanomolar concentrations of O_2, this respiratory chain can function in energy metabolism, indicating that the response of this bacterium to O_2 is more complex than was previously thought.

1. Introduction

Members of the *Bacteroides* class of eubacteria are classified as obligate anaerobes, incapable of growth in the presence of atmospheric oxygen (Loesche, 1969). Depending upon the strain and growth conditions used, the growth-inhibitory pO_2 for these bacteria can range from 0.1% (Baughn and Malamy, 2004) to 4% (Loesche, 1969). The requirement for a low pO_2 restricts the growth of these bacteria to environments such as wastewater sludge (Shoop *et al.*, 1990), the lower gastrointestinal tract of animals (Lee *et al.*, 1968; Berchtold *et al.*, 1999), and the gingival crevice of mammals (Tanner *et al.*, 1979; Clark *et al.*, 1988). Since transit to new colonization sites often involves exposure to growth-restricting amounts of O_2, the ability to withstand an aerobic challenge (aerotolerance) is critical to the fitness of this class of bacteria (Tang *et al.*, 1999). Moreover, many environments in which these obligate anaerobes thrive have a regular influx of small amounts of O_2. Thus, the ability to tolerate low-level exposure to O_2 is important for persistence at these sites.

Bacteroides spp. account for greater than 30% of the cultivatable bacterial species from the human colon (Holdeman *et al.*, 1976; Wang *et al.*, 2003). In this environment, *Bacteroides spp.* thrive on complex sugars, such as starch and cellulose (Kotarski and Salyers, 1984; Berchtold *et al.*, 1999). This commensal relationship benefits *Bacteroides spp.* by providing a nutrient rich, low-pO_2 growth environment; and benefits the host by providing fermentation products from otherwise undigestible oligosaccharides. Despite this commensalism, some of these *Bacteroides spp.* are also important opportunistic pathogens. In fact, *Bacteroides fragilis* is among the most frequently isolated bacterial species from intra-abdominal and intra-peritoneal infections, even though this species is only a minor constituent of the human feces (Tally and Ho, 1987; Namavar *et al.*, 1989; Brook, 2002). *Bacteroides* infections commonly arise following spillage of colonic contents into the normally sterile abdominal cavity (Brook, 2002). Often times, fibrotic abscesses are formed around material that cannot be cleared by the host immune response. If untreated, these abscesses can suppurate, resulting in dissemination of bacteria to adjacent sites. In addition to the opportunistic pathogens of the intestinal microflora, other *Bacteroides spp.*, such as *Porphyromonas gingivalis*, *Prevotella intermedia*, and *Tannerella forsythensis* are associated with the development of periodontal disease, as well as oral-facial and neck abscesses (Brook, 2002). Since mammalian tissues are essentially aerobic (Bornside *et al.*, 1976), the ability to both survive transient exposure to O_2 and subsequently establish a low-pO_2 growth environment has been regarded as critical for the virulence of *Bacteroides spp.* (Tang *et al.*, 1999).

*For correspondence email: michael.malamy@tufts.edu

2. Response of *Bacteroides spp.* to Oxidative Stress

Studies on *B. fragilis* indicate that O_2 inhibits growth in a manner that is independent of the environmental reduction-oxidation (redox) potential (Walden and Hentges, 1975; Onderdonk *et al.*, 1976). When this bacterium is exposed to ambient air, there is a rapid arrest of cell division (Schumann *et al.*, 1983). Within the first hour of air-exposure DNA replication ceases (Schumann *et al.*, 1983), and there is an increase in expression of at least 26 peptides, some of which are known to be critical for survival during periods of oxidative stress (Rocha *et al.*, 1996). Following several hours of air-exposure, the closely related bacterium *B. thetaiotaomicron* suffers oxidative inactivation of many iron-sulfur cluster-containing dehydratases of central metabolism, such as aconitase and fumarase (Pan and Imlay, 2001). This observation suggests that O_2 might inhibit growth of these bacteria by interfering with energy metabolism. However, while growth arrest occurs almost immediately upon air-exposure, iron-sulfur cluster inactivation requires several hours of air-exposure (Pan and Imlay, 2001). Further, *B. fragilis* continues to consume glucose and synthesize RNA and protein for several hours, albeit at slower rates, indicating that these cells maintain metabolic activity for some time in the presence of O_2 (Onderdonk *et al.*, 1976; Schumann *et al.*, 1983). Failure to re-establish an anaerobic environment ultimately results in death of these bacteria, likely due to the irreversible DNA damage caused by toxic O_2 derivatives (Takeuchi *et al.*, 1999).

3. Formation of Reactive Oxygen Species

Molecular oxygen is not usually harmful to cells, rather, it is the toxic derivatives of O_2, known as reactive oxygen species (ROS), that are the major cause of oxidative damage in cells (Halliwell and Gutteridge, 1984; Fridovich, 1998, 1999; Imlay, 2003). Two common forms of ROS encountered by air-exposed cells are the superoxide radical (O_2^-) and hydrogen peroxide (H_2O_2). These oxidants are inadvertently formed via the univalent transfer of electrons from a number of electron transfer complexes, such as flavin-containing oxidoreductases, to O_2 (Massey, 1994; Imlay, 1995, 2003; Messner and Imlay, 1999, 2002; Huycke *et al.*, 2001). Enzymes of the succinate: quinone oxidoreductase (SQR) family often transfer a single electron to O_2 forming O_2^-, which is rapidly released from the complex (Massey, 1994; Imlay, 1995, 2003; Messner and Imlay, 2002). The semi-oxidized flavin (flavosemiquinone) is then available to transfer its second electron to another incoming molecule of O_2. These reactions are summarized by eq. 1.

$$2O_2 + \text{flavin}_{red} \rightarrow 2O_2^- + 2H^+ + \text{flavin}_{ox} \qquad (1)$$

Many anaerobes rely on the SQR fumarate reductase for anaerobic energy metabolism (Baughn and Malamy, 2003), which might increase the risk for O_2^- production upon exposure to O_2 (Pan and Imlay, 2001). However, fumarate reductase deficient strains of *B. fragilis* do not show increased resistance to air-exposure, relative to the wild-type strain (Baughn and Malamy, unpublished). Thus, fumarate reductase-mediated O_2^- production does not contribute significantly to *B. fragilis* aerosensitivity.

Unlike fumarate reductase, several other flavoprotein complexes (such as NADH dehydrogenase and sulfite reductase) do not readily release O_2^-, rather, the O_2^- forms a hydroperoxide adduct with the active site flavin (Massey, 1994; Imlay, 2003). Upon protonation of the unstable flavin-hydroperoxide, H_2O_2 is released from the complex (Massey, 1994; Messner and Imlay, 1999; Imlay, 2003). These reactions are summarized by eq. 2.

$$O_2 + \text{flavin}_{red} \rightarrow H_2O_2 + \text{flavin}_{ox} \qquad (2)$$

In addition, H_2O_2 can also be formed by the dismutation of O_2^-. This reaction (eq. 3) can be catalyzed by reduced transition metals, such as Fe^{2+}, Mn^{2+} or Cu^{2+}, or by metalloproteins, such as iron-sulfur cluster proteins (Halliwell and Gutteridge, 1984).

$$2O_2^- + 2H^+ \xrightarrow{\text{reduced transition metal}} H_2O_2 + O_2 \qquad (3)$$

During periods of oxidative stress, cells are also challenged by the extremely reactive hydroxyl radical ($HO^\cdot$), which is generated upon interaction of H_2O_2 with reduced transition metals (eq. 4), such as Fe^{2+} and Cu^{2+} (Halliwell and Gutteridge, 1984).

$$H_2O_2 + Fe^{2+} \rightarrow HO^\cdot + HO^- + Fe^{3+} \qquad (4)$$

Since the reactivity of $HO^\cdot$ with organic molecules is diffusion-rate limited (Czapski, 1984), it is assumed that the only way to prevent damage by this oxidant is to inhibit its formation.

4. ROS Induced Damage

ROS can cause damage to multiple cellular components including proteins, DNA and lipids. In organisms capable of aerobic growth, such as *Escherichia coli* and *Saccharomyces cerevisiae*, an increase in the intracellular O_2^- concentration has been linked to defects in multiple metabolic pathways and elevated mutation frequencies (Carlioz and Touati, 1986; Farr *et al.*, 1986; Gralla and Valentine, 1991). Proteins containing solvent-exposed iron-sulfur clusters are especially prone to damage by O_2^- (Flint *et al.*, 1993). In a study of enzymes of central metabolism in *B. thetaiotaomicron*, many iron-sulfur cluster proteins were found to be inactivated in aerated cells, while other non-iron-sulfur enzymes were unaffected by air-exposure (Pan and Imlay, 2001). Further, when cell extracts were incubated with different forms of ROS, fumarase and aconitase were rapidly inactivated by O_2^-, while pyruvate:ferredoxin oxidoreductase (PFOR) was found to be more sensitive to O_2 than it was to O_2^-

(Pan and Imlay, 2001). Interestingly, fumarate reductase was less affected by these oxidants, presumably because the iron-sulfur clusters of this enzyme are buried within the complex, inaccessible to ROS (Pan and Imlay, 2001). With the exception of PFOR, these enzymes could be reactivated *in vitro* by the addition of thiol-containing reductant and reduced iron salt, and *in vivo* by incubating the cells under anaerobic conditions (Pan and Imlay, 2001). Several factors involved in iron-sulfur cluster assembly have been identified in aerobes and facultative anaerobes (Zheng *et al.*, 1998a; Mühlenhoff *et al.*, 2003; Outten *et al.*, 2003; Roy *et al.*, 2003). Though such factors have not been described for obligate anaerobes, it is possible that an analogous system might play a role in reactivation of some ROS sensitive enzymes.

ROS-mediated mutagenesis has been observed in obligate anaerobes, such as *Porphyromonas gingivalis*, as well as in facultative anaerobes and aerobes (Farr *et al.*, 1986; Gralla and Valentine, 1991; Lynch and Kuramitsu, 1999; Won Kim *et al.*, 2003). The most common ROS-associated DNA lesion is the 8-hydroxydeoxyguanosine adduct (8-oxo-Gua), which frequently causes G to C and G to T transversions *in vitro* and *in vivo* (Valentine *et al.*, 1998; Takeuchi *et al.*, 1999; Tuo *et al.*, 2000). Such damage frequently occurs in cells under O_2^- stress, however, O_2^- is not directly associated with DNA damage *in vitro* (Lesko *et al.*, 1980). Rather, 8-oxo-Gua and other lesions form under conditions that favor the generation of reactive hydroxyl radicals, such as the presence of Fe^{2+} and H_2O_2 (Tuo *et al.*, 2000). Since Fe^{2+} can bind to DNA, the presence of H_2O_2 results in formation of damaging hydroxyl radicals proximate to the DNA (Rai *et al.*, 2001). Given these observations, it is likely that the mutagenic effect of O_2^- *in vivo* is indirect, as well. Consistent with this prediction, oxidation of iron-sulfur clusters by O_2^- can significantly raise the concentration of free intracellular Fe^{2+} (Keyer and Imlay, 1996), and high levels of Fe^{2+} can promote DNA damage in cells regardless of the presence of O_2^- (Keyer *et al.*, 1995). Furthermore, this defect is exacerbated by excess H_2O_2, and can be suppressed by the over-production of intracellular iron chelators, such as picolinic acid (Keyer and Imlay, 1996; Maringanti and Imlay, 1999). Thus, it is likely that *in vivo*, O_2^--mediated mutagenesis occurs through Fe^{2+}-dependent sensitization of DNA to intracellular H_2O_2.

ROS are responsible for several other forms of oxidative damage, such as peroxidation of lipids, peptides and other organic molecules. In addition to disrupting the cellular functions of these biomolecules, these lesions can be propagated to adjacent molecules, resulting in wide-spread oxidative damage (Halliwell and Gutteridge, 1984; Fridovich, 1999). In addition, peroxides can also oxidize peptidyl methionine residues to form methionine sulfoxide and methionine sulfone, and cysteine residues to form cysteine-sulfenic acid and cystine disulfides (Kido and Kassell, 1975; Claiborne *et al.*, 1999; Agbas *et al.*, 2002). If these target residues are important for function of a specific protein, its activity will be altered by this change in oxidation state. Because of the deleterious affects of ROS, organisms that are able to survive even transient exposure to O_2 must possess mechanisms for ROS tolerance. Further, since ROS are elaborated as part of the innate immune response of many animals, successful pathogens must be capable of mounting an effective oxidative stress response.

5. Mechanisms for ROS Tolerance in *Bacteroides spp.*

5.1. The Oxidative Stress Response

When *B. fragilis* is exposed to conditions of oxidative stress, the expression of at least 28 proteins is increased, indicating that this bacterium has an adaptive oxidative stress response (OSR) (Rocha *et al.*, 1996). Since at least 22 of these OSR proteins are coordinately regulated during exposure of this bacterium to O_2, paraquat (an O_2^- generating reagent), and H_2O_2, there are likely overlapping mechanisms that are important for regulation of gene expression under these different stress conditions (Rocha *et al.*, 1996). However, since there are also some notable differences in these OSR protein profiles, this bacterium also has distinct responses to these different conditions of oxidative stress (Rocha *et al.*, 1996). Furthermore, since neither elevation of the environmental redox potential, nor induction of a general stress response can induce a similar protein profile, the OSR is specific to ROS-mediated stress (Rocha *et al.*, 1996).

As is the case for *E. coli*, some genes involved in the *B. fragilis* OSR are regulated by the redox-sensitive transcriptional activator OxyR (Rocha *et al.*, 2000; Smalley *et al.*, 2002; Herren *et al.*, 2003). The majority of the OxyR-induced genes play important roles in resistance to peroxide. In *E. coli*, the OxyR regulon includes genes for catalase (*katG*), alkyl hydroperoxide reductase (*ahpCF*), an iron-binding DNA-protective protein (*dps*), glutaredoxin (*grxA*), glutathione reductase (*gorA*), the ferric-iron uptake regulatory protein (*fur*), and at least 30 others (Zheng and Storz, 2000; Pomposiello and Demple, 2001; Zheng *et al.*, 2001). Likewise, in *B. fragilis*, transcription of genes encoding catalase (*katB*), alkyl hydroperoxide reductase (*ahpCF*), and DNA protective protein (*dps*) are positively regulated by OxyR (Rocha *et al.*, 2000). In addition, the *B. fragilis* OxyR regulon includes genes for thioredoxin peroxidase (*tpx*) and a putative RNA-binding protein (*rbpA*) (Herren *et al.*, 2003). Recently, it was shown that OxyR also regulates Dps in *P. gingivalis* (Ueshima *et al.*, 2003), indicating that this is a common theme for regulation of the peroxide response in *Bacteroides spp.*

When *E. coli* is cultivated aerobically in the absence of added peroxide, the homotetrameric protein OxyR represses its own expression and is found in the reduced (inactive) state (Christman *et al.*, 1989). When peroxide levels increase, OxyR is activated through oxidation of cysteine residues 199 and 208, which form an intramolecular disulfide bond (Zheng *et al.*, 1998b). This reversible oxidation is thought to cause a conformational change that allows OxyR to bind to target promoters via an amino-terminal LysR-like helix-turn-helix motif (Christman *et al.*, 1989; Åslund *et al.*, 1999). Subsequently, transcription of the target genes is activated when promoter-bound OxyR makes contact with the α-subunit of RNA polymerase (Tao *et al.*, 1993). Once the intracellular level of peroxide decreases, OxyR is inactivated by glutaredoxin-mediated reduction (Zheng *et al.*, 1998b; Åslund *et al.*, 1999). Since oxidized OxyR activates expression of glutaredoxin, these factors constitute an autoregulatory network (Zheng *et al.*, 1998b; Åslund *et al.*, 1999;).

Unlike the *E. coli* OxyR, the *B. fragilis* OxyR is constitutively expressed, and the protein can be activated upon exposure of cells to either molecular oxygen or peroxide (Rocha *et al.*, 2000). It is unclear if this difference in OxyR activation is due to a difference in the level of peroxide produced in air-exposed cells, or if the *B. fragilis* OxyR has different oxidation properties. However, since the *B. fragilis* OxyR shows conservation of cysteine residues 199 and 208, as well as the amino-terminal helix-turn-helix motif, this protein most likely activates transcription in the same way as its *E. coli* counterpart (Rocha *et al.*, 2000). Accordingly, an aspartate to glycine mutation at residue 202 of the *B. fragilis* OxyR gives this protein constitutive (redox-independent) activity (Rocha *et al.*, 2000), similar to the H198R and R201C mutations that result in constitutive activation of the *E. coli* OxyR (Kullik *et al.*, 1995).

Due to the essential role of OxyR in activation of the peroxide response, *B. fragilis* strains that express the constitutively active form of this protein show heightened resistance to H_2O_2 and organic peroxides (Herren *et al.*, 2003). However, OxyR-constitutive and *oxyR*-null strains of *B. fragilis* and *P. gingivalis* are unaffected in aerotolerance (Rocha *et al.*, 2000; Ueshima *et al.*, 2003), indicating that the harmful effects of O_2 on *Bacteroides spp.* might not involve peroxide toxicity.

Many genes of the *B. fragilis* OSR are regulated in an OxyR-independent manner (Smalley *et al.*, 2002; Herren *et al.*, 2003). These include genes for superoxide dismutase (*sod*), an aerobic-type ribonucleotide reductase (*nrdAB*), and cytochrome-*c* peroxidase (*ccp*) (Smalley *et al.*, 2002; Herren *et al.*, 2003). In addition, genes for the DNA-protective protein Dps (*dps*), and the predicted RNA-binding protein Rbp (*rbp*) show both OxyR-dependent and OxyR-independent regulation (Rocha *et al.*, 2000; Herren *et al.*, 2003). In *E. coli*, the iron-sulfur cluster containing transcriptional activator SoxR indirectly controls expression of several OxyR-independent OSR genes that are important for survival in the presence of O_2^- (Zheng and Storz, 2000; Pomposiello and Demple, 2001). This transcription factor contains an O_2^--labile iron-sulfur cluster. Oxidation of this cluster causes SoxR to activate transcription of *soxS* encoding a second regulatory protein, which binds to target promoters and activates transcription by recruitment of RNA polymerase (Zheng and Storz, 2000; Pomposiello and Demple, 2001). Although a portion of the *B. fragilis* OSR is specific for response to O_2^- stress (Rocha *et al.*, 1996), candidate regulatory proteins, such as SoxR or SoxS, have yet to be identified.

5.2. Prevention of ROS Formation

Since *Bacteroides spp.* are sensitive to O_2, as well as to reactive derivatives of this molecule, one "first line of defense" for avoiding oxidative damage is to maintain a low pO_2 environment. Such a strategy is employed by the obligate aerobe *Azotobacter vinelandii* to maintain the low cytoplasmic pO_2 required for the oxygen-sensitive process of nitrogen fixation (Kelly *et al.*, 1990). This rapid O_2-depletion system requires the activity of a high affinity cytochrome *bd* oxidase that couples the reduction of O_2 to the oxidation of reducing equivalents such as NADH (Kelly *et al.*, 1990). While mutant strains that are defective for this electron transport system (ETS) are capable of fixing nitrogen in the presence of <1.5% O_2, these strains are unable to fix nitrogen under fully aerobic conditions (Kelly *et al.*, 1990). A similar O_2 consumption system has been described in the obligately anaerobic ε-proteobacterium *Desulfovibrio gigas* (Lemos *et al.*, 2001). Although it has been proposed that this O_2-respiratory chain is critical for the establishment of an anaerobic environment, as well as for energy conservation during exposure to O_2 (Cypionka, 2000), the importance of aerobic respiration in the response of obligate anaerobes to atmospheric O_2 has yet to be confirmed by genetic or biochemical means. However, it is important to note that many *Bacteroides spp.* express O_2-dependent NADH oxidase activity that seems to correlate with the degree of aerotolerance of these species (Amano *et al.*, 1988).

We have recently identified a cytochrome *bd* oxidase-dependent ETS in *B. fragilis* that is capable of consuming O_2 at the expense of NADH (Baughn and Malamy, 2004). During growth in the presence of nanomolar concentrations of O_2, this ETS can functionally replace the anaerobic (fumarate reductase) respiratory chain in energy metabolism (Baughn and Malamy, 2004). Since *B. fragilis* can grow in, and benefit from the presence of nanomolar concentrations of O_2, this bacterium should no longer be considered an obligate anaerobe, rather, it should be classified as a nanaerobe. Further, since the genes for cytochrome *bd* oxidase are present in many other "obligate" anaerobes, such as other *Bacteroides spp.*, *Desulfovibrio spp.*, *Geobacter sulfurreducens*, *Archaeoglobus fulgidus* and *Methanosarcina barkerii*, it is likely that these organisms are nanaerobes as well.

Since ROS are often generated when O_2 interacts with electron transfer complexes, another means by which organisms can prevent the formation of ROS is by the regulation of electron flow. If oxidoreductase complexes, such as SQR, are unable to pass electrons to O_2, the formation of O_2^- and peroxide can be prevented. Thus, the oxidative inactivation of central metabolism that is observed in air-exposed *B. thetaiotaomicron* might represent a strategy for limiting the production of ROS (Pan and Imlay, 2001). However, in *E. coli*, the oxidative inactivation of some metabolic enzymes, such as fumarase and aconitase, is associated with an increase in the concentration of free intracellular Fe^{2+} (Keyer and Imlay, 1996). Since free Fe^{2+} sensitizes cells to ROS (Keyer and Imlay, 1996), iron sequesteration might be equally important for limiting the formation of $HO^{\cdot}$ in the presence of peroxide (Maringanti and Imlay, 1999). Consistent with this model, *P. gingivalis* strains that are defective for synthesis of the iron-storage protein ferritin are slightly compromised for survival in the presence of exogenous peroxide (Ueshima *et al.*, 2003). However, it is important to note that these strains do no show increased sensitivity to air-exposure, relative to the wild-type strain (Ueshima *et al.*, 2003).

5.3. ROS Detoxification

5.3.1. Superoxide Dismutase

Superoxide dismutase (Sod), which catalyzes the conversion of O_2^- to O_2 and H_2O_2 (eq. 3), has been detected in several obligate anaerobes, including many *Bacteroides spp.* (Gregory *et al.*, 1977; Tally *et al.*, 1977). Upon exposure to air, Sod activity in *P. gingivalis* and *B. fragilis* increases by approximately 2 to 10-fold (Privalle and Gregory, 1979; Amano *et al.*, 1990), which is coincident with an increase in the level of *sod* mRNA (Smalley *et al.*, 2002). In these bacteria, Sod is cambialistic, meaning that the enzyme can contain either Mn^{2+} or Fe^{2+} with little effect on specific activity (Gregory, 1985; Amano *et al.*, 1990). Interestingly, Sod purified from anaerobically grown *P. gingivalis* or *B. fragilis* contained mostly Fe, whereas Mn was the predominant metal found in Sod from aerated cells (Gregory, 1985; Amano *et al.*, 1990). This observation is consistent with the suggestion that *Bacteroides spp.* might have a mechanism for sequesteration of Fe during exposure to O_2 (Ueshima *et al.*, 2003).

For most organisms, Sod is important for preventing ROS-mediated damage in the presence of O_2 (Carlioz and Touati, 1986; Farr *et al.*, 1986; Gralla and Valentine, 1991). Accordingly, the degree of aerotolerance of many obligate anaerobes seems to correlate with Sod activity levels (McCord *et al.*, 1971; Tally *et al.*, 1977; Amano *et al.*, 1988). Studies with *P. gingivalis* indicate that Sod-deficient strains are more sensitive to air-exposure than the wild-type parental strain (Nakayama, 1994). As such, Sod-deficient strains also show increased mutation rates, indicating that O_2-lethality might result, from ROS-mediated DNA damage (Lynch and Kuramitsu, 1999). In addition, we have found that iron-sulfur cluster containing dehydratases, such as fumarase and aconitase, are rapidly inactivated in *B. fragilis* Sod-deficient strains when exposed to atmospheric concentrations of O_2 (Baughn and Malamy, unpublished). Thus, O_2^- also interferes with central metabolism during air-exposure, as has been suggested for *B. thetaiotaomicron* (Pan and Imlay, 2001). Since Sod is essential for maintenance of enzymes of central metabolism, and for genome stability, the ability to quench O_2^- is a major determinant for aerobic survival of *Bacteroides spp.*

5.3.2. Peroxidases

H_2O_2 and other peroxides are readily generated in the presence of O_2, therefore, aerotolerant organisms must possess means for the degradation of these oxidants. Some *Bacteroides spp.*, such as *B. fragilis*, express a heme-containing catalase (Gregory *et al.*, 1977), an iron-containing cytochrome-*c* peroxidase (Herren *et al.*, 2003), and at least two thiol-specific peroxidases (Rocha and Smith, 1998; Herren *et al.*, 2003). Since catalase is the major peroxidase involved in the degradation of H_2O_2 (eq. 5), Δ*katG* strains of

$$2H_2O_2 \rightarrow 2H_2O + O_2 \quad (5)$$

B. fragilis show heightened sensitivity to H_2O_2 exposure (Rocha *et al.*, 1996). However, it is important to note that the catalase-deficient strains are not compromised for aerotolerance (Rocha *et al.*, 1996). Unlike catalase, cytochrome-*c* peroxidase and thiol-specific peroxidases reduce peroxides, such as ROOH, to ROH and H_2O (eq. 6). These

$$ROOH + 2H^+ + 2e^- \rightarrow ROH + H_2O \quad (6)$$

peroxidases have less substrate-specificity than catalase. Due to this property, constitutive expression of the thiol-specific peroxidase alkyl hydroperoxide reductase can functionally replace catalase for resistance to H_2O_2 (Rocha and Smith, 1998). Strains lacking cytochrome-*c* peroxidase and thiol-specific peroxidases show wild-type resistance to H_2O_2 and heightened sensitivity to organic peroxides (Rocha and Smith, 1999; Herren *et al.*, 2003). Although mutation frequencies have not been reported, these peroxidase-deficient strains do not appear to be compromised for aerotolerance, indicating that the production of peroxide during air-exposure is not overwhelming for these bacteria (Rocha *et al.*, 1996; Rocha and Smith, 1999). However, these enzymes might prove to be important for survival in the presence of the host-mediated respiratory burst.

Since *P. gingivalis* does not produce a heme-containing catalase, this bacterium must use alternative strategies for H_2O_2 detoxification (Amano *et al.*, 1988). It has recently been demonstrated that the porphyrin-derivative μ-oxybis heme, as well as the NADH peroxidase rubrerythrin, are involved in the degradation of H_2O_2 in *P. gingivalis* (Smalley *et al.*, 2000; Sztukowska *et al.*, 2002). Indeed, strains lacking either of these peroxidases are compromised for survival in the presence of exogenous H_2O_2 (Smalley *et al.*, 2000; Sztukowska *et al.*, 2002). In addition, *P. gingivalis* encodes a thiol-specific peroxidase homologue (Nelson *et al.*, 2003), however, the role for this enzyme in peroxide resistance has not been assessed. Despite the involvement of some of these peroxidases in the *P. gingivalis* OSR, their role in organic peroxide resistance and aerotolerance has yet to be described.

5.4. Maintenance of ROS Sensitive Targets

Several lines of evidence indicate that the prevention of ROS-mediated DNA damage is crucial for aerotolerance of *Bacteroides spp.* This point is illustrated by the observation that Sod-deficient strains of *P. gingivalis* suffer increased rates of mutagenesis during exposure to O_2 (Lynch and Kuramitsu, 1999). In addition, studies comparing oxidative DNA damage in *Bacteroides spp.* indicate that aerosensitive species, such as *P. melaninogenica*, accumulate far more 8-oxo-Gua lesions during air-exposure than aerotolerant species, such as *B. fragilis* (Takeuchi *et al.*, 1999; Takeuchi *et al.*, 2000). This damage was found to coincide with the accumulation of intracellular ROS, such as H_2O_2 and O_2^- (Takeuchi *et al.*, 1999; 2000).

In addition to the prevention of damage, repair of oxidative DNA damage is also important for *Bacteroides* aerotolerance. In fact, mutant strains of *B. thetaiotaomicron* that are deficient for recombination repair are extremely sensitive to air-exposure in the presence of ROS generating compounds, such as resazurin (Cooper *et al.*, 1997). Further, it has been shown that *B. fragilis* expresses an aerobic ribonucleotide reductase that is important for aerotolerance and recovery from an aerobic challenge (Smalley *et al.*, 2002). Thus, repair of DNA lesions during and following periods of oxidative stress is important for aerotolerance in *Bacteroides spp.*

An additional component of *Bacteroides* aerotolerance is the Bat complex (Tang *et al.*, 1999; Tang, 2000). This predicted integral-membrane complex is encoded by a large operon, *batFGHABCDE*, that is highly conserved within *Bacteroides spp.* (Tang *et al.*, 1999; Tang, 2000; Nelson *et al.*, 2003; Xu *et al.*, 2003). The *batF* gene product shows strong homology to MoxR, a protein that is required for assembly of the large methanol oxidase complex in *Paracoccus denitrificans* (VanSpanning *et al.*, 1991; Tang, 2000). The other products of the *bat* operon do not show homology to any proteins of known function (Tang *et al.*, 1999; Tang, 2000). However, BatC and BatE contain tetratricopeptide repeats that might be involved in protein-protein interactions (Lamb *et al.*, 1995; Tang *et al.*, 1999; Tang, 2000). Strains of *B. fragilis* with transposon insertions in the *bat* operon were identified in a screen for mutants that were unable to grow on tissue culture monolayers (Tang *et al.*, 1999; Tang, 2000). Since this defect could not be reversed by the presence of wild-type *B. fragilis*, the Bat function is required *in cis* (Tang *et al.*, 1999; Tang, 2000). Further, these strains were unable to grow in a rat abscess model, indicating that the Bat complex is essential for virulence of *B. fragilis* (Tang *et al.*, 1999; Tang, 2000). In addition, it was found that these strains were compromised for growth in environments with an elevated redox potential, and were also severely defective for aerotolerance (Tang *et al.*, 1999; Tang, 2000). These phenotypes could be partially reversed in the presence of reducing agents, such as cysteine and thiodiglycol, suggesting that the Bat complex might be important for maintaining some process that is sensitive to oxidative conditions (Tang *et al.*, 1999; Tang, 2000). It is important to note that the non-sulfur-containing reductant tris-(2-carboxylethyl)phosphine also partially restored aerotolerance, indicating that the Bat mutants are not simply defective for sulfur metabolism during oxidative stress (Tang *et al.*, 1999; Tang, 2000). It has been proposed that this complex might be involved in the export of reducing equivalents, such as cysteine or glutathione (Tang *et al.*, 1999; Tang, 2000), however, clarification of the Bat mechanism awaits further study.

6. Concluding Remarks

Since *Bacteroides spp.* face exposure to ROS when traversing to new colonization sites, and these sites might have a constant influx of small amounts of O_2, ROS defense mechanisms are probably critical for the virulence and fitness of these bacteria. Studies of ROS tolerance have focused mainly on the gut commensal/opportunistic pathogen *B. fragilis*, and on the oral pathogen *P. gingivalis*. These studies have demonstrated that the *Bacteroides* OSR is comprised of several factors that are common to other bacterial OSR systems as well as some novel factors. It is clear that these systems are important for survival of *Bacteroides spp.* in many *in vitro* assays; however, the role of these systems in host colonization and infection requires further investigation. With the recent and ongoing sequencing of the genomes of several *Bacteroides spp.*, it is apparent that many of these mechanisms for ROS tolerance are conserved within this class of obligate anaerobes. These sequence data will permit genome-wide expression analyses that will enable the identification of additional candidate aerotolerance determinants.

Acknowledgements

This study was supported by Public Health Grant AI-19497 from the National Institute of Allergy and Infectious Disease of the National Institutes of Health.

References

Agbas, A., Chen, X., Hong, O., Kumar, K.N., and Michaelis, E.K. 2002. Superoxide modification and inactivation of a neuronal receptor-like complex. Free Rad. Biol. Med. 32: 512-524.

Amano, A., Tamagawa, H., Takagaki, M., Murakami, Y., Shizukuishi, S., and Tsunemitsu, A. 1988. Relationship between enzyme activities involved in oxygen metabolism and oxygen tolerance in black-pigmented *Bacteroides*. J. Dent. Res. 67: 1196-1199.

Amano, A., Shizukuishi, S., Tamagawa, H., Iwakura, K., Tsunasawa, S., and Tsunemitsu, A. 1990. Characterization of superoxide dismutase purified from either anaerobically maintained or aerated *Bacteroides gingivalis*. J. Bacteriol. 172: 1457-1463.

Åslund, F., Zheng, M., Beckwith, J., and Storz, G. 1999. Regulation of the OxyR transcription factor by hydrogen peroxide and the cellular thiol-disulfide status. Proc. Natl. Acad. Sci. USA. 96: 6161-6165.

Baughn, A.D., and Malamy, M.H. 2003. The essential role of fumarate reductase in haem-dependent growth stimulation of *Bacteroides fragilis*. Microbiology. 149: 1503-1511.

Baughn, A.D., and Malamy, M.H. 2004. The strict anaerobe *Bacteroides fragilis* grows in and benefits from nanomolar concentrations of oxygen. Nature. 427: 441-444.

Berchtold, M., Chatzinotas, A., Schönhuber, W., Brune, A., Amann, R., Hahn, D., and König, H. 1999. Differential enumeration and *in situ* localization of microorganisms in the hindgut of the lower termite *Mastotermes darwiniensis* by hybridization with rRNA-targeted probes. Arch. Microbiol. 172: 407-416.

Bornside, G.H., Donovan, W.E., and Myers, M.B. 1976. Intracolonic tensions of oxygen and carbon dioxide in germfree, conventional, and gnotobiotic rats. Proc. Soc. Exp. Biol. Med. 151: 437-441.

Brook, I. 2002. Microbiology of polymicrobial abscesses and implications for therapy. J. Antimicrob. Chemother. 50: 805-810.

Carlioz, A., and Touati, D. 1986. Isolation of superoxide dismutase mutants in *Escherichia coli*: is superoxide dismutase necessary for aerobic life? EMBO J. 5: 623-630.

Christman, M.F., Storz, G., and Ames, B.N. 1989. OxyR, a positive regulator of hydrogen peroxide-inducible genes in *Escherichia coli* and *Salmonella typhimurium*, is homologous to a family of bacterial regulatory proteins. Proc. Natl. Acad. Sci. USA. 86: 3484-3488.

Claiborne, A., Yeh, J.I., Mallett, C., Luba, J., Crane III, E.J., Charrier, V., and Parsonage, D. 1999. Protein-sulfenic acids: Diverse roles for an unkikely player in enzyme catalysis and redox regulation. Biochemistry. 38: 15407-15416.

Clark, W.B., Magnusson, I., Abee, C., Collins, B., Beem, J.E., and McArthur, W.P. 1988. Natural occurrence of black-pigmented *Bacteroides* species in the gingival crevice of the squirrel monkey. Infect. Immun. 56: 2392-2399.

Cooper, A.J., Kalinowski, A.P., Shoemaker, N.B., and Salyers, A.A. 1997. Construction and characterization of a *Bacteroides thetaiotaomicron recA* mutant: Transfer of *Bacteroides* integrated conjugative elements is RecA independent. J. Bacteriol. 179: 6221-6227.

Cypionka, H. 2000. Oxygen respiration by *Desulfovibrio* species. Annu. Rev. Microbiol. 54: 827-848.

Czapski, G. 1984. Reaction of ·OH. Methods Enzymol. 105: 209-214.

Farr, S.B., D' Ari, R., and Touati, D. 1986. Oxygen-dependent mutagenesis in *Escherichia coli* lacking superoxide dismutase. Proc. Natl. Acad. Sci. USA. 83: 8268-8272.

Flint, D.H., Tuminello, J.F., and Emptage, M.H. 1993. The inactivation of Fe-S cluster containing hydro-lyases by superoxide. J. Biol. Chem. 268: 22369-22376.

Fridovich, I. 1998. Oxygen toxicity: a radical explanation. J. Exp. Biol. 201: 1203-1209.

Fridovich, I. 1999. Fundamental aspects of reactive oxygen species, or what's the matter with oxygen? Ann. NY Acad. Sci. 893: 13-18.

Gralla, E.B., and Valentine, J.S. 1991. Null mutants of *Saccharomyces cerevisiae* Cu, Zn superoxide dismutase: Characterization and spontaneous mutation rates. J. Bacteriol. 173: 5918-5920.

Gregory, E.M., Kowalski, J.B., and Holdeman, L.V. 1977. Production and some properties of catalase and superoxide dismutase from the anaerobe *Bacteroides distasonis*. J. Bacteriol. 129: 1298-1302.

Gregory, E.M. 1985. Characterization of the O_2-induced manganese-containing superoxide dismutase from *Bacteroides fragilis*. Arch. Biochem. Biophys. 238: 83-89.

Halliwell, B., and Gutteridge, J.M.C. 1984. Oxygen toxicity, oxygen radicals, transition metals and disease. Biochem. J. 219: 1-14.

Herren, C.D., Rocha, E.R., and Smith, C.J. 2003. Genetic analysis of an important oxidative stress locus in the anaerobe *Bacteroides fragilis*. Gene. 316: 167-175.

Holdeman, L.V., Good, I.J., and Moore, W.E.C. 1976. Human fecal flora: Variation in bacterial composition within individuals and a possible effect of emotional stress. Appl. Environ. Microbiol. 31: 359-375.

Huycke, M.M., Moore, D., Joyce, W., Wise, P., Shepard, L., Kotake, Y., and Gilmore, M.S. 2001. Extracellular superoxide production by *Enterococcus faecalis* requires demethylmenaquinone and is attenuated by functional terminal quinol oxidases. Mol. Microbiol. 42: 729-740.

Imlay, J.A. 1995. A metabolic enzyme that rapidly produces superoxide, fumarate reductase of *Escherichia coli*. J. Biol. Chem. 270: 19767-19777.

Imlay, J.A. 2003. Pathways of oxidative damage. Annu. Rev. Microbiol. 57: 395-418.

Kelly, M.J.S., Poole, R.K., Yates, M.G., and Kennedy, C. 1990. Cloning and mutagenesis of genes encoding cytochrome *bd* terminal oxidase complex in *Azotobacter vinelandii*:

Mutants deficient in the cytochrome d complex are unable to fix nitrogen in air. J. Bacteriol. 172: 6010-6019.

Keyer, K., Strohmeier Gort, A., and Imlay, J.A. 1995. Superoxide and the production of oxidative DNA damage. J. Bacteriol. 177: 6782-6790.

Keyer, K., and Imlay, J.A. 1996. Superoxide accelerates DNA damage by elevating free-iron levels. Proc. Natl. Acad. Sci. USA. 93: 13635-13640.

Kido, K., and Kassell, B. 1975. Oxidation of methionine residues of porcine and bovine pepsins. Biochemistry. 11: 631-635.

Kotarski, S.F., and Salyers, A.A. 1984. Isolation and characterization of outer membranes of *Bacteroides thetaiotaomicron* grown on different carbohydrates. J. Bacteriol. 158: 102-109.

Kullik, I., Toledano, M.B., Tartaglia, L.A., and Storz, G. 1995. Mutational analysis of the redox-sensitive transcriptional regulator OxyR: Regions important for oxidation and transcriptional activation. J. Bacteriol. 177: 1275-1284.

Lamb, J.R., Tugendreich, S., and Hieter, P. 1995. Tetratricopeptide repeat interaction: To TPR or not to TPR? Trends Biochem. Sci. 20: 257-259.

Lee, A., Gordon, J., and Dubos, R. 1968. Enumeration of the oxygen sensitive bacteria usually present in the intestine of healthy mice. Nature. 220: 1137-1139.

Lemos, R.S., Gomes, C.M., Santana, M., LeGall, J., Xavier, A.V., and Teixeira, M. 2001. The 'strict' anaerobe *Desulfovibrio gigas* contains a membrane-bound oxygen-reducing respiratory chain. FEBS Lett.. 496: 40-43.

Lesko, S.A., Lorentzen, R.J., and Ts'o, P.O.P. 1980. Role of superoxide in deoxyribonucleic acid strand scission. Biochemistry. 19: 3023-3028.

Loesche, W.J. 1969. Oxygen sensitivity of various anaerobic bacteria. Appl. Microbiol. 18: 723-727.

Lynch, M.C., and Kuramitsu, H.K. 1999. Role of superoxide dismutase activity in the physiology of *Porphyromonas gingivalis*. Infect. Immun. 67: 3367-3375.

Maringanti, S., and Imlay, J.A. 1999. An intracellular iron chelator pleiotropically suppresses enzymatic and growth defects of superoxide dismutase-deficient *Escherichia coli*. J. Bacteriol. 181: 3792-3802.

Massey, V. 1994. Activation of molecular oxygen by flavins and flavoproteins. J. Biol. Chem. 269: 22459-22462.

McCord, J.M., Keele Jr., B.B., and Fridovich, I. 1971. An enzyme-based theory of obligate anaerobiosis: The physiological function of superoxide dismutase. Proc. Natl. Acad. Sci. USA. 68: 1024-1027.

Messner, K.R., and Imlay, J.A. 1999. The identification of primary sites of superoxide and hydrogen peroxide formation in the aerobic respiratory chain and sulfite reductase complex of *Escherichia coli*. J. Biol. Chem. 274: 10119-10128.

Messner, K.R., and Imlay, J.A. 2002. Mechanism of superoxide and hydrogen peroxide formation by fumarate reductase, succinate dehydrogenase, and aspartate oxidase. J. Biol. Chem. 277: 42563-42571.

Mühlenhoff, U., Gerber, J., Richhardt, N., and Lill, R. 2003. Components involved in assembly and dislocation of iron-sulfur clusters on the scaffold protein Isu1p. EMBO J. 22: 4815-4825.

Nakayama, K. 1994. Rapid viability loss on exposure to air in a superoxide dismutase-deficient mutant of *Porphyromonas gingivalis*. J. Bacteriol. 176: 1939-1943.

Namavar, F., Theunissen, E.B.M., Verweij-van Vught, A.M.J.J., Peerbooms, P.G.H., Bal, M., Hoitsma, H.F.W., and MacLaren, D.M. 1989. Epidemiology of the *Bacteroides fragilis* group in the colonic flora in 10 patients with colonic cancer. J. Med. Microbiol. 29: 171-176.

Nelson, K., Fleishmann, R., DeBoy, R., Paulsen, I., Fouts, D., Eisen, J., Daugherty, S., Dodson, R., Durkin, A., Gwinn, M., Haft, D., Kolonay, J., Nelson, W., White, O., Mason, T., Tallon, L., Gray, J., Granger, D., Tettlin, H., Dong, H., Galvin, J., Duncan, M., Dewhirst, F., and Fraser, C. 2003. Complete genome sequence of the oral pathogenic bacterium *Porphyromonas gingivalis* strain W83. J. Bacteriol. 185: 5591-5601.

Onderdonk, A.B., Johnston, J., Mayhew, J.W., and Gorbach, S.L. 1976. Effect of dissolved oxygen and Eh on *Bacteroides fragilis* during continuous culture. Appl. Environ. Microbiol. 31: 168-172.

Outten, F.W., Wood, M.J., Muñoz, F.M., and Storz, G. 2003. The SufE protein and the SufBCD complex enhance SufS cysteine desulfurase activity as part of a sulfur transfer pathway for Fe-S cluster assembly in *Escherichia coli*. J. Biol. Chem. 278: 45713-45719.

Pan, N., and Imlay, J.A. 2001. How does oxygen inhibit central metabolism in the obligate anaerobe *Bacteroides thetaiotaomicron*? Mol. Microbiol. 39: 1562-1571.

Pomposiello, P.J., and Demple, B. 2001. Redox-operated genetic switches: the SoxR and OxyR transcription factors. Trends Biotechnol. 19: 109-114.

Privalle, C.T., and Gregory, E.M. 1979. Superoxide dismutase and O_2 lethality in *Bacteroides fragilis*. J. Bacteriol. 138: 139-145.

Rai, P., Cole, T.D., Wemmer, D.E., and Linn, S. 2001. Localization of Fe^{2+} at an RTGR sequence within a DNA duplex explains preferential cleavage by Fe^{2+} and H_2O_2. J. Mol. Biol. 312: 1089-1101.

Rocha, E.R., Selby, T., Coleman, J.P., and Smith, C.J. 1996. Oxidative stress response in an anaerobe, *Bacteroides fragilis*: a role for catalase in protection against hydrogen peroxide. J. Bacteriol. 178: 6895-6903.

Rocha, E.R., and Smith, C.J. 1998. Characterization of a peroxide-resistant mutant of the anaerobic bacterium *Bacteroides fragilis*. J. Bacteriol. 180: 5906-5912.

Rocha, E.R., and Smith, C.J. 1999. Role of the alkyl hydroperoxide reductase (*ahpCF*) gene in oxidative stress defense of the obligate anaerobe *Bacteroides fragilis*. J. Bacteriol. 181: 5701-5710.

Rocha, E.R., Owens Jr., G., and Smith, C.J. 2000. The redox-sensitive transcriptional activator OxyR regulates the peroxide response regulon in the obligate anaerobe *Bacteroides fragilis*. J. Bacteriol. 182: 5059-5069.

Roy, A., Solodovnikova, N., Nicholson, T., Antholine, W., and Walden, W.E. 2003. A novel eukaryotic factor for cytosolic Fe-S cluster assembly. EMBO J. 22: 4826-4835.

Schumann, J.P., Jones, D.T., and Woods, D.R. 1983. Effect of oxygen and UV irratiation on nucleic acid and protein syntheses in *Bacteroides fragilis*. J. Bacteriol. 156: 1366-1368.

Shoop, D.S., Myers, L.L., and LeFever, J.B. 1990. Enumeration of enterotoxigenic *Bacteroides fragilis* in municipal sewage. Appl. Environ. Microbiol. 56: 2243-2244.

Smalley, D., Rocha, E.R., and Smith, C.J. 2002. Aerobic-type ribonucleotide reductase in the anaerobe *Bacteroides fragilis*. J. Bacteriol. 184: 895-903.

Smalley, J.W., Birss, A.J., and Silver, J. 2000. The periodontal pathogen *Porphyromonas gingivalis* harnesses the chemistry of the μ-oxo bishaem of iron protoporphyrin IX to protect against hydrogen peroxide. FEMS Microbiol. Lett. 183: 159-164.

Sztukowska, M., Bungo, M., Potempa, J., Travis, J., and Kurtz Jr., D.M. 2002. Role of ruberythrin in the oxidative stress response of *Porphyromonas gingivalis*. Mol. Microbiol. 44: 479-488.

Takeuchi, T., Nakaya, Y., Kato, N., Watanabe, K., and Morimoto, K. 1999. Induction of oxidative DNA damage in anaerobes. FEBS Lett. 450: 178-180.

Takeuchi, T., Kato, N., Watanabe, K., and Morimoto, K. 2000. Mechanism of oxidative DNA damage induction in a strict anaerobe, *Prevotella melaninogenica*. FEMS Microbiol. Lett. 192: 133-138.

Tally, F.P., Goldin, B.R., Jacobus, N.V., and Gorbach, S.L. 1977. Superoxide dismutase in anaerobic bacteria of clinical significance. Infect. Immun. 16: 20-25.

Tally, F.P., and Ho, J.L. 1987. Management of patients with intraabdominal infections due to colonic perforation. Curr. Topics Infect. Dis. 8: 266-295.

Tang, Y.P., Dallas, M.M., and Malamy, M.H. 1999. Characterization of the BatI (Bacteroides aerotolerance) operon in *Bacteroides fragilis*: Isolation of a *B. fragilis* mutant with reduced aerotolerance and impaired growth in *in vivo* model systems. Mol. Microbiol. 32: 139-149.

Tang, Y.P. 2000. Identification and characterization of genes in *Bacteroides fragilis* involved in aerotolerance. PhD Thesis, Dept. Mol. Biol. & Microbiol. Boston: Tufts University.

Tanner, A.C., Haffer, C., Bratthall, G.T., Visconti, R.A., and Socransky, S.S. 1979. A study of the bacteria associated with advancing periodontitis in man. J. Clin. Periodontol. 6: 278-307.

Tao, K., Fujita, N., and Ishihama, A. 1993. Involvement of the RNA polymerase alpha subunit C-terminal region in co-operative interaction and transcriptional activation with OxyR protein. Mol. Microbiol. 7: 859-864.

Tuo, J., Liu, L., Poulsen, H.E., Weimann, A., Svendsen, O., and Loft, S. 2000. Importance of guanine nitration and hydroxylation in DNA *in vitro* and *in vivo*. Free Rad. Biol. Med. 29: 147-155.

Ueshima, J., Shoji, M., Ratnayake, D.B., Abe, K., Yoshida, S., Yamamoto, K., and Nakayama, K. 2003. Purification, gene cloning, gene expression, and mutants of Dps from the obligate anaerobe *Porphyromonas gingivalis*. Infect. Immun. 71: 1170-1178.

Valentine, M.R., Rodriquez, H., and Termini, J. 1998. Mutagenesis by peroxy radical is dominated by transversions at deoxyguanosine: Evidence for the lack of involvement of 8-oxo-dG and/or abasic site formation. Biochemistry. 37: 7030-7038.

VanSpanning, R.J.M., Wansell, C.W., DeBoer, T., Hazelaar, M.J., Anazawa, H., Harms, N., Oltmann, L.F., and Stouthamer, A.H. 1991. Isolation and characterization of the *moxJ*, *moxG*, *moxI*, and *moxR* genes of *Paracoccus denitrificans*: Inactivation of *moxJ*, *moxG*, and *moxR* and the resultant effect on methylotrophic growth. J. Bacteriol. 173: 6948-6961.

Walden, W.C., and Hentges, D.J. 1975. Differential effects of oxygen and oxidation-reduction potential on the multiplication of three species of anaerobic intestinal bacteria. Appl. Microbiol. 30: 781-785.

Wang, X., Heazlewood, S.P., Krause, D.O., and Florin, T.H.J. 2003. Molecular characterization of the microbial species that colonize human ileal and colonic mucosa by using 16S rDNA sequence analysis. J. Appl. Microbiol. 95: 508-520.

Won Kim, H., Murakami, A., Williams, M.V., and Ohigashi, H. 2003. Mutagenecity of reactive and nitrogen species as detected by co-culture of activated inflammatory leukocytes and AS52 cells. Carcinogenesis. 24: 235-241.

Xu, J., Bjursell, M.K., Himrod, J., Deng, S., Carmichael, L.K., Chiang, H.C., Hooper, L.V., and Gordon, J.I. 2003. A genomic view of the human-*Bacteroides thetaiotaomicron* symbiosis. Science. 299: 2074-2076.

Zheng, L., Cash, V.L., Flint, D.H., and Dean, D.R. 1998a. Assembly of iron-sulfur clusters. J. Biol. Chem. 273: 13264-13272.

Zheng, M., Åslund, F., and Storz, G. 1998b. Activation of the OxyR transcription factor by reversible disulfide bond formation. Science. 279: 1718-1721.

Zheng, M., and Storz, G. 2000. Redox sensing by prokaryotic transcription factors. Biochem. Pharmacol. 59: 1-6.

Zheng, M., Wang, X., Templeton, L.J., Smulski, D.R., LaRossa, R.A., and Storz, G. 2001. DNA microarray-mediated transcriptional profiling of the *Escherichia coli* response to hydrogen peroxide. J. Bacteriol. 183: 4562-4570.

Chapter 9

Clostridial and Bacteroides Toxins: Structure and Mode of Action

Michel R. Popoff*

Abstract

Clostridia are the bacteria, which produce the highest number of toxins, and are involved in severe diseases in man and animals. Most of the clostridial toxins are pore-forming toxins responsible for gangrenes and gastro-intestinal diseases. Among them, perfringolysin has been largely studied and it is the paradigm of the cholesterol binding cytotoxins, whereas *Clostridium septicum* alpha toxin, which is related to aerolysin, is the prototype of several clostridial toxins forming small pores. Other toxins active on the cell surface possess an enzymatic activity such as phospholipase C and collagenase and are involved in the degradation of specific cell membrane or extracellular matrix components. Three groups of clostridial toxins have the ability to enter cells: large clostridial toxins, binary toxins, and neurotoxins. The binary and large clostridial toxins alter the actin cytoskeleton by enzymatically modifying the actin monomers and the regulatory proteins from the Rho family, respectively. Clostridial neurotoxins proteolyse key components of the neuroexocytosis system. Botulinum neurotoxins inhibit neurotransmission at the neuromuscular junctions, whereas tetanus toxin targets the inhibitory interneurons of the central nervous system. The high potency of clostridial toxins result from their specific targets, which have an essential cellular function, and from the type of modification that they induce.

1. Introduction

Certain anaerobic bacteria are pathogenic for man and animals. Their main virulence determinants consist of secreted toxins, which are essentially produced by species from the *Clostridium* genus, and more rarely by anaerobic Gram-negative bacteria. The virulence factors of pathogenic anaerobic Gram-negative bacteria are multiple and include capsule, resistance to phagocytosis, adherence factors, products of metabolism, hydrolytic enzymes, and in some cases toxins. The best characterized toxin from this group of bacteria is the *Bacteroides fragilis* enterotoxin. *Clostridium* are anaerobic Gram-positive bacteri, whose main habitat is the environment. These fermentative bacteria secrete numerous hydrolytic enzymes, which degrade an extremely wide range of organic substrates to provide the nutrients and energy required for their growth. Recent genome sequencing of *Clostridium perfringens* and *Clostridium tetani* shows that these microorganisms lack many genes that encode enzymes necessary for amino acid biosynthesis, and therefore they are dependent on appropriate substrates (Shimizu *et al.*, 2002; Brüggemann *et al.*, 2003). Among more than 150 *Clostridium* species, 15 produce potent toxins, which are responsible for severe diseases in man and animals. Pathogenic *Clostridium* spp. are not invasive bacteria, but they secrete active molecules, which act at a distance from the bacteria and, which cause all the symptoms and lesions of the clostridial diseases. *Clostridium* can enter an organism by two ways. The oral route permits, upon the presence of risk factors, their growth and toxin production in some compartments of the digestive tract, leading to the onset of intestinal or foodborne acquired diseases (Figure 1). In the case of botulism, preformed toxin in food is sufficient to cause the disease. The second possibility is the entry of *Clostridium* by the tegument due to a wound permitting anaerobic bacterial growth in the presence of necrotic tissues. The local production of toxins generates gangrene or induces a specific neurological disease (tetanus) (Figure 2).

Clostridial toxins include a large variety of proteins differing in their size, structure and mechanism of action. They also differ by their ability to diffuse in the organism and by their site of action. Some toxins act locally, others can pass through the mucosal barrier and disseminate to several tissues through the blood circulation. For example botulinum neurotoxins pass through the intestinal mucosa and bind to neuronal endings, tetanus toxin is transported along the axons to the nervous central system, and epsilon-toxin can cross the blood brain barrier. Some toxins interact with various types of cells and are generally involved in gangrene (Figure 2), others act on specific cells such as intestinal cells leading to enteritis, necrotic and/or hemorrhagic enterocolitis, and certain others are specific to neurons causing neurological

*For correspondence email: mpopoff@pasteur.fr

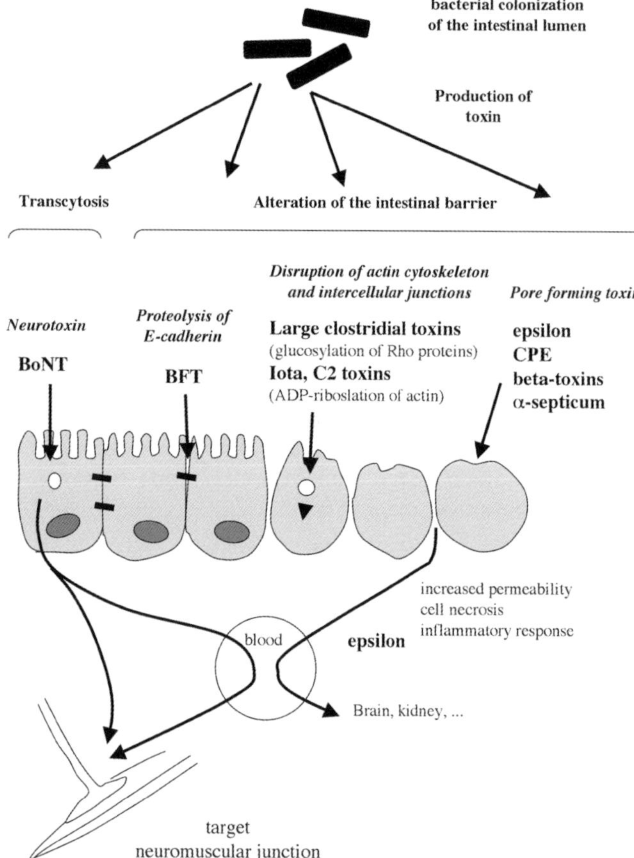

Figure 1. Toxins from anaerobes involved in intestinal diseases or digestive acquired diseases.

disorders such as botulism and tetanus (Figure 1). Numerous clostridial toxins, approximately one third, are pore-forming toxins. This could represent a basic system to release useful nutrients from cells. The other toxins exhibit an enzymatic activity, and can be considered as highly specific extracellular hydrolytic enzymes, which modify a crucial target in eukaryotic cells leading to a rapid death or to an irreversible alteration of an essential cellular function. Some toxins possessing an enzymatic activity act at the cell surface, but many of them have acquired the machinery to translocate across the cell membrane and to recognize an intracellular target.

According to their mechanism of action, toxins can be divided into those acting at the cell membrane (pore formation, phospholipase activity) (Table 1), and toxins which enter the cells and interact with an intracellular target such as actin monomers, small GTPases, or proteins of the exocytosis machinery (Table 2).

2. Membrane- and Extracellular Matrix-Damaging Toxins

2.1. Pore Forming Toxins

2.1.1. Theta Toxin (or Perfringolysin, PFO) and Other Cholesterol Binding Cytotoxins

PFO is produced by all *C. perfringens* strains and the *pfo* gene is located on the chromosomal DNA near the origin of replication (Katayama *et al.*, 1995). PFO is synthesized with a 27 amino acid signal peptide, and the mature protein consists of 472 amino acids (53 kDa) (Tweten, 1988). PFO belongs to the cholesterol-binding cytotoxins. This family encompasses toxins which are

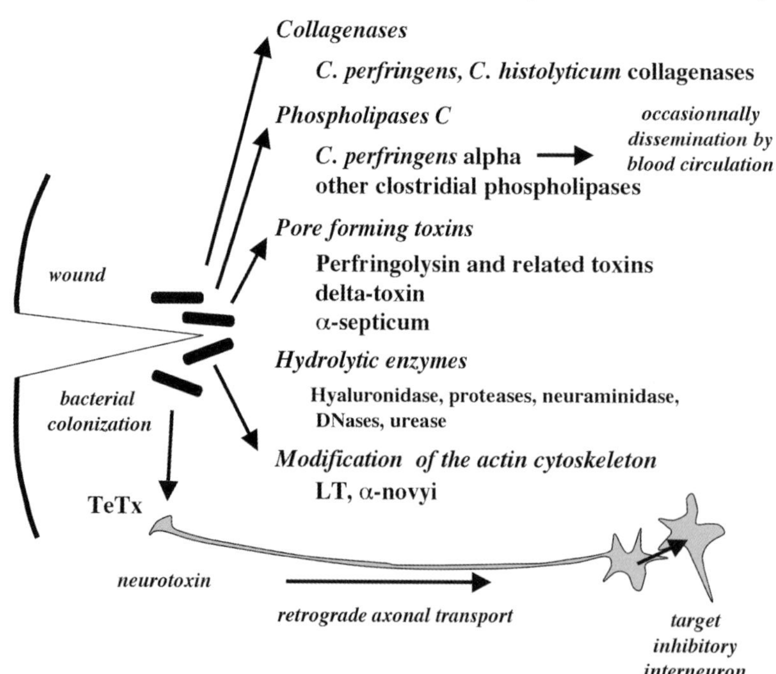

Figure 2. Toxins from anaerobes involved in myonecrosis, gangrene, or diseases acquired from contaminated wounds.

Table 1. Toxins from anaerobic bacteria acting on cell membrane or on extracellular matrix

Species	Toxin	Activity
Cholesterol binding cytotoxins (Pore forming toxins)		
C. bifermentans	bifermentolysin	
C. botulinum	botunolysin	
C. chauvoei	chauveolysin	
C. histolyticium	histolycolysin	
C. novyi	novyilysin	Pore formation
C. perfringens	perfringolysin	
C. septicum	septicolysin	
C. sordellii	sordellilysin	
C. tetani	tetanolysin	
Other Pore forming toxins		
C. septicum	alpha-toxin	
C. perfringens	beta1-, beta2-toxins	
C. perfringens	delta-toxin	Pore formation
C. perfringens	epsilon-toxin	
C. perfringens	enterotoxin	
Phospholipase activity		
C. absonum	PLC	
C. baratii	PLC	
C. bifermentans	PLC	Phospholipase C
C. novyi	β-toxin	
C. perfringens	α-toxin	
C. sordellii	PLC	
Proteases		
C. histolyticum	collagenases	Collagenase
C. perfringens	collagenase (kappa-toxin)	
Bacteroides fragilis	enterotoxin (fragilysin)	Protease (E-cadherin)

produced by numerous Gram-positive bacteria such as listeriolysin O of *Listeria monocytogenes*, pneumolysin of *Streptococcus pneumoniae*, and streptolysin O (SLO) from *Streptococcus pyogenes*. Various *Clostridia* produce related toxins (*C. botulinum*, *C. chauvoei*, *C. tetani*, and so on) (Table 1). Members of this toxin family exhibit 40-80% identity at the primary structure level and share common biological properties and structural characteristics (Alouf, 1997; 2003).

These toxins bind to membrane cholesterol in a wide variety of mammalian cells, and possess a single, sensitive and highly conserved Cys. They form large oligomeric structures, which insert into the membrane leading to lysis. The unique Cys has been found to be critical for activity (Alouf, 1997).

2.1.1.1. Structure

PFO has an unusual elongated rod shape. The molecule is rich in β-sheet and it is hydrophilic without significant patches of hydrophobic residues on the surface. Four domains can be distinguished in the PFO molecule. Domain 1 has a seven-stranded antiparallel β-sheet and is connected to domain 4 by the elongated domain 2. Domain 3 consists of β-sheets and α-helices. The C-terminal part (domain 4) folds into a separate and compact β-sandwich domain (Rossjohn *et al.*, 1997). Cys is located in a conserved 11 amino acid sequence (ECTGLAWEWWR) near the C-terminus in domain 4, which is involved in binding to cholesterol. Thiol modification of the Cys residue with a bulky reagent such as 5,5'-dithiobis (2-nitrobenzoic acid) or substitution of Trp436, Trp438 or Trp439 of this segment with Phe reduces significantly the binding step and the subsequent hemolytic activity (Sekino-Suzuki *et al.*, 1996). However, replacement of Cys with an amino acid of similar size retains an active conformation. Molecular modeling shows that cholesterol binding to this region induces a displacement of the Trp rich loop. It is proposed that the high affinity of PFO and also of the cholesterol binding cytotoxins (Kd 10^{-9}M) to the cholesterol receptor is involved in concentrating the toxin in cholesterol molecules organized in arcs on the target membrane, promoting oligomerization and membrane insertion (Rossjohn *et al.*, 1997). Cholesterol is clustered in membrane microdomains or rafts, and PFO is a useful tool to identify such membrane rafts (Waheed *et al.*, 2001).

2.1.1.2. Mode of Action

The proposed model of PFO pore formation includes the binding of water-soluble PFO monomers to cholesterol

Table 2. Toxins from anaerobic bacteria active intracellularly (UDP-*N*-AGlu, UDP-*N*-acetylglucosamine)

Species	Toxin	Enzymatic activity	Substrate
Toxins modifying the actin cytoskeleton			
Binary toxins			
C. botulinum C	C2 toxin		cellular G-actin
C. botulinum D	C2 toxin	ADP-ribosylation	cellular G-actin
C. difficile	*C. difficile* transferase		cellular and muscular G-actin
C. perfringens	iota toxin		
C. spiroforme	*C. spiroforme* toxin		
C. botulinum C	C3 enzyme		
C. botulinum D	C3 enzyme	ADP-ribosylation	RhoA, B, C
C. limosum	C3 enzyme		
Large clostridial toxins			
C. difficile	ToxA	Glucosylation	Rho, Rac, Cdc42
	ToxB(VPI10463)	(UDP-glucose)	Rho, Rac, Cdc42
	ToxB(8864)		Rac
C. novyi	α-toxin	(UDP-*N*-AGlu)	Rho, Rac, Cdc42
C. sordellii	LT82	(UDP-glucose)	Rac, Ras, Rap, Ral
	LT9048		Rac, Cdc42, Ras, Rap
	HT		Rho, Rac, Cdc42
Clostridial neurotoxins			
C. botulinum A	BoNT/A		SNAP25
C. botulinum B	BoNT/B		VAMP
C. botulinum C	BoNT/C1		Syntaxin, SNAP25
C. botulinum D	BoNT/D		VAMP
C. botulinum E	BoNT/E	Zinc dependent Protease	SNAP25
C. butyricum E	BuNT/E		SNAP25
C. botulinum F	BoNT/F		VAMP
C. baratii F	BaNT/F		VAMP
C. argentinense	BoNT/G		VAMP
C. tetani	TeTx		VAMP

of the lipid bilayer, mediated by domain 4. Only a short hydrophobic loop at the tip of the domain 4 β-sandwich is involved in cholesterol binding (Ramachandran *et al.,* 2002). PFO monomers bound to cholesterol assemble and oligomerize to form a prepore complex (Shepard *et al.,* 2000). Oligomers consist of 40 to 50 monomers forming large arcs and rings on the membrane surface, leading to large pores between 300 Å and 450 Å in diameter. Domains 1, 2 and 4 fit into L-shaped repeating units connected to the corresponding domains of the neighboring partners, thus forming a cylindrical structure. Two binding sites have been identified in domain 1-domain 1 interactions. Interaction of domain 4 with cholesterol induces a conformational change of domains 3, which are rotated from domains 2 and form a belt in the outside face of the cylinder. This promotes the exposure of hydrophobic residues and the insertion of a transmembrane β-barrel into the lipid bilayer. A bundle of three α-helices of domain 3 unfolds forming two amphipathic β-sheets. Each monomer contributes two amphipathic β-hairpins to the formation of the transmembrane β-barrel (Shatursky *et al.,* 1999; Heuck *et al.,* 2001). Monomers do not insert their transmembrane hairpins individually, but cooperation between PFO monomers is required to drive the insertion of the prepore complex, which appears to be an all or none process (Hotze *et al.,* 2002). The charged face of the domain 4 amphipathic β-hairpin forms the inner lining of the pore, while the other face is protected from the hydrophobic part of the lipid bilayer by cholesterol molecules (Rossjohn *et al.,* 1997). The role of cholesterol in pore formation is still speculative. Alternatively, a monomer bound to a cholesterol molecule could oligomerize and insert into the membrane without requirement for additional cholesterol molecules (Bayley, 1997). The cholesterol binding toxins exhibit a novel mode of action in which the toxin receptor has multiple roles in binding, pore formation and stabilization.

2.1.1.3. Cellular Effects

PFO participates with alpha-toxin in the production of local lesions of gangrene which are characterized by a rapid extension of myonecrosis and a polymorphonuclear leucocyte absence at the site of infection. Experimental study with mutant strains of *C. perfringens* showed that PFO induces tissue destruction and an anti-inflammatory response. However, the effects were less pronounced than those elicited by alpha-toxin, and both toxins could have a synergistic action in gangrenous lesions (Stevens *et al.,* 1997). PFO and alpha-toxin stimulate leucocyte adherence probably by increasing vascular leucostasis and local ischemia. PFO is leukocytotoxic at high doses, and at a sublethal concentration in endothelial cells, it significantly stimulates the production of intercellular leukocyte adherence module-1 (ICAM-1) and adherent glycoprotein CD11b/CD18, which contribute to leukostasis in vessels adjacent to gangrenous lesions (Bryant *et al.,* 1993; Bryant and Stevens, 1996). The toxin effects on endothelial cells participate in the disruption of endothelial integrity resulting in local edema and ultimately in systemic shock and multiorgan failure. Thus, PFO and alpha-toxin exhibit synergistic effects (Awad *et al.,* 2001; Stevens and Bryant. 2002).

2.1.2. Clostridium Septicum α-Toxin

C. septicum α-toxin is the main virulence factor of this bacterium. α-Toxin is hemolytic, dermonecrotic and lethal, and belongs to the pore forming toxin family, which includes *S. aureus* α-toxin (27% amino acid sequence identity), aerolysin, and the *Pseudomonas aeruginosa* cytotoxins

α-Toxin is secreted by means of a signal peptide as an inactive prototoxin, which is proteolytically activated by removing 45 C-terminal amino acids (propeptide). Various proteases can activate α-toxin such as trypsin and proteinase K, but furin, a cell surface protease, seems to be the main activator *in vivo*.

α-Toxin precursor binds to a glycosylphosphatidylinositol (GPI)-anchored membrane protein, activity resembling that of aerolysin. But, the receptors for the two toxins are different. In contrast to aerolysin, the *N*-glycan of anchored proteins is not important for the binding to α-toxin (Hong *et al.,* 2002). After binding to its receptor, the prototoxin is cleaved to the active form (41 kDa) by furin and possibly by other cell proteases. When the toxin is activated in solution, the monomers aggregate into non-functional complexes. But when the toxin is activated after its binding to membrane, it forms very active oligomers. The propeptide remains associated to the toxin by non-covalent interactions even after the proteolytic cleavage. The propeptide is probably displaced when the monomers associate in oligomers by strong interactions. Thus, the propeptide probably stabilizes an unstable conformation and facilitates the correct assembly of oligomers on the cell membrane. Oligomerization occurs prior to the insertion of the resulting complexes into the membrane. Activated monomers bound to their GPI receptor diffuse laterally in the membrane permitting interaction with other monomers and the formation of oligomers. The oligomers (6-7 monomers) show a structure of an amphipathic β-barrel. α-Toxin is a hydrophilic molecule with no obvious transmembrane domains. The hydrophobic amino acid side chains of each β-strand are probably located on the outside of the β-barrel and interact with the lipid bilayer, while the alternating hydrophilic residues are directed toward the fluid-filled-channel. The α-toxin pores are estimated to be 1-2 nm in diameter, which is similar to those formed by other related toxins (Ballard *et al.,* 1993; Gordon *et al.,* 1999; Tweten, 2001).

C. septicum is involved in gas gangrene and also in non traumatic myonecrosis of the intestinal mucosa, which occurs in patients with intestinal malignancy, neutropenia, leukemia or diarrhea. This infection is accompanied by a profound shock and it is fulminant and often fatal. In animals, *C. septicum* is responsible for gangrene and enterotoxemia (braxy).

2.1.3. C. perfringens Enterotoxin

Enterotoxigenic *C. perfringens* strains are responsible for food poisoning in humans. Most of the *C. perfringens* ingested with food die upon exposure to the stomach acidity. But, when ingested in high number, some bacteria can pass into the small intestine. Food involved in *C. perfringens* poisoning contain at least 10^5 bacteria/g. *C. perfringens* multiplies rapidly in the intestine and sporulates. Enterotoxigenic *C. perfringens* are also involved in antibiotic associated diarrhea, in chronic non-food borne diarrhea, and in infant death syndrome in humans, and in diarrhea in foals and piglets (Petit *et al.,* 1999).

C. perfringens enterotoxin (CPE) is a 319 amino acid protein (35.5 kDa), which is synthesized during the sporulation phase and does not contain any signal peptide. The toxin is released from the bacteria during mother cell lysis.

2.1.3.1. Mode of Action

CPE is cytotoxic for Vero cells and intestinal epithelial cells including CaCo-2, I407 and Hep3b. The first step consists of CPE binding to a cell membrane receptor which is only present in the CPE-sensitive cells. The 30 C-terminal amino acids of CPE are involved in the binding to receptor (Hanna *et al.,* 1991; Kokai-Kun and McClane, 1997). CPE binds first to a receptor identified as claudin (isoform 3, 4, 6, 7, 8, or 14), which is an essential component of tight junctions (Katahira *et al.,* 1997; Sonoda *et al.,* 1999; Fujita *et al.,* 2000). This results in the formation of a 90-100 kDa complex (small complex) in the plasma membrane. At this step, the toxin is largely accessible to antibodies and proteases such as pronase, indicating that it is exposed at the cell surface (Wieckowski *et al.,* 1998). Subsequently, a postbinding maturation occurs when cells are incubated at 37°C, consisting of the formation of an intermediate (135 kDa) and then a large complex (160 or 200 kDa) by association with a membrane protein. The large complex is significantly resistant to SDS and pronase, probably a result of CPE insertion into the membrane. Occludin, a major structural protein of tight junctions, is part of the 200 kDa complex (Singh *et al.,* 2000). CPE large complex increases the membrane permeability to small molecules (<200 Da) possibly by pore formation. The loss of small molecules (ions, amino acids, nucleotides etc.) causes an inhibition of macromolecule synthesis and cytopathic effects leading to morphological changes, permeability alterations for large molecules, and cell lysis. In addition, 200 kDa complex formation in polarized CaCo-2 cells results in the removal of occludin from tight junctions and increased paracellular permeability (McClane, 2001; Singh *et al.,* 2001). This mechanism is probably responsible for the intestinal fluid accumulation and epithelium desquamation observed *in vivo*.

In the small intestine of experimental animals, CPE induces a desquamation of the intestinal cells, particularly those of the villi tips, and a rapid loss of fluid and electrolytes. The ileum is the most sensitive segment of the intestine, whereas little or no effects are observed in the colon. Histopathological changes probably have a major role in fluid and electrolyte perturbations (McClane, 1996).

2.1.4. *C. perfringens* Delta Toxin

Delta–toxin is an extracellular hemolysin which is active only on the erythrocytes of even-toed ungulates (sheep, ox, pig, goat) whereas the erythrocytes from other species are resistant. In addition, this toxin lyses other eukaryotic cells such as human monocytes, rabbit macrophages and platelets from human, rabbit, goat and guinea pigs (Jolivet-Reynaud *et al.,* 1988).

Delta-toxin is a 40 kDa basic protein (pI 9.1) produced by some *C. perfringens* type B and C strains. The toxin binds to ganglioside GM2 on the cell surface which is responsible for cell specificity. GM2 is undetectable on cells insensitive to δ-toxin. The toxin can bind to liposomes containing GM2. The cytolysis is a multistep process which is initiated by a rapid and tight irreversible binding of the toxin by its C-terminal domain to GM2 receptor. The toxin remains on the cell surface without insertion into the lipid bilayer and induces cell swelling and membrane disruption by an unknown mechanism (Jolivet-Reynaud and Alouf, 1983; Jolivet-Reynaud *et al.,* 1986; Jolivet-Reynaud *et al.,* 1989).

2.1.5 *Clostridium perfringens* Beta Toxins

Beta1-toxin is produced by *C. perfringens* type B and C, which are involved in necrotic enteritis in young animals, in sheep with enterotoxemia, and in humans (Pigbel and Darmbrand). The disease is characterized by necrosis and inflammation of the intestinal mucosa with bleeding to the lumen. Beta1-toxin is very labile and sensitive to protease degradation. For this reason, the beta1-induced pathology is only observed in particular circumstances such as in newborns in which the protease activity of the digestive tract is low. The risk factors involved in human disease are low-protein diet inducing low trypsic activity in the intestine and consumption of sweet potatoes, which contain a trypsin inhibitor. The low-protease activity permits a high level of active toxin into the intestinal lumen.

Beta1-toxin is synthesized as a 336 amino acid protein, the first 27 amino acids of which constitute a signal peptide. The secreted protein has a molecular weight of 34,861 Da and a pI of 5.5 (Hunter *et al.,* 1993). The beta1-toxin gene is localized on large plasmids in *C. perfringens* which also contain the insertion sequence IS*1151* (Katayama *et al.,* 1996; Gibert *et al.,* 1997).

In addition, beta1-toxin is thermolabile and highly sensitive to oxidizing and thiol group reagents (Sakurai *et al.,* 1980). Therefore, the high lability of this toxin has hindered attempts to study it. The toxin is dermonecrotic and lethal, but it is not hemolytic. When intravenously injected in rat, beta1-toxin induces an elevated blood pressure and vascular contraction, possibly by interacting with the autonomic nervous system and stimulating catecholamine release (Sakurai, 1995).

Beta1-toxin is closely related, at the amino acid level, to pore forming cytolysins produced by *Staphylococcus aureus*: 28% similarity with α-toxin, 18 to 28% with the A, B, and C components of gamma toxin, 17 and 28% with the S and F components of leukocidin and leukocidin R (Hunter *et al.,* 1993). *S. aureus* α-toxin is a single component toxin, which forms hexamers and which inserts into cell membrane leading to pore formation. In contrast to *S. aureus* toxins, beta1-toxin is cytotoxic but it is neither cytolytic nor hemolytic (Hunter *et al.,* 1993). Beta1-toxin probably acts by pore formation. It has been reported that it associates with human umbilical vein endothelial cell membranes in multimeric complexes (Steinthorsdottir *et al.,* 2000). In addition, recombinant beta1-toxin forms cation selective channels in planar bilayers of artificial phospholipids (Shatursky *et al.,* 2000).

Certain amino acids from beta1-toxin, which are conserved in *S. aureus* α-toxin have been mutated (Y134, Y203, R212 and D167). The substitutions R212E and Y203F induced a 12.5- and 2.5-fold decrease in lethal activity, respectively. The residue counterparts in α-toxin are involved in binding to substrate and oligomerization and have been found to be important for the *S. aureus* α-toxin activity. However, the replacement of His (four residues in the beta1-sequence) to Leu did not alter activity. No strict functional homology can be deduced between the two proteins (Nagahama *et al.*, 1999).

Previous biochemical data showed that thiol groups are essential for beta1-toxin activity (Sakurai *et al.*, 1980). The replacement of the unique Cys at position 265 of the beta1-sequence by Ser or Ala did not change the activity level (Steinthorsdottir *et al.*, 1998; Nagahama *et al.*, 1999). However, the substitution of Cys by a residue with a large side chain or possessing functional groups and mutation of residues (Y266, L268 and W275) close to Cys induced a complete loss of activity. The site encompassing the Cys residue seems to be essential for this activity (Nagahama *et al.*, 1999). Although the structure-function of beta1-toxin remains to be determined, the recent results indicate that beta1-toxin forms cation selective pores resulting from oligomerization of the toxin into susceptible membranes.

Beta2-toxin has been identified from a *C. perfringens* strain isolated in a piglet that died of necrotic enteritis. The Beta2-toxin gene is localized on plasmid, but it is not linked to the beta1-toxin gene. *C. perfringens* can produce beta1- and beta2-toxin, separately or in combination. beta2-toxin strains are mainly involved in piglet necrotic enteritis and in horse typhlocolitis (Gibert *et al.*, 1997; Herholz *et al.*, 1999).

The toxin is a secreted protein containing a signal peptide (30 N-terminal residues). The mature protein has a molecular weight of 27,670 Da and a pI of 5.01. The beta2-toxin sequence is not significantly related to that of beta1-toxin, and both proteins show no cross immunological reactions. What Beta1- and Beta2-toxin have in common are that they are highly sensitive to trypsin, are cytotoxic for the I407 cell line, and they both experimentally induce necrotic enteritis (Jolivet-Reynaud *et al.*, 1986; Gibert *et al.*, 1997). Beta2-toxin is not related to any other protein in the Data Bank and its mode of action remains unknown.

2.1.6. Clostridium perfringens Epsilon-Toxin

Epsilon-toxin is the major virulence factor of *C. perfringens* types B and D. Fatal enterotoxemia, which is common in sheep and goat, and more rarely in cattle is caused by *C. perfringens* D. Overgrowth of *C. perfringens* D in the intestine of susceptible animals, generally as a consequence of overeating of foods containing a large proportion of starch or sugars, produce large amounts of epsilon-toxin. The toxin is absorbed through the intestinal mucosa and spread in the different organs by the blood circulation causing blood pressure elevation, vascular permeability increase, lung edema and kidney necrosis (pulpy kidney disease in lambs).

The major pathological changes are observed in the brain: congestion and edema of the meninges, perivascular and intercellular edema, and necrotic foci of the nervous tissue. Epsilon-toxin can pass through the blood-brain barrier and accumulates specifically in the brain (Nagahama and Sakurai, 1991; Nagahama and Sakurai, 1992). The neurological disorders (retraction of the head, opithotonus, convulsions, agonal struggling, hazard roaming) seem to result from epsilon-toxin action on the hippocampus leading to an excessive release of glutamate (Miyamoto *et al.*, 1998).

The epsilon-toxin gene (*etx*) is localized on plasmids of various sizes and is flanked by insertion sequence and transposon genes. IS*1151* and a gene related to the transposase (*tnpa*) gene from Tn*3* lie upstream of *etx*, and a region with similarity to transposases from *S. aureus* and *Lactococcus* is located downstream from *etx* (Rood, 1998).

2.1.6.1. Structure

Epsilon-toxin is synthesized as a protein containing a signal peptide (32 N-terminal amino acids). The secreted protein (32,981 Da) is inactive and it is called prototoxin (Hunter *et al.*, 1992). The prototoxin is activated by proteases such as trypsin, α-chymotrypsin, and λ-protease, which is produced by *C. perfringens*. Activation by λ-protease is comparable to that obtained with trypsin plus α-chymotrypsin. The λ-protease removes 11 N-terminal and 29 C-terminal residues whereas trypsin plus α-chymotrypsin cleaves 13 N-terminal residues and the same C-terminal amino acids (Figure 1). This results in a reduction of size (28.6 kDa) and an important decrease of pI from 8.02 to 5.36 accompanied probably by a conformational change. The charged C-terminal residues could prevent the interaction of the protein with its substrate or receptor (Minami *et al.*, 1997).

Epsilon-toxin contains three domains, which are mainly composed of β-sheets. The overall structure is significantly related to that of the pore forming toxin aerolysin from *Aeromonas hydrophila* (Cole, 2003). Essential amino acids for the lethal activity have been identified by biochemistry and mutagenesis. A previous work with chemical modifications shows that His residues are required for the active site, and Trp and Tyr residues are necessary for the binding to target cells (Sakurai, 1995). The molecule contains a unique Trp and two His residues. Amino acid substitutions showed that His106 is important for biological activity, whereas His149 and Trp190 probably are needed to maintain the structure of epsilon-toxin, but they are not essential for the activity (Oyston *et al.*, 1998).

2.1.6.2. Cellular Effects

Specific activity of epsilon-toxin is also observed in cultured cells. Among the cell lines which have been tested, Madin Darby Canine Kidney (MDCK) and, to a lesser extent, the human leiomyoblastoma (G-402) cells are sensitive to epsilon-toxin. Marked swelling is observed in the first phase of intoxication, followed by blebbing and membrane disruption. The cytotoxicity can be monitored by using an indicator of lysosomal integrity (neutral red) or mitochondrial integrity (3-(4,5-dimethylthiazol-2-yl)-2,5-diphenyltetrazolium bromide MTT) (Payne *et al.*, 1994; Lindsay *et al.*, 1995; Petit *et al.*, 1997; Shortt *et al.*, 2000).

Epsilon-toxin binds to the MDCK cell surface, preferentially to the apical site, and recognizes a specific membrane receptor which is not present in insensitive cells. Binding of the toxin to its receptor leads to the formation of large membrane complexes which are very stable when incubation is performed at 37°C. In contrast, the complexes formed at 4°C are dissociated by SDS and heating. This suggests that a maturation process, probably involving fluidity of the lipid bilayer, is required. Endocytosis and internalization of the toxin into the cell was not observed, and the toxin remains associated with the cell membrane throughout the intoxication process (Petit *et al.*, 1997). The epsilon-toxin large membrane complex in MDCK cells and synaptosomes correspond to the heptamerization of toxin molecules within the membrane and pore formation (Petit *et al.*, 1997; Miyata *et al.*, 2001). In contrast to CPE which forms a heterotrimeric complex, epsilon-toxin seems to interact with only one class of membrane protein (Petit *et al.*, 1997).

The epsilon-toxin receptors have been found to be a 34 kDa protein or glycoprotein in MDCK cells and a 26 kDa sialyglycoprotein in rat brain. Lipid environment seems to be important for the binding of epsilon-toxin to the cell surface, since detergents prevent the toxin-cell surface interaction (Payne *et al.*, 1997; Petit *et al.*, 1997). In synaptosomes and MDCK cells, the epsilon-toxin receptor has been localized in lipid raft microdomains (Miyata *et al.*, 2002).

The cytotoxicity is associated with a rapid loss of intracellular K^+, and an increase of Cl^- and Na^+, whereas an increase of Ca^{++} occurs later. In addition, the loss of viability correlates with the entry of propidium iodide, indicating that the epsilon-toxin forms large pores in the cell membrane. Pore formation was evident in artificial lipid bilayers. Epsilon-toxin induces the formation of water-filled channels that are permeable to hydrophilic solutes up to a molecular mass of 1 kDa, which represent general diffusion pores slightly selective for anions (Petit *et al.*, 2001). In polarized MDCK cells, epsilon-toxin induces a rapid and dramatic increase in permeability. Pore formation in the cell membrane is likely responsible for the permeability change of cell monolayers. Actin cytoskeleton and organization of tight and adherens junctions are not altered, and the paracellular permeability to macromolecules is not significantly increased upon epsilon-toxin treatment (Petit *et al.*, 2003). Therefore, epsilon-toxin is a very potent toxin, which alters the permeability of cell monolayers such as epithelium and endothelium, causing edema.

Intravenous injection of epsilon-toxin in experimental animals leads to damage and permeability increase of the blood brain barrier, accumulation of the toxin in the brain and formation of perivascular edema (Buxton, 1978; Nagahama and Sakurai, 1991; Finnie *et al.*, 1999; Zhu *et al.*, 2001). Endothelial cells are possibly the primary target cells, but epsilon-toxin also alters the blood brain barrier and binds specifically to the brain (Nagahama and Sakurai, 1991; Zhu *et al.*, 2001). Epsilon-toxin could directly interact with hippocampus neurons leading to an excessive release of glutamate, which is probably responsible for the nervous symptoms of excitation (Petit *et al.*, 1997; Miyamoto *et al.*, 1998; 2000; Borrmann *et al.*, 2001). The precise mechanism of epsilon-toxin action on neuronal cells remains to be elucidated.

2.2. Toxins Modifying Cell Membrane and/or Connective Tissue Proteins

2.2.1. C. perfringens Alpha-Toxin and Related Clostridial Phospholipases

C. perfringens alpha-toxin is a zinc dependent phospholipase C which degrades phosphatidylcholine (PC, or lecithin) and sphingomyelin. It is dermonecrotic, hemolytic, and is lethal for mice. The toxin causes membrane damage to a variety of erythrocytes and cultured mammalian cells. It is the major *C. Perfringens* toxin involved in gangrene, which is characterized by extensive local tissue destruction and necrosis progressing to profound shock and death (Stevens *et al.*, 1988). Trauma-induced gas gangrene was common in soldiers from World War I. Alpha-toxin is produced by all *C. perfringens* strains. The toxinotype A strains usually produce the toxin in higher amounts than the other toxinotypes. The alpha-toxin gene has been characterized from several *C. perfringens* strains originating from mammals, and the deduced protein sequences are highly conserved. However, an avian *C. perfringens* strain was found to contain a divergent alpha-toxin (57-58 amino acid differences from the mammal strains) (Justin *et al.*, 2002). Phospholipases C is also produced by several *Clostridium* species (Table 1).

2.2.1.1. Structure and Mode of Action

C. perfringens alpha-toxin is secreted by means of a signal peptide (28 first amino acids). The mature protein is 370 amino acids long (43 kDa), and contains two domains, an alpha-helical N-terminal domain (residues 1-246) harboring the active site, and a β-sandwich C-terminal domain (residues 256-370), which is involved in membrane binding. Alpha-toxin is related at the amino acid level (more than 50% identity) and/or on the basis of antigenic cross-reactivity, to the phospholipases C from *Clostridium bifermentans*, *Clostridium novyi* (γ-toxin), *C. baratii* and *C. absonum*. The N-terminal domain is related (29% identity) to the non-toxic *Bacillus cereus* phosphatidylcholine phospholipase C (PC-PLC) which does not possess the C-terminal domain. Both proteins have similar tertiary structure (Naylor *et al.*, 1998; Titball *et al.*, 1999; Justin *et al.*, 2002).

The N-terminal domain of alpha-toxin (residues 1-249) supports the hydrolysis of PC and sphingomyelin, but it is not hemolytic and lethal (Titball *et al.*, 1991). The double bond of PC has been found to be the binding site of alpha-toxin (Nagahama *et al.*, 1998). Crystallographic studies and site-directed mutagenesis shows that the active site includes two tightly bound zinc ions and one loosely bound zinc ion, which are coordinated to five histidine, one tryptophane, two aspartic and one glutamate residues (Nagahama *et al.*, 1995; Guillouard *et al.*, 1996; Nagahama *et al.*, 1997; Naylor *et al.*, 1998). A hydrophobic groove on the enzyme surface is involved in binding to phospholipid (Naylor *et al.*, 1998). Two loops (residues 55-93 and 132-149) on either side of the active site cleft and close to the third zinc-binding site, undergo a change of conformation between active and inactive configuration. Alpha-toxin adopts two conformations. In the open or active structure, the active site contains three zinc ions and is accessible to substrate binding, whereas in the closed or inactive conformation, the two loops occlude the active site and one ion binding site is lost. The binding of alpha-toxin to its membrane receptor through its C-terminal domain probably induces a conformational change of the N-terminal domain from the closed to open configuration (Eaton *et al.*, 2002).

The C-terminal domain has no enzymatic activity, but it is required for hemolytic, lethal and sphingomyelinase activities (Titball, 1993b). This domain is an eight-stranded antiparallel β-sandwich containing a Ca^{++} binding site. Other putative Ca^{++} binding sites could be localized on the protein surface (Naylor *et al.*, 1998). Calcium ions have been found to be essential for the binding of alpha-toxin to phospholipid monolayers (Moreau *et al.*, 1988).

The importance of the C-terminal domain in alpha-toxin activity is supported by the comparison with *Clostridium bifermentans* PLC, which is weakly hemolytic and non lethal. *C. bifermentans* PLC shows an overall 47% sequence identity with alpha-toxin. The homology increases to 75% in the active site and membrane binding domains. The loop (residues 329-337) of alpha-toxin, close to the Ca^{++} binding site contains higher residue changes compared to *C. bifermentans* PLC and could be crucial for the toxicity by reducing binding to membrane (Naylor *et al.*, 1998). The C-terminal domain of *C. bifermentans* PLC lacks Tyr-331 and Phe-334, which are present in *C. perfringens* alpha-toxin. The restitution of these residues by constructing hybrid proteins between *C. bifermentans* PLC and *C. perfringens* alpha-toxin, restores the lethal and hemolytic activities. Thus, the C-terminal domain, which controls the interaction with membrane lipids, greatly potentiates the biological activities of the clostridial phospholipases (Jepson *et al.*, 1999).

The alpha-toxin C-terminal domain is structurally similar to the C2 domain of eukaryotic proteins which are involved in Ca^{++} dependent phospholipid binding such as pancreatic lipase, soybean lipoxygenase, synaptotagmin, phosphoinositide-specific phospholipase C (PI-PLC) and arachidonate lipoxygenase. The Ca^{++} binding site of synaptotagmin is conserved in the alpha-toxin C-terminal domain. Alpha-toxin binds to membrane in a manner similar to that of eukaryotic proteins. A binding surface has been identified in the C-terminal domain, whose charge-charge interaction with the phospholipid headgroups is mediated by calcium ions. The calcium ions confer positive charges on the phospholipid membrane allowing the binding to the negatively charged toxin domain. Three calcium-binding sites are coordinated partially by side-chains of the C-terminal domain, and partially by phospholipid phosphate groups (Guillouard *et al.*, 1997; Naylor *et al.*, 1998). It is assumed that the C-terminal domain of alpha-toxin preferentially binds to phospholipids in intact membrane and triggers conformational changes and activation of the enzyme. The active site would close after detachment from the membrane. Therefore, alpha-toxin could discriminate between phospholipids from intact and lysed cells, resulting in an increase of cell lysis (Eaton *et al.*, 2002).

2.2.1.2. Cellular Effects

Alpha-toxin is the major *C. perfringens* toxin responsible for gangrene. Indeed, immunization with recombinant alpha-toxin protected mice against gangrene, and alpha-toxin negative *C. perfringens* mutant obtained by allelic exchange shows a reduced virulence (Williamson and Titball, 1993; Awad *et al.*, 1995). The role of alpha-toxin in non-gangrenous diseases such as enteritis and/or enterotoxemia in man and animals is still questionable. Necrotic enteritis in chickens has been associated with *C. perfringens* alpha-toxin (reviewed in Songer, 1996; Titball *et al.*, 1999). Interestingly, alpha-toxin from an avian *C. perfringens* strain differs from that of other mammalian strains, with a structure that remains in an active form, and having different substrate specificity (Justin *et al.*, 2002). This suggests that specific alpha-toxin variants could be involved in intestinal lesions. Alpha-toxin from a *C. perfringens* strain isolated from gas gangrene differs from that of a calf intestinal strain by 3 amino acid changes. Both toxins displayed the same levels of enzymatic and toxic activities, but alpha-toxin from the calf strain was more resistant to α-chymotrypsin (Ginter *et al.*, 1996). The implication in intestinal diseases of alpha-toxin variant resistant to protease degradation is not yet established.

Alpha-toxin cleaves PC containing at least one unsaturated fatty acyl chain, and the resulting products are diacylglycerol (DG) and phosphorylcholine (Nagahama *et al.*, 1998). Sphingomyelin is cleaved to yield ceramide and phosphorylcholine. Alpha-toxin from an avian *C. perfringens* strain is more active towards PC versus sphingomyelin, in contrast to alpha-toxin from mammal strains. This could possibly represent an adaptation to host species (Justin *et al.*, 2002). Alpha-toxin is a hot-cold hemolysin. The hemolysis is obtained by incubating the erythrocytes with the toxin at 37°C, and then by chilling the suspension below 10°C.

Oligomerization of the toxin and pore formation seem not to occur (Nagahama *et al.*, 1996). Alpha-toxin hydrolyzes membrane PC and sphingomyelin, preferentially from the outer leaflet (Titball, 1993a). It has been shown that alpha-toxin damages membranes of PC-cholesterol liposomes and that membrane fluidity plays an important role in binding and/or activity of the toxin (Nagahama *et al.*, 1996; 1998). Large amounts of toxin immediately lyse cells probably by degradation of membrane phospholipid (Sakurai *et al.*, 1993). However, the hemolysis does not only result from the alpha-toxin phospholipase C activity but from activation of endogenous phospholipases A_2, C and D. The hemolytic activity is correlated with the formation of phosphatidic acid (PA), DG, IP3 and degradation of PIP2. It has been shown that alpha-toxin stimulates endogenous phospholipases C and D (PLC and PLD) in erythrocyte membranes through a GTP-binding protein. Endogenous PLC induces a rapid PA increase, and activation of PLD is responsible for late PA formation that is closely linked to hemolysis (Sakurai *et al.*, 1993; Ochi *et al.*, 1996; Titball *et al.*, 1999). The products of phospholipid hydrolysis (DG, ceramide) also activate the arachidonic cascade resulting in the formation of thromboxanes, leukotrienes and prostaglandins, which generate local inflammation and vasoconstriction. This contributes to the anoxia of infected tissues promoting the local growth of *C. perfringens* (reviewed in Titball *et al.*, 1999).

In contrast to other bacterial myonecrosis, *C. perfringens* gas gangrene is characterized by the absence of inflammatory cells in affected tissues and the accumulation of leukocytes within vessels at the periphery of the lesions. Alpha-toxin is cytolytic at high concentration, and in sublytic amounts, it impairs the migration of leukocytes and promotes their aggregation. The toxin strongly stimulates the expression of endothelial cell-leukocyte adherence molecule-1 (ELAM-1) and ICAM-1 causing a hyperadhesion of leukocytes to vessel endothelium. Alpha-toxin is also a potent activator of IL-8 in endothelial cells, which is a neutrophil chemoattractant. The recruitment of leukocytes coupled to toxin-induced hyperadhesion contributes to the leucostasis at the periphery of the lesions, and could enhance respiratory burst activity and neutrophil-mediated vascular injury (Bryant and Stevens, 1996). The rapid destruction of tissues also results from a drastic attenuation of local and regional blood flow. Alpha-toxin promotes both platelet/platelet and platelet/neutrophil aggregates. The toxin activates the platelet fibrinogen receptor (gpIIbIIIa) and stimulates homotypic and heterotypic platelet aggregation in a soluble fibrinogen-dependent manner. The PLC-mediated activation of gpIIbIIIa is correlated with a depletion of intracellular Ca stores, a subsequent opening of plasma-membrane calcium channels, and a transient increase of cytosolic calcium. Therefore, intracellular Ca^{++} could be the mediator of gpIIbIIIa activation. Leukocyte aggregates are observed intravascularly and associated with vascular endothelium. Free intravascular aggregates are responsible for the blockade of blood flow in small vessels, and subsequently for the anoxia and necrosis of tissues (Bryant *et al.*, 2000b; 2000a; Bryant *et al.*, 2003).

In addition, alpha-toxin stimulates the production of platelet activating factor (PAF) and prostacyclin in endothelial cells. PAF synthesis is dependent on protein kinase C activation, possibly in response to DG, resulting from PC hydrolysis by alpha-toxin. The vasoactive lipids (PAF and prostacyclin) may contribute to myocardial dysfunction and to the marked edema observed in gangrenous tissues (Bunting *et al.*, 1997; Stevens *et al.*, 1997). Clostridial gangrene is commonly associated with profound shock accompanied by hypotension, and alpha-toxin directly impairs myocardial contractility (Stevens and Bryant, 2002). When injected intravenously, alpha-toxin suppresses the myocardial contractility and induces a profound hypotension, bradychardia, shock, and multiorgan failure. The myotoxic effects could result from the direct activity of alpha-toxin on muscle cell membrane, from alteration of local blood flow, and/or inflammatory mediators as a consequence of endothelial cells stimulated by alpha-toxin (Stevens *et al.*, 1988).

All the mechanisms of physiological disorders induced by alpha-toxin are not yet fully understood. It has been found that UDP-glucose deficient cells are highly sensitive to lysis by alpha-toxin resulting, possibly, in an alteration of membrane glycophospholipid composition or glycoproteins favoring the interaction between the toxin and the membrane. Low UDP-glucose level that is found in diabetic and ischemic tissues, probably promotes extensive tissue destruction (Flores-Diaz *et al.*,1998).

2.2.2. Clostridial Collagenases

Clostridia are anaerobic bacteria, which ferment a wide range of substrates for their nutritional requirements. This is achieved by production and secretion of hydrolytic enzymes. Clostridial hydrolytic enzymes include polysaccharide-, lipid-degrading enzymes, proteases including collagenases and gelatinases, and nucleases (reviewed in Matsushita and Okabe, 2001). Many clostridial hydrolytic enzymes act as virulence factors or contribute to the pathogenicity of *Clostridia*. They are mainly involved in gangrene lesions, where they participate in tissue destruction and spreading *Clostridium*.

C. histolyticum is responsible for dramatic myonecrosis and gangrene. The main secreted toxins are collagenases (also called α-clostripain), which exist in at least six different forms with molecular masses ranging from 68 to 125 kDa, and are divided into two classes ColG and ColH based on amino acid sequence similarities and specificities toward peptide substrates. They are encoded by two genes, *colG* and *colH*, which are probably derived from duplication of an ancestral gene. Both genes encode a 116 kDa precursor, which is processed in the C-terminal part to yield the multiple forms (Matsushita *et al.,* 1999). ColH consists of four segments S1, S2a, S2b, and S3, of which S2a and S2b are homologous, and ColG also possesses four segments S1, S2, S3a and S3b. All collagenase forms contain in the N-terminal part (segment 1) the consensus active site HExxH of zinc metalloproteases (Jung *et al.,* 1993). The other segments are involved in substrate recognition and binding. Thus S3 is required for binding to collagen (Matsushita *et al.,* 1998; Matsushita and Okabe, 2001). *C. histolyticum* collagenases are specific for collagen and gelatin. They cleave peptide bonds on the amino side of the glycine residue in the PxGP sequence, and they degrade their substrates to small dialyzable peptides. Collagenases play an important role in the degradation of the connective tissue.

C. perfringens produces a collagenase, ColA or κ-toxin, which is related to ColG from *C. histolyticum* and contains the characteristic motif HExxH (Matsushita *et al.,* 1994; Matsushita *et al.,* 1999). ColA has an additional effect on phospholipase alpha-toxin in the destruction of tissues.

A gene encoding a collagenase (Col-T) related to *C. histolyticum* collagenases, has been identified on the same plasmid, which contains the Tetanus toxin (TeTx) gene. ColT could be involved in the destruction of tissue permitting invasion by *C. tetani* (Brüggemann *et al.,* 2003).

2.2.3. Hydrolytic Enzymes

In addition to toxins, several pathogenic *Clostridium* produce potent hydrolytic enzymes including collagenases, proteases, lipases, hyaluronidase, neuraminidases, DNase, and urease, which are additional virulence factors mainly involved in *Clostridium*-mediated gangrenes. The best example is *C. perfringens*, which secretes a large number of hydrolytic enzymes (Figure 1). These enzymes degrade extracellular substrates and components resulting from cell lysis, and thus participate in the destruction of tissues synergistically with the membrane damaging toxins (Petit *et al.,* 1999). Furthermore, bacterial extracellular proteases activate toxins, which are secreted as inactive precursors. This is the case of *C. perfringens* λ-protease, which activates epsilon- and iota-toxin precursors, and *C. botulinum* type A protease, which activates the neurotoxin.

2.2.4. Bacteroides fragilis Enterotoxin

Bacteroides fragilis is a commensal bacterium of the intestine, but is occasionally responsible for abdominal abscesses and septicemia. Some strains produce an enterotoxin (BFT) and can cause diarrhea in humans, lambs, and more rarely in other young animals. The *bft* gene is localized on a 6 kbp pathogenicity island in the chromosome of enterotoxigenic *B. fragilis* strains. BFT is synthesized as a 397 amino acid protein and is secreted by means of a signal peptide (18 amino acids) as an inactive precursor. The 193 N-terminal residues (propeptide) are linked to the 186 C-terminal residues (mature protein) by a trypsin cleavage site. The mature protein contains a characteristic zinc-binding metalloprotease motif (HExxHxxGxxH). *In vitro* BFT is able to proteolyze actin, gelatin, casein, and azocoll. Three highly related (92-96% identity) isoforms have been characterized. BFT2, but not BFT1 and BFT3, contains an additional C-terminal extension of 20 amino acids forming an amphipathic structure. This suggests that BFT2 could oligomerize and insert into the cell membrane forming a pore.

BFT induces morphological changes in cultured intestinal and renal cells (HT29, T84, MDCK, CaCo-2) that are able to form intercellular junctions. Cell rounding, increased cell volume, and effacement of microvilli and apical junctional complexes are observed. BFT does not enter cells, but cleaves the extracellular domain of E-cadherin in an ATP-independent manner. This step is followed by the degradation of the intracellular domain of E-cadherin probably by ATP-dependent cellular proteases. A reorganization of actin filaments occurs without a decrease in the content of actin filaments. In polarized T84 cells treated with BFT, the apical actin ring and microvilli actin filaments are lost, whereas actin filaments accumulate in the basal pole. The permeability of polarized cell monolayers increases in a time- and concentration-dependent manner. BFT is more active when applied to the basolateral side of polarized cells. The cell effects are reversible, since cells regain a normal aspect by 2-3 days after toxin treatment that correlates with resynthesis of E-cadherin. In addition, BFT stimulates the secretion of interleukin-8 by intestinal cells, which promotes an inflammatory response with recruitment of polymorphonuclear leucocytes to the intestinal mucosa. The inflammatory response is supposed to contribute to the intestinal secretory effects. BFT is the only known bacterial toxin, which modifies the actin cytoskeleton by cleaving a cell surface molecule. The BFT-dependent proteolysis of the extracellular domain of E-cadherin leads to the loss of the intracellular domain. Since the actin filaments are connected to E-cadherin via catenins, BFT induces a relocalization of the actin filaments from the apical to the basal side, which is accompanied by the loss of microvilli and disorganization of the adherens and tight junctions. This is followed by a decrease in intestinal cell barrier function (reviewed in Sears 2001).

3. Toxins Modifying an Intracellular Target

3.1. Toxins Modifying the Actin Cytoskeleton

3.1.1. Clostridial Actin ADP-Ribosylating Toxins (Clostridial Binary Toxins)

The clostridial binary toxins have a common structure consisting of two independent protein components that are not covalently linked, one being the binding component (100 kDa), and the other the enzymatic component (45 kDa). Both components are required for the biological activity. Two families can be distinguished. Iota family encompasses iota-toxin produced by *C. perfringens* type E, *C. spiroforme* toxin and ADP-ribosyltransferase (CDT) synthesized by some strains of *C. difficile*. The C2 family corresponds to the C2 toxins from *C. botulinum* C and D. At the amino acid sequence level, the components of the Iota family are highly related (80-85% identity), whereas C2 toxins show 31-40% identity with the Iota family proteins. Members of the Iota and C2 families show no immunological cross-reaction and no functional cross-complementation (reviewed in Popoff, 2000). These toxins are also related to the vegetative insecticidal proteins (VIP) produced by *Bacillus cereus* and *Bacillus thuringiensis* (Warren *et al.*, 1996).

3.1.1.1. Structure

The binary toxin components are encoded by two genes that are closely linked and are organized in an operon. The Iota family components are synthesized during the exponential growth phase and are secreted by means of a signal peptide as inactive proteins. They are proteolytically activated by removing a 20 kDa N-terminal peptide from the binding component (80 kDa for the active form) and 9 to 11 N-terminal residues from the enzymatic component (Gibert *et al.*, 2000). In contrast, C2 toxin components are produced during the sporulation phase and are released upon mother cell lysis. Signal peptide is lacking in C2 toxin proteins. The C2 toxin-binding component also requires proteolytic processing resulting in the removal of a 20 kDa N-terminal peptide (Ohishi, 2000).

The binding components of actin-ADP-ribosylating toxins show an overall sequence identity of 31-43% with the binding component (protective antigen or PA) of the *Bacillus anthracis* toxin. The PA structure has been solved and shows 4 distinct domains essentially composed of hydrophilic β-strands (Petosa *et al.*, 1997). The N-terminal domain contains the binding site for the enzymatic components (edema factor and lethal factor) (Cunningham *et al.*, 2002), and the C-terminal domain (domain 4) is involved in the recognition of the cell surface receptor (Petosa *et al.*, 1997). No specific function has been attributed to domain 3, which is the smallest one. Domain 2 contains long β-strands and plays a central role in oligomerization. An amphipathic flexible loop (amino acids 302-325) forms a β-hairpin. In the heptameric structure, the β-hairpins of each monomer assemble to form a β-barrel. The transmembrane pore is thus composed of 14 antiparalell β-strands, which insert into the membrane allowing the translocation of the partially unfolded enzymatic components into the cytosol (Cunningham *et al.*, 2002; Mogridge *et al.*, 2002a; 2002b; Abrami *et al.*, 2003). The PA structure and domain organization are probably conserved in the binding components of the actin-ADP-ribosylating toxins. The C-terminus of the binding components of iota toxin (Ib) and C2 toxin (C2-II), corresponding to domain 4 of PA, has been identified as the receptor-binding domain (Blöcker *et al.*, 2000; Marvaud *et al.*, 2001). The lower amino-acid conservation (less than 10%) between these domains suggests that each toxin recognizes distinct receptors, which include a membrane protein for iota toxin (Stiles *et al.*, 2000), and a *N*-glycan containing the α-D-mannoside-β1, 2-*N*-acetyglucosamine motif for C2 toxin (Eckhardt *et al.*, 2000). The N-terminal parts of mature Ib and C2-II are required for the docking of the enzymatic components Ia and C2-I, respectively (Barth *et al.*, 1998; Marvaud *et al.*, 2001).

The structure of the enzymatic components of VIP (VIP-II) and iota toxin (Ia) shows two distinct but related structural domains that are probably derived from a duplication of an ancestral gene. Each domain core consists of a perpendicular packing of a five-stranded mixed β-sheet against a three-stranded antiparallel β-sheet flanked by four consecutive helices. The C-terminal domain which is homologous to C3 contains the catalytic site (see below), whereas the N-terminus interacts with the binding component allowing cell internalization of the enzymatic component (Han *et al.*, 1999; Tsuge *et al.*, 2003). The Ia segment (129-257aa) corresponding to the compact core of the N-domain mediates the interaction with Ib (Marvaud *et al.*, 2002), whereas the C2-I domain that binds to C2-II, has been mapped to the segment 1-87 corresponding to the 4 α-helices upstream from the N-domain core of VIP-II (Barth *et al.,* 2002).

3.1.1.2. Mode of Action

The mature binding components (about 80 kDa) trigger the internalization of the respective enzymatic components into the cell by receptor-mediated endocytosis (Simpson *et al.,* 1987; Perelle *et al.,* 1993). Mature binding components recognize a specific cell membrane receptor (Stiles *et al.,* 2000), heptamerize to form small ion permeable channels, and trap the enzymatic components into endocytic vesicles which are subsequently translocated into the cytosol upon an acidic pulse from early endosomes (Schmid *et al.,* 1994; Barth *et al.,* 2000; Bachmeyer *et al.,* 2001; Knapp *et al.,* 2002). Like anthrax toxin, the enzymatic components only bind to oligomerized binding components (Mogridge *et al.,* 2002a). But in contrast to PA which is proteolytically activated by furin once bound to the cell surface receptor, Ib and C2-II need to be processed in solution by trypsin or α-chymotrypsin. Unprocessed Ib and C2-II can bind to cell surface receptor, but they do not oligomerize and do not mediate the entry of the enzymatic components (Ohishi. 1987; Gibert *et al.,* 2000; Stiles *et al.,* 2002). The binding component heptamers insert into the membrane and form pores permitting the passage of molecules such as ions. As shown with the inhibition of C2 toxin by chloroquine, which binds to a region close to the pore-forming segment, the enzymatic component is likely directly delivered through the pore of the binding component heptamers (Bachmeyer *et al.,* 2001). Cytosolic chaperones such as Hsp90 seem to facilitate the transport of C2-I across the vesicle membrane (Haug *et al.,* 2003), possibly in a similar manner as found for diphtheria toxin (Ratts *et al.,* 2003). In addition, iota toxin is able to transcytose across polarized intestinal cells such as CaCo-2 cells. Ib can either bind to apical or basolateral surfaces and is recycled to the opposite cell side, where it is able to trap and internalize Ia. Thus Ib remains on the cell surface up to 3 h and mediates the endocytosis of a large number of Ia molecules (Richard *et al.,* 2002).

In cells, the actin cytoskeleton is in dynamic equilibrium between actin monomers and filaments. The enzymatic components catalyze the ADP-ribosylation of actin monomers at Arg-177 but not of polymerized F-actin. Those of the Iota family modify all the actin isoforms including cellular and muscular isoforms, whereas C2 toxin only interacts with cytoplasmic and smooth muscle γ-actin (Aktories *et al.,* 1986; Ohishi and Tsuyama, 1986; Vandekerckhove *et al.,* 1987; Mauss *et al.,* 1990). The target residue, Arg-177, is located in the actin-actin binding site, which is buried in the actin filament. Therefore, actin ADP-ribosylating toxins only modify monomeric actin. Actin complexed to gelsolin is also a substrate of ADP-ribosylation. The cumbersome ADP-ribose on the actin-binding site prevents the nucleation and polymerization of ADP-ribosylated actin monomers. Moreover, ADP-ribosylated actin acts as capping protein by binding to the barbed end of the actin filament and inhibits the further addition of unmodified actin monomers. Actin filament depolymerization at the pointed end is not modified and the released actin monomers are immediately ADP-ribosylated. In addition, ADP-ribosylation inhibits the intrinsic ATPase activity of actin (Reuner *et al.,* 1996). Microinjection of ADP-ribosylated actin monomers to cells induces the same effects as C2 or iota toxin (Kiefer *et al.,* 1996; Reuner *et al.,* 1996). This results in a complete disassembly of actin filaments and accumulation of actin in monomeric form (reviewed in Aktories and Wegner, 1992; Aktories, 1994; Aktories and Koch, 1995; Aktories, 2000).

3.1.1.3. Cellular Effects

The most prominent effects of Clostridial ADP-ribosylating toxins are the depolymerization of the actin filaments and the increase of the pool of G-actin. The intermediate filaments are also disorganized, whereas the microtubules are not affected. The cells are rounded, lose the staining of actin filaments, detach from the support, and die. Cells treated with C2 stop proliferating but remain viable for about two days. C2 toxin causes a delay of the G2/M transition in Hela cells by decreasing the activation of cdc25-C phosphatase and the subsequent activation of $p34^{cdc2}$ protein kinase, which is required for cell cycle progression. The link between the C2 toxin-dependent breakdown of actin cytoskeleton and the enzymatic mechanism of mitosis control is unsolved (Barth *et al.,* 1999).

The main effects induced by Clostridial ADP-ribosylating toxins include: change in morphology (rounding), inhibition of migration and activation of leucocytes (Vershueren *et al.,* 1995), inhibition of smooth muscle contraction, impairment of endocytosis, exocytosis, and cytokinesis. Contraction of ileum muscle, which is rich in β-actin, is inhibited by C2 toxin. Since this toxin specifically modifies monomeric actin, a dynamic equilibrium between monomeric and polymerized actin is required for smooth muscle contraction in this type of cell (Mauss *et al.,* 1989).

The depolymerization of actin cytoskeleton by iota toxin induces a disorganization of tight and basolateral intercellular junctions and a subsequent increase in permeability of cultured intestinal cell monolayers (Richard *et al.,* 2002). In rabbit lungs, C2 toxin causes edema and alteration of the endothelial cell membranes (Ermert *et al.,* 1997). Massive edema and hypotonic shock observed in animals injected with C2 toxins probably result from the breakdown of the vascular endothelium barrier (Aktories and Koch, 1995).

The binary toxin-producing *Clostridium* species are involved in necrotizing enteritis and diarrhea in animals and occasionally in humans. *C. perfringens* E causes enterotoxemia in calves and other young animals. *C. spiroforme* is responsible for enteritis and death in rabbits and rarely in humans. *C. botulinum* C2 toxin induces intestinal hemorrhagic lesions in avians.

3.1.2 Rho-GTPase ADP-Ribosylating Enzyme (C3 Enzyme)

In addition to C2 toxin, *C. botulinum* C and D produce another ADP-ribosyltransferase that is termed C3 enzyme. C3-related enzymes are also produced by *Clostridium limosum, Bacillus cereus, Bacillus thuringiensis* and *Staphylococcus aureus* (epidermal cell differentiation inhibitor, EDIN), and are specific to Rho proteins.

3.1.2.1. Structure

C3 enzyme (28 kDa) only possesses a catalytic domain and lacks the binding and translocation domains permitting the entry of toxins into cells. The crystal structure shows that C3 consists of a core structure of five antiparallel β-strands packed against a three-stranded antiparallel β-sheet, flanked by four consecutive α-helices (Han *et al.*, 2001). Interestingly, the C3 structure is similar to that of the catalytic domain of the enzymatic components VIP2 and Ia (Han *et al.*, 1999; Tsuge *et al.*, 2003).

Although no overall sequence homology has been found with the other ADP-ribosylating toxins, C3 retains the conserved NAD binding site and catalytic pocket that consists of an α-helix (α3 in C3) bent over a β-sheet (Domenighini and Rappuoli, 1996). The two perpendicular antiparallel β-sheets form a central cleft. The amino acids, which have an essential role in the ADP-ribosylation mechanism, are conserved: a hydrophobic segment containing the motif Ser-Thr-Ser flanked by an Arg (Arg88) about 50 residues upstream and a Glu (Glu174) about 40 residues downstream from this motif (Han *et al.*, 2001). The hydrophobic pocket is involved in the binding of the nicotinamide ring, Gly89 interacts with the NAD carboxamide group and Arg88 and Arg146 bind to the NAD phosphate groups (Han *et al.*, 2001). Glu174 has the pivotal role in transfer of the adenine diphosphate ribose group from NAD to the Asn41 of Rho. Mutational and photo-affinity analyses have confirmed the importance of Glu174 (Jung *et al.*, 1993; Just *et al.*, 1995a; Saito *et al.*, 1995; Böhmer *et al.*, 1996).

The molecular basis of substrate recognition by C3 exoenzymes is still poorly understood. The N-terminal half of RhoA, and mainly Arg5, Lys6, Val43, Glu47 and Glu54 are involved in the interaction with C3. The exchange of these residues in Rac1 renders this GTPase susceptible to C3 (Wilde *et al.*, 2000). Unlike the other ADP-ribosyltransferases, C3 does not contain an active site loop, but possesses two adjacent protruding turns (ADP-ribosylating toxin turn-turn motif, ARTT) involved in the specific recognition of the substrate. Two ARTT motifs of C3 encompassing Ser207-Ala210 (turn 1) and Gly211-Glu214 (turn 2) have been shown to mediate the specific recognition of RhoA. Phe209 in turn 1, which is the only solvent exposed hydrophobic residue near the active site, interacts with the hydrophobic patches adjacent to Asn41 of RhoA, and Gln212 in turn 2 probably contacts Asn41 (Han *et al.*, 2001). Gln212 is conserved in all C3 enzymes and EDIN, whereas binary toxins that specifically recognize actin, contain a conserved Glu at this position (Han *et al.*, 1999; Han *et al.*, 2001). The ARTT loop adopts a conformational change upon NAD binding, permitting the transition from a buried to a solvent-exposed position for Gln212. This indicates that C3 acts in three steps, NAD binding, Rho binding, and Rho ADP-ribosylation (Ménétrey *et al.*, 2002).

3.1.2.2. Mode of Action

C3 enzyme ADP-ribosylates specifically Rho protein, which is a G-protein from the Ras superfamily. These proteins are active in the GTP-bound form and inactive in the GDP-bound form. The transition from GDP- (inactive) to GTP-bound form (active) is stimulated by an exchange factor (GEF) in response to an external signal transduced by a membrane receptor. These proteins undergo a conformational change in two regions called switch I and switch II, when they are in GTP- or in GDP-bound form. Switch I interacts with the downstream effectors and switch II is involved in the GTPase activity that is stimulated by a GTPase-activating protein (GAP). GTPases act as molecular switches between a membrane receptor activated by an external factor and a downstream effector. Ras proteins are involved in the control of proliferation-differentiation, and Rho proteins (Rho, Rac, Cdc42) regulate the actin cytoskeleton. In contrast to Ras, which is permanently bound to membrane, Rho proteins bound to GDP are in the cytosol in association with a protein called guanine dissociation inhibitor (GDI), and they translocate to the membrane after dissociation of the Rho-GDI complex, possibly by ERM (ezrin/radixin/moesin) proteins. At the membrane, Rho is activated by GEFs and interacts with its effectors.

C3 ADP-ribosylates Rho at Asn41 which is localized on an extended stretch close to the switch I. Rho-GDP is a preferred substrate, and Rho associated with GDI is not ADP-ribosylated by C3, since Asn41 is masked in the Rho-GDI complex (Bourmeyster *et al.*, 1992). ADP-ribosylation of Rho-Asn41 by C3 does not impair the exchange of GDP with GTP, the intrinsic and GAP-stimulated GTPase activity, or the interaction with its effectors (Ren *et al.,* 1996; Sehr *et al.,* 1998). ADP-ribosylation of Rho by C3 increases the stability of the Rho-GDI complexes preventing its translocation to the membrane, its subsequent activation by GEFs, and interaction with its effectors (Fujihara *et al.,* 1997; Genth *et al.,* 2003). Thus C3 traps Rho in a permanent inactive form in the cytosol.

3.1.2.3. Cellular Effects

C3 has been used in many studies to investigate the functions of Rho. Although, C3 ADP-ribosylates the three isoforms RhoA, B and C, most of the cellular effects described with this enzyme are related to RhoA. The first evidence that Rho is involved in actin cytoskeleton organization comes from the initial study of C3 on Vero cells in which the effects are characterized by a cell rounding up and destruction of actin filaments (Chardin *et al.,* 1989). Phalloidin staining shows the disappearance of actin stress fibers whereas the cortical actin is only partially disorganized. Patches of aggregated short filaments are still visible, in contrast to the actin ADP-ribosylating toxins, which completely depolymerize the actin cytoskeleton. Experiments with constitutively active RhoA demonstrate that this GTPase stimulates stress fiber formation and focal adhesion complex assembly (Ridley and Hall, 1992). C3 exoenzyme inhibits the Rho-dependent actin organization induced by growth factors, lysophosphatidic acid (LPA), and thrombin in various cell lines. Upon stimulation by LPA, Rho activates protein kinase N (PKN), whose substrate is still unclear. A well-documented Rho effector is Rho-kinase (RhoK or Rock), which phosphorylates myosin light chains and inhibits myosin phosphatase. RhoK is involved in smooth muscle and endothelial cell contraction, formation of stress fibers and focal adhesion, aggregation of platelets, neurite retraction, tumor cell invasion, and growth factor-induced cell mobility (reviewed in Fukata *et al.,* 2001). Other effectors of Rho include citron-kinase involved in cytokinesis, PI4P5-kinase, p140mDia, involved in actin polymerization, rhophilin, and rhotekin the functions of which are still unknown. All the cellular Rho effects can be blocked by C3.

C3 induces a disorganization of the actin stress fibers, cell morphology change, loss of intercellular junctions (mainly tight junctions), impairment of endocytosis, exocytosis, cytokinesis, and neuronal plasticity, inhibition of cell cycle progression and movement, and induction of apoptosis (reviewed in Aktories *et al.,* 2000).

The effects of C3 in natural disease are not known. However, this enzyme represents a valuable tool to dissect cell molecular mechanisms that are dependent on actin regulation and Rho protein.

3.1.3 Rho-GTPase Glucosylating Toxins (Large Clostridial Toxins)

Large clostridial toxins are 250-300 kDa proteins encompassing *Clostridium difficile* toxins A and B (ToxA and ToxB), *Clostridium sordellii* lethal toxin (LT), hemorrhagic toxin (HT), and *Clostridium novyi* α-toxin (α-novyi). ToxB and LT are closely related (76% amino acid sequence identity), but are more distantly related to ToxA and α-novyi (48-60% identity).

3.1.3.1. Structure

These toxins are single protein chains containing three functional domains. The C-terminal part one third contains multiple repeated sequences and it is involved in the recognition of a cell surface receptor. A trisaccharide (Gal-α1-3Gal-β1-4GlcNac) has been found to be the motif recognized by ToxA. The cell receptor in rabbit ileal brush border has been identified as the glycoprotein sucrase-isomaltase. However, this receptor is not expressed in human colon. The receptors for other large clostridial toxins have not been characterized. The central part contains a hydrophobic segment and probably mediates the translocation of the toxin across the membrane. The enzymatic site characterized by the DxD motif surrounded by a hydrophobic region, and the substrate recognition domain are localized within the 548 N-terminal residues (Hofmann *et al.,* 1997; 1998). The aspartate residues seem to be involved in the coordination of the divalent cation (mainly Mn^{++}) which increases the hydrolase activity and/or the binding of UDP-glucose (reviewed in Just *et al.,* 2000). The region from amino acid 133 to 517 is implicated in the nucleotide-sugar specificity (Busch *et al.,* 2000b), and a conserved hydrophobic residue (Trp102) plays a critical role in the binding of the cosubstrate UDP-glucose (Busch *et al.,* 2000a). Chimeric molecules between ToxB and LT were used to identify the sites of substrate recognition. Amino acids 408 to 468 of ToxB ensure the specificity for Rho, Rac and Cdc42, whereas in LT, the recognition of Rac and Cdc42 is mediated by residues 364 to 408, and that of Ras proteins by residues 408 to 516 (Hofmann *et al.,* 1998). Amino acids 22-27 of Rho and Ras GTPases which are part of the transition of the α1-helix to the switch I region, constitute the major domain recognized by the glucosylating toxins (Müller *et al.,* 1999).

3.1.3.2. Mode of Action

Large clostridial toxins enter cells by receptor-mediated endocytosis. The cytotoxic effects are blocked by endosomal and lysosomal acidification inhibitors (monensin, bafilomycin A1, ammonium chloride) and the inhibiting effects can be by-passed by an extracellular acidic pulse (Fiorentini and Thelestam, 1991; Popoff *et al.,* 1996; Qa'dan *et al.,* 2000; Barth *et al.,* 2001; Qa'dan *et al.*, 2001). This indicates that the large clostridial toxins translocate from early endosomes upon an acidification step. At low pH, ToxB and LT induce channel formation in cell membranes and artificial lipid bilayers, and show an increase of hydrophobicity as determined with fluorescence methods (Qa'dan *et al.*, 2000; 2001). This probably involves a conformational change and insertion of the toxin into the membrane, possibly mediated by the hydrophobic segment of the central domain. Only the proteolytically cleaved N-terminal fragment is translocated into the cytosol (Rupnik *et al.*, 2003).

Large clostridial toxins catalyze the glucosylation of 21 kDa G-proteins from UDP-glucose, except α–novyi which uses UDP-*N*-acetylglucosamine as cosubstrate. ToxA and ToxB glucosylate Rho, Rac and Cdc42 at Thr-37, whereas LT glucosylates Ras at Thr-35, Rap, Ral and Rac at Thr-37 (Table 1). The large clostridial toxins cleave the cosubstrate and transfer the glucose moiety to the acceptor amino acid of the Rho proteins (Just *et al.*, 1995b; Just *et al.*, 1995c; Popoff *et al.*, 1996). The conserved Thr, which is glucosylated, is located in switch I. Thr37/35 is involved in the coordination of Mg^{++} and subsequently to the binding of the β and γ phosphates of GTP. The hydroxyl group of Thr37/35 is exposed to the surface of the molecule in its GDP-bound form, which is the only accessible substrate of glucosylating toxins. The nucleotide binding of glucosylated Ras by LT is not grossly altered, but the GEF activation of GDP forms is decreased (Hermann *et al.*, 1998). Glucosylation of Thr35 completely prevents the recognition of the downstream effector, keeping the G-protein in an inactive form (Hermann *et al.*, 1998). The crystal structure of Ras modified by LT shows that glucosylation prevents the formation of the GTP binding conformation of the Ras effector loop, which is required for the interaction with the effector Raf (Vetter *et al.*, 2000). Similar results were found with RhoA when it is glucosylated by ToxB (Sehr *et al.*, 1998). In addition, glucosylation of the GTPase slightly reduces the intrinsic GTPase activity, completely inhibits GAP-stimulated GTP hydrolysis (Hermann *et al.*, 1998), and leads to accumulation of the GTP-bound form of Rho in the membrane where it is tightly bound (Genth *et al.*, 1999).

3.1.3.3. Cellular Effects

Large clostridial toxins, by modifying Rho proteins, induce cell rounding, with loss of actin stress fibers, reorganization of the cortical actin, and disruption of the intercellular junctions. ToxA and ToxB disrupt apical and basal actin filaments and subsequently disorganize the ultrastructure and component distribution (ZO-1, ZO-2, occludin, claudin) of tight junctions, whereas E-cadherin junctions show little alteration (Nusrat *et al.*, 2001; Chen *et al.*, 2002). In contrast, LT, which only modifies Rac among the Rho proteins, alters the permeability of intestinal cell monolayers causing a redistribution of E-cadherin whereas ZO-1 labeling is not significantly affected (Richard *et al.*, 1999).

ToxA and ToxB have been reported to induce apoptosis as a consequence of Rho glucosylation. In addition to the effects on the cytoskeleton, the inactivation of Rho proteins impairs many other cellular functions such as endocytosis, exocytosis, lymphocyte activation, immunoglobulin-mediated phagocytosis in macrophages, NADPH oxidase regulation, smooth muscle contraction, phospholipase D activation, and transcriptional activation mediated by JNK, and/or p38 (Just *et al.*, 2000).

ToxA produces a severe inflammatory response in mammalian intestine characterized by an epithelial cell necrosis and massive infiltration with inflammatory cells (Pothoulakis and Lamont, 2001). In monocyte cells, ToxA stimulates cytokine (TNF-α, IL-1β, IL-6 and IL-8) release and activation of p38 MAP kinase, whereas the activation of ERK and JNK is only transient (Warny *et al.*, 2000; Pothoulakis and Lamont, 2001). p38 activation is required for IL-8 production, IL-1b release, monocyte necrosis and intestinal mucosa inflammation. ToxA-induced p38 activation could be mediated by toxin binding to a membrane receptor independently of Rho-GTPase glucosylation (Warny *et al.*, 2000). Other Rho-independent cellular effects induced by ToxA include the activation of NF-kB, and subsequent release of IL-8 and possibly other inflammatory cytokines (Wershil *et al.*, 1998), mitochondrial damage, apoptosis (He *et al.*, 2000), and activation of a neuroimmune pathway (reviewed in Pothoulakis and Lamont, 2001).

Unlike ToxA and ToxB, LT modifies Ras and blocks the MAP-kinase cascade and phospholipase D regulation. However, implication of this cellular pathway blockage in cytotoxicity has not been demonstrated (Schmidt *et al.*, 1998; Ben El Hadj *et al.*, 1999). PLD inhibition by these toxins is restored by RalA (Schmidt *et al.*, 1998). It has been shown that RalA and ARF directly interact with PLD1, but the role of RalA in PLD activity is still unknown (Luo *et al.*, 1997; Kim *et al.*, 1998).

C. difficile is the etiological agent of pseudomembranous colitis and about 30% of the postantibiotic diarrhea. ToxA which experimentally induces necrotic and hemorrhagic intestinal lesions seems to be the main virulence factor in natural disease. In addition, ToxB and ToxA could participate in the recruitment of inflammatory cells, which are abundant in the lesions. *C. sordellii* and *C. novyi* are involved in gangrene, and *C. sordellii* is also an agent of hemorrhagic enteritis and enterotoxemia in cattle.

3.2 Toxins Blocking Neurosecretion (Clostridial Neurotoxins)

C. botulinum and *C. tetani* secrete very potent neurotoxins, which are responsible for the neurological disorders, botulism and tetanus, in humans and animals. Several recent reviews detail the mode of action of neurotoxins (Poulain *et al.*, 1997; Herreros *et al.*, 1999; Bigalke and Shoer, 2000; Humeau *et al.*, 2000; Schiavo *et al.*, 2000; Meunier, 2002a; Meunier *et al.*, 2002b).

C. tetani forms a homogeneous bacterial species which produces only one type of TeTx, whereas botulinum neurotoxin (BoNT)-producing strains are heterogeneous. *C. botulinum* is divided into 4 groups, which on the basis of phenotypic and genotypic parameters, correspond to different species. In addition, some strains of other species, such as *C. butyricum* and *C. baratii*, can produce a related BoNT type E and F respectively. Seven toxinotypes of BoNT (A, B, C1, D, E, F and G) are distinguished according to their antigenic properties.

BoNTs are associated with non-toxic proteins (ANTPs) to form large complexes. ANTPs encompass a non-toxic and non-hemagglutinin component (NTNH) and several hemagglutinin components (HA34, HA17 and HA70 in *C. botulinum* A) (reviewed in Popoff and Marvaud, 1999). The function of ANTPs is still unclear; they could protect the neurotoxin from the acidic pH of the stomach and from digestive proteases.

3.2.1. Structure

BoNTs and TeTx share a common structure. They are synthesized as a precursor protein (about 150 kDa), which is inactive or weakly active. The precursor which does not contain signal peptide, is released from the bacteria possibly by a cell-wall exfoliation mechanism (Call *et al.*, 1995). The precursor is proteolytically activated in the extra-bacterial medium either by *Clostridium* proteases or by exogenous proteases such as digestive proteases in the intestinal tract. The active neurotoxin consists of a light chain (L, about 50 kDa) and a heavy chain (H, about 100 kDa), which remain linked by a disulfide bridge. The structure of BoNT shows three distinct domains: L-chain containing α-helices and β-strands including the catalytic zinc binding motif, the N-terminal part of the H-chain forming two unusually long and twisted α-helices, and the C-terminal part of the H-chain consisting of two distinct subdomains involved in the recognition of the receptor (Umland *et al.*, 1997; Lacy *et al.*, 1998; Lacy and Stevens, 1999; Emsley *et al.*, 2000).

3.2.2. Mode of Action

Although BoNTs and TeTx have different routes of entry, they display a similar intracellular mechanism of action. BoNTs enter by the oral route and transcytosis across the digestive mucosa (Maksymowych *et al.*, 1999). After diffusion into the extracellular fluid and blood stream, BoNTs target motoneuron endings. In contrast, TeTx is formed in wounds colonized by *C. tetani*. TeTx diffuses in the extracellular fluid and can be internalized in all types of nerve endings (sensory, adrenergic and motoneurons).

Each type of BoNT and TeTx recognize specific receptors on demyelinated terminal nerve endings. The nature of the receptors is still controversial. They probably consist of two parts, a ganglioside of the G1b series (Gd1b, GT1b) an associated protein. Experimental data suggests that synaptotagmin isoforms, which are transmembrane proteins from the synaptic vesicles, are the protein receptors of BoNTs. It has been found that synaptotagmin II associated with ganglioside G_{T1b} is involved in the binding of BoNT/B (Nishiki *et al.*, 1994; 1996). GT1b is required in neuroblastoma expressing synaptotagmin I for binding and subsequent activity of BoNT/A. However, synaptotagmin I in the absence of ganglioside does not support BoNT/A activity (Yowler *et al.*, 2002). Polysialoganglioside-binding sites have been characterized in the C domain of the heavy chain (Umland *et al.*, 1997; Lacy *et al.*, 1998). TeTx exhibits two carbohydrate-binding sites, whereas BoNT/A and B show only one (Rummel *et al.*, 2003). The protein part of the TeTx receptor has been identified as a glycosylphos phatidylinositol-anchored protein of 15 kDa (Herreros *et al.*, 2000; 2001; Munro *et al.*, 2001). The localization of TeTx receptor on lipid rafts probably favors the mobility of the toxin bound to its receptor and its subsequent oligomerization and cell entry. The high affinity of BoNTs and TeTx for presynaptic membranes probably results from multiple interactions with the ganglioside and protein parts of the receptor.

Neurotoxin bound to its receptor is internalized by receptor-mediated endocytosis. An essential difference between both types of neurotoxins is that BoNTs are directly endocytosed in clathrin-coated vesicles, which, when acidified, trigger the translocation of the L chain into the cytosol. Therefore, BoNT L chain is delivered in the peripheral nervous system, to the neuromuscular junctions where it blocks the release of acetylcholine leading to a flaccid paralysis. In contrast, TeTx enters different endocytic vesicles, which are not acidified. The vesicles retrogradely transport the toxin, in a microtubule-dependent manner, to the cell body of neurons in the spinal cord. Like nerve growth factors, TeTx is transported by tubulo-vesicular organelles characterized by the presence of the neurotoxin receptor $p75^{NTR}$ (Lalli and Schiavo, 2002). The C-terminal fragment of TeTx drives the retrograde transport of the toxin, and can be used to transport heterologous protein in the same way (Li *et al.*, 2001; Maslos *et al.*, 2002). Then, TeTx carries out a transynaptic migration and reaches the target neurons, which are the inhibitory interneurons involved in the regulation of the motoneurons. TeTx enters target inhibitory interneurons via vesicles that are acidified permitting the delivery of L chain into the cytosol, where it inhibits the regulated release of glycine and GABA. TeTx seems to enter synaptic vesicles directly. Acidification of the vesicle lumen triggers a conformational change of the neurotoxin, which is required for cytosolic translocation of the L chain. H chains form tetramers and insert into lipid membranes, thus forming cation selective channels permeable to small molecules (<700 Da). The mechanism of translocation is not completely understood. Neurotoxins probably use a mechanism similar to that of diphtheria toxin. In this model, the H chain forms a hydrophylic cleft. The hydrophobic residues of H and partially unfolded L chains face the lipids and the hydrophilic segment of L chain glides upon one of the H chains. Then, the L chain refolds within the neutral pH of the cytosol. Protein chaperones are possibly involved in this mechanism, as it has been found for diphtheria toxin (Humeau *et al.*, 2000; Schiavo *et al.*, 2000; Meunier *et al.*, 2002b; Ratts *et al.*, 2003).

The L chains of all clostridial neurotoxins contain a zinc-binding motif (HExxH) characteristic of zinc-endopeptidases. They cleave one of the three members of the SNARE (soluble NSF attachment protein receptor) proteins. TeTx and BoNT/B, D, F and G cut synaptobrevin (or VAMP for vesicular associated membrane protein), BoNT/A and E cut SNAP25 (25kDa synaptosome associated protein), and BoNT/C1 cuts both SNAP25 and syntaxin. The cleavage sites are different for each neurotoxin except BoNT/B and TeTx, which cut synaptobrevin at the same site. Thus, TeTx and BoNT/B share the same molecular mechanism. However, they induce opposite symptoms. This clearly indicates that the different clinical signs result from the site of intoxication and not from a different mechanism of action. Synaptobrevin is a transmembrane protein anchored in the synaptic vesicle membrane. Palmytoilation of SNAP25 in the middle of the protein mediates its binding to the presynaptic membrane. Syntaxin is essentially composed of α-helices and associates with SNAP25 and VAMP to form the SNARE complex, in which the cytosolic domains of the three proteins form a highly twisted four-helix bundle, that lies parallel to the membrane surface (Poirier *et al.*, 1998; Sutton *et al.*, 1998). Syntaxin also interacts in a calcium-dependent manner with synaptotagmin which is probably the Ca^{++} sensor triggering neurotransmitter release. The SNARE complex is very stable to various denaturing agents, and recruits soluble cytosolic proteins such as NSF (*N*-ethymaleimide-sensitive factor) and SNAPs (soluble NSF accessory proteins) forming the 20S SNARE complex, which has been recognized as essential in vesicle targeting and fusion (Söllner *et al.*, 1993a; Söllner *et al.*, 1993b; Hayashi *et al.*, 1995). The 20S SNARE complex is rapidly disassembled by NSF-dependent hydrolysis of ATP (Barnard *et al.*, 1997). Thus cycling assembly of SNARE proteins to regenerate the SNARE complex and disassembly is essential in the exocytosis process. Synaptic vesicles containing the neurotransmitter bind to the presynaptic membrane at active zones. In the resting state, VAMP and SNAP25 are free and syntaxin is associated with a chaperone protein (n-sec1). In the primed state induced by Mg-ATP and temperature, which renders the synaptic vesicles available for exocytosis, SNARE proteins partially and reversibly assemble. Syntaxin bound to a calcium sensor protein (synaptotagmin) does not strongly associate with VAMP and SNAP25. Upon calcium entry, the calcium sensor detaches from syntaxin allowing the SNARE complex to form. The complex joins the synaptic vesicle and presynaptic membranes leading to membrane fusion (Chen *et al.*, 2001). Clostridial neurotoxins only cleave SNARE proteins when disassembled, but not in complex form. When an individual SNARE protein is cleaved by a clostridial neurotoxin, SNARE complex formation is not inhibited but its stability is reduced and the release of neurotransmitter is blocked (Hayashi *et al.*, 1995; Pellegrini *et al.*, 1995). Even the physiological effects induced by the cleavage of either VAMP, SNAP25 or syntaxin are not equivalent at the neuromuscular junctions. All the clostridial neurotoxins cause blockage of regulated neurotransmission, which varies in intensity and duration according to each neurotoxin type. BoNTs act at the neuromuscular junctions by preventing the regulated release of acetylcholine, leading to flaccid paralysis. The blockage of glycine or GABA liberation in the inhibitory interneurons by TeTx disrupts the balance between stimulation from the sensitive nerve and from the motoneuron, thus leading to a permanent excitation of the muscles (spastic paralysis) (Humeau *et al.*, 2000; Meunier *et al.*, 2002a).

The remodeling of neuromuscular junction after clostridial neurotoxin treatment has been mainly studied with BoNT/A. After neuromuscular junction paralysis induced by BoNT/A, outgrowth of intramuscular axons spreads out from the node of Ranvier (nodal sprouting) and at motor nerve terminals (terminal sprouting). Sprouts are observed within 24 hours of muscle inactivity and increases in number and complexity for about 50 days after BoNT/A treatment (Juzans *et al.*, 1996). Newly formed sprouts are able to cause spontaneous and regulated release of acetyl-choline and permit the recovery of neuromuscular transmission after a sublethal BoNT/A intoxication. Sprouts persist even when muscle recovers normal activity, but they are eliminated when the originally intoxicated motor nerve terminal becomes functional (de Paiva *et al.*, 1999). Muscle derived signaling factors such as growth factors IGF-1 and IGF-2 are possibly involved in the induction of sprouting (reviewed in Meunier *et al.,* 2002b). The negative control of sprouting by the original endplate has not been characterized, but neural-cell adhesion molecule (N-CAM) and tenascin-C are possible regulators (reviewed in Meunier *et al.,* 2002a).

The duration of intoxication depends on the BoNT types. Indeed, the half-lives of exocytosis blockage in rat cerebellum neurons is more than 31 days for BoNT/A, more than 25 days for BoNT/C1, about 10 days for BoNT/B, about 2 days for BoNT/F and less than one day for BoNT/E (Foran *et al.,* 2003). The intoxication process is much longer with BoNT/A than with BoNT/E, even though both toxins recognize the same intracellular target, SNAP25. Removal of the 9 C-terminal residues from SNAP25 by BoNT/A does not prevent the formation of SNARE complex, but its stability is reduced. In contrast, the removal of 17-C-terminal residues from SNAP25 by BoNT/E totally inhibits the SNARE complex assembly (Hayashi *et al.,* 1995; Eleopra *et al.,* 1998; Raciborska and Charlton, 1999). A likely hypothesis is that the most BoNT/E-modified form of SNAP25 is rapidly degraded and replaced by newly synthesized SNAP25 allowing rapid recovery, whereas the limited truncated SNAP25 by BoNT/A is more slowly recognized by the cell reparation mechanism. However, it was recently shown that the long lasting effect of BoNT/A, BoNT/B and BoNT/C1 is due to the persistence of the toxin within the neurons, whereas the short paralysis induced by BoNT/E and BoNT/F correlates with the turnover of newly synthesized SNAP25 and VAMP, respectively (Foran *et al.,* 2003).

3.3.3. Therapeutic Use

The paralytic properties of BoNTs are used in the treatment of several human diseases characterized by the hyperfunction of cholinergic terminals such as blepharospasm, spasmodic torticolli, hemifacial paralysis and many other dystonias. Local injection of BoNT into a hyperactive muscle causes paralysis of the affected muscle for a few months. BoNT/A is the most widely used since this toxinotype induces paralysis for a long period. Therefore, the most potent toxin is also an efficient drug (Jankovic and Brin, 1991; 1997).

4. Concluding Remarks

Clostridium is the bacterial genus that produces the highest number and the most potent toxins. The large and diverse group of toxins, their structure, their modes of action, and the interactions with their targets are most intriguing. What advantage is served by the seemingly accidental interactions with host organisms, which are not obligately required for survival? Why do some clostridial toxins recognize such specific target molecules in evolved organisms, such as crucial proteins required for neuroexocytosis? Whether or not these toxins represent extremely frightening threats, they also constitute very useful tools in cell biology, and powerful therapeutic agents such as BoNTs, or potential useful vectors to transport active molecules in target cells.

References

Abrami, L., Liu, S., Cosson, P., Leppla, S.H., and van der Goot, F.G. 2003. Anthrax toxin triggers endocytosis of its receptor via a lipid raft-mediated clathrin-dependent process. J. Cell Biol. 160: 321-328.

Aktories, K. 1994. Clostridial ADP-ribosylating toxins; effects on ATP and GTP-binding proteins. Mol. Cell Biochem. 138: 167-176.

Aktories, K. 2000. Bacterial protein toxins as tools in cell Biology and pharmacology. In: Cellular Microbiology, P. Cossart, P. Boquet, S. Normark, and R. Rappuoli, eds. ASM Press, Washington, D.C. p. 221-237.

Aktories, K., Bärmann, M., Ohishi, I., Tsuyama, S., Jakobs, K.H., and Habermann, E. 1986. Botulinum C2 toxin ADP-ribosylates actin. Nature. 322: 390-392.

Aktories, K., Barth, H., and Just, I. 2000. *Clostridium botulinum* C3 exoenzyme and C3-like transferases. In: Bacterial Protein Toxins, K. Aktories, and I. Just, eds. Springer, Berlin. p. 207-233.

Aktories, K., and Koch, G. 1995. Modification of actin and Rho proteins by Clostridial ADP-ribosylating toxins. In: Bacterial Toxins and Virulence Factors in Disease, J. Moss, B. Iglewski, M. Vaughan, and A. T. Tu, eds. Marcel Dekker, New York, New York. p. 491-520.

Aktories, K., and Wegner, A. 1992. Mechanisms of the cytopathic action of actin-ADP-ribosylating toxins. Mol. Microbiol. 6: 2905-2908.

Alouf, J. 1997. Cholesterol binding toxins (*Streptococcus, Bacillus, Clostridium, Listeria*). In: Guidebook to Protein Toxins and their Use in Cell Biology, R. Rappuoli, and C. Montecucco, eds. Sambrook & Tooze Publications, Oxford. p. 7-10.

Alouf, J.E. 2003. Molecular features of the cytolytic pore forming bacterial protein toxins. Fol. Microbiol. 48: 5-16.

Awad, M.M., Bryant, A.E., Stevens, D.L., and Rood, J.L. 1995. Virulence studies on chromosomal α-toxin and θ-toxin mutants constructed by allelic exchange provide genetic evidence for the essential role of α-toxin in *Clostridium perfringens*-mediated gas gangrene. Mol. Microbiol. 15: 191-202.

Awad, M.M., Ellenor, D.M., Bod, R.L., Emmins, J.J., and Rood, J.I. 2001. Synergistic effects of alpha-toxin and perfringolysin O in *Clostridium perfringens*-mediated gas gangrene. Infect. Immun. 69: 7904-7910.

Bachmeyer, C., Benz, R., Barth, H., Aktories, K., Gibert, M., and Popoff, M.R. 2001. Interaction of *Clostridium botulinum* C2 toxin with lipid bilayer membranes and Vero cells: inhibition of channel function in chloroquine and related compounds *in vitro* and toxin action *in vivo*. FASEB J. 15:1658-1660.

Ballard, J., Sokolov, Y., Yuan, W.L., Kagan, B.L., and Tweten, R.K. 1993. Activation and mechanism of *Clostridium septicum* alpha toxin. Mol. MIcrobiol. 10: 627-634.

Barnard, R.J.O., Morgan, A., and Burgoyne, R.D. 1997. Stimulation of NSF ATPase activity by alpha-SNAP is required for SNARE complex disassembly and exocytosis. J. Cell Biol. 139: 875-883.

Barth, H., Blöcker, D., Behlke, J., Bergsma-Schutter, W., Brisson, A., Benz, R., and Aktories, K. 2000. Cellular uptake of *Clostridium botulinum* C2 toxin requires oligomerization and acidification. J. Biol. Chem. 275: 18704-18711.

Barth, H., Hofmann, F., Olenik, C., Just, I., and Aktories, K. 1998. The N-terminal part of the enzyme component (C2I) of the binary *Clostridium botulinum* C2 toxin interacts with the binding component C2II and functions as a carrier system for a Rho ADP-ribosylating C3-like fusion toxin. Infect. Immun. 66: 1364-1369.

Barth, H., Klinger, M., Aktories, K., and Kinzel, V. 1999. *Clostridium botulinum* C2 toxin delays entry into mitosis and activation of p34^{cdc2} kinase and cdc25-C phosphatase in Hela cells. Infect. Immun. 67: 5083-5090.

Barth, H., Pfeifer, G., Hofmann, F., Maier, E., Benz, R., and Aktories, K. 2001. Low pH-induced formation of ion channels by *Clostridium difficile* toxin B in target cells. J. Biol. Chem. 276: 10670-10676.

Barth, H., Roebling, R., Fritz, M., and Aktories, K. 2002. The binary *Clostridium botulinum* C2 toxin as a protein delivery system. J. Biol. Chem. 277: 5074-5081.

Bayley, H. 1997. Toxin structure: part of a hole. Curr. Biol. 7: R763-R767.

Ben El Hadj, N., Popoff, M.R., Marvaud, J.C., Payrastre, B., Boquet, P., and Geny, B. 1999. G-protein-stimulated phospholipase D activity is inhibited by lethal toxin from *Clostridium sordellii* in HL-60 cells. J. Biol. Chem. 274: 14021-14031.

Bigalke, H., and Shoer, L.F. 2000. Clostridial neurotoxins. In: Bacterial Protein Toxins, K. Aktories, and I. Just, eds. Springer, Berlin. p. 407-443.

Blöcker, D., Barth, H., Maier, E., Benz, R., Barbieri, J.T., and Aktories, K. 2000. The C terminus of component C2II of *Clostridium botulinum* C2 toxin is essential for receptor binding. Infect. Immun. 68: 4566-4573.

Böhmer, J., Jung, M., Sehr, P., Fritz, G., Popoff, M.R., Just, I., and Aktories, K. 1996. Active site mutation of the C3-like ADP-ribosyltransferase from *Clostridium limosum* - Analysis of glutamic acid 174. Biochemistry. 35: 282-289.

Borrmann, E., Günther, H., and Köhler, H. 2001. Effect of *Clostridium perfringens* epsilon toxin on MDCK cells. FEMS Immunol. Med. Microbiol. 31: 85-92.

Bourmeyster, N., Strasia, M.J., Garin, J., Gagnon, J., Boquet, P., and Vignais, P. 1992. Copurification of Rho protein and the Rho GDP dissociation inhibitor from bovine neutrophil cytosol. Effects of phosphoinositides on Rho ADP-ribosylation by the C3 exoenzyme of *Clostridium botulinum*. Biochemistry. 31: 12863-12869.

Brüggemann, H., Bäumer, S., Fricke, W.F., Wiezr, A., Liesagang, H., Decker, I., Herzberg, C., Martinez-Arias, R., Henne, A., and Gottschalk, G. 2003. The genome sequence of *Clostridium tetani*, the causative agent of tetanus disease. Proc. Natl. Acad. Sci. USA. 100: 1316-1321.

Bryant, A.E., Bayer, C.R., Hayes-Schroer, S.M., and Stevens, D. 2003. Activation of platelet gpIIIa by phospholipase C from *Clostridium perfringens* involves store-operated calcium entry. J. Infect. Dis.187: 408-417.

Bryant, A.E., Bergstrom, R., Zimmerman, G.A., Salyer, J.L., Hill, H.R., Tweten, R.K., Sato, H., and Stevens, D.L. 1993. *Clostridium perfringens* invasiveness is enhanced by effects of theta toxin upon PMNL structure and function: The roles of leukocytotoxicity and expression of CD11/CD18 adherence glycoprotein. FEMS Immunol. Med. Microbiol. 7: 321-326.

Bryant, A.E., Chen, R.Y.Z., Nagata, Y., Wang, Y., Lee, C.H., Finegold, S., Guth, P.H., and Stevens, D.L. 2000a. Clostridial gas gangrene. I. Cellular and molecular mechanisms of microvascular dysfunction induced by exotoxins of *Clostridium perfringens*. J. Infect. Dis. 182: 799-807.

Bryant, A.E., Chen, R.Y.Z., Nagata, Y., Wang, Y., Lee, C.H., Finegold, S., Guth, P.H., and Stevens, D.L. 2000b. Clostridial gas gangrene. II. Phospholipase C-induced activation of platelet gpIIbIIIa mediates vascular occlusion and myonecrosis in *Clostridium perfringens* gas gangrene. J. Infect. Dis. 182: 808-815.

Bryant, A.E., and Stevens, D.L. 1996. Phospholipase C and perfringolysin O from *Clostridium perfringens* upregulate endothelial cell-leukocyte adherence molecule 1 and intercellular leukocyte adherence molecule 1 expression and induce interleukin-8 synthesis on cultured human umbilical vein endothelial cells. Infect. Immun. 64: 358-362.

Bunting, M., Lorant, D.E., Bryant, A.E., Zimmerman, G.A., Mcintyre, T.M., Stevens, D.L., and Prescott, S.M. 1997. Alpha toxin from *Clostridium perfringens* induces proinflammatory changes in endothelial cells. J. Clin. Invest. 100: 565-574.

Busch, C., Hofmann, F., Gerhard, R., and Aktories, K. 2000a. Involvement of a conserved tryptophan residue in the UDP-glucose binding of large clostridial cytotoxin glycosyltransferases. J. Biol. Chem. 275: 13228-13234.

Busch, C., Schömig, K., Hofmann, F., and Aktories, K. 2000b. Characterization of the catalytic domain of *Clostridium novyi* alpha toxin. Infect. Immun. 68: 6378-6383.

Buxton, D. 1978. The use of an imunoperoxidase technique to investigate by light and electron microscopy the sites of binding of *Clostridium welchii* type D ε-toxin in mice. J. Med. Microbiol. 11: 289-292.

Call, J.E., Cooke, P.H., and Miller, A.J. 1995. *In situ* characterization of *Clostridium botulinum* neurotoxin synthesis and export. J. Appl. Bacteriol. 79: 257-263.

Chardin, P., Boquet, P., Madaule, P., Popoff, M.R., Rubin, E.J., and Gill, D.M. 1989. The mammalian G protein *rhoC* is ADP-ribosylated by *Clostridium botulinum* exoenzyme C3 and affects actin microfilaments in Vero cells. EMBO J. 8: 1087-1092.

Chen, M.L., Pothoulakis, C., and LaMont, J.T. 2002. Protein kinase C signaling regulates ZO-1 translocation and increased paracellular flux of T84 colonocytes exposed to *Clostridium difficile* toxin A. J. Biol. Chem. 277: 4247-4254.

Chen, Y.A., Scales, S.J., and Scheller, R.H. 2001. Sequential SNARE assembly underlies priming and triggering of exocytosis. Neuron. 30: 161-170.

Cole, A. 2003. Structural studies on epsilon toxin from *Clostridium perfringens*. In: Protein Toxins of the Genus

Clostridium and Vaccination, C. Duchesnes, J. Mainil, M. R. Popoff, and R. Titball, eds. Presses de la Faculté de Médecine Vétérinaire, Liege. p. 95.

Cunningham, K., Lacy, D.B., Mogridge, J., and Collier, R.J. 2002. Mapping the lethal factor and edema factor binding sites on oligomeric anthrax protective antigen. Proc. Natl. Acad. Sci USA. 99: 7049-7053.

de Paiva, A., Meunier, F.A., Molgo, J., Aoki, K.R., and Dolly, O. 1999. Functional repair of motor endplates after botulinum neurotoxin type A poisoning: biphasic switch of synaptic activity between nerve sprouts and their parent terminals. Proc. Natl. Acad. Sci. USA. 96: 3200-3205.

Domenighini, M., and Rappuoli, R. 1996. Three conserved consensus sequences identify the NAD-binding site of ADP-ribosylating enzymes, expressed by eukaryotes, bacteria and T-even bacteriophages. Mol. Microbiol. 21: 667-674.

Eaton, J.T., Naylor, C.E., Howells, A.M., Moss, D.S., Titball, R.W., and Basak, A.K. 2002. Crystal structure of the *C. perfringens* alpha-toxin with the active site closed by a flexible loop region. J. Mol. Biol. 319: 275-281.

Eckhardt, M., Barth, H., Blöcker, D., and Aktories, K. 2000. Binding of *Clostridium botulinum* C2 toxin to asparagine-linked complex and hybrid carbohydrates. J. Biol. Chem. 275: 2328-2334.

Eleopra, R., Tugnoli, V., Rossetto, O., De Grandis, D., and Montecucco, C. 1998. Different time courses of recovery after poisoning with botulinum neurotoxin serotypes A and E in humans. Neurosci. Lett. 256: 135-138.

Emsley, P., Fotinou, C., Black, I., Fairweather, N.F., Charles, I.G., Watts, C., Hewitt, E., and Isaacks, N.W. 2000. The structures of the Hc fragment of tetanus toxin with carbohydrate subunit complexes provide insight into ganglioside binding. J. Biol. Chem. 275: 8889-8894.

Ermert, L., Duncker, H.R., Brückner, H., Grimminger, F., Hansen, T., Rössig, R., Aktories, K., and Seeger, W. 1997. Ultrastructural changes of lung capillary endothelium in response to botulinum C2 toxin. J. Appl. Physiol. 82: 382-388.

Finnie, J.W., Blumbergs, P.C., and Manavis, J. 1999. Neuronal damage produced in rat brains by *Clostridium perfringens* type D epsilon-toxin. J. Comp. Path. 120: 415-420.

Fiorentini, C., and Thelestam, M. 1991. *Clostridium difficile* toxin A and its effects on cells. Toxicon. 29: 543-567.

Flores-Diaz, M., Alape-Giron, A., Titball, R.W., Moos, M., Guillouard, I., Cole, S., Howells, A.M., von Eichel-Streiber, C., Florin, I., and Thelestam, M. 1998. UDP-glucose deficiency causes hypersensitivity to the cytotoxic effect of *Clostridium perfringens* phospholipase C. J. Biol. Chem. 273: 24433-24438.

Foran, P.G., Mohammed, N., Lisk, G.O., Nagwaney, S., Lawrence, G.W., Johnson, E., Smith, L., Aoki, K.R., and Dolly, J.O. 2003. Evaluation of the therapeutic usefulness of botulinum neurotoxin B, C1, E, and F compared with the long lasting type A. J. Biol. Chem. 278: 1363-1371.

Fujihara, H., Walker, L.A., Gong, M.C., Lemichez, E., Boquet, P., Somlyo, A.V., and Somlyo, A.P. 1997. Inhibition of RhoA translocation and calcium sensitization by *in vivo* ADP-ribosylation with the chimeric toxin DC3B. Mol. Biol. Cell. 8: 2437-2447.

Fujita, K., Katahira, J., Horiguchi, Y., Sonoda, N., Furuse, M., and Tsukita, S. 2000. *Clostridium perfringens* enterotoxin binds to the second extracellualr loop of claudin-3, a tight junction integral membrane protein. FEBS Lett. 476: 258-261.

Fukata, Y., Amano, M., and Kaibuchi, K. 2001. Rho-Rho-kinase pathway in smooth muscle contraction and cytoskeletal reorganization of non-muscle cells. Trends Pharmacol. Sci. 22: 32-39.

Genth, H., Aktories, K., and Just, I. 1999. Monoglucosylation of RhoA at threonine 37 blocks cytosol membrane recycling. J. Biol. Chem. 274: 29050-29056.

Genth, H., Gerhard, R., Maeda, A., Amano, M., Kaibuchi, K., and Aktories, K., Just, I. 2003. Entrapment of Rho ADP-ribosylated by *Clostridium botulinum* C3 exoenzyme in the Rho-Guanine nucleotide dissociation inhibitor-1 complex. J. Biol. Chem. 278: 28523-28527.

Gibert, M., Jolivet-Reynaud, C., and Popoff, M.R. 1997. Beta2 toxin, a novel toxin produced by *Clostridium perfringens*. Gene. 203: 65-73.

Gibert, M., Petit, L., Raffestin, S., Okabe, A., and Popoff, M.R. 2000. *Clostridium perfringens* iota-toxin requires activation of both binding and enzymatic components for cytopathic activity. Infect. Immun. 68: 3848-3853.

Ginter, A., Williamson, E.D., Dessy, F., Coppe, P., Bullifent, H., Howells, A., and Titball, R.W. 1996. Molecular variation between α-toxins from the type strain (NCTC8237) and clinical isolates of *Clostridium perfringens* associated with disease in man and animals. Microbiology. 142: 191-198.

Gordon, V.M., Nelson, K.L., Buckley, J.T., Stevens, V.L., Tweten, R.K., Elwood, P.C., and Leppla, S.H. 1999. *Clostridium septicum* alpha-toxin uses glycosylphosphatid ylinositol-anchored protein receptors. J. Biol. Chem. 274: 27274-27280.

Guillouard, I., Alzari, P.M., Saliou, B., and Cole, S.T. 1997. The carboxy-terminal C2-like domain of the α-toxin from *Clostridium perfringens* mediates calcium-dependent membrane recognition. Mol. Microbiol. 26: 867-876.

Guillouard, I., Garnier, T., and Cole, S. 1996. Use of site-directed mutagenesis to probe structure-function relationship of alpha-toxin from *Clostridium perfringens*. Infect. Immun. 64: 2440-2444.

Han, S., Arvai, A.S., Clancy, S.B., and Tainer, J.A. 2001. Crystal structure and novel recognition motif of Rho ADP-ribosylating C3 exoenzyme from *Clostridium botulinum*. Structural insights for recognition specifcity and catalysis. J. Mol. Biol. 305: 95-107.

Han, S., Craig, J.A., Putnam, C.D., Carozzi, N.B., and Tainer, J.A. 1999. Evolution and mechanism from structures of an ADP-ribosylating toxin and NAD complex. Nature Struct. Biol. 6: 932-936.

Hanna, P.C., Mietzner, T.A., Schoolnik, G.K., and McClane, B.A. 1991. Localization of the receptor-binding region of *Clostridium perfringens* enterotoxin utilizing cloned toxin fragments and synthetic peptides. The 30 C-terminal amino acids define a functional binding region. J. Biol. Chem. 266: 11037-11043.

Haug, G., Leemhuis, J., Tiemann, D., Meyer, D.K., Aktories, K., and Barth, H. 2003. The host cell chaperonne Hsp90 is essential for translocation of the binary *Clostridium* botulinum C2 toxin into the cytosol. J. Biol. Chem. 278: 32266-32274.

Hayashi, T., Yamasaki, S., Nauenburg, S., Binz, T., and Niemann, H. 1995. Disassembly of the reconstituted synaptic vesicle membrane fusion complex in vitro. EMBO J. 14: 2317-2325.

He, D., Hagen, S.J., Pothoulakis, C., Chen, M., Medina, N.D., Warny, M., and Lamont, J.T. 2000. *Clostridium difficile* Toxin A causes early damage to mitochondria in cultured cells. Gastroenterol. 119: 139-150.

Herholz, C., Miserez, R., Nicolet, J., Frey, J., Popoff, M.R., Gibert, M., Gerber, H., and Straub, R. 1999. Prevalence of β2-toxigenic *Clostridium perfringens* in horses with intestinal disorders. J. Clin. Microbiol. 37: 358-361.

Hermann, C., Ahmadian, M.R., Hofmann, F., and Just, I. 1998. Functional consequences of monoglucosylation of Ha-Ras at effector domain amino acid threonine 35. J. Biol. Chem. 273: 16134-16139.

Herreros, J., Lalli, G., Montecucco, C., and Schiavo, G. 1999. Pathophysiological properties of clostridial neurotoxins. In: The Comprehensive Sourcebook of Bacterial Protein Toxins, J. E. Alouf, and J. H. Freer, eds. Academic Press, London. p. 202-228.

Herreros, J., Lalli, G., Montecucco, C., and Schiavo, G. 2000. Tetanus toxin fragment C binds to a protein present in neuronal cell lines and motoneurons. J. Neurosci. 74: 1941-1950.

Herreros, J., Ng, T., and Schiavo, G. 2001. Lipid rafts act as specialized domains for tetanus toxin binding and internalization into neurons. Mol. Biol. Cell. 12: 2947-2960.

Heuck, A.P., Tweten, R.K., and Johnson, A.E. 2001. β-barrel pore-forming toxins: intriguing dimorphic proteins. Biochemistry. 40: 9065-9073.

Hofmann, F., Busch, C., and Aktories, K. 1998. Chimeric clostridial cytotoxins: identification of the N-terminal region involved in protein substrate recognition. Infect. Immun. 66: 1076-1081.

Hofmann, F., Busch, C., Prepens, U., Just, I., and Aktories, K. 1997. Localization of the glucosyltransferase activity of *Clostridium difficile* Toxin B to the N-terminal part of the holotoxin. J. Biol. Chem. 272: 11074-11078.

Hong, Y., Ohishi, K., Inoue, N., Kang, J.Y., Shime, H., Horiguchi, Y., van der Goot, F.G., Sugimoto, N., and Kinoshita, T. 2002. Requirement of *N*-glycan on GPI-anchored proteins for efficient binding of aerolysin but not *Clostridium speticum* α-toxin. EMBO J. 21: 5047-5056.

Hotze, E.M., Heuck, A.P., Czajkowsky, D.M., Shao, Z., Johnson, A.E., and Tweten, R.K. 2002. Monomer-monomer interactions drive the prepore to pore conversion of a β-barrel-forming cholesterol-dependent cytolysin. J. Biol. Chem. 277: 11597-11605.

Humeau, Y., Doussau, F., Grant, N.J., and Poulain, B. 2000. How botulinum and tetanus neurotoxins block neurotransmitter release. Biochimie. 82: 427-446.

Hunter, S.E., Brown, E., Oyston, P.C.F., Sakurai, J., and Titball, R.W. 1993. Molecular genetic analysis of beta-toxin of *Clostridium perfringens* reveals sequence homology with alpha-toxin, gamma-toxin, and leukocidin of *Staphylococcus aureus*. Infect. Immun. 61: 3958-3965.

Hunter, S.E., Clarke, I.N., Kelly, D.C., and Titball, R.W. 1992. Cloning and nucleotide sequencing of the *Clostridium perfringens* epsilon-toxin gene and its expression in *Escherichia coli*. Infect. Immun. 60: 102-110.

Jankovic, J., and Brin, M.F. 1991. Therapeutic uses of botulinum toxin. N. Engl. J. Med. 324: 1186-1194.

Jankovic, J., and Brin, M.F. 1997. Botulinum toxin: historical perspective and potential new indications. Muscle Nerve Suppl. 6: S129-S145.

Jepson, M., Howells, A., Bullifent, H.L., Bolgiano, B., Crane, D., Miller, J., Holley, J., Jayasekera, P., and Titball, R.W. 1999. Differences in the carboxy-terminal (putative phospholipid binding) domains of *Clostridium perfringens* and *Clostridium bifermentans* phospholipases C influence the hemolytic and lethal properties of these enzymes. Infect. Immun. 67: 3297-3301.

Jolivet-Reynaud, C., and Alouf, J.E. 1983. Binding of *Clostridium perfringens* ^{125}I-labeled Δ-toxin to erythrocytes. J. Biol. Chem. 258: 1871-1877.

Jolivet-Reynaud, C., Hauttecoeur, B., and Alouf, J. 1989. Interaction of *Clostridium perfringens* δ toxin with erythrocyte and liposome membranes and relation with the specific binding to the ganglioside GM2. Toxicon. 27: 1113-1126.

Jolivet-Reynaud, C., Launay, J.M., and Alouf, J.E. 1988. Damaging effects of *Clostridium perfringens* delta toxin on blood platelets and their relevance to ganglioside GM2. Arch. Biochem. Biophys. 262: 59-66.

Jolivet-Reynaud, C., Popoff, M.R., Vinit, M.A., Ravisse, P., Moreau, H., and Alouf, J.E. 1986. Enteropathogenicity of *Clostridium perfringens* beta toxin and other clostridial toxins. Zbl. Bakteriol. S15: 145-151.

Jung, M., Just, I., Van Damme, J., Vandekerckhove, J., and Aktories, K. 1993. NAD-binding site of the C3-like ADP-ribosyltransferase from *Clostridium limosum*. J. Biol. Chem. 268: 23215-23218.

Just, I., Hofmann, F., and Aktories, K. 2000. Molecular mechanism of action of the large clostridial cytotoxins. In Bacterial Protein Toxins, K. Aktories, I. Just, ed. Springer, Berlin. p. 307-331.

Just, I., Selzer, J., Jung, M., Van Damme, J., Vandekerckhove, J., and Aktories, K. 1995a. Rho-ADP-ribosylating exoenzyme from *Bacillus cereus*. Purification, characterization, and identification of the NAD-binding site. Biochemistry. 34: 334-340.

Just, I., Selzer, J., Wilm, M., von Eichel-Streiber, C., Mann, M., and Aktories, K. 1995b. Glucosylation of Rho proteins by *Clostridium difficile* toxin B. Nature. 375: 500-503.

Just, I., Wilm, M., Selzer, J., Rex, G., von Eichel-Streiber, C., Mann, M., and Aktories, K. 1995c. The enterotoxin from *Clostridium difficile* (ToxA) monoglucosylates the Rho proteins. J. Biol. Chem. 270: 13932-13936.

Justin, N., Walker, N., Bullifent, H.L., Songer, G., Bueschel, D.M., Jost, H., Naylor, C., Miller, J., Moss, D.S., Titball, R.W., and Basak, A.K. 2002. The first strain of *Clostridium perfringens* isolated from an avian source has an alpha-toxin with divergent structural and kinetics properties. Biochemistry. 41: 6253-6262.

Juzans, P., Comella, J.X., Molgo, J., Faille, L., and Angaut-Petit, D. 1996. Nerve terminal sprouting in botulinum type-A treated mouse *Levator auris Longus* muscle. Neuromusc. Disord. 6: 177-185.

Katahira, J., Inoue, N., Horiguchi, Y., Matsuda, M., and Sugimoto, N. 1997. Molecular cloning and functional characterization of the receptor for *Clostridium perfringens* enterotoxin. J. Cell Biol. 136: 1239-1247

Katayama, S., Dupuy, B., and Cole, S.T. 1995. Rapid expansion of the physical and genetic map of the chromosome of

Clostridium perfringens CPN50. J. Bacteriol. 177: 5680-5685.

Katayama, S., Dupuy, B., Daube, G., China, B., and Cole, S. 1996. Genome mapping of *Clostridium perfringens* strains with I-*Ceu*I shows many virulence genes to be plamsid-borne. Mol. Gen. Genet. 251: 720-726.

Kiefer, G., Lerner, M., Sehr, P., Just, I., and Aktories, K. 1996. Cytotoxic effects by microinjection of ADP-ribosylated skeletal muscle G-actin in PtK2 cells in the absence of *Clostridium perfringens* iota toxin. Med. Microbiol. Immunol. 184: 175-180.

Kim, J.H., Lee, S.D., Han, J.M., Lee, T.G., Kim, Y., Park, J.B., Lambeth, J.D., Suh, P.G., and Ryu, S.H. 1998. Activation of phospholipase D1 by direct interaction with ADP-ribosylation factor 1 and RalA. FEBS Lett. 430: 231-235.

Knapp, O., Benz, R., Gibert, M., Marvaud, J.C., and Popoff, M.R. 2002. Interaction of *Clostridium perfringens* iota-toxin with lipid bilayer membranes. J. Biol. Chem. 277: 6143-6152.

Kokai-Kun, J.F., and McClane, B.A. 1997. Deletion analysis of the *Clostridium perfringens* enterotoxin. Infect. Immun. 65: 1014-1022.

Lacy, D.B., and Stevens, R.C. 1999. Sequence homology and structural analysis of the clostridial neurotoxins. J. Mol. Biol. 291: 1091-1104.

Lacy, D.B., Tepp, W., Cohen, A.C., Das Gupta, B.R., and Stevens, R.C. 1998. Crystal structure of botulinum neurotoxin type A and implications for toxicity. Nature Struct. Biol. 5:898-902.

Lalli, G., and Schiavo, G. 2002. Analysis of retrograde transport in motor neurons reveals common endocytic carriers for tetanus toxin and neutrophin receptor $p75^{NTR}$. J. Cell Biol. 156: 233-239.

Li, Y., Foran, P., Lawrence, G., Mohammed, N., Chan-Kwo-Chion, C., Lisk, G., Aoki, R., and Dolly, O. 2001. Recombinant forms of tetanus toxin engineered for examining and exploiting neuronal trafficking pathways. J. Biol. Chem. 276: 31394-31401.

Lindsay, C.D., Hambrook, J.L., and Upshall, D.G. 1995. Examination of toxicity of *Clostridium perfringens* ε-toxin in the MDCK cell line. Toxic. In Vitro. 9: 213-218.

Luo, J.Q., Liu, X., Hammond, S.M., Colley, W.C., Feig, L.A., Frohman, M.A., Morris, A.J., and Foster, D.A. 1997. RalA interacts directly with the Arf-responsive, PIP2-dependent phospholipase D1. Biochem. Biophys. Res. Commun. 235: 854-859.

Maksymowych, A.B., Rienhard, M., Malizio, C.J., Goodnough, M.C., Johnson, E.A., and Simpson, L.L. 1999. Pure botulinum neurotoxin is absorbed from the stomach and small intestine and produces peripheral neuromuscular blockade. Infect. Immun. 67: 4708-4712.

Marvaud, J.C., Smith, T., Hale, M.L., Popoff, M.R., Smith, L.A., and Stiles, B.G. 2001. *Clostridium perfringens* iota-toxin:mapping of receptor binding and Ia docking domains on Ib. Infect. Immun. 69: 2435-2441.

Marvaud, J.C., Stiles, B.G., Chenal, A., Gillet, D., Giber, M., Smith, L.A., and Popoff, M.R. 2002. *Clostridium perfringens* iota toxin: mapping of the Ia domain involved in docking with Ib and cellular internalization. J. Biol. Chem. 277: 43659-43666.

Maslos, U., Kissa, K., St Cloment, C., and Brûlet, P. 2002. Retrograde trans-synaptic transfer of green fluorescent protein allows the genetic mapping of neuronal circuits in transgenic mice. Proc. Ntl Acad. Sci. USA. 99: 10120-10125.

Matsushita, O., Jung, C.M., Katayama, S., Minami, J., Takahashi, Y., and Okabe, A. 1999. Gene duplication and multiplicity of collagenases in *Clostridium histolyticum*. J. Bacteriol. 181: 923-933.

Matsushita, O., Jung, C.M., Minami, J., Katayama, S., Nishi, N., and Okabe, A. 1998. A study of the collagen-binding domain of a 116 kDa *Clostridium histolyticum* collagenase. J. Biol. Chem. 273: 3643-3648.

Matsushita, O., and Okabe, A. 2001. Clostridial hydrolytic enzymes degrading extracellualr components. Toxicon. 39: 1769-1780.

Matsushita, O., Yoshihara, K., Katayama, S., Minami, J., and Okabe, A. 1994. Purification and characterization of a *Clostridium perfringens* 120-kilodalton collagenase and nucleotide sequence of the corresponding gene. J. Bacteriol. 176: 149-156.

Mauss, S., Chaponnier, C., Just, I., Aktories, K., and Gabbiani, G. 1990. ADP-ribosylation of actin isoforms by *Clostridium botulinum* C2 toxin and *Clostridium perfringens* iota toxin. Eur. J. Biochem. 194: 237-241.

Mauss, S., Koch, G., Kreye, V.A.W., and Aktories, K. 1989. Inhibition of the contraction of the isolated longitudinal muscle of the guinea-pig ileum by botulinum C2 toxin: evidence for a role of G/F -actin transition in smooth muscle contraction. Naunyn-Schmiedebergs Arch. Pharmacol. 340: 345-351.

McClane, B.A. 1996. An overview of *Clostridium perfringens* enterotoxin. Toxicon. 34: 1335-1343

McClane, B.A. 2001. The complex interaction between *Clostridium perfringens* enterotoxin and epithelial tight junctions. Toxicon. 39: 1781-1791.

Ménétrey, J., Flatau, G., Stura, E.A., Charbonnier, J.B., Gas, F., Teulon, J.M., Le Du, M.H., Boquet, P., and Ménez, A. 2002. NAD binding induces conformational changes in Rho ADP-ribosylating *Clostridium botulinum* C3 exoenzyme. J. Biol. Chem. 277: 30950-30957.

Meunier, F.A., Herreros, J., Schiavo, G., Poulain, B., and Molgo, J. 2002a. Molecular mechanism of action of botulinal neurotoxins and the synaptic remodeling they induce in vivo at the skeletal neuromuscular junction. In: Handbook of Neurotoxicology, J. Massaro, ed. Humana Press, Totowa. p. 305-347.

Meunier, F.A., Schiavo, G., and Molgo, J. 2002b. Botulinum neurotoxins: from paralysis to recovery of functional neuromuscular transmission. J. Physiol. 96: 105-113.

Minami, J., Katayama, S., Matsushita, O., Matsushita, C., and Okabe, A. 1997. Lambda-toxin of *Clostridium perfringens* activates the precursor of epsilon-toxin by releasing its N- and C-terminal peptides. Microbiol. Immunol. 41: 527-535.

Miyamoto, O., Minami, J., Toyoshima, T., Nakamura, T., Masada, T., Nagao, S., Negi, T., Itano, T., and Okabe, A. 1998. Neurotoxicity of *Clostridium perfringens* epsilon-toxin for the rat hipocampus via glutamanergic system. Infect. Immun. 66: 2501-2508.

Miyamoto, O., Sumitami, K., Nakamura, T., Yamagani, S., Miyatal, S., Itano, T., Negi, T., and Okabe, A. 2000. *Clostridium perfringens* epsilon toxin causes excessive release of glutamate in the mouse hippocampus. FEMS Microbiol. Lett. 189: 109-113.

Miyata, S., Matsushita, O., Minami, J., Katayama, S., Shimamoto, S., and Okabe, A. 2001. Cleavage of C-terminal peptide is essential for heptamerization of *Clostridium perfringens* ε-toxin in the synaptosomal membrane. J. Biol. Chem. 276: 13778-13783.

Miyata, S., Minami, J., Tamai, E., Matsushita, O., Shimamoto, S., and Okabe, A. 2002. *Clostridium perfringens* ε-toxin forms a heptameric pore within the detergent-insoluble microdomains of Madin-Darby Canine Kidney Cells and rat synaptosomes. J. Biol. Chem. 277: 39463-39468.

Mogridge, J., Cuningham, K., Lacy, D.B., Mourez, M., and Collier, R.J. 2002a. The lethal and edema factors of anthrax toxin bind only to oligomeric forms of the protective antigen. Proc. Natl. Acad. Sci. USA. 99: 7045-7048.

Mogridge, J., Cunningham, K., and Collier, R.J. 2002b. Stoichiometry of anthrax toxin complexes. Biochemistry. 41: 1079-1082.

Moreau, H., Pieroni, G., Jolivet-Reynaud, C., Alouf, J.E., and Verger, R. 1988. A new kinetic approach for studying phospholipase C (*Clostridium perfringens* α toxin) activity on phospholipid monolayers. Biochemistry. 27: 2319-2323.

Müller, S., von Eichel-Streiber, C., and Moos, M. 1999. Impact of amino acids 22-27 of rho subfamily GTPases on glucosylation by the large clostridial cytotoxins TcsL-1522, TcdB-1470 and tcdB-8864. Eur. J. Biochem. 266: 1073-1080.

Munro, P., Kojima, H., Dupont, J.L., Bossu, J.L., Poulain, B., and Boquet, P. 2001. High sensitivity of mouse neuronal cells to tetanus toxin requires a GPI-anchored protein. Biochem. Biophys. Res. Comm. 289: 623-629.

Nagahama, M., Michiue, K., Mukai, M., Ochi, S., and Sakurai, J. 1998. Mechanism of membrane damage by *Clostridium perfringens* alpha-toxin. Microbiol. Immunol. 42: 533-538.

Nagahama, M., Michiue, K., and Sakurai, J. 1996. Membrane-damaging action of *Clostridium perfringens* alpha-toxin on phospholipid liposomes. Bioch. Biophys. Acta. 1280: 120-126.

Nagahama, M., Miyawaki, T., Kihara, A., Mukai, M., Sakaguchi, Y., Ochi, S., and Sakurai, J. 1999. Thiol group reagent-sensitive *Clostridium perfringens* beta-toxin does not require a thiol group for lethal activity. Biochem. Biophys. Acta. 1454: 97-105.

Nagahama, M., Nakayama, T., K., M., and Sakurai, J. 1997. Site-specific mutagenesis of *Clostridium perfringens* alpha-toxin: replacement of Asp-56, Asp-130, or Glu-152 causes loss of enzymatic and hemolytic activites. Infect. Immun. 65: 3489-3492.

Nagahama, M., and Sakurai, J. 1991. Distribution of labeled *Clostridium perfringens* epsilon toxin in mice. Toxicon. 29: 211-217.

Nagahama, M., and Sakurai, J. 1992. High-affinity binding of *Clostridium perfringens* epsilon-toxin to rat brain. Infect. Immun. 60: 1237-1240.

Nagahama, M.Y., Okagawa, T., Nakayama, T., Nishioka, E., and Sakurai, J. 1995. Site directed mutagenesis of histidine residues in *Clostridium perfringens* alpha-toxin. J. Bacteriol. 177: 1179-1185.

Naylor, C.E., Eaton, J.T., Howells, A., Justin, N., Moss, D.S., Titball, R.W., and Basak, A.K. 1998. Structure of the key toxin in gas gangrene. Nature Struct. Biol. 5: 738-746.

Nishiki, T., Kamata, Y., Nemoto, Y., Omori, A., Ito, T., Takahashi, M., and Kozaki, S. 1994. Identification of protein receptor for *Clostridium botulinum* type B neurotoxin in rat brain synaptosomes. J. Biol. Chem. 269: 10498-10503.

Nishiki, T., Tokuyama, Y., Kamata, Y., Nemoto, Y., Yoshida, A., Sato, K., Sekigichi, M., Taakahashi, M., and Kozaki, S. 1996. The high-affinity of *Clostridium botulinum* type B neurotoxin to synaptotagmin II associated with gangliosides G_{T1B}/G_{D1a}. FEBS Lett. 378: 253-257.

Nusrat, A., von Eichel-Streiber, C., Turner, J.R., Verkade, P., Madara, J.L., and Parkos, C.A. 2001. *Clostridium difficile* toxins disrupt epithelial barrier function by altering membrane microdomain localization of tight junction proteins. Infect. Immun. 69: 1329-1336.

Ochi, S., Hashimoto, K., Nagahama, M., and Sakurai, J. 1996. Phospholipd metabolism induced by *Clostridium perfringens* alpha-toxin elicits a hot cold type of hemolysis in rabbit erythrocytes. Infect. Immun. 64: 3930-3933.

Ohishi, I. 1987. Activation of botulinum C2 toxin by trypsin. Infect. Immun. 55: 1461-1465.

Ohishi, I. 2000. Structure and function of actin-adenosine-diphosphate-ribosylating toxins. In: Bacterial Protein Toxins, K. Aktories, and I. Just, eds. Springer, Berlin. p. 253-273.

Ohishi, I., and Tsuyama, S. 1986. ADP-ribosylation of nonmuscle actin with component I of C2 toxin. Biochem. Biophys. Res. Commun. 136: 802-806.

Oyston, P.C.F., Payne, D.W., Havard, H.L., Williamson, E.D., and Titball, R.W. 1998. Production of a non-toxic site-directed mutant of *Clostridium perfringens* ε-toxin which induces protective immunity in mice. Microbiology. 144: 333-341.

Payne, D., Williamson, E.D., and Titball, R.W. 1997. The *Clostridium perfringens* epsilon-toxin. Rev. Med. Microbiol. 8: S28-S30.

Payne, D.W., Williamson, E.D., Havard, H., Modi, N., and Brown, J. 1994. Evaluation of a new cytotoxicity assay for *Clostridium perfringens* type D epsilon toxin. FEMS Microbiol. Lett. 116: 161-168.

Pellegrini, L.L., O'Connor, V., Lottspeich, F., and Betz, H. 1995. Clostridial neurotoxins compromise the stability of a low energy SNARE complex mediating NSF activation of synaptic vesicle fusion. EMBO J. 14: 4705-4713.

Perelle, S., Gibert, M., Boquet, P., and Popoff, M.R. 1993. Characterization of *Clostridium perfringens* iota-toxin genes and expression in *Escherichia coli*. Infect. Immun. 61: 5147-5156 (Author's correction, 5163:4967, 1995).

Petit, L., Gibert, M., Gillet, D., Laurent-Winter, C., Boquet, P., and Popoff, M.R. 1997. *Clostridium perfringens* epsilon-toxin acts on MDCK cells by forming a large membrane complex. J. Bacteriol. 179: 6480-6487.

Petit, L., Gibert, M., Gourch, A., Bens, M., Vandewalle, A., and Popoff, M.R. 2003. *Clostridium perfringens* Epsilon Toxin Rapidly Decreases Membrane Barrier Permeability of Polarized MDCK Cells. Cell. Microbiol. 5: 155-164.

Petit, L., Gibert, M., and Popoff, M.R. 1999. *Clostridium perfringens*: toxinotype and genotype. Trends Microbiol. 7: 104-110.

Petit, L., Maier, E., Gibert, M., Popoff, M.R., and Benz, R. 2001. *Clostridium perfringens* epsilon-toxin induces a rapid change in cell membrane permeability to ions and forms channels in artificial lipid bilayers. J. Biol. Chem. 276: 15736-15740.

Petosa, C., Collier, J.R., Klimpel, K.R., Leppla, S.H., and Liddington, R.C. 1997. Crystal structure of the anthrax toxin protective antigen. Nature. 385: 833-838.

Poirier, M.A., Xiao, W.Z., MasOsko, J.C., Chan, C., Shin, Y.K., and Bennett, M.K. 1998. The synaptic SNARE complex is a parallel 4-stranded helical bundle. Nat. Struct. Biol. 5: 765-769.

Popoff, M.R. 2000. Molecular Biology of Actin-ADP-Ribosylating Toxins. In: Bacterial Protein Toxins, K. Aktories, and I. Just, eds. Springer, Berlin. p. 275-302.

Popoff, M.R., Chaves-Olarte, E., Lemichez, E., Von Eichel-Streiber, C., Thelestam, M., Chardin, P., Cussac, D., Antonny, B., Chavrier, P., Flatau, G., Giry, M., de Gunzburg, J., and Boquet, P. 1996. Ras, Rap, and rac small GTP-binding proteins are targets for *Clostridium sordellii* lethal toxin glucosylation. J. Biol. Chem. 271: 10217-10224.

Popoff, M.R., and Marvaud, J.C. 1999. Structural and genomic features of clostridial neurotoxins. In: The Comprehensive Sourcebook of Bacterial Protein Toxins, J. E. Alouf, and J. H. Freer, eds. Academic Press, London. p. 174-201.

Pothoulakis, C., and Lamont, J.T. 2001. Microbes and microbial toxins: paradigms for microbial-mucosa interactions II. The integrated response of the intestine to *Clostridium difficile* toxins. Am. J. Physiol. Gastrointest. Liver Physiol. 280: G178-G183.

Poulain, B., Dousseau, F., Colasante, C., Deloye, F., and Molgo, J. 1997. Cellular and molecular mode of action of botulinum and tetanus neurotoxins. Adv. Organ Biol. 2: 285-313.

Qa'dan, M., Spyres, L.M., and Ballard, J.D. 2000. pH-induced conformational changes in *Clostridium difficile* toxin B. Infect. Immun. 68: 2470-2474.

Qa'dan, M., Spyres, L.M., and Ballard, J.D. 2001. pH-induced cytopathic effects of *Clostridium sordellii* lethal toxin. Infect. Immun. 69: 5487-5493.

Raciborska, D., and Charlton, M. 1999. Retention of cleaved synaptosome-associated protein of 25 kDa (SNAP25) in neurouscular junctions: a new hypothesis to explain persistence of botulinum A poisoning. Can. J. Physiol. Pharmacol. 77: 679-688.

Ramachandran, R., Heuck, A.P., Tweten, R.K., and Johnson, A.E. 2002. Structural insights into the membrane-anchoring mechanism of a cholesterol-dependent cytolysin. Nat. Struct. Biol. 9: 823-827.

Ratts, R., Zeng, H., Berg, E.A., Blue, C., McCom, M.E., Costello, C.E., vanderSpek, J.C., and Murphy, J.R. 2003. The cytosolic entry of diphtheria toxin catalytic domain requires a host cell cytosolic translocation factor complex. J. Cell Biol. 160: 1139-1150.

Ren, X.D., Bokoch, G.M., Traynor-Kaplan, A., Jenkins, G.H., Anderson, R.A., and Schwartz, M.A. 1996. Physical association of the small GTPase Rho with a 68-kDa phosphatidylinositol 4-phosphate 5-kinase in swiss 3T3 cells. Mol. Biol. Cell. 7: 435-442.

Reuner, K.H., Dunker, P., van der does, A., Wiederhold, M., Just, I., Aktories, K., and Katz, N. 1996. Regulation of actin synthesis in rat hepatocytes by cytoskeleton rearrangements. Eur. J. Cell Biol. 69: 189-196.

Richard, J.F., Mainguy, G., Gibert, M., Marvaud, J.C., Stiles, B., and Popoff, M.R. 2002. Transcytosis of iota toxin across polarized CaCo-2 cell monolayers. Mol. Microbiol. 43: 907-917.

Richard, J.F., Petit, L., Gilbert, M., Marvaud, J.C., Bouchaud, C., and Popoff, M.R. 1999. Bacterial toxins modifying the actin cytoskeleton. Internatl. Microbiol. 2: 185-194.

Ridley, A.J., and Hall, A. 1992. The small GTP-binding protein Rho regulates the assembly of focal adhesions and actin stress fibers in response to growth factors. Cell. 70: 389-399.

Rood, J.I. 1998. Virulence genes of *Clostridium perfringens*. Annu. Rev. Microbiol. 52: 333-360.

Rossjohn, J., Feil, S.C., McKinstry, W.J., Tweten, R.K., and Parker, M.W. 1997. Structure of a cholesterol-binding thiol-activated cytolysin and a model of its membrane form. Cell. 89: 685-692.

Rummel, A., Bade, S., Alves, J., Bigalke, H., and Binz, T. 2003. Two carbohydrate binding sites in the H_{CC}-domain of tetanus neurotoxin are required for toxicity. J. Mol. Biol. 326: 835-847.

Rupnik, M., von Eichel-Streiber, C., Popoff, M.R., and Söling, H.D. 2003. Limited proteolysis of large clostridial cytotoxins results in a functional catalytic domain. In: ETOX11. p. Abstratc n°. Praha: P82.

Saito, Y., Nemoto, Y., Ishizaki, T., Watanabe, N., Morii, N., and Narumiya, S. 1995. Identification of Glu^{173} as the critical amino acid residue for the ADP-ribosyltransferase activity of *Clostridium botulinum* C3 exoenzyme. FEBS Lett. 371: 105-109

Sakurai, J. 1995. Toxins of *Clostridium perfringens*. Rev Med. Microbiol. 6: 175-185.

Sakurai, J., Fujii, Y., and Matsuura, M. 1980. Effect of oxidizing agents and sulfhydryl group reagents on beta toxin from *Clostridium perfringens* type C. Microbiol. Immunol. 24: 595-601.

Sakurai, J., Ochi, S., and Tanaka, H. 1993. Evidence for coupling of *Clostridium perfringens* alpha-toxin-induced hemolysis to stimulated phosphatidic acid formation in rabbit erythrocytes. Infect. Immun. 61: 3711-3718.

Schiavo, G., Matteoli, M., and Montecucco, C. 2000. Neurotoxins affecting neuroexocytosis. Physiol. Rev. 80: 717-766.

Schmid, A., Benz, R., Just, I., and Aktories, K. 1994. Interaction of *Clostridium botulinum* C2 toxin with lipid bilayer membranes. J. Biol. Chem. 269: 16706-16711.

Schmidt, M., Vos, M., Thiel, M., Bauer, B., Grannas, A., Tapp, E., Cool, R.H., de Gunzburg, J., von Eichel Streiber, C., and Jakobs, K.H. 1998. Specific inhibition of phorbol ester-stimulated phospholipase D by *Clostridium sordellii* lethal toxin and *Clostridium difficile* toxin B-1470 in HEK-293 cells. J. Biol. Chem. 273: 7413-7422

Sears, C.L. 2001. The toxins of *Bacteroides fragilis*. Toxicon 39: 1737-1746.

Sehr, P., Gili, J., Genth, H., Just, I., Pick, E., and Aktories, K. 1998. Glucosylation and ADP ribosylation of Rho proteins: effects on nucleotide binding, GTPase activity, and effector coupling. Biochemistry. 37: 5296-5304.

Sekino-Suzuki, N., Nakamura, M., Mitsui, K., and Ohno-Iwashita, O. 1996. Contribution of individual tryptophan residues to the structure and activity of θ-toxin (pefringolysinb O), a cholesterol-binding cytolysin. Eur. J. Biochem. 241: 941-947.

Shatursky, O., Bayles, R., Rogers, M., Jost, B.H., Songer, J.G., and Tweten, R.K. 2000. *Clostridium perfringens* beta-toxin forms potential-dependent, cation-selective channels in lipid bilayers. Infect. Immun. 68: 5546-5551.

Shatursky, O., Heuck, A., Shepard, L., Rossjhon, J., Parker, M., Johnson, A., and Tweten, R. 1999. The mechanism of membrane insertion of a cholesterol-dependent cytolysin: a novel paradigm for pore-forming toxins. Cell. 99: 293-299.

Shepard, L., Shatursky, O., Johnson, A., and Tweten, R. 2000. The mechanism of pore assembly for a cholesterol-dependent cytolysin: formation of a large prepore complex precedes the insertion of hte transmembrane b-hairpins. Biochemistry. 39: 10284-10293.

Shimizu, T., Ohtani, K., Hirakawa, H., Ohshima, K., Yamashita, A., Shiba, T., Ogasawara, N., Hattori, M., Kuhara, S., and Hayashi, H. 2002. Complete genome sequence of *Clostridium perfringens,* an anaerobic flesh-eater. Proc. Natl. Acad. Sci. USA. 99: 996-1001.

Shortt, S.J., Titball, R.W., and Lindsay, C.D. 2000. An assessment of the in vitro toxicology of *Clostridium perfringens* type D epsilon-toxin in human and animal cells. Hum. Experim. Toxicol. 19: 108-116.

Simpson, L.L., Stiles, B.G., Zepeda, H.H., and Wilkins, T.D. 1987. Molecular basis for the pathological actions of *Clostridium perfringens* iota toxin. Infect. Immun. 55: 118-122.

Singh, U., Mitic, L.L., Wieckowski, E.U., Anderson, J.M., and McClane, B.A. 2001. Comparative biochemical and immunocytochemical studies reveal differences in the effects of *Clostridium perfringens* enterotoxin on polarized CaCo-2 cells *versus* Vero cells. J. Biol. Chem. 276: 33402-33412.

Singh, U., Van Itallie, C.M., Mitic, L.L., Anderson, J.M., and McClane, B.A. 2000. CaCo-2 cells treated with *Clostridium perfringens* enterotoxin form multiple complex species, one of which contains the tight junction protein occludin. J. Biol. Chem. 275: 18407-18417.

Söllner, T., Bennett, M.K., Whiteheart, S.W., Schelleer, R.H., and Rothmann, J.E. 1993a. A protein assembly-disassembly pathway in vitro that may correspond to sequential steps of synaptic vesicle docking, activation, and fusion. Cell. 75: 409-418.

Söllner, T., Whiteheart, S.W., Brunner, M., Erdjument-Bromage, H., Geromanos, S., Tempst, P., and Rothman, J.E. 1993b. SNAP receptors implicated in vesicle targeting and fusion. Nature. 362: 318-324.

Songer, J.G. 1996. Clostridial enteric diseases of domestic animals. Clin. Microbiol. Rev. 9: 216-234.

Sonoda, N., Furuse, M., Sasaki, H., Yonemura, S., Katahira, J., Horiguchi, Y., and Tsukita, S. 1999. *Clostridium perfringens* enterotoxin fragment removes specific claudin from tight junction strands: evidence for direct involvement of claudin in tight junction barrier. J. Cell Biol. 147: 195-204.

Steinthorsdottir, V., Fridiksdottir, V., Gunnarson, E., and Andresson, O. 1998. Site-directed mutagenesis of *Clostridium perfringens* beta-toxin expression of wild type and mutant toxins in *Bacillus subtilis*. FEMS Microbiol. Lett. 158: 17-23.

Steinthorsdottir, V., Halldorson, H., and Andresson, O. 2000. *Clostridium perfringens* beta-toxin forms multimeric transmembrane pores in human endothelial cells. Microb. Pathog. 28: 45-50.

Stevens, D.L., and Bryant, A.E. 2002. The role of Clostridial toxins in the pathogenesis of gas gangrene. Clin. Infect. Dis. 35: S93-S100.

Stevens, D.L., Troyer, B.E., Merrick, D.T., Mitten, J.E., and Olson, R.D. 1988. Lethal effects and cardiovascular effects of purified α- and Θ-toxins from *Clostridium perfringens*. J. Infect. Dis. 157: 272-279.

Stevens, D.L., Tweten, R.K., Awad, M.M., Rood, J.I., and Bryant, A.E. 1997. Clostridial gas gangrene: evidence that α and τ toxin differentially modulate the immune response and induce acute tissue necrosis. J. Infect. Dis. 176: 189-195.

Stiles, B., Hale, M.L., Marvaud, J.C., and Popoff, M.R. 2000. *Clostridium perfringens* iota toxin: binding studies and characterization of cell surface receptor by fluorescence-activated cytometry. Infect. Immun. 68: 3475-3484.

Stiles, B.G., Hale, M.L., Marvaud, J.C., and Popoff, M.R. 2002. *Clostridium perfringens* iota toxin: characterization of the cell-asociated iota b complex. Biochem. J. 367: 801-808.

Sutton, R.B., Fasshauer, D., Jahn, R., and Brunger, A.T. 1998. Crystal structure of a SNARE complex involved in synaptic exocytosis at 2.4 angstrom resolution. Nature. 395: 347-353.

Titball, R.W. 1993a. Bacterial phospholipases C. Microbiol. Rev. 57: 347-366.

Titball, R.W. 1993b. Biochemical and immunological properties of the C-terminal domain of the alpha-toxin of *Clostridium perfringens*. FEMS Microbiol. Lett. 110: 45-50.

Titball, R.W., Leslie, D.L., Harvey, S., and Kelly, D. 1991. Hemolytic and sphingomyelinase activities of *Clostridium perfringens* alpha-toxin are dependent on a domain homologous to that of an enzyme from the human arachidonic acid pathway. Infect. Immun. 59: 1872-1874.

Titball, R.W., Naylor, C.E., and Basak, A.K. 1999. The *Clostridium perfringens* α-toxin. Anaerobe. 5: 51-64.

Tsuge, H., Nagahama, M., Nishimura, H., Hisatsune, J., Sakaguchi, Y., Itogawa, Y., Katunuma, N., and Sakurai, J. 2003. Crystal structure and site-directed mutagenesis of enzymatic components from *Clostridium perfringens* iota-toxin. J. Mol. Biol. 325: 471-483.

Tweten, R.K. 1988. Nucleotide sequence of the gene for perfringolysin O (theta toxin) from *Clostridium perfringens*: significant homology with the genes for streptolysin and pneumolysin. Infect. Immun. 56: 3235-3240.

Tweten, R.K. 2001. *Clostridium perfringens* beta toxin and *Clostridium septicum* alpha toxin: their mechanisms and possible role in pathogenesis. Vet. Microbiol. 82: 1-9.

Umland, T.C., Wingert, L.M., Swaminathan, S., Furey, W.F., Schmidt, J.J., and Sax, M. 1997. The structure of the receptor binding fragment H_C of tetanus neurotoxin. Nature Struct. Biol. 4: 788-792.

Vandekerckhove, J., Schering, B., Bärmann, M., and Aktories, K. 1987. *Clostridium perfringens* iota toxin ADP-ribosylates skeletal muscle actin in Arg-177. FEBS Lett. 255: 48-52.

Vershueren, H., van der Taelen, I., Dewit, J., De Braekeleer, J., De Baetselier, P., Aktories, K., and Just, I. 1995. Effects of *Clostridium botulinum* C2 toxin and cytochalasin D on in vitro invasiveness, motility and F-actin content of a murine T-lymphoma cell line. Eur. J. Cell Biol. 66: 335-341.

Vetter, I.R., Hofmann, F., Wohlgemuth, S., Hermann, C., and Just, I. 2000. Structural consequences of monoglucosylation of Ha-Ras by *Clostridium sordellii* lethal toxin. J. Mol. Biol. 301: 1091-1095.

Waheed, A.A., Shimada, Y., Heijnen, H.F.G., Nakamura, M., Inomata, M., Hayashi, M., Iwashita, S., Slot, J.W., and Ohno-Iwashita, Y. 2001. Selective binding of perfringolysin O derivative to cholesterol-rich membrane microdomùains (rafts). Proc. Natl. Acad. Sci. USA. 98: 4926-4931.

Warny, M., Keates, A.C., Keates, S., Castagliuolo, I., Zacks, J.K., Aboudola, A., Pothoulakis, C., LaMont, J.T., and Kelly, C.P. 2000. p38 MAP kinase activation by *Clostridium difficile* toxin A mediates monocyte necrosis, IL-8 production, and enteritis. J. Clin. Invest. 105: 1147-1156.

Warren, G., Koziel, M., Mullins, M.A., Nye, G., Carr, B., Desai, N., Kostichka, K., Duck, N., and Estruch, J.J. 1996. Novel pesticidal proteins and strains. World Intellectual Property Organization. Patent application. WO 96/10083.

Wershil, B., Castagliuolo, I., and Pothoulakis, C. 1998. Mast cell involvement in *Clostridium difficile* toxin A-induced intestinal fluid secretion and neutrophil recruitment in mice. Gastroenterology. 114: 956-964.

Wieckowski, E.U., Kokai-Kun, J.F., and McClane, B.A. 1998. CHaracterization of membrane-associated *Clostridium perfringens* enterotoxin following pronase treatment. Infect. Immun. 66: 5897-5905.

Wilde, C., Genth, H., Aktories, K., and Just, I. 2000. Recognition of RhoA by *Clostridium botulinum* C3 exoenzyme. J. Biol. Chem. 275: 16478-16483.

Williamson, E.D., and Titball, R.W. 1993. A genetically engineered vaccine against the alpha-toxin of *Clostridium perfringens* also protects mice against experimental gas gangrene. Vaccine. 11: 1253-1258.

Yowler, B.C., Kensinger, R.D., and Schengrund, C.L. 2002. Botulinum neurotoxin A activity is dependent upon the presence of specific gangliosides in neuroblastoma cells expressing synaptotagmin I. J. Biol. Chem. 27: 32815-32819.

Zhu, C., Ghabriel, M.N., Blumbergs, P.C., Reilly, P.L., Manavis, J., Youssef, J., Hatami, S., and Finnie, J.W. 2001. *Clostridium perfringens* prototoxin-induced alteration of endothelial barrier antigen (EBA) immunoreactivity at the blood brain barrier (BBB). Exp. Neurol. 169: 72-82.

Chapter 10

Toxin Gene Regulation in *Clostridium*

Michiko M. Nakano*, Peter Zuber, and Abraham L. Sonenshein

Abstract

Recent findings have greatly increased our understanding of toxin gene regulation in *Clostridium*. The VirR/VirS two-component regulatory proteins regulate transcription of genes encoding extracellular toxins in *Clostridium perfringens*. Some genes are directly activated by the VirR response regulator that interacts with the target DNA. In other cases, a VirR/VirS-dependent regulatory RNA is required for gene expression via an as yet unidentified mechanism. The transcription of the enterotoxin gene in *C. perfringens* appears to be regulated by mother cell-specific sigma factors, σ^E and σ^K, which are present and active only during sporulation. Transcriptional regulation utilizing alternative sigma factors has also been uncovered in recent studies of toxin gene regulation in *Clostridium difficile, Clostridium tetani*, and *Clostridium botulinum*. Continued genome sequence determination of *Clostridium* spp., as well as transcriptome and proteome analyses, will greatly contribute to furthering our understanding of how toxin genes are regulated by various regulators and environmental factors.

1. Introduction

The anaerobic, spore-forming, Gram-positive bacteria of the genus *Clostridium* include species that have a serious impact on human health. Gas gangrene, tetanus, and botulism, for example, are caused by potent toxins produced by *Clostridium perfringens*, *Clostridium tetani*, and *Clostridium botulinum*, respectively. The structures, modes of action, and cellular effects of these toxins have been well-studied and are described in the accompanying chapter (Chapter 9). In contrast, a limited number of studies focusing on the regulation of the toxin genes have been reported. Until very recently toxin gene regulation has been studied primarily in *C. perfringens* because of its oxygen tolerance, relatively fast growth, and amenability to genetic manipulation due, in part, to the availability of shuttle plasmids for cloning. Tn*916* mutagenesis and reporter systems to monitor gene expression have been developed as well (reviewed by Rood and Cole, 1991; Rood, 1997). However, methods of genetic manipulation, such as transformation by electroporation, conjugation, development of vector systems, and reporter constructs, are now becoming available for other *Clostridium* spp. (see the reviews and the references cited below).

The complete genome sequences of *C. perfringens* (Shimizu *et al.*, 2002a) and *C. tetani* (Bruggemann *et al.*, 2003) have very recently become available and sequencing projects for the *C. botulinum* and *Clostridium difficile* genomes are nearly complete (NCBI). Comparative analyses of genome sequences have identified potential virulence factors and provided important clues about the physiology of the bacteria. Mobile genetic elements and genes likely to have been acquired by horizontal transfer are relatively uncommon in the genomes of *C. perfringens* and *C. tetani*, indicating that their genomes are more stable than those of the enteric bacteria. The virulence genes in these organisms are scattered around their chromosomes and pathogenicity islands appear to be absent. Only the genes specifying surface layer proteins, which are absent in *C. perfringens*, are clustered in the genome of *C. tetani*. Consistent with its growth physiology, *C. tetani* carries genes encoding various sodium ion-dependent symporters, most of which likely transport amino acids, and enzymes that function in the degradation of amino acids to pyruvate. In contrast, *C. perfringens* undergoes sugar fermentation and genes for utilization of a variety of sugars as fermentation substrates were found in its genome.

This chapter focuses on recent studies of toxin gene regulation in *C. perfringens*, *C. botulinum, C. tetani*, and *C. difficile*. There are excellent reviews available that describe the genetics of toxin production, as well as analyses of toxin gene regulation in *Clostridium* spp. (Rood, 1997; McClane, 1998; Rood, 1998; Marvaud *et al.*, 2000; Johnson and Bradshaw, 2001; Raffestin *et al.*, 2004; Sonenshein *et al.*, 2004).

*For correspondence email: mnakano@ebs.ogi.edu

2. VirR/VirS-Dependent Toxin Gene Regulation in *C. perfringens*

C. perfringens is a human pathogen that causes gas gangrene and food poisoning. It produces numerous toxins and extracellular enzymes (Chapter 9). Because of its ease of manipulation in the laboratory, *C. perfringens* is the *Clostridium* species most often used for genetic studies of toxin production. Recent studies have shown that a two-component regulatory system composed of a histidine sensor kinase, VirS, and a response regulator, VirR, controls transcription of the *plc* gene, which encodes alpha-toxin (phospholipase C), *pfoA,* encoding theta-toxin (perfringolysin O), and *colA,* encoding kappa-toxin (collagenase). VirR/VirS is also required for production of sialidase, protease, and haemoglutinin (Lyristis *et al.*, 1994; Shimizu *et al.*, 1994).

2.1. VirR/VirS Two-Component Signal Transduction System

The *virR* gene was independently isolated by two groups using different approaches. In one study, the cloned *pfoA* gene was found to be unable to complement certain perfringolysin O-nonproducing strains obtained by chemical mutagenesis. The *virR* gene was isolated from members of a chromosomal clone library that restored toxin production to this mutant (Shimizu *et al.*, 1994). In the other study, Tn*916* mutagenesis was used to isolate a pleiotropic mutant that showed reduced phospholipase C, protease, and sialidase production, and did not produce perfringolysin O. The Tn*916* insertion was located within *virS*, which was subsequently found to reside downstream from the *virR* gene (Lyristis *et al.*, 1994). The *virR* and *virS* genes constitute an operon and their products positively regulate the production of alpha-toxin, theta-toxin, and kappa-toxin at the transcriptional level. The *pfoA* gene is primarily transcribed from a promoter strictly dependent on VirR/VirS. The *plc* gene is transcribed from a single VirR/VirS-dependent promoter, but some residual transcription is observed in a *virS* mutant. The *colA* gene is transcribed from two promoters, one of which is dependent on VirR/VirS. These results explain why the *virS*::Tn*916* mutant does not produce perfringolysin O, but still produces a detectable level of phospholipase C and collagenase (Ba-Thein *et al.*, 1996).

In vitro experiments suggested that *pfoA* is directly activated by VirR/VirS, whereas the VirR/VirS-dependent regulation of *plc* and *colA* is indirect. Gel mobility shift assays using purified His-tagged VirR showed that VirR binds to the regulatory region of *pfoA*, but not to the regulatory regions of other VirR-regulated genes, such as *plc* and *colA* (Cheung and Rood, 2000b). DNase I footprinting analysis of the *pfoA* promoter identified the VirR binding region, which contains two imperfect direct repeats of the sequence CCCAGTTNTNCAC. Site-directed mutagenesis, together with DNA-binding

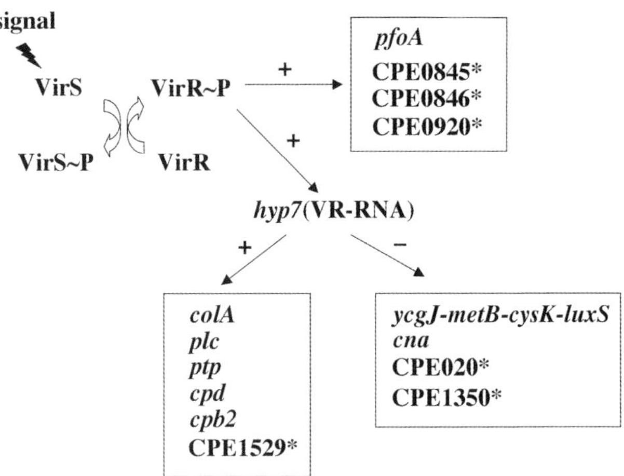

Figure 1. Regulatory pathway of extracellular toxin genes in *C. perfringens*. VirR phosphorylated by VirS activates transcription of toxin genes directly or indirectly through VR-RNA. It also functions as a negative regulator through VR-RNA. The product encoded by each gene is listed in Table 1. Direct or indirect regulation by VirR for genes marked with asterisks is characterized only by the presence or absence of the consensus VirR-binding sequence. Details as well as references are described in the text.

and DNase I protection experiments, revealed that VirR binds independently to the two repeated sequences. Consistent with the inability of VirR to bind to *plc* and *colA*, no version of the repeated sequence is found in the upstream regions of these two genes.

The predicted VirR/VirS-dependent regulator of *plc* and *colA* was identified by a search for new VirR/VirS-regulated genes, using a differential display method with cDNA probes prepared from total RNA of wild type and a *virR* mutant (Banu *et al.*, 2000). A mutation in one of the VirR/VirS-dependent genes, *hyp7*, blocked transcription of *colA* and *plc*, but not *pfoA*, suggesting that the *hyp7* gene encodes a regulatory factor specifically required for the transcription of *colA* and *plc*.

Other VirR/VirS-dependent genes identified in this study include *ptp* (protein tyrosine phosphatase) and *cpd* (2′, 3′-cyclic nucleotide 2′-phosphodiesterase). In addition, an operon that contains *ycgJ* (encoding a hypothetical protein), *metB* (encoding cystathionine gamma-synthase), *cysK* (encoding cysteine synthase), and *ygaG* (later identified as *luxS*, see below) appeared to be negatively regulated by VirR/VirS (Ohtani *et al.*, 2000). Transcription of the operon was only observed in early to mid-exponential growth phase (Banu *et al.*, 2000).

2.2. VirR-Regulated RNA (VR-RNA)

The aforementioned *hyp7*, which is required for the transcription of *colA* and *plc*, was also shown to activate *ptp* and *cpd* and repress *ycgJ-metB-cysK-ygaG* (*luxS*) (Shimizu *et al.*, 2002c). The VirR/VirS-regulated promoter region of *hyp7* gene contains the sequence CCAGTTACGCAC, which is similar to the proposed

Table 1. Genes regulated by the VirR/VirS signal transduction system

Gene	Product	Regulation[1]	Reference[2]
pfoA	perfringolysin O (theta-toxin)	VirRS / P	1, 3
colA	collagenase (kappa-toxin)	VR-RNA / P	1, 2
plc	phospholipase C (alpha toxin)	VR-RNA / P	1, 2
hyp7	VR-RNA	VirRS / P	2
ptp	protein tyrosine phosphatase	VR-RNA / P	2, 6
cpd	2′, 3′-cyclic nucleotide 2'-phosphodiesterase	VR-RNA / P	2, 6
ycgJ	hypothetical protein	VR-RNA / N	2, 6
metB	cystathionine gamma-synthase	VR-RNA / N	2, 6
cysK	cysteine synthase	VR-RNA / N	2, 6
luxS	autoinducer-2 production protein	VR-RNA / N	2, 6
cpb2	beta 2-toxin	VR-RNA / P	4
cna	collagen adhesin	VR-RNA / N	4
CPE0846	α-clostripain	VirRS* / P	5, 7
CPE0202	probable cell wall binding protein	VR-RNA* / N	5
CPE0845	hypothetical protein	VirRS* / P	7
CPE0920	hypothetical protein	VirRS* / P	7
CPE1350	fructose bisphosphate aldolase	VR-RNA* / N	5
CPE1529	hypothetical protein	VR-RNA* / P	5

[1] Expression was regulated positively (P) or negatively (N) by VirR/S or VR-RNA. Asterisks indicate probable regulation predicted by sequences in the promoter regions.
[2] References: 1, (Ba-Thein *et al.*, 1996); 2, (Banu *et al.*, 2000); 3, (Cheung and Rood, 2000); 4, (Ohtani *et al.*, 2003); 5, (Shimizu *et al.*, 2002b); 6, (Shimizu *et al.*, 2002c); 7, (Shimizu *et al.*, 2002a).

VirR-binding sequence (CCCAGTTNTNCAC). The model that emerges suggests that VirR directly activates the transcription of *hyp7*, the product of which in turn controls the target genes listed above. Surprisingly, the activity of *hyp7* as a transcription regulator is attributed not to a *hyp7*-encoded protein, but rather to the *hyp7* transcript (Shimizu *et al.*, 2002c). The entire *hyp7* transcript, known as the VirR-regulated RNA (VR-RNA), is predicted to form a tight and compact secondary structure with base pairing between its 5′- and 3′- ends that creates an axis of symmetry. However, deletion analysis showed that only a small 3′-proximal region of VR-RNA is needed for its regulatory activity. Whether the 5′-region plays any role in regulation has not been determined. The lack of sequence similarity between the 3′ region of VR-RNA and the promoter regions of VR-RNA-regulated genes implies that VR-RNA does not regulate the expression of its target genes by annealing to complementary DNA or RNA. The authors suggest instead that VR-RNA forms ribonucleoprotein complex with unidentified proteins, as in the case of RNAIII, which regulates expression of virulence factors in *Staphylococcus aureus* (Novick *et al.*, 1993). A proposed VirR/VirS regulatory pathway in toxin production is shown in Figure 1.

Another regulatory RNA (*virX*) was reported to act as a positive regulator of *plc*, *colA*, and *pfoA* (Ohtani *et al.*, 2002a). The detailed regulatory mechanism for this RNA and its relation with VR-RNA remain to be determined.

2.3. The VirR/VirS Regulon

Two-dimensional gel electrophoresis of culture fluids of wild-type and *virR* mutant cells, followed by matrix-assisted laser adsorption ionization-time of flight/mass spectrometry, confirmed the previous VirR/VirS-dependent genes and revealed four additional secreted proteins whose synthesis is VirR/VirS-dependent (Table 1) (Shimizu *et al.*, 2002b). Eight previously unrecognized target proteins were produced in higher concentration in the *virR* mutant, suggesting that their genes are negatively regulated by VirR/VirS. However, transcriptome analysis showed that only seven of the fifteen genes that encode the various secreted proteins are regulated at the transcriptional level by VirR/VirS. Other proteins are processed by a VirR/VirS-dependent extracellular cysteine protease that is similar to α-clostripain of *Clostridium histolyticum* (Dargatz *et al.*, 1993). Similarly, iota-toxin produced in *C. perfringens* type E strains appears to be processed by a VirR/VirS-dependent protease (Gibert *et al.*, 2000).

Scanning the complete *C. perfringens* genome sequence for the VirR binding site identified three more genes that are likely to be directly regulated by VirR. These include the α-clostripain gene described above, and two open-reading frames encoding hypothetical proteins (Shimizu *et al.*, 2002a).

The complete nucleotide sequence of the 54.3-kb plasmid pCP13 found in certain strains of *C. perfringens* indicates that the plasmid carries genes encoding beta2-toxin (*cpb2*) and a possible collagen adhesin (*cna*), which may contribute to the pathogenicity of the bacterium (Shimizu *et al.*, 2002a). The transcription of *cpb2* and *cna* is under the positive and negative control of VR-RNA, respectively (Ohtani *et al.*, 2003). VirR/VirS-regulated genes, thus far identified or proposed, are listed in Table 1.

2.4. Characterization of VirR and VirS

Two-component regulatory proteins are utilized by prokaryotes and lower eukaryotes to sense and respond to environmental changes. The N-terminal domain of a sensor kinase recognizes an environmental or metabolic signal, leading to autophosphorylation at a conserved histidine residue near the C-terminus. The phosphoryl group is then transferred to the invariant aspartate residue in the N-terminal domain of the cognate response regulator, which results in a conformational change that activates the C-terminal effector domain (reviewed by Stock *et al.*, 2000).

The VirS sensor kinase contains six putative transmembrane regions. An amino acid substitution of His255 showed that this residue is likely to be the auto-phosphorylation site that is essential for VirS function (Cheung and Rood, 2000a). DXGXG and GXGL motifs in histidine kinases have been proposed to be nucleotide-binding sites and are usually located in close proximity to one another (Parkinson and Kofoid, 1992). However, the DXGXG motif of VirS, which was shown to be essential for activity, is predicted to be in the cytoplasmic loop between transmembrane regions 4 and 5. Two glutamate residues in the putative transmembrane domains are also indispensable for VirS function (Cheung and Rood, 2000a).

The N-terminal receiver domain of VirR has strong sequence similarity to that of other response regulators. The C-terminal domain does not have a canonical helix-turn-helix motif, but is similar to the C-terminal domain of the LytTR family of response regulators, which includes AgrA and LytR from *S. aureus* and AlgR from *Pseudomonas aeruginosa* (Nikolskaya and Galperin, 2002). Most of the transcriptional regulators in this class are involved in production of extracellular polysaccharides and secreted proteins including toxins and bacteriocins. The most conserved sequence FhRhHRS (where "h' indicates hydrophobic amino acids) is located at the beginning of the second α helix (residues 184-190 in VirR). Site-directed mutagenesis demonstrated that R186, H188, and S190 are essential for VirR function. The purified VirR mutant proteins, R186K, S190A, and S190C, were shown to have much lower binding affinity to the target DNA (McGowan *et al.*, 2002). The C-terminus of the LytTR domain contains a positively charged cluster of Lys and Arg residues. A SKHR motif within this cluster was also shown to be required for binding to the target DNA, and is, thus, indispensable for VirR activity (McGowan *et al.*, 2003). Structural studies of VirR should provide significant insights into the mechanism by which the LytTR family of response regulators functions in transcriptional activation.

2.5. Cell-Cell Signaling for Toxin Production

Extracellular signaling molecules mediate cell density-dependent gene regulation, also known as quorum sensing (reviewed by Bassler, 2002). An early report described a dialyzable substance (called substance A) produced by certain theta-toxin non-producing strains during exponential growth phase that restored toxin production to another class of non-producing strains of *C. perfringens*, suggesting that cell-cell communication is involved in toxin production (Imagawa and Higashi, 1992).

Cell-cell communication systems in Gram-negative bacteria are usually mediated by homoserine lactones, which serve as signaling molecules, whereas secreted peptides are the primary factors in intercellular communication in Gram-positive bacteria (Kleerebezem *et al.*, 1997). However, autoinducer 2 (AI-2), the production of which requires the *luxS* gene, was recently shown to mediate interspecies communication (Schauder *et al.*, 2001; Chen *et al.*, 2002). As described above, *C. perfringens luxS* resides within an operon that is negatively regulated by the VirR/VirS system (Banu *et al.*, 2000). Cell-free culture fluid from the wild-type strain of *C. perfringens*, but not from the *luxS* mutant, greatly stimulates the luminescence of *Vibrio harveyi* tester strains in which production of luminescence is dependent on AI-2. This result indicates that the *luxS* gene functions in AI-2 production in *C. perfringens* (Ohtani *et al.*, 2002b).

The production of alpha-, kappa-, and theta-toxins was reduced in the *luxS* mutant, although reduced transcription was only seen in the case of the theta-toxin gene (*pfoA*). Conditioned medium prepared from a wild-type culture stimulates toxin production and *pfoA* transcription in the *luxS* mutant. In contrast the supernatant prepared from cultures of the *luxS* mutant has less of a stimulatory effect. The *luxS*-mediated stimulation requires VirR/VirS. However, overexpression of *virR* results in increased toxin gene transcription in the *luxS* mutant, which may indicate that VirR/VirS is able to regulate the transcription even in the absence of AI-2, suggesting that AI-2 may not be the signal for VirR/VirS (Ohtani *et al.*, 2002b). Alternatively, overexpression of *virR* might bypass the requirement for VirS. If so, one could not eliminate the possibility that AI-2 is indeed the signal to activate VirS. The possibility that substance A described above is AI-2 and that either AI-2 or substance A is the signal for VirR/VirS remains to be tested. AI-2-mediated cell-cell signaling might play a role in the pathogenicity of *C. perfringens* during multiple infection with other AI-2-producing bacteria (Ohtani *et al.*, 2002b).

2.6. Other Possible Regulatory Systems Affecting Expression of Extracellular Toxin Genes in *C. perfringens*

As described above, the expression of the theta toxin gene, *pfoA*, is directly activated by the VirR/VirS system. The *pfoR* gene, which is located upstream of *pfoA*, was shown to positively affect the expression of *pfoA* when the genes were cloned in *Escherichia coli* (Shimizu *et al.*, 1991). However, a recent study showed that *pfoA* expression is independent of *pfoR* in *C. perfringens* (Awad and Rood, 2002). Therefore the relationship between PfoR and VirR/VirS-dependent regulation, if any, remains unclear.

The *plc* gene carries, immediately upstream from its promoter, three poly(A) tracts on the same face of the DNA helix, which were shown to be required for the promoter activity *in vivo*. *In vitro* studies have also shown that the curved DNA containing the poly(A) tracts stimulates transcription *in vitro*, the stimulatory effect being more prominent at lower temperature (Matsushita *et al.*, 1996). It was proposed that low temperature-dependent stimulation is due to an increase in the bending angle, which facilitates the formation of a closed complex (Katayama *et al.*, 1999). Hydroxyl radical footprinting and fluorescence polarization assays indicated that the C-terminal domain of the α subunit binds to the minor grooves of the three phased (A)-tracts and that the binding affinity is increased at lower temperature (Katayama *et al.*, 2001). Whether these poly(A) tracts are required for VR-RNA-dependent regulation of the *plc* gene was not determined.

3. Gene Regulation of Enterotoxin Production in *C. perfringens*

Enterotoxin, which causes food poisoning in humans and animals, is the only known *C. perfringens* toxin that is not secreted from cells (Chapter 9). The *cpe* gene encoding enterotoxin is located on the chromosome in food-borne gastrointestinal disease isolates but on a plasmid in non-food-poisoning isolates. A *cpe*-carrying plasmid was shown to be conjugative, suggesting a possible means of transferring the *cpe* gene between cells of the intestinal flora (Brynestad *et al.*, 2001).

3.1. Sporulation-Dependent CPE Production

Enterotoxin is produced only in sporulating cells (Duncan *et al.*, 1972) and mutants blocked at stage 0 of sporulation are unable to produce CPE (Duncan, 1973). C2 toxin production in *C. botulinum* C2 also appears to be associated with sporulation (Nakamura *et al.*, 1978). Expression of the enterotoxin gene is regulated at the transcriptional level (Melville *et al.*, 1994). When the *cpe* gene was introduced by transformation into natural isolates of *cpe*-negative *C. perfringens*, all strains were able to produce CPE in a sporulation-associated manner and the *cpe* mRNA in the transformants was only detected during sporulation, indicating that the *cpe*-negative isolates produce the regulatory factors required for *cpe* transcriptional regulation (Czeczulin *et al.*, 1996).

The *cpe* promoter regions are classified into two types. One class contains a 45-bp insertion not found in the other class, but *cpe* transcription initiates in *C. perfringens* at the same transcriptional start site for either type of promoter (Melville *et al.*, 1994). The expression of *cpe* was measured both in *C. perfringens* and *Bacillus subtilis* by using translational fusions of each type of promoter to *E. coli gusA* encoding β-glucuronidase. Both types of promoter exhibited higher activity in stationary phase, especially in the fusion-bearing *C. perfringens* cells.

3.2. SigE- and SigK-Dependent Transcription of the *cpe* Gene

Deletion analysis of the *cpe* promoter region identified three promoters, P1 and P2, which are separated by 15-bp, and an upstream promoter P3 (Zhao and Melville, 1998). Transcription from a DNA fragment carrying both P1 and P2, as measured using fusions to the *E. coli* β-glucuronidase (*gusA*) gene, is similar to that from the full promoter and P3 exhibits about 50% of the full activity. Transcription from either P1/P2 or P3 is strongly induced during stationary phase. Alignment of the sequences of P1, P2, and P3 with those of the consensus promoters recognized by different sigma factors in *B. subtilis* (reviewed by Haldenwang, 1995) suggested that P1 is a SigK-dependent promoter and P2 and P3 are

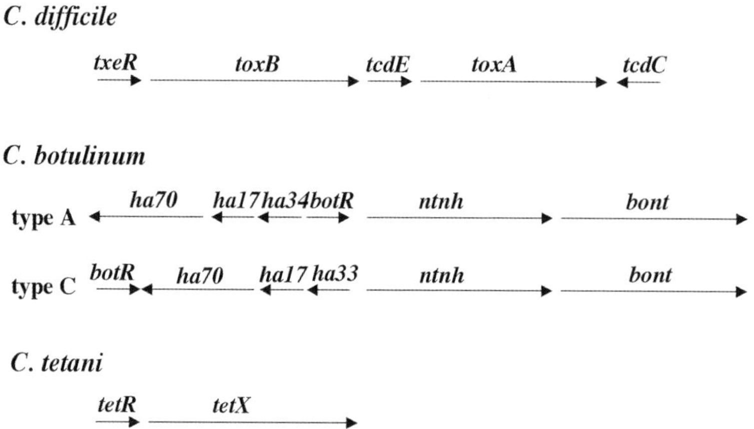

Figure 2. Organization of the toxin gene clusters from *C. difficile*, *C. botulinum*, and *C. tetani*. The gene clusters in *C. botulinum* are shown in only type A and C strains.

SigE-dependent promoters (Zhao and Melville, 1998). Although these assignments remain to be tested using more direct approaches such as *in vitro* transcription with RNA polymerase reconstituted from core enzyme and each sigma, a previous report that CPE is produced mostly in the mother cell compartment rather than the forespore (Roper *et al.*, 1976) supports the hypothesis that the *cpe* gene is transcribed by the mother-cell specific SigE and SigK holoenzymes (Zhao and Melville, 1998). Genes likely encoding SigE and SigK were detected in the genome of *C. perfringens* (Shimizu *et al.*, 2002a). Unlike the *sigK* genes in *B. subtilis* (Stragier *et al.*, 1989) and *C. difficile* (Haraldsen and Sonenshein, 2003), which are disrupted by a prophage-like element termed *skin* (*sigK* intervening sequence), the *C. perfringens sigK* lacks the *skin* element and is intact.

4. Toxin Gene Regulation by Alternative Sigma Factors

Recent studies have revealed a new global regulatory system involving alternative sigma factors that control expression of toxin genes at the level of transcription initiation. These sigma factors are TxeR in *C. difficile*, TetR in *C. tetani*, and BotR in *C. botulinum*. As described below these sigma factors share sequence similarity and are functionally interchangeable (Marvaud *et al.*, 1998a).

4.1. TxeR in *C. difficile*

C. difficile is a causative agent of pseudomembranous colitis and antibiotic-associated diarrhea (Chapter 9). The toxin A and B genes, *toxA* (or *tcdA*) and *toxB* (*tcdB*), are located in a 19-kb pathogenicity locus, together with three additional genes, *txeR* (*tcdD*), *tcdE* (*txe2*), and *tcdC* (*txe3*) (Figure 2). In early exponential growth phase, a high level of *tcdC* transcription and low levels of *toxA*, *toxB*, *txeR*, and *tcdE* transcription are seen, but this expression pattern is reversed in stationary phase, suggesting that TcdC might be a negative regulator of the other genes within the cluster (Hundsberger *et al.*, 1997), a hypothesis that has not been experimentally tested.

The presence of *toxA*- and *toxB*-specific promoters with sequence elements in common was shown by primer extension analysis and RNase T2 protection assays (Dupuy and Sonenshein, 1998). Induction of transcription from these promoters during stationary phase is blocked if the culture medium contains glucose or other rapidly metabolizable carbon sources (Dupuy and Sonenshein, 1998).

The *txeR* gene product is a 22-kDa protein that contains a helix-turn-helix motif, which is characteristic of DNA binding proteins (Hundsberger *et al.*, 1997). To determine whether TxeR is involved in transcriptional regulation of *toxA* and *toxB*, their putative promoter regions were fused to a reporter gene carrying the *C. difficile* toxin A repeating units (ARU). The expression of ARU in *E. coli* was strongly enhanced when *txeR* was expressed in trans (Moncrief *et al.*, 1997). To rule out the possibility that the apparent regulation seen in *E. coli* is an artifact of using a Gram-negative host, fusions of *toxA* and *toxB* to the *gusA* gene were also studied in *C. perfringens*. (*C. difficile* itself could not be used at the time because no method for introducing engineered DNA into that organism was then known.) TxeR is indeed required for transcription of *toxA* and *toxB* in *C. perfringens* (Dupuy and Sonenshein, 1998).

Such recapitulation of *tox* gene regulation in *C. perfringens* indicates that TxeR is likely to be a direct regulator of toxin gene expression. Surprisingly, however, TxeR does not bind to the *toxA* or *toxB* promoter regions. This mystery was resolved by the demonstration that TxeR is an alternative RNA polymerase sigma factor (Mani and Dupuy, 2001). Addition of TxeR greatly enhances the binding affinity of RNA polymerase to *toxA* and *toxB* promoter DNA and stimulates transcription from these promoters *in vitro* by RNA polymerase holoenzyme from *E. coli*, *B. subtilis* or *C. difficile* (Mani and Dupuy, 2001). More convincingly, TxeR enables *E. coli* and *B. subtilis* core

RNA polymerases to bind to the *tox* promoters and to initiate transcription from these promoters (Mani and Dupuy, 2001). TxeR also activates its own transcription, whether in *E. coli, C. perfringens,* or *C. difficile*. The expression of *txeR* in these organisms, like that of the *tox* genes, increases when cells enter stationary phase and expression is repressed in medium containing glucose (Mani *et al.*, 2002). Sequence alignment of TxeR revealed that it is distantly related to certain sigma factors of the extracytoplasmic function (ECF) sigma family but defines a new sub-family of sigma factors (reviewed by Helmann, 2002).

The induction of *tox* gene expression seen in stationary phase cells and its repression by glucose can now be attributed to the effects of growth state on transcription of *txeR*. But what controls expression of the *txeR* gene? CodY, a protein found almost ubiquitously in the low G+C Gram-positive bacteria, represses during rapid exponential growth phase dozens of *B. subtilis* genes that are induced when cells make the transition to stationary phase (Slack *et al.*, 1995; Molle *et al.*, 2003). CodY is activated as a repressor by two types of effector molecules, which generally have additive effects. First, repression is dependent on the concentration of GTP, which interacts directly with CodY (Ratnayake-Lecamwasam *et al.*, 2001). Second, the affinity of CodY for target DNAs is increased substantially in the presence of the amino acids isoleucine and valine (R. P. Shivers and A. L. Sonenshein, unpublished). *C. difficile* encodes a close homolog of *B. subtilis* CodY, which has been purified after cloning and overexpression in *E. coli*. The purified protein binds tightly to the *txeR* promoter region and the binding is stimulated separately and additively by GTP and isoleucine (A. Villapakkam, J. Nordman, and A. L. Sonenshein, unpublished). Thus, CodY is likely to be the regulatory protein that ties toxin gene expression to the growth state of the cells.

Carbon catabolite repression in many Gram-positive bacteria including *B. subtilis* is mediated by CcpA, a member of the *lac* repressor family (reviewed by Stulke and Hillen, 2000). The apparent *C. difficile* homolog of CcpA was overexpressed in *E. coli* and purified. This protein was able to bind to the *txeR* promoter region *in vitro* (E. Miller, J. Nordman, and A. L. Soneneshein, unpublished). Therefore, it is likely that CcpA mediates catabolite repression of the toxin genes by controlling synthesis of TxeR. For both CodY and CcpA, their hypothesized roles in toxin gene regulation will only be proved when knock-out mutations in their respective genes have been constructed and shown to cause derepression of *txeR*. It is not clear that it will be possible to create such mutations using currently available methods.

4.2. BotR in *C. botulinum*

Botulinum neurotoxins (BoNT) are the most potent toxins known and cause botulism in both animals and humans (Chapter 9). Therapeutic use of BoNT has also attracted considerable attention (reviewed by Johnson, 1999). Different strains of *C. botulinum* produce structurally similar but immunologically distinct types of BoNT that fall into seven characterized serotypes, A to G.

Botulism toxin genes (*bont*) are always found in an operon with and downstream of the *ntnh* (non-toxic, non-haemaglutinin component) gene (Henderson *et al.*, 1996). A divergently transcribed operon encodes one or more haemaglutinin (HA) components. The two operons are sometimes separated by and sometimes bordered by the *botR* gene (Figure 2), although *botR* has not been found in the vicinity of *bont* in *C. botulinum* E, nor has it been found elsewhere in the genome as determined by PCR or DNA/DNA hybridization (Raffestin *et al.*, 2004). The *botR* gene encodes a product with significant similarity to TxeR. A type A *C. botulinum* strain that overexpresses *botR/A* under the control of its own promoter produces higher-than-expected levels of BoNT/A and associated non-toxic proteins, as a result of increased transcription. Conversely, partial inhibition of *botR/A* expression by antisense RNA results in reduced transcription of genes encoding BoNT/A, NTNH, and HA70. Gel mobility shift assays and immunoprecipitation experiments showed that crude extracts containing BotR/A contain proteins that can bind to the regions upstream of the two operon promoters, suggesting that BotR/A directly activates transcription (Marvaud *et al.*, 1998b). A recent study showed that BotR/A does not bind directly to the promoters unless it forms a complex with core RNA polymerase (Raffestin *et al.*, 2004).

4.3. TetR in *C. tetani*

A gene (*tetR*) homologous to *txeR* and *botR* is found immediately upstream of the tetanus toxin (TeTx) gene, both of which are carried on a plasmid (Eisel *et al.*, 1986; Fairweather and Lyness, 1986). TetR exhibits 50 to 65% identity with BotR proteins. Overproduction of TetR by amplifying its gene on a high-copy number plasmid leads to an increase in the production of TetTx, which can be attributed to higher *tetX* transcription. Similarly, overproduction of BotR/A (60% identity with TetR) in *C. tetani* increases TetTx production, whereas overproduction of BotR/C, which shows less identity (50%) with TetR, is less effective in stimulating toxin production. These results demonstrate that TetR positively regulates TeTx gene expression and that BotR and TetR are functionally interchangeable (Marvaud *et al.*, 1998a). Binding of TetR to the *tetX* promoter requires core RNA polymerase, and furthermore, BotR/A-RNAP and TetR-RNAP are interchangeable to direct transcription *in vitro*, confirming that TetR and BotR/A, like TxeR, are alternative sigma factors involved in toxin gene regulation (Raffestin *et al.*, 2004).

4.4. Global Gene Regulation by Alternative Sigma Factors

Newly identified TxeR-family sigma factors have been shown to play a pivotal role in clostridial toxin production. TxeR shares significant similarity, not only with BotR and TetR, but also with *C. perfringens* UviA (Garnier and Cole, 1988), the putative positive regulator of *bcn*, the UV-inducible bacteriocin locus (Hundsberger *et al.*, 1997; Dupuy and Sonenshein, 1998). Consistent with this similarity among the regulators, the TxeR-recognized promoters, particularly in their -35 regions, exhibit high similarity with each other as well as with the *bcn* promoter (Mani *et al.*, 2002). The complete genome of *C. tetani* contains two more putative σ factors that show sequence similarity to TxeR (Bruggemann *et al.*, 2003). Furthermore, the transcription of *toxA* and *toxB* exhibits a maximal level at 37°C and low levels at 22 and 42°C (Karlsson *et al.*, 2003). This temperature-control also requires TxeR and *txeR* expression itself is autoinduced at 37°C, suggesting that TxeR-class sigma factors respond to multiple environmental signals. As described above, TxeR shows some similarity to ECF sigma factors, whose activity is controlled by anti-sigma factors. Future studies are needed to determine whether a similar mechanism controls the activity of TxeR.

5. Nutritional Regulation

Toxin gene expression in *C. difficile* is subject to complex nutritional regulation, but mechanistic studies of such regulation are still at an early stage. The presence of rapidly metabolizable carbon sources represses toxin gene expression in *C. difficile* cultured in complex media as described above (Dupuy and Sonenshein, 1998) and may be attributable to CcpA, but the repression is not observed in defined media (Karlsson *et al.*, 1999). Toxin production is also suppressed by certain amino acids (Karasawa *et al.*, 1997; Yamakawa *et al.*, 1998; Karlsson *et al.*, 1999; 2000) and the production is greatly enhanced under biotin-limited conditions (Yamakawa *et al.*, 1996; Karlsson *et al.*, 1999; Maegawa *et al.*, 2002). The production of BoNT in *C. botulinum* is also affected by certain amino acids (Patterson-Curtis and Johnson, 1989; Leyer and Johnson, 1990). The effect of biotin limitation was proposed to be related to a block in purine biosynthesis (Maegawa *et al.*, 2002) or to the essential role of biotin as a co-factor in carboxylation reactions (Karlsson *et al.*, 1999). Reductive carboxylation of propionyl-CoA generates α-ketoglutarate, which is a precursor in isoleucine biosynthesis (Karlsson *et al.*, 1999). (It is noteworthy that isoleucine increases the affinity of CodY for target DNAs as described above.) Surprisingly, an increase in bicarbonate concentration also enhances toxin production (Karlsson *et al.*, 1999). A similar stimulatory effect by CO_2 on type *B botulinum* neurotoxin (*bont/B*) gene expression was recently reported (Lövenklev *et al.*, 2004). The relationship between the effects of biotin and bicarbonate is unclear. Butyric acid in growth media induces toxin production in *C. difficile*, whereas butanol suppresses production, suggesting a link between the induction of metabolic pathways involved in butyric acid production and regulation of toxin gene expression at least in certain media (Karlsson *et al.*, 2000; 2003).

6. Future Prospects

This review has summarized recent discoveries related to toxin gene regulation in *Clostridium* spp.. The field has clearly seen substantial progress in the last several years and prospects for future advances are highly promising. The identification of the signal that activates the VirR/VirS system will be key to furthering our understanding of how the bacteria decide when to synthesize virulence factors and how the signal transduction system operates. Elucidation of the regulatory mechanism of VR-RNA remains to be carried out. An important question that needs to be answered rigorously is how different environmental signals are sensed, culminating in activation of the TxeR-class of sigma factors. On-going genome sequencing and post-genomic analysis will undoubtedly provide valuable information that will drive future studies in the field of toxin gene regulation. In the short term, these studies will require better tools for genetic manipulation. In the long run, the results obtained will help us to develop strategies for preventing, diagnosing, and treating the serious diseases caused by the toxin-producing *Clostridium* species.

Acknowledgements

We thank Michel Popoff for valuable discussions and M. Popoff and Peter Rådström for communicating unpublished results. We thank Shunji Nakano for comments on the manuscript. Unpublished work from the laboratory of A. L. Sonenshein was made possible by a research grant (GM042219) from the U. S. Public Health Service.

References

Awad, M.M., and Rood, J.I. 2002. Perfringolysin O expression in *Clostridium perfringens* is independent of the upstream *pfoR* gene. J. Bacteriol. 184: 2034-2038.

Ba-Thein, W., Lyristis, M., Ohtani, K., Nisbet, I.T., Hayashi, H., Rood, J.I., and Shimizu, T. 1996. The *virR/virS* locus regulates the transcription of genes encoding extracellular toxin production in *Clostridium perfringens*. J. Bacteriol. 178: 2514-2520.

Banu, S., Ohtani, K., Yaguchi, H., Swe, T., Cole, S.T., Hayashi, H., and Shimizu, T. 2000. Identification of novel VirR/VirS-regulated genes in *Clostridium perfringens*. Mol. Microbiol. 35: 854-864.

Bassler, B.L. 2002. Small talk. Cell-to-cell communication in bacteria. Cell. 109: 421-424.

Bruggemann, H., Baumer, S., Fricke, W.F., Wiezer, A., Liesegang, H., Decker, I., Herzberg, C., Martinez-Arias, R., Merkl, R., Henne, A., and Gottschalk, G. 2003. The genome sequence of *Clostridium tetani*, the causative agent of tetanus disease. Proc. Natl. Acad. Sci .USA. 100: 1316-1321.

Brynestad, S., Sarker, M.R., McClane, B.A., Granum, P.E., and Rood, J.I. 2001. Enterotoxin plasmid from *Clostridium perfringens* is conjugative. Infect. Immun. 69: 3483-3487.

Chen, X., Schauder, S., Potier, N., Van Dorsselaer, A., Pelczer, I., Bassler, B.L., and Hughson, F.M. 2002. Structural identification of a bacterial quorum-sensing signal containing boron. Nature. 415: 545-549.

Cheung, J.K., and Rood, J.I. 2000a. Glutamate residues in the putative transmembrane region are required for the function of the VirS sensor histidine kinase from *Clostridium perfringens*. Microbiology. 146: 517-525.

Cheung, J.K., and Rood, J.I. 2000b. The VirR response regulator from *Clostridium perfringens* binds independently to two imperfect direct repeats located upstream of the *pfoA* promoter. J. Bacteriol. 182: 57-66.

Czeczulin, J.R., Collie, R.E., and McClane, B.A. 1996. Regulated expression of *Clostridium perfringens* enterotoxin in naturally *cpe*-negative type A, B, and C isolates of *C. perfringens*. Infect. Immun. 64: 3301-3309.

Dargatz, H., Diefenthal, T., Witte, V., Reipen, G., and von Wettstein, D. 1993. The heterodimeric protease clostripain from *Clostridium histolyticum* is encoded by a single gene. Mol. Gen. Genet. 240: 140-145.

Duncan, C.L., Strong, D.H., and Sebald, M. 1972. Sporulation and enterotoxin production by mutants of *Clostridium perfringens*. J. Bacteriol. 110: 378-391.

Duncan, C.L. 1973. Time of enterotoxin formation and release during sporulation of *Clostridium perfringens* type A. J. Bacteriol. 113: 932-936.

Dupuy, B., and Sonenshein, A.L. 1998. Regulated transcription of *Clostridium difficile* toxin genes. Mol. Microbiol. 27: 107-120.

Eisel, U., Jarausch, W., Goretzki, K., Henschen, A., Engels, J., Weller, U., Hudel, M., Habermann, E., and Niemann, H. 1986. Tetanus toxin: primary structure, expression in *E. coli*, and homology with botulinum toxins. EMBO J. 5: 2495-2502.

Fairweather, N.F., and Lyness, V.A. 1986. The complete nucleotide sequence of tetanus toxin. Nucleic Acids Res. 14: 7809-7812.

Garnier, T., and Cole, S.T. 1988. Studies of UV-inducible promoters from *Clostridium perfringens in vivo* and *in vitro*. Mol. Microbiol. 2: 607-614.

Gibert, M., Petit, L., Raffestin, S., Okabe, A., and Popoff, M.R. 2000. *Clostridium perfringens* iota-toxin requires activation of both binding and enzymatic components for cytopathic activity. Infect. Immun. 68: 3848-3853.

Haldenwang, W.G. 1995. The sigma factors of *Bacillus subtilis*. Microbiol. Rev. 59: 1-30.

Haraldsen, J.D., and Sonenshein, A.L. 2003. Efficient sporulation in *Clostridium difficile* requires disruption of the σ^K gene. Mol. Microbiol. 48: 811-821.

Helmann, J.D. 2002. The extracytoplasmic function (ECF) sigma factors. Adv. Microb. Physiol. 46: 47-110.

Henderson, I., Whelan, S.M., Davis, T.O., and Minton, N.P. 1996. Genetic characterisation of the botulinum toxin complex of *Clostridium botulinum* strain NCTC 2916. FEMS Microbiol. Lett. 140: 151-158.

Hundsberger, T., Braun, V., Weidmann, M., Leukel, P., Sauerborn, M., and von Eichel-Streiber, C. 1997. Transcription analysis of the genes *tcdA-E* of the pathogenicity locus of *Clostridium difficile*. Eur. J. Biochem. 244: 735-742.

Imagawa, T., and Higashi, Y. 1992. An activity which restores theta toxin activity in some theta toxin-deficient mutants of *Clostridium perfringens*. Microbiol. Immunol. 36: 523-527.

Johnson, E.A. 1999. Clostridial toxins as therapeutic agents: benefits of nature's most toxic proteins. Annu. Rev. Microbiol. 53: 551-575.

Johnson, E.A., and Bradshaw, M. 2001. *Clostridium botulinum* and its neurotoxins: a metabolic and cellular perspective. Toxicon. 39: 1703-1722.

Karasawa, T., Maegawa, T., Nojiri, T., Yamakawa, K., and Nakamura, S. 1997. Effect of arginine on toxin production by *Clostridium difficile* in defined medium. Microbiol. Immunol. 41: 581-585.

Karlsson, S., Burman, L.G., and Åkerlund, T. 1999. Suppression of toxin production in *Clostridium difficile* VPI 10463 by amino acids. Microbiology. 145: 1683-1693.

Karlsson, S., Lindberg, A., Norin, E., Burman, L.G., and Åkerlund, T. 2000. Toxins, butyric acid, and other short-chain fatty acids are coordinately expressed and down-regulated by cysteine in *Clostridium difficile*. Infect. Immun. 68: 5881-5888.

Karlsson, S., Dupuy, B., Mukherjee, K., Norin, E., Burman, L.G., and Åkerlund, T. 2003. Expression of *Clostridium difficile* toxins A and B and their sigma factor TcdD is controlled by temperature. Infect. Immun. 71: 1784-1793.

Katayama, S., Matsushita, O., Jung, C.M., Minami, J., and Okabe, A. 1999. Promoter upstream bent DNA activates the transcription of the *Clostridium perfringens* phospholipase C gene in a low temperature-dependent manner. EMBO J. 18: 3442-3450.

Katayama, S., Matsushita, O., Tamai, E., Miyata, S., and Okabe, A. 2001. Phased A-tracts bind to the alpha subunit of RNA polymerase with increased affinity at low temperature. FEBS Lett. 509: 235-238.

Kleerebezem, M., Quadri, L.E., Kuipers, O.P., and de Vos, W.M. 1997. Quorum sensing by peptide pheromones and two-component signal-transduction systems in Gram-positive bacteria. Mol. Microbiol. 24: 895-904.

Leyer, G.J., and Johnson, E.A. 1990. Repression of toxin production by tryptophan in *Clostridium botulinum* type E. Arch. Microbiol. 154: 443-447.

Lövenklev, M., Artin, I., Hagberg, O., Borch, E., Holst, E., and Rådström, P. 2004. Quantitative interaction effects of carbon dioxide, sodium chloride and sodium nitrite on neurotoxin gene expression in nonproteolytic *Clostridium botulinum* type B. Appl. Environ. Microbiol. 70: 2928-2934.

Lyristis, M., Bryant, A.E., Sloan, J., Awad, M.M., Nisbet, I.T., Stevens, D.L., and Rood, J.I. 1994. Identification and molecular analysis of a locus that regulates extracellular toxin production in *Clostridium perfringens*. Mol. Microbiol. 12: 761-777.

Maegawa, T., Karasawa, T., Ohta, T., Wang, X., Kato, H., Hayashi, H., and Nakamura, S. 2002. Linkage between toxin production and purine biosynthesis in *Clostridium difficile*. J. Med. Microbiol. 51: 34-41.

Mani, N., and Dupuy, B. 2001. Regulation of toxin synthesis in *Clostridium difficile* by an alternative RNA polymerase sigma factor. Proc. Natl. Acad. Sci. USA. 98: 5844-5849.

Mani, N., Lyras, D., Barroso, L., Howarth, P., Wilkins, T., Rood, J.I., Sonenshein, A.L., and Dupuy, B. 2002. Environmental response and autoregulation of *Clostridium difficile* TxeR, a sigma factor for toxin gene expression. J. Bacteriol. 184: 5971-5978.

Marvaud, J.C., Eisel, U., Binz, T., Niemann, H., and Popoff, M.R. 1998a. TetR is a positive regulator of the tetanus toxin gene in *Clostridium tetani* and is homologous to *botR*. Infect. Immun. 66: 5698-5702.

Marvaud, J.C., Gibert, M., Inoue, K., Fujinaga, Y., Oguma, K., and Popoff, M.R. 1998b. *botR/A* is a positive regulator of botulinum neurotoxin and associated non-toxin protein genes in *Clostridium botulinum* A. Mol. Microbiol. 29: 1009-1018.

Marvaud, J.C., Raffestin, S., Gibert, M., and Popoff, M.R. 2000. Regulation of the toxinogenesis in *Clostridium botulinum* and *Clostridium tetani*. Biol. Cell. 92: 455-457.

Matsushita, C., Matsushita, O., Katayama, S., Minami, J., Takai, K., and Okabe, A. 1996. An upstream activating sequence containing curved DNA involved in activation of the *Clostridium perfringens plc* promoter. Microbiology. 142: 2561-2566.

McClane, B.A. 1998. New insights into the genetics and regulation of expression of *Clostridium perfringens* enterotoxin. Curr. Top. Microbiol. Immunol. 225: 37-55.

McGowan, S., Lucet, I.S., Cheung, J.K., Awad, M.M., Whisstock, J.C., and Rood, J.I. 2002. The FxRxHrS motif: a conserved region essential for DNA binding of the VirR response regulator from *Clostridium perfringens*. J. Mol. Biol. 322: 997-1011.

McGowan, S., O'Connor, J.R., Cheung, J.K., and Rood, J.I. 2003. The SKHR motif is required for biological function of the VirR response regulator from *Clostridium perfringens*. J. Bacteriol. 185: 6205-6208.

Melville, S.B., Labbe, R., and Sonenshein, A.L. 1994. Expression from the *Clostridium perfringens cpe* promoter in *C. perfringens* and *Bacillus subtilis*. Infect. Immun. 62: 5550-5558.

Molle, V., Nakaura, Y., Shivers, R.P., Yamaguchi, H., Losick, R., Fujita, Y., and Sonenshein, A.L. 2003. Additional targets of the *Bacillus subtilis* global regulator CodY identified by chromatin immunoprecipitation and genome-wide transcript analysis. J. Bacteriol. 185: 1911-1922.

Moncrief, J.S., Barroso, L.A., and Wilkins, T.D. 1997. Positive regulation of *Clostridium difficile* toxins. Infect. Immun. 65: 1105-1108.

Nakamura, S., Serikawa, T., Yamakawa, K., Nishida, S., Kozaki, S., and Sakaguchi, G. 1978. Sporulation and C2 toxin production by *Clostridium botulinum* type C strains producing no C1 toxin. Microbiol. Immunol. 22: 591-596.

Nikolskaya, A.N., and Galperin, M.Y. 2002. A novel type of conserved DNA-binding domain in the transcriptional regulators of the AlgR/AgrA/LytR family. Nucleic Acids. Res. 30: 2453-2459.

Novick, R.P., Ross, H.F., Projan, S.J., Kornblum, J., Kreiswirth, B., and Moghazeh, S. 1993. Synthesis of staphylococcal virulence factors is controlled by a regulatory RNA molecule. EMBO J. 12: 3967-3975.

Ohtani, K., Takamura, H., Yaguchi, H., Hayashi, H., and Shimizu, T. 2000. Genetic analysis of the *ycgJ-metB-cysK-ygaG* operon negatively regulated by the VirR/VirS system in *Clostridium perfringens*. Microbiol. Immunol. 44: 525-528.

Ohtani, K., Bhowmik, S.K., Hayashi, H., and Shimizu, T. 2002a. Identification of a novel locus that regulates expression of toxin genes in *Clostridium perfringens*. FEMS Microbiol. Lett. 209: 113-118.

Ohtani, K., Hayashi, H., and Shimizu, T. 2002b. The *luxS* gene is involved in cell-cell signalling for toxin production in *Clostridium perfringens*. Mol. Microbiol. 44: 171-179.

Ohtani, K., Kawsar, H.I., Okumura, K., Hayashi, H., and Shimizu, T. 2003. The VirR/VirS regulatory cascade affects transcription of plasmid-encoded putative virulence genes in *Clostridium perfringens* strain 13. FEMS Microbiol. Lett. 222: 137-141.

Parkinson, J.S., and Kofoid, E.C. 1992. Communication modules in bacterial signaling proteins. Annu. Rev. Genet. 26: 71-112.

Patterson-Curtis, S.I., and Johnson, E.A. 1989. Regulation of neurotoxin and protease formation in *Clostridium botulinum* Okra B and Hall A by arginine. Appl. Environ. Microbiol. 55: 1544-1548.

Raffestin, S., Marvaud, J.C., Cerrato, R., Dupuy, B., and Popoff, M.R. 2004. Organization and regulation of the neurotoxin genes in *Clostridium botulinum* and *Clostridium tetani*. Anaerobe. 10: 93-100.

Ratnayake-Lecamwasam, M., Serror, P., Wong, K.W., and Sonenshein, A.L. 2001. *Bacillus subtilis* CodY represses early-stationary-phase genes by sensing GTP levels. Genes Dev. 15: 1093-1103.

Rood, J.I., and Cole, S.T. 1991. Molecular genetics and pathogenesis of *Clostridium perfringens*. Microbiol. Rev. 55: 621-648.

Rood, J.I. 1997. Genetic analysis of *Clostridium perfrigens*. In: The Clostridia: Molecular Biology and Pathogenesis. J.I. Rood, B.A. McClane, J.G. Songer, and R.W. Titball, eds. Academic Press, London. p. 65-71.

Rood, J.I. 1998. Virulence genes of *Clostridium perfringens*. Annu. Rev. Microbiol. 52: 333-360.

Roper, G., Short, J.A., and Walker, D.A. 1976. The ultrastructure of *Clostridium perfringens* spores. Academic Press, London.

Schauder, S., Shokat, K., Surette, M.G., and Bassler, B.L. 2001. The LuxS family of bacterial autoinducers: biosynthesis of a novel quorum-sensing signal molecule. Mol. Microbiol. 41: 463-476.

Shimizu, T., Okabe, A., Minami, J., and Hayashi, H. 1991. An upstream regulatory sequence stimulates expression of the perfringolysin O gene of *Clostridium perfringens*. Infect. Immun. 59: 137-142.

Shimizu, T., Ba-Thein, W., Tamaki, M., and Hayashi, H. 1994. The *virR* gene, a member of a class of two-component response regulators, regulates the production of perfringolysin O, collagenase, and hemagglutinin in *Clostridium perfringens*. J. Bacteriol. 176: 1616-1623.

Shimizu, T., Ohtani, K., Hirakawa, H., Ohshima, K., Yamashita, A., Shiba, T., Ogasawara, N., Hattori, M., Kuhara, S., and Hayashi, H. 2002a. Complete genome sequence of *Clostridium perfringens*, an anaerobic flesh-eater. Proc. Natl. Acad. Sci. USA. 99: 996-1001.

Shimizu T, Shima K, Yoshino K, Yonezawa K, Shimizu T, Hayashi H. 2002b. Proteome and transcriptome analysis of the virulence genes regulated by the VirR/VirS system in *Clostridium perfringens*. J. Bacteriol.184: 2587-2594.

Shimizu, T., Yaguchi, H., Ohtani, K., Banu, S., and Hayashi, H. 2002c. Clostridial VirR/VirS regulon involves a regulatory RNA molecule for expression of toxins. Mol. Microbiol. 43: 257-265.

Slack, F.J., Serror, P., Joyce, E., and Sonenshein, A.L. 1995. A gene required for nutritional repression of the *Bacillus subtilis* dipeptide permease operon. Mol. Microbiol. 15: 689-702.

Sonenshein, A.L., Haraldsen, J.D., and Dupuy, B. 2004. RNA polymerase and alternative sigma factors. In: Handbook on Clostridia. P. Dürre, ed. CRC Press, Boca Raton, Florida. In press.

Stock, A.M., Robinson, V.L., and Goudreau, P.N. 2000. Two-component signal transduction. Annu. Rev. Biochem. 69: 183-215.

Stragier, P., Kunkel, B., Kroos, L., and Losick, R. 1989. Chromosomal rearrangement generating a composite gene for a developmental transcription factor. Science. 243: 507-512.

Stulke, J., and Hillen, W. 2000. Regulation of carbon catabolism in *Bacillus* species. Annu. Rev. Microbiol. 54: 849-880.

Yamakawa, K., Karasawa, T., Ikoma, S., and Nakamura, S. 1996. Enhancement of *Clostridium difficile* toxin production in biotin-limited conditions. J. Med. Microbiol. 44: 111-114.

Yamakawa, K., Karasawa, T., Ohta, T., Hayashi, H., and Nakamura, S. 1998. Inhibition of enhanced toxin production by *Clostridium difficile* in biotin-limited conditions. J. Med. Microbiol. 47: 767-771.

Zhao, Y., and Melville, S.B. 1998. Identification and characterization of sporulation-dependent promoters upstream of the enterotoxin gene (*cpe*) of *Clostridium perfringens*. J. Bacteriol. 180: 136-142.

Chapter 11

Use of Anaerobic Bacteria for Cancer Therapy

J. Martin Brown* and Shie-chau Liu

Abstract

This chapter describes the power of genetically engineered bacteria in cancer therapy. In the applications we will be considering, the bacteria are genetically engineered to carry a specific gene into tumors. This is not what is classically thought of as gene therapy, which is defined as the introduction of a gene, or part of a gene, into the cancer cells (or normal cells). In other words bacteria are not vectors for the introduction of genes into mammalian cells. However, anaerobic bacteria (and other types of bacteria) can and do concentrate in tumors by various means and can carry a gene of interest to produce a protein of choice in tumors. This can be a powerful adjunct to cancer therapy. In this chapter we will consider necrosis-targeted therapy, of which species of the obligate anaerobe *Clostridium* are the prototypical agent.

1. Hypoxia and Necrosis in Solid Tumors

Necrotic regions are a common, if not a universal, feature of human solid tumors, although this is rarely quantitated. Dang and colleagues (2001) recently reported that all 20 liver metastases of colorectal carcinoma ≥ 1cm^3 in size had 25-75% of their volume occupied by necrosis. These necrotic regions typically occur at a distance from functioning blood vessels beyond the diffusion distance of oxygen (Thomlinson and Gray, 1955), as demonstrated also in the *in vitro* spheroid model (Groebe and Vaupel, 1988). Because these necrotic regions typically develop due to prolonged lack of oxygen, they are usually intimately associated with hypoxic but viable cells. These hypoxic regions in tumors are best detected by polarographic oxygen electrodes, and measurements with such electrodes have shown that the majority of primary tumors or metastases of the head and neck, cervix, brain and melanomas, sarcomas, anal prostate and pancreatic cancers have regions that are severely hypoxic (Rampling *et al.*, 1994; Brizel *et al.*, 1995; Vaupel and Hockel, 1995; Sundfor *et al.*, 1997; Becker *et al.*, 1998; Brizel *et al.*, 1999; Movsas *et al.*, 1999; Koong *et al.*, 2000). Moreover, the fact that hypoxic cells are resistant to killing by ionizing radiation, and are also more slowly proliferating than well oxygenated cells makes them resistant to treatment by radiotherapy and chemotherapy (Brown and Giaccia, 1998; Wouters *et al.*, 2002).

2. Certain Species of Clostridia Colonize Solid Tumors.

The genus *Clostridium* comprises a large and heterogeneous group of Gram-positive, spore-forming bacteria that become vegetative and grow only in the absence (or at very low levels) of oxygen. Malmgren and Flanagan were the first to demonstrate that intravenous injection of spores of *C. tetani* colonized solid tumors by observing that tumor-bearing mice died of tetanus within 48 hr of intravenous injection of *C. tetani* spores, whereas non tumor-bearing animals were unaffected (Malmgren and Flanigan, 1955). Möse and Möse (1959; 1964) later reported that a nonpathogenic clostridial strain, *C. butyricum* M-55, localized and germinated in solid Ehrlich tumors, causing extensive lysis without any concomitant effect on normal tissues. Such observations were soon confirmed and extended by a number of investigators using tumors in mice, rats, hamsters, and rabbits (Engelbart and Gericke, 1964; Thiele *et al.*, 1964) and were followed by clinical studies with cancer patients (Carey *et al.*, 1967; Heppner and Mose, 1978; Heppner *et al.*, 1983). While the anaerobic bacteria did not significantly alter tumor control or eradication, these clinical reports established the safety of this approach as well as the fact that colonization of human tumors occurs following intravenous injection of clostridial spores.

3. Clostridia-Directed Enzyme Prodrug Therapy (CDEPT)

The above studies showing the specific colonization of tumors by non-pathogenic clostridial species prompted the suggestion first made in 1994 that they could be used as a tumor specific gene delivery system (Lemmon *et al.*, 1994). In essence, the tumor-targeting concept we proposed was to inject spores of a nonpathogenic

*For correspondence email: mbrown@stanford.edu

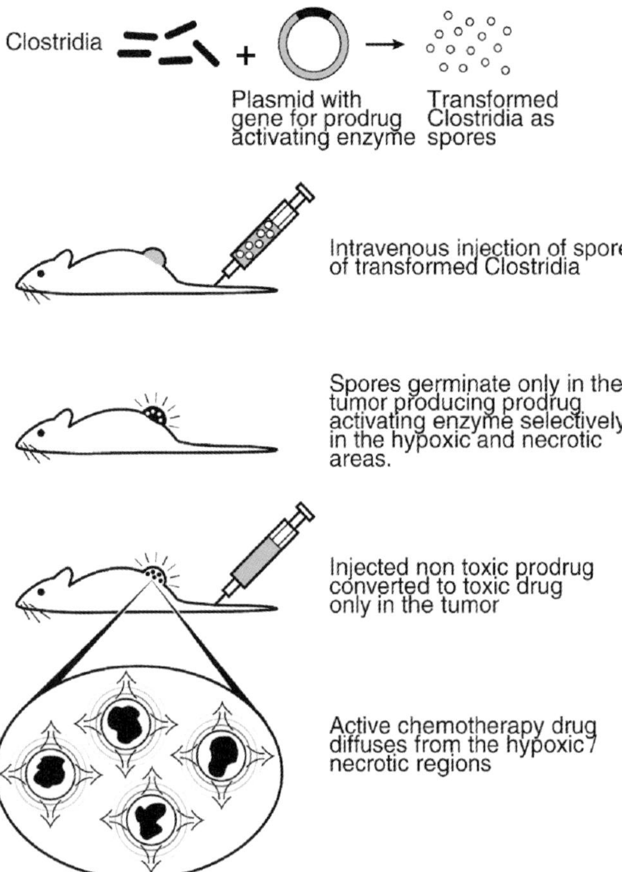

Figure 1. Diagrammatic representation of tumor-specific targeting of chemotherapy using obligate anaerobes

clostridial strain, genetically engineered to produce a specific protein potentially harmful to the tumor when the spores germinate and grow in the hypoxic/necrotic areas of the tumor. As such germination to the vegetative form will occur only in the hypoxic/necrotic regions, the protein should be produced only in the tumor. We suggested that control of the cytotoxicity could be achieved by a two-step process similar to that used by the antibody-directed enzyme prodrug therapy (ADEPT) approach. Rather than using antibodies, however, our proposal was to have the clostridia produce the enzyme in the tumor following which the animal would be treated systemically with a non-toxic prodrug, which would be metabolized to a highly cytotoxic species only by the enzyme. Since the enzyme is localized to the tumor, the generation of the toxic metabolite should also be localized to the tumor (Figure 1). More details are given in the following sections.

3.1. Choice of Clostridial Species

When we first began to develop the clostridial system for targeting of anticancer therapy the most important requirement was to choose a nonpathogenic strain of clostridia that would allow genetic manipulations. Because of the long history, dating back to World War I (Jones and Woods, 1986) of use of the strain *C. acetobutylicum* (later renamed *C. beijerinckii*) in producing the solvents acetone and butanol, and the more recent molecular manipulations of this species to exploit this biotechnological potential (Minton *et al.*, 1990; Narberhaus and Bahl, 1992; Walter *et al.*, 1992), this strain was an obvious candidate. It was also important that this strain be able to germinate in tumor tissue, and indeed tumor lysis has been reported for this strain (though to a lesser extent than the original *C. butyricum* strain (later renamed *C. oncolyticum*) (Mose and Mose, 1964).

This early work enabled us to demonstrate that we could genetically engineer *C. beijerinckii* to produce *Escherichia coli* cytosine deaminase (CD) that was active in metabolizing 5-fluorocytosine to 5-fluorouracil (Fox *et al.*, 1996). Importantly, we also showed that genetically engineered clostridia expressed the protein (in this case from the *E. coli*-derived *nitroreductase* gene) in experimental tumors in mice, whereas no protein was expressed in any normal tissues including that of the GI tract (Lemmon *et al.*, 1997). These studies established the proof of principle that the expression of an active foreign enzyme can be produced selectively in tumors with no expression of the exogenous enzyme in any normal tissues.

However, our preclinical studies with genetically engineered *C. beijerinckii* failed to produce antitumor activity when the relevant prodrug was injected systemically (5-fluorocytosine for CD-expressing tumors and CB1954 for nitroreductase-expressing tumors). We reasoned that this lack of antitumor activity was a result of insufficient levels of prodrug-activating enzymes because of low levels of viable clostridia in the tumors. Indeed, our experiments showed that *C. beijerinckii* produced only 10^5 to 10^6 bacteria per gram of tumor (Lemmon *et al.*, 1997) whereas experiments with *C. oncolyticum* and a closely related species *C. sporogenes* gave 10^8 to 2 x 10^8 bacteria/g of tumor in experimental tumors. Unfortunately, until very recently all attempts to transform *C. sporogenes* were unsuccessful. Very recently our laboratory found that the reason for the failure was high levels of DNAase produced by the organism, and we have now successfully transformed C. *sporogenes* using PEG-based transfection buffer containing DNAase inhibitors (see below).

3.2. Transformation of *C. sporogenes*

An electroporation technique developed initially in our lab [modified from a method of Oultram *et al.* (1988) was used with the clostridial strain *C. beijerinckii* NCIMB 8052, with the pAMβ1 plasmids derivatives (e.g. pMTL500E and pMTL540FT) encoding resistance

Table 1. Transformation of Clostridial species using electroporation

A.	***C. beijerinckii***
1.	Grow bacteria to mid-log phase (until OD_{600} reaches about 0.6) in 100 ml of 2xYTG broth (per liter: 16 g Bacto tryptone, 10 g Bacto yeast extract, 5 g NaCl).
2.	Harvest and wash cells once in 10 ml ice-cold electroporation buffer (270 mM sucrose, 1 mM $MgCl_2$, 7 mM Na phosphate, pH 7.4),
3.	Resuspend cells in 2 ml ice-cold electroporation buffer.
4.	Incubate the cell suspension on ice for 10 min.
5.	Add DNA (0.5 – 1 µg) to 0.4 ml of the cell suspension held in a Bio-Rad electroporation cuvette and keep the cells on ice for an additional 4 min
6.	Deliver an electric pulse with the conditions set at 1.25 kV (voltage 6.25 kV/cm in 0.2-cm inter-electrode cuvettes), capacitance 25 µF, resistance 100 ohm and time constant about 3.5 ms using a Bio-Rad *E. coli* Pulser^{TM}.
7.	Keep cells on ice for a further 4 min.
8.	Dilute 0.4 ml of the cell suspension with 3.6 ml 2 x YTG.
9.	Incubate the cell suspension at 37°C for 3 hrs for phenotypic expression.
10.	Plate cells on a selective medium containing erythromycin at 20 µg/ml.
B.	***C. sporogenes***
1.	Pick a single colony from a freshly streaked plate to 5 ml TPGY broth [per liter: trypticase peptone 20 g, Bacto peptone 5 g, yeast extract 5 g, cysteine-HCl 1 g, add glucose to 0.1%, (wt./vol.)] and set up serial dilutions (up to 10^5 time dilution).
2.	Incubate overnight at 37°C under hypoxic conditions.
3.	Use the two lowest dilutions of the overnight culture to inoculate 300 ml of fresh TPGY broth.
4.	Incubate at 37°C until OD_{600} reaches 0.7 to 0.8.
5.	Harvest cells by centrifugation at 8,000 rpm at 4°C for 10 min in a Beckman JA-14 rotor and wash 2 times with 10 ml of ice-cold electroporation buffer consisting of 10 % polyethylene glycol (PEG 8000) and 1 mg/ml of polyanetholesulfonic acid.
6.	Centrifuge the cell suspension at 10,000 rpm at 4°C for 10 min in a Beckman JA-20 rotor.
7.	Suspend cells in 3 ml of ice-cold electroporation buffer containing additional DNase inhibitor, aurintricarboxylic acid at final concentration of 0.1 mg/ml.
8.	Keep cell suspension in a dry-ice bath for an additional 10 min.
9.	After a freeze-thaw cycle, transfer 400 µl of cell suspension into a 0.2 cm electroporation cuvette (Bio-Rad Laboratories, Richmond, CA, USA).
10.	Mix with plasmid DNA by pipetting and apply a single electric pulse (capacitance 25 µF, voltage 6.25 kV/cm, and time constant approximately 3.0 ms) with an *E. coli* Pulser^{TM} (Bio-Rad Laboratories, Richmond, CA, USA).
11.	Dilute the pulsed cells with 4.6 ml of TPGY broth containing 25 mM $MgCl_2$, incubate at 37°C for 4 hrs.
12.	Plate the cells onto TPGY agar supplemented with erythromycin at 10 µg/ml.

to erythromycin (Em^R) (Table 1A). Em^R transformants appeared 2-3 days after incubation with a maximum of 2 x 10^3 transformants obtained per µg DNA. All manipulations (including removal of supernatant from cell pellets and delivery of electric pulse) were performed under anaerobic conditions.

This electroporation procedure provides a rapid, reproducible, and efficient means for transforming intact cells of *C. beijerinckii* with plasmid DNA. However our early attempts to extend this protocol to *C. sporogenes* were unsuccessful. As we have reported (Liu *et al.*, 2002) the major obstacle in transforming *C. sporogenes* is high-level extracellular DNase secretion by this organism. Thus, a novel electroporation procedure had to be developed to transform this class of nuclease producing clostridia. Within the past 2 years we have succeeded in transforming *C. sporogenes* with enzyme-producing plasmid constructs.

A summary of the successful transformation procedure for *C. sporogenes* is shown in Table 1B. Em^R transformants should appear after incubation for 3-4 days at 37°C under hypoxic conditions. It has been reported that a freeze-thaw cycle enhances the transformation efficiency and sometimes is essential to the transformation of many Gram–negative and Gram-positive bacteria (Dityatkin *et al.*, 1974; Holsters *et al.*, 1978; Selvaraj and Iyer, 1981; Merrick *et al.*, 1987). While the freeze-thaw cycle is thought to weaken the cell wall sufficiently to permit DNA to enter, the exact role of divalent metal ions of Mg^{+2} remains unclear. However, it is thought to be involved in the process of cell regeneration.

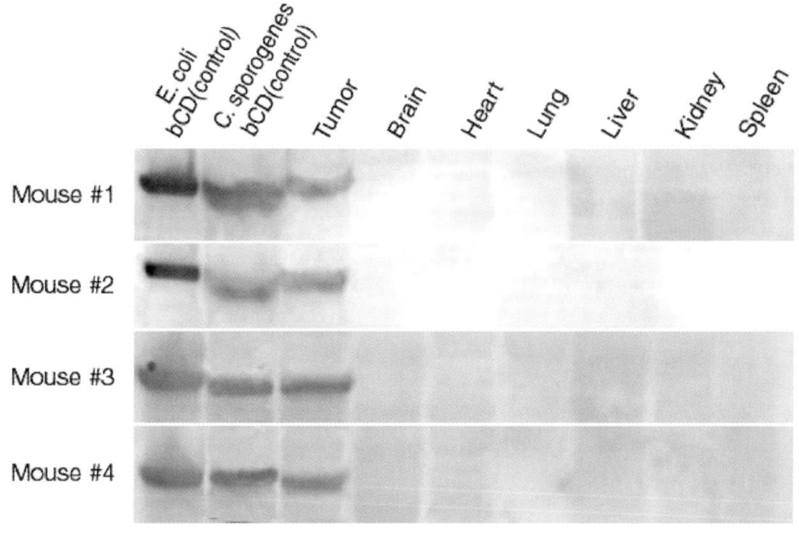

Figure 2. Immunoblot for CD in cell extracts from SCCVII tumors and from several normal tissues from the same mice injected i.v. 7 days prior to animal sacrifice with spores of CD-transformed *C. sporogenes*. The tumors were approximately 100 mg at the time of i.v. injection. 100μg of cell extract from the tumor and normal mouse tissues was loaded on the gel. Shown also are cell extracts from CD-transformed *E. coli* and *C. sporogenes* with 1 and 20 μg cell extract loaded respectively. From (Liu *et al.*, 2002) with permission.

The triphenylmethane dye aurintricarboxylic acid (ATA) is a polyanionic, polyaromatic compound and is a general inhibitor of nucleases. ATA has been shown to inhibit a wide variety of exo and endonucleases (Hallick *et al.*, 1977; Benchokroun *et al.*, 1995). It was postulated that this compound would inhibit the association of any nucleic acid binding protein with nucleic acid. The ATA molecule carries three carboxylic acid moieties that are negatively charged at physiological pH and appears to be a non-specific enzyme inhibitor by virtue of its structure, with effects on many systems (Bina-Stein and Tritton, 1976). It has been suggested that these groups facilitate extensive ionic binding of the molecule to positively charged groups of cellular proteins (Bina-Stein and Tritton, 1976). A nuclease is known to be more sensitive to inhibition by ATA than many other enzymes, and endonuclease activity has been demonstrated to be particularly sensitive to inhibition by ATA (Bina-Stein and Tritton, 1976). The action of ATA as a nuclease inhibitor can be used advantageously in the transformation of highly active nuclease- producing clostridial strains.

Polyethylene glycol (PEG) has long been used to facilitate the chemically-induced transformation of yeast, Gram-negative and Gram-positive bacteria (Klebe *et al.*, 1983; Takahashi *et al.*, 1983; Smith, 1985) and even clostridial species (Soutschek-Bauer and Staudenbauer, 1987). While the exact mechanism by which the PEG promotes DNA uptake has not been determined, it is believed that in addition to protecting incoming DNA from nucleases, it could also mediate fusion of the DNA to the recipient cell membrane. The molecular weight of the PEG seems less critical for the effectiveness of transformation as an average molecular weight from 1000 to 8000 usually yields equally effective results.

In conclusion, electroporation offers perhaps the simplest, fastest, and most promising alternative to establish a transformation system for introducing DNA into formerly untransformable clostridia. However, not every clostridial strain can be readily electroporated and when successful, frequencies and efficiencies vary. Optimal conditions differ from strain to strain. The culture medium and the growth phase affect frequency, as do the addition of agents that weaken the cell wall (*e.g.*, cycloserine, penicillin, DL-threonine, glycine and muramidases). Transformation efficiency can be enhanced by optimizing these factors. The ionic strength and the pH of the electroporation buffer, and the addition of metal ions and DNase inhibitors can have a significant influence on electroporetic success. The most important parameters for bacteria, however, appear to be field strength, duration of the pulse, and the shape of the pulse. Thus, the system will benefit from the use of new types of power supplies and electroporation-cell designs leading to lower cell lethality and permitting higher electric discharges. The system will also benefit from a better understanding of the actual mechanism of how DNA transverses the murein (or peptidoglycan) layer of the cell wall and what influence membrane composition has on electroporetic success. However, the underlying mechanism of electroporation remains elusive and the physical chemistry of what happens to a bacterial membrane during a high-voltage electric pulse is equally vague. What is clear is that electroporation provides a viable alternative method for enabling DNA to enter bacterial cells and it has a significant advantage over most current methods. Our data lead us to believe that electroporation will be the method of choice for the transformation of the formerly untransformable clostridial strains.

3.3. Antitumor Activity of *C. sporogenes* Transformed with Cytosine Deaminase.

We have recently shown that intravenous injection into tumor-bearing mice of spores of *C. sporogenes*

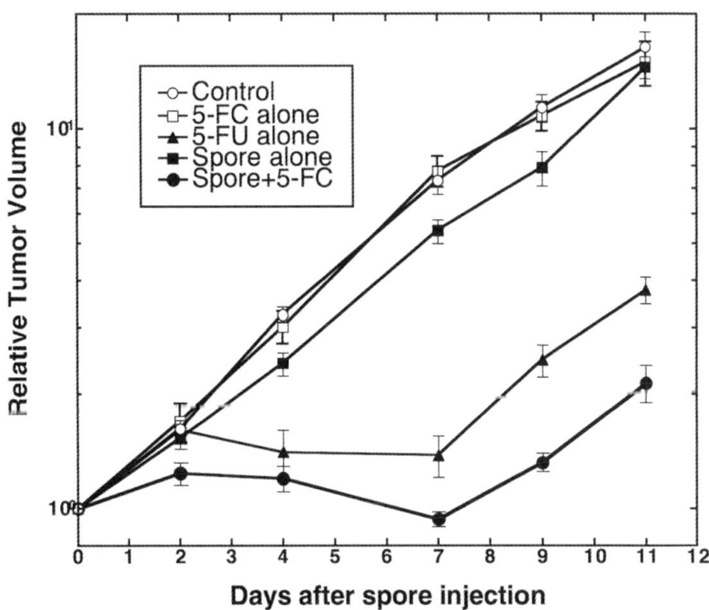

Figure 3. Growth delay produced by daily (5x/week) injections of 5-FC (500 mg/kg/day) in SCCVII tumors injected on day 0 with 10^8 spores of cytosine deaminase-transformed *C. sporogenes*. Shown also is the growth delay produced by a similar injection schedule of the MTD of 5-FU in the same experiment. Data from 5 mice per group with the error bars showing SEM for each group. From (Liu *et al.*, 2002) with permission

transformed with an *E. coli* cytosine deaminase (CD)-expressing plasmid resulted in tumor-specific expression of CD (Figure 2). Importantly, we found significant anti-tumor efficacy of systemically injected 5-fluorocytosine (5FC) following IV injection of these recombinant spores (Liu *et al.*, 2002) (Figure 3). This was at least as great as that achieved with the maximum tolerated doses of 5-fluorouracil (5-FU) given systemically. We have also shown that the 5-FC in the tumor is actively converted to 5-FU and that the levels achieved are at least 10 times greater than those achievable with systemic drug administration. Furthermore, there was no detectable 5-FU in the blood stream with these high levels in the tumor. These data demonstrate that systemically applied recombinant clostridial spores with a non-toxic prodrug can produce significant and specific antitumor activity.

3.4. CDEPT has Significant Advantages Over Antibody and Gene-Directed Enzyme Prodrug Therapy

Both ADEPT (antibody directed) and GDEPT (gene directed enzyme prodrug therapy) are viable approaches to achieve locally high concentrations of anticancer drugs in tumors and are appropriately receiving significant effort both in preclinical and clinical studies. However, we believe that the clostridial approach to enzyme/prodrug therapy (CDEPT) has a number of intrinsic advantages as outlined below.

3.4.1. Lack of an Immune Response

An immune response against the antibody/enzyme conjugates has been a major problem with ADEPT (Melton and Sherwood, 1996). However, clostridial spores elicit an undetectable immune response, and when the spores germinate into vegetative bacteria, they are present in necrotic areas of the tumor, which are immune-privileged sites. In fact, it has been found both in clinical (Carey *et al.*, 1967) and preclinical studies (Gericke *et al.*, 1979) that multiple treatments are effective. In studies with intravenous (i.v.) injection of *C. sporogenes* and *C. acetobutylicum* we have found no loss of numbers of bacteria in experimental tumors for up to 14-17 days respectively following a single i.v. injection of spores (Liu *et al.*, 2002; Nuyts *et al.*, 2002).

3.4.2. An Advantageous Intratumor Distribution

The necessity of delivering enzyme/antibody conjugates, or vectors for GDEPT, through the bloodstream makes it likely that the highest concentrations of enzyme will be perivascular. However, the resistant cells in the tumor are likely to be the nonproliferating hypoxic cells distant from the vasculature and often close to necrosis. In contrast the prodrug-activating enzyme from clostridia will be at its highest concentration in areas adjacent to necrosis and far from blood vessels. Not only does this guarantee that the enzyme from clostridia will be at its highest level adjacent to the distant cells, it also minimizes the problem of leakage of activated drug back into the blood vessels, a problem that has been reported for ADEPT (Martin *et al.*, 1997).

3.4.3. The Same Construct will be Universally Applicable to all Cancer Patients

As opposed to ADEPT, which requires either tumor-specific or possibly patient-specific enzyme/antibody conjugates, the clostridial approach, because it depends only on tumor necrosis, will have universal applicability.

In a recent publication on the potential of clostridia to target human tumors, Dang and colleagues reported that all 20 liver metastases of colorectal carcinoma ≥ 1cm^3 in size had 25-75% of their volume occupied by necrosis (Dang *et al*., 2001).

3.4.4. Higher Tumor to Normal Tissue-Targeting Ratios

Major problems with ADEPT and GDEPT are that tumor targeting cannot be 100% effective. In most cases the majority of the injected material will not be localized in the tumor but in the reticular endothelial system. This necessitates efforts to clear such nonbound material. On the other hand, with clostridia it appears that 100% of the novel protein expressed from the recombinant clostridia is within the tumor.

3.5. Combination of CDEPT with Vascular-Targeting Agents

Because of the need for hypoxic necrotic areas in tumors for vegetative growth of obligate anaerobes it would seem that the CDEPT approach would not be effective in small tumors lacking necrosis. However, recent studies with vascular-targeting agents promise not only to enhance the efficacy of clostridial therapy in medium sized tumors but also raise the exciting possibility that the minimum tumor size for efficacy can be substantially reduced. Though the prototype vascular-targeting agent flavone acetic acid (FAA) has been studied in preclinical work for over 12 years, it is only recently that this approach is showing promise in the clinic. Two classes of compounds show both preclinical and clinical activity: DMXAA (5,6-dimethylxanthenone-4-acetic acid), which acts primarily by inducing TNF in tumors (Joseph *et al*., 1999; Zhao *et al*., 2002), and tubulin-binding agents such as combretastatin 4A and an analog ZD 6126 currently being developed by AstraZeneca (Tozer *et al*., 1999; Goto *et al*., 2002). The mechanism of action of these agents is a rapid selective occlusion of tumor blood vessels leading to necrosis within 16-24 hrs.

These effects of vascular targeting agents have been shown to increase both the tumor colonization efficiency of i.v. injected clostridial spores (Theys *et al*., 2001) and antitumor activity of nonrecombinant clostridia (Dang *et al*., 2001). The reason for the increased tumor colonization by clostridia is the large increase in tumor necrosis caused by these vascular-targeting agents. Not only does this increase the number of active clostridia per tumor but also reduces the volume of viable tumor tissue that needs to be exposed to the product of the vegetative clostridia.

Thus, vascular-targeting agents could be an important adjunct to clostridia-directed enzyme prodrug therapy (CDEPT), not only in potentiating its effectiveness in medium to large tumors but in extending its range into very small tumors.

3.6. Will CDEPT Work in the Clinic?

It was the lack of any adverse effects in preclinical testing of *C. butyricum* that led Möse and Möse to inject themselves with the spores, thereby demonstrating the lack of human pathogenicity. Clinical studies with cancer patients followed using i.v. injection of 10^9 to 10^{10} spores per individual (Carey *et al*., 1967; Heppner and Mose, 1978; Heppner *et al*., 1983). Typically, a low-grade fever occurs from 1-3 days following injection, with further increase in temperature from days 5-8, which coincides with lysis in the center of the tumor. In subsequent clinical studies with 49 patients with malignant gliomas Heppner and Mose injected clostridial spores into the carotid artery on the side of the tumor (Heppner *et al*., 1983). All tumors showed lysis, following which the tumors were removed surgically. However, there was no prolongation of the recurrence-free interval. In more recent studies with the nonpathogenic clostridial strain *C. butyricum*, CNRZ 528, Fabricius and colleagues (Fabricius *et al*., 1993) injected 68 clinical volunteers with 3.5 to 7.0 x 10^9 spores per patient in order to determine the half-life of blood clearance of the spores. This was found to be 1-2 days in man. No adverse effects were noted.

These clinical reports demonstrate that spores of non-pathogenic strains of clostridia can be given safely, that the spores germinate in the necrotic regions of tumors, and that in the case of *C. oncolyticum*, lysis of the tumors can occur. Also in preclinical studies with mice and rabbits, a similar lack of *C. oncolyticum* pathogenicity has been seen (Thiele *et al*., 1964). In our own studies we have seen no adverse effects following i.v. injections in control and tumor bearing mice with up to 10^9 genetically engineered *C. sporogenes* per mouse.

An important question is whether there will be an immune response against the clostridial spores, the vegetative bacteria, or the secreted protein that might preclude treatment of patients on multiple occasions. However, both the clinical (Carey *et al*., 1967) and experimental data (Gericke *et al*., 1979) so far suggest that multiple treatments are effective. In our own studies we have found undiminished expression of cytosine deaminase and no loss of numbers of bacteria in experimental tumors for up to 14 days following a single intravenous injection.

In summary, we believe that clostridia-directed enzyme/prodrug therapy has established its likelihood of being effective in the clinic.

4. *Bifidobacterium longum*: An Alternative to *Clostridium*?

Another anaerobic bacterium that can selectively grow in the hypoxic regions of solid tumors after i.v. injection is *Bifidobacterium longum* (Yazawa *et al.*, 2000). These are Gram-positive anaerobes found in the lowest small intestine and large intestine of humans and other animals. Studies with mice with B16 melanoma or Lewis lung tumors have shown tumor-specific proliferation with minimal levels in normal tissues from 24 hr after injection. As with clostridia the proliferation of the bacteria is in the necrotic areas of the tumor. *B. longum* has been successfully transformed with a shuttle vector described by Matsumura *et al.*, (1997) and these transformed bacteria colonize tumors similarly to the wild-type bacterium. However, results published to date show that levels of only 1-4 x 10^6 bacteria per gram of tumor are obtained. In addition, recent work of Dang and colleagues has shown that following i.v. injection of *B. longum*, the growing bacteria were tightly clustered within colonies in the necrotic areas rather than being distributed throughout the necrotic regions (Dang *et al.*, 2001). Thus, the rather low colonization efficiency and the tendency to clump rather than distribute within necrotic areas would appear to make *B. longum* inferior to the optimized strains of clostridia for enzyme prodrug therapy.

5. Future Directions

To date, antitumor efficacy has been demonstrated using one enzyme (cytosine deaminase) with the prodrug 5-FC, which it converts into the active anticancer 5-fluorouracil (5-FU). However, 5-FU is a drug primarily active against proliferating cells, so there is a strong need for work with other enzymes that can convert prodrugs into more active anticancer agents. Thus, one avenue for future work is exploration of other enzyme/prodrug combinations. Another avenue that should be explored is the combination of CDEPT with antivascular agents. Since these agents cause massive vascular shutdown and necrosis and have already been shown to increase the colonization of clostridia in experimental tumors, further work is needed to explore the potential of such combinations. Not only should this be performed in experimental tumors of standard size but also should be expanded to include small metastases, which normally do not have necrosis and so would not normally be colonized by obligate anaerobes. If systemic administration of antivascular agents such as DMXAA and combretastatin can produce necrosis in small metastases, they would then be susceptible to treatment with CDEPT. Finally, Phase I clinical trials need to be performed with the CDEPT approach. These have already been performed with attenuated *Salmonella*, but the lack of robust colonization by this organism has lessened enthusiasm for this approach. There is reason to believe, however, that genetically modified clostridia will not suffer the same problems since studies with non-genetically modified clostridia in the '70s and '80s demonstrated high levels of colonization and tumor lysis following i.v. injection of clostridial spores. Colonization and measurement of levels of novel enzymes and conversion of prodrug to active drug (for example, 5-FC to 5-FU) could readily be performed in a relatively small Phase I clinical study. Positive results from such a study would augur well for larger scale applications of CDEPT.

References

Becker, A., Hansgen, G., Bloching, M., Weigel, C., Lautenschlager, C., and Dunst, J. 1998. Oxygenation of squamous cell carcinoma of the head and neck: comparison of primary tumors, neck node metastases, and normal tissue. Int. J. Radiat. Oncol. Biol. Phys. 42: 35-41.

Benchokroun, Y., Couprie, J., and Larsen, A.K. 1995. Aurintricarboxylic acid, a putative inhibitor of apoptosis, is a potent inhibitor of DNA topoisomerase II *in vitro* and in Chinese hamster fibrosarcoma cells. Biochem. Pharmacol. 49: 305-313.

Bina-Stein, M., and Tritton, T.R. 1976. Aurintricarboxylic acid is a nonspecific enzyme inhibitor. Mol. Pharmacol. 12: 191-193.

Brizel, D.M., Dodge, R.K., Clough, R.W., and Dewhirst, M.W. 1999. Oxygenation of head and neck cancer: changes during radiotherapy and impact on treatment outcome. Radiother. Oncol. 53: 113-117.

Brizel, D.M., Rosner, G.L., Prosnitz, L.R., and Dewhirst, M.W. 1995. Patterns and variability of tumor oxygenation in human soft tissue sarcomas, cervical carcinomas, and lymph node metastases. Int. J. Radiat. Oncol. Biol. Phys. 32: 1121-1125.

Brown, J.M., and Giaccia, A.J. 1998. The unique physiology of solid tumors: Opportunities (and problems) for cancer therapy. Cancer Res. 58: 1408-1416.

Carey, R.W., Holland, J.F., Whang, H.Y., Neter, E., and Bryant, B. 1967. Clostridial oncolysis in man. Europ. J. Cancer 3: 37-46.

Dang, L.H., Bettegowda, C., Huso, D.L., Kinzler, K.W., and Vogelstein, B. 2001. Combination bacteriolytic therapy for the treatment of experimental tumors. Proc. Natl. Acad. Sci. USA. 98: 15155-15160.

Dityatkin, S.Y., Il'yashenko, B.N., and Lisovskaya, K.V. 1974. Infection of frozen and thawed bacteria with isolated DNA of phage 1phi7. Sov. Genet. 8: 1158-1162.

Engelbart, K., and Gericke, D. 1964. Oncolysis by clostridia V. Transplanted tumors of the hamster. Cancer Res. 24: 239-243.

Fabricius, E.M., Schneeweiss, U., Schau, H.P., Schmidt, W., and Benedix, A. 1993. Quantitative investigations into the elimination of *in vitro*-obtained spores of the non-pathogenic *Clostridium butyricum* strain CNRZ 528, and their persistence in organs of different species following intravenous spore administration. Res. Microbiol. 144: 741-753.

Fox, M.E., Lemmon, M.J., Mauchline, M.L., Davis, T.O., Giaccia, A.J., Minton, N.P., and Brown, J.M. 1996. Anaerobic bacteria as a delivery system for cancer gene therapy: activation of 5-fluorocytosine by genetically engineered clostridia. Gene Therapy. 3: 173-178.

Gericke, D., Dietzel, F., Konig, W., Ruster, I., and Schumacher, L. 1979. Further progress with oncolysis due to apathogenic clostridia. Zentralbl. Bakteriol. [Orig A]. 243: 102-112.

Goto, H., Yano, S., Zhang, H., Matsumori, Y., Ogawa, H., Blakey, D.C., and Sone, S. 2002. Activity of a new vascular targeting agent, ZD6126, in pulmonary metastases by human lung adenocarcinoma in nude mice. Cancer Res. 62: 3711-3715.

Groebe, K., and Vaupel, P. 1988. Evaluation of oxygen diffusion distances in human breast cancer xenografts using tumor-specific in vivo data: role of various mechanisms in the development of tumor hypoxia. Int. J. Radiat. Oncol. Biol. Phys. 15: 691-697.

Hallick, R.B., Chelm, B.K., Gray, P.W., and Orozco, E.M., Jr. 1977. Use of aurintricarboxylic acid as an inhibitor of nucleases during nucleic acid isolation. Nucleic Acids Res. 4: 3055-3064.

Heppner, F., Mose, J., Ascher, P.W., and Walter, G. 1983. Oncolysis of malignant gliomas of the brain. 13th Int. Cong. Chemother. 226: 38-45.

Heppner, F., and Mose, J.R. 1978. The liquefaction (oncolysis) of malignant gliomas by a non pathogenic clostridium. Acta Neuro. 12: 123-125.

Holsters, M., de Waele, D., Depicker, A., Messens, E., van Montagu, M., and Schell, J. 1978. Transfection and transformation of *Agrobacterium tumefaciens*. Mol. Gen. Genet. 163: 181-187.

Jones, D.T., and Woods, D.R. 1986. Acetone-butanol fermentation revisited. Microbiol. Rev. 50: 484-524.

Joseph, W.R., Cao, Z., Mountjoy, K.G., Marshall, E.S., Baguley, B.C., and Ching, L.M. 1999. Stimulation of tumors to synthesize tumor necrosis factor-alpha in situ using 5,6-dimethylxanthenone-4-acetic acid: a novel approach to cancer therapy. Cancer Res. 59: 633-638.

Klebe, R.J., Harriss, J.V., Sharp, Z.D., and Douglas, M.G. 1983. A general method for polyethylene-glycol-induced genetic transformation of bacteria and yeast. Gene. 25: 333-341.

Koong, A.C., Mehta, V.K., Le, Q.T., Fisher, G.A., Terris, D.J., Brown, J.M., Bastidas, A.J., and Vierra, M. 2000. Pancreatic tumors show high levels of hypoxia. Int. J. Radiat. Oncol. Biol. Phys. 48: 919-922.

Lemmon, M.J., Elwell, J.H., Brehm, J.K., Mauchline, M.L., N, M., Minton, N.P., Giaccia, A.J., and Brown, J.M. 1994. Anaerobic bacteria as a gene delivery system to tumors. Proc. Am. Assoc. Cancer Res. 35: 374.

Lemmon, M.L., Van Zijl, P., Fox, M.E., Mauchline, M.L., Giaccia, A.J., Minton, N.P., and Brown, J.M. 1997. Anaerobic bacteria as a gene delivery system that is controlled by the tumor microenvironment. Gene Therapy. 4: 791-796.

Liu, S.C., Minton, N.P., Giaccia, A. J., and Brown, J.M. 2002. Anticancer efficacy of systemically delivered anaerobic bacteria as gene therapy vectors targeting tumor hypoxia/necrosis. Gene Therapy. 9: 291-296.

Malmgren, R.A., and Flanigan, C.C. 1955. Localization of the vegetative form of *Clostridium tetani* in mouse tumors following intravenous spore administration. Cancer Res. 15: 473-478.

Martin, J., Stribbling, S.M., Poon, G.K., Begent, R.H., Napier, M., Sharma, S.K., and Springer, C.J. 1997. Antibody-directed enzyme prodrug therapy: pharmacokinetics and plasma levels of prodrug and drug in a phase I clinical trial. Cancer Chemother. Pharmacol. 40: 189-201.

Matsumura, H., Takeuchi, A., and Kano, Y. 1997. Construction of *Escherichia coli-Bifidobacterium longum* shuttle vector transforming *B. longum* 105-A and 108-A. Biosci. Biotechnol. Biochem. 61: 1211-1212.

Melton, R.G., and Sherwood, R.F. 1996. Antibody-enzyme conjugates for cancer therapy. J. Natl. Cancer Inst. 88: 153-65.

Merrick, M.J., Gibbins, J.R., and Postgate, J.R. 1987. A rapid and efficient method for plasmid transformation of *Klebsiella pneumoniae* and *Escherichia coli*. J. Gen. Microbiol. 133: 2053-2057.

Minton, N.P., Brehm, J.K., Oultram, J.D., Thompson, D.E., Swinfield, T.-J., Pennock, A., Schimming, S., Whelan, S.M., Vetter, U., Young, M., and Staudenbauer, W.L. 1990. Vector systems for the genetic analysis of *Clostridium acetobutylicum*. In: Clinical and Molecular Aspects of Anaerobes. S. P. Borriello, ed. Wrightson Biomedical Publishing Ltd., p. 187-201.

Mose, J.R., and Mose, G. 1959. Onkolyseversuche mit apathogenen anaeroben Sporenbildern am Ehrlich Tumor des Maus. Z. Krebsforsch. 63: 63-74.

Mose, J.R., and Mose, G. 1964. Oncolysis by clostridia. I. Activity of *Clostridium butyricum* (M-55) and other nonpathogenic clostridia against the Ehrlich carcinoma. Cancer Res. 24: 212-216.

Movsas, B., Chapman, J.D., Horwitz, E.M., Pinover, W.H., Greenberg, R.E., Hanlon, A.L., Iyer, R., and Hanks, G.E. 1999. Hypoxic regions exist in human prostate carcinoma. Urology. 53: 11-18.

Narberhaus, F., and Bahl, H. 1992. Cloning, sequencing, and molecular analysis of the *gro*ESL operon of *Clostridium acetobutylicum*. J. Bacteriol. 174: 3282-3289.

Nuyts, S., Van Mellaert, L., Theys, J., Landuyt, W., Lambin, P., and Anne, J. 2002. Clostridium spores for tumor-specific drug delivery. Anticancer Drugs. 13: 115-125.

Oultram, J.D., Loughlin, M., Swinfield, T.J., Brehm, J.K., Thompson, D.E., and Minton, N.P. 1988. Introduction of plasmids into whole cells of *Clostridium acetobutylicum* by electroporation. FEMS Microbiol. Letts. 56: 83-88.

Rampling, R., Cruickshank, G., Lewis, A.D., Fitzsimmons, S.A., and Workman, P. 1994. Direct measurement of pO_2 distribution and bioreductive enzymes in human malignant brain tumors. Int. J. Radiat. Oncol. Biol. Phys. 29: 427-431.

Selvaraj, G., and Iyer, V.N. 1981. Genetic transformation of *Rhizobium meliloti* by plasmid DNA. Gene. 15: 279-83.

Smith, C.J. 1985. Polyethylene glycol-facilitated transformation of *Bacteroides fragilis* with plasmid DNA. J. Bacteriol 164: 466-469.

Soutschek-Bauer, E., and Staudenbauer, W.L. 1987. Synthesis and secretion of a heat-stable carboxymethylcellulose from *Clostridium thermocellum* in *Bacillus subtilis* and *Bacillus stearothermophilus*. Mol. Gen. Genet. 208: 537-541.

Sundfor, K., Lyng, H., Kongsgard, U.L., Trope, C., and Rofstad, E.K. 1997. Polarographic measurement of pO_2 in cervix carcinoma. Gynecol. Oncol. 64: 230-236.

Takahashi, W., Yamagata, H., Yamaguchi, K., Tsukagoshi, N., and Udaka, S. 1983. Genetic transformation of *Bacillus brevis* 47, a protein-secreting bacterium, by plasmid DNA. J. Bacteriol. 156: 1130-1134.

Theys, J., Landuyt, W., Nuyts, S., Van Mellaert, L., Bosmans, E., Rijnders, A., Van Den Bogaert, W., van Oosterom, A., Anne, J., and Lambin, P. 2001. Improvement of Clostridium tumour targeting vectors evaluated in rat rhabdomyosarcomas. FEMS Immunol. Med. Microbiol. 30: 37-41.

Thiele, E.H., Arison, R.N., and Boxer, G.E. 1964. Oncolysis by clostridia. III. Effects of clostridia and chemotherapeutic agents on rodent tumors. Cancer Res. 24: 222-233.

Thiele, E.H., Arison, R.N., and Boxer, G.E. 1964. Oncolysis by clostridia. IV. Effect of nonpathogenic clostridial spores in normal and pathological tissues. Cancer Research. 24: 234-238.

Thomlinson, R.H., and Gray, L.H. 1955. The histological structure of some human lung cancers and the possible implications for radiotherapy. Brit. J. Cancer. 9: 539-549.

Tozer, G.M., Prise, V.E., Wilson, J., Locke, R.J., Vojnovic, B., Stratford, M.R., Dennis, M.F., and Chaplin, D.J. 1999. Combretastatin A-4 phosphate as a tumor vascular-targeting agent: early effects in tumors and normal tissues. Cancer Res. 59: 1626-1634.

Vaupel, P.W., and Hockel, M. 1995. Oxygenation status of human tumors: A reappraisal using computerized pO_2 histography. In: Tumor Oxygenation. P.W. Vaupel, D.K. Kelleher, and M. Gunderoth, eds. Stuttgart: Gustav Fischer Verlag. p. 219-232.

Walter, K.A., Bennett, G.N., and Papoutsakis, E.T. 1992. Molecular characterization of two *Clostridium acetobutylicum* ATCC 824 butanol dehydrogenase isozyme genes. J. Bacteriol. 174: 7149-7158.

Wouters, B.G., Weppler, S.A., Koritzinsky, M., Landuyt, W., Nuyts, S., Theys, J., Chiu, R.K., and Lambin, P. 2002. Hypoxia as a target for combined modality treatments. Eur. J. Cancer 38: 240-257.

Yazawa, K., Fujimori, M., Amano, J., Kano, Y., and Taniguchi, S. 2000. *Bifidobacterium longum* as a delivery system for cancer gene therapy: selective localization and growth in hypoxic tumors. Cancer Gene Ther. 7: 269-274.

Zhao, L., Ching, L.M., Kestell, P., and Baguley, B.C. 2002. The antitumour activity of 5,6-dimethylxanthenone-4-acetic acid (DMXAA) in TNF receptor-1 knockout mice. Br. J. Cancer 87: 465-470.

Chapter 12

Bioremediation Processes Coupled to the Microbial Reduction of Fe(III)oxides in Sedimentary Environments

Robert T. Anderson*

Abstract

In recent years the contribution of anaerobic processes to *in situ* contaminant transformation has been recognized as an increasingly important if not dominant process during the natural attenuation of groundwater contaminants. Microbial processes coupled to Fe(III) reduction have been viewed as of particular importance because upon the depletion of dissolved oxygen, Fe(III) is generally the most abundant potential electron acceptor within subsurface sediments. A wide diversity of Fe(III)-reducing bacteria are known but relatively few are found to be important for *in situ* contaminant transformation. Members of the *Geobacteraceae* have been detected *in situ* in association with contaminant degradation consistent with the known physiology of these organisms. *Geobacteraceae* are known to utilize aromatic hydrocarbons as growth substrates, remove important metal contaminants from groundwater and have been associated with halogenated solvent degradation. *Geobacter* species are also known to utilize electrodes as electron acceptors, a consequence of respiring solid phase Fe(III). Recent insights into the mechanism of metal reduction in *Geobacter* species may explain the prevalence of this group in sediments. Further genome-enabled studies of *Geobacteraceae* physiology are likely to lead to gene expression assays for monitoring *in situ* metabolism and suggest ways to promote contaminant degradation *in situ*.

1. Introduction

The potential for *in situ* bioremediation techniques to restore contaminated subsurface environments is well known. For sites contaminated with organics, the stimulation and maintenance of an aerobic bacterial community provides conditions suitable for the relatively rapid removal of contaminants via microbial oxidation to carbon dioxide. Thus, many engineered bioremediation techniques seek to provide oxygen and nutrients *in situ* to sustain an aerobic contaminant-degrading population of bacteria (Flathman *et al.*, 1994; Norris *et al.*, 1994; Chapelle, 1999). However, many contaminated sites contain large areas dominated by anaerobic processes (Lyngkilde *et al.*, 1992b; Baedecker *et al.*, 1993; Bjerg *et al.*, 1995; Borden *et al.*, 1995; Bekins *et al.*, 2001; Christensen *et al.*, 2001). Anaerobic conditions develop within the subsurface due to an overwhelming supply of organic carbon in the form of contaminants relative to the small amount of oxygen available in the subsurface immediately following a contaminant spill (Chapelle, 1993; Anderson *et al.*, 1997). Aerobic bacteria coupling the oxidation of contaminants to the reduction of oxygen rapidly consume all the available oxygen. Upon the depletion of oxygen, anaerobic conditions develop where contaminant transformation reactions are coupled to the reduction of anaerobic electron acceptors such as nitrate, Fe(III), sulfate and carbon dioxide (methanogenesis) (Chapelle, 1993; Lovley *et al.*, 1995a; Anderson *et al.*, 1997). Over time, these anaerobic areas within the contaminated subsurface are characterized by high concentrations of reduced chemical species such as Fe(II) and sulfides. Injection of oxygen into such highly reduced areas can be problematic due to consumption via abiotic reactions with reduced metals and sulfides found in abundance in such environments. Several anaerobic bioremediation techniques take advantage of the fact that anaerobic electron acceptors such as nitrate and sulfate can be injected at far greater concentration than oxygen (Reinhard *et al.*, 1997; Hutchins *et al.*, 1998; Anderson *et al.*, 2000; Cunningham *et al.*, 2001). Recent emphasis on the use of intrinsic bioremediation or monitored natural attenuation for organic contamination and the use of bioremediation for metal-contaminated sites have drawn attention to the capabilities of Fe(III)-reducing bacteria because many sites have very large areas dominated by microbial Fe(III) reduction (Lovley, 1995a; Anderson *et al.*, 1997; Madsen, 2001). Unlike other anaerobic electron acceptors (nitrate, sulfate, carbon dioxide), Fe(III) is insoluble and addition to the subsurface is, at present, technically difficult. However, the benefits of maintaining an Fe(III)-reducing population are

*For correspondence email: rtanders@microbio.umass.edu

substantial. This chapter will summarize the contaminant transformation capabilities of Fe(III)-reducing bacteria, their prevalence in subsurface environments and some of the unique features of this metabolism that may lead to new *in situ* treatment techniques.

2. Availability of Fe(III)oxides in Sediments for Microbial Reduction

When discussing the potential for contaminant degradation coupled to Fe(III) reduction, such as would be appropriate for an intrinsic or engineered bioremediation strategy, the form of Fe(III) available to microorganisms for use as an electron acceptor is of paramount importance. Soils and sediments contain a variety of Fe(III)oxide minerals, many in crystalline form, but not all of these Fe(III) forms are available to microorganisms for use as electron acceptors (Schwertmann *et al.*, 1977; 1992). Several recent reviews on the iron metabolism in soils and sediments have examined the overall thermodynamics of Fe(III) reduction for some of the more common Fe(III)oxides relative to other terminal electron accepting processes (TEAPs) likely to be encountered in anaerobic environments (Thamdrup, 2000; Straub *et al.*, 2001). Organic matter oxidation coupled to Fe(III) reduction, depending on the form (amorphous Fe, lepidocrocite, goethite, hematite), is a thermodynamically favorable process and the energy of reaction generally decreases in the order amorphous Fe(III) > lepidocrocite > goethite > hematite, with goethite and hematite reduction less energetically favorable than sulfate reduction when coupled to acetate oxidation (Thamdrup, 2000). This order of decreasing energy correlates with experimental data indicating an increase in the extent of microbial Fe(III) reduction in laboratory cultures containing model Fe(III) minerals of decreasing crystallinity; amorphous Fe(III) being the least crystalline, hematite the most crystalline (Ottow, 1969; Lovley, 1987). Further studies examining the effect of Fe(III) oxides on methane production in aquatic sediments revealed a substantial inhibition of methane production in the presence of amorphous Fe(III) oxides but not hematite (Lovley *et al.*, 1986a; 1986b). Quantitative measurements of methane and Fe(II) revealed no change in the overall reducing equivalents indicating the addition of Fe(III) did not affect the overall metabolism within the sediments. The data were interpreted as a shift in electron flow from methane production to amorphous Fe(III) reduction, consistent with amorphous Fe(III) serving as a source of electron acceptor (Lovley *et al.*, 1986a; 1986b). Hematite addition had no effect on methane production indicating no shift in electron flow and therefore was not utilized as an electron acceptor (Lovley *et al.*, 1986b). Detailed examinations of the same aquatic sediments revealed a persistent fraction of Fe(III) in sediment, later shown to be crystalline Fe(III), in which methane production was the dominant process that persisted with depth, confirming the results of the laboratory studies and demonstrating that crystalline Fe(III) is largely unavailable for microbial reduction in the environment (Lovley *et al.*, 1986a; 1986b; Phillips *et al.*, 1993). More recent data also demonstrates the lowering of steady state hydrogen levels and inhibition of methane production in the presence of amorphous Fe(III) whereas goethite and hematite addition did not impede methanogenesis nor lower steady state hydrogen concentrations and were not reduced in laboratory incubations of sediment consistent with thermodynamic calculations under conditions likely to be found *in situ* (Thamdrup, 2000; Anderson *et al.*, 2004). These results are consistent with examinations of natural organic matter degradation in sediments which correlates with amorphous Fe(III) reduction indicating amorphous Fe(III)oxides as the likely dominant source of microbial reducible Fe(III) in sediments (Lovley *et al.*, 1986b; Roden *et al.*, 2003).

In contrast to the detailed studies of aquatic sediment, much work has been performed indicating a potential for direct crystalline Fe(III) oxide reduction (Roden *et al.*, 1996; Urrutia *et al.*, 1998 ; Urrutia *et al.*, 1999; Roden *et al.*, 2000; 2002; Roden, 2003a). However, few of these examples are of relevance to processes occurring in the environment. Crystalline Fe(III) reduction has primarily been evaluated in cell suspension studies with unrealistic inocula of laboratory-grown (aerobically) cultures (Roden *et al.*, 1996; 2000), an electron donor (lactate) rarely detected in the environment (Roden *et al.*, 1996; Urrutia *et al.*, 1998) and with Fe(III)oxide minerals that cannot be verified as pure crystalline Fe(III) forms (Roden, 2003a; 2003b). Despite optimal culturing conditions, which may favor crystalline Fe(III) reduction in the laboratory (Glasauer *et al.*, 2003), only a small fraction (few percent) of the crystalline Fe(III) is reduced (Roden *et al.*, 2002; Roden, 2003a). The inferred mechanism of crystalline Fe(III) preservation in sediment is the pacification of reducible surface sites by Fe(II) preventing further crystalline Fe(III) reduction (Roden *et al.*, 2002; Roden, 2003a). This mechanism implies that crystalline Fe(III) oxide reduction, if possible, is likely to be found only in environments devoid of amorphous Fe(III), or other reducible Fe(III) forms such as clay-associated Fe(III) (Kostka *et al.*, 1996; 1999; Shelobolina *et al.*, 2003), or in areas where the amount of crystalline Fe(III) is present at orders of magnitude greater abundance than non-crystalline forms (Roden, 2003a). The arguments against the potential for direct use of crystalline Fe(III) as an electron acceptor in the environment do not exclude the possibility that chelation of Fe(III) by organic or inorganic ligands may be an important Fe(III) dissolution mechanism increasing the availability of crystalline Fe(III) and contributing to a loss of crystalline Fe(III) from sediments in close proximity to contamination such as landfill leachate (Heron *et al.*, 1995).

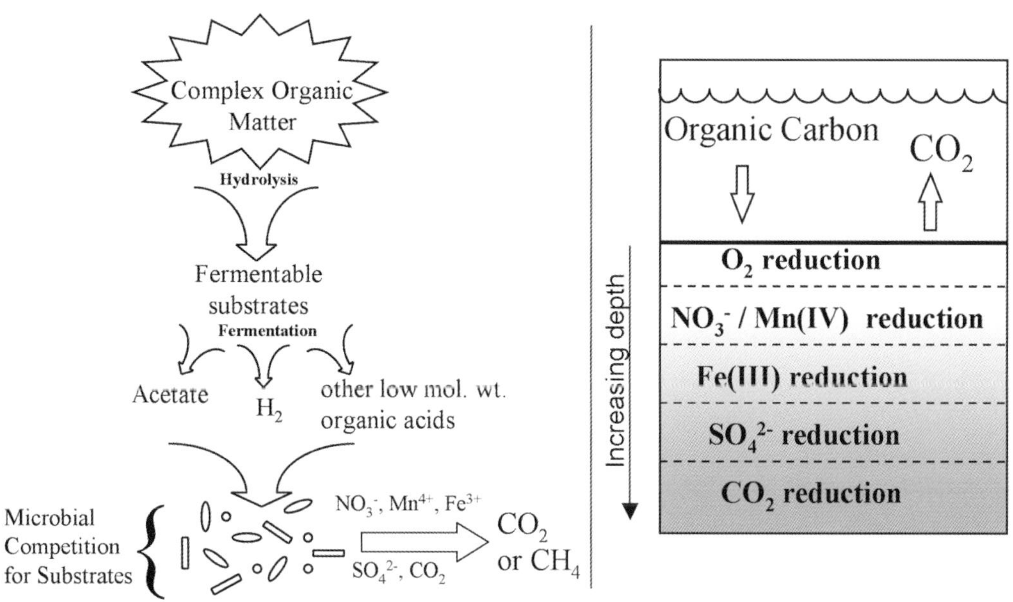

Figure 1. Anaerobic degradation of organic matter and the distribution of terminal electron accepting processes (TEAPs) with depth in aquatic sediments.

2.1. Assessment of the Amorphous Fe(III) Content of Sediments

Acid extraction of sediments allows the evaluation of the potential and progression of Fe(III) reduction in sediments. When evaluating the bioremediation potential of sediments where Fe(III) reduction is a dominant process, a simple 0.5N HCl extraction procedure provides the best estimate of the amount of Fe(III) available for use as an electron acceptor (Lovley *et al.*, 1987b; Heron *et al.*, 1994; Roden *et al.*, 2003). The basis for this procedure stems from a variety of laboratory and field observations. For example, laboratory evaluations of sediments to which amorphous Fe(III) was added produced Fe(II) as long as 0.5N HCl-extractable Fe(III) could be detected in sediment (Lovley *et al.*, 1987b). Once 0.5N HCl-extractable Fe(III) was depleted, Fe(II) production stopped (Lovley *et al.*, 1987b). When portions of a previously anaerobic sediment were oxidized to produce "native" amorphous Fe(III) and then added to methanogenic sediments a similar effect was observed, methane production ceased during production of Fe(II) and resumed once the added amorphous Fe(III) was depleted (Lovley *et al.*, 1986b). Other studies report the progression of Fe(III) reduction with time as monitored by the 0.5N HCl-extraction procedure, generally reported as a ratio of acid extractable Fe(II) to total acid extractable Fe, which correlates with shifts in the sediment microbial community to organisms known for Fe(III) reduction (Snoeyenbos-West *et al.*, 2000; Holmes *et al.*, 2002). Upon the depletion of 0.5N HCl extractable Fe(III), comparable decreases in the prevalence of known Fe(III)-reducing organisms within the sediment microbial community decreases (Holmes *et al.*, 2002). These data and several examples from the field correlating a lack of 0.5N HCl extractable Fe(III) with the lack of Fe(III)-reducing conditions and vice versa indicate the 0.5N HCl extractable Fe(III) analysis as the best available technique yet developed for determining the amount of Fe(III) in sediments available for use as a electron acceptor by anaerobic bacteria (Lovley *et al.*, 1987b; Heron *et al.*, 1994; Anderson *et al.*, 1998; 1999; Lovley *et al.*, 2000; Roden *et al.*, 2003).

3. Distribution of Microbial Processes in Sedimentary Environments

Subsurface contamination results in the development of distinct zones of microbial metabolism within the subsurface based on the availability of potential electron acceptors. While the physiology and metabolism of subsurface microbial communities is not known in great detail, the progression of stimulated anaerobic processes within the subsurface can be inferred from better-studied examples of processes occurring in aquatic sediments (Lovley *et al.*, 1995a; Madsen, 1995). Anaerobic degradation processes, unlike aerobic degradation, require a succession of processes to breakdown complex organic matter (Figure 1) (Ponnamperuma, 1972; Reeburgh, 1983; Lovley, 1987; Lovley *et al.*, 1995a). Upon the depletion of oxygen, degradation reactions coupled to nitrate reduction are followed by Mn(IV) reduction, Fe(III) reduction, sulfate reduction and methanogenesis in this order as each electron acceptor is successively depleted (Lovley *et al.*, 1995a; Madsen, 1995). Distinct zones develop with depth in aquatic sediments dominated by these electron accepting

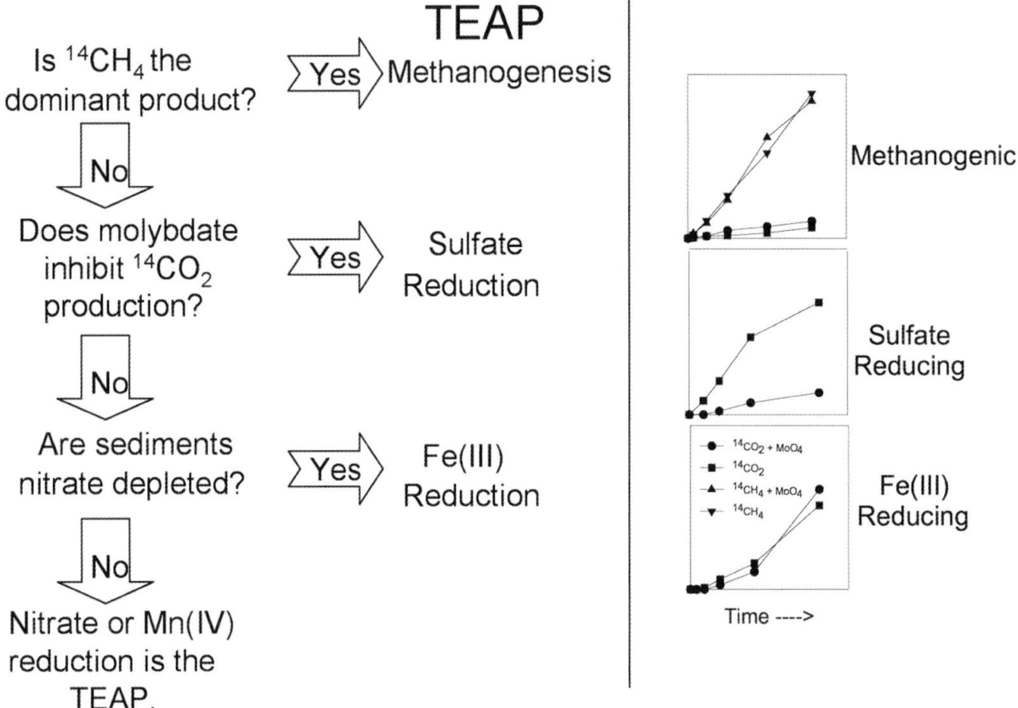

Figure 2. TEAP determination in sediment samples using [2-^{14}C]acetate and examples of raw data indicating the predominant deduced electron accepting process.

processes (Figure 1). The observed zonation with depth is often interpreted in the context of a thermodynamic argument whereby the depletion of electron acceptors correlates with the depletion of the more energetically favorable electron acceptor first (Stumm *et al.*, 1981; Bouwer *et al.*, 1993). However, there is no thermodynamic reason why a reaction of less favorable energy should not also proceed (McCarty, 1972). A more complete description of the progression of microbial processes stems from a consideration of the physiological controls on microbial metabolism. As an example, many sedimentary environments contain little nitrate and manganese but most sedimentary environments contain Fe(III). Therefore Fe(III) reduction becomes a prevalent anaerobic microbial process upon the introduction of organic matter and depletion of oxygen. For porewaters that also contain sulfate it is not thermodynamically apparent why Fe(III) reduction tends to be a dominant process prior to the stimulation of sulfate reduction, also an energetically favorable process. The reason for this succession is physiological. In the presence of microbial reducible Fe(III), Fe(III)-reducing bacteria are able to metabolize substrates such as hydrogen and acetate to levels too low to support sulfate reduction (Lovley *et al.*, 1987a; Cord-Ruwisch *et al.*, 1988; Lovley *et al.*, 1988a). Likewise, sulfate reducers can metabolize substrates (hydrogen and acetate) to levels too low to support methanogenesis (Lovley *et al.*, 1983; Cord-Ruwisch *et al.*, 1988; Lovley *et al.*, 1988a; Achtnich *et al.*, 1995). For sites where nitrate is an abundant electron acceptor, nitrate-reducers would be expected to outcompete Fe(III)-reducers by the same mechanism which explains the onset of Fe(III) reduction only upon the depletion of nitrate (Finneran *et al.*, 2002). In aquatic sediments the net result of this competition is a stratification of terminal electron accepting processes (TEAPs) with depth that correlates with a decreasing energy of reaction (Lovley *et al.*, 1995a; Anderson *et al.*, 1997).

A similar progression of processes is observed within plumes of contaminated groundwater downgradient from a source of organic contamination (Lyngkilde *et al.*, 1992b; 1992a; Patterson *et al.*, 1993; Vroblesky *et al.*, 1994; Bjerg *et al.*, 1995; Borden *et al.*, 1995; Chapelle *et al.*, 1995; Albrechtsen *et al.*, 1999; Anderson *et al.*, 1999; Bekins *et al.*, 2001; Roling *et al.*, 2001; Anderson *et al.*, 2003). The distinction between TEAPs is not absolute and is subject to the heterogeneities of subsurface groundwater flow but the overall model provides a workable framework from which to evaluate subsurface microbial processes (Lovley *et al.*, 1988a; 1994a; Bjerg *et al.*, 1995; Jakobsen *et al.*, 1998; Anderson *et al.*, 2003). In this model the least energetically favorable processes tend to be found in areas nearest the source of contamination in sediments where contamination has existed for the longest period of time and consequently where the greatest depletion of potential electron acceptors has occurred. Therefore, in "older" contaminant plumes methanogenic processes would tend to be found within and immediately downgradient of source areas while more energetically favorable TEAPs (sulfate reduction, Fe(III) reduction, nitrate reduction) would tend to occur within contaminant plumes located further downgradient (Lovley, 1995a; Lovley *et al.*, 1995a). While exact estimates on how fast TEAP zones develop *in situ* likely depends on a variety of factors including groundwater flow rate and the nature of the contaminants, there are indications that these can occur relatively rapidly (Lovley *et al.*, 1995b; Chapelle *et al.*, 2002).

3.1. TEAP Determination in Sediments

In order to evaluate the progression of subsurface microbial processes it is frequently necessary to collect sediment samples from the subsurface. Assessing subsurface microbial ecology from ground water samples alone can be misleading (Chapelle *et al.*, 1995; Anderson *et al.*, 1997). Many of the products of anaerobic metabolism such as methane, Fe(II), hydrogen sulfide, Mn(II) are soluble and therefore mobile within groundwater (Christensen *et al.*, 1994; Anderson *et al.*, 1997). Detection of these end products far from the point of production tends to blur the spatial detection of TEAPs within the subsurface (Christensen *et al.*, 1994; Anderson *et al.*, 1997). Dissolved hydrogen concentrations have been employed to deduce subsurface microbial processes (Chapelle, 1993; Lovley *et al.*, 1994a; Vroblesky *et al.*, 1994; Chapelle *et al.*, 1995; Chapelle *et al.*, 1997; Jakobsen *et al.*, 1998). However, care must be taken when employing this technique *in situ* (Chapelle *et al.*, 1997). A central assumption regarding the use of dissolved hydrogen as a TEAP indicator is the assumption that assessed processes are at steady state which may not be applicable at many sites (Watson *et al.*, 2003). A particularly clear TEAP determination, although quite specialized and not widely employed, is the evaluation of radiolabeled acetate oxidation in sediment samples in the presence of microbial inhibitors (Winfrey *et al.*, 1977; Lovley *et al.*, 1994a; Anderson *et al.*, 1999; 2000). The oxidation of [2-^{14}C] acetate in the presence and absence of molybdate can provide conclusive data on the dominant microbial processes occurring in sediments when coupled with data on the availability of potential electron acceptors gathered from the point of sediment collection (Figure 2). Acetate oxidation that produces a large proportion of ^{14}C-methane relative to ^{14}C-carbon dioxide indicates methanogenesis as the dominant TEAP. Production of ^{14}C carbon dioxide that can be inhibited in the presence of molybdate, a specific inhibitor of sulfate reduction (Oremland *et al.*, 1988), indicates a microbial community dominated by sulfate reduction. Production of ^{14}C carbon dioxide in Fe(II)-containing sediments that is not inhibited by molybdate indicates Fe(III)-reduction as a dominant TEAP. Finally, production of ^{14}C carbon dioxide in the presence of Mn(II) or nitrate indicates the presence of a community dominated by manganese reduction or nitrate reduction, respectively. These analyses, while labor intensive, provide reasoned, data-supported documentation of the dominant TEAP occurring in collected sediments allowing an analysis of the distribution of TEAPs *in situ* and/or a prerequisite analysis for further laboratory sediment studies.

4. Dissimilatory Fe(III)-Reducing Bacteria

Many contaminated aquifers contain large areas dominated by microbial Fe(III)-reduction (Borden *et al.*, 1995; Bekins *et al.*, 2001; Christensen *et al.*, 2001; Roling *et al.*, 2001; Anderson *et al.*, 2003). This suggests that significant microbially-mediated, contaminant degradation processes are occurring within this zone (Lovley *et al.*, 1989; Lu *et al.*, 1999; Lovley *et al.*, 2000; Christensen *et al.*, 2001; Roling *et al.*, 2001). Consequently there has been increased interest in the physiology of Fe(III)-reducing organisms and the bioremediation potential of this novel form of metabolism.

Dissimilatory Fe(III) reduction is widespread throughout the phylogenetic spectrum (Lovley, 2000c). The ability to conserve energy for growth coupled to the reduction of Fe(III) spans several major groups within the *Bacteria* including the β, γ, δ and ε groups within the *Proteobacteria*, members of Fibrobacteres/Acidobacteria and Firmicutes, and also extends to several thermophilic and hyperthermophilic members of the *Archaea* (Vargas *et al.*, 1998; Lovley, 2000b; 2000c; Tor *et al.*, 2001; Kashefi *et al.*, 2002). The wide diversity of organisms capable of Fe(III) reduction and examples of this metabolism within deeply branching members of the *Archaea* suggest that Fe(III) reduction may be an ancient respiratory process and a remnant metabolism of an early anoxic Earth (Vargas *et al.*, 1998; Lovley, 2000b). The largest described group of dissimilatory Fe(III) reducers are of the family *Geobacteraceae* (Lovley, 2000c; 2000b). Members of this family include the *Geobacter*, *Pelobacter*, *Desulfuromonas* and *Desulfuromusa* species. All of the organisms with the exception of the *Pelobacter* species are capable of oxidizing multicarbon compounds such as acetate (among others) completely to carbon dioxide and coupling this process to growth and the reduction of Fe(III). Many also utilize hydrogen as an electron donor. All of these organisms are strict anaerobes but some may tolerate exposure to low levels of oxygen (D.R. Lovley unpublished). Members of γ-*Proteobacteria* including the *Shewanella*, *Aeromonas* and *Ferrimonas* groups are also a well known group of Fe(III)-reducers (Lovley, 2000b; 2000c). These organisms are known to incompletely oxidize multi-carbon compounds such as lactate and glucose to acetate, coupling this process to growth and the reduction of Fe(III) (Coates *et al.*, 1998a;

Lovley, 2000c). Members of these groups are remarkably metabolically diverse in terms of the ability to use a great variety of electron acceptors including oxygen.

5. Prevalence of *Geobacteraceae* in Sediments

Within Fe(III)-reducing sediments it is generally the *Geobacteraceae* that are the dominant group of Fe(III) reducers detected. This has been demonstrated in several laboratory experiments in which the composition of the microbial community has been evaluated using 16S-rRNA-based techniques during stimulated Fe(III) reduction (Snoeyenbos-West *et al.*, 2000; Holmes *et al.*, 2002). In these studies Fe(III) reduction was stimulated upon the addition of a suitable electron donor such as acetate. *Geobacteraceae,* coincident with Fe(II) production, were detected and become selectively enriched comprising as much as 40% of the detected microbial community (Snoeyenbos-West *et al.*, 2000; Holmes *et al.*, 2002). *Geobacteraceae* are also the dominant detected Fe(III)-reducing group when other electron donors such as lactate, benzoate and formate are added to sediments (Snoeyenbos-West *et al.*, 2000). Several examinations of contaminated sedimentary environments and bioremediation test plots also have noted the prevalence of *Geobacteraceae*, and to a lesser extent *Geothrix* species (Fibrobacteres/Acidobacteria), within sediments dominated by Fe(III) reduction (Anderson *et al.*, 1998; Roling *et al.*, 2001; Anderson *et al.*, 2003; Cummings *et al.*, 2003). In an example of an *in situ* bioremediation test of stimulated uranium reduction *Geobacteraceae* comprised up to 89% of the detected groundwater microbial community downgradient from an acetate injection gallery (Anderson *et al.*, 2003). An *in situ* test of stimulated halogenated solvent degradation noted the detection of *Desulfuromonas* species that may play an important role in environments where reductive dehalogenation is occurring (He *et al.*, 2002). While many Fe(III)-reducing organisms can potentially be isolated from a variety of soils and sediments (Caccavo Jr *et al.*, 1996; Coates *et al.*, 1996; 1998b; Fredrickson *et al.*, 1998; Cummings *et al.*, 1999; Coates *et al.*, 2001), the available evidence gathered to date from molecular characterizations of aquifer sediments both in the lab and from the field indicates *Geobacteraceae* as a dominant and widely detected group of Fe(III)-reducing bacteria within Fe(III)-reducing sediments at mesophilic temperatures (Lovley, 2000b; 2000c; Snoeyenbos-West *et al.*, 2000).

6. Potential Competitive Advantages of *Geobacter* species within sediments

6.1. Acetate and Hydrogen Utilization

There are several potential reasons why *Geobacteraceae* might be expected to be a dominant group within Fe(III)-reducing sediments beginning with the difference in types of electron donors utilized by known Fe(III)-reducing bacteria. In anaerobic environments the predominant intermediates produced during anaerobic degradation processes are hydrogen and acetate (Lovley *et al.*, 1982; Lovley, 1987). All the *Geobacteraceae* (except *Pelobacter sp.*) are capable of completely oxidizing acetate to carbon dioxide while coupling this process to Fe(III) reduction (Lovley, 2000c). Many are also capable of utilizing hydrogen and completely oxidizing other organic acids, including aromatic compounds (Lovley, 2000c). This capacity to utilize a wide variety of low molecular weight organic acids, in particular acetate, as electron donors is a likely reason why these organisms are readily detected in sedimentary environments (Coates *et al.*, 1996; 1998b; 2001). Unlike the *Geobacteraceae*, other known Fe(III)-reducing bacteria couple the incomplete oxidation of electron donors such as lactate and pyruvate, which are not typically found as intermediates during anaerobic degradation processes in sediments (Lovley *et al.*, 1982; Lovley, 2000b; 2000c). Lactate and pyruvate metabolism coupled to Fe(III) reduction results in the production of acetate by these organisms (Lovley, 2000b; 2000c). While these incomplete oxidizers can still be isolated from subsurface sediments using the appropriate electron donors (Pedersen *et al.*, 1996; Fredrickson *et al.*, 1998), the lack of ability to utilize the most common degradation intermediate, acetate, may be a reason limiting detection of these organisms by 16S rRNA-based techniques in aquifer sediments (Snoeyenbos-West *et al.*, 2000; Roling *et al.*, 2001; Holmes *et al.*, 2002).

6.2. Mechanisms of Enzymatic Fe(III) Reduction

Differences in the mechanism of Fe(III) reduction among the more well-studied Fe(III)-reducing organisms may also confer a competitive advantage on *Geobacteraceae* detected within sediments. While *Geobacter* species must contact Fe(III)oxides in order to reduce them, other Fe(III)-reducing species have been shown to produce extracellular, soluble Fe(III) chelators and/or electron shuttles and need not physically contact Fe(III) oxides for reduction (Nevin *et al.*, 2000a; Newman *et al.*, 2000; Nevin *et al.*, 2002). Production of extracellular Fe(III) chelators or electron shuttles for a one electron transfer is energetically expensive unless these exported components can be utilized many times (Nevin *et al.*, 2002). This may be most appropriate for Fe(III)-reducing organisms living in high density populations such as biofilms near a source of Fe(III) in a nutrient rich environment where the energy expended in producing these compounds may be recovered (Nevin *et al.*, 2002). Direct reduction of Fe(III) oxides may be most efficient in low density populations such as aquifer sediments where production of Fe(III) chelating or electron shuttling compounds are likely to diffuse away from the organism and thus be of little benefit (Nevin *et al.*, 2002). The energetic cost of synthesizing such compounds would quickly become a net energy loss to the organism. Consideration of the differences in the mechanism of Fe(III) reduction among Fe(III)-reducers is of importance because it is likely to have an impact on the way Fe(III) reduction is modeled in the subsurface depending on the dominant mechanism present, inferred from the dominant organisms detected.

6.3. Nonenzymatic Fe(III) Reduction: Humics Reduction

The redox cycling of humic materials offers a potential mechanism of abiotic Fe(III) reduction in sediments. Humic materials are widespread in the environments and the observation that humic materials can be utilized as electron acceptors by metal-reducing bacteria suggests this may be a widespread form of bacterial metabolism. Humic materials contain quinone moieties susceptible to enzymatic reduction by metal-reducing bacteria (Lovley *et al.*, 1996a; Coates *et al.*, 1998b; Lovley *et al.*, 1998; Scott *et al.*, 1998). Enzymatically reduced quinone structures are capable of reducing Fe(III) oxides abiotically (Lovley *et al.*, 1996a; Scott *et al.*, 1998; Nevin *et al.*, 2002). This potential for humics respiration coupled with the ubiquity of humic materials in the environment indicates that bacterial respiration of Fe(III) oxides need not be a direct enzymatic reduction in the presence of humic materials. During humics respiration the quinone groups within humic materials are enzymatically reduced (Lovley *et al.*, 1996a; Scott *et al.*, 1998). As the soluble humic materials diffuse away from the cell they are oxidized upon contact with Fe(III)oxides, where Fe(III) is reduced. The re-oxidized humic materials are then available once again for use as an electron acceptor during humics respiration. This redox cycling of humic materials between metal-reducing bacteria and Fe(III) oxides has been termed "electron shuttling" and greatly increases the rate of Fe(III) reduction in sediments presumably by alleviating the need for bacteria to directly contact Fe(III) oxides (Nevin *et al.*, 2000b; Lovley, 2001; Nevin *et al.*, 2002). Also, analyses of steady state hydrogen concentrations in Fe(III)-reducing sediment indicate that electron shuttling may provide a competitive advantage to humics reducers relative to Fe(III) reducers or methanogens (Cervantes *et al.*, 2000; Anderson *et al.*, 2003). Laboratory incubations of sediments to which varying concentrations of the humics analog anthraquinone-2,6-disulfonate (AQDS) were added lowered steady state hydrogen concentrations to levels lower than those typically found under Fe(III)-reducing conditions (Anderson *et al.*, 2004). This effect was observed in sediments in the presence of as little as 1μM AQDS (Anderson *et al.*, 2004). These data suggest that in the presence of sufficient humic materials and Fe(III), humics reducers ought to be able to outcompete Fe(III)-reducers for substrates in a manner similar to the way in which Fe(III)-reducers outcompete sulfate reducers for substrates, mentioned above, and in the manner in which quinone reducers have been shown to outcompete methanogenic processes (Cervantes *et al.*, 2000).

In addition to reduced humic materials shuttling electrons to Fe(III) oxides, reduced humics may also serve as electron donors for other bacteria catalyzing the reduction of electron acceptors such as nitrate, selenate, arsenate and fumarate (Lovley *et al.*, 1999; Lovley, 2001). These data indicate a potential for "interspecies electron transfer" between organisms but the environmental relevance of this process or humics reduction in general has yet to be investigated in detail (Lovley, 2001).

7. Contaminant Transformation and Fe(III) Reduction

Contaminant degradation under Fe(III)-reducing conditions is an important component of the natural attenuation of groundwater contaminants at many sites and has potential applications during engineered site remediation. At many contaminated sites where dissolved oxygen has been depleted in the subsurface Fe(III) is generally the most abundant potential electron acceptor. The known physiological capabilities of Fe(III)-reducing bacteria including petroleum aromatic hydrocarbon degradation, contaminant metal transformation and reductive dehalogenation implies a potential for a wide range of contaminant transformations under Fe(III)-reducing conditions *in situ*.

7.1. Aromatic Hydrocarbon Oxidation Coupled to Microbial Fe(III) Reduction

Early reports on contaminant transformations under Fe(III)-reducing conditions focused on the oxidation of aromatic hydrocarbons commonly found in petroleum contaminated environments. Monoaromatic hydrocarbons such as those of the BTEX group (benzene, toluene, ethylbenzene and the xylenes) are common groundwater components found downgradient of petroleum hydrocarbon-contaminated areas and landfill leachate plumes (Baedecker *et al.*, 1989; Salanitro, 1993; Borden *et al.*, 1995; Christensen *et al.*, 2001; Madsen, 2001). Some of the first indications that aromatic

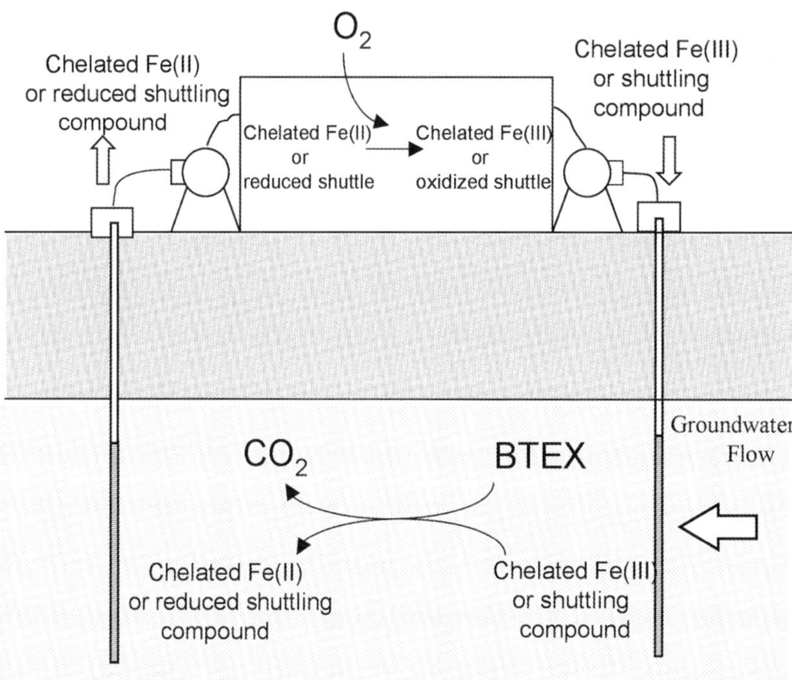

Figure 3. Conceptualized plan for the addition of chelated Fe(III) and/or electron shuttling compounds to sediments in an *in situ* bioremediation scheme.

hydrocarbons may be degraded under Fe(III)-reducing conditions found *in situ* came from the site of a crude oil pipeline spill near Bemidji, Minnesota. Sediments collected from downgradient portions of the aquifer dominated by Fe(III)-reducing conditions produced $^{14}CO_2$ from ^{14}C-labeled benzoate when incubated under *in situ* conditions (Lovley *et al.*, 1989). The results correlated with laboratory results detailing the oxidation of aromatic compounds, including the BTEX component toluene, by the Fe(III)-reducing organism *Geobacter metallireducens* (Lovley *et al.*, 1988b; 1989; Lonergan *et al.*, 1991; Lovley *et al.*, 1993). Similar radiotracer studies later demonstrated a potential for the anaerobic oxidation of toluene, benzene and even naphthalene in Fe(III)-reducing sediments from the Bemidji site (Anderson *et al.*, 1998; 1999; Rooney-Varga *et al.*, 1999). While toluene appears to be rapidly degraded in Fe(III)-reducing sediments from many aquifers (Anderson *et al.*, 1999) and is well known to the oxidized under anaerobic conditions in general (Evans *et al.*, 1988; Colberg *et al.*, 1995; Lovley, 1997), benzene tends to be more recalcitrant in aquifers. However, Fe(III)-reducing sediments collected from within the contaminant plume of the Bemidji site readily oxidized benzene to carbon dioxide (Anderson *et al.*, 1998). This is of importance because benzene is the primary contaminant of concern at petroleum-contaminated sites due to its toxicity. The downgradient location within the aquifer of sediments exhibiting a potential for anaerobic benzene oxidation correlated with an observed decrease in benzene concentrations within the groundwater and the detection of Fe(III)-reducing bacteria of the family *Geobacteraceae*, known oxidizers of aromatic compounds (Anderson *et al.*, 1998; Rooney-Varga *et al.*, 1999; Lovley *et al.*, 2000). A potential for naphthalene, but not phenanthrene, degradation was also observed at this location (Anderson *et al.*, 1999). These data suggest that active areas of aromatic hydrocarbon degradation under Fe(III)-reducing conditions may exist at specific downgradient locations within contaminated aquifers and could be related to the availability of oxidized Fe(III) for use as a terminal electron acceptor (Lovley *et al.*, 2000).

Increasing the availability of Fe(III) in aquifer sediments may increase the rate of aromatic hydrocarbon degradation. Addition of Fe(III) chelators to sediments has been shown to stimulate the rate of degradation of important groundwater contaminants such as benzene (Lovley *et al.*, 1994b; 1996b). Strong Fe(III) chelators such as nitrilotriacetic acid (NTA) solubilize Fe(III) within porewaters of sediment incubations and presumably increase the availability of Fe(III) for use as an electron acceptor by aromatic oxidizing, Fe(III)-reducing bacteria. These laboratory observations indicate that, should the potential for aromatic hydrocarbon oxidation, including benzene oxidation, exist within such sediments the rate of degradation can be enhanced by increasing the availability of Fe(III). The solubilization of Fe(III) upon the addition of an Fe(III) chelator alleviates the requirement of Fe(III)-reducing bacteria to contact Fe(III) oxides in order to reduce them. Soluble electron acceptors are generally more available to microorganisms than insoluble acceptors and the increase in the rate of degradation reflects the increase in Fe(III) availability. A similar mechanism of stimulated aromatic hydrocarbon degradation is also observed in sediments to which humic materials and/or AQDS have been added (Lovley *et al.*, 1996b; Anderson *et al.*, 1999). In addition to the capacity for humic materials to solubilize Fe(III), humic materials can also participate in electron shuttling reactions as described above. Both mechanisms stimulate the reduction of Fe(III) by increasing the accessibility of solid phase Fe(III). The observed increases in rate of aromatic hydrocarbon degradation upon the addition of these compounds to sediments has led to the proposal of several *in situ* bioremediation schemes aimed at promoting *in situ* Fe(III) accessibility in the subsurface in order to stimulate the removal contaminant hydrocarbons (Figure 3). The bioremediation schemes focus on providing and/or solubilizing Fe(III) within the subsurface to promote organic contaminant oxidation coupled to microbial Fe(III) reduction. These strategies have yet to be demonstrated in the field.

7.2. Halogenated Solvent Degradation Under Fe(III)-Reducing Conditions

Halogenated organics are some of the most widely detected groundwater contaminants (Westrick *et al.*, 1984; Hardman, 1991; McCarty *et al.*, 1994). While a great variety of halogenated contaminants are found in the environment, the result of spills or other releases, much attention has been focused on the chlorinated solvents PCE (tetrachloroethene) and TCE (trichloroethene) and their transformation products (Vogel *et al.*, 1987; McCarty *et al.*, 1991; Semprini *et al.*, 1992; Hopkins *et al.*, 1993a; McCarty *et al.*, 1994; Norris *et al.*, 1994). In addition to the potential for PCE and TCE transformation under co-metabolic methanotrophic (Semprini *et al.*, 1994), phenol-oxidizing (Hopkins *et al.*, 1993a; 1993b; 1995), toluene-oxidizing (Hopkins *et al.*, 1995; McCarty *et al.*, 1998) conditions more recent investigations have centered on evaluating the potential for stimulating chlorinated solvent respiration under natural or stimulated anaerobic conditions *in situ* (Lollar *et al.*, 2001; He *et al.*, 2002; Lendvay *et al.*, 2003). Hydrogen has generally been perceived as the preferred electron donor for reductive dehalogenation primarily because many of the chloro-respirers (*Desulfitobacterium*, *Dehalococcoides*, *Dehalospirillum* and *Dehalobacter* species) available in pure culture utilize hydrogen as an electron donor during this process (Holliger *et al.*, 1999) and these organisms have been detected *in situ* (He *et al.*, 2002). However, acetate appears to also stimulate reductive dehalogenation (He *et al.*, 2002). This observation correlates with the known metabolism of *Desulfuromonas* isolates capable of coupling reductive dechlorination of PCE to the oxidation of acetate (Loffler *et al.*, 2000; Sung *et al.*, 2003) and the detection of *Desulfuromonas* species *in situ* (Lendvay *et al.*, 2003). This is of importance in terms of Fe(III)-reduction because *Desulfuromonas* species are of the family *Geobacteraceae*, known acetate-oxidizing, dissimilatory Fe(III)-reducers widely detected in subsurface environments (Lovley, 2000c). The observation that acetate addition to the subsurface can result in complete reductive dehalogenation of PCE (He *et al.*, 2002) and that PCE reductive dehalogenation is not inhibited in the presence of Fe(III) (both are reduced) in pure cultures of acetate-oxidizing , PCE-reducing bacteria (Sung *et al.*, 2003) suggests a potential for chlorinated solvent transformation under natural or stimulated Fe(III)-reducing conditions *in situ*. *Desulfuromonas* species in pure culture reduce PCE to DCE (dichloroethene) (Sung *et al.*, 2003). Earlier investigations of sediment noted the potential for degradation of important intermediates of reductive dehalogenation, such as DCE and VC (vinyl chloride), under Fe(III)-reducing conditions (Bradley *et al.*, 1996; Bradley *et al.*, 1998a; Bradley *et al.*, 1998b; Bradley *et al.*, 1998c). The demonstrated potential for *Desulfuromonas* species to reduce PCE to DCE, the detection of *Desulfuromonas* species *in situ*, and the demonstrated degradation potential of intermediates of reductive dechlorination under Fe(III)-reducing conditions suggests that several Fe(III)-reducing organisms may be involved in the complete reduction of PCE to ethane *in situ*. More detailed investigations on the importance of acetate during reductive dehalogenation processes are sure to be forthcoming because acetate is soluble and acetate oxidation provides up to four times more reducing equivalents than hydrogen.

7.3. *In Situ* Contaminant Metal Remediation

A comprehensive review of the interactions of microorganisms with contaminant metals is beyond the scope of this section and has been recently reviewed in detail (Lovley, 2000a; Keith-Roach *et al.*, 2002; Lloyd *et al.*, 2002). Many organisms are capable of reducing metals and many metals upon reduction become much less soluble (Lovley, 1993; 1995a; 1995b; Lovley *et al.*, 1997). Fe(III)-reducing bacteria, in particular, are known to reduce a wide variety of contaminant metals including uranium (Lovley *et al.*, 1991; Gorby *et al.*, 1992; Lovley, 1995b). The reduction of soluble U(VI) to insoluble U(IV) by Fe(III)-reducing bacteria upon the stimulation of Fe(III) reduction within the contaminated subsurface has been proposed as a method to remove soluble U(VI) from the groundwater of uranium-contaminated sites (Lovley, 1995a; Lloyd *et al.*, 2000a; Anderson *et al.*, 2002; Lloyd *et al.*, 2002). An example of this strategy was recently investigated at a uranium mill tailings site in Rifle, CO (USA) (Anderson *et al.*, 2003). Acetate addition to the subsurface stimulated metal-reducing conditions resulting in an observed loss of soluble

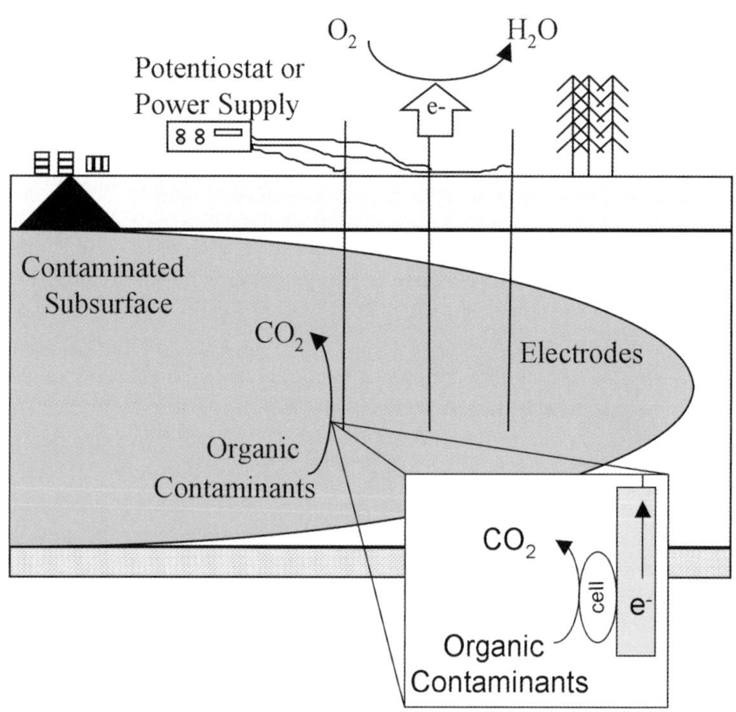

Figure 4. Conceptualized plan for using electrodes placed within the contaminated subsurface as a source of electron acceptor for aromatic-oxidizing, electrode-reducing bacteria.

U(VI) from the groundwater. Up to 70% removal of U(VI) was observed relative to initial concentrations and in some wells the final concentrations were close to or below the established maximum contaminant levels (MCL) for the site. The loss of U(VI) occurred coincident with the stimulation of Fe(III)-reducing conditions and enrichment of known Fe(III)- and U(VI)-reducing bacteria of the *Geobacteraceae* family in the groundwater (Anderson *et al.*, 2003). Further loss of soluble U(VI) was not observed upon a shift in microbial processes from Fe(III)-reducing conditions to sulfate reduction indicating that the activity of Fe(III)-reducing bacteria was a necessary requirement for the removal of U(VI) from groundwater. Loss of vanadium, a common co-contaminant associated with uranium mill tailings, was also enhanced upon the stimulation of Fe(III)-reducing conditions consistent with the demonstrated potential for bacterially-mediated vanadium reduction and consequent precipitation (Lyalikova *et al.*, 1992; Carpentier *et al.*, 2003; Ortiz-Bernad *et al.*, 2004). However, vanadium is susceptible to abiotic reduction by Fe(II) and the stimulated removal of soluble vanadium was likely not solely due to enzymatic reduction within the subsurface but also abiotic reduction by Fe(II) (Ortiz-Bernad *et al.*, 2004).

Technetium, an important radionuclide contaminant, is also susceptible to both enzymatic reduction by metal-reducing bacteria (and sulfate reducing bacteria) and reduction by Fe(II) (Lovley, 1995a; Lloyd *et al.*, 2000b; 2002). It is anticipated that stimulation of Fe(III)-reducing conditions within the subsurface will stimulate the removal of technetium from groundwater via a combination of enzymatic and abiotic reduction (Lovley, 1995a; Lloyd *et al.*, 2002).

Chromium is susceptible to removal from solution under stimulated Fe(III)-reducing conditions (Wielinga *et al.*, 2001). A wide variety of metal-reducing bacteria will enzymatically reduce Cr(VI) to Cr(III) thereby precipitating chromium from solution (Lovley, 1995a; Wielinga *et al.*, 2001). However, Cr(VI) is also susceptible to abiotic reduction by Fe(II) and the stimulation of Fe(III)-reducing conditions *in situ* is likely to promote chromium removal from groundwater via enzymatic and abiotic reduction by Fe(II). An *in situ* test of abiotic Cr(VI) removal from groundwater revealed the persistence of anoxic conditions and removal of soluble chromium years (3.5) after installation suggesting that reduction of the subsurface and production of Fe(II) whether via chemical or biological means can promote long term removal of Cr(VI) from groundwater (Fruchter *et al.*, 2000).

8. Electrode Reduction

One of the more recent and novel aspects of solid phase electron acceptor respiration is the demonstration of direct electrode respiration by Fe(III)-reducing bacteria (Kim *et al.*, 1999; Bond *et al.*, 2002; Tender *et al.*, 2002; Bond *et al.*, 2003; Chaudhuri *et al.*, 2003; Holmes *et al.*, 2004). Laboratory cultures of Fe(III)-reducing bacteria have been shown to couple the oxidation of simple organics to the reduction of a graphite electrode suspended in the culture media and electrically coupled to another electrode suspended in aerobic media (Kim *et al.*, 1999; Bond *et al.*, 2002). Production of current coincident with the depletion of electron donor and increases in

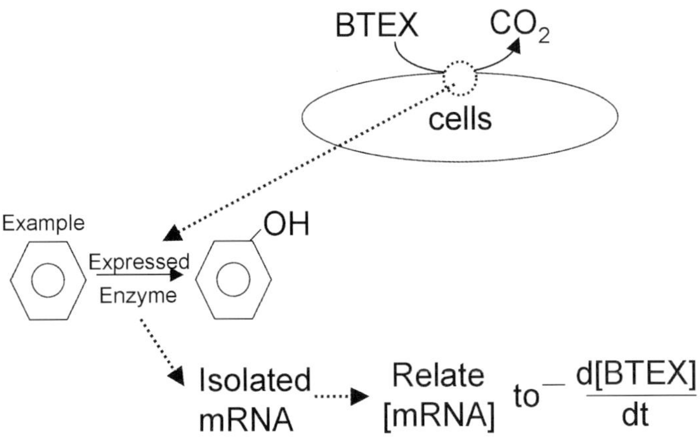

Figure 5. Genomic information from environmentally relevant microorganisms leads to the possibility of evaluating rates of subsurface contaminant degradation processes from expressed mRNA of specific enzyme-catalyzed reactions.

cell protein content indicated that the organisms were conserving energy for growth by electrode reduction (Bond *et al.*, 2002). Electrodes placed into marine sediments electrically connected to other electrodes placed in the overlying, aerobic water produced current and selected for a microbial community capable of Fe(III) reduction relative to control electrodes in an open circuit configuration (Tender *et al.*, 2002; Holmes *et al.*, 2004). These results indicate that conducting surfaces in anaerobic environments when in electrical contact with oxygenated water can serve as a source of electron acceptor for Fe(III)-reducing bacteria found in the environment. This discovery has sparked interest not only for the obvious applications to microbial fuel cells but also as a potential method of electron acceptor supply to contaminated subsurface areas (Bond *et al.*, 2002; Chaudhuri *et al.*, 2003). The demonstration of benzoate degradation coupled to electrode reduction in cell cultures of *Geobacter metallireducens* hints at the possibility of using electrodes to stimulate aromatic hydrocarbon degradation within heavily contaminated portions of petroleum-contaminated aquifers (Figure 4) (Bond *et al.*, 2002). However, the full potential of this application has yet to be studied in detail.

9. Genome-Enabled Bioremediation: A Glimpse of the Future

In situ bioremediation practices are largely empirical undertakings because much of the physiology and community dynamics of subsurface microorganisms, the catalytic agents of bioremediation, are poorly understood. With the advent of genomic studies of organisms of environmental relevance the possibility of bringing bioremediation into the realm of quantitative understanding is proceeding rapidly (Madsen, 1998; Lovley, 2003). Subsurface remediation requires knowledge of the types of microorganisms present, their metabolic potential and how these organisms respond to environmental changes. Often, this type of information is gathered from costly and time consuming laboratory studies, the results of which may bear little semblance to *in situ* processes. However, gene expression-based studies may be the key to understanding bioremediation processes to the degree that chemical processes in the environment are now understood and modeled with sophisticated geochemical and contaminant transport models (Lovley, 2003). Gene expression may also be useful for understanding the nutritional needs and/or stresses of subsurface communities. As the genomes of sequenced environmentally relevant organisms becomes ever more annotated and as the genomes of other bacteria become available, the possibility of conducting gene expression assays for specific, enzymatically catalyzed contaminant transformations become possible. Relating levels of expressed mRNA coding for a specific enzyme involved in a reaction pathway to the overall rate of microbially-mediated contaminant transformation within that reaction pathway provides a way to precisely quantify contaminant degradation from a mechanistic perspective (Figure 5). This is a fundamentally different approach to understanding microbially-catalyzed contaminant degradation, different than Monod-based or kinetics-based models which largely infer contaminant transformation from growth or assumed kinetics of reaction (Watson *et al.*, 2003). The advent of gene expression to environmental studies holds the promise of evaluating the identity and status of subsurface microbial communities and providing estimates of rates of microbially-catalyzed contaminant transformation reactions that are based on the physiological expression of contaminant-transforming enzymes (Lovley, 2003).

10. Summary

Microbial processes coupled to the reduction of Fe(III)oxides contribute to the removal of contaminants from the environment. This is especially true of contaminated subsurface environments where Fe(III) reduction can be a dominant process within contaminated aquifers. While Fe(III)-reducing organisms are known to oxidize important groundwater contaminants such as aromatic hydrocarbons, immobilize contaminant metals such as uranium and may be associated with reductive dehalogenation *in situ*, much remains to be discovered about metal reduction metabolism. A few genomes of Fe(III)-reducing bacteria have been completely sequenced (Heidelberg *et al.*, 2002; Methe *et al.*, 2003). Several more are planned and/or in progress. These descriptions will ultimately provide a glimpse of the genetic potential of Fe(III)-reducing organisms, their contaminant degradation potential, nutritional requirements, dynamics of growth and the physiological mechanism of solid phase electron acceptor reduction. These studies of Fe(III)-reducing bacteria and similar analyses of other sequenced organisms will provide the basis for understanding microbial processes in the environment from a more fundamental perspective, perhaps revolutionizing the field of *in situ* bioremediation.

References

Achtnich, C., Bak, F., and Conrad, R. 1995. Competition for electron donors among nitrate reducers, ferric iron reducers, sulfate reducers, and methanogens in anoxic paddy soil. Biol. Fertil. Soils. 19: 65-72.

Albrechtsen, H.-J., Bjerg, P.L., Ludvigsen, L., Rugge, K., and Christensen, T.H. 1999. An anaerobic field injection experiment in a landfill leachate plume, Grindsted, Denmark 2. Deduction of anaerobic (methanogenic, sulfate- and Fe(III)-reducing) redox conditions. Wat. Resour. Res. 35: 1247-1256.

Anderson, R.T., Ciufo, S.A., Housewright, M.E., and Lovley, D.R. 2004a. Assessment of extent, mechanisms and rates of microbial Fe(III) reduction in subsurface environments. In: The Biogeochemistry of Iron Cycling in Natural Environments. C. Zhang, and J.D. Coates, eds. Kluwer Academic/Plenum Publishers, New York, New York. In press.

Anderson, R.T., and Lovley, D.R. 1997. Ecology and biogeochemistry of *in situ* groundwater bioremediation. In: Advances in Microbial Ecology. Vol.15 J.G. Jones, ed. Plenum Press, New York. p. 289-350.

Anderson, R.T., and Lovley, D.R. 1999. Naphthalene and benzene degradation under Fe(III)-reducing conditions in petroleum-contaminated aquifers. Bioremediation J. 3: 121-135.

Anderson, R.T., and Lovley, D.R. 2000. Anaerobic bioremediation of benzene under sulfate-reducing conditions in a petroleum-contaminated aquifer. Environ. Sci. Technol. 34: 2261-2266.

Anderson, R.T., and Lovley, D.R. 2002. Microbial redox interactions with uranium: an environmental perspective. In: Interactions of Microorganisms with Radionuclides. M. Keith-Roach, and F. Livens, eds. Elsevier Science Limited, Amsterdam. p. 205-223.

Anderson, R.T., Rooney-Varga, J.N., Gaw, C.V., and Lovley, D.R. 1998. Anaerobic benzene oxidation in the Fe(III) reduction zone of petroleum-contaminated aquifers. Environ. Sci. Technol. 32: 1222-1229.

Anderson, R.T., Vrionis, H.A., Ortiz-Bernad, I., Resch, C.T., Long, P.E., Dayvault, R., Karp, K., Marutzky, S., Metzler, D.R., Peacock, A., White, D.C., Lowe, M., and Lovley, D.R. 2003. Stimulating the *in situ* activity of *Geobacter* species to remove uranium from the groundwater of a uranium-contaminated aquifer. Appl. Environ. Microbiol. 69: 5884-5891.

Baedecker, M.J., Cozzarelli, I.M., Siegel, D.I., Bennett, P.C., and Eganhouse, R.P. 1993. Crude oil in a shallow sand and gravel aquifer: 3. Biogeochemical reactions and mass balance modeling in anoxic ground water. Appl. Geochem. 8: 569-586.

Baedecker, M.J., Siegel, D.I., Bennett, P., and Cozzarelli, I.M. 1989. The fate and effects of crude oil in a shallow aquifer I. The distribution of chemical species and geochemical facies. In: U. S. Geological Survey Water Resources Division Report 88-4220. G.E. Mallard and S.E. Ragone, eds. U. S. Geological Survey, Reston, Virginia. p. 13-20.

Bekins, B.A., Cozzarelli, I.M., Godsy, E.M., Warren, E., Essaid, H.I., and Tuccillo, M.E. 2001. Progression of natural attenuation processes at a crude oil spill site: II. Controls on spatial distribution of microbial populations. J. Contam. Hydrol. 53: 387-406.

Bjerg, P.L., Rugge, K., Pedersen, J.K., and Christensen, T.H. 1995. Distribution of redox-sensitive groundwater quality parameters downgradient of a landfill (Grindsted, Denmark). Environ. Sci. Technol. 29: 1387-1394.

Bond, D.R., Holmes, D.E., Tender, L.M., and Lovley, D.R. 2002. Electrode-reducing microorganisms that harvest energy from marine sediments. Science. 295: 483-485.

Bond, D.R., and Lovley, D.R. 2003. Electricity production by *Geobacter sulfurreducens* attached to electrodes. Appl. Environ. Microbiol. 69: 1548-1555.

Borden, R.C., Gomez, C.A., and Becker, M.T. 1995. Geochemical indicators of intrinsic bioremediation. Ground Water. 33: 180-189.

Bouwer, E.J., and Zehnder, A.J.B. 1993. Bioremediation of organic compounds - putting microbial metabolism to work. TIBTECH. 11: 360-367.

Bradley, P.M., and Chapelle, F.H. 1996. Anaerobic mineralization of vinyl chloride in aquifer sediments. Environ. Sci. Technol. 30: 2084-2086.

Bradley, P.M., and Chapelle, F.H. 1998a. Microbial mineralization of VC and DCE under different terminal electron accepting conditions. Anaerobe. 4: 81-87.

Bradley, P.M., Chapelle, F.H., and Lovley, D.R. 1998b. Humics acids as electron acceptors for anaerobic microbial oxidation of vinyl chloride and dichloroethene. Appl. Environ. Microbiol. 64: 3102-3105.

Bradley, P.M., Chapelle, F.H., and Wilson, J.T. 1998c. Field and laboratory evidence for intrinsic biodegradation of vinyl chloride contamination in a Fe(III)-reducing aquifer. J. Contam. Hydrol. 31: 111-127.

Caccavo Jr, F., Coates, J.D., Rossello-Mora, R.A., Ludwig, W., Schleifer, K.H., Lovley, D.R., and McInerney, M.J. 1996. *Geovibrio ferrireducens*, a phylogenetically distinct dissimilatory Fe(III)-reducing bacterium. Arch. Microbiol. 165: 370-376.

Carpentier, W., Sandra, K., De Smet, I., Brige, A., De Smet, L., and Van Beeumen, J. 2003. Microbial reduction and precipitation of vanadium by *Shewanella oneidensis*. Appl. Environ. Microbiol. 69: 3636-3639.

Cervantes, F.J., van der Velde, S., Lettinga, G., and Field, J.A. 2000. Competition between methanogenesis and quinone respiration for ecologically important substrates in anaerobic consortia. FEMS Microbiol. Ecol. 34: 161-171.

Chapelle, F.H. 1993. Ground-water microbiology and geochemistry. John Wiley & Sons, New York, New York.

Chapelle, F.H. 1999. Bioremediation of petroleum hydrocarbon-contaminated ground water: the perspectives of history and hydrology. Ground Water. 37: 122-132.

Chapelle, F.H., Bradley, P.M., Lovley, D.R., O'Neill, K., and Landmeyer, J.E. 2002. Rapid evolution of redox processes in a petroleum hydrocarbon-contaminated aquifer. Ground Water. 40: 353-360.

Chapelle, F.H., McMahon, P.B., Dubrovsky, N.M., Fujii, R.F., Oaksford, E.T., and Vroblesky, D.A. 1995. Deducing the distribution of terminal electron-accepting processes in hydrologically diverse groundwater systems. Water Resour. Res. 31: 359-371.

Chapelle, F.H., Vroblesky, D.A., Woodward, J.C., and Lovley, D.R. 1997. Practical considerations for measuring hydrogen concentrations in groundwater. Environ. Sci. Technol. 31: 2873-2877.

Chaudhuri, S.K., and Lovley, D.R. 2003. Electricity generation by direct oxidation of glucose in mediatorless microbial fuel cells. Nature Biotechnol. 21: 1229-1232.

Christensen, T., H., Kjeldsen, P., Albrechtsen, H.-J., and Heron, G. 1994. Attenuation of pollutants in landfill leachate polluted aquifers. Critical Reviews in Environ. Sci. Technol. 24: 119-202.

Christensen, T.H., Kjeldsen, P., Bjerg, P.L., Jensen, D.L., Christensen, J.B., Baun, A., Albrechtsen, H., and Heron, G. 2001. Biogeochemistry of landfill leachate plumes. Appl. Geochem. 16: 659-718.

Coates, J.D., Bhupathiraju, V.K., Achenbach, L.A., McInerney, M.J., and Lovley, D.R. 2001. *Geobacter hydrogenophilus*, *Geobacter chapellei*, *Geobacter grbiciae*, three new, strictly anaerobic, dissimilatory Fe(III)-reducers. Int. J. Syst. Evol. Microbiol. 51: 581-588.

Coates, J.D., Councell, T., Ellis, D.J., and Lovley, D.R. 1998a. Carbohydrate oxidation coupled to Fe(III) reduction, a novel form of anaerobic metabolism. Anaerobe. 4: 277-282.

Coates, J.D., Ellis, D.J., Blunt-Harris, E.L., Gaw, C.V., Roden, E.E., and Lovley, D.R. 1998b. Recovery of humics-reducing bacteria from a diversity of environments. Appl. Environ. Microbiol. 64: 1504-1509.

Coates, J.D., Lonergan, D.J., Jenter, H., and Lovley, D.R. 1996. Isolation of *Geobacter* species from a variety of sedimentary environments. Appl. Environ. Microbiol. 62: 1531-1536.

Colberg, P.J.S., and Young, L.L. 1995. Anaerobic degradation of nonhalogenated homocyclic aromatic compounds coupled with nitrate, iron or sulfate reduction. In: Microbial Transformation and Degradation of Toxic Organic Chemicals. L.L. Young, and C.E. Cerniglia, eds. Wiley-Liss, Inc., New York, New York. p. 307-330.

Cord-Ruwisch, R., Seitz, H., and Conrad, R. 1988. The capacity of hydrogenotrophic anaerobic bacteria to compete for traces of hydrogen depends on the redox potential of the terminal electron acceptor. Arch. Microbiol. 149: 350-357.

Cummings, D.E., Caccavo Jr, F., Spring, S., and Rosenzweig, R.F. 1999. *Ferribacter limneticum*, gen. nov., sp. nov., an Fe(III)-reducing microorganism isolated from mining-impacted freshwater lake sediments. Arch. Microbiol. 171: 183-188.

Cummings, D.E., Snoeyenbos-West, O.L., Newby, D.T., Niggemyer, A.M., Lovley, D.R., Achenbach, L.A., and Rosenzweig, R.F. 2003. Diversity of Geobacteraceae species inhabiting metal-polluted freshwater lake sediments ascertained by 16S rDNA analyses. Microb. Ecol. 46: 257-269.

Cunningham, J.A., Rahme, H., Hopkins, G.D., Lebron, C., and Reinhard, M. 2001. Enhanced *in situ* bioremediation of BTEX-contaminated groundwater by combined injection of nitrate and sulfate. Environ. Sci. Technol. 35: 1663-1670.

Evans, W.C., and Fuchs, G. 1988. Anaerobic degradation of aromatic compounds. Ann. Rev. Microbiol. 42: 289-317.

Finneran, K.T., Housewright, M.E., and Lovley, D.R. 2002. Multiple influences of nitrate on uranium solubility during bioremediation of uranium-contaminated subsurface sediments. Environ. Microbiol. 4: 510-516.

Flathman, P., E., Jerger, D., E., and Exner, J., H. 1994. Bioremediation Field Experience. CRC Press, Inc., Boca Raton, Florida.

Fredrickson, J.K., Zachara, J.M., Kennedy, D.W., Dong, H., Onstott, T.C., Hinman, H.W., and Li, S. 1998. Biogenic iron mineralization accompanying the dissimilatory reduction of hydrous ferric oxide by a groundwater bacterium. Geochim. Cosmochim. Acta. 62: 3239-3257.

Fruchter, J.S., Cole, C.R., Williams, M.D., Vermeul, V.R., Amonette, J.E., Szecsody, J.E., Istok, J.D., and Humphrey, M.D. 2000. Creation of a subsurface permeable treatment zone for aqueous chromate contamination using *in situ* redox manipulation. Groundwater Monit. Rem. 20: 66-77.

Glasauer, S., Weidler, P.G., Langley, S., and Beveridge, T.J. 2003. Controls on Fe reduction and mineral formation by a subsurface bacterium. Geochim. Cosmochim. Acta. 67: 1277-1288.

Gorby, Y.A., and Lovley, D.R. 1992. Enzymatic uranium precipitation. Environ. Sci. Technol. 26: 205-207.

Hardman, D., J. 1991. Biotransformation of halogenated compounds. Critical Reviews in Biotechnology. 11: 1-40.

He, J., Sung, Y., Dollhope, M.E., Fathepure, B.Z., Tiedje, J.M., and Loffler, F.E. 2002. Acetate versus hydrogen as direct electron donors to stimulate the microbial reductive dechlorination process at chloroethene-contaminated sites. Environ. Sci. Technol. 36: 3945-3952.

Heidelberg, J.F., Paulsen, I.T., Nelson, K.E., Gaidos, E.J., Nelson, W.C., Read, T.D., Eisen, J.A., Seshadri, R., Ward, N., Methe, B., Clayton, R.A., Meyer, T., Tsapin, A., Scott, J., Beanan, M., Brinkac, L., Daugherty, S., DeBoy, R.T., Dodson, R.J., Durkin, A.S., Haft, D.H., Kolonay, J.F., Madupu, R., Petersen, J.D., Umayam, L.A., White, O., Wolf, A.M., Vamathevan, J., Weidman, J., Impraim, M., Lee, K., Berry, K., Lee, C., Mueller, J., Khouri, H., Gill, J., Utterback, T.R., McDonald, L.A., Feldblyum, T.V., Smith, H.O., Venter, J.C., Nealson, K.H., and Fraser, C.M. 2002. Genome sequence of the dissimilatory metal ion-reducing bacterium *Shewanella oneidensis*. Nature Biotechnol. 20: 1118-1123.

Heron, G., and Christensen, T.H. 1995. Impact of sediment-bound iron on redox buffering in a landfill leachate polluted aquifer (Vejen, Denmark). Environ. Sci. Technol. 29: 187-192.

Heron, G., Crouzet, C., Bourg, A.C.M., and Christensen, T.H. 1994. Speciation of Fe(II) and Fe(III) in contaminated aquifer sediments using chemical extraction techniques. Environ. Sci. Technol. 28: 1698-1705.

Holliger, C., Wohlfarth, G., and Diekert, G. 1999. Reductive dechlorination in the energy metabolism of anaerobic bacteria. FEMS Microbiol. Ecol. 22: 383-398.

Holmes, D.E., Bond, D.R., O'Neill, R.A., Reimers, C.E., and Lovley, D.R. 2004. Microbial communities associated with electrodes harvesting electricity from a variety of aquatic sediments. Microb. Ecol. in press.

Holmes, D.E., Finneran, K.T., O'Neil, R.A., and Lovley, D.R. 2002. Enrichment of *Geobacteraceae* associated with stimulation of dissimilatory metal reduction in uranium-contaminated aquifer sediments. Appl. Environ. Microbiol. 68: 2300-2306.

Hopkins, G., D., Semprini, L., and McCarty, P., L. 1993a. Microcosm and *in situ* field studies of enhanced biotransformation of trichloroethylene by phenol-utilizing microorganisms. Appl. Environ. Microbiol. 59: 2277-2285.

Hopkins, G.D., and McCarty, P.L. 1995. Field evaluation of *in situ* aerobic cometabolism of trichloroethylene and three dichloroethylene isomers using phenol and toluene as the primary substrates. Environ. Sci. Technol. 29: 1628-1637.

Hopkins, G.D., Munakata, J., Semprini, L., and McCarty, P.L. 1993b. Trichloroethylene concentration effects on pilot field-scale *in-situ* groundwater bioremediation by phenol-oxidizing microorganisms. Environ. Sci. Technol. 27: 2542-2547.

Hutchins, S.R., Miller, D.E., and Thomas, A. 1998. Combined laboratory/field study on the use of nitrate for *in situ* bioremediation of a fuel-contaminated aquifer. Environ. Sci. Technol. 32: 1832-1840.

Jakobsen, R., Albrechtsen, H., Rasmussen, M., Bay, H., Bjerg, P.L., and Christensen, T.H. 1998. H_2 concentrations in a landfill leachate plume (Grindsted, Denmark): *In situ* energetics of terminal electron acceptor processes. Environ. Sci. Technol. 32: 2142-2148.

Kashefi, K., Tor, J.M., Holmes, D.E., VanPraagh, C.V.G., Reysenbach, A.-L., and Lovley, D.R. 2002. *Geoglobus ahangari*, gen. nov., sp. nov., a novel hyperthermophilic Archaeum capable of oxidizing organic acids and growing autotrophically on hydrogen with Fe(III) serving as the sole electron acceptor. Int. J. Syst. Evol. Microbiol. 52: 719-728.

Keith-Roach, M.J., and Livens, F.R., eds. 2002. Interactions of Microorganisms with Radionuclides. Radioactivity in the Environment. Elsevier, Amsterdam.

Kim, B.H., Kim, H.J., Hyun, M.S., and Park, D.H. 1999. Direct electrode reaction of Fe(III)-reducing bacterium *Shewanella putrefaciens*. J. Microbiol. Biotechnol. 9: 127-131.

Kostka, J.E., Wu, J., Nealson, K.H., and Stucki, J.W. 1999. The impact of structural Fe(III) reduction by bacteria on the surface chemistry of smectite clay minerals. Geochim. Cosmochim. Acta. 63: 3705-3713.

Kostka, J.M., Nealson, K.H., Wu, J., and Stucki, J.W. 1996. Reduction of structural Fe(III) in smectite by a pure culture of *Shewanella putrefaciens* strain MR-1. Clays Clay Miner. 44: 522-529.

Lendvay, J.M., Loffler, F.E., Dollhopf, M., Aiello, M.R., Daniels, G., Fathepure, B.Z., Gebhard, M., Heine, R., Helton, R., Shi, J., Krajmalnik-Brown, R., Major Jr., C.L., Barcelona, M.J., Petrovskis, E., Hickey, R., Tiedje, J.M., and Adriaens, P. 2003. Bioreactive barriers: a comparison of bioaugmentation and biostimulation for chlorinated solvent remediation. Environ. Sci. Technol. 37: 1422-1431.

Lloyd, J.R., Chesnes, J., Glasauer, S., Bunker, D.J., Livens, F.R., and Lovley, D.R. 2002. Reduction of actinides and fission products by Fe(III)-reducing bacteria. Geomicrobiol. J. 19: 103-120.

Lloyd, J.R., and Macaskie, L.E. 2000a. Bioremediation of radionuclide-containing wastewaters. In: Enviromental Microbe-Metal Interactions. D.R. Lovley, ed. ASM Press, Washington, D.C. p. 277-327.

Lloyd, J.R., Sole, V.A., Van Praagh, C.V.G., and Lovley, D.R. 2000b. Direct and Fe(II)-mediated reduction of technetium by Fe(III)-reducing bacteria. Appl. Environ. Microbiol. 66: 3743-3749.

Loffler, F.E., Sun, Q., Li, J., and Tiedje, J.M. 2000. 16S rRNA gene-based detection of tetrachloroethene-dechlorinating *Desulfuromonas* and *Dehalococcoides* species. Appl. Environ. Microbiol. 66: 1369-1374.

Lollar, S.B., Slater, G.F., Sleep, B., Witt, M., Klecka, G.M., Harkness, M., and Spivak, J. 2001. Stable carbon isotope evidence for intrinsic bioremediation of tetrachloroethene and trichloroethene at area 6, Dover Air Force Base. Environ. Sci. Technol. 35: 261-269.

Lonergan, D.J., and Lovley, D.R. 1991. Microbial oxidation of natural and anthropogenic aromatic compounds coupled to Fe(III) reduction. In: Organic Substances and Sediments in Water. R.A. Baker, ed. Lewis Publishers, Inc., Chelsea, Michigan. p. 327-338.

Lovley, D.R. 1987. Organic matter mineralization with the reduction of ferric iron: A review. Geomicrobiol. J. 5: 375-399.

Lovley, D.R. 1993. Anaerobes into heavy metal: dissimilatory metal reduction in anoxic environments. Trends Ecol. Evol. 8: 213-217.

Lovley, D.R. 1995a. Bioremediation of organic and metal contaminants with dissimilatory metal reduction. J. Industr. Microbiol. 14: 85-93.

Lovley, D.R. 1995b. Microbial reduction of iron, manganese, and other metals. Adv. Agron. 54: 175-231.

Lovley, D.R. 1997. Potential for anaerobic bioremediation of BTEX in petroleum-contaminated aquifers. J. Industr. Microbiol. 18: 75-81.

Lovley, D.R., ed. 2000a. Environmental Microbe-Metal Interactions. ASM Press, Washington, D.C.

Lovley, D.R. 2000b. Fe(III) and Mn(IV) reduction. In: Environmental microbe-metal interactions. D.R. Lovley, ed. ASM Press, Washington, D.C. p. 3-30.

Lovley, D.R. 2000c. Dissimilatory Fe(III)- and Mn(IV)-reducing Prokaryotes. In: The Procaryotes. M. Dworkin, S. Falkow, E. Rosenberg, K.-H. Schleifer and E. Stackebrandt, eds. Springer-Verlag, New York, New York. http://link.springer-ny.com/link/service/books/10125/.

Lovley, D.R. 2001. Reduction of iron and humics in subsurface environments. In: Subsurface Microbiology and Biogeochemistry. J.K. Frederickson and M. Fletcher, eds. Wiley-Liss, Inc., New York, New York. p. 193-217.

Lovley, D.R. 2003. Cleaning up with genomics: Applying molecular biology to bioremediation. Nature Reviews – Microbiology. 1: 35-44.

Lovley, D.R., and Anderson, R.T. 2000. Influence of dissimilatory metal reduction on fate of organic and metal contaminants in the subsurface. Hydrogeology J. 8: 77-88.

Lovley, D.R., Baedecker, M.J., Lonergan, D.J., Cozzarelli, I.M., Phillips, E.J.P., and Siegel, D.I. 1989. Oxidation of aromatic contaminants coupled to microbial iron reduction. Nature. 339: 297-299.

Lovley, D.R., and Chapelle, F.H. 1995a. Deep subsurface microbial processes. Rev. Geophys. 33: 365-381.

Lovley, D.R., and Chapelle, F.H. 1995b. A modeling approach to elucidating the distribution and rates of microbially catalyzed redox reactions in anoxic groundwater. In: Mathematical Models in Microbial Ecology. A.L. Koch, J.A. Robinson, and G.A. Milliken, eds. Chapman and Hall, New York, New York. p. 196-209.

Lovley, D.R., Chapelle, F.H., and Woodward, J.C. 1994a. Use of dissolved H_2 concentrations to determine the distribution of microbially catalyzed redox reactions in anoxic ground water. Environ. Sci. Technol. 28: 1205-1210.

Lovley, D.R., and Coates, J.D. 1997. Bioremediation of metal contamination. Curr. Opin. Biotechnol. 8: 285-289.

Lovley, D.R., Coates, J.D., Blunt-Harris, E.L., Phillips, E.J.P., and Woodward, J.C. 1996a. Humic substances as electron acceptors for microbial respiration. Nature. 382: 445-448.

Lovley, D.R., Fraga, J.L., Blunt-Harris, E.L., Hayes, L.A., Phillips, E.J.P., and Coates, J.D. 1998. Humic substances as a mediator for microbially catalyzed metal reduction. Acta Hydrochimica et Hydrobiologica. 26: 152-157.

Lovley, D.R., Fraga, J.L., Coates, J.D., and Blunt-Harris, E.L. 1999. Humics as an electron donor for anaerobic respiration. Environ. Microbiol. 1: 89-98.

Lovley, D.R., Giovannoni, S.J., White, D.C., Champine, J.E., Phillips, E.J.P., Gorby, Y.A., and Goodwin, S. 1993. *Geobacter metallireducens* gen. nov. sp. nov., a microorganism capable of coupling the complete oxidation of organic compounds to the reduction of iron and other metals. Arch. Microbiol. 159: 336-344.

Lovley, D.R., and Goodwin, S. 1988a. Hydrogen concentrations as an indicator of the predominant terminal electron-accepting reactions in aquatic sediments. Geochimica et Cosmochimica Acta. 52: 2993-3003.

Lovley, D.R., and Klug, M.J. 1982. Intermediary metabolism of organic carbon in the sediment of a eutrophic lake. Appl. Environ. Microbiol. 43: 552-560.

Lovley, D.R., and Klug, M.J. 1983. Sulfate reducers can out compete methanogens at freshwater sulfate concentrations. Appl. Environ. Microbiol. 45: 187-192.

Lovley, D.R., and Phillips, E.J.P. 1986a. Availability of ferric iron for microbial reduction in bottom sediments of the freshwater tidal Potomac River. Appl. Environ. Microbiol. 52: 751-757.

Lovley, D.R., and Phillips, E.J.P. 1986b. Organic matter mineralization with reduction of ferric iron in anaerobic sediments. Appl. Environ. Microbiol. 51: 683-689.

Lovley, D.R., and Phillips, E.J.P. 1987a. Competitive mechanisms for inhibition of sulfate reduction and methane production in the zone of ferric iron reduction in sediments. Appl. Environ. Microbiol. 53: 2636-2641.

Lovley, D.R., and Phillips, E.J.P. 1987b. Rapid assay for microbially reducible ferric iron in aquatic sediments. Appl. Environ. Microbiol 53: 1536-1540.

Lovley, D.R., and Phillips, E.J.P. 1988b. Novel mode of microbial energy metabolism: organic carbon oxidation coupled to dissimilatory reduction of iron or manganese. Appl. Environ. Microbiol. 54: 1472-1480.

Lovley, D.R., Phillips, E.J.P., Gorby, Y.A., and Landa, E.R. 1991. Microbial reduction of uranium. Nature. 350: 413-416.

Lovley, D.R., and Woodward, J.C. 1996b. Mechanisms for chelator stimulation of microbial Fe(III)- oxide reduction. Chem. Geol. 132: 19-24.

Lovley, D.R., Woodward, J.C., and Chapelle, F.H. 1994b. Stimulated anoxic biodegradation of aromatic hydrocarbons using Fe(III) ligands. Nature. 370: 128-131.

Lu, G., Clement, T.P., Zheng, C., and Wiedemeier, T.H. 1999. Natural attenuation of BTEX compounds: model development and field application. Ground Water. 37: 707-717.

Lyalikova, N.N., and Yurkova, N.A. 1992. Role of microorganisms in vanadium concentration and dispersion. Geomicrobiol. J. 10: 15-26.

Lyngkilde, J., and Christensen, T.H. 1992a. Fate of organic contaminants in the redox zones of a landfill leachate pollution plume (Vejen, Denmark). J. Contam. Hydrol. 10: 291-307.

Lyngkilde, J., and Christensen, T.H. 1992b. Redox zones of a landfill leachate pollution plume (Vejen, Denmark). J. Contam. Hydrol. 10: 273-289.

Madsen, E., L. 1995. Impacts of Agricultural Practices on Subsurface Microbial Ecology. In: Advances in Agronomy. Vol. 54 D.L. Sparks, ed. Academic Press, Inc., San Diego, California. p. 1-67.

Madsen, E.L. 1998. Epistemology of environmental microbiology. Environ. Sci. Technol. 32: 429-439.

Madsen, E.L. 2001. Intrinsic bioremediation of organic subsurface contaminants. In: Subsurface Microbiology and biogeochemistry. J.K. Frederickson, and M. Fletcher, eds. John Wiley & Sons, Inc., New York, New York. p. 249-278.

McCarty, P.L. 1972. Energetics of organic matter degradation. In: Water Pollution Microbiology. R. Mitchell, ed. John Wiley & Sons, New York, New York. p. 91-118.

McCarty, P.L., Goltz, M.N., Hopkins, G.D., Dolan, M.E., Allan, J.P., Kawakami, B.T., and Carrothers, T.J. 1998. Full-scale evaluation of *in situ* cometabolic degradation of trichloroethylene in groundwater through toluene injection. Environ. Sci. Technol. 32: 88-100.

McCarty, P.L., and Semprini, L. 1994. Ground-water treatment for chlorinated solvents. In: Handbook of Bioremediation. R.D. Norris, R.E. Hinchee, R. Brown, P.L. McCarty, L. Semprini, J.T. Wilson, D.H. Kampbell, M. Reinhard, E.J. Bouwer, R.C. Borden, T.M. Vogel, J.M.Thomas, and C.H..Ward, eds. Lewis Publishers, Boca Raton, Florida. p. 87-116.

McCarty, P.L., Semprini, L., Dolan, M.E., Harmon, T.C., Tiedeman, C., and Gorelick, S.M. 1991. *In Situ* Methanogenic Bioremediation for Contaminated Groundwater at St. Joseph, Michigan. In: On-Site Bioreclamation: Processes for Xenobiotic and HydrocarbonTreatment. R.E. Hinchee, and R.F. Olfenbuttel, eds. Butterworth-Heinemann, Boston, Massachusetts. p. 16-40.

Methe, B.A., Nelson, K.E., Eisen, J.A., Paulsen, I.T., Nelson, W., Heidelberg, J.F., Wu, D., Wu, M., Ward, N., Beanan, M.J., Dodson, R.J., Madupu, R., Brinkac, L.M., Daugherty, S.C., DeBoy, R.T., Durkin, A.S., Gwinn, M., Kolonay, J.F., Sullivan, S.A., Haft, D.H., Selengut, J., Davidsen, T.M., Zafar, N., White, O., Tran, B., Romero, C., Forberger, H.A., Weidman, J., Khouri, H., Feldblyum, T.V., Utterback, T.R., Van Acken, S.E., Lovley, D.R., and Fraser, C.M. 2003. Genome of *Geobacter sulfurreducens*: metal reduction in subsurface environments. Science. 302: 1967-1969.

Nevin, K.P., and Lovley, D.R. 2000a. Lack of production of electron-shuttling compounds or solubilization of Fe(III) during reduction of insoluble Fe(III) by *Geobacter metallireducens*. Appl. Environ. Microbiol. 66: 2248-2251.

Nevin, K.P., and Lovley, D.R. 2000b. Potential for nonenzymatic reduction of Fe(III) via electron shuttling in subsurface sediments. Environ. Sci. Technol. 34: 2472-2478.

Nevin, K.P., and Lovley, D.R. 2002. Mechanisms for Fe(III) oxide reduction in sedimentary environments. Geomicrobiol. J. 19: 141-159.

Newman, D.K., and Kolter, R. 2000. A role for excreted quinones in extracellular electron transfer. Nature 405: 94-97.

Norris, R., D., Hinchee, R., E., Brown, R., McCarty, P., L., Semprini, L., Wilson, J., T., Kampbell, D., H., Reinhard, M., Bouwer, E., J., Borden, R., C., Vogel, T., M., Thomas, J., M., and Ward, C., H. 1994. Handbook of Bioremediation. CRC Press, Inc., Boca Raton, Florida.

Oremland, R.S., and Capone, D.G. 1988. Use of "specific" inhibitors in biogeochemistry and microbial ecology. Adv. Microb. Ecol. 10: 285-383.

Ortiz-Bernad, I., Anderson, R.T., Vrionis, H.A., and Lovley, D.R. 2004. Vanadium respiration by *Geobacter metallireducens*: a novel strategy for the *in situ* removal of vanadium from groundwater. Appl. Environ. Microbiol. 70: 3091-3095.

Ottow, J.C.G. 1969. Der Einfluss von Nitrat, Chlorat, Sulfat, Eisenoxidform und Wachstumbedinungen auf das Ausmass der bakteriellen Eisenreduktion. Z. Pflanzenernaehr. Bodenkd. 124: 238-253.

Patterson, B., M., Pribac, F., Barber, C., Davis, G., B., and Gibbs, R. 1993. Biodegradation and retardation of PCE and BTEX compounds in aquifer material from Western Australia using large-scale columns. J. Contam. Hydrol. 14: 261-278.

Pedersen, K., Arlinger, J., Ekendahl, S., and Hallbeck, L. 1996. 16S rRNA gene diversity of attached and unattached bacteria in boreholes along the access tunnel to the Apso hard rock laboratory. FEMS Microbiol. Ecol. 19: 249-262.

Phillips, E., Lovley, D.R., and Roden, E.E. 1993. Composition of non-microbially reducible Fe(III) in aquatic sediments. Appl. Environ. Microbiol. 59: 2727-2729.

Ponnamperuma, F.N. 1972. The chemistry of submerged soils. Adv. Agron. 24: 29-96.

Reeburgh, W.S. 1983. Rates of biogeochemical processes in anoxic sediments. Ann. Rev. Earth Planet. Sci. 11: 269-298.

Reinhard, M., Shang, S., Kitanidis, P.K., Orwin, E., Hopkins, G.D., and Lebron, C.A. 1997. *In situ* BTEX biotransformation under enhanced nitrate- and sulfate-reducing conditions. Environ. Sci. Technol. 31: 28-36.

Roden, E.E. 2003a. Diversion of electron flow from methanogenesis to crystalline Fe(III) oxide reduction in carbon-limited cultures of wetland sediment microorganisms. Appl. Environ. Microbiol. 69: 5702-5706.

Roden, E.E. 2003b. Fe(III) oxide reactivity toward biological versus chemical reduction. Environ. Sci. Technol. 37: 1319-1324.

Roden, E.E., and Urrutia, M.M. 2002. Influence of biogenic Fe(II) on bacterial crystalline Fe(III) oxide reduction. Geomicrobiol. J. 19: 209-251.

Roden, E.E., Urrutia, M.M., and Mann, C.J. 2000. Bacterial reductive dissolution of crystalline Fe(III) oxide in continuous-flow column reactors. Appl. Environ. Microbiol. 66: 1062-1065.

Roden, E.E., and Wetzel, R.G. 2003. Competition between Fe(III)-reducing and methanogenic bacteria for acetate in iron-rich freshwater sediments. Microb. Ecol. 45: 252-258.

Roden, E.E., and Zachara, J.M. 1996. Microbial reduction of crystalline iron(III) oxides: influence of oxide surface area and potential for cell growth. Environ. Sci. Technol. 30: 1618-1628.

Roling, W.F.M., Van Breukelen, B.M., Braster, M., Lin, B., and H.W., V.V. 2001. Relationships between microbial community structure and hydrochemistry in a landfill leachate-polluted aquifer. Appl. Environ. Microbiol. 67: 4619-4629.

Rooney-Varga, J., Anderson, R.T., Fraga, J.L., Ringleberg, D., and Lovley, D.R. 1999. Microbial communities associated with anaerobic benzene degradation in a petroleum-contaminated aquifer. Appl. Environ. Microbiol. 65: 3056-3063.

Salanitro, J., P. 1993. The role of bioattenuation in the management of aromatic hydrocarbon plumes in aquifers. Ground Water Monit. Rev. 13: 150-161.

Schwertmann, U., and Fitzpatrick, R.W. 1992. Iron minerals in surface environments. In: Biomineralization Processes of Iron and Manganese. H.C.W. Skinner, and R.W. Fitzpatrick, eds. Catena Verlag, Germany. p. 7-30.

Schwertmann, U., and Taylor, R.M. 1977. Iron oxides. In: Minerals in soil environments. J.B. Dixon, and S.B. Weed, eds. Soil Science Society of America, Madison, Wisconsin. p. 145-180.

Scott, D.T., McKnight, D.M., Blunt-Harris, E.L., Kolesar, S.E., and Lovley, D.R. 1998. Quinone moieties act as electron acceptors in the reduction of humic substances by humics-reducing microorganisms. Environ. Sci. Technol. 32: 2984-2989.

Semprini, L., Grbic-Galic, D., McCarty, P., L., and Roberts, P., V. 1992. Methodologies for evaluating *in-situ* bioremediation of chlorinated solvents. Ada, OK, Robert S. Kerr Environmental Research Laboratory.

Semprini, L., Hopkins, G., D., Grbic-Galic, D., McCarty, P., L., and Roberts, P., V. 1994. A Laboratory and Field Evaluation of Enhanced *In Situ* Bioremediation of Trichoroethylene, *cis-* and *trans*-Dichloroethylene, and Vinyl Chloride by Methanotrophic Bacteria. In: Bioremediation Field Experience. P.E. Flathman, D.E. Jerger, and J.H. Exner,, eds. Lewis Publishers, Boca Raton, Florida. p. 383-412.

Shelobolina, E.S., Anderson, R.T., Vodyanitskii, Y.N., and Lovley, D.R. 2003. Importance of clay minerals for Fe(III) respiration in a petroleum-contaminated aquifer. Geobiology. in press.

Snoeyenbos-West, O.L., Nevin, K.P., Anderson, R.T., and Lovley, D.R. 2000. Enrichment of Geobacters species in response to stimulation of Fe(III) reduction in sandy aquifer sediments. Microb. Ecol. 39: 153-167.

Straub, K.L., Benz, M., and Schink, B. 2001. Iron metabolism in anoxic environments at near neutral pH. FEMS Microbiol. Ecol. 34: 181-186.

Stumm, W., and Morgan, J.J. 1981. Aquatic Chemistry. John Wiley & Sons, New York.

Sung, Y., Ritalahti, K.M., Sanford, R.A., Urbance, J.W., Flynn, S.J., Tiedje, J.M., and Loffler, F.E. 2003. Characterization of two tetrachloroethene-reducing, acetate-oxidizing anaerobic bacteria and their description as *Desulfuromonas michiganensis* sp. nov. Appl. Environ. Microbiol. 69: 2964-2974.

Tender, L.M., Reimers, C.E., Stecher III, H.A., Holmes, D.E., Bond, D.R., Lowy, D.A., Pilobello, K., Fertig, S.J., and Lovley, D.R. 2002. Harnessing microbially generated power on the seafloor. Nature Biotechnology. 20: 821-825.

Thamdrup, B. 2000. Bacterial manganese and iron reduction in aquatic sediments. In: Advances in Microbial Ecology. Vol. 16 B. Schink, ed. Kluwer Academic/Plenum Publishers, New York, New York. p. 41-84.

Tor, J.M., Kashefi, K., and Lovley, D.R. 2001. Acetate oxidation coupled to Fe(III) reduction in hyperthermophilic microorganisms. Appl. Environ. Microbiol. 67: 1363-1365.

Urrutia, M.M., Roden, E.E., Fredrickson, J.K., and Zachara, J.M. 1998. Microbial and surface chemistry controls on reduction of synthetic Fe(III) oxide minerals by the dissimilatory iron-reducing bacterium *Shewanella alga*. Geomicrobiol. J. 15: 269-291.

Urrutia, M.M., Roden, E.E., and Zachara, J.M. 1999. Influence of aqueous and solid-phase Fe(II) complexants on microbial reduction of crystalline oxides. Environ. Sci. Technol. 33: 4022-4028.

Vargas, M., Kashefi, K., Blunt-Harris, E.L., and Lovley, D.R. 1998. Microbiological evidence for Fe(III) reduction on early Earth. Nature. 395: 65-67.

Vogel, T.M., Criddle, C.S., and McCarty, P.L. 1987. Transformations of halogenated aliphatic compounds. Environ. Sci. Technol. 21: 722-736.

Vroblesky, D.A., and Chapelle, F.H. 1994. Temporal and spatial changes of terminal electron-accepting processes in a petroleum hydrocarbon-contaminated aquifer and the significance for contaminant biodegradation. Water Resour. Res. 30: 1561-1570.

Watson, I.A., Oswald, S.E., Mayer, K.U., Wu, Y., and Banwart, S.A. 2003. Modeling kinetic processes controlling hydrogen and acetate concentrations in an aquifer-derived microcosm. Environ. Sci. Technol. 37: 3910-3919.

Westrick, J., J., Mello, J., W., and Thomas, R., F. 1984. The groundwater supply survey. J. Am. Water Works Assoc. 76: 52-59.

Wielinga, B., Mizuba, M.M., Hansel, C.M., and Fendorf, S. 2001. Iron promoted reduction of chromate by dissimilatory iron-reducing bacteria. Environ. Sci. Technol. 35: 522-527.

Winfrey, M.R., and Zeikus, J.G. 1977. Effect of sulfate on carbon and electron flow during microbial methanogenesis in freshwater sediments. Appl. Environ. Microbiol. 33: 275-281.

Chapter 13

Prokaryotic Arsenate and Selenate Respiration

Joanne M. Santini* and John F. Stolz

Abstract

The oxyanions of arsenic and selenium can be used by prokaryotes as terminal electron acceptors in anaerobic respiration. Prokaryotes that respire these semi-metals are phylogenetically diverse and have been isolated from both pristine and contaminated environments. A total of 33 different species of prokaryotes have been isolated representing eight different prokaryotic phyla. These organisms are not obligate arsenate or selenate respirers as they can utilise a number of different terminal electron acceptors and also use a variety of different electron donors. The mechanisms by which these organisms respire with arsenate and selenate are slowly becoming uncovered. The enzymes responsible for the reduction are either periplasmic or membrane-bound terminal reductases linked to electron transport chains involved in energy generation. To date, all the respiratory arsenate and selenate reductases characterised are members of the DMSO reductase family of molybdenum-containing enzymes.

1. Introduction

Arsenic and selenium are both semi-metals that lie adjacent to one another in the Periodic table in groups 15 (pnictogen) and 16 (chalcogen), respectively. Both elements are naturally occurring in aquatic and terrestrial environments and the crustal abundance for both is about 0.0001% (Stolz and Oremland, 1999). Arsenic is commonly found in weathered volcanic and sedimentary rocks, fossil fuels and a number of minerals (e.g., arsenopyrite, realgar, loellingite, orpiment) whereas selenium is found in fossil fuels, shales, alkaline soils and minerals (e.g., ferroselite, schmeiderite, challomenite) (Stolz and Oremland, 1999). The most common chemical oxidation states for arsenic and selenium differ. The primary oxidation states for arsenic are arsenate (V), arsenite (III), elemental (0) and arsine (-III), while those for selenium are selenate (VI), selenite (IV), elemental (0) and selenide (-II) (Stolz *et al.*, 2002). Arsenate, a phosphate analogue, can enter cells via the phosphate transport system and is toxic because it can interfere with normal phosphorylation processes by replacing phosphate. Arsenite has recently been demonstrated to enter cells at neutral pH by aqua-glyceroporin (glycerol transport proteins) in bacteria, yeast and mammals (Mukhopadhyay *et al.*, 2002; Wysocki *et al.*, 2001) and its toxicity lies in its ability to bind sulfhydryl groups of cysteine residues in proteins thereby inactivating them. Selenium is an analogue of sulfur and substitutes for sulfur in thiols. The oxyanions of arsenic and selenium can be used by prokaryotes for metabolism and their toxicity has prompted much interest in these oxidation/reduction mechanisms.

This chapter will concentrate on oxyanions of arsenic and selenium that can be used as terminal electron acceptors for anaerobic respiration. Prior to discussing the mechanisms of arsenate and selenate respiration it is worth noting that arsenate can also be reduced to arsenite by a resistance mechanism that requires energy in the form of ATP (Mukhopadhyay *et al.*, 2002). At least three different but structurally related arsenate reductases have evolved convergently in prokaryotes and yeast (Rosen, 2002). The genes that encode the proteins involved in the resistance are either plasmid- or chromosomally-borne. The most well studied system is that of the *Escherichia coli* plasmid R773 that comprises five genes, *arsRDABC* organised in an operon (Chen *et al.*, 1986). The *arsC* gene encodes a small (16 kDa) soluble arsenate reductase, ArsC, which mediates the reduction of arsenate to arsenite in the cytoplasm using reducing equivalents from reduced glutaredoxin. The products of the *arsA* and *arsB* genes, act as an arsenite efflux pump and the *arsR* and *arsD* gene products are involved in regulation of the *ars* operon. A second system has been described for *Staphylococcus aureus*. The *ars* operon in plasmid pI258 of *S. aureus* contains only *arsB*, *arsC*, and *arsD* (Rosen, 2002). In this case, reduced thioredoxin provides the electrons to reduce As(V), and As(III) is expelled from the cell via an ATP-independent ArsB. In all cases of ArsC-based arsenate reduction, energy is expended.

A substantial amount of free energy is available from the reduction of arsenate to arsenite and selenate to

*For correspondence email: j.santini@latrobe.edu.au

Table 1. Arsenate and selenate-respiring prokaryotes

Prokaryote	Isolated from	Phylogeny	Respiration with		Reference
			As(V)	Se(VI)	
Bacillus sp. str. JMM-4	gold-mine	low G+C	+	-	Santini *et al.* (2002)
Bacillus selenitireducens	hypersaline lake	low G+C	+	-[a]	Switzer Blum *et al.* (1998)
Bacillus arsenicoselenatis	hypersaline lake	low G+C	+	+	Switzer Blum *et al.* (1998)
Desulfitobacterium sp. str. GBFH	As-contaminated sediments, Lake	low G+C	+	+	Niggemyer *et al.* (2001)
Desulfosporosinus auripigmenti	lake surface sediments	low G+C	+	-	Newman *et al.* (1997); Stackebrandt *et al.* (2003)
Bacillus sp. str. HT-1	hamster stool	low G+C	+	NK	Herbel *et al.* (2002)
Clostridium sp. str. OhilAs	river sediment	low G+C	+	-	Stolz *et al.* (2002)
SF-1	Se-polluted sediment	NK[b]	-	+	Fujita *et al.* (1997)
Salana multivorans	river sediment	high G+C	-	+	von Wintzingerode *et al.* (2001a)
Chrysiogenes arsenatis	gold-mine	Chrysiogenetes	+	-	Macy *et al.*(1996); Macy *et al.* (2001)
HGM-K1	mine	Aquificales	+	+	Takai *et al.* (2002)
Deferribacter desulfuricans	hydrothermal vent	Deferribacteres	+	-	Takai *et al.* (2003)
Selenihalanaerobacter shriftii	Dead Sea	Halanaerobacter	-	+	Switzer Blum *et al.* (2001)
Citrobacter sp. str. TSA-1	termite hindgut	γ-Proteobacteria	+	+	Herbel *et al.* (2001)
Shewanella sp. str. ANA-3	As-treated wood	γ-Proteobacteria	+	-	Saltikov *et al.* (2003)
AK4OH1	Se-contaminated water	γ-Proteobacteria	-	+	Knight *et al.* (2002)
Ke4OH1	Se-contaminated water	γ-Proteobacteria	-	+	Knight *et al.* (2002)
Aeromonas hydrophila	Great Bay Estuary	γ-Proteobacteria	NK	+	Knight and Blackmore (1998)
Thauera selenatis	Se-contaminated water	β-Proteobacteria	-	+	Macy *et al.* (1993)
Bordetella petrii	river sediment	β-Proteobacteria	NK	+	von Wintzingerode *et al.* (2001b)
Wolinella sp. str. BRA-1	bovine rumen fluid	ε-Proteobacteria	+	NK	Herbel *et al.* (2001)
Sulfurospirillum barnesii	Se-contaminated water	ε-Proteobacteria		+ +	Oremland *et al.* 1994; Stolz *et al.* (1999)
Sulfurospirillum arsenophilum	As-contaminated sediments	ε-Proteobacteria		+ -	Ahmann *et al.* 1994; Stolz *et al.* (1999)
Desulfomicrobium sp. str. Ben-RB	gold mine	δ-Proteobacteria		+ -	Macy *et al.* (2000)
Pyrobaculum aerophilum	hot spring	Crenarchaeota		+ +	Völkl *et al.* (1993); Huber *et al.* (2000)
Pyrobaculum arsenaticum	hot spring	Crenarchaeota		+ +	Huber *et al.* (2000)

[a] respires selenite

[b] Has been classified as a *Bacillus* sp. based on morphological and physiological properties.

NK - Not Known

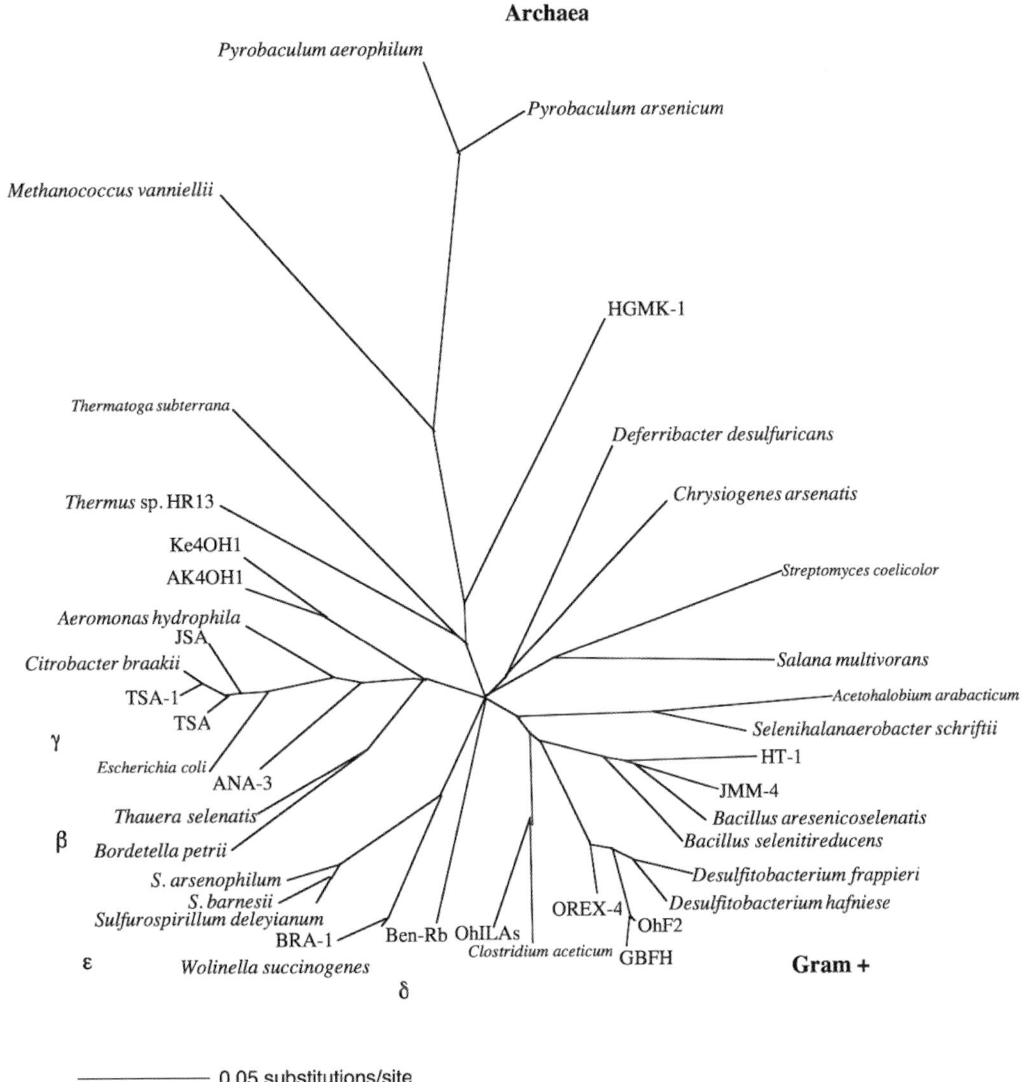

Figure 1. Phylogeny of prokaryotes that respire oxyanions of arsenic and selenium. The 16S rRNA gene sequences were aligned using Clustal X (Jeannmougin *et al.*, 1998) and the unrooted neighbor joining tree was constructed using PAUP (Swafford, 1998). See Table 1 for the description of the species. (N.B.: The species in smaller font have been included to help define the branching and do not respire oxyanions of arsenic or selenium).

selenite, which is coupled to the oxidation of a variety of electron donors, and many prokaryotes that respire these compounds have been isolated (Table 1). In this chapter we will outline the diversity of arsenate and selenate respiring prokaryotes. Focus, however, will be on the mechanisms of arsenate and selenate respiration.

2. Diversity of Arsenate and Selenate Respirers

The prokaryotes isolated to date that can respire oxyanions of arsenic and selenium are physiologically and phylogenetically distinct (Table 1) (Figure 1). They are not confined to any particular prokaryotic domain (with representatives from both the Bacteria and Archaea) or phylum, and many are classified as members of established genera (i.e., *Pyrobaculum*, *Bacillus*, *Sulfurospirillum*, *Desulfosporosinus*) (Figure 1). The majority are environmental strains, having been isolated from freshwater sediments, estuaries, soda lakes, hot springs and gold mines (Table 1). Recently, however, arsenate and selenate respiring bacteria have been isolated from the gastrointestinal tract of animals (Herbel *et al.*, 2002). These organisms range from mesophiles to extremophiles, and can be adapted to extremes of temperature, pH and salinity. At the writing of this chapter, a total of 18 species respire arsenate, 15 are known to respire selenate, and at least two species respire selenite. Among these organisms, seven are capable of using arsenate and selenate, and one (*Bacillus selenitireducens*) can use both arsenate and selenite as terminal electron acceptors (Switzer Blum *et al.*, 1998). While the different species may share the ability to use these oxyanions for respiration they may need different concentrations for growth. For

Table 2. Comparison of the respiratory arsenate reductase from *Chyrsiogenes arsenatis* and *Bacillus selenitireducens*

	Chrysiogenes arsenatis	*Bacillus selenitireducens*
Cellular location	Periplasm	Membranes
pH optimum	6.5	9.5
Subunit composition & size	ArrA, 87 kDa; ArrB, 29 kDa	ArrA, 110 kDa; ArrB, 34 kDa
Native molecular weight (kDa)	123	150
Cofactors	Mo, Fe, S	Mo, Fe
V_{max} (U/mg)	7013	2.5
K_m (μM)	300	34

example, concentrations of arsenate above 5 mM inhibit the growth of *Sulfurospirillum* species (Stolz *et al.*, 1999), whereas the *Clostridium* species strain OhILAs can grow in medium with excess of 40 mM arsenate. In addition, these organisms may have even greater metabolic versatility, being able to utilise other electron acceptors. For example, *Sulfurospirillum barnesii* can respire nitrate, nitrite, fumarate, thiosulfate, Fe(III), dimethylsulfoxide, and trimethylamine oxide in addition to selenate and arsenate (Oremland *et al.*, 1994; Stolz *et al.*, 1999). These organisms may also use a variety of different electron donors such as acetate, hydrogen, glucose, glycerol and even aromatic hydrocarbons (Oremland *et al.*, 1994; Switzer Blum *et al.*, 1998; Niggemyer *et al.*, 2001; Knight *et al.*, 2002; Santini *et al.*, 2002). The combination of phylogenetic conformity (i.e., the existence of closely related species that don't respire arsenate or selenate) and the aforementioned metabolic versatility compromises the utility of 16S rRNA based molecular probes for identifying these organisms in the environment. Thus, it is important that a multifaceted approach, employing enrichment cultures utilising different electron donors and varying concentrations of electron acceptor in consort with molecular approaches (i.e., functional probes) targeting expressed proteins, be taken when attempting to identify these organisms in the environment.

3. Mechanisms of Arsenate and Selenate Respiration

Dissimilatory arsenate and selenate reduction are associated with terminal reductases that are either periplasmic or membrane-bound, and linked to electron transport chains involved in energy generation. To date, two arsenate and one selenate reductase have been purified and characterised (Schröder *et al.*, 1997; Krafft and Macy, 1998; Afkar *et al.*, 2003). Interestingly, they all appear to be members of the DMSO reductase family of molybdoenzymes (Hille *et al.*, 1999; Kisker *et al.*, 1999).

3.1. Dissimilatory Arsenate Reduction

Three arsenate reductases (Arr) from phylogenetically distant bacteria have been studied, two of these, from *Chrysinogenes arsenatis* (Krafft and Macy, 1998) and *B. selenitireducens* (Afkar *et al.*, 2003), have been purified and characterised whereas the third from *Shewanella* sp. str. ANA-3 has only been studied through the cloning and sequencing of the genes encoding the enzyme (Saltikov and Newman, 2003). The enzyme purified from *C. arsenatis* was found to be periplasmic (Krafft and Macy, 1998), whereas the enzyme purified from *B. selenitireducens* was membrane-bound (Afkar *et al.*, 2003). Table 2 shows the comparisons between the two purified Arr's. The optimum pH for enzyme activity varies and in the case of the *B. selenitireducens* Arr the optimum pH is similar to that for growth of the organism (pH 9.8). This enzyme also requires a NaCl concentration of 150 g l^{-1} for optimal activity. Both Arr's are heterodimers consisting of two heterologous subunits, ArrA and ArrB (Table 2). The greatest variation between the enzymes appears to be their affinity for arsenate. The K_m for arsenate of the *C. arsenatis* and *B. selenitireducens* Arr's are 300 μM and 34 μM, respectively. The V_{max} for the enzymes are 7013 and 2.5 μmol arsenate reduced min^{-1} mg $protein^{-1}$, respectively. This suggests that although the *B. selenitireducens* has a higher affinity for arsenate the *C. arsenatis* Arr has a higher enzyme turnover. The enzymes also vary in their ability to use different electron acceptors. When reduced benzyl viologen is used as the artificial electron donor the *C. arsenatis* Arr appears to be specific for arsenate, as nitrate, sulfate, selenate, fumarate and arsenite do not serve as alternative electron acceptors. The *B. selenitireducens* Arr however can also use arsenite (24% of specific activity for arsenate), selenate (48%) and selenite (29%) as electron acceptors when reduced methyl viologen is provided as the artificial electron donor. Nitrate, nitrite, fumarate, thiosulfate or phosphate, however, cannot be used as alternative electron acceptors. The reason for use of these alternative electron acceptors is not clear although the use of arsenite and selenate does not appear to be physiologically relevant, as the organism does not respire with these electron acceptors (Switzer Blum *et al.*, 1998). *B. selenitireducens* does respire with selenite but whether the Arr is responsible for selenite respiration is not presently known. Mutations in the *arr* genes may prove useful in determining whether the Arr performs a dual function as an arsenate and selenite reductase.

Both Arr's contain molybdenum and iron as cofactors and the Arr from *C. arsenatis* has also been demonstrated to contain acid-labile sulfur. The presence of molybdenum suggests that they are both members of the family of mononuclear molybdenum enzymes (Hille *et al.*, 1999; Kisker *et al.*, 1999). The presence of iron and sulfur suggests that these enzymes also contain several iron-sulfur clusters as prosthetic groups. Mononuclear molybdoenzymes have been divided into three families based on the structure of their molybdenum centers (Hille, 1999). The xanthine oxidase family is the largest and most diverse family (the molybdenum hydroxylases) and includes enzymes that catalyse the hydroxylation of a broad range of aldehydes and aromatic heterocyclic compounds. The sulfite oxidase family (the eukaryotic oxo transferases) also includes the eukaryotic assimilatory nitrate reductases, which catalyse oxygen atom transfer to or from a substrate. The molybdenum in both the xanthine oxidase and sulfite oxidase families is bound by a single pyranopterin cofactor that contains a phosphate group at position 2 (Stolz and Basu, 2002). The DMSO reductase family, which is comprised of molybdoenzymes exclusively found in prokaryotes, catalyse either oxygen atom transfer or other oxidation-reduction reactions. It is this latter family that includes a wide range of enzymes including the three types of prokaryotic nitrate reductases Nar, Nas, and Nap (Stolz and Basu, 2002). In this case the phosphate is replaced by guanine diphosphate and the molybdenum is bound by two pyranopterin cofactors (MGD).

Although the Arr of ANA-3 has not been purified and characterised, it was shown to function in arsenate respiration through gene knock-out experiments (Saltikov and Newman, 2003). While ANA-3 possesses both a gene for ArsC-resistance and a respiratory arsenate reductase gene only a few kb apart, their products apparently function independently (Saltikov *et al.*, 2003, Saltikov and Newman, 2003). The *arr* operon in ANA-3 lies immediately downstream of the *ars* operon and contains only two genes *arrA* and *arrB*. They encode proteins of predicted molecular masses of 95.2 kDa (ArrA) and 25.7 kDa (ArrB), respectively, which are similar to those of the purified Arr's (Table 2). The N-terminal region of ArrA displays characteristics associated with the Tat (twin-arginine translocation) signal sequences (Berks, 1996; Berks *et al.*, 2000; Sargent *et al.*, 2002). The Tat pathway functions to transport folded proteins, predominantly those containing redox cofactors involved in respiratory and photosynthetic electron transport chains across the cytoplasmic membrane. The conserved motif S/T-R-R-X-F-L-K is found at the N-terminus of these signal sequences as is also found in ArrA. A predicted cleavage site (Alanine 42) also resides in this sequence. No signal sequence was found in ArrB. Although the cellular location of the ANA-3 Arr has not been determined, prediction from the sequences suggests that the enzyme is either periplasmic or membrane-bound and that the ArrA signal sequence is used to export ArrA/ArrB as a fully folded complex across the cytoplasmic membrane.

A BLAST search of the current genome databases using the inferred protein sequences from ANA-3 and *Bacillus selenitireducens* resulted in the discovery of the putative *arr* operon in *Desulfitobacterium hafniense* (Afkar *et al.*, 2003). This freshwater low G+C gram-positive bacterium had been previously shown to respire arsenate, however, no biochemical analysis has been done (Niggemyer *et al.*, 2001). A DNA sequence contig from *D. hafniense* (http://www.jgi.doe.gov) has at least seven genes including genes encoding a two component regulatory system (sensor kinase and response regulator) and an ORF of unknown function, immediately upstream of the putative *arrA*. Immediately downstream of *arrA* are two genes, *arrB* and *torD*, that encode an iron sulfur protein (ArrB) and a chaperone protein homolog, respectively. A putative Tat signal sequence was found in ArrA, similar to that found in *Shewanella* strain ANA-3, indicating that the enzyme is exported across the cytoplasmic membrane.

The finding that the purified Arr's of *C. arsenatis* and *B. selenitireducens* are in fact molybdoenzymes was further confirmed by determining the N-terminal sequences of each subunit. The N-terminal sequences of the ArrA and ArrB subunits are aligned with the molybdenum-containing subunits of other molybdoenzymes (Figures 2 and 3). The N-terminal regions of the putative ArrA and ArrB subunits of the ANA-3 Arr are also shown. The N-terminal sequences of the ArrA subunits contain a conserved Cys-X_2-Cys-X_3-Cys motif present in the molybdenum-containing subunits of molybdoenzymes. This motif may bind a [3Fe-4S] or [4Fe-4S] cluster. All three ArrA sequences were found to be most similar to each other but also shared similarities to the catalytic subunit of the polysulfide reductase (PsrA) of *Wolinella succinogenes* and *Shewanella oneidensis*, and the catalytic subunit of thiosulfate reductase (PhsA) of *Salmonella typhimurium* (Figure 2, see also Table 3). Interestingly, the only residues conserved in the N-terminal sequence of arsenite oxidase from "*Alcaligenes faecalis*" were the cysteine residues involved in binding of the iron-sulfur cluster. Each of these is the catalytic subunit of the respective enzymes and all contain molybdopterin guanosine dinucleotide as the organic component of their molybdenum cofactor. In fact, based on the corresponding amino acid sequences of the ArrA subunits of ANA-3 and *B. selenitireducens,* the conserved cysteine motif is Cys-X_2-Cys-X_3-Cys-X_{27}-

```
B.s. ArrA    1   ------SQENKEQG--EWIASVCQGCTAWCAVQ
C.a. ArrA    1   QTGTGASAMGEAEG--KWIPSTCQGCTTWCP–
D.h. ArrA   29   ALAQGENTAPSAEG--KWMSTLCQGCTTWCAIQ
ANA-3ArrA   40   AAELPAPLRRTGVG--EWLATTCQGCTSWCAKQ
W.s. PsrA   28   PGTLGALEKQEIKGSAKFVPSICEMCTSSCTIE
Sa.t.TsrA   25   PGALARNPIAGINGKTTLTPSLCEMCSFRCPIQ
S.o. PsrA   29   SLAALESKQLKGMG--KEIASICEMCSTRCPIS
A.f. AoxA    1   GCPNDRITLPPANA--QRTNMTCHFCIVGCGYH
E.c. FdhH    1   ---------------MKKVVTVCPYCASGCKIN
T.s. SerA   54   EDLYRKEWTWDSTG---FIT-HSNGCVAGCAWR
```

Figure 2. Sequence alignment of the N termini of ArrA and related molybdenum-containing proteins. The sequences are the arsenate reductase from *B. selenitireducens* (*B.s.* ArrA), *C. arsenatis* (*C.a.* ArrA), *D. hafniense* (*D.h.* ArrA), and *Shewanella* sp. strain ANA-3 (ANA-3 ArrA),; the polysulfide reductase from *Wolinella succinogenes* (*W.s.* PsrA), and *S. oneidensis* (*S.o.* PsrA), the thiosulfate reductase from *Salmonella typhimurium* (*Sa.t.* TsrA); the arsenite oxidase from "*Alcaligenes faecalis*" (*A.f.* AoxA); the formate dehydrogenase from *E. coli* (*E.c.* FdhH), and the selenate reductase from *Thauera selenatis* (*T.s.* SerA). The numbers prior to the sequence indicate the number of additional amino acids (including signal sequence) at the N terminus. Conserved residues are boxed in light grey, the cysteine residues of the [Fe-S] cluster-binding site are boxed in dark grey.

Cys which defines a [4Fe-4S] cluster. The N-terminal sequences of the ArrB subunits share a conserved Cys-X_2-Cys-X_2-Cys-X_3-Cys motif that functions in binding an iron-sulfur cluster (Figure 3). The ArrB sequences are more similar to each other than to other iron-sulfur proteins of molybdenum-containing enzymes (Figure 3). These latter proteins include the PsrB of the polysulfide reductase of *W. succinogenes* and TtrB of the *S. typhimurium* tetrathionate reductase (Afkar *et al.*, 2003). These proteins all function in electron transfer and are are components of different electron transport chains. Sequence analysis of the ANA-3 putative ArrB indicates that it may contain four iron-sulfur clusters (Saltikov and Newman, 2003). The arrangement of these putative clusters is similar to that of the DMSO reductase DmsB subunit of *E. coli*, which ligates four [4Fe-4S] clusters (Trieber *et al.*, 1996).

Phylogenetic analysis of the molybdenum-containing catalytic subunit of the DMSO reductase family indicates that this family can be further subdivided into four distinct clades based on sequence homology (Figure 4). The Nar clade is comprised of the membrane-bound respiratory nitrate reductase (NarG), but also includes chlorate and perchlorate reductase, dimethylsulfide dehydrogenase, and selenate reductase (see below). The Nas/Nap clade contains assimilatory and periplasmic nitrate reductases, as well as formate dehydrogenases. A small, but nevertheless distinctive clade is comprised solely of arsenite oxidase. The last clade includes the family's namesake, DMSO reductase, as well as TMAO reductase and biotin sulfoxide reductase. ArrA is a sub-branch of this group, clustering with polysulfide reductase and thiosulfate reductase (Figure 4). Whether this group further resolves into its own subfamily with the addition of more sequences remains to be seen. Nevertheless, the variations in function of the enzymes in the DMSO reductase family are truly remarkable.

```
B.s. ArrB    --AKKNYAMTIDLQACIGCAGCAVTCKN-ENSTS-
C.s. ArrB    ----AKYGMAIDLHKCAGCGACGLACKT-QNNTDD
D.h. ArrB    --------MVINLQKCVGCGACGIACKN-ENNVDQ
ANA-3ArrB    ----MRLGMVIDLQKCVGCGGCSLACKT-ENNTND
W.s. PsrB    --MAKKYGMIHDENLCIGCQACNIACRS-ENKIPD
Sa.t.TtrB    GSPRHRYAMLIDLRRCIGCQSCTVSCTIENQTPQG
St.a.NarH    MKIKAQVAMVLNLDKCIGCHTCSVTCKNTWTNRPG
B.su.NarH    MKIKAQIGMVMNLDKCIGCHACSVTCKNTWTNRSG
T.s. SerB    M-SQRQLAYVFDLNKCIGCHTCTMACKQLWTNRDG
```

Figure 3. Sequence alignment of the N termini of ArrB and related iron-sulfur proteins. The sequences are for the iron sulfur containing subunits of the arsenate reductase from *B. selenitireducens* (*B.s.* ArrB), *C. arsenatis* (*C.a.* ArrB), *D. hafniense* (*D.h.* ArrB), and *Shewanella* sp. strain ANA-3 (ANA-3 ArrB); the polysulfide reductase from *Wolinella succinogenes* (*W.s.* PsrB), the tetrathionate reductase from *Salmonella typhimurium* (*Sa.t.* TsrB; this sequence starts at the 45th amino acid residue); the nitrate reductase from *Staphylococcus aureus* (*St.a.* NarH), *B. subtilis* (*B.su.* NarH), and the selenate reductase from *Thauera selenatis* (*T.s.* SerB). Conserved residues are boxed in light grey, the cysteine residues of the [Fe-S] cluster-binding site are boxed in dark grey.

Table 3. Name, abbreviations, prokaryotic species, and accession numbers of the molybdenum-containing enzymes of the DMSO reductase family used in Figure 2

Protein	Abbr.	Species	Accession
Arsenate reductase	ArrA	*Desulfitobacterium hafniense*	NZAAAW02000205
Arsenate reductase	ArrA	*Bacillus selenitireducens*	AY283639
Arsenate reductase	ArrA	*Shewanella trabarsenatis*	AAQ01672
Polysulfide reductase	PsrA	*Wolinella succinogenes*	P31075
Thiosulfate reductase	PhsA	*Salmonella typhimurium*	AAC36934
DMSO reductase	DorA	*Escherichia coli*	P18775
DMSO reductase	DorA	*Rhodobacter sphaeroides*	AAC13660
Biotin sulfoxide reductase	BisC	*Escherichia coli*	AAB18528
TMAO reductase	TorA	*Escherichia coli*	P58360
Arsenite oxidase	AoxA	*Alicaligenes faecalis*	1G8KA
Arsenite oxidase	AoxB	*Cenibacterium arsenoxidans*	AAN05581
Assimilatory nitrate reductase	NasA	*Pseudomonas putida*	AF203789
Assimilatory nitrate reductase	NasA	*Oscillatoria chalybea*	X89445
Assimilatory nitrate reductase	NasA	*Klebsiella pneumoniae*	Q06457
Assimilatory nitrate reductase	NasA	*Bacillus subtilis*	D30689
Periplasmic nitrate reductase	NapA	*Campylobacter jejuni*	Cj0780
Periplasmic nitrate reductase	NapA	*Rhodobacter sphaeroides*	AF06954
Periplasmic nitrate reductase	NapA	*Paracoccus pantatrophus*	Z36773
Periplasmic nitrate reductase	NapA	*Escherichia coli*	P33937
Periplasmic nitrate reductase	NapA	*Desulfovibrio desulfuricans*	Y18045
Respiratory nitrate reductase	NarG	*Aeropyrum pernyx*	AP000061
Respiratory nitrate reductase	NarG	*Thermus thermophilus*	Y10124
Respiratory nitrate reductase	NarG	*Haloarculum marimortuii*	AJ277440
Respiratory nitrate reductase	NarG	*Escherichia coli*	X16181
Respiratory nitrate reductase	NarG	*Pseudomonas fluorescences*	U71398
Respiratory nitrate reductase	NarG	*Bacillus subtilis*	Z49884

3.2. Dissimilatory Selenate Reduction

To date, *Thauera selenatis* is the only organism where selenate reduction has been studied in detail. The enzyme, selenate reductase, has been purified and characterised (Schröder *et al.*, 1997) and the genes cloned and sequenced (Krafft *et al.*, 2000). The selenate reductase (Ser), like the Arr of *C. arsenatis*, is periplasmic. Ser is a heterotrimer consisting of three heterologous subunits, SerA (96 kDa), SerB (40 kDa) and SerC (23 kDa) with a native molecular weight of 180 kDa. The K_m for selenate is 16 µM and V_{max} 40 µmol selenate reduced min^{-1} mg $protein^{-1}$. Cofactor constituents include molybdenum, iron, acid-labile sulfur and heme *b*. The heme *b* of Ser displays a reduced spectrum of 425 (soret), 528 (β) and 558 (α) nm. The oxidised spectrum results in a shift of the soret peak to 415 nm indicative of a cytochrome *b* spectrum. Ser, like Arr, also appears to be a molybdenum-containing enzyme. The only other molybdoenzymes known to contain a periplasmic cytochrome *b* are the dimethylsulfide dehydrogenase (Ddh) of *Rhodovulum sulfidophilum* (Hanlon *et al.*, 1996) and the ethylbenzene dehydrogenase (Ebd) of *Azoarcus* strain EbN1 (Kniemeyer and Heider, 2001); similar UV/Visible spectra were observed with the cytochrome *b*'s of these enzymes. Both of these latter enzymes, like Ser, are heterotrimers consisting of three subunits of 94, 38 and

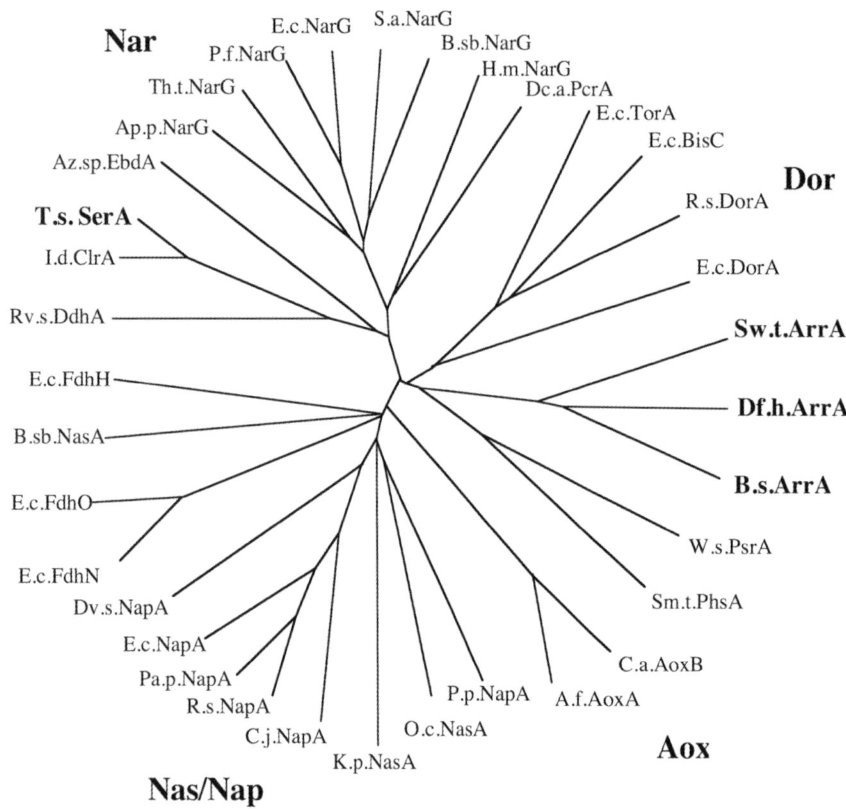

Figure 4. Phylogeny of ArrA and SerA as related to other members of the DMSO reductase family of mononuclear molybdenum-containing enzymes. The sequences were aligned using Clustal X (Jeannmougin *et al.*, 1998) and the unrooted neighbor joining tree was constructed using PAUP (Swafford, 1998). See Table 3 for list of enzymes, prokaryotic species, and abbreviations.

32 kDa for Ddh and 96, 43 and 32 kDa for Ebd. Recently, another enzyme in this class, chlorate reductase (Clr) was isolated from *Ideonella dechloratans* (Thorell *et al.*, 2003). The cellular location has not been determined but it also consists of three heterologous subunits of 94, 36 and 27 kDa and, like the others, contains heme-*b* (Thorell *et al.*, 2003). All four enzymes carry out quite different reactions, Ser and Clr are terminal reductases reducing selenate and chlorate, respectively, and both Ddh and Ebd oxidise DMS and ethylbenzene, respectively. The latter two enzymes do not reduce selenate. Clr however has been demonstrated to reduce Se(VI) but the rate was only 7% of that observed with chlorate (Thorell *et al.*, 2003) and Ser has been observed to reduce chlorate (S. Bydder and J.M. Santini, unpublished results).

The *T. selenatis* selenate reductase (*ser*) genes appear to reside in the same transcriptional unit as *serABDC*. The *serA*, *serB* and *serC* genes encode the Ser subunits and the fourth open reading frame between *serB* and *serC* encodes a putative protein of unknown function (see below). The arrangement of these genes is similar to the genes of the Ddh (*ddhABDC*) (McDevitt *et al.*, 2002b), Ebd (*ebdABDC*) (Rabus *et al.*, 2002), *E. coli* respiratory nitrate reductase (*narGHJI*) (Blasco *et al.*, 1989) and Clr (*clrABDC*) (Thorell *et al.*, 2003).

The first gene in the putative *serABDC* operon encodes SerA (882 aa), which contains at its N-terminus a conserved twin-arginine motif that directs the transport of this protein to the periplasm using the Tat pathway (Berks 1996; Berks *et al.*, 2000). SerA is similar to the molybdenum-containing subunits of members of the DMSO reductase family of molybdoenzymes. In fact, SerA is most similar to the molybdenum-containing catalytic subunits of Clr (ClrA), Ddh (DdhA), Ebd (EbdA) and Nar (NarG) (Figure 4). As stated above, these enzymes form a distinct clade in the DMSO reductase family (Figure 4) (McDevitt *et al.*, 2002b; McEwan *et al.*, 2002). The N-terminus of SerA contains a conserved motif composed of a histidine and three cysteine residues that may define a [4Fe-4S] cluster. This motif also occurs in ClrA, DdhA, EbdA and NarG. It has recently been shown from the structure of the *E. coli* nitrate reductase that this motif in NarG binds a [4Fe-4S] cluster (Bertero *et al.*, 2003). The coordination of this cluster by the residues outlined is relatively uncommon, with only two other examples, the [Ni-Fe] hydrogenase from *Desulfovibrio gigas* and the Fe hydrogenase from *Clostridium pasteurianum* (Bertero *et al.*, 2003). Based on the nitrate reductase crystal structure the molybdenum atom is coordinated by an aspartate residue (Asp222). This may be yet another distinguishing feature of the Nar clade. In the other four subgroups of this family the Mo atom is coordinated by a side chain of a serine, cysteine, selenocysteine, or alanine residue; in the latter subgroup which includes the arsenite oxidase of "*A. faecalis*" there is no covalent bond between the Mo atom and the protein (Ellis *et al.*, 2001). This Asp residue is conserved in SerA, ClrA, DdhA and EbdA at the corresponding position. Whether this residue coordinates the Mo atom in these proteins is at present unknown, but the close phylogenetic association of these proteins is compelling.

The *serB* gene encodes a protein (326 aa) that does not contain a leader sequence. This protein is probably exported to the periplasm using the leader sequence of SerA for export. The SerAB complex is folded in the cytoplasm where the molybdenum is inserted and exported together as a fully folded complex. SerB shares similarities to the Fe-S proteins of Clr, Ddh, Ebd and Nar. SerB contains four cysteine motifs that may function in binding 4[Fe-S] clusters. These motifs are also conserved in ClrB, DdhB, EbdB and NarH. The third motif (from the N-terminus) in SerB, ClrB, and DdhB, instead of containing four cysteines, contains three and a tyrosine in place of cysteine; in EbdB this cysteine is replaced by a histidine. NarH contains a tryptophan in place of the tyrosine and this cluster has been shown to coordinate a [3Fe-4S] cluster and the other three coordinate 3[4Fe-4S] clusters (Bertero *et al.*, 2003).

The *serC* gene encodes a protein (211 aa) that shares similarity with ClrC, DdhC and EbdC. The N-terminus of SerC contains a Sec-dependent leader sequence suggesting that SerC is exported to the periplasm in an unfolded state using the general secretory (Sec) apparatus (Pugsley, 1993; Rapoport *et al.*, 1996). SerC is probably the heme-*b* containing subunit as it, like ClrC, DdhC and EbdC, contains two conserved amino acids, histidine and methionine, that may act as ligands for the heme iron. In fact, the *ddhC* gene has been shown to encode the *b*-type cytochrome in Ddh (McDevitt *et al.*, 2002a). Once exported to the periplasm by the Sec pathway the heme *b* is inserted into the apoprotein, which results in correct folding of the protein.

A fourth open-reading frame resides between *serB* and *serC*, namely *serD*. This gene encodes a putative protein (193 aa) that is not part of the purified Ser. SerD shares sequence similarities with ClrD, DdhD and EbdD, which are also not components of the mature holoenzymes. The function of these proteins is not presently known. It is possible to gauge some information based on the position of *serD* in the putative *ser* operon, which is similar to the position of *narJ* in the *narGHJI* operon. The NarJ protein is also not part of the active Nar, but has been shown to act as a molecular chaperone during biogenesis of the molybdoenzyme and/or has a role in insertion of the molybdenum cofactor into the immature polypeptide (Blasco *et al.*, 1992; 1998).

An electron transport chain for the reduction of selenate to selenite can be proposed for *T. selenatis*. A soluble periplasmic cytochrome *c* has recently been isolated from *T. selenatis* that can act as the physiological electron donor to Ser (S. Bydder and J. M. Santini, unpublished results). Electrons may pass from this cytochrome to SerC (heme-*b*-containing subunit). The cytochrome *c* could accept electrons from either a membrane-bound quinol-oxidising cytochrome bc_1 complex or directly from mobile electron carriers in the respiratory chain such as quinol. The electrons are then shuttled from SerC to the Fe-S clusters of SerB and then onto the Fe-S cluster of SerA. From here the electrons are transferred to the catalytic site (Mo site) of SerA where selenate is reduced to selenite.

Synthesis of Ser is regulated. Maximum Ser activity (100%) was observed when *T. selenatis* was grown with selenate as the sole terminal electron acceptor. Ser activity decreased 2.4- and 5-fold when the organism was grown with selenate/nitrate (42%) and nitrate (20%) as terminal electron acceptors, respectively (Macy, 1994). No Ser activity was detected when the organism was grown under aerobic conditions suggesting that anaerobic conditions are essential for Ser expression. Although the Ser of *S. barnesii* has not been purified, preliminary experiments suggest that enzyme synthesis is also regulated (Stolz *et al.*, 1997). Maximum Ser activity was found when the organism was grown with selenate as sole terminal electron acceptor. Activity decreased by 2-, 3.1- and 6.3-fold when the organism was grown with nitrate (51%), thiosulfate (32%) and fumarate (16%) as terminal electron acceptors, respectively. Whether anaerobic conditions are also required for Ser expression remains to be determined. Interestingly, membrane fractions of *S. barnesii* grown with selenate as the sole terminal electron acceptor contained a reduced absorbance spectrum indicative of *b*-type cytochromes (Stolz *et al.*, 1997). It is not known whether this cytochrome is in fact part of Ser. It is noteworthy that this cytochrome was found in the membrane fractions and not in the periplasm suggesting that the Ser of *S. barnesii* is membrane-bound.

4. Conclusions

Dissimilatory arsenate and selenate reduction are two processes by which prokaryotes can obtain energy. Both elements are toxic at elevated concentrations yet these organisms have found ways that not only allow them to cope with the toxicity of these compounds but to respire with them. They are widespread in the environment and represent several different phylogenetic groups. They are physiologically unique using different electron donors for growth and can use other terminal electron acceptors such as oxygen, nitrate and sulfate. The mechanisms by which they reduce arsenate or selenate appear to be conserved in that they all involve molybdenum-containing enzymes that also contain other redox cofactors such as iron and sulfur. Further research on the mechanisms of arsenate and selenate respiration will allow for the evolution of these enzymes to be discerned. Whether these processes have come about by convergent or divergent evolution remains an intriguing and open question.

References

Afkar, E., Lisak, J., Saltikov, C., Basu, P., Oremland, R.S., and Stolz, J.F. 2003. The respiratory arsenate reductase from *Bacillus selenitireducens* strain MLS10. FEMS Microbiol. Lett. 206:107-112.

Ahmann, D., Roberts, A.L., Krumholz, L.R., and More, F.M.M. 1994. Microbe grows by reducing arsenic. Nature. 371: 750.

Berks, B.C. 1996. A common export pathway for proteins binding complex redox cofactors? Mol. Microbiol. 22: 393-404.

Berks, B.C., Sergent, F., and Palmer, T. 2000. The Tat export pathway. Mol. Microbiol. 35: 670-274.

Bertero, MG., Rothery, R.A., Palak, M., Hou, C., Lim, C., Lim, D., Blasco, F., Weiner, J.H., and Strynadka, N.C.J. 2003. Insights into the respiratory electron transfer pathway from the structure of nitrate reductase A. Nature Struct. Biol. 10: 681-687.

Blasco, F., Iobbi, C., Giordano, G., Chippaux, M., and Bonnefoy, V. 1989. Nitrate reductase of *Escherichia coli*: completion of the nucleotide sequence of the *nar* operon and reassessment of the role of the alpha and beta subunits in iron binding and electron transfer. Mol. Gen. Genet. 218: 249-256.

Blasco, F., Pommier, J., Augier, V., Chippaux, M., and Giordano, G. 1992. Involvement of the *narJ* or *narW* gene product in the formation of active nitrate reductase in *Escherichia coli*. Mol. Microbiol. 6: 221-230.

Blasco, F., Dos Santos, J.P., Magalon, A., Frixon, C., Guigliarelli, B., Santini, C.L., and Giordano, G. 1998. NarJ is a specific chaperone required for molybdenum cofactor assembly in nitrate reductase A of *Escherichia coli*. Mol. Microbiol. 28: 435-447.

Chen, C.M., Misra, T.K., Silver, S., and Rosen. B.P. 1986. Nucleotide sequence of the structural genes for an anion pump. The plasmid encoded arsenical resistance operon. J. Biol. Chem. 261: 15030-15038.

Ellis, P.J., Conrads, T., Hille, R., and Kuhn, P. 2001 Crystal structure of the 100 kDa arsenite oxidase from *Alcaligenes faecalis* in two crystal forms at 1.64 Å and 2.03 Å. Structure. 9: 125-132.

Fujita, M., Ike, M., Nishimoto, S., Takahashi, K., and Kashiwa, M. 1997. Isolation and characterisation of a new selenate-reducing bacterium, *Bacillus* sp. SF-1. J. Ferment. Bioeng. 83: 517-522.

Hanlon, S.P., Holt, R.A., Moore, G.R., and McEwan, A.G. 1996. Dimethylsulfide: acceptor oxidoreductase from *Rhodobacter sulfidophilus*. The purified enzyme contains *b*-type heme and a pterin molybdenum cofactor. Eur. J. Biochem. 239: 391-396.

Herbel, M.J., Switzer Blum, J., Hoeft, S.E., Cohen, S.M., Arnold, L.L., Lisak, J., Stolz, J.F., and Oremland, R.S. 2002. Dissimilatory arsenate reductase activity and arsenate-respiring bacteria in bovine rumen fluid, hamster faeces, and the termite hindgut. FEMS Microbiol. Ecol. 41: 59-67.

Hille, R., Rétey, J., Bartlewski-Hof, U., Reichenbecher, W., and Schink, B. 1999. Mechanistic aspects of molybdenum-containing enzymes. FEMS Microbiol. Rev. 22: 489-501.

Huber, R., Sacher, M., Vollmann, A., Huber, H., and Rose, D. 2000. Respiration of arsenate and selenate by hyperthermophilic Archaea. Syst. Appl. Microbiol. 23: 305-314.

Jeannmougin, F., Thompson, J.D., Gouy, M., Higgins,D.G., and Gibson, T.J. 1998. Multiple sequence alignment with Clustal X. Trends Biochem. Sci. 23: 403-405.

Kisker, C., Schindelin, H., Baas, D., Rétey, J., Mechenstock, R.U., and Kroneck, P.M.H. 1999. A structural comparison of molybdenum cofactor-containing enzymes. FEMS Microbiol. Rev. 22: 503-521.

Kniemeyer, O., and Heider, J. 2001. Ethylbenzene dehydrogenase, a novel hydrocarbon-oxidising molybdenum/iron-sulfur/heme enzyme. J. Biol. Chem. 276: 21381-21386.

Knight, V., and Blakemore, R. 1998. Reduction of diverse electron acceptors by *Aeromonas hydrophila*. Arch. Microbiol. 169: 239-248.

Knight, V.K., Nijenhuis, I., Kerkhof, L.J. Häggblom, M.M. 2002. Degradation of aromatic compounds coupled to selenate reduction. Geomicrobiol. J. 19: 77-86.

Krafft, T., and Macy, J.M. 1998. Purification and characterisation of the respiratory arsenate reductase of *Chrysiogenes arsenatis*. Eur. J. Biochem. 255: 647-653.

Krafft, T., Bowen, A., Theis, F., and Macy, J.M. 2000. Cloning and sequencing of the genes encoding the periplasmic-cytochrome *b*-containing selenate reductase of *Thauera selenatis*. DNA Sequence. 10: 365-377.

Macy, J.M., Rech, S., Dorsch, A.M., Stackebrandt, E., and Sly, L.I. 1993. *Thauera selenatis* gen. nov., sp. nov., a member of the beta subclass of Proteobacteria with a novel type of anaerobic respiration. Int. J. Syst. Bacteriol. 43: 135-142.

Macy, J.M. 1994. Biochemistry of selenium metabolism by *Thauera selenatis* gen. nov. sp. nov. and use of the organism for bioremediation of selenium oxyanions in San Joaquin valley drainage water. In: Selenium in the Environment. W.T. Frankenberger, Jr. and S. Benson, eds. Marcel Dekker, Inc., New York, New York. p. 421-444.

Macy, J.M., Nunan, K., Hagen, K.D., Dixon, D.R., Harbour, P.J., Cahill, M., and Sly, L.I. 1996. *Chrysiogenes arsenatis* gen. nov., sp., nov., a new arsenate-respiring bacterium isolated from gold mine wastewater. Int. J. Syst. Bacteriol. 46: 1153-1157.

Macy, J.M., Santini, J.M., Pauling, B.V., O'Neill, A.H., and Sly, L.I. 2000. Two new arsenate-sulfate-reducing bacteria: mechanisms of arsenate reduction. Arch. Microbiol. 173: 49-57.

Macy, J.M., Krafft, T., and Sly, L.I. 2001. Genus I. *Chrysiogenes* Macy, Nunan, Hagen, Dixon, Harbour, Cahill, and Sly 1996, 1156^{VP}. In: Bergey's Manual of Systematic Bacteriology, 2nd ed. Volume 1. The Archaea and the Deeply Branching and Phototrophic Bacteria. D.R. Boone, R.W. Castenholz, and G.M. Garrity, eds. Springer-Verlag, New York, New York. p. 412-415.

McDevitt, C.A., Hanson, G.R., Noble, C.J., Cheesman, M.R., and McEwan, A.G. 2002a. Characterization of the redox centers in dimethyl sulfide dehydrogenase from *Rhodovulum sulfidophilum*. Biochemistry. 41: 15234-15244.

McDevitt, C.A., Hugenholtz, P., Hanson, G.R., and McEwan, A.G. 2002b. Molecular analysis of dimethylsulfide dehydrogenase from *Rhodovulum sulfidophilum*: its place in the dimethyl sulfoxide reductase family of microbial molybdopterin-containing enzymes. Mol. Microbiol. 44: 1575-1587.

McEwan, A.G., Ridge, J.P., and McDevitt, C.A. 2002. The DMSO reductase family of microbial molybdenum enzymes; molecular properties and role in the dissimilatory reduction of toxic elements. Geomicrobiol. J. 9: 3-21.

Mukhopadhyay, R., Rosen, B.P., Phung, L.T., and Silver, S. 2002. Microbial arsenic: from geocycles to genes and enzymes. FEMS Microbiol. Rev. 26: 311-325.

Newman, D.K., Kennedy, E.K., Coates, J.D., Ahmann, D., Ellis, D.J., Lovley, D.R., and Morel, F.M.M. 1997. Dissimilatory arsenate and sulfate reduction in *Desulfotomaculum auripigmentum* sp. nov. Arch. Microbiol. 168: 380-388.

Niggemyer, A., Spring, S., Stackebrandt, E., and Rosenzweig, R.F. 2001. Isolation and characterization of a novel As(V)-reducing bacterium: implications for arsenic mobilisation and the genus *Desulfitobacterium*. Appl. Environ. Microbiol. 67: 5568-5580.

Oremland, R.S., Switzer Blum, J., Culbertson, C.W., Visscher, P.T., Miller, L.G., Dowdle, P., and Strohmaier, F.E. 1994. Isolation, growth, and metabolism of an obligately anaerobic selenate-reducing bacterium, strain SES-3. Appl. Environ. Microbiol. 60: 3011-3019.

Pugsley, A.P. 1993. The complete general secretory pathway in Gram-negative bacteria. Microbiol. Rev. 57: 50-108.

Rabus, R., Kube, M., Beck, A., Widdel, F., and Reinhardt, R. 2002. Genes involved in the anaerobic degradation of ethylbenzene in a dentifying bacterium, strain EbN1. Arch. Microbiol. 178: 506-516.

Rapoport, T.A., Jungnickel, B., and Kutay, U. 1996. Protein transport across the eukaryotic endoplasmic reticulum and bacterial inner membranes. Annu. Rev. Biochem. 65: 271-303.

Rosen, B.P. 2002. Biochemistry of arsenic detoxification. FEBS Lett. 529: 86-92.

Saltikov, C.W., Cifuentes, A., Venkateswaran, K., and Newman, D.K. 2003. The *ars* detoxification system is advantageous but not required for As(V) respiration by the genetically tractable *Shewanella* species strain ANA-3. Appl. Environ. Microbiol. 69: 2800-2809.

Saltikov, C.W., and Newman, D.K. 2003. Genetic identification of a respiratory arsenate reductase. Proc. Natl. Acad. Sci. USA. 100:10983-10988.

Sargent, F., Bercks, B.C., Palmer, T. 2002. Assembly of membrane-bound respiratory complexes by the Tat protein-transport system. Arch. Microbiol. 178: 77-84.

Santini, J.M., Stolz, J.F., and Macy, J.M. 2002. Isolation of a new arsenate-respiring bacterium – physiological and phylogenetic studies. Geomicrobiol. J. 19: 41-52.

Schröder, I., Rech, S., Krafft, T., and Macy, J.M. 1997. Purification and characterisation of the selenate reductase from *Thauera selenatis*. J. Biol. Chem. 272: 23765-23768.

Stackebrandt, E., Schumann, P., Schüler, E., and Hippe, H. 2003. Reclassification of *Desulfotomaculum auripigmentum* as *Desulfosporosinus auripigmenti* corrig., comb.nov. Int. J. Syst. Evol. Microbiol. 53: 1439-1443.

Stolz, J.F., Gugliuzza, T., Switzer Blum, J., Oremland, R., and Martinez Murillo, F. 1997. Differential cytochrome content and reductase activity in *Geospirillum barnesii* strain SeS3. Arch. Microbiol. 167: 1-5.

Stolz, J.F., Ellis, D.J., Switzer Blum, J., Ahmann, D., Lovley, D.R., and Oremland, R.S. 1999. *Sulfurospirillum barnesii* sp. nov. and *Sulfurospirillum arsenophilum* sp. nov., new members of the *Sulfurospirillum* clade of the ε *Proteobacteria*. Int. J. Syst. Bacteriol. 49: 1177-1180.

Stolz, J.F., and Oremland, R.S. 1999. Bacterial respiration of arsenic and selenium. FEMS Microbiol. Rev. 23: 615-627.

Stolz, J.F., and Basu, P. 2002. Evolution of nitrate reductase: molecular and structural variations on a common function. Chembiochem. 3: 198-206

Stolz, J.F., Basu, P., and Oremland, R.S. 2002. Microbial transformation of elements: the case of arsenic and selenium. Int. Microbiol. 5: 207-207.

Swafford, D.L. 1998. Phylogenetic analysis using Parsimony (*and other methods). Version 4. Sinauer Associates, Sunderland, Massachusetts.

Switzer Blum, J., Burns Bindi, A., Buzzelli, J., Stolz, J.F., and Oremland, R.S. 1998. *Bacillus arsenicoselenatis*, sp. nov., and *Bacillus selenitireducens*, sp. nov.: two haloalkaliphiles from Mono Lake, California that respire oxyanions of selenium and arsenic. Arch. Microbiol. 171: 19-30.

Switzer Blum, J., Stolz, J.F., Oren, A., and Oremland, R.S. 2001. *Selenihalanaerobacter shriftii* gen. nov., sp. nov., a halphilic anaerobe from Dead Sea sediments that respires selenate. Arch. Microbiol. 175: 208-219.

Takai, K., Hirayama, H., Sakihama, Y., Inagaki, F., Yamato, Y., and Horikoshi, K. 2002. Isolation and metabolic characteristics of previously uncultured members of the order *Aquificales* in a subsurface gold mine. Appl. Environ. Microbiol. 68: 3046-3054.

Takai, K., Kobayashi, H., Nealson, K.H., and Horikoshi, K. 2003. *Deferribacter desulfuricans* sp. nov., a novel sulfur-, nitrate- and arsenate-reducing thermophile isolated from a deep-sea hydrothermal vent. Int. J. Syst. Evol. Microbiol. 53: 839-846.

Thorell, H.D., Stenklo, K., Karlsson, J., and Nilsson, T. 2003. A gene cluster for chlorate metabolism in *Ideonella dechloratans*. Appl. Environ. Microbiol. 69: 5585-5592.

Trieber, C.A., Rothery, R.A., and Weiner, J.H. 1996. Engineering a novel iron-sulfur cluster into the catalytic subunit of *Escherichia coli* dimethyl-sulfoxide reductase. J. Biol. Chem. 271: 4620-4626.

Völkl, P., Huber, R., Drobner, E., Rachel, R., Burggraf, S., Trincone, A., and Stetter, K.O. 1993. *Pyrobaculum aerophilum* sp. nov., a novel nitrate-reducing hyperthermophilic Archaeum. Appl. Environ. Microbiol. 59: 2918-2926.

von Wintzingerode, F., Göbel, U.B, Siddiqui, R.A., Rösick, U., Schumann, P., Frühling, A., Rohde, M., Pukall, R., Stackebrandt, E. 2001a. *Salana multivorans* gen. nov., sp. nov., a novel actinobacterium isolated from an anaerobic bioreactor and capable of selenate reduction. Int. J. Syst. Evol. Microbiol. 51: 1653-1661.

von Wintzingerode, F., Schattke, A., Siddiqui, R.A., Rösick, U., Göbel, U.B., and Gross, R. 2001b. *Bordetella petrii* sp. nov., isolated from an anaerobic bioreactor, and emended description of the genus *Bordetella*. Int. J. Syst. Evol. Microbiol. 51: 1257-1265.

Wysocki, R., Chéry, C.C., Wawrzycka, D., Van Hulle, M., Cornelis, R., Thevelein, J.M., and Tamás, M.J. 2001. The glycerol channel Fps1p mediates the uptake of arsenite and antimonite in *Saccharomyces cerevisiae*. Mol. Microbiol. 40: 1391-1401.

Chapter 14

Molecular and Cellular Biology of Acetogenic Bacteria

Volker Müller*, Frank Imkamp, Andreas Rauwolf, Kirsten Küsel, and Harold L. Drake*

Abstract

Acetogenic bacteria are acetate-producing anaerobes that utilize CO_2 as a terminal electron acceptor. The reductive pathway by which acetogens reduce CO_2 is termed the acetyl-coenzyme A (acetyl-CoA) "Wood-Ljungdahl" pathway and yields acetate as a catabolic end product. In addition to being a terminal electron-accepting process, the acetyl-CoA pathway also provides the cell with a mechanism for the fixation of CO_2 under autotrophic conditions. Pathways that are biochemically very similar to the acetyl-CoA pathway are utilized by other prokaryotes for the autotrophic fixation of CO_2 and the oxidation of acetate. Thus, the acetyl-CoA pathway and processes that are biochemically very similar to it serve a variety of functions in nature. The main objectives of this chapter are to examine the (a) diverse metabolic features of acetogens that allow them to colonize diverse habitats, (b) regulatory and molecular aspects of specialized processes by which acetogens reduce CO_2, synthesize acetate, and conserve energy, and (c) *in situ* consequences of the physiological capabilities of acetogens.

1. Introduction to Acetogenic Bacteria and Acetogenesis

Acetogens are defined as anaerobic prokaryotes that utilize the acetyl-CoA pathway (see Section 2) for the reduction of CO_2 to acetate (Wood and Ljungdahl, 1991; Diekert, 1992; Drake, 1994; Drake *et al.*, 2004). (Note: No distinction between acetate and acetic acid is made in this article.) This pathway is a terminal electron-accepting process that conserves energy and provides a basis for the autotrophic fixation of CO_2. The ability of an organism to produce acetate at the expense of H_2-CO_2 is evidence that the organism is an acetogen; the ability of an organism to convert a hexose to three molecules of acetate is likewise evidence that the organism is an acetogen. As will be outlined in this chapter, acetogens are metabolically very versatile and, hence, colonize diverse habitats.

*For correspondence email: vmueller@em.uni-frankfurt.de or harold.drake@bitoek.uni-bayreuth.de

Acetogenesis is a term that denotes the ability of acetogens to produce acetate via the acetyl-coenzyme A (acetyl-CoA) pathway. Under certain conditions, acetate can be the sole product formed by acetogens, and the term homoacetogenesis has been used to denote this capability. Although the ability of acetogens to reduce CO_2 to acetate is their main hallmark, this capability is conditional, i.e., is dependent on the organism, growth conditions, and substrate utilized. Indeed, under certain conditions, acetate might not even be formed during the growth of an acetogen (Section 5).

It should be noted that the terms acetogen, acetogenesis, and homoacetogenesis have been used in association with organisms that form acetate by mechanisms that do not involve the acetyl-CoA pathway, i.e., cannot reduce CO_2 to acetate (e.g., see Brulla and Bryant, 1989; Stams and Dong, 1995; Tholen *et al.*, 1997; Causey *et al.*, 2003). However, this usage is less than ideal, as it does not differentiate between acetogens and acetate-forming non-acetogens, and creates confusion in the literature regarding the nature of acetogens.

1.1. Discovery of Acetogenesis and Acetogenic Bacteria

In 1932, Fischer and co-workers reported that microorganisms in wastewater catalyzed the H_2-dependent reduction of CO_2 to acetate (Fischer *et al.*, 1932). This report was the first time acetogenesis was mentioned in the literature. In 1936 and subsequent years, Wieringa reported on the properties of a spore-forming bacterium, *Clostridium aceticum*, that synthesized acetate from H_2 and CO_2 according to the following reaction (Wieringa, 1936; 1939-1940; 1941):

$$4\,H_2 + 2\,CO_2 \rightarrow CH_3COOH + 2\,H_2O \qquad (1)$$

The reaction catalyzed by *C. aceticum* was essentially identical to that observed by Fischer *et al.* (1932) with wastewater and constituted a new autotrophic process, i.e., a process by which organic carbon is produced via the reduction of CO_2.

In 1942, Fontaine and co-workers isolated a thermophilic, heterotrophic bacterium, *Clostridium thermoaceticum*, that converted glucose to acetate in a stoichiometry that approximated the following reaction (Fontaine *et al.*, 1942):

$$C_6H_{12}O_6 \rightarrow 3\ CH_3COOH \qquad (2)$$

As with Reaction 1, this process constituted a biological novelty in the early 1940's, and it was proposed that a one-carbon compound, i.e., CO_2, was reabsorbed and reduced to acetate during the oxidation of glucose (Fontaine *et al.*, 1942). Although there is no obvious overlap between Reactions 1 and 2, subsequent work with *C. thermoaceticum*, including the first published biological experiments with ^{14}C (see Kamen, 1963), demonstrated that the conversion of glucose to three molecules of acetate involved a unique reductive process by which CO_2 was reduced to acetate (Barker, 1944; Barker and Kamen, 1945):

oxidation:

$$C_6H_{12}O_6 + 2\ H_2O \rightarrow 2\ CH_3COOH + 2\ CO_2 + 8\ H \qquad (3)$$

reduction: $8\ H + 2\ CO_2 \rightarrow CH_3COOH + 2\ H_2O$ (4)

net: $C_6H_{12}O_6 \rightarrow 3\ CH_3COOH$ (5)

Note that Reactions 1 and 4 are essentially identical. Thus, it became apparent that the reductive synthesis of acetate from CO_2 was a process centrally important to both the autotrophic synthesis of acetate (i.e., H_2-CO_2-dependent acetogenesis) by *C. aceticum* and the heterotrophic synthesis of acetate (i.e., glucose-dependent acetogenesis) by *C. thermoaceticum.*

Subsequent work by Wood, which included historically important ^{13}C- and ^{14}C-labelling experiments, conclusively proved that both the methyl and carboxyl carbons of one of the three acetates formed during glucose-dependent acetogenesis were derived from CO_2 (Wood, 1952a, 1952b; see Singleton, 1997a, 1997b for excellent historical treatments of the early studies of Wood). The process by which CO_2 is reduced to acetate is now referred to as the acetyl-CoA pathway. Although the ability of acetogens to reduce CO_2 to acetate was firmly established in the 1940's, it took nearly 5 decades of research before the enzymology of this process, i.e., the acetyl-CoA pathway, was fully resolved (see the following references for reviews of the studies centrally important to the resolution of the pathway: Ljungdahl, 1986; Wood and Ljungdahl, 1991; Ragsdale, 1991, 1997; Drake, 1994; Drake, *et al.* 2004).

1.2. Genera and Phylogeny

Acetogens have only been found in the domain Bacteria and have been isolated from very diverse habitats, ranging from the termite gut to deep subsurface ecosystems. There are 21 bacterial genera that contain acetogenic species, i.e., species that have been termed acetogens; approximately 100 acetogenic species have been reported in the literature (Drake *et al.,*, 2004). These 21 genera and the first species of each genus to be classified as an acetogen are (quotation marks indicate that the name has not been validated; a question mark indicates that it is less than certain that the organism uses the acetyl-CoA pathway [see (Drake *et al.*, 2004)]:

1. *Acetitomaculum ruminis* [ATCC 43876^T (Greening and Leedle, 1989)]
2. *Acetoanaerobium noterae* [ATCC 35199^T (Sleat *et al.*, 1985)]
3. *Acetobacterium woodii* [ATCC 29683^T (Balch *et al.*, 1977)]
4. *Acetohalobium arabaticum* [DSM 5501^T (Zhilina and Zavarzin, 1990)]
5. *Acetonema longum* [DSM 6540^T (Kane and Breznak, 1991)]
6. "*Bryantella formatexigens*" [DSM 14469^T (Wolin *et al.*, 2003)]
7. "*Butyribacterium methylotrophicum*" [ATCC 33266^T (Zeikus *et al.*, 1980)]
8. Caloramator fervidus (?) [ATCC 43204^T (Patel *et al.*, 1987)]
9. *Clostridium aceticum* [DSM 1496^T (Wieringa, 1936)]
10. *Eubacterium limosum* [ATCC 8486^T (Sharak-Genthner *et al.*, 1981)]
11. *Holophaga foetida* [DSM 6591^T (Bak *et al.*, 1992)]
12. *Moorella thermoacetica* [ATCC 35608^T (Fontaine *et al.*, 1942)]
13. *Natroniella acetigena* [DSM 9952^T (Zhilina *et al.*, 1996)]
14. *Natronincola histidinovorans* [DSM 11416^T (Zhilina *et al.*, 1998)]
15. *Oxobacter pfennigii* [DSM 3222^T (Krumholz and Bryant, 1985)]
16. *Ruminococcus productus* [ATCC 35244^T (Lorowitz and Bryant, 1984)]
17. *Sporomusa ovata* [DSM 2662^T (Möller *et al.*, 1984)]
18. *Syntrophococcus sucromutans* [DSM 3224^T (Krumholz and Bryant, 1986)]
19. *Thermoacetogenium phaeum* [DSM 12270^T (Hattori *et al.*, 2000)]
20. *Thermoanaerobacter kivui* [ATCC 33488^T (Leigh *et al.*, 1981)]
21. "*Treponema primitia*" [DSM 12427^T (Graber *et al.*, 2004)]

The genus *Clostridium* was taxonomically reorganized in 1994 (Collins *et al.*, 1994), and many clostridial acetogens were placed in new genera at that time. For example, the acetogen *C. thermoaceticum* (Fontaine *et*

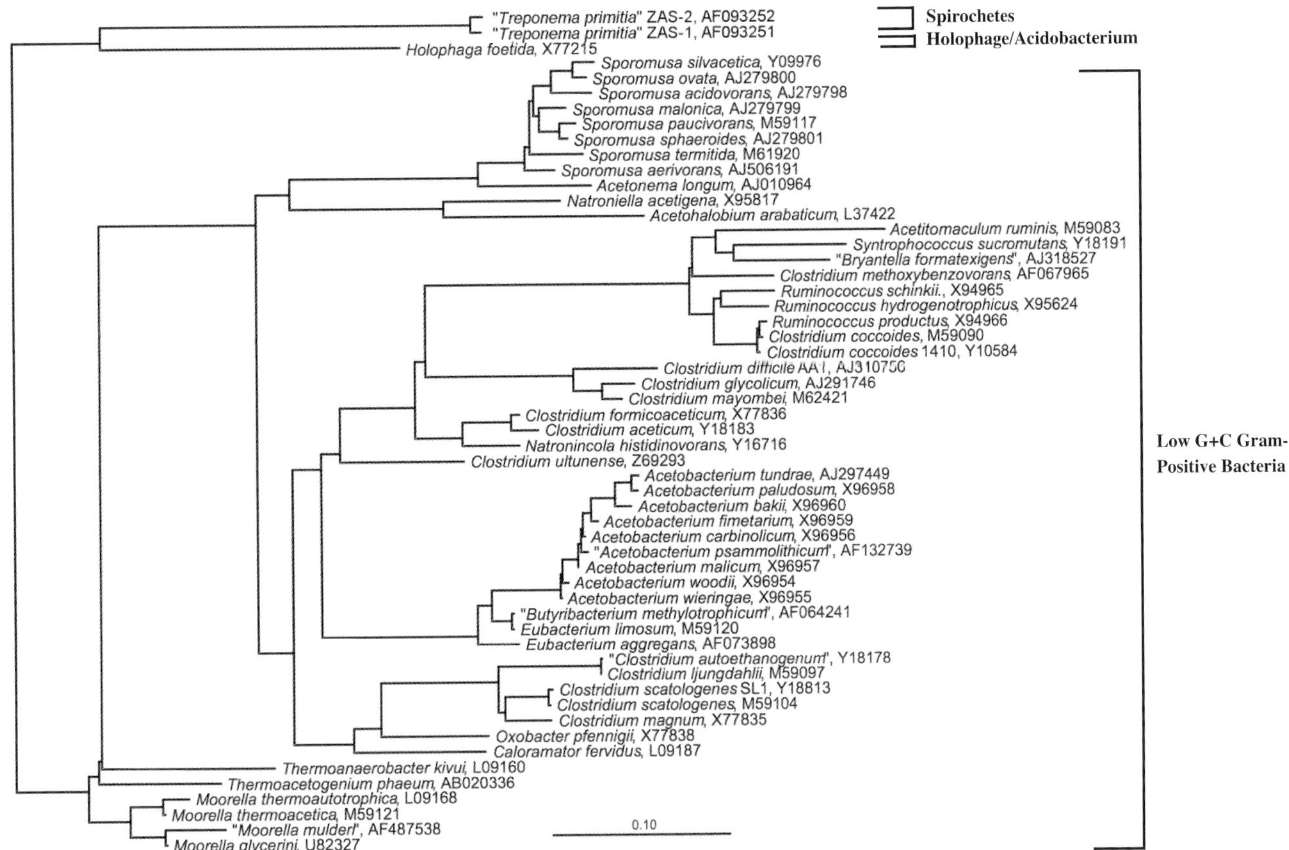

Figure 1. Parsimony tree containing 20 genera of acetogenic bacteria. The tree is based on the 16S rRNA sequences of the indicated species (ARB-release June 2002). *Acetoanaerobium* is absent because a sequence for a species of this genus is not available. The numbers in parentheses are the sequence accession numbers. Bar corresponds to 10 nucleotide substitutions per 100 sequence positions. (Used with permission from Drake *et al*., 2004.)

al., 1942) was reclassified as *Moorella thermoacetica* (this organism will be referred to as *M. thermoacetica* hereafter). The reader is directed to a recent review that presents a complete description of all acetogens and their taxonomic classsification (Drake *et al*., 2004).

An organism's ability to engage the acetyl-CoA pathway is a unifying physiological feature of acetogens; however, acetogens display extreme genetic diversity. For example, the G + C content of the genomes of *Clostridium ljungdahlii* and *Holophaga foetida* are 22 mol% and 62 mol%, respectively (Tanner *et al*., 1993; Liesack *et al*., 1994). Thus, acetogens do not form a taxonomic group of phylogenically closely related organisms (Tanner and Woese, 1994; Drake *et al*., 2004).

Within the phylogenic tree of the Bacteria, most acetogens are found in one phylum, namely, the Gram-positive bacteria with low DNA G+C-content; acetogens are also affiliated with the phyla Spirochetes and *Holophaga/Acidobacterium* (Figure 1). Some genera (e.g., *Acetobacterium* and *Sporomusa*) are exclusively occupied by acetogens. However, many acetogens are phylogenically dispersed among non-acetogens, and several genera (e. g., *Clostridium*, *Ruminococcus*, *Eubacterium*, *Thermoanaerobacter*, *Treponema*) contain both acetogenic and non-acetogenic species. For example, the closest phylogenic relative of the acetogen *Clostridium formicoaceticum* is the non-acetogen *Clostridium felsineum* (99.3% 16S rRNA gene sequence similarity). In the case of *Clostridium glycolicum*, some strains are acetogenic, even though the type strain is not (Ohwaki and Hungate, 1977; Küsel *et al*., 2001). Thus, the phylogenic position (i.e., the 16S rRNA gene sequence) of a bacterium cannot be taken as absolute evidence that the organism is an acetogen. This fact has made it impossible to develop broad-based, acetogen-specific oligonucleotide probes and primers that are based on 16S rRNA gene sequences.

Prolonged cultivation in the laboratory appears to cause certain acetogens to lose their ability to engage the acetyl-CoA pathway, which might explain why nearly identical phylogenic species can display different acetogenic capabilities (Drake *et al*., 2004). Likewise, the true metabolic abilities of an acetogen may go undetected upon initial isolation. For example, *M. themoacetica* was isolated as an obligate heterotroph (Fontaine *et al*., 1942) but was shown to be capable of chemolithoautotrophic

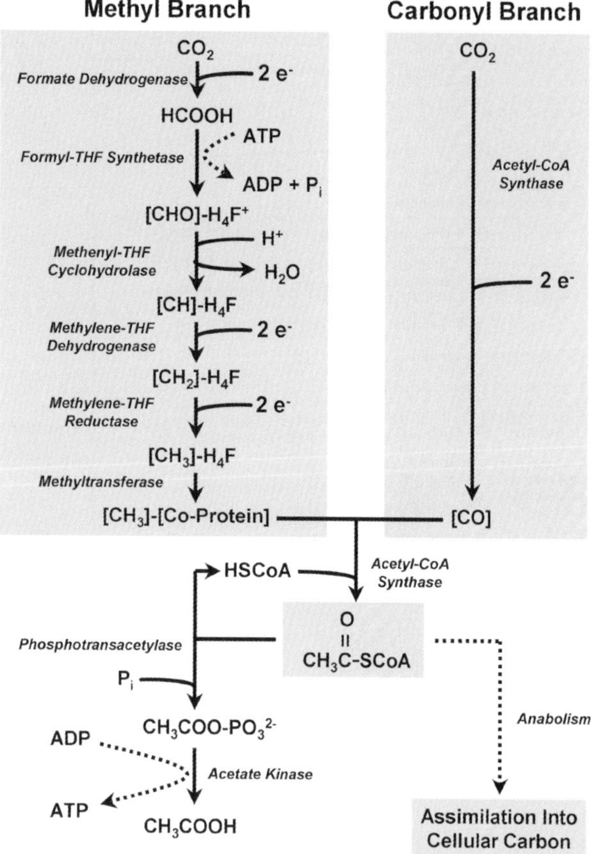

Figure 2. The acetyl-CoA pathway. Abbreviations: H_4F, tetrahydrofolate; HSCoA, coenzyme A; P_i, inorganic phosphate; e^-, electron; Co-Protein, corrinoid enzyme; ATP, adenosine 5'-triphosphate. Brackets indicate that a particular C_1 unit is bound to a cofactor or structurally associated with an enzyme. (Modified from Drake *et al.*, 2004.)

growth nearly 60 years later (Daniel *et al.*, 1990) and has since proven to be an organism with enormous physiological versatility (Drake and Daniel, 2004). Thus, it is not surprising that some organisms were originally described as non-acetogens but later found to be capable of acetogenesis [e.g., *Clostridium coccoides* (Kamlage *et al.*, 1997) and *Clostridium scatologenes* (Küsel *et al.*, 2000)]. It is likely that future studies will disclose the occurrence of acetogens in genera and phyla that currently do not contain acetogenic species.

2. The Acetyl-CoA Pathway

The acetyl-CoA pathway is a terminal electron-accepting process that reduces and fixes CO_2. Acetate is the catabolic end product of this pathway; anabolically, the acetyl moiety of acetyl-CoA, an intermediate in the pathway, is assimilated into cellular carbon. The acetyl-CoA pathway is often referred to as the Wood-Ljungdahl pathway in recognition of Harland G. Wood and Lars G. Ljungdahl, the two biochemists who were responsible for elucidating most of its enzymological features.

2.1. Reduction of CO_2 to Acetate

The acetyl-CoA pathway is composed of two reductive branches (Figure 2). On the methyl branch, CO_2 is reduced to the methyl level, while the other branch reduces CO_2 to the carbonyl level. Adenosine 5'-triphosphate (ATP) is required for the activation of formate, an intermediate in the pathway, and is produced in the terminal reaction of the pathway via acetate kinase. Thus, the acetyl-CoA pathway does not yield any net ATP_{SLP} (ATP formed via substrate-level phosphorylation). However, the standard Gibb's free energy ($\Delta G^{\circ\prime}$) for the overall reductive synthesis of acetate from CO_2 is approximately –100 kJ per mol acetate formed, and energy is conserved via chemiosmotic processes that are coupled to the pathway (see Section 4). Although the pathway can serve as a terminal electron-accepting process, it also provides acetogens with a mechanism for the assimilation of CO_2 and other one-carbon molecules (e.g., formate and methanol) into biomass (Ljungdahl and Wood, 1965; Eden and Fuchs, 1982; 1983; Ljungdahl, 1986). Thus, the pathway functions in both dissimilatory and assimilatory capacities during the growth of acetogens.

Although the acetyl-CoA pathway can be presented in a somewhat cyclic form (e.g., Wood and Ljungdahl, 1991; Ragsdale, 1997), one should note that it is a linear "one-carbon" process that does not involve a recycled multi-carbon intermediate to which CO_2 is fixed (e.g., the Calvin cycle, the reductive tricarboxylic acid cycle, and the hydroxypropionate cycle are CO_2-fixing processes that are dependent upon ribulose biphosphate, oxalacetate, and acetyl-CoA, respectively, for the initial fixation of CO_2). The electron carriers and cofactors that are utilized in the acetyl-CoA pathway cycle between different states, but the flow of carbon in the pathway is linear (Figure 2).

2.2. Interface to Catabolism

Under acetogenic conditions, the recycling of reduced electron carriers (NAD, ferredoxin, etc.) that are generated during catabolism is strictly dependent upon the reduction of CO_2 to acetate. The importance of this redox coupling is illustrated in Figure 3. During the homoacetogenic conversion of hexoses, the electron carriers that are reduced during glycolysis (i.e., the Embden-Merhof-Parnas pathway) and by pyruvate-ferredoxin oxidoreductase are reoxidized via the acetyl-CoA pathway. This specialized anaerobic catabolism yields 4 moles of ATP_{SLP} per mole hexose, which is twice the number of ATP_{SLP} formed via ethanol or homolactate fermentation of hexoses.

The acetyl-CoA pathway (Figure 2) and the pathway by which glucose is catabolized to acetate (Figure 3) were resolved with the model acetogen *M. thermoacetica* (Wood and Ljungdahl, 1991). Relatively few acetogens have been examined in detail. It is therefore uncertain

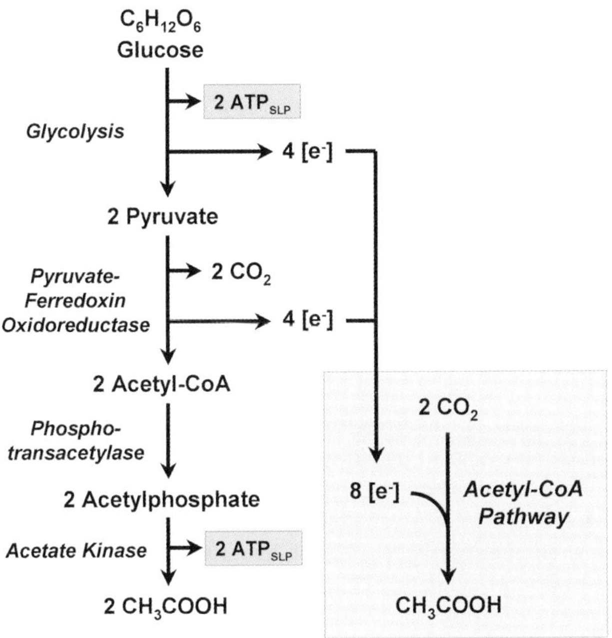

Figure 3. Glucose-dependent homoacetogenesis. During the overall conversion of glucose to acetate, most of the CO_2 that is reduced to acetate via the acetyl-CoA pathway is likely derived from exogenous CO_2. Abbreviations: ATP_{SLP}, adenosine 5'-triphosphate (ATP) that is produced by substrate-level phosphorylation; CoA, coenzyme A; [e^-], electron. (Modified from Drake, 1994.)

if all acetogens utilize the same catabolic pathways. The ability of an acetogen to grow heterotrophically on hexoses may not always be dependent upon glycolysis (Figure 3), and autotrophic growth might involve reductive processes that are not identical to those displayed in Figure 2. As noted below, acetogens conserve energy via more than one chemiosmotic process, and it is likely that slight variations of the acetyl-CoA pathway exist among acetogens.

2.3. Importance of Exogenous CO_2

Depending on the acetogen and growth substrate(s), exogenous CO_2 can be essential to growth (Andreesen *et al.*, 1970; O'Brien and Ljungdahl, 1972; Braun and Gottschalk, 1981; Savage *et al.*, 1987). For example, even though the stoichiometry of hexose-dependent acetogenesis (Reaction 2) does not indicate that exogenous CO_2 is involved in the metabolism of hexoses, growth on hexoses can be significantly impaired, and even blocked entirely, unless exogenous CO_2 is supplied in the growth medium. The reason why exogenous CO_2 is required for optimal growth can be envisioned by examining Figure 3. Glycolysis generates reducing equivalents prior to the production of CO_2 via the decarboxylation of pyruvate, and recycling of the electron carriers involved in glycolysis cannot occur unless CO_2 is readily available. Indeed, based on ^{14}C-labeling studies with *M. thermoacetica*, carbons 3 and 4 of glucose enter the free pool of CO_2 (Wood, 1952b; O'Brien and Ljungdahl, 1972), and only two-thirds of the carbon from uniformly-labeled glucose is recovered in acetate (Martin and Drake, unpublished data; see also Martin *et al.*, 1985). Thus, even though oxidative metabolism of acetogens is often depicted as being CO_2-independent in metabolic schemes, the cytoplasmic oxidation of substrates can be tightly linked to the availability of CO_2 (Drake, 1994).

The importance of exogenous CO_2 to acetogens is exemplified by (a) their inability to grow on substrates [e.g., CO (Savage *et al.*, 1987)] that generate excess CO_2 during acetogenesis unless supplemental CO_2 is supplied in the growth medium, (b) their ability to generate growth-essential CO_2-equivalents from carboxylated lignin derivatives in the absence of supplemental CO_2 (Hsu *et al.*, 1990a; 1990b), and (c) the occurrence of high levels of carbonic anhydrase in acetogenic cells (Braus-Stromeyer *et al.*, 1997). Carbonic anhydrase is widespread in nature and catalyzes the following reaction (Lindskog *et al.*, 1971; Karrasch *et al.*, 1989; Albers and Ferry, 1994; Vandenberg *et al.*, 1996):

$$CO_2 + H_2O \leftrightarrow HCO_3^- + H^+ \qquad (6)$$

Specific activities of this enzyme in certain acetogens are among the highest in biological systems (Braus-Stromeyer *et al.*, 1997). Given the importance of CO_2 in the acetyl-CoA pathway, carbonic anhydrase might increase intracellular levels of CO_2 in acetogens. Acetogens are sensitive to acidic conditions (Baronofsky *et al.*, 1984; Wiegel, 1994), and a secondary function of carbonic anhydrase might be in maintaining cytoplasmic pH homeostasis.

3. Enzymological and Molecular Features of the Acetyl-CoA Pathway

The main enzymological features of the acetyl-CoA pathway that is utilized by acetogens were resolved during decades of research that used *M. thermoacetica* as the model acetogen; many of the physical and catalytic properties of the enzymes of the pathway have been summarized in several excellent reviews (Ljungdahl, 1986, 1994; Wood and Ljungdahl, 1991; Ragsdale, 1991; 1997; 2004).

3.1. The Methyl Branch of the Pathway: Reductive Synthesis of a Methyl Group

The initial reaction on the methyl branch of the pathway (Figure 2) is the reduction of CO_2 to formate by formate dehydrogenase; this reaction is reversible, in that the enzyme can also rapidly oxidize formate. The enzyme from *M. thermoacetica* is an $\alpha_2\beta_2$ tetramer (340 kDa) that contains tungsten, selenium, and iron sulfur centers;

NADP is used as an electron acceptor. Formate is then activated by a tetrahydrofolate (H_4F)-dependent reaction that is catalyzed by formyl-H_4F synthase. The synthesis of 10-formyl-H_4F is endergonic and is driven by the hydrolysis of ATP. Formyl-H_4F synthase has been purified, and the corresponding gene has been cloned and sequenced (Lovell *et al.*, 1988; 1990). Water is split off by 5,10-methenyl-H_4F-cyclohydrolase in the next reaction on the methyl branch, and the resulting methenyl group is reduced to methylene-H_4F by methylene-H_4F dehydrogenase. These activities are catalyzed by two monofunctional (*A. woodii*) or one bifunctional (*M. thermoacetica*) enzymes (Ljungdahl *et al.*, 1980; Ragsdale and Ljungdahl, 1984). Methylene-H_4F is then reduced by 5,10-methylene-H_4F reductase, an enzyme that contains FAD, zinc, and iron-sulfur centers. The enyzme can use NADH and methylene blue (*R. productus*) or $FADH_2$ and ferredoxin (*Clostridium formicoaceticum*) as electron donors (Wohlfahrt *et al.*, 1990). This enzyme might be localized on the membrane, suggesting that it might interact with an unknown membranous electron carrier (see Section 4).

The methyl group that is bound to H_4F is the precursor of the methyl group of acetate and is initially transferred to the enzyme acetyl-CoA synthase. There are different forms of acetyl-CoA synthase in different prokarotes, and these enzymes have unusual and interesting properties. The enzymology of this reaction has been studied intensively (Ragsdale, 1991; Ragsdale and Kumar, 1996; Barondeau and Lindahl, 1997; Russell *et al.*, 1998). The methyl group of methyl-H_4F is abstracted as a methyl cation, but transfer from a tertiary amine requires an activation that is facilitated by protonation of the N5 of methyl-H_4F (Seravalli *et al.*, 1999). The activated methyl group is then attacked by a supernucleophile and undergoes a nucleophilic displacement reaction. This reaction is catalyzed by a methyltransferase, a small, dimeric protein of 66 kDa (Drake *et al.*, 1981). The methyltransferase has been cloned and sequenced, and produced in an active form in *Escherichia coli* (Roberts *et al.*, 1994). High-resolution structures revealed a pterin-binding site within a negatively charged region of the protein and identified residues specifically involved in pterin binding (Doukov *et al.*, 2000). This binding motif is also found in methionine synthase. The initial methyl-group acceptor is a corrinoid-containing protein that contains the supernucleophile cobalt. Co(I) of the corrinoid with its free electron pair attacks the methyl group and abstracts the methyl cation, yielding methyl-Co(III) (Banerjee and Ragsdale, 2003). The Co(I)/Co(II) half-cell reaction has a standard redox potential below -500 mV, and the corrinoid protein is subject to inactivation by oxidation. Therefore, activating enzymes that convert Co(II) to Co(I) are present in corrinoid-containing organisms (Banerjee and Ragsdale, 2003).

Corrinoids are very abundant in acetogens, methanogens, and other anaerobes (Dangel *et al.*, 1987), and catalyze isomerisations, methyl-transfer reactions, and dehalogenations (Banerjee and Ragsdale 2003). A role for methyl corrinoids in acetogenesis was discovered in the 1960's (Poston *et al.*, 1964; 1966). Corrinoids contain a planar corrin ring, an upper cobalt ligand, and a lower cobalt ligand. Thus, cobalt is complexed by six substituents, four from the corrin ring and two axial ligands. The upper ligand is a methyl, adenosyl, hydroxyl, or cyanosyl group. The methyl ligand occurs in methyltransferases, and the adenosyl ligand occurs in C-C rearrangement mutases or enzymes catalyzing elimination reactions. The hydroxyl ligand is a degradation product, and the cyanosyl ligand is an artefact that occurs during preparation. The lower axial ligand varies in organisms but is typically a benzimidazole (as in vitamin B_{12}) or an adenine derivative (Kräutler, 1990). For example, *A. woodii* contains vitamin B_{12}, whereas *Sporomusa ovata* contains para-cresolyl-cobamide as a lower axial ligand (Stupperich *et al.*, 1990; Stupperich, 1994). The lower ligand can adopt two conformations, a base-on form, where it is coordinated to the central cobalt atom, and a base-off form, that is not coordinated. A given corrinoid can be in the base-on or base off-form, and the different forms change the redox potential considerably. For example, the removal of the lower axial ligand in the corrinoid from *M. thermoacetica* increases the standard redox potential of the Co(II)/Co(I) half-cell reaction by 150 mV (Harder *et al.*, 1989).

The corrinoid protein also has iron-sulfur centers; the enzyme from *M. thermoacetica* contains 5-methox ybenzimidazolylcobamide, and a $[4Fe\text{-}4S]^{2+/1+}$ center, and neither the 5-methoxybenzimidazolylcobamide nor a histidine residue coordinates to the cobalt atom (base-off, His-off configuration) (Ragsdale *et al.*, 1987; Wirt *et al.*, 1995). The gene encoding the corrinoid protein has been cloned from *M. thermoacetica*, sequenced, and expressed in *E. coli*; unfortunately, the heterologously produced enzyme was inactive (Roberts *et al.*, 1989).

3.2. The Carbonyl Branch of the Pathway: Acetyl-CoA Synthase

The methyl group of the methylated corrinoid protein is then transferred to acetyl-CoA synthase. Acetyl-CoA synthase has been purified from *M. thermoacetica* and *A. woodii*, and is an $\alpha_2\beta_2$ tetramer (Ragsdale *et al.*, 1983a; 1983b). Acetyl-CoA synthase was historically discovered as carbon monoxide (CO) dehydrogenase, i.e., an oxidoreductase that can oxidize CO (Diekert and Thauer, 1978). The function of the enzyme in the synthesis of acetyl-CoA was resolved later (Hu *et al.*, 1982). The designations acetyl-CoA synthase and CO dehydrogenase will be used in the following discussion to highlight the dual reactions catalyzed by this enzyme. The CO

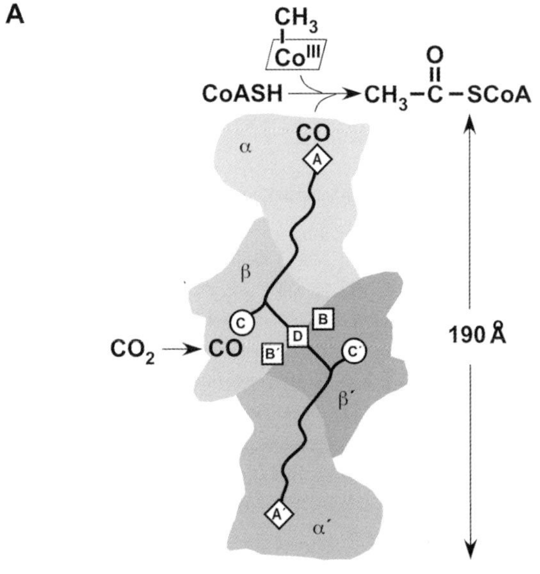

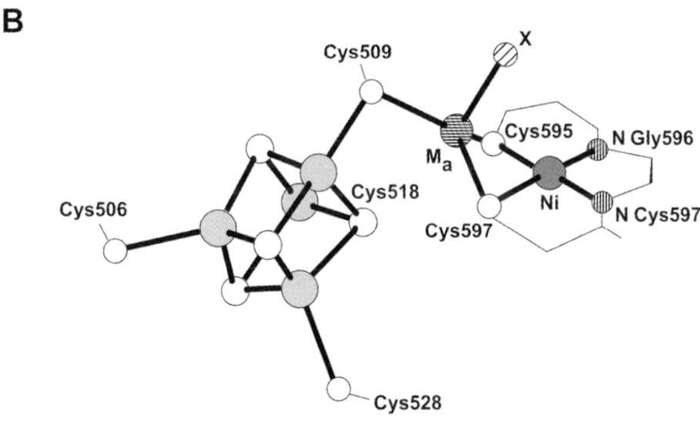

Figure 4. Model of the structure of the acetyl-CoA synthase. Panel A: Localization of metal centers and subunit composition. The reactive "A clusters" (α subunits) are connected with the "C cluster" (β subunits) via the CO channel. The "D cluster" is an Fe_4S_4 center located at the interface of the β subunits. Panel B: Schematic presentation of the reactive "A cluster", consisting of an Fe_4S_4 cluster and a binuclear metal center (proximal M_a: either Cu or Ni; distal: Ni). X represents an unidentified non-protein group. Shadings: Fe_4S_4 cluster Fe atoms are shown in light grey; sulfur atoms are in white; Ni is dark grey; M_a has horizontal hatching; N has vertical hatching; X has angular hatching. (Republished with permission from Grahame, 2003.)

dehydrogenase reaction is catalyzed by the β subunits, while the acetyl-CoA synthase reaction is catalyzed by the α subunits (Ragsdale and Kumar, 1996). Recently, high resolution structures of the acetyl-CoA synthase from *M. thermoacetica* were obtained independently by two groups (Doukov *et al.*, 2002; Darnault *et al.*, 2003). Acetyl-CoA synthase from *M. thermoacetica* has an α-ββ-α configuration with a major axis of 190 Å. Each α subunit harbors an "A cluster"; the "A clusters" are 148 Å apart from each other. However, the "A clusters" are connected with a narrow interior channel that has two branches to the "C clusters" of the β subunits. This channel presumably concentrates the CO and prevents this poisonous gas from interacting with other cellular constituents (Grahame 2003) (Figure 4). The "A cluster" catalyzes the synthesis of acetyl-CoA from CO, CoA, and methylated corrinoid protein (CH_3-Co-Protein) according to the following reaction:

$$CO + CH_3\text{-Co-Protein} + CoA\text{-}SH \rightarrow CH_3\text{-}CO\text{-}S\text{-}CoA + \text{Co-Protein} \quad (7)$$

The "C cluster" in the β subunits catalyze the reversible oxidation of CO to CO_2:

$$CO + H_2O \rightarrow CO_2 + 2\,H^+ + 2\,e^- \quad (8)$$

Crystallographic data revealed the presence of different metal centers in acetyl-CoA synthase and yielded insights on the reaction mechanism(s) of the enzyme. The "C cluster" contains a distorted cubane that is constructed from a nickel atom, three iron atoms, and four sulfur atoms that are bridged to an additional iron atom. Slightly different structures were observed for the "C clusters" of CO dehydrogenases from *Carboxydothermus hydrogenoformans* (Dobbek *et al.*, 2001) and *Rhodospirillum rubrum* (Drennan *et al.*, 2001). The two structures resolved for the acetyl-CoA synthase from *M. thermoacetica* (Doukov *et al.*, 2002; Darnault *et al.*, 2003) and that from *C. hydrogenoformans* (Svetlitchnyi *et al.*, 2004) are very similar; however, different proposals have been made relative to the metal content and geometry of the metal clusters. All structures have a [4Fe-4S] cubane that is bridged to a binuclear metal site. The metal site that is distal to the cubane contains nickel. The nature of the metal site that is proximal to the cubane is less certain; this site has been interpreted to contain Cu (Doukov *et al.*, 2002), Zn in the closed form or Ni in the open form (Darnault *et al.*, 2003), or Ni (Svetlitchnyi *et al.*, 2004). These interpretations suggest the presence of Cu-Ni, Zn-Ni, and Ni-Ni binuclear sites. The finding of copper in the active site (Doukov *et al.*, 2002) was unexpected and is controversial since (a) copper is apparently not required for catalysis by, or synthesis of, the enzyme, (b) a metalloenzyme having an active site with three different transition metals had not been previously reported, and (c) such a biological role for copper is unprecedented. The most recent analyses indicate that the occurrence of copper in the active site is an artifact; in fact, copper is an inhibitor of acetyl-CoA synthase (Bramlett *et al.*, 2003; Seravalli *et al.*, 2003). A nickel insertase is required for the biosynthesis of acetyl-CoA synthase (Loke and Lindahl, 2003).

Three consecutive processes/reactions for the synthesis of acetyl-CoA by acetyl-CoA synthase have been proposed: (a) a mononuclear reaction in which the proximal Ni site binds both CO and CH_3, (b) a condensation reaction that yields an acetyl moiety, and (c) a displacement reaction by the thiolate anion of CoA. The binuclear mechanisms at a Cu-Ni or at a Ni-Ni site suggest the binding of CO at the proximal transition metal and CH_3 at the distal Ni site (Grahame, 2003). Acetyl-CoA is subsequently converted to acetate by the sequential reactions catalyzed by phosphotransacetylase and acetate kinase (Schaupp and Ljungdahl, 1974; Drake *et al.*, 1981).

3.3. Additional Oxidoreductases Centrally Important to Acetogenesis

The reducing equivalents required in the acetyl-CoA pathway are obtained from (a) the hydrogenase-dependent oxidation of molecular hydrogen during autotrophic growth or (b) the NADH and reduced ferredoxin that are generated during heterotrophic growth, i.e., produced from the oxidation of organic substrates. Despite the importance of hydrogenase in the metabolism of acetogens, little is known about this oxidoreductase in these anaerobes. Two hydrogenases have been observed in *M. thermoacetica* (Kellum and Drake, 1984). One hydrogenase is expressed when cells are cultivated on glucose, and another hydrogenase is expressed when glucose and CO are provided as co-substrates. A membrane bound hydrogenase was also reported (Ljungdahl, 1994). A partially purified hydrogenase from the cytoplasm of fructose-grown *A. woodii* was devoid of nickel but contained iron-sulfur clusters (Ragsdale and Ljungdahl, 1984). Ferredoxin, flavodoxin, and CO dehydrogenase served as electron acceptors, and CO strongly inhibited the enyzme. A membrane-bound hydrogenase was solubilized and purified from *Sporomusa sphaeroides* (Kamlage and Blaut, 1993; Dobrindt and Blaut, 1996). The enzyme had two subunits of 65 and 37 kDa, and was an $\alpha\beta$ dimer. The native protein contained iron, sulfur, and nickel and was inactivated by O_2. Cytochrome *b* and viologen dyes were reduced, whereas FAD, NAD, and FMN were not. *T. kivui* displays exceptionally fast growth at the expense of H_2 and likewise has the highest hydrogenase specific activity of all known acetogens (Leigh *et al.*, 1981; Daniel *et al.*, 1990). Membranes from autotrophically cultivated cells of this acetogen have formate dehydrogenase, as well as hydrogenase and acetyl-CoA synthase activities, and a complex by which these oxidoreductases might interact has been proposed; significantly, growth conditions influenced the localization of these enzymes (Braus-Stromeyer and Drake, 1996; Drake *et al.*, 1997).

4. Bioenergetics

As noted in Section 2.1, the net gain of ATP_{SLP} in the acetyl-CoA pathway is zero (Figure 2). Thus, a chemiosmotic process for the conservation of energy must occur for the chemolithoautotrophic growth of an acetogen. Experimental evidence indicates that two different mechanisms are used by acetogens for ion gradient-driven phosphorylation via ATP synthase (ATPase) (Müller, 2003). One mechanism is proton (H^+)-dependent and the other is sodium ion (Na^+)-dependent; *M. thermoacetica* (Ljungdahl, 1994) and *A. woodii* (Müller and Gottschalk, 1994) have served as the primary models for the study of these two energy-conserving processes, respectively.

4.1. Energy Conservation in Proton-Dependent Acetogens

The occurrence of membraneous electron carriers is correlated with the ability of certain acetogens to conserve energy via a proton-dependent, chemiosmotic process. The pioneering work of the group of L. G. Ljungdahl identified several membrane-integrated electron carriers in *M. thermoacetica* and the closely related *Moorella thermoautotrophica* (formerly, *Clostridium thermoautotrophicum*). Menaquinone MK-7 (2 methyl-3-heptaprenyl-1.4-naphtoquinone; E_o' = - 74 mV) and two b-type cytochromes (cyt b_{559}, E_o' = - 215 mV; cyt b_{554}, E_o' = - 57 mV) were identified in 1975; subsequent studies demonstrated that a flavoprotein co-purified with cyt b_{559} (Gottwald *et al.*, 1975; Das *et al.*, 1989). These membranous components are very likely involved in electron transport processes (Ljungdahl, 1994; Das and Ljungdahl, 2003). However, the nature of the electron donor and electron acceptor systems is not known, mainly because all of the enyzmes of the acetyl-CoA pathway occur in the cytoplasmic fraction when cells are disrupted with a French press. This problem has necessitated the preparation of inside out membrane vesicles under mild conditions that prevent or reduce the loss of membrane-bound enzymes during the disruption of cells.

Using inside out membrane vesicles, it has been possible to show that certain enzymes of the acetyl-CoA pathway may be at least loosely associated with the cytoplasmic membrane. These enzymes include: hydrogenase, CO dehydrogenase, and NADH dehydrogenase (all enzymes that can generate reducing equivalents), and also the electron accepting methylene-H_4F reductase (Hugenholtz and Ljungdahl, 1989). Inverted membrane vesicles catalyze the oxidation of CO (Hugenholtz *et al.*, 1987) and establishment of a CO-dependent transmembrane electrochemical potential ($\Delta\mu_{H}+$) that drives the uptake of amino acids (Hugenholtz and Ljungdahl, 1990). Although these observations proved that a transmembrane electrochemical potential can be formed via the membranous oxidation of CO, they did not reveal the specific membranous components and mechanisms involved.

Thauer *et al.* (1977) suggested that methylene-H_4F reductase (i.e., the reduction of methylene-H_4F) might constitute the last stage of a membranous electron transport chain that could be engaged by electrons generated from hydrogenase, CO dehydrogenase, or NADH dehydrogenase. Such an electron transport chain could result in the generation of a transmembrane proton

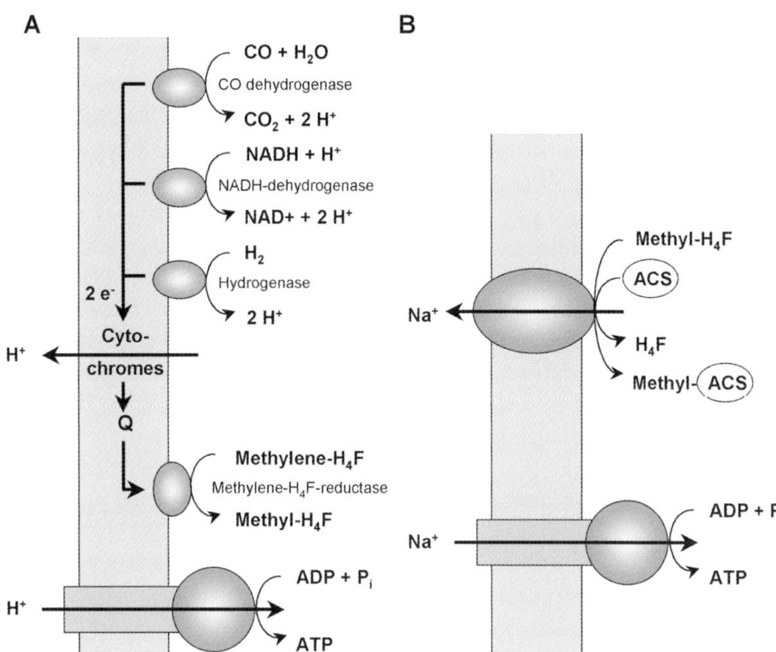

Figure 5. Chemiosmotic mechanisms by which acetogens conserve energy. Panel A: Proton-dependent conservation of energy with the hypothetical involvement of different oxidoreductases that activate reducing equivalents for electron transport. Panel B: Na^+-dependent conservation of energy; the Na^+ gradient might be generated by a methyltransfer reaction that is facilitated by several enzymes (see text).

potential that drives the synthesis of ATP (Figure 5A). Although this is a likely scenario, experimental evidence for such an energy-conserving process has not been obtained.

The transmembrane electrochemical potential is utilized by a proton-dependent F_1F_O ATPase that synthesizes ATP. The enzyme and subcomplexes thereof have been isolated from membranes of *M. thermoacetica* and *M. thermoautotrophica* (Ivey and Ljungdahl, 1986; Das and Ljungdahl, 2003). In contrast to any other bacterial F_1F_O ATPase, the enzyme contained only six subunits (αβγδε and c), being devoid of subunits *a* and *b*. It was thus speculated that the architecture of the F_1F_O ATPase from these acetogens is simpler than that of other bacteria (Das and Ljungdahl, 1997; Das *et al.*, 1997). However, this possibility was hard to explain because subunits *a* and *b* are essential for catalysis (i.e., the synthesis or hydrolysis of ATP). More recently, it has been determined that the operon encoding the F_1F_O ATPase is identical to that of *E. coli* and thus contains the genes for subunits *a* and *b* (Das *et al.*, 1997). The current proposal is that the enzyme from *M. thermoacetica* and *M. thermoautotrophica* has an $\alpha_3\beta_3\gamma\delta\varepsilon ab_2c_X$ architecture. In earlier studies, subunits *a* and *b* were apparently not solubilized from the cytoplasmic membrane, were degraded, or simply lost during the purification procedure. Loss of subunits during the purification of F_1F_O ATPase is not an uncommon problem when purifying this enzyme.

4.2. Energy Conservation in Na^+-Dependent Acetogens

Methanogenic archaea use a modified version of the acetyl-CoA pathway. Methanogens are dependent on Na^+ for the reduction of CO_2 (Perski *et al.*, 1981; 1982), and methyltetrahydromethanopterin:coenzyme M methyltransferase (essential to methanogenesis) is a membrane-bound, Na^+-dependent enzyme that uses the energy that is released during methyl transfer from methyl-tetrahydromethanopterin to coenzyme M for the vectorial transport of Na^+ across the cytoplasmic membrane (Müller *et al.*, 1988; 1993; Lienard *et al.*, 1996; Gottschalk and Thauer, 2001). Because the acetogenic and methanogenic pathways for the reduction of CO_2 are biochemically very similar, it was likely that Na^+ could be essential to certain acetogens. Indeed, *A. woodii* (Heise *et al.*, 1989), *T. kivui* (Yang and Drake, 1990), and *R. productus* (Geerligs *et al.*, 1989) were found to be dependent on Na^+ for growth and acetogenesis. It is significant to note that heterotrophic (e.g., glucose-dependent) growth was not dependent on Na^+; only autotrophic (i.e., H_2-CO_2-dependent) growth, which is strictly dependent upon the acetyl-CoA pathway for conservation of energy and carbon assimilation, displayed a requirement for Na^+.

Studies with cell suspensions of *A. woodii* made it apparent that the acetyl-CoA pathway in this acetogen indeed requires Na^+. Furthermore, an important function of the acetyl-CoA pathway is to generate a Na^+ motive force across the cytoplasmic membrane via a primary, electrogenic Na^+ pump. The Na^+-dependent reaction sequence leading from methylene-H_4F to a methylated intermediate was resolved using substrates that feed into the acetyl-CoA pathway at different levels (Heise *et al.*,

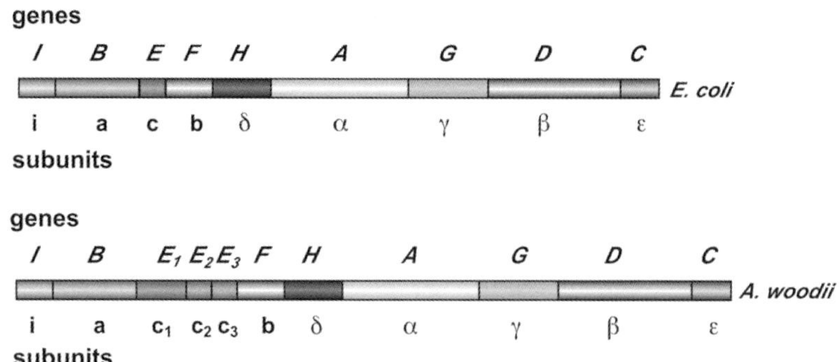

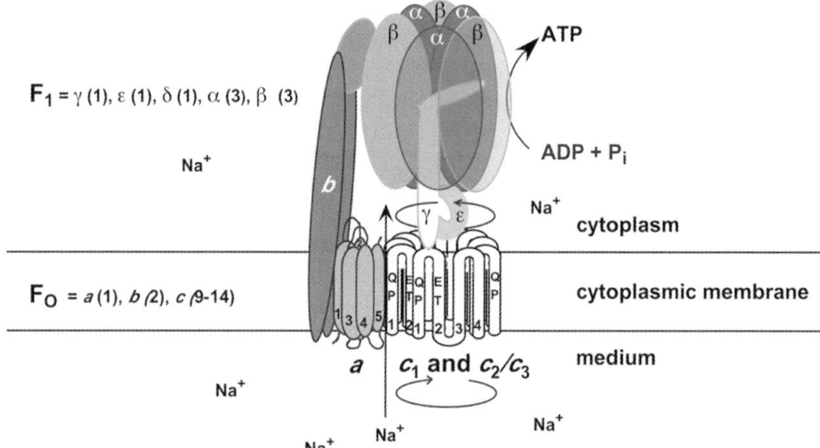

Figure 6. The Na^+ F_1F_O ATPase of *A. woodii*. Panel A: The genomic organization of the *atp* operons of *A. woodii* and *E. coli*. Panel B: A hypothetical model of the enzyme. The stoichiometries of subunits c_1 and $c_{2/3}$ are unknown.

1989). This reaction sequence is catalyzed by methylene-H_4F reductase and one or two methyltransferases according to the following reactions (ACS is acetyl-CoA synthase):

Methylene-H_4F + 2 H → Methyl-H_4F (9)

Methyl-H_4F + Co-Protein → Methyl-Co-Protein + H_4F (10)

Methyl-Co-Protein + ACS → Methyl-ACS + Co-Protein (11)

Reaction 9 is catalyzed by methylene-H_4F reductase. Its cytoplasmic localization and NADH dependence, as well as thermodynamic considerations, made the involvement of this enzyme in the translocation of Na^+ unlikely (Wohlfarth and Diekert, 1991; Heise *et al.*, 1992). This leaves Reactions 10 and 11 as the most likely candidates for the translocation of Na^+. This reaction sequence has never been studied with Na^+-dependent acetogens, and our understanding of the associated enzymes is based on results from studies with proton-dependent acetogens in which the entire sequence is catalyzed by soluble, cytoplasmic enzymes (Ragsdale, 1991). However, in Na^+-dependent acetogens, the situation might be different. As outlined above, methyltransferases require corrinoids as cofactors, and the occurrence of membrane-bound corrinoids in Na^+-dependent acetogens (Dangel *et al.*, 1987) suggested that they might contain membrane-bound methyltransferases. Together with the analogy of the pathways in acetogens and methanogens, this finding led to the hypothesis that Na^+ extrusion is catalyzed by a membrane-bound, corrinoid-containing multi-subunit methyltransferase in Na^+-dependent acetogens (Müller and Gottschalk, 1994) (Figure 5B). However, this hypothesis has not been verified by experimental analyses.

4.3. ATPase of *A. woodii* is a Na^+-F_1F_O-Type Enzyme with Unusual Subunit Composition

One of the most important functional goals of catabolism is the production of adenosine 5'-triphosphate (ATP). Most of the ATP in eukaryotes (chloroplasts and mitochondria) and prokaryotes is synthesized by the enzyme F_1F_O ATPase. This enzyme can transform energy from a gradient of ions across the cytoplasmic membrane to synthesize ATP (Mitchell, 1961) and can

A

B

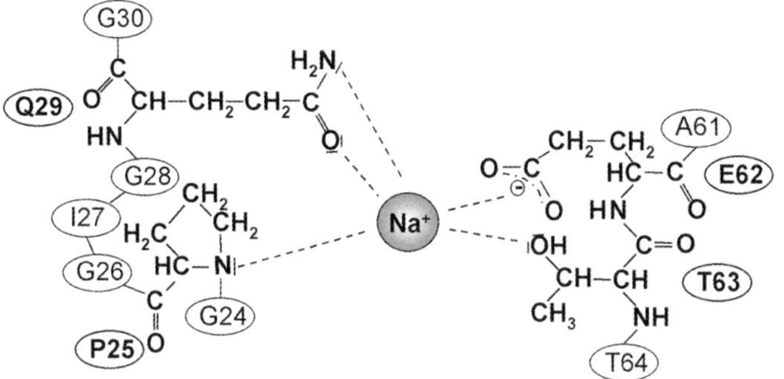

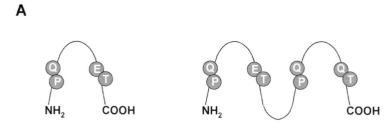

Figure 7. The Na^+-binding site of the Na^+ F_1F_O ATPase of *A. woodii*. Panel A: The Na^+-binding site in subunit c_1 and hairpin 1 of subunit $c_{2/3}$. Hairpin 2 of subunit c_1 does not contain the conserved glutamate and, therefore, cannot bind either Na^+ or a proton. Panel B: Detailed close-up of the Na^+-binding site.

also catalyze the reverse reaction, i.e., the hydrolysis of ATP coupled to the efflux of cations (H^+ or Na^+) through the membrane (Müller *et al.*, 1999; Müller and Grüber, 2003). F_1F_O ATPase in its simplest bacterial form is composed of eight subunits ($\alpha_3\beta_3\gamma\delta\varepsilon ab_2c_X$). The F_1F_O ATPase has three morphological components: (a) a membrane-bound domain, F_O, that contains the ion channel, (b) an approximately spherical cytoplasmic domain, F_1, that contains the catalytic sites (Stock *et al.*, 2000; Grüber *et al.*, 2001; Nishi and Forgac, 2002), and (c) two stalks that connect the F_1 and F_O domains.

The F_1 domain is constructed from the $\alpha_3\beta_3\gamma\delta\varepsilon$ subunit complex. Three copies of subunits α and β alternate around a central stalk, i.e., around subunit γ. Subunit ε is located at the bottom of the central stalk and connects it to the F_O domain. Subunits a, b, and c in a stoichiometry of 1 : 2 : 9-14 (Stock *et al.*, 2000) constitute the F_O domain. Subunit b contains two transmembrane α helices (Dmitriev *et al.*, 1999) and a rather large cytoplasmic domain that is proposed to be a part of the peripheral stalk (Dunn, 1992; Wilkens and Capaldi, 1998; Böttcher and Gräber, 2000) (Figure 6). Ion transport is catalyzed by subunits *a* and *c*. Subunit *a* has five or six predicted transmembrane helices, whereas subunit *c* has two transmembrane α helices. Subunit *c* carries the active carboxylate (aspartate or glutamate, Asp-61 in *E. coli*) that is involved in the translocation of protons and undergoes protonation/deprotonation cycles during translocation of protons (Rastogi and Girvin, 1999). Structural data show the *c* polypeptides are organized in a ring with a stoichiometry of 10, 11, and 14 *c*'s in yeast (Stock *et al.*, 1999), *Ilyobacter tartaricus* (Stahlberg *et al.*, 2001), and chloroplasts (Seelert *et al.*, 2000), respectively. The structure of this ring is unknown but it is assumed that the subunit *c* monomers are arranged front-to-back, yielding two concentric rings. It is still a matter of debate whether helix one or two makes the outer ring of the oligomer (Groth and Walker, 1997; Jones *et al.*, 1998; Fillingame *et al.*, 2000; Schnick *et al.*, 2000).

F_1F_O ATPases are the smallest membranous rotary devices known in biology and are of considerable interest in biochemistry and nanotechnology. Ion flow through the membrane along the *a*-*c*-interface is coupled to a rotation of the ring of proteolipids (Sambongi *et al.*, 1999) which drives the rotation of the central stalk (i.e., subunit γ) (Yoshida *et al.*, 2001). Rotation of subunit γ within the $\alpha_3\beta_3$ headpiece results in the liberation of ATP from the β-subunit. Such a mechanism requires a stator, which is most likely composed of subunits *b* and δ.

ATPase from *A. woodii* uses Na^+ as the coupling ion (Heise *et al.*,1991; 1993; Müller *et al.*, 2001). Biochemical, immunological, and molecular studies identified the Na^+ ATPase of *A. woodii* as a member of the F_1F_O ATPases (Reidlinger and Müller, 1994; Reidlinger *et al.*, 1994; Forster *et al.*, 1995; Rahlfs *et al.*, 1999). The ATPase was solubilized from the cytoplasmic membrane and purified to apparent homogeneity, but, like the enzymes purified from *M. thermoacetica* and *M. thermoautotrophica*, subunits *a* and *b* were missing (Reidlinger and Müller, 1994). Preparative electrophoretic methods later yielded a purified enzyme that contained three additional subunits (*b*, *a*, c_1) (Aufurth *et al.*, 2000). The Na^+-binding site resides in the rotory component of ATPase (Kaim and Dimroth, 1993). Subunit *c* employs at least a triade of residues [Glu62, Thr63, and Gln29 (numbering per the *A. woodii* subunit $c_{2/3}$)] for the binding of Na^+ (Zhang and Fillingame, 1995; Kaim *et al.*, 1997). Pro25 (*A. woodii* numbering) of subunit *c* might also be involved in Na^+ binding. The Na^+-binding site is depicted in Figure 7.

The *atp* operon of *A. woodii* that encodes the Na^+ F_1F_O ATPase contains homologues of the nine genes present in the *E. coli atp* operon (Forster *et al.*, 1995; Rahlfs *et al.*, 1999) (Figure. 7). The order of the genes is *atpIBE*$_1$*E*$_2$*E*$_3$*FHAGDC*, and the genes constitute one polycistronic message. In contrast to other known F_1F_O ATPase operons, the *atp* operon from *A. woodii* contains three tandemly organized genes (*atpE*$_1$, *atpE*$_2$, *atpE*$_3$) encoding subunit *c*. AtpE_2 (subunit c_2) and AtpE_3 (subunit c_3) are 100% identical at the amino acid level; only 18 base substitutions occur at the DNA level (Rahlfs *et al.*, 1999). This pattern is strong evidence for a duplication of an ancestral gene. The deduced molecular mass of the polypeptides c_2 and c_3 is 8.18 kDa. As with their bacterial homologues, they are likely organized in the membrane in a hairpin-like structure, having two transmembrane helices that are connected by a polar loop. Most interestingly, *atpE*$_1$ with 546 base pairs is more than double the size of *atpE*$_{2/3}$. The first and second halves are 66% identical at the DNA level, indicating a duplication of a precursor and subsequent fusion of the two gene copies. The deduced molecular mass of subunit c_1 is 18.37 kDa with four predicted transmembrane helices arranged in two hairpins. However, the membrane-buried ion-binding residue (Glu62 in Atp$E_{2/3}$; Glu79 in hairpin one of AtpE_1) is substituted by a glutamine residue in hairpin two. Because proteolipids with four transmembrane helices but only one ion-binding site occur in only V_1V_O ATPases from eukarya (Müller and Grüber, 2003), they are often referred to as "eukaryal" or "16-kDa" *c* subunits. The subunit *c* oligomer of the F_1F_O ATPase from *A. woodii* indeed comprises a mixture of "bacterial-like" 8- and "eukaryal-like" 16-kDa polypeptides, and is the first of its kind found in nature (Aufurth *et al.*, 2000). The stoichiometry of the different polypeptides in the *c*-oligomer has not yet been determined.

What selective pressure could account for the multiplication of subunit *c*-encoding genes? One must note that the subunits of ATPase are present in different stoichiometries ($a_1b_2c_{9-14}\delta\alpha_3\gamma\beta_3\varepsilon$) and that subunit *c* has by far the highest copy number (9-14 copies of the "bacterial" 8 kDa *c* subunit, depending on the species) in the complex. Most information on the regulation of *c* subunit synthesis is derived from the *E. coli* paradigm. In *E. coli*, the subunit *c*-encoding gene is part of a polycistronic message, and enhanced synthesis of subunit *c* is achieved by enhancement of translation (McCarthy *et al.*, 1985). In addition, but to a lesser extent, regulation by differential mRNA stability can contribute to differential gene expression (McCarthy *et al.*, 1991). Apparently, multiplication of the *atpE* gene and embedding the copies into the operon is another way to increase the concentration of subunit *c*. Although this strategy is apparently applicable to *A. woodii*, additional mechanisms cannot be excluded.

The duplication of the proteolipid encoding gene *per se* has no obvious importance to the function of ATPase, since the products of genetically engineered duplicated proteolipid genes from *E. coli* are fully functional in transporting protons and synthesizing ATP (Jones and Fillingame, 1998). The striking feature, however, of subunit c_1 of *A. woodii* is not its size but rather the fact that the ion-translocating residue is not conserved in helix two. A loss of one ion-translocating residue is also encountered in the eukaryal "16-kDa *c* subunits" from V_1V_O ATPases and this loss has the dramatic consequence that the V_1V_O ATPases are not able to synthesize ATP *in vivo* (Müller and Grüber, 2003).

The synthesis of ATP by the ATPase is directly dependent on the number of ions translocated per ATP synthesized. Thermodynamic predictions based on the equation $\Delta G_p = -nF\Delta p$ indicate that a phosphorylation potential (ΔG_p) of 50 to 70 kJ/mol is sustained by the use of 3 - 4 ions (n)/ATP at a physiological electrochemical ion potential of -180 mV (Δp) (F denotes the Faraday constant). However, if the number of ions is lower than this value, ATP can no longer be synthesized. Although the number of monomers in the *c*-ring may vary from 9 to 14 (Stock *et al.*, 1999; Seelert *et al.*, 2000; Stahlberg *et al.*, 2001), it is assumed that 24 transmembrane helices are present per oligomer for the following calculation. Three ATP-synthesizing or hydrolyzing centers would give a stoichiometry of 4 ions per ATP synthesized if 12 copies of subunit $c_{2/3}$ with one ion-binding site each constitute the *c*-ring. In contrast, six copies of the "16-kDa subunit *c*" with four transmembrane helices constitute the *c*-ring of V_1V_O ATPases. Since the ion-translocating group is lost in the first pair of transmembrane helices, the stoichiometry is only 2 ions per ATP, which is too low for synthesis of ATP. On the other hand, if the number of ions is low, the same phosphorylation potential can account for a much higher electrochemical ion potential, making the enzyme a more thermodynamically efficient proton pump. In general, a lower number of carboxylates per *c*-ring yields a less efficient coupling.

Taking the above considerations into account, it is reasonable to assume that an organism could, depending on its cellular needs, alter the function of the ATPase between ATP synthesis and ATP hydrolysis by varying the number of ion-translocating residues. This possibility

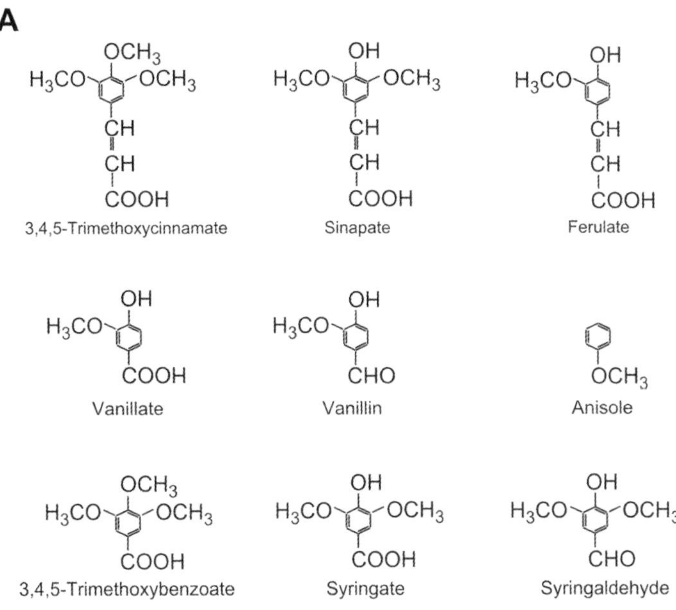

Figure 8. Aromatic substrates that are utilized by acetogens. Panel A: Structures of representative phenolic compounds. Panel B: Reduction of caffeate to hydrocaffeate as carried out by *A. woodii*. Reduction of the acrylate group is coupled to the efflux of Na^+ from the cytoplasm into the medium.

would be very attractive for a substrate-dependent regulation of the function of the ATPase of *A. woodii*. The F_1F_O ATPase could work as an ATP-dependent ion pump during fermentation and generate the transmembrane potential; in contrast, it could work as an ATP synthase during autotrophic growth on $H_2 + CO_2$. The switch from pump to synthase could be performed by changing the $c_1/c_{2/3}$ ratio. Testing this hypothesis is a challenging task for future experiments.

5. Novel Electron Donors and Acceptors: Enzymology and Regulation

The first isolated acetogens were characterized as H_2-utilizing autotrophs or hexose-utilizing heterotrophs (Section 1.1), and, for the most part, the broad substrate range of acetogens went unnoticed for many decades. Studies during the last two decades have revealed a number of different electron donors that are used by acetogens. These substrates include, but are not restricted to, one-carbon compounds (e.g., CO, formate, methyl groups), alcohols (e.g., methanol, acetoin, ethanol, glycerol), aldehydes (e.g., glyoxylate, benzaldehyde), a large variety of hexoses, pentoses, and short-chain carbohydrates (e.g., cellobiose), carboxylates (e.g., pyruvate, lactate), and even dicarboxylates like oxalate and fumarate (Drake *et al.*, 2004). Furthermore, CO_2 is not the only terminal electron acceptor that can be used by acetogens; several alternative energy-conserving electron acceptors can also be utilized. Combinations thereof will lead to situations where acetate is no longer formed. For example, a growth-supportive, non-acetogenic redox reaction that is carried out by some acetogens is the oxidation of a methyl group that is coupled to the reduction of nitrate (Section 5.2).

5.1. Novel Electron Donors

5.1.1. Utilization of Methyl-Level Substrates

Given the function of the methyl branch of the acetyl-CoA pathway, it is not surprising that methyl-containing substrates such as methanol, methyl chloride, and methoxylated aromatic compounds (Figure 8A) are utilized by acetogens as both electron donors and preformed methyl groups. Methyl groups are metabolized in a manner that simulates the acetogenic disproportionation of methanol:

$$1\ CH_3OH + H_2O \rightarrow 1\ CO_2 + 6\ H \quad (12)$$

$$3\ CH_3OH + 3\ CO_2 + 6\ H \rightarrow 3\ CH_3COOH + 3\ H_2O \quad (13)$$

Sum of 12 and 13:

$$4\ CH_3OH + 2\ CO_2 \rightarrow 3\ CH_3COOH + 2\ H_2O \quad (14)$$

One methyl group is oxidized to CO_2 via a reversal of the acetyl-CoA pathway. This oxidation generates six reducing equivalents that are used for the reduction of three molecules of CO_2 to the carbonyl (i.e., CO) level. Condensation of the three enzyme-associated, carbonyl-level molecules with three methyl groups yields three acetates. It should be noted that acetogenesis from methyl groups requires exogenous CO_2 (Reaction 14).

The exact route by which acetogens metabolize methyl groups is unresolved. Methyl groups might be converted directly to methyl-H_4F (Figure 9A). An enzyme-bound methylated corrinoid might also be involved as an intermediate. Alternatively, the methyl group could be oxidized to free formaldehyde. Formaldehyde could then be oxidized to CO_2 via formate, or could react spontaneously in a chemical reaction with H_4F and yield methylene-H_4F, which would then be oxidized to CO_2 via a reversal of the methyl branch of

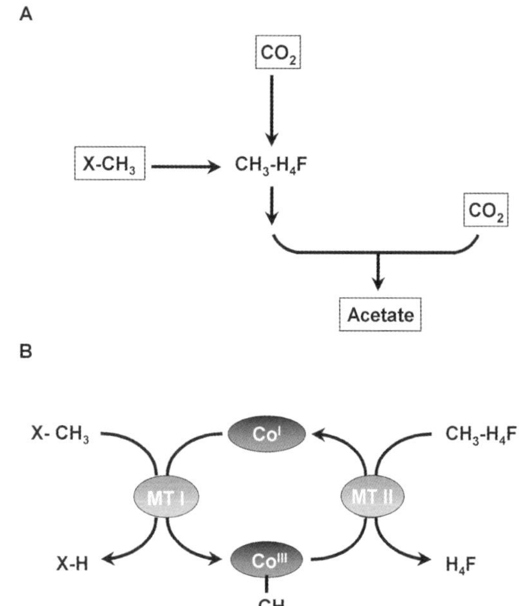

Figure 9. Methyltransferase reactions in acetogens. Panel A: Entrance of a preformed methyl group from substrates like methanol or methylchloride into the acetyl-CoA pathway (shown in an abbreviated form). Panel B: Enzymological mechanism by which a preformed methyl group is transferred to tetrahydrofolate (H_4F) (see text). Abbreviations: MT I, methyltransferase I; MT II, methyltransferase II; Co, corrinoid protein; X-CH_3, a hypothetical, bound intermediate form of a methyl group.

the acetyl-CoA pathway (Figure 2). For the vanillate O-demethylase system from *M. thermoacetica* (Naidu and Ragsdale, 2001) and *Acetobacterium dehalogenans* (Kaufmann *et al.*, 1997), as well as the O-demethylase system for methoxylated aromatic compounds from *Holophaga foetida* (Kreft and Schink, 1994), H_4F is methylated by one component of the O-demethylase system.

Methyl transfer appears to be more complicated in the Na^+-dependent *A. woodii*. Current evidence suggests that a methyltransferase of this acetogen is membrane bound and transfers the methyl group of methyl-H_4F to the acetyl-CoA synthase complex. The methyltransferase translocates Na^+ out of the cell concomitant with methyl transfer, thereby generating a Na^+ motive force across the cytoplasmic membrane (Section 4). Thus, such a methyltransferase would be fundamental to both carbon flow and the conservation of energy.

One might predict that methanol-dependent acetogenesis by *A. woodii* would be Na^+ dependent if the methyl group was first transferred to H_4F and then further metabolized by a membranous methyltransferase. However, this is not always the case. The growth of *A. woodii* is independent of Na^+ when both methanol (a preformed methyl group, thus by-passing the need to reduce CO_2 to the methyl level on the carbonyl branch of the acetyl-CoA pathway) and CO (a preformed carbonyl group, thus by-passing the need to reduce CO_2 to CO on the carbonyl branch of the acetyl-CoA pathway) are supplied as growth substrates (Heise *et al.*, 1989). It thus appears that alternative routes exist for channeling methyl groups into the acetyl-CoA pathway (e.g., via direct transfer of methyl groups to the corrinoid protein of acetyl-CoA synthase). All acetogens tested so far have significantly lower growth yields with methanol than with methoxylated aromatic compounds (Tschech and Pfennig, 1984; Schuppert and Schink, 1990; Daniel *et al.*, 1991), a result that is consistent with the direct transfer of the methyl group of methanol into the corrinoid pool.

The mechanism of the methyl transfer reaction has been analyzed in detail with different O-demethylase systems. In *A. dehalogenans*, methyltransferase I demethylates the substrate and transfers the methyl group to a corrinoid protein. The methyl group is then transferred to H_4F by methyltransferase II (Kaufmann *et al.*, 1998) (Figure 9B). The corrinoid cofactor contains Co(I), a supernucleophile able to abstract the methyl group from methanol. Spontaneous oxidation of Co(I) to Co(II) renders the enzyme inactive, and a fourth protein in the enzyme complex reductively reestablishes the Co(I) status of the corrinoid.

Methyltransferases with similar subunit composition catalyze the demethylation of veratrol by *A. dehalogenans* (Engelmann *et al.*, 2001) and vanillate by *M. thermoacetica* (Naidu and Ragsdale, 2001). Methyltransferases with similar catalytic mechanisms are also present in methanogenic archaea where they catylaze the transfer of methyl groups to coenzyme M (van der Meijden *et al.*, 1983; Ferguson *et al.*, 1996; Harms and Thauer 1996; Sauer and Thauer, 1997; Sauer *et al.*, 1997; Sauer *et al.*, 2000; Tallant *et al.*, 2001).

The demethylation reactions in *H. foetida* involve additional, unique processes. Methyl substituent groups of methoxylated aromatic compounds can be transferred to H_4F by a system similar to that described above or, alternatively, be used to methylate sulfide to methanethiol and dimethylsulfide (Bak *et al.*, 1992; Kreft and Schink, 1993; Liesack *et al.*, 1994; Kreft and Schink, 1994). After *O*-demethylation, the aromatic moiety is degraded to acetate via the phloroglucinol pathway (Kreft and Schink, 1993). It is unclear whether the methylation of sulfide is a metabolic or co-metabolic process. The ability of *H. foetida* to metabolize aromatic rings to acetate is not shared with other known acetogens, but highlights the diverse abilities of this bacteriological group. The capacity to *O*-demethylate methoxylated aromatic compounds is a very common feature of numerous acetogens and is likely of ecological importance relatively to the overall *in situ* biodegradation of phenylmethylethers, i.e., numerous degradation products of lignin. The dechlorination of methyl chloride by *A. dehalogenans* is another process of ecological relevance (Traunecker *et al.*, 1991). Little information is available about this process, but the mechanism appears

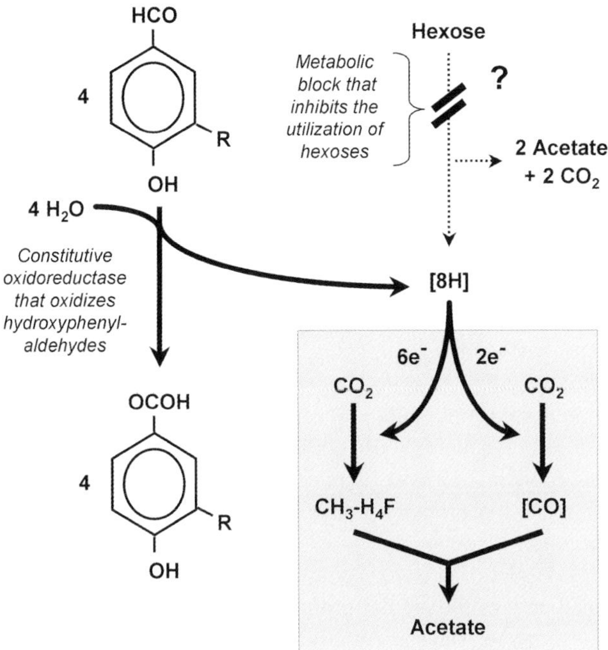

Figure 10. Effect of hydroxyphenylaldehydes on the flow of reductant to the acetyl-CoA pathway (shaded box) in *C. formicoaceticum*. Abbreviations: H_4F, tetrahydrofolate; R, a substituent group; H, reducing equivalent; e$^-$, electron. (Modified from Drake and Küsel, 2003.)

to be similar to that used for methoxylated compounds and results in the methylation of H_4F (Diekert and Wohlfahrt, 1994).

O-demethylating and methyltransferase activities are not constitutive, but are induced by their methyl-containing substrates (Wu *et al.*, 1988; Daniel *et al.*, 1991; Häggblom *et al.*, 1993; Misoph *et al.*, 1996; Kreft and Schink, 1997; Kaufmann *et al.*, 1997; Naidu and Ragsdale, 2001; Engelmann *et al.*, 2001). Few data are available on the organization of the genes encoding the methyltransferase systems. A 723 bp fragment encoding a part of the corrinoid protein and methyltransferase I of the vanillate *O*-demethylase system from *A. dehalogenans* was cloned, and the clustering of the two genes indicate that the genes coding for the methyltransferase system are organized in an operon (Kaufmann *et al.*, 1998b). An operon structure would facilitate the efficient regulation of the expression of these genes.

The existence of different demethylase systems in the same organism indicates that different demethylases have different substrate specificities. Unfortunately, the lack of genomic information makes it difficult to ascertain how many different methyltransferase systems might exist in one acetogen. Nonetheless, an acetogen might harbor broad demethylating potentials, as is exemplified by the ability of *M. thermoacetica* to grow on and *O*-demethylate at least 20 methoxylated aromatic compounds (Daniel *et al.*, 1991); this capacity appears to be due to one or a few broad specificity *O*-demethylase systems rather than numerous separate ones. Although it is obvious that efficient regulatory mechanisms must be present for these systems, information on how they are regulated at the gene level is unavailable.

5.1.2. Additional Oxidative Processes

It is beyond the scope of this article to evaluate all of the diverse substrates that different acetogens can oxidize and thereby use as a source of reductant and energy. For the most part, information on these processes at the molecular level does not exist. Nonetheless, a few of these substrates and their physiological importance are noteworthy. Some of these processes appear to have features that are unique to acetogens.

Relatively few known anaerobes metabolize oxalate. *M. thermoacetica* is arguably the most metabolically robust acetogen thus far isolated (Drake and Daniel, 2004) and readily oxidizes oxalate, and also glyoxylate and glycolate, with the concomitant synthesis of acetate via the acetyl-CoA pathway (Daniel and Drake, 1993; Drake and Daniel, 2004; Seifritz *et al.*,1999; 2003; Daniel *et al.*, 2004). The biochemical mechanisms by which these two-carbon molecules are catabolized are unknown; evidence to date suggests that the pathway used for the oxidation of oxalate does not involve nicotinamide adenine dinucleotide (NAD^+), nicotinamide adenine dinucleotide phosphate ($NADP^+$), or CoA-level intermediates. Several species of the acetogenic genus *Acetobacterium* also harbor very robust metabolic capabilities. For example, *Acetobacterium carbinolicum* (Eichler and Schink, 1984; Schuppert and Schink, 1990) and *Acetobacterium malicum* (Tanaka and Pfennig, 1988), both isolated from freshwater sediments, metabolize a large variety of substrates, including betaine, glycol ethers, and various alcoxyethanols.

Clostridium formicoaceticum can utilize hydroxyphenylaldehydes (e.g., 4-hydroxybenzaldehyde) as a sole source of reductant for the synthesis of acetate (Frank *et al.*, 1998; Drake and Küsel, 2003). The aldehyde group is oxidized to the carboxyl level, and the reducing equivalents that are generated by this oxidation are used preferentially in the reduction of CO_2 to acetate (Figure 10). Thus, the use of hydroxyphenylaldehydes as a source of reductant has a profound effect on the catabolism of other substrates, including those normally conceived to be ideal acetogenic substrates (e.g., hexoses). The mechanism for this hydroxyphenylaldehyde-dependent effect on the catabolism of other substrates has not been resolved; inhibition could occur at the level of transport or at another catabolic event in glycolysis. Nonetheless, this finding is unusual in that (a) a hexose is not catabolized preferentially to the substituent group of an aromatic compound, and (b) it demonstrates that acetogens can regulate reductant flow at the level of substrate utilization. Likewise, the ability to grow at the expense of hydroxyphenylaldehydes reinforces the importance of chemiosmotic processes that acetogens use to conserve energy, because the engagement of a chemiosmotic process is obligatory for hydroxyphenylaldehyde-dependent growth (i.e., as noted in Section 2.1, the reduction of CO_2 to acetate does not yield any net ATP_{SLP}).

The use of polymers such as cellulose is not a common feature of known acetogens. However, the acetogen *B. formatexigens* was isolated on amorphous cellulose and carboxymethylcellulose, and initially had the capacity to degrade these cellulose polymers (Wolin *et al.*, 2003). Unfortunately, *B. formatexigens* lost its ability to use these polymers after prolonged cultivation in the laboratory. As with other bacteria, acetogens can lose some of their catabolic abilities after prolonged cultivation; indeed, they can even lose their ability to engage the acetyl-CoA pathway. Another recent isolate that is phylogenically related to *M. thermoacetica* is cellulolytic (Karita *et al.*, 2003). This organism has not been well characterized yet, and its acetogenic nature is not fully known. However, these observations argue for the existence of cellulolytic acetogens.

5.2. Alternative Terminal Electron Acceptors

5.2.1. Reductive Processes

Many acetogens can use alternative, energy-conserving terminal electron acceptors for the recycling of reduced intracellular electron carriers; thus, acetogens are not always strictly dependent on the acetyl-CoA pathway, i.e., the use of CO_2 as a terminal electron acceptor. The alternative terminal electron acceptors that can be used by acetogens are diverse and include aromatic acrylate groups, fumarate, nitrate, nitrite, thiosulfate, and dimethylsulfoxide.

M. thermoacetica and *M. thermoautotrophica* preferentially use nitrate as a terminal electron acceptor, which is reduced to nitrite and ammonium (Seifritz *et al.*, 1993; Fröstl *et al.*, 1996). Cells preferentially use nitrate even in the presence of CO_2 during growth on numerous substrates, including vanillin, vanillate, methanol, glucose, CO, and H_2. Analogous observations have been made when cells, growing on glyoxylate, were supplemented with nitrite (Seifritz *et al.*, 2003), thus demonstrating that the dissimilation of nitrite can also be growth supportive. *M. thermoautotrophica* and *M. thermoacetica* are also able to use sulfur compounds (e.g., thiosulfate and dimethylsulfoxide) as terminal electron acceptors (Beaty and Ljungdahl, 1990). Although little is known of these processes, electrons derived from substrates like ethanol, butanol, or propanol can be transferred to certain inorganic sulfur compounds when these acetogens are faced with a limitation for CO_2.

C. formicoaceticum grows with fumarate as the sole carbon and energy source. Fumarate is disporportionated as follows:

$$\text{Fumarate} + 2\ H_2O \rightarrow \text{Acetate} + 2\ CO_2 + 4\ H \quad (15)$$

$$2\ \text{Fumarate} + 4\ H \rightarrow 2\ \text{Succinate} \quad (16)$$

Sum of 15 and 16:

$$3\ \text{Fumarate} + 2\ H_2O \rightarrow 2\ \text{Succinate} + \text{Acetate} + 2\ CO_2 \quad (17)$$

The reducing equivalents that are generated during the oxidation of fumarate to acetate and CO_2 are not transferred to another acceptor, but are instead used for the reduction of additional fumarate (Dorn *et al.*, 1978a; 1978b). Under CO_2-limiting conditions, *C. formicoaceticum* is able to use fumarate as an electron acceptor independently of dismutation (Matthies *et al.*, 1993). Electrons generated from the oxidation of methanol or the methyl group of vanillate are transferred to fumarate. The same holds true for *C. aceticum* that can transfer H_2-derived electrons to fumarate under CO_2-limiting conditions.

A very significant discovery was made when it was observed that compounds derived from the degradation of lignin were used as terminal electron acceptors by *A. woodii* (Bache *et al.*, 1981). This capability occurs in several acetogens and is exemplified by the ability of *A. woodii* to reduce the double bond of various phenylacrylates [e.g., 3,4,5-trimethoxycinnamate (Figure 8A)]. A phenylacrylate rather than CO_2 is the preferred electron acceptor when cells are grown on methanol or formate (Tschech and Pfennig, 1984). When methyl groups are oxidized in the presence of caffeate, hydrocaffeate is produced and acetate is not formed. *R. productus* is another acetogen that conserves energy during the reduction of phenylacrylates (Misoph *et al.*, 1996). Under certain conditions when reductant is derived from methoxyl groups, this redox coupling can be utilized as the sole energy-conserving process. *R. productus* also engages a wide range of fermentation processes and can produce succinate, lactate, ethanol, and formate as reduced end products from the catabolism of fructose, xylose, pyruvate, and glycerol (Misoph and Drake, 1996).

5.2.2. Regulatory Effects

The regulatory mechanisms that control the channeling of substrate-derived electrons away from the acetyl-CoA pathway and towards an alternative terminal electron acceptor remain obscure. Consistent with the observation that nitrate-supplemented cultures of *M. thermoacetica* do not synthesize acetate (Seifritz *et al.*, 1993), it has been shown that (a) the capacity to use nitrate as a terminal electron acceptor is not constitutive and (b) nitrate represses the acetyl-CoA pathway. A first hint for the basis of this nitrate-dependent regulation was the observation that membranes of *M. thermoautotrophica* and *M. thermoacetica* grown in the presence of nitrate or nitrite do not contain cytochrome b, a crucial membranous electron carrier in the reduction of methylene- to methyl-H_4F on the methyl branch of the acetyl-CoA pathway (Fröstl *et al.*, 1996; Seifritz *et al.*, 2003).

There are conflicting reports on the effect of nitrate on the synthesis and activities of the enzymes of the acetyl-CoA pathway. Fröstl *et al.* (1996) found that nitrate-dissimilating cells contained relatively normal levels (i.e., specific activities) of the central enzymes of the pathway. In contrast, Arendsen *et al.* (1999) observed that nitrate decreased the activities of some enzymes (e.g., acetyl-CoA synthase and methyltransferase). Northern blot analyses revealed that this regulation took place at the level of transcription. At the same time, genes coding for components of nitrate respiration were activated. The nitrate-dependent block on the formation of acetyl-CoA is overcome with preformed methyl and carbonyl groups (Fröstl *et al.*, 1996), demonstrating that (a) the catalytic ability to synthesize acetyl-CoA (assumed to be at the level of acetyl-CoA synthase) is retained during nitrate dissimilation (as this ability is required for anabolism under basal conditions with methyl- and carbonyl-level precursors as the sole source of carbon), and (b) control of reductant flow is the main reason why the catabolic function of the acetyl-CoA pathway is repressed when nitrate is dissimilated (Drake *et al.*, 1997; Drake and Küsel, 2003; Drake *et al.*, 2004). In this regard, the activity of hydrogenase, an oxidoreductase that might have an essential function in the flow of reductant during heterotophic growth, is significantly inhibited/repressed in nitrate-dissimilating cells (Fröstl *et al.*, 1996).

The dismutation of fumarate, and also fumarate respiration, are regulated by CO_2 in *C. formicoaceticum* (Matthies *et al.*, 1993). As outlined above, *C. formicoaceticum* can use fumarate as a terminal electron acceptor when growing on methanol or vanillate. Substrate-derived electrons are preferentially used for acetogenesis when both CO_2 and fumarate are available. H_2-dependent fumarate respiration is inhibited by CO_2 in *C. aceticum*, acetogenesis is the preferred electron-accepting process when CO_2 is readily available (Matthies *et al.*, 1993).

The utilization of caffeate by *A. woodii* is not inhibited by CO_2 (Tschech and Pfennig, 1984). Electrons from substrates like methanol or formate are preferentially transferred to caffeate even in the presence of CO_2. In contrast, when *R. productus* is provided with both ferulate and CO_2 as terminal electron acceptors, the reduction of both acceptors occur simultaneously (Misoph *et al.*,1996). An acrylate oxidoreductase is induced by a utilizable phenylacrylate, and reduction of the phenylacrylate becomes the primary energy-conserving process when CO_2 becomes limiting. The capacity of *A. woodii* to use caffeate as a terminal electron acceptor is also not constitutive, but induced by caffeate (Imkamp and Müller, unpublished data). Although little information on the reduction of phenylacrylates by actogens is available, these findings indicate that the process is regulated at the gene-level; in certain cases, the availability of CO_2 might be important to this regulation. CO_2 also influences the ability of *R. productus* to engage alternative fermentation processes, such as the ethanolic fermentation of glycerol (Misoph and Drake, 1996).

5.3. Coupling the Reduction of Alternative Electron Acceptors to the Conservation of Energy

The reduction of CO_2 to acetate via the acetyl-CoA pathway is coupled to H^+- and Na^+-dependent chemiosmotic processes that conserve energy (Section 4). The use of an alternative terminal electron acceptor not only serves as an electron sink but is also coupled to the conservation of energy. However, unlike the detailed information on how energy is conserved when CO_2 is reduced to acetate, very little is understood about the physiological and molecular processes by which acetogens conserve energy when alternative electron acceptors are used.

Growth experiments with *M. thermoacetica* and *M. thermoautotrophica* demonstrated that the reduction of nitrate to nitrite and ammonium is coupled to an energy conserving mechanism (Seifritz *et al.*, 1993; Fröstl *et al.*, 1996; Seifritz *et al.*, 2003). Cultures grown in the presence of nitrate produced several times more biomass than cultures that were not supplemented with nitrate, and certain alcohols that were not utilizable for acetogenesis were oxidized and growth supportive during the dissimilation of nitrate. That the oxidation of H_2 and CO can be coupled to the conservation of energy when nitrate is reduced strongly indicates that the reduction of nitrate can be coupled to a chemiosmotic process (i.e., there is no known process by which ATP_{SLP} could be formed under such conditions). The simplest case would be the positioning of an oxidoreductase (e.g., hydrogenase) on the exterior side of the membrane that releases protons outwardly while delivering electrons towards nitrate via associated electron carriers (the redox coupling to nitrate being either a membrane-associated or cytoplasmic process). Such a redox couple has precedence in the nitrate-dependent respiratory pathway of *E. coli* (Richardson *et al.*, 2001; Simon, 2002; Jormakka *et al.*, 2003). This possibility is also supported by the occurrence of membranous electron carriers and a proton-dependent ATPase in *M. thermoacetica* and *M. thermoautotrophica* (Section 4.1).

Fumarate supports the growth of *C. formicaceticum* under basal conditions, i.e., conditions that demand the fumarate-dependent conservation of energy (Matthies *et al.*, 1993). A fumarate reductase is located at the outer side of the cytoplasmatic membrane (Dorn *et al.*, 1978b), and *b*-type cytochromes and menaquinone occur in this acetogen as well (Gottwald *et al.*, 1975). These findings suggest that *C. formicaceticum* utilizes an energy-conserving mechanism that is similar to those described for H_2-dependent fumarate respiration in *Wolinella succinogenes* and *E. coli* (Cecchini *et al.*, 2002; Kröger *et al.*, 2002).

The cell yields of *A. woodii* increase during the reduction of phenylacrylates (Tschech and Pfennig, 1984). Thus, utilization of this unique half-cell redox reaction must be coupled to the conservation of energy. Resting cells of *A. woodii* couple the H_2-dependent reduction of caffeate to the synthesis of ATP (Hansen *et al.*, 1988), and this synthesis of ATP occurs by a Na^+-dependent chemiosmotic mechanism (Figure 8B) (Imkamp and Müller, 2002). This dependence on Na^+ is reminiscent of the requirement of this ion during the reduction of CO_2 to acetate. The reduction of caffeate is accompanied by the generation of a transmembrane electrochemical Na^+ gradient that drives the synthesis of ATP via Na^+-dependent F_1F_O ATPase (Imkamp and Müller, 2002). However, the mechanism by which Na^+ is extruded during the caffeate-dependent respiration has not been elucidated.

6. *In Situ* Processes and Environmental Impact

The ability of acetogens to use a wide range of electron donors and electron acceptors, and to withstand transient oxidative conditions, indicate that they have numerous trophic links in diverse habitats. However, the magnitudes of the *in situ* activities of this bacterial group remain mostly unresolved. Three reasons why it is difficult to assess acetogens under *in situ* conditions are: (a) the end product of acetogenesis, acetate, is difficult to measure in complex habitats and is subject to rapid consumption by other organisms; (b) their ability to utilize a wide range of electron donors and acceptors indicates that they have functional diversity and are not strictly dependent upon acetogenesis; and (c) they do not form an exclusive phylogenic group, thus making it impossible to develop acetogen-specific molecular probes that are based on 16S rRNA gene sequences. Despite these limitations in evaluating the *in situ* activities of acetogens, acetogens are ubiquitous and can have robust ecological effects in their habitats. Likewise, acetyl-CoA synthase and the acetyl-CoA pathway have far reaching biological significance.

6.1. Functional Diversity and Evolution of the Acetyl-CoA Pathway

Metabolic pathways that are biochemically similar to the acetyl-CoA pathway are utilized by non-acetogenic prokaryotes (e.g., sulfate-reducing bacteria and methanogens) for the autotrophic fixation of CO_2 or the oxidation of acetate (Fuchs, 1990; 1994; Ferry, 1994; Whitman, 1994). Although these metabolic processes are not identical and make use of dissimilar co-factors and enzymes, the general biochemical features of these different forms of the acetyl-CoA pathway have considerable overlap. Thus, acetyl-CoA synthase-related processes serve more that one function and are widely distributed in nature. Indeed, the acetyl-CoA pathway appears to be the dominant process by which acetate is anaerobically oxidized (Fuchs, 1990).

Life on earth originated under anoxic conditions, and the first autotrophs may have been ancestors of methanogens (Schopf *et al.*, 1983; Brock, 1989). Given the occurrence of acetyl-CoA synthase in methanogens, an acetyl-CoA synthase-dependent pathway may have constituted the first autotrophic process (Fuchs, 1986; Wood and Ljungdahl, 1991; Lindahl and Chang, 2001). The acetyl-CoA pathway is a linear process (i.e., no intermediates are recycled) and arguably more simple than the Calvin cycle, the reductive tricarboxylic acid cycle, and the hydroxypropionate cycle. For example, the Calvin cycle requires more energy to fix a molecule of CO_2 than does the acetyl-CoA pathway. The acetyl-CoA pathway is multi-functional, and it is uncertain if the pathway first evolved for catabolism (i.e., the oxidation of acetate) or anabolism (i.e., the assimilation of CO_2). Although the evolutionary linkage is uncertain, microorganisms that have acetyl-CoA synthase-related enzymes appear to have had a common ancestor (Lindahl and Chang, 2001).

6.2. Competitiveness and Vigor of Acetogens

In anoxic habitats that lack inorganic terminal electron acceptors (e.g., sulfate and nitrate), acetogens must compete with fermentative microorganisms for compounds that are derived from the initial breakdown of cellulose and lignin (McInerney and Bryant, 1981). Two reasons why acetogens might be considered ecologically noncompetitive are (a) the CO_2/acetate half-cell reaction (-290 mV) is one of the most negative terminal electron-accepting reactions in biology (Diekert and Wohlfarth, 1994a; 1994b), and (b) acetogens have low affinity for H_2 when CO_2 is utilized as a terminal electron acceptor (Cord-Ruwisch, *et al.*, 1988). Despite the thermodynamic limitations of acetogenesis, acetogens can be quite competitive. For example, acetogenesis appears to be the most competitive anaerobic glucose-consuming process in anoxic paddy soils and certain fresh water sediments (Krumböck and Conrad, 1991). Although it is not possible to determine how much carbon is processed globally via acetogens and metabolic pathways that make use of acetyl-CoA synthase, numerous observations indicate that acetogens and processes associated with the acetyl-CoA pathway have enormous impact in various

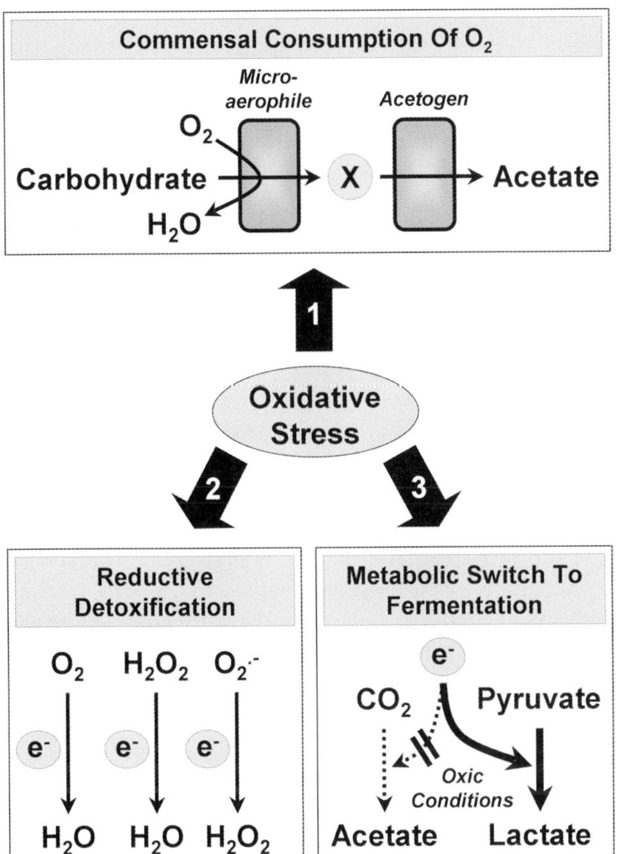

Figure 11. Three mechanisms by which acetogens cope with oxidative stress. Abbreviations: X, products (e.g., H_2, formate, lactate) that are derived from the partial oxidation of carbohydrates; e^-, electron.

habitats. Indeed, although most acetogens have been obtained from habitats that are anoxic (e.g., sewage and gastrointestinal tracts), they can be easily isolated from very diverse environments, including oxic soils, plant roots, hypersaline and alkaline habitats, aquifers and deep subsurface materials, and marine sediments.

Many factors enhance the survival and *in situ* effectiveness of acetogens. Numerous acetogens form spores, and the ability to sporulate is likely an important factor to the competitiveness of acetogens. The durability of the spores of acetogens is exemplified in the revival of *C. aceticum* spores after decades of dessication (Braun *et al.*, 1981). Spores of *M. thermoacetica* are among the most heat resistant spores known and can have decimal reduction times (i.e., the time required to achieve a 90% reduction in the number of viable spores) of almost 2 hours at 121° C (Byrer *et al.*, 2000).

Acetogens compete with methanogens for H_2. The simultaneous utilization of multiple substrates can increase the competitiveness of acetogens for H_2 and other substrates (Breznak *et al.*, 1991; Liu and Suflita, 1993; Peters *et al.*, 1998). Because methanogens are less competitive for H_2 under acidic and psychrophilic conditions, acetogens may become more competitive under certain *in situ* conditions (Phelps and Zeikus, 1984; Conrad *et al.*, 1989; Nozhevnikova *et al.*, 1994; Zavarzin *et al.*, 1994). Sodium-proton antiporters (Terracciano *et al.*, 1987) might enable acetogens to cope with acidic environments (Schink, 1994). Acetogens capable of growth under mildly acidic (Küsel *et al.*, 2000) and low temperature conditions (Conrad *et al.*, 1989; Kotsyurbenko *et al.*, 1995; 1996) have been isolated.

Acetogens appear to have a low affinity (i.e., are non-competitive) for H_2 during H_2-dependent acetogenesis (Cord-Ruwisch *et al.*, 1988). However, the close association of acetogens with H_2-producing microbes [e.g., protozoa (Leadbetter *et al.*, 1999)] can significantly increase the *in situ* success/effectiveness of H_2-dependent acetogenesis. The use of alternative electron acceptors can significantly increase the affinity of acetogens for H_2 (Cord-Ruwisch *et al.*, 1988), and can likewise increase H_2-dependent growth yields eight fold (Fröstl *et al.*, 1996). Alternative electron acceptors can also influence the oxidation of other substrates by acetogens. For example, ethanol and propanol are not acetogenic substrates for *M. thermoacetica*, but are rapidly utilized and growth supportive when nitrate serves as a terminal electron acceptor (Fröstl *et al.*, 1996).

Transient amounts of O_2 will theoretically influence the competitiveness of acetogens and other anaerobes in habitats that are subject to fluctuating oxic-anoxic conditions. Acetogens are less sensitive to oxic conditions than are methanogens, as is evidenced by the occurrence of acetogens and relative absence of methanogens in well-aerated soils and litter; indeed, acetogens appear to be a dominant group of anaerobes in such habitats (Küsel and Drake, 1994; 1995; 1996, 1999; Peters and Conrad, 1995; 1996; Wagner *et al.*, 1996; Küsel *et al.*, 1999b, 2002; Reith *et al.*, 2002).

Acetogens remain competitive in habitats that are subject to transient fluxes of O_2 by their (a) commensal association with O_2-consuming microorganisms, (b) ability to catabolize O_2 and its toxic by-products, and (c) ability to switch to classic fermentative metabolism (e.g., ethanol or lactate fermentation) when conditions become too oxic (Figure 11). The commensal interaction of acetogens with a microaerophilic partner (Figure 11, Arrow 1) is exemplified by the commensal growth of *Thermicanus aegyptius* with *M. thermoacetica* (Gößner *et al.*, 1999). Numerous oxidative stress enzymes (i.e., enzymes that remove/metabolize O_2 or its toxic side products) or their genes have been detected in acetogens; these include NADH-oxidase, peroxidase, superoxide dismutase, catalase, flavoprotein/rubredoxin (functions as an NADH oxidase), rubredoxin oxidoreductase (functions as a superoxide reductase), and rubrerythrin (functions as a peroxide reductase) (Das *et al.*, 2001; Küsel *et al.*, 2001; Drake *et al.*, 2002; 2004; Karnholz *et al.*, 2002; Boga and Brune, 2003; Kurtz, 2003; Silaghi-Dumitrescu *et al.*, 2003). As depicted in Figure 11 (Arrow 2), the majority of these catalysts are reductive. The acetyl-CoA pathway contains numerous enzymes that are highly sensitive to O_2 (i.e., oxidation), and engagement of an alternative terminal electron-accepting process that operates at a higher redox potential than that of the acetyl-CoA pathway and is facilitated by enzymes that are less sensitive of oxic conditions can provide an acetogen with an excellent alternative to acetogenesis if conditions become too oxic (Figure 11, Arrow 3). This metabolic switch is exemplified in the ability of the acetogen *Clostridium glycolicum* RD-1 to use both lactate and ethanol fermentation under oxic conditions (Küsel *et al.*, 2001).

6.3. Acetogenesis in Diverse Habitats

6.3.1. Gastrointestinal Ecosystems

Mammalian gastrointestinal tracts are heavily colonized with acetogens (Prins and Lankhorst, 1977; Wolin and Miller, 1983; 1994; Breznak and Kane, 1990; Mackie and Bryant, 1994; Leedle *et al.*, 1995). ^{13}C- and ^{14}C-labelling studies have confirmed that the acetyl-CoA pathway is responsible for a part of the acetate formed in the human intestine, and it is estimated that 10^{10} kg of acetate are synthesized from H_2-CO_2 each year in the collective colons of the human population (Lajoie *et al.*, 1988; Wolin and Miller, 1994; Doré *et al.*, 1995; Bernalier *et al.*, 1996a; 1996b; Miller and Wolin, 1996; Wolin *et al.*, 1999). Microbially-produced acetate is absorbed across the intestinal wall and constitutes a source of carbon and energy for the mammalian host. Numerous H_2-utilizing acetogens have been isolated from human feces (Wolin and Miller, 1994; Bernalier *et al.*, 1996a; 1996b; 1996c; Kamlage *et al.*, 1997; Leclerc *et al.*, 1997a; 1997b; Wolin *et al.*, 2003).

Mammals that do not excrete methane have greater numbers of H_2-utilizing fecal acetogens than do methane-excreting mammals (Prins and Lankhorst, 1977; Doré *et al.*, 1995), thus suggesting that fecal acetogens become more significant in the absence of methanogens. However, methanogenesis is the primary terminal event in the gastrointestinal system of ruminants, yet cultured values of acetogens in the rumen can be as great as those of methanogens (Krumholz and Bryant, 1986; Leedle and Greening, 1988; Mackie and Bryant, 1994; Rieu-Lesme *et al.*, 1995; 1996a; 1996b,;1998). Dietary changes of ruminants can affect the relative distribution of ruminal acetogens and methanogens (Leedle and Greening, 1988). Although numerous ruminal acetogens have been isolated, the *in situ* importance of acetogens in ruminal ecosystems remains poorly resolved.

Acetogens are fundamental constituents of the termite gut (Breznak, 1994). 10^{12} kg of acetate are produced each year via the acetyl-CoA pathway of acetogens that colonize the hindgut of termites; although this microbial niche has a volume of only 1 μl, this amount of acetate is 5-fold greater than the global reduction of CO_2 to methane via methanogenesis (Breznak and Kane, 1990). Microbially-produced acetate, one-third of which is synthesized from the reduction of CO_2, is the primary source of carbon and energy of wood-feeding termites (Breznak, 1994). Cellulolytic protozoa convert wood polysaccharides (e.g., cellulose) to acetate, H_2, and CO_2, and acetogens subsequently convert H_2-CO_2 to acetate in the hindgut of the wood-feeding "lower" termites (Breznak and Switzer, 1986; Brauman *et al.*, 1992). The symbiotic association between H_2-producing protozoa and H_2-consuming acetogenic spirochetes [e.g., *Treponema primitia* (Graber *et al.*, 2004)] is enhanced by the direct attachment of the acetogen on the exterior of the protozoan (Leadbetter *et al.*, 1999). The spacial orientation of acetogens relative to *in situ* trophic/chemical gradients determine their effectiveness in the termite gut (Schmitt-Wagner and Brune, 1999; Tholen and Brune, 1999; 2000).

6.3.2. Terrestrial, Subsurface, and Aquatic Ecosystems

Acetate is a dominate organic compound in soil solution (Tani *et al.*, 1993), and numerous observations indicate that acetate is centrally important to the flux of carbon in soils. The formation of acetate in soils is theoretically maximized in anoxic microzones; in this regard, 25% of the organic carbon of soil can be converted to acetate under anoxic conditions, a value that approximates 40 g acetate per kg dry wt. soil (Küsel and Drake, 1994). The capacity to form acetate in soils is concomitant with acetogenic activities and the occurrence of both H_2- and CO-utilizing acetogens (Küsel and Drake, 1995; Wagner *et al.*, 1996; Kuhner *et al.*, 1997; Küsel *et al.*, 1999b). It has been estimated that 10^{12} kg of acetate is present in the first meter of the terrestrial surface at any one moment (Drake *et al.*, 2004). The occurrence and activity of acetogens in soils make it probable that they play a role in the production of this acetate. Indeed, the first acetogen isolated was obtained from soil [i.e., ditch mud (Wieringa, 1936)], and 10% of the 10^{13} kg of acetate that are formed and further metabolized annually in terrestrial soils and sediments are attributed to acetogens and the acetyl-CoA pathway (Wood and Ljungdahl, 1991). The terrestrial subsurface harbors 10-fold more prokaryotes than does the terrestrial surface (Whitman *et al.*, 1998), and occurrence of acetogens in the subsurface (Wellsbury *et al.*, 1997; 2002; Krumholz *et al.*, 1999) indicate that acetogenesis is an active process during the turnover of carbon in the subsurface biosphere.

Acetogens are easily isolated from water-saturated sediments, and acetogenesis can become an important process under certain conditions in such habitats. For example, when sulfate becomes depleted in marine sediments, the flow of carbon and reductant shifts towards methanogenesis, and acetogenesis (i.e., the reduction of CO_2 to acetate) can occur at rates equivalent to those of the peak rates of methanogenesis or sulfate reduction during this transitional period (Hoehler *et al.*, 1999). Although the reduction of sulfate is a dominant, anaerobic terminal process in marine habitats, H_2-utilizing acetogens occur in significant numbers in marine sediments (Küsel *et al.*, 1999a). Acetogens also occur in oceanic sediments (Wellsbury *et al.*, 1997, 2002), hypersaline habitats (Zavarzin *et al.*, 1994; Zhilina *et al.*, 1996; 1998), and subsurface aquifers (Kotelnikova and Pedersen, 1997; 1998; Krumholz *et al.*, 1999; Kotelnikova, 2002; Haveman and Pedersen, 2002), thus indicating that that the reductive synthesis of acetate from CO_2 occurs in diverse water-logged habitats.

Anaerobic bacteria can be important rhizosphere colonists (Großkopf *et al.*, 1998; Hines *et al.*, 1999). Seagrass rhizosphere is colonized by acetogens, and *O*-demethylatng acetogens are tightly associated with seagrass roots (Küsel *et al.*, 1999a). The hybridization of seagrass root thin sections with ^{33}P-labeled probes that are specific for acetogens indicate that acetogens can occur within plant tissues (Küsel *et al.*, 1999a). The detection of phylogenically diverse formyltetrahydrofolate synthetase sequences in salt marsh plant roots indicates that such roots are colonized by phylogenically diverse acetogens (Lovell, 1994; Leaphart *et al.*, 2001; 2003). Acetogenesis might also occur in the rhizophere of rice (Conrad *et al.*, 1989; Rosencrantz *et al.*, 1999).

The fluctuating gradient of O_2 that is generated around plant roots (Gilbert and Frenzel, 1995; Revsbech *et al.*, 1999) makes the occurrence of anaerobes in the rhizosphere somewhat surprising and indicates that rhizosphere anaerobes have mechanisms for coping with transient gradients of O_2. In this regard, acetogens isolated from the plant roots of different marine habitats display a high tolerance to O_2 (Küsel *et al.*, 2001; 2003).

7. Conclusions

Acetogens are a metabolically robust group of anaerobic bacteria, and the literature on the 21 genera of acetogens has reached enormous proportions. Although the mechanism by which acetogens reduce CO_2 to acetate has been resolved, our understanding of the molecular processes by which their diverse catabolic processes are regulated is very limited. Likewise, the abundance of acetogens in nature is well established, but their *in situ* activities are poorly understood. Resolving the molecular and regulatory aspects of the diverse metabolic capabilities of acetogens represents an important challenge of future studies. Resolving the magnitude of the diverse *in situ* trophic links of acetogens will be equally challenging and important.

Acknowledgements

The authors thank John Breznak and Steve Ragsdale for supplying unpublished information. Current support for the authors' laboratories is derived in part from funds from the German Research Society (DFG) to HD and VM, and the German Ministry of Education, Research and Technology (BMBF) to HD, and is gratefully acknowledged.

References

Albers, B.E., and Ferry, J.G. 1994. A carbonic anhydrase from the archaeon Methanosarcina thermophila. Proc. Natl. Acad. Sci. USA. 91: 6909-6913.

Andreesen, J.R., Gottschalk, G., and Schlegel, H.G. 1970. *Clostridium formicoaceticum* nov. spec., isolation, description and distinction from *C. aceticum* and *C. thermoaceticum*. Arch. Microbiol. 72: 154-174.

Arendsen, A.F., Soliman, M.Q., and Ragsdale, S.W. 1999. Nitrate-dependent regulation of acetate biosynthesis and nitrate respiration by *Clostridium thermoaceticum*. J. Bacteriol. 181: 1489-1495.

Aufurth, S., Schägger, H., and Müller, V. 2000. Identification of subunits *a*, *b*, and c_1 from *Acetobacterium woodii* Na^+-F_1F_O -ATPase. Subunits c_1, c_2, and c_3 constitute a mixed c-oligomer. J. Biol. Chem. 275: 33297-33301.

Bache, R., and Pfennig, N. 1981. Selective isolation of *Acetobacterium woodii* on methoxylated aromatic acids and determination of growth yields. Arch. Microbiol. 130: 255-261.

Bak, F., Finster, K., and Rothfuß, F. 1992. Formation of dimethylsulfide and methanethiol from methoxylated aromatic compounds and inorganic sulfide by newly isolated anaerobic bacteria. Arch. Microbiol. 157: 529-534.

Balch, W.E., Schoberth, S., Tanner, R.S., and Wolfe, R.S. 1977. *Acetobacterium*, a new genus of hydrogen-oxidizing, carbon dioxide-reducing, anaerobic bacteria. Int. J. Sys. Bacteriol. 27: 355-361.

Banerjee, R., and Ragsdale, S.W. 2003. The many faces of vitamin B_{12}: catalysis by cobalamin-dependent enzymes. Annu. Rev. Biochem. 72: 209-247.

Barker, H.A. 1944. On the role of carbon dioxide in the metabolism of *Clostridium thermoaceticum*. Proc. Nat. Acad. Sci. USA. 30: 88-90.

Barker, H.A., and Kamen, M.D. 1945. Carbon dioxide utilization in the synthesis of acetic acid by *Clostridium thermoaceticum*. Proc. Natl. Acad. Sci. USA. 31: 219-225.

Barondeau, D.P., and Lindahl, P.A. 1997. Methylation of carbon monoxide dehydrogenase from *Clostridium thermoaceticum* and mechanism of acetyl coenzyme A synthesis. J. Am. Chem. Soc. 119: 3959-3970.

Baronofsky, J.J., Schreurs, W.J.A., and Kashket, E.R. 1984. Uncoupling by acetic acid limits growth of and acetogenesis by *Clostridium thermoaceticum*. Appl. Environ. Microbiol. 48: 1134-1139.

Beaty, P.S., and Ljungdahl, L.G. 1990. Thiosulfate reduction by *Clostridium thermoaceticum* and *Clostridium thermoautotrophicum* during growth on methanol. Abstr. Ann. Meet. Am. Soc. for Microbiol. Abstract I-7, p. 199.

Bernalier, A., Lelait, M., Rochet, V., Grivet, J.-P., Gibson, G.R., and Durand, M. 1996a. Acetogenesis from H_2 and CO_2 by methane- and non-methane-producing human colonic bacterial communities. FEMS Microbiol. Ecol. 19: 193-202.

Bernalier, A., Rochet, V., Leclerc, M., Doré, J., and Pochart, P. 1996b. Diversity of H_2/CO_2-utilizing acetogenic bacteria from feces of non-methane-producing humans. Curr. Microbiol. 33: 94-99.

Bernalier, A., Willems, A., Leclerc, M., Rochet, V., and Collins, M.D. 1996c. *Ruminococcus hydrogenotrophicus* sp. nov., a new H_2/CO_2-utilizing acetogenic bacterium isolated from human feces. Arch. Microbiol. 166: 176-183.

Boga, H., and Brune, A. 2003. Hydrogen-dependent oxygen reduction by homoacetogenic bacteria isolated from termite guts. Appl. Environ. Microbiol. 69: 779-786.

Böttcher, B., and Gräber, P. 2000. The structure of the H^+-ATP synthase from chloroplasts and its subcomplexes as revealed by electron microscopy. Biochim. Biophys. Acta. 1458: 404-416.

Bramlett, M.R., Tan, X., and Lindahl, P.A. 2003. Inactivation of acetyl-CoA synthase/carbon monoxide dehydrogenase by copper. J. Am. Chem. Soc. 125: 9316-9317.

Brauman, A., Kane, M.D., Labat, M., and Breznak, J.A. 1992. Genesis of acetate and methane by gut bacteria of nutritionally diverse termites. Science. 257:1384-1387.

Braun, K., and Gottschalk, G. 1981. Effect of molecular hydrogen and carbon dioxide on chemo-organotrophic growth of *Acetobacterium woodii* and *Clostridium aceticum*. Arch. Microbiol. 128: 294-298.

Braun, M., Mayer, F., and Gottschalk, G. 1981. *Clostridium aceticum* (Wieringa), a microorganism producing acetic acid from molecular hydrogen and carbon dioxide. Arch. Microbiol. 128: 288-293.

Braus-Stromeyer, S.A., and Drake, H.L. 1996. Expression and localization of CO_2-fixing enzymes during autotrophic growth by the acetogen *Acetogenium kivuii*. Abstr. Ann. Meet. Am. Soc. for Microbiol. Abstract K-162, p. 563.

Braus-Stromeyer, S.A., Schnappauf, G., Braus, G.H., Gößner, A.S., and Drake, H.L. 1997. Carbonic anhydrase in *Acetobacterium woodii* and other acetogenic bacteria. J. Bacteriol. 179: 7197-7200.

Breznak, J.A. 1994. Acetogenesis from carbon dioxide in termite guts. In: Acetogenesis. H.L. Drake, ed. Chapman and Hall, New York, New York. p. 303-330.

Breznak, J.A., and Blum, J.S. 1991. Mixotrophy in the termite gut acetogen, *Sporomusa termitida*. Arch. Microbiol. 156: 105-110.

Breznak, J.A., and Kane, M.D. 1990. Microbial H_2/CO_2 acetogenesis in animal guts: nature and nutritional significance. FEMS Microbiol. Rev. 87: 309-314.

Breznak, J.A., and Switzer, J.M. 1986. Acetate synthesis from H_2 plus CO_2 by termite gut microbes. Appl. Environ. Microbiol. 52: 623-630.

Brock, T.D. 1989. Evolutionary relationships of the autotrophic bacteria. In: Autotrophic Bacteria. H.G. Schlegel and B. Bowien, eds. Science Tech. Publishers, Madison, Wisconsin. p. 499-512.

Brulla, W.J., and Bryant, M.P. 1989. Growth of the syntrophic anaerobic acetogen, strain PA-1, with glucose or succinate as energy source. Appl. Environ. Microbiol. 55: 1289-1290.

Byrer, D.E., Rainey, F.A., and Wiegel, J. 2000. Novel strains of *Moorella thermoacetica* form unusually heat-resistant spores. Arch. Microbiol. 174: 334-339.

Causey, T.B., Zhou, S., Shanmugam, K.T., and Ingram, L.O. 2003. Engineering the metabolism of *Escherichia coli* W3110 for the conversion of sugar to redox-neutral and oxidized products: homoacetate production. Proc. Natl. Acad. Sci. USA. 100: 825-832.

Cecchini, G., Schröder, I., Gunsalus, R.P., and Maklashina, E. 2002. Succinate dehydrogenase and fumarate reductase from *Escherichia coli*. Biochim. Biophys. Acta. 1553: 140-157.

Collins, M.D., Lawson, P.A., Willems, A., Cordoba, J.J., Fernandez-Garayzabal, J., Garcia, P., Cai, J., Hippe, H., and Farrow, J.A.E. 1994. The phylogeny of the genus *Clostridium*: Proposal of five new genera and eleven new species combinations. Int. J. Syst. Bacteriol. 44: 812-826.

Conrad, R., Bak, F., Seitz, H.-J., Thebrath, B., Mayer, H.P., and Schütz, H. 1989. Hydrogen turnover by psychrotrophic homoacetogenic and mesophilic methanogenic bacteria in anoxic paddy soil and lake sediment. FEMS Microbiol. Ecol. 62: 285-294.

Cord-Ruwisch, R., Seitz, H.-J., and Conrad, R. 1988. The capacity of hydrogenotrophic anaerobic bacteria to compete for traces of hydrogen depends on the redox potential of the terminal electron acceptor. Arch. Microbiol. 149: 350-357.

Dangel, W., Schulz, H., Diekert, G., König, H., and Fuchs, G. 1987. Occurence of corrinoid-containing membrane proteins in anaerobic bacteria. Arch. Microbiol. 148: 52-56.

Daniel, S.L., and Drake, H.L. 1993. Oxalate- and glyoxylate-dependent growth and acetogenesis by *Clostridium thermoaceticum*. Appl. Environ. Microbiol. 59: 3062-3069.

Daniel, S.L., Hsu, T., Dean, S.I., and Drake, H.L. 1990. Characterization of the H_2- and CO-dependent chemolithotrophic potentials of the acetogens *Clostridium thermoaceticum* and *Acetogenium kivui*. J. Bacteriol. 172: 4464-4471.

Daniel, S.L., Keith, E.S., Yang, H.C., Lin, Y.S., and Drake, H.L. 1991. Utilization of methoxylated aromatic compounds by the acetogen *Clostridium thermoaceticum*: expression and specificity of the Co-dependent *O*-demethylating activity. Biochem. Biophys. Res. Commun. 180: 416-422.

Daniel, S.L., Pilsl, C., and Drake, H.L. 2004. Oxalate metabolism by the acetogenic bacterium *Moorella thermoacetica*. FEMS Microbiol. Lett. In press.

Darnault, C., Volbeda, A., Kim, E.J., Legrand, P., Vernede, X., Lindahl, P.A., and Fontecilla-Camps, J.C. 2003. Ni-Zn-[Fe4-S4] and Ni-Ni-[Fe4-S4] clusters in closed and open subunits of acetyl-CoA synthase/carbon monoxide dehydrogenase. Nat. Struct. Biol. 10: 271-279.

Das, A., and Ljungdahl, L.G. 1997. Composition and primary structure of the F_1F_O ATP synthase from the obligately anaerobic bacterium *Clostridium thermoaceticum*. J. Bacteriol. 179: 3746-3755.

Das, A., and Ljungdahl, L.G. 2003. Electron-transport systems in acetogens. In: Biochemistry and physiology of anaerobic bacteria. L.G. Ljungdahl, M.W. Adams, L.L Barton, J.G. Ferry and M.K. Johnson, eds. Springer-Verlag, New York, New York. p. 191-204.

Das, A., Coulter, E.D., Kurtz, Jr., D.M., and Ljungdahl, L.G. 2001. Five-gene cluster in *Clostridium thermoaceticum* consisting of two divergent operons encoding rubredoxin oxidoreductase - rubredoxin and rubrerythrin-type flavodoxin - high-molecular-weight rubredoxin. J. Bacteriol. 183: 1560-1567.

Das, A., Hugenholtz, J., Van Halbeek, H., and Ljungdahl, L.G. 1989. Structure and function of a menaquinone involved in electron transport in membranes of *Clostridium thermoautotrophicum* and *Clostridium thermoaceticum*. J. Bacteriol. 171: 5823-5829.

Das, A., Ivey, D.M., and Ljungdahl, L.G. 1997. Purification and reconstitution into proteoliposomes of the F_1F_O ATP synthase from the obligately anaerobic gram-positive bacterium *Clostridium thermoautotrophicum*. J. Bacteriol. 179: 1714-1720.

Diekert, G. 1992. The acetogenic bacteria. In: The Prokaryotes, 2nd edition. A. Balows, H.G. Trüper, M. Dworkin, W. Harder, and K.-H. Schleifer, eds. Springer-Verlag, New York, New York. p. 517-533.

Diekert, G., and Wohlfarth, G. 1994. Metabolism of homoacetogens. Antonie van Leeuwenhoek. 66: 209-221.

Diekert, G., and Wohlfarth, G. 1994a. Energetics of acetogenesis from C_1 units. In: Acetogenesis. H.L. Drake, ed. Chapman and Hall, New York, New York. p. 157-179.

Diekert, G., and Wohlfarth, G. 1994b. Metabolism of homoacetogens. Antonie van Leeuwenhoek. 66: 209-221.

Diekert, G., and Thauer, R.K. 1978. Carbon monoxide oxidation by *Clostridium thermoaceticum* and *Clostridium formicoaceticum*. J. Bacteriol. 136: 597-606.

Dmitriev, O., Jones, P.C., Jiang, W., and Fillingame, R.H. 1999. Structure of the membrane domain of subunit *b* of the *Escherichia coli* F_OF_1 ATP synthase. J. Biol. Chem. 274: 15598-15604.

Dobbek, H., Svetlitchnyi, V., Gremer, L., Huber, R., and Meyer, O. 2001. Crystal structure of a carbon monoxide dehydrogenase reveals a [Ni-4Fe-5S] cluster. Science. 293: 1281-1285.

Dobrindt, U., and Blaut, M. 1996. Purification and characterization of a membrane-bound hydrogenase from *Sporomusa sphaeroides* involved in energy-transducing electron transport. Arch. Microbiol. 165: 141-147.

Doré, J., Pochart, P., Bernalier, A., Goderel, I., Morvan, B., and Rambaud, J.C. 1995. Enumeration of H_2-utilizing methanogenic archaea, acetogenic and sulfate-reducing bacteria from human feces. FEMS Microbiol. Ecol. 17: 279-284.

Dorn, M., Andreesen, J.R., and Gottschalk, G. 1978a. Fermentation of fumarate and L-malate by *Clostridium formicoaceticum*. J. Bacteriol. 133: 26-32.

Dorn, M., Andreesen, J.R., and Gottschalk, G. 1978b. Fumarate reductase of *Clostridium formicoaceticum*. A peripheral membrane protein. Arch. Microbiol. 119: 7-11.

Doukov, T., Seravalli, J., Stezowski, J.J., and Ragsdale S.W. 2002. Crystal structure of a methyltetrahydrofolate- and corrinoid-dependent methyltransferase. Structure Fold. Des. 8: 817-830.

Doukov, T.I., Iverson, T.M., Seravalli, J., Ragsdale, S.W., and Drennan, C.L. 2003. A Ni-Fe-Cu center in a bifunctional carbon monoxide dehydrogenase/acetyl-CoA synthase. Science. 28: 567-572.

Drake, H.L. 1994. Acetogenesis, acetogenic bacteria, and the acetyl-CoA «Wood-Ljungdahl» pathway: past and current perspectives. In: Acetogenesis. H.L. Drake, ed. Chapman and Hall, New York. p. 3-60.

Drake, H.L., and Daniel, S.L. 2004. Physiology of the thermophilic acetogen *Moorella thermoacetica*. Res. Microbiol. In press.

Drake, H.L., Daniel, S.L., Küsel, K., Matthies, C., Kuhner, C., and Braus-Strohmeyer, S. 1997. Acetogenic bacteria: what are the *in situ* consequences for their diverse metabolic versatilities? Biofactors. 6: 13-24.

Drake, H.L., Hu, S.I., and Wood, H.G. 1981. Purification of five components from *Clostridium thermoaceticum* which catalyze synthesis of acetate from pyruvate and methyltetrahydrofolate. J. Biol. Chem. 56: 11137-11144.

Drake, H.L., and Küsel, K. 2003. How the diverse physiological potentials of acetogens determine their *in situ* realities. In: Biochemistry and Physiology of Anaerobic Bacteria. L.G. Ljungdahl, M.W. Adams, L.L Barton, J.G. Ferry and M.K. Johnson, eds. Springer-Verlag, New York, New York. p. 171-190.

Drake, H.L., Küsel, K., and Matthies, C. 2002. Ecological consequences of the phylogenetic and physiological diversities of acetogens. Antonie van Leeuwenhoek. 81: 203-213.

Drake, H.L., Küsel, K., and Matthies, C. 2004. Acetogenic prokaryotes. In: The Prokaryotes, 3rd Edition. M. Dworkin, S. Falkow, E. Rosenberg, K.-H. Schleifer, and E. Stackebrandt, eds. Springer-Verlag, New York, New York. In press.

Drennan, C.L., Heo, J., Sintchak, M.D., Schreiter, E., and Ludden, P.W. 2001. Life on carbon monoxide: X-ray structure of *Rhodospirillum rubrum* Ni-Fe-S carbon monoxide dehydrogenase. Proc. Natl. Acad. Sci. USA. 98: 11973-11978.

Dunn, S.D. 1992. The polar domain of the *b*-subunit of *Escherichia coli* F_1F_O-ATPase forms an elongated dimer that interacts with the F_1 sector. J. Biol. Chem. 267: 7630-7636.

Eden, G., and Fuchs, G. 1982. Total synthesis of acetyl coenzyme A involved in autotrophic CO_2 fixation in *Acetobacterium woodii*. Arch. Microbiol. 133: 66-74.

Eden, G., and Fuchs, G. 1983. Autotrophic CO_2 fixation in *Acetobacterium woodii*. II. Demonstration of enzymes involved. Arch. Microbiol. 135: 68-73.

Eichler, B., and Schink, B. 1984. Oxidation of primary aliphatic alcohols by *Acetobacterium carbinolicum* sp. nov., a homoacetogenic anaerobe. Arch. Microbiol. 140: 147-152.

Engelmann, T., Kaufmann F., and Diekert, G. 2001. Isolation and characterization of a veratrol:corrinoid protein methyl transferase from *Acetobacterium dehalogenans*. Arch. Microbiol. 175: 376-383.

Ferguson, D.J., Krzycki, J.A., and Grahame, D.A. 1996. Specific roles of methylcobamide:coenzyme M methyltransferase isozymes in metabolism of methanol and methylamines in *Methanosarcina barkeri*. J. Biol. Chem. 271: 5189-5194.

Ferry, J.G. 1994. CO dehydrogenase of methanogens. In: Acetogenesis. H.L. Drake, ed. Chapman and Hall, New York, New York. p. 539-556.

Fillingame, R.H., Jiang, W., Dmitriev, O.Y., and Jones, P.C. 2000. Structural interpretations of F_O rotary function in the *Escherichia coli* F_1F_O ATP synthase. Biochim. Biophys. Acta. 1458: 387-403.

Fischer, F., Lieske, R., and Winzer, K. 1932. Biologische Gasreaktionen. II. Über die Bildung von Essigsäure bei der biologischen Umsetzung von Kohlenoxyd und Kohlensäure mit Wasserstoff zu Methan. Biochem. Z. 245: 2-12.

Fontaine, F.E., Peterson, W.H., McCoy, E., Johnson, M.J., and Ritter, G.J. 1942. A new type of glucose fermentation by *Clostridium thermoaceticum* n. sp. J. Bacteriol. 43: 701-715.

Forster, A., Daniel, R., and Müller, V. 1995. The Na^+-translocating ATPase of *Acetobacterium woodii* is a F_1F_O-type enzyme as deduced from the primary structure of its β, γ and ε subunits. Biochim. Biophys. Acta. 1229: 393-397.

Frank, C., Schwarz, U., Matthies, C., and Drake, H. L. 1998. Metabolism of aromatic aldehydes as co-substrates by the acetogen *Clostridium formicoaceticum*. Arch. Microbiol. 170: 427-434.

Fröstl, J.M., Seifritz, C., and Drake, H.L. 1996. Effect of nitrate on the autotrophic metabolism of the acetogens *Clostridium thermoautotrophicum* and *Clostridium thermoaceticum*. J. Bacteriol. 178: 4597-4603.

Fuchs, G. 1986. CO_2 fixation in acetogenic bacteria: variations on a theme. FEMS Microbiol. Rev. 39: 181-213.

Fuchs, G. 1990. Alternatives to the Calvin cycle and the Krebs cycle in anaerobic bacteria: pathways with carbonylation chemistry. In: The Molecular Basis of Bacterial Metabolism. G. Hauska and R. Thauer, eds. Springer, Berlin. p. 13-20.

Fuchs, G. 1994. Variations of the acetyl-CoA pathway in diversely related microorganisms that are not acetogens. In: Acetogenesis. H.L. Drake, ed. Chapman and Hall, New York, New York. p. 507-520.

Geerligs, G., Schönheit, P., and Diekert, G. 1989. Sodium dependent acetate formation from CO_2 in *Peptostreptococcus productus* (strain Marburg). FEMS Microbiol. Lett. 57: 253-258.

Gilbert, B., and Frenzel, P. 1995. Methanotrophic bacteria in the rhizosphere of rice microcosms and their effect on porewater methane concentration and methane emission. Biol. Fertil. Soils. 20: 93-100.

Gößner, A., Devereux, R., Ohnemüller, N., Acker, G., Stackebrandt, E., and Drake, H.L. 1999. *Thermicanus aegyptius* gen. nov., sp. nov., isolated from oxic soil, a facultative microaerophile that grows commensally with the thermophilic acetogen *Moorella thermoacetica*. Appl. Environ. Microbiol. 65: 5124-5133.

Gottschalk, G., and Thauer, R.K. 2001. The Na^+-translocating methyltransferase complex from methanogenic archaea. Biochim. Biophys. Acta. 1505: 28-36.

Gottwald, M., Andreesen, J.R., LeGall, J., and Ljungdahl, L.G. 1975. Presence of cytochrome and menaquinone in *Clostridium formicoaceticum* and *Clostridium thermoaceticum*. J. Bacteriol. 122: 325-328.

Graber, J.R., Leadbetter, J.R., and Breznak, J. 2004. Description of *Treponema azotonutricium* sp. nov., and *Treponema primitia* sp. nov., the first spirochetes isolated from termite guts. Appl. Environ. Microbiol. 70: 1315-1320.

Grahame, D.A. 2003. Acetate C-C bond formation and decomposition in the anaerobic world: the structure of a central enzyme and its key active-site metal cluster. Trends Biochem. Sci. 28: 221-224.

Greening, R.C., and Leedle, J.A.Z. 1989. Enrichment and isolation of *Acetitomaculum ruminis*, gen. nov., sp. nov.: acetogenic bacteria from the bovine rumen. Arch. Microbiol. 151: 399-406.

Großkopf, R., Stubner, S., and Liesack, W. 1998. Novel euryarchaeotal lineages detected on rice roots and in the anoxic bulk soil of flooded rice microcosms. Appl. Environ. Microbiol. 64: 4983-4989.

Groth, G., and Walker, J.E. 1997. Model of the *c*-subunit oligomer in the membrane domain of F-ATPases. FEBS Lett. 410: 117-123.

Grüber, G., Wieczorek, H., Harvey, W.R., and Müller, V. 2001. Structure-function relationships of A-, F- and V-ATPases. J. Exp. Biol. 204: 2597-2605.

Häggblom, M.M., Berman, M.H., Frazer, A.C., and Young, L.Y. 1993. Anaerobic O-demethylation of chlorinated guaiacols by *Acetobacterium woodii* and *Eubacterium limosum*. Biodegradation. 4: 107-114.

Hansen, B., Bokranz, M., Schönheit, P., and Kröger, A. 1988. ATP formation coupled to caffeate reduction by H_2 in *Acetobacterium woodii* NZva16. Arch. Microbiol. 150: 447-451.

Harder, S., Lu, W.P., Feinberg, B.A., and Ragsdale, S.W. 1989. Spectroelectrochemical studies of the corrinoid/iron-sulfur protein involved in acetyl coenzyme A synthesis by *Clostridium thermoaceticum*. Biochemistry. 28: 9080-9087.

Harms, U., and Thauer, R.K. 1996. Methylcobalamin: coenzyme M methyltransferase isoenzymes MtaA and MtbA from *Methanosarcina barkeri*. Cloning, sequencing and differential transcription of the encoding genes, and functional overexpression of the *mtaA* gene in *Escherichia coli*. Eur. J. Biochem. 235: 653-659.

Hattori, S., Kamagata, Y., Hanada, S., and Shoun, H. 2000. *Thermoacetogenium phaeum* gen. nov., sp. nov., a strictly anaerobic, thermophilic, syntrophic acetate-oxidizing bacterium. Int. J. Syst. Evol. Microbiol. 50: 1601-1609.

Haveman, S.A., and Pedersen, K. 2002. Distribution of culturable microorganisms in Fennoscandian Shield groundwater. FEMS Microbiol. Ecol. 39: 129-137.

Heise, R., Müller, V., and Gottschalk, G. 1989. Sodium dependence of acetate formation by the acetogenic bacterium *Acetobacterium woodii*. J. Bacteriol. 171: 5473-5478.

Heise, R., Müller, V., and Gottschalk, G. 1992. Presence of a sodium-translocating ATPase in membrane vesicles of the homoacetogenic bacterium *Acetobacterium woodii*. Eur. J. Biochem. 206: 553-557.

Heise, R., Müller, V., and Gottschalk, G. 1993. Acetogenesis and ATP synthesis in *Acetobacterium woodii* are coupled via a transmembrane primary sodium ion gradient. FEMS Microbiol. Lett. 112: 261-268.

Heise, R., Reidlinger, J., Müller, V., and Gottschalk, G. 1991. A sodium-stimulated ATP synthase in the acetogenic bacterium *Acetobacterium woodii*. FEBS Lett. 295: 119-122.

Hines, M.E., Evans, R.S., Sharak Genthner, B.R., Willis, S.G., Friedman, S., Rooney-Varga, J.N., and Devereux, R. 1999.

Molecular phylogenetic and biogeochemical studies of sulfate-reducing bacteria in the rhizosphere of Spartina alterniflora. Appl. Environ. Microbiol. 65: 2209-2216.

Hoehler, T.M., Albert, D.B., Alperin, M.J., and Martens, C.S. 1999. Acetogenesis from CO_2 in an anoxic marine sediment. Limnol. Oceanogr. 44:662-667.

Hsu, T., Daniel, S.L., Lux, M.F., and Drake, H.L. 1990a. Biotransformations of carboxylated aromatic compounds by the acetogen *Clostridium thermoaceticum*: generation of growth-supportive CO_2 equivalents under CO_2-limited conditions. J. Bacteriol. 172: 212-217.

Hsu, T., Lux, M.F., and Drake, H.L. 1990b. Expression of an aromatic-dependent decarboxylase which provides growth-essential CO_2 equivalents for the acetogenic (Wood) pathway of *Clostridium thermoaceticum*. J. Bacteriol. 172: 5901-5907.

Hugenholtz, J., and Ljungdahl, L.G. 1989. Electron transport and electrochemical proton gradient in membrane vesicles of *Clostridium thermoaceticum*. J. Bacteriol. 171: 2873-2875.

Hugenholtz, J., and Ljungdahl, L.G. 1990. Amino acid transport in membrane vesicles of *Clostridium thermoautotrophicum*. FEMS Microbiol. Lett. 69: 117-122.

Hugenholtz, J., Ivey, D.M., and Ljungdahl, L.G. 1987. Carbon monoxide-driven electron transport in *Clostridium thermoautotrophicum* membranes. J. Bacteriol. 169: 5845-5847.

Hu, S.-I., Drake, H.L., and Wood, H.G. 1982. Synthesis of acetyl coenzym A from carbon monoxide, methytetrahydrofolate, and coenzyme A by enzyme from *Clostridium thermoaceticum*. J. Bacteriol. 149: 440-448.

Imkamp, F., and Müller, V. 2002. Chemiosmotic energy conservation with Na^+ as the coupling ion during hydrogen-dependent caffeate reduction by *Acetobacterium woodii*. J. Bacteriol. 184: 1947-1951.

Ivey, D.M., and Ljungdahl, L.G. 1986. Purification and characterization of the F_1-ATPase from *Clostridium thermoaceticum*. J. Bacteriol. 165: 252-257.

Jones, P.C., and Fillingame, R.H. 1998. Genetic fusions of subunit *c* in the F_O sector of H^+-transporting ATP synthase. Functional dimers and trimers and determination of stoichiometry by cross-linking analysis. J. Biol. Chem. 273: 29701-29705.

Jones, P.C., Jiang, W., and Fillingame, R.H. 1998. Arrangement of the multicopy H^+-translocating subunit *c* in the membrane sector of the *Escherichia coli* F_1F_O ATP synthase. J. Biol. Chem. 273: 17178-17185.

Jormakka, M., Byrne, B., and Iwata, S. 2003. Protonmotive force generation by a redox loop mechanism. FEBS Lett. 545: 25-30.

Kaim, G., and Dimroth, P. 1993. Formation of a functionally active sodium-translocating hybrid F_1F_O ATPase in *Escherichia coli* by homologous recombination. Eur. J. Biochem. 218: 937-944.

Kaim, G., Wehrle, F., Gerike, U., and Dimroth, P. 1997. Molecular basis for the coupling ion selectivity of F_1F_O ATP synthases: probing the liganding groups for Na^+ and Li^+ in the *c* subunit of the ATP synthase from *Propionigenium modestum*. Biochemistry. 36: 9185-9194.

Kamen, M.D. 1963. The early history of carbon-14. J. Chem. Ed. 40: 234-242.

Kamlage, B., and Blaut, M. 1993. Isolation of a cytochrome-deficient mutant strain of *Sporomusa sphaeroides* not capable of oxidizing methyl groups. J. Bacteriol. 175: 3043-3050.

Kamlage, B., Gruhl, B., and Blaut, M. 1997. Isolation and characterization of two new homoacetogenic hydrogen-utilizing bacteria from the human intestinal tract that are closely related to *Clostridium coccoides*. Appl. Environ. Microbiol. 63: 1732-1738.

Kane, M.D., and Breznak, J.A. 1991. *Acetonema longum* gen. nov. sp. nov., an H_2/CO_2 acetogenic bacterium from the termite, *Pterotermes occidentis*. Arch. Microbiol. 156: 91-98.

Karita, S., Nakayama, K., Goto, M., Sakka, K., Kim, W.-J., and Ogawa, S. 2003. A novel cellulolytic, anaerobic, and thermophilic bacterium, Moorella sp. strain F21. Biosci. Biotechnol. Biochem. 67: 183-185.

Karnholz, A., Küsel, K., Gößner, A., Schramm, A., and Drake, H.L. 2002. Tolerance and metabolic response of acetogenic bacteria toward oxygen. Appl. Environ. Microbiol. 68: 1005-1009.

Karrasch, M., Bott, M., and Thauer, R.K. 1989. Carbonic anhydrase activity in acetate grown *Methanosarcina barkeri*. Arch. Microbiol. 151: 137-142.

Kaufmann, F., Wohlfarth, G., and Diekert, G. 1998a. *O*-demethylase from *Acetobacterium dehalogenans* - substrate specificity and function of the participating proteins. Eur. J. Biochem. 253: 706-711.

Kaufmann, F., Wohlfarth, G., and Diekert, G. 1998b. *O*-demethylase from *Acetobacterium dehalogenans*. Cloning, sequencing and active expression of the gene encoding the corrinoid protein. Eur. J. Biochem. 257: 515-521.

Kaufmann, F., Wohlfarth, G., and Diekert, G. 1997. Isolation of *O*-demethylase, an ether-cleaving enzyme system of the homoacetogenic strain MC. Arch Microbiol. 168: 136-142.

Kellum, R., and Drake, H.L. 1984. Effects of cultivation gas phase on hydrogenase of the acetogen *Clostridium thermoaceticum*. J. Bacteriol. 160: 466-469.

Kotelnikova, S. 2002. Microbial production and oxidation of methane in deep subsurface. Earth Sci. Rev. 58: 367-395.

Kotelnikova, S., and Pedersen, K. 1997. Evidence for methanogenic Archaea and homoacetogenic Bacteria in deep granitic rock aquifers. FEMS Microbiol. Rev. 20: 339-349.

Kotelnikova, S., and Pedersen, K. 1998. Distribution and activity of methanogens in deep granitic aquifers at Äspö Hard Rock Laboratory, Sweden. FEMS Microbiol. Ecol. 26: 121-134.

Kotsyurbenko, O.R., Nozhevnikova, A.N., Soloviova, T.I., and Zavarin, G.A. 1996. Methanogenesis at low temperatures by microflora of tundra wetland soil. Antonie van Leeuwenhoek. 69: 75-86.

Kotsyurbenko, O.R., Simankova, M.V., Nozhevnikova, A.N., Zhilina, T.N., Bolotina, N.P., Lysenko, A.M., and Osipov, G.A. 1995. New species of psychrophilic acetogens: *Acetobacterium bakii* sp. nov., *A. paludosum* sp. nov., *A. fimetarium* sp. nov. Arch. Microbiol. 163: 29-34.

Kräutler, B. 1990. Chemistry of methylcorrinoids related to their roles in bacterial C_1 metabolism. FEMS Microbiol. Rev. 87: 349-354.

Kreft, J.U., and Schink, B. 1993. Demethylation and degradation of phenylmethylethers by the sulfide-methylating homoacetogenic bacterium strain TMBS-4. Arch. Microbiol. 159: 308-315.

Kreft, J.U., and Schink, B. 1994. *O*-demethylation by the homoacetogenic anaerobe *Holophaga foetida* studied by a new photometric methylation assay using electrochemically produced cob(I)alamin. Eur. J. Biochem. 226: 945-951.

Kreft, J.U., and Schink, B. 1997. Specificity of O-demethylation in extracts of the homoacetogenic *Holophaga phoetida* and demethylation kinetics measured by a coupled photometric assay. Arch. Microbiol. 167: 363-368.

Kröger, A., Biel, S., Simon, J., Gross, R., Unden, G., and Lancaster, C.R. 2002. Fumarate respiration of *Wolinella succinogenes*: enzymology, energetics and coupling mechanism. Biochim. Biophys. Acta. 1553: 23-38.

Krumböck, M., and Conrad, R. 1991. Metabolism of position-labelled glucose in anoxic methanogenic paddy soil and lake sediment. FEMS Microbiol. Ecol. 85: 247-256.

Krumholz, L.R., and Bryant, M.P. 1985. *Clostridium pfennigii* sp. nov. uses methoxyl groups of monobenzenoids and produces butyrate. Int. J. Sys. Bacteriol. 35: 454-456.

Krumholz, L.R., and Bryant, M.P. 1986. *Syntrophococcus sucromutans* sp. nov. gen. nov. uses carbohydrates as electron donors and formate, methoxymonobenzenoids or Methanobrevibacter as electron acceptor systems. Arch. Microbiol. 143: 313-318.

Krumholz, L.R., Harris, S.H., Tay, S.T., and Suflita, S.M. 1999. Characterization of two subsurface H_2-utilizing bacteria, *Desulfomicrobium hypogeium* sp. nov. and *Acetobacterium psammolithicum* sp. nov., and their ecological roles. Appl. Environ. Microbiol. 65: 2300-2306.

Kuhner, C.H., Frank, C., Grießhammer, A., Schmittroth, M., Acker, G., Gößner, A., and Drake, H.L. 1997. *Sporomusa silvacetica* sp. nov., an actogenic bacterium isolated from aggregated forest soil. Int. J. Syst. Bacteriol. 47: 352-358.

Kurtz, D.M., Jr. 2003. Oxygen and anaerobes. In: Biochemistry and Physiology of Anaerobic Bacteria. L.G. Ljungdahl, M.W. Adams, L.L Barton, J.G. Ferry and M.K. Johnson, eds. Springer-Verlag, New York, New York. p. 128-142.

Küsel, K., and Drake, H.L. 1994. Acetate synthesis in soil from a Bavarian beech forest. Appl. Environ. Microbiol. 60: 1370-1373.

Küsel, K., and Drake, H.L. 1995. Effects of environmental parameters on the formation and turnover of acetate by forest soils. Appl. Environ. Microbiol. 61: 3667-3675.

Küsel, K., and Drake, H.L. 1996. Anaerobic capacities of leaf litter. Appl. Environ. Microbiol. 62: 4216-4219.

Küsel, K, and Drake, H.L. 1999. Microbial turnover of low molecular weight organic acids during leaf litter decomposition. Soil Biol. Biochem. 31: 107-118.

Küsel, K., Dorsch, T., Acker, G., Stackebrandt, E., and Drake, H.L. 2000. *Clostridium scatologenes* strain SL1 isolated as an acetogenic bacterium from acidic sediments. Int. J. Syst. Evol. Microbiol. 50: 537-546.

Küsel, K., Gößner, A., Lovell, C.R., and Drake, H.L. 2003. Ecophysiology of an aerotolerant acetogen, *Sporomusa* ST-1, isolated from Juncus roots. Abstr. Ann. Meet. Am. Soc. for Microbiol. Abstract Q-375, p. 582.

Küsel, K., Karnholz, A., Trinkwalter, T., Devereux, R., Acker, G., and Drake, H. L. 2001. Physiological ecology of *Clostridium glycolicum* RD-1, an aerotolerant acetogen isolated from sea grass roots. Appl. Environ. Microbiol. 67: 4734-4741.

Küsel, K., Pinkart, H.C., Drake, H.L., and Devereux, R. 1999a. Acetogenic and sulfate-reducing bacteria inhabiting the rhizoplane and deep cortex cells of the sea grass *Halodule wrightii*. Appl. Environ. Microbiol. 65: 5117-5123.

Küsel, K., Wagner, C., and Drake, H.L. 1999b. Enumeration and metabolic product profiles of the anaerobic microflora in the mineral soil and litter of a beech forest. FEMS Microbiol. Ecol. 29: 91-103.

Küsel, K., Wagner, C., Trinkwalter, T., Gößner, A. S., Bäumler, R., and Drake, H.L. 2002. Microbial reduction of Fe(III) and turnover of acetate in Hawaiian soils. FEMS Microbiol. Ecol. 40: 73-81.

Lajoie, S.F., Bank, S., Miller, T.L., and Wolin, M.J. 1988. Acetate production from hydrogen and [^{13}C]carbon dioxide by the microflora of human feces. Appl. Environ. Microbiol. 54: 2723-2727.

Leadbetter, J.R., Schmidt, T.M., Graber, J.R., and Breznak, J.A. 1999. Acetogenesis from H_2 plus CO_2 by sprirochetes from termite guts. Science. 283: 686-689.

Leaphart, A., and Lovell, C.R. 2001. Recovery and analysis of formyltetrahydrofolate synthetase gene sequences from natural populations of acetogenic bacteria. Appl. Environ. Microbiol. 67: 1392-1395.

Leaphart, A.B., Friez, M.J., and Lovell, C.R. 2003. Formyltetrahydrofolate synthetase sequences from salt marsh plant roots reveal a diversity of acetogenic bacteria and other bacterial functional groups. Appl. Environ. Microbiol. 69: 693-696.

Leclerc, M., Bernalier, A., Donadille, G., and Lelait, M. 1997a. H_2/CO_2 metabolism in acetogenic bacteria isolted from the human colon. Anaerobe. 3: 307-315.

Leclerc, M., Bernalier, A., Lelait, M., and Grivet., J.-P. 1997b. ^{13}C-NMR study of glucose and pyruvate catabolism in four acetogenic species isolated from the human colon. FEMS Microbiol. Lett. 146: 199-204.

Leedle, J.A.Z., and Greening, R.C. 1988. Postprandial changes in methanogenic and acidogenic bacteria in the rumens of steers fed high- or low-forage diets once daily. Appl. Environ. Microbiol. 54: 502-506.

Leedle, J.A.Z., Lotrario, J., Hovermale, J., and Craig, A.M. 1995. Forestomach anaerobic microflora of the bowhead whale (*Balaena mysticetus*). Abstr. Ann. Meet. Am. Soc. Microbiol. Abstract N-8. p. 334.

Leigh, J.A., Mayer, F., and Wolfe, R.S. 1981. *Acetogenium kivui*, a new thermophilic hydrogen-oxidizing, acetogenic bacterium. Arch. Microbiol. 129: 275-280.

Lienard, T., Becher, B., Marschall, M., Bowien, S., and Gottschalk, G. 1996. Sodium ion translocation by N^5-methyltetrahydromethanopterin: coenzyme M methyl-transferase from *Methanosarcina mazei* Gö1 reconstituted in ether lipid liposomes. Eur. J. Biochem. 239: 857-864.

Liesack, W., Bak, F., Kreft, J.U., and Stackebrandt, E. 1994. *Holophaga foetida* gen. nov., sp. nov., a new, homoacetogenic bacterium degrading methoxylated aromatic compounds. Arch. Microbiol. 162: 85-90.

Lindahl, P.A., and Chang, B. 2001. The evolution of acetyl-CoA synthase. Origins of Life and Evolution of the Biosphere. 31: 403-434.

Lindskog, S., Henderson, L.E., Kannan, K.K., Liljas, A., and Strandberg, P.O.B. 1971. Carbonic anhydrase. The Enzymes. 5: 587-665.

Liu, S., and Suflita, J.M. 1993. H_2/CO_2-dependent anaerobic *O*-demethylation activity in subsurface sediments and by an isolated bacterium. Appl. Environ. Microbiol. 59: 1325-1331.

Ljungdahl, L., and Wood, H.G. 1965. Incorporation of C^{14} from carbon dioxide into sugar phosphates, carboxylic acids, and amino acids by *Clostridium thermoaceticum*. J. Bacteriol. 89: 1055-1064.

Ljungdahl, L.G. 1986. The autotrophic pathway of acetate synthesis in acetogenic bacteria. Ann. Rev. Microbiol. 40: 415-450.

Ljungdahl, L.G. 1994. The acetyl-CoA pathway and the chemiosmotic generation of ATP during acetogenesis. In: Acetogenesis. H.L. Drake, ed. Chapman and Hall, New York, New York. p. 63-87.

Ljungdahl, L.G., Irion, E., and Wood, H.G. 1966. Role of corrinoids in the total synthesis from CO_2 by *Clostridium thermoaceticum*. Fed. Proc. 25: 1642-1648.

Ljungdahl, L.G., O´Brien, W.E., Moore, M.R., and Liu, M.-T. 1980. Methylenetetrahydrofolate dehydrogenase from *Clostridium formicoaceticum* and methylenete trahydrofolate dehydrogenase, methylenetetrahydro folate cyclohydrolase (combined) from *Clostridium thermoaceticum*. Methods Enzymol. 66: 599-609.

Loke, H.K., and Lindahl, P.A. 2003. Identification and preliminary characterization of AcsF, a putative Ni-insertase used in the biosynthesis of acetyl-CoA synthase from *Clostridium thermoaceticum*. J. Inorg. Biochem. 93: 33-40.

Lorowitz, W.H., and Bryant, M.P. 1984. *Peptostreptococcus productus* strain that grows rapidly with CO as the energy source. Appl. Environ. Microbiol. 47: 961-964.

Lovell, C.R. 1994. Development of DNA probes for the detection and identification of acetogenic bacteria. In: Acetogenesis. H.L. Drake, ed. Chapman and Hall, New York, New York. p. 236-253.

Lovell, C.R., Przybyla, A., and Ljungdahl, L.G. 1988. Cloning and expression in *Escherichia coli* of the *Clostridium thermoaceticum* gene encoding thermostable formyltetrahydrofolate synthetase. Arch. Microbiol. 149: 280-285.

Lovell, C.R., Przybyla, A., and Ljungdahl, L.G. 1990. Primary structure of the thermostable formyltetrahydrofolate synthetase from *Clostridium thermoaceticum*. Biochemistry. 29: 5687-5694.

Mackie, R.I., and Bryant, M.P. 1994. Acetogenesis and the rumen: syntrophic relationships. In: Acetogenesis. H.L. Drake, ed. Chapman and Hall, New York, New York. p. 331-364.

Martin, D.R., Misra, A., and Drake, H.L. 1985. Dissimilation of carbon monoxide to acetic acid by glucose-limited cultures of *Clostridium thermoaceticum*. Appl. Environ. Microbiol. 49: 1412-1417.

Matthies, C., Freiberger, A. and Drake, H.L. 1993. Fumarate dissimilation and differential reductant flow by *Clostridium formicoaceticum* and *Clostridium aceticum*. Arch. Microbiol. 160: 273-278.

McCarthy, J.E., Gerstel, B., Surin, B., Wiedemann, U., and Ziemke, P. 1991. Differential gene expression from the *Escherichia coli atp* operon mediated by segmental differences in mRNA stability. Mol. Microbiol. 10: 2447-2258.

McCarthy, J.E., Schairer, H.U., and Sebald, W. 1985. Translational initiation frequency of *atp* genes from *Escherichia coli*: identification of an intercistronic sequence that enhances translation. EMBO J. 4: 519-526.

McInerney, M.J., and Bryant, M.P. 1981. Basic principles of bioconversions in anaerobic digestion and methanogenesis. In: Biomass conversion processes for energy and fuels. S.S. Sofer and O.R. Zaborsky, eds. Plenum, New York, New York. p. 277-296.

Miller, T., and Wolin, M.J. 1996. Pathways of acetate, propionate, and butyrate formation by the human fecal microbial flora. Appl. Environ. Microbiol. 62: 1589-1592

Misoph, M., and Drake, H. L. 1996. Effect of CO_2 on the fermentation capacities of the acetogen *Peptostreptococcus productus* U-1. J. Bacteriol. 178: 3140-3145.

Misoph, M., Daniel, S.L., and Drake, H.L. 1996. Bidirectional usage of ferulate by the acetogen *Peptostreptococcus productus* U-1: CO_2 and aromatic acrylate groups as competing electron acceptors. Microbiology. 142: 1983-1988.

Mitchell, P. 1961. Coupling of phosphorylation to electron and hydrogen transfer by a chemiosmotic mechanism. Nature. 191: 144-148.

Möller, B., Oßmer, R., Howard, B.H., Gottschalk, G., and Hippe, H. 1984. *Sporomusa*, a new genus of gram-negative anaerobic bacteria including *Sporomusa sphaeroides* spec. nov. and *Sporomusa ovata* spec. nov. Arch. Microbiol. 139: 388-396.

Müller, V. 2003. Energy conservation in acetogenic bacteria. Appl. Environ. Microbiol. 69: 6345-6353.

Müller, V. and Grüber, G. 2003. ATP synthases: structure, function and evolution of unique energy converters. Cell. Mol. Life Sci. 60: 474-494.

Müller, V., and Gottschalk, G. 1994. The sodium ion cycle in acetogenic and methanogenic bacteria: generation and utilization of a primary electrochemical sodium ion gradient. In: Acetogenesis. H.L. Drake, ed. Chapman and Hall, New York, New York. p. 127-156.

Müller, V., Aufurth, S., and Rahlfs, S. 2001. The Na^+ cycle in *Acetobacterium woodii*: identification and characterization of a Na^+ translocating F_1F_O-ATPase with a mixed oligomer of 8 and 16 kDa proteolipids. Biochim. Biophys. Acta. 1505: 108-120.

Müller, V., Blaut, M., and Gottschalk, G. 1993. Bioenergetics of methanogenesis. In: Methanogenesis. J.G. Ferry, ed. Chapman and Hall, New York, New York. p. 360-406.

Müller, V., Ruppert, C., and Lemker, T. 1999. Structure and function of the A_1A_O ATPases from methanogenic archaea. J. Bioenerg. Biomembrane. 31: 15-28.

Müller, V., Winner, C., and Gottschalk, G. 1988. Electron transport-driven sodium extrusion during methanogenesis from formaldehyde + H_2 by *Methanosarcina barkeri*. Eur. J. Biochem. 178: 519-525.

Naidu, D., and Ragsdale, S.W. 2001. Characterization of a three-component vanillate *O*-demethylase from *Moorella thermoacetica*. J. Bacteriol. 183: 3276-3281.

Nishi, T., and Forgac, M. 2002. The vacuolar H^+-ATPases - nature's most versatile proton pumps. Nat. Rev. Mol. Cell. Biol. 3: 94-103.

Nozhevnikova, A., Kotsyurbenko, O.R., and Simankova, M.V. 1994. Acetogenesis at low temperature. In: Acetogenesis. H.L. Drake, ed. Chapman and Hall, New York, New York. p. 416-431.

O'Brien, W.E., and Ljungdahl, L.G. 1972. Fermentation of fructose and synthesis of acetate from carbon dioxide by *Clostridium formicoaceticum*. J. Bacteriol. 109: 626-632.

Ohwaki, K., and Hungate, R.E. 1977. Hydrogen utilization by clostridia in sewage sludge. Appl. Environm. Microbiol. 33: 1270-1274.

Patel, B.K.C., Monk, C., Littleworth, H., Morgan, H.W., and Daniel, R.M. 1987. *Clostridium fervidus* sp. nov., a new chemoorganotrophic acetogenic thermophile. Int. J. Sys. Bacteriol. 37: 123-126.

Perski, H.J., Moll, J., and Thauer, R.K. 1981. Sodium dependence of growth and methane formation in *Methanobacterium thermoautotrophicum*. Arch. Microbiol. 130: 319-321.

Perski, H.J., Schönheit, P., and Thauer, R.K. 1982. Sodium dependence of methane formation in methanogenic bacteria. FEBS Lett. 143: 323-326.

Peters, V., and Conrad, R. 1995. Methanogenic and other strictly anaerobic bacteria in desert soil and other oxic soils. Appl. Environ. Microbiol. 61: 1673-1676.

Peters, V., and Conrad, R. 1996. Sequential reduction processes and initiation of CH_4 production upon flooding of oxic upland soils. Soil Biol. Biochem. 28: 371-382.

Peters, V., Janssen, P.H., and Conrad, R. 1998. Efficiency of hydrogen utilization during unitrophic and mixotrophic growth of *Acetobacterium woodii* on hydrogen and lactate in the chemostat. FEMS Microbiol. Ecol. 26: 317-324.

Phelps, T.J., and Zeikus, J.G. 1984. Influence of pH on terminal carbon metabolism in anoxic sediments from a mildly acidic lake. Appl. Environ. Microbiol. 48: 1088-1095.

Poston, J.M., Kuratomi, K., and Stadtman, E.R. 1964. Methyl-vitamin B_{12} as a source of methyl groups for the synthesis of acetate by cell free extracts of *Clostridium thermoaceticum*. Ann. NY Acad. Sci. 121: 804-806.

Poston, J.M., Kuratomi, K., and Stadtman, E.R. 1966. The conversion of carbon dioxide to acetate. I. The use of cobalt methylcobalmin as a source of methyl groups for the synthesis of acetate by cell free extracts of *Clostridium thermoaceticum*. J. Biol. Chem. 241: 4209-4216.

Prins, R.A., and Lankhorst, A. 1977. Synthesis of acetate from CO_2 in the cecum of some rodents. FEMS Microbiol. Lett. 1: 255-258.

Ragsdale, S.W. 1991. Enzymology of the acetyl-CoA pathway of autotrophic CO_2 fixation. CRC Crit. Rev. Biochem. Mol. Biol. 26: 261-300.

Ragsdale, S.W. 1997. The Eastern and Western branches of the Wood/Ljungdahl pathway: how the East and West were won. Biofactors. 6: 3-11.

Ragsdale, S.W. 2004. Life with carbon monoxide. CRC Crit. Rev. Biochem. Mol. Biol. (in press).

Ragsdale, S.W., and Kumar, M. 1996. Ni containing carbon monoxide dehydrogenase/acetyl-CoA synthase. Chem. Rev. 96: 2515-2539.

Ragsdale, S.W., and Ljungdahl, L.G. 1984. Hydrogenase from *Acetobacterium woodii*. Arch. Microbiol. 139: 361-365.

Ragsdale, S.W., and Ljungdahl, L.G. 1984. Purification and properties of NAD-dependent 5,10-methylenetetrahydrofolate dehydrogenase from *Acetobacterium woodii*. J. Biol. Chem. 259: 3499-3503.

Ragsdale, S.W., Clark, J.E., Ljungdahl, L.G., Lundi, L.L., and Drake, H.L. 1983a. Properties of purified carbon monoxide dehydrogenase from *Clostridium thermoaceticum*, a nickel, iron-sulfur protein. J. Biol. Chem. 258: 2364-2369.

Ragsdale, S.W., Lindahl, P.A., and Münck, E. 1987. Mössbauer, EPR, and optical studies of the corrinoid/iron-sulfur protein involved in the synthesis of acetyl-coenzyme A by *Clostridium thermoaceticum*. J. Biol. Chem. 26: 1489-14297.

Ragsdale, S.W., Ljungdahl, L.G., and DerVartanian, D.V. 1983b. Isolation of carbon monoxide dehydrogenase from *Acetobacterium woodii* and comparison of its properties with those of the *Clostridium thermoaceticum* enzyme. J. Bacteriol. 255: 1224-1237.

Rahlfs, S., Aufurth, S., and Müller, V. 1999. The Na^+-F_1F_O-ATPase operon from *Acetobacterium woodii*. Operon structure and presence of multiple copies of *atpE* which encode proteolipids of 8- and 18-kDa. J. Biol. Chem. 274: 33999-34004.

Rastogi, V.K., and Girvin, M.E. 1999. Structural changes linked to proton translocation by subunit *c* of the ATP synthase. Nature. 402: 263-268.

Reidlinger, J., and Müller, V. 1994. Purification of ATP synthase from *Acetobacterium woodii* and identification as a Na^+-translocating F_1F_O-type enzyme. Eur. J. Biochem. 223: 275-283.

Reidlinger, J., Mayer, F., and Müller, V. 1994. The molecular structure of the Na^+-translocating F_1F_O-ATPase of *Acetobacterium woodii*, as revealed by electron microscopy, resembles that of H^+-translocating ATPases. FEBS Lett. 356: 17-20.

Reith, F., Drake, H.L., and Küsel, K. 2002. Anaerobic activities of bacteria and fungi in moderately acidic conifer and leaf litter. FEMS Microbiol. Ecol. 41: 27-35.

Revsbech, N.P., Pedersen, O., Reichardt, W., and Briones, A. 1999. Microsensor analysis of oxygen and pH in the rice rhizosphere under field and laboratory conditions. Biol. Fertil. Soils 29: 379-385.

Richardson, D.J., Berks, B.C., Russell, D.A., Spiro, S., and Taylor, C.J. 2001. Functional, biochemical and genetic diversity of prokaryotic nitrate reductases. Cell Mol. Life Sci. 58: 165-178.

Rieu-Lesme, F., Dauga, C., Fonty, G., and Doré, J. 1998. Isolation from the rumen of a new acetogenic bacterium phylogenetically closely related to *Clostridium difficile*. Anaerobe. 4: 89-94.

Rieu-Lesme, F., Dauga, C., Morvan, B., Bouvet, O.M.M., Grimont, P.A.D., and Doré, J. 1996a. Acetogenic coccoid spore-forming bacteria isolated from the rumen. Res. Microbiol. 147: 753-764.

Rieu-Lesme, F., Fonty, G., and Doré, J. 1995. Isolation and characterization of a new hydrogen-utilizing bacterium from the rumen. FEMS Microbiol. Lett. 125: 77-82.

Rieu-Lesme, F., Morvan, B., Collins, M.D., Fonty, G., and Willems, A. 1996b. A new H_2/CO_2-using acetogenic bacterium from the rumen: description of *Ruminococcus schinkii* sp. nov. FEMS Microbiol. Lett. 140: 281-286.

Roberts, D.L., James-Hagstrom, J.E., Garvi, D.K., Gorst, C.M., Runquist, J.A., Baur, J.R., Haase, F.C., and Ragsdale, S.W. 1989. Cloning and expression of the gene cluster encoding key proteins involved in acetyl-CoA synthesis in *Clostridium thermoaceticum*: CO dehydrogenase, the corrinoid/Fe-S rotein, and methyltransferase. Proc. Natl. Acad. Sci. USA. 86: 32-36.

Roberts, D.L., Zhao, S.Y., Doukov, T., and Ragsdale, S.W. 1994. The reductive acetyl coenzyme A pathway: sequence and heterologous expression of active methyltetrahydrofolate:

corrinoid/ iron-sulfur protein methyltransferase from *Clostridium thermoaceticum*. J. Bacteriol. 176: 6127-6130.

Rosencrantz, D., Rainey, F.A., and Janssen, P.H. 1999. Culturable populations of *Sporomusa* spp. and *Desulfovibrio* spp. in the anoxic bulk soil of flooded rice microcosms. Appl. Environ. Microbiol. 65: 3526-3533.

Russell, W.K., Stahlhandske, C.M.V, Jinqiang, X., Scott, R.A., and Lindahl, P.A. 1998. Spectroscopic, redox and structural characterization of the Ni-labile and nonlabile forms of the acetyl-CoA synthase active site of carbon monoxide dehydrogenase. J. Am. Chem. Soc. 120: 7502-7510.

Sambongi, Y., Iko, Y., Tanabe, M., Omote, H., Iwamoto-Kihara, A., Ueda, I., Yanagida, T., Wada, Y., and Futai, M. 1999. Mechanical rotation of the *c* subunit oligomer in ATP synthase (F_1F_O): direct observation. Science. 286: 1722-1724.

Sauer, K., and Thauer, R.K. 2000. Methyl-coenzyme M formation in methanogenic archaea. Involvement of zinc in coenzyme M activation. Eur. J. Biochem. 267: 2498-2504.

Sauer, K., and Thauer, R.K. 1997. Methanol:coenzyme M methyltransferase from *Methanosarcina barkeri* - zinc dependence and thermodynamics of the methanol: cob(I)alamin methyltransferase reaction. Eur. J. Biochem. 249: 280-285.

Sauer, K., Harms, U., and Thauer, R.K. 1997. Methanol: coenzyme M methyltransferase from *Methanosarcina barkeri* - purification, properties and encoding genes of the corrinoid protein MT1. Eur. J. Biochem. 243: 670-677.

Savage, M.D., Wu, Z., Daniel, S.L., Lundie, L.L., Jr., and Drake, H.L. 1987. Carbon monoxide-dependent chemolithotrophic growth of *Clostridium thermoautotrophicum*. Appl. Environ. Microbiol. 53: 1902-1906.

Schaub, A., and Ljungdahl, L.G. 1974. Purification and properties of acetate kinase from *Clostridium thermoaceticum*. Arch. Microbiol. 10: 121-129.

Schink, B. 1994. Diversity, ecology, and isolation of acetogenic bacteria. In: Acetogenesis. H.L. Drake, ed. Chapman and Hall, New York, New York. p. 197-235.

Schmitt-Wagner, D., and Brune, A. 1999. Hydrogen profiles and localization of methanogenic activities in the highly compartmentalized hindgut of soil-feeding higher termites (Cubitermes spp.). Appl. Environ. Microbiol. 65: 4490-4496.

Schnick, C., Forrest, L.R., Sansom, M.S., and Groth, G. 2000. Molecular contacts in the transmembrane *c*-subunit oligomer of F-ATPases identified by tryptophan substitution mutagenesis. Biochim. Biophys. Acta. 1459: 49-60.

Schopf, J.W., Hayes, J.M., and Walter, M.R. 1983. Evolution of the earth's earliest ecosystems: recent progress and unsolved problems. In: Earth's Earliest Biosphere. J.W. Schopf, ed. Princeton University Press, Princeton, New Jersey. p. 361-384.

Schuppert, B., and Schink, B. 1990. Fermentation of methoxyacetate to glycolate and acetate by newly isolated strains of *Acetobacterium* sp. Arch. Microbiol. 153: 200-204.

Seelert, H., Poetsch, A., Dencher, N.A., Engel, A., Stahlberg, H., and Müller, D.J. 2000. Structural biology. Proton-powered turbine of a plant motor. Nature. 405: 418-419.

Seifritz, C., Daniel, S.L., Gossner, A., and Drake, H.L. 1993. Nitrate as a preferred electron sink for the acetogen *Clostridium thermoaceticum*. J. Bacteriol 175: 8008-8013.

Seifritz, C., Fröstl, J.M., Drake, H.L., and Daniel, S.L. 1999. Glycolate as a metabolic substrate for the acetogen *Moorella thermoacetica*. FEMS Microbiol. Lett. 170: 399-405.

Seifritz, C., Drake, H.L., and Daniel. S.L. 2003. Nitrite as an energy-conserving electron sink for the acetogenic bacterium *Moorella thermoacetica*. Curr. Microbiol. 46: 329-333.

Seravalli, J., Gu, W., Tam, A., Strauss, E., Begley, T.P., and Cramer, S.P. 2003. Functional copper at the acetyl-CoA synthase active site. Proc. Natl. Acad. Sci. USA. 100: 3689-3694.

Seravalli, J, Shoemaker, R.K, Sudbeck M.J., and Ragsdale, S.W. 1999. Binding of (6R,S)-methyltetrahydrofolate to methyltransferase from *Clostridium thermoaceticum*: role of protonation of methyltetrahydrofolate in the mechanism of methyl transfer. Biochemistry. 38: 5736-5745.

Sharak-Genthner, B.R., Davies, C.L., and Bryant, M.P. 1981. Features of rumen and sewage sludge strains of *Eubacterium limosum*, a methanol- and H_2-CO_2-utilizing species. Appl. Environ. Microbiol. 42: 12-19.

Silaghi-Dumitrescu, R., Coulter, E.D., Das, A., Ljungdahl, L.G., Jameson, G.N.L., Huynh, B.H., and Kurtz, Jr., D.M. 2003. A flavodiiron protein and high molecular weight rubredoxin from *Moorella thermoacetica* with nitric oxide reductase activity. Biochemistry. 42: 2806-2815.

Simon, J. 2002. Enzymology and bioenergetics of respiratory nitrite ammonification. FEMS Microbiol. Rev. 26: 285-309.

Singleton, Jr., R. 1997a. Harland Goff Wood: an American biochemist. In: Comprehensive Biochemistry: History of Biochemistry, vol. 40. G. Semenza and R. Jaenicke, eds. Elsevier Science B.V., Amsterdam. p. 333-382.

Singleton, Jr., R. 1997b. Heterotrophic CO_2-fixation, mentors, and students: the Wood-Werkman reactions. J. History of Biol. 30: 91-120.

Sleat, R., Mah, R.A., and Robinson, R. 1985. *Acetoanaerobium noterae* gen. nov., sp. nov.: an anaerobic bacterium that forms acetate from H_2 and CO_2. Int. J. Sys. Bacteriol. 35: 10-15.

Stahlberg, H., Müller, D.J., Suda, K., Fotiadis, D., Engel, A., Meier, T., Matthey, U., and Dimroth, P. 2001. Bacterial Na^+-ATP synthase has an undecameric rotor. EMBO Rep. 2: 229-233.

Stams, A. J. M., and Dong, X. 1995. Role of formate and hydrogen in the degradation of propionate and butyrate by defined suspended cocultures of acetogenic and methanogenic bacteria. Antonie van Leeuwenhoek. 68: 281-284.

Stock, D., Gibbons, C., Arechaga, I., Leslie, A.G., and Walker, J.E. 2000. The rotary mechanism of ATP synthase. Curr. Opin. Struct. Biol. 10: 672-679.

Stock, D., Leslie, A.G. and Walker, J.E. 1999. Molecular architecture of the rotary motor in ATP synthase. Science. 286: 1700-1705.

Stupperich, E. 1994. Corrinoid-dependent mechanism of acetogenesis from methanol. In: Acetogenesis. H.L. Drake, ed. Chapman and Hall, New York, New York. p. 180-191.

Stupperich, E., Eisinger, H.J., and Schurr, S. 1990. Corrinoids in anaerobic bacteria. FEMS Microbiol. Rev. 87: 355-359.

Svetlitchnyi, V., Dobbek, H., Meyer-Klaucke, W., Meins, T., Thiele, B., Römer, P., Huber, R., and Meyer, O. 2004. A functional Ni-Ni-[4Fe-4S] cluster in the monomeric acetyl-CoA synthase from *Carboxydothermus hydrogenoformans*. Proc. Natl. Acad. Sci. USA. 101: 446-451.

Tallant, T.C., Paul, L., and Krzycki, J. 2001. The MtsA subunit of the methylthiol:coenzyme M methyltransferase of *Methanosarcina barkeri* catalyses both half-reactions of corrinoid-dependent dimethylsulfide: coenzyme M methyl transfer. J. Biol. Chem. 276: 4485-4493.

Tani, M., Higashi, T., and Nagatsuka, S. 1993. Dynamics of low-molecular weight aliphatic carboxylic acids (LACAs) in forest soils. I. Amount and composition of LACAs in different types of forest soils. Soil Sci. Plant Nutr. 39: 485-495.

Tanaka, K., and Pfennig, N. 1988. Fermentation of 2-methoxyethanol by *Acetobacterium malicum* sp. nov. and *Pelobacter venetianus*. Arch. Microbiol. 149: 181-187.

Tanner, R.S., and Woese, C.R. 1994. A phylogenetic assessment of the acetogens. In: Acetogenesis. H.L. Drake, ed. Chapman and Hall, New York, New York. p. 254-269.

Tanner, R.S., Miller, L.M., and Yang, D. 1993. *Clostridium ljungdahlii* sp. nov., and acetogenic species in clostridial rRNA homology group I. Int. J. Sys. Bacteriol. 43: 232-236.

Terracciano, J.S., Schreurs, W.J.A., and Kashket, E.R. 1987. Membrane H^+ conductance of *Clostridium thermoaceticum* and *Clostridium acetobutylicum*: evidence for electrogenic Na^+/H^+ antiport in *Clostridium thermoaceticum*. Appl. Environ. Microbiol. 53: 782-786.

Thauer, R.K., Jungermann, K., and Decker, K. 1977. Energy conservation in chemotrophic anaerobic bacteria. Bact. Review. 41: 100-180.

Tholen, A., and Brune, A. 1999. Localization and *in situ* activities of homoacetogenic bacteria in the highly compartmentalized hindgut of soil-feeding higher termites (*Cubitermes* spp.). Appl. Environ. Microbiol. 65: 4497-4505.

Tholen, A., and Brune, A. 2000. Impact of oxygen on metabolic fluxes and *in situ* rates of reductive acetogenesis in the hindgut of the wood-feeding termite *Reticulitermes flavipes*. Environ. Microbiol. 2: 436-449.

Tholen, A., Schink, B., and Brune, A. 1997. The gut microflora of *Reticulitermes flavipes*, its relation to oxygen, and evidence for oxygen-dependent acetogenesis by the most abundant *Enterococcus* sp. FEMS Microbiol. Ecol. 24: 137-149.

Traunecker, J., Preuß, A., and Diekert, G. 1991. Isolation and characterization of a methyl chloride utilizing, strictly anaerobic bacterium. Arch. Microbiol. 156: 416-421.

Tschech, A., and Pfennig, N. 1984. Growth yield increase linked to caffeate reduction in *Acetobacterium woodii*. Arch. Microbiol. 137: 163-167.

Valiyaveetil, F.I., and Fillingame, R.H. 1997. On the role of Arg-210 and Glu-219 of subunit *a* in proton translocation by the *Escherichia coli* F_1F_O-ATP synthase. J. Biol. Chem. 272: 32635-32641.

van der Meijden, P., Heythuysen, H.J., Pouwels, F.P., Houwen, F.P., van der Drift, C., and Vogels, G.D. 1983. Methyltransferase involved in methanol conversion by *Methanosarcina barkeri*. Arch. Microbiol. 134: 238-242.

Vandenberg, J.I., Carter, N.D., Bethell, H.W.L., Nogradi, A., Ridderstrale, Y., Metcalfe, J.C., and Grace, A.A. 1996. Carbonic anhydrase and cardiac pH regulation. Am. J. Physiol. 40: 1838-1846.

Wagner, C., Grießhammer, A., and Drake, H.L. 1996. Acetogenic capacities and the anaerobic turnover of carbon in a Kansas prairie soil. Appl. Environ. Microbiol. 62: 494-500.

Wellsbury, P., Goodman, K., Barth, T., Cragg, B.A., Barnes, S.P., and Parkes, R.J. 1997. Deep marine biosphere fuelled by increasing organic matter availability during burial and heating. Nature. 388: 573-576.

Wellsbury, P., Mather, I., and Parkes, R.J. 2002. Geomicrobiology of deep, low organic carbon sediments in the Woodlark Basin, Pacific Ocean. FEMS Microbiol. Ecol. 42: 59-70.

Whitman, W.B. 1994. Autotrophic acetyl coenzyme A biosynthesis in methanogens. In: Acetogenesis. H.L. Drake, ed. Chapman and Hall, New York, New York. p. 521-538.

Whitman, W.B., Coleman, D.C., and Wiebe, W.J. 1998. Prokaryotes: the unseen majority. Proc. Natl. Acad. Sci. USA. 95: 6578-6583.

Wiegel, J. 1994. Acetate and the potential of homoacetogenic bacteria for industrial applications. In: Acetogenesis. H.L. Drake, ed. Chapman and Hall, New York. p. 484-504

Wieringa, K.T. 1936. Over het verdwijnen van waterstof en koolzuur onder anaerobe voorwaarden. Antonie van Leeuwenhoek. 3: 263-273.

Wieringa, K.T. 1939-1940. The formation of acetic acid from carbon dioxide and hydrogen by anaerobic spore-forming bacteria. Antonie van Leeuwenhoek. 6: 251-262.

Wieringa, K.T. 1941. Über die Bildung von Essigsäure aus Kohlensäure und Wasserstoff durch anaerobe Bazillen. Brennstoff-Chemie. 22: 161-164.

Wilkens, S., and Capaldi, R.A. 1998. Electron microscopic evidence of two stalks linking the F_1 and F_O parts of the *Escherichia coli* ATP synthase. Biochim. Biophys. Acta 1365: 93-97.

Wirt, M.D., Kumar, M., Wu, J.J., Scheuring, E.M., Ragsdale, S.W. and Chance, M.R. 1995. Structural and electronic factors in heterolytic cleavage: formation of the Co(I) intermediate in the Corrinoid/iron-sulfur protein from *Clostridium thermoaceticum*. Biochemistry. 34: 5269-5273.

Wohlfarth, G., and Diekert, G. 1991. Thermodynamics of meth ylenetetrahydrofolate reduction to methyltetrahydrofolate and its implications for the energy metabolism of homoacetogenic bacteria. Arch. Microbiol. 155: 378-381.

Wohlfarth, G., Geerligs, G., and Diekert, G. 1990. Purification and properties of a NADH-dependent 5,10-methylenetetra hydrofolate reductase from *Peptostreptococcus productus*. Eur. J. Biochem. 12: 411-417.

Wolin, M.J., and Miller, T.L. 1983. Carbohydrate fermentation. In: Human Intestinal Flora in Health and Disease. D.A. Hentges, ed. Academic Press, New York, New York. p. 147-165.

Wolin, M.J., and Miller, T.L. 1994. Acetogenesis from CO_2 in the human colonic ecosystem. In: Acetogenesis. H.L. Drake, ed. Chapman and Hall, New York, New York. p. 365-385.

Wolin, M.J., Miller, T.L., Collins, M.D., and Lawson, P.A. 2003. Formate-dependent growth and homoacetogenic fermentation by a bacterium from human feces: description of *Bryantella formatexigens* gen. nov., sp. nov. Appl. Environ. Microbiol. 69: 6321-6326.

Wolin, M.J., Miller, T.L., Yerry, S., Zhang, Y., Bank, S., and Weaver, G.A. 1999. Changes of fermentation pathways of fecal microbial communities associated with a drug treatment that increases dietary starch in the human colon. Appl. Environ. Microbiol. 65: 2807-2812.

Wood, H.G. 1952a. A study of carbon dioxide fixation by mass determination on the types of C^{13}-acetate. J. Biol. Chem. 194:905-931.

Wood, H.G. 1952b. Fermentation of 3,4-C^{14}- and 1-C^{14} -labeled glucose by *Clostridium thermoaceticum*. J. Biol. Chem. 199:579-583.

Wood, H.G., and Ljungdahl, L.G. 1991. Autotrophic character of the acetogenic bacteria. In: Variations in Autotrophic Life. J.M. Shively, L.L. Barton, eds. Academic Press, San Diego, California. p. 201-250.

Wu, Z.R., Daniel, S.L., and Drake, H.L. 1988. Characterization of a CO-dependent *O*-demethylating enzyme system from *Clostridium thermoaceticum*. J. Bacteriol. 170: 5747-5750.

Yang, H. and Drake, H.L. 1990. Differential effects of sodium on hydrogen- and glucose-dependent growth of the acetogenic bacterium *Acetogenium kivui*. Appl. Environ. Microbiol. 56: 81-86.

Yoshida, M., Muneyuki, E., and Hisabori, T. 2001. ATP synthase - a marvellous rotary engine of the cell. Nat. Rev. Mol. Cell Biol. 2: 669-677.

Zavarzin, G.A., Zhilina, T.N., and Pusheva, M.A. 1994. Halophilic acetogenic bacteria. In: Acetogenesis. H.L. Drake, ed. Chapman and Hall, New York, New York. p 432-444.

Zeikus, J.G., Lynd, L.H., Thompson, T.E., Krzycki, J.A., Weimer, P.J., and Hegge, P.W. 1980. Isolation and characterization of a new, methylotrophic, acidogenic anaerobe, the Marburg strain. Curr. Microbiol. 3: 381-386.

Zhang, Y., and Fillingame, R.H. 1995. Changing the ion binding specificity of the *Escherichia coli* H^+-transporting ATP synthase by directed mutagenesis of subunit *c*. J. Biol. Chem. 270: 87-93.

Zhilina, T.N., and Zavarzin, G.A. 1990. Extremely halophilic, methylotrophic, anaerobic bacteria. FEMS Microbol. Rev. 87: 315-322.

Zhilina, T.N., Detkova, E.N., Rainey, F.A., Osipov, G.A., Lysenko, A.M., Kostrikina, N.A., and Zavarzin, G.A. 1998. *Natronoincola histidinovorans* gen. nov., sp. nov., a new alkaliphilic acetogenic anaerobe. Curr. Microbiol. 37: 177-185.

Zhilina, T.N., Zavarzin, G.A., Detkova, E.N., and Rainey, F.A. 1996. *Natroniella acetigena* gen. nov. sp. nov., an extremely halophilic, homoacetogenic bacterium: a new member of Haloanaerobiales. Curr. Microbiol. 32: 320-326.

Chapter 15

Anaerobic Oxidation of Inorganic Nitrogen Compounds

Ingo Schmidt* and Mike S. M. Jetten

Abstract

The biological nitrogen cycle is a complex interplay between many microorganisms catalyzing different reactions. The ammonia and nitrite oxidizing bacteria that were thought to be chemolithoautotrophic were placed into the family *Nitrobacteraceae*. For a long time, the oxidation of the inorganic nitrogen compounds ammonia and nitrite by nitrifiers was thought to be restricted to oxic environments, and the metabolic flexibility of these organisms seemed to be limited. The discovery of a novel pathway for anaerobic ammonia oxidation by *Planctomyces* (anammox) and the finding of an anoxic metabolism by "classical" nitrifiers showed that these assumptions are no longer valid.

1. Nitrifier

Microorganisms participating in nitrification have been characterized as obligatory chemolithoautotrophic ammonia- or nitrite-oxidizing bacteria. The heterotrophic aerobic nitrifiers and their metabolism will not be discussed in this chapter. Lithotrophic nitrifiers belong to the family *Nitrobacteraceae* (Watson *et al.*, 1989), although they are not necessarily related phylogenetically. Although they were thought to be restricted in their metabolic potentialities, they have been found in many ecosystems like fresh water, salt water, sewage systems, soils, and on/in rocks as well as in masonry (Mansch *et al.*, 1998; Bothe *et al.*, 2000), in extreme habitats at high temperatures (Egorova, 1975), in Antarctic soils (Arrigo *et al.*, 1995; Wilson *et al.*, 1997), and in environments with pH values of about 4 like acid tea and forest soils (de Boer *et al.*, 2001; Burton *et al.*, 2001) as well as with pH values of about 10 like soda lakes (Sorokin *et al.*, 1998, 2001). Growth under suboptimal conditions seems to be possible by ureolytic activity, aggregate formation (de Boer *et al.*, 1991), or in biofilms on the surfaces of substrata (Allison *et al.*, 1993). It is even more interesting that aerobic nitrifiers were also found in anoxic environments (Weber *et al.*, 2001; Abeliovich *et al.*, 1992). For many years there was no consistent explanation for this observation. The idea of a drift from oxic layers was not convincing, since the cell numbers of ammonia oxidizers were (too) high even deep in the anoxic layers, and they responded almost immediately to exposure to ammonia and oxygen with increasing ammonia oxidation activities (Abeliovich *et al.*, 1992; Weber *et al.*, 2001) indicating an active metabolism in this anoxic ecosystem. First evidence for the survival strategy of nitrifiers was provided by identifying the anoxic denitrification capabilities of both ammonia and nitrite oxidizers. *Nitrosomonas* and *Nitrobacter* were shown to grow with hydrogen or organic compounds as electron donor and nitrite or nitrate, respectively, the products of their oxic energy generation, as electron acceptor (Freitag *et al.*, 1987; Bock *et al.*, 1995). *Nitrosomonas* has further been shown to oxidize ammonia in an oxygen-independent process (Schmidt *et al.*, 2001c). N_2O_4 is used as oxidant for ammonia oxidation leading to the production of nitrite that is then used as electron acceptor (denitrification). Furthermore, a new group of anaerobic nitrite-dependent ammonia oxidizers (anammox) was discovered (Mulder *et al.*, 1995; Strous, 1999). These organisms, through a complex process, take advantage of the free energy of reactions between oxidized and reduced nitrogen compounds. As a rare compound in biological processes, hydrazine was shown to be an intermediate of the anammox pathway.

The aim of this chapter is to summarize these novel findings in anoxic nitrogen conversion and to evaluate their ecological importance.

2. Ammonia Oxidizers

Ammonia oxidizers gain energy using three different pathways, one oxic and two anoxic. The mechanism of the aerobic ammonia oxidation was the first pathway discovered (Winogradsky, 1949; Hooper, 1969). The aerobic ammonia oxidation had been thought to be the only energy generation reaction until Abeliovich *et al.* (1992) and Bock *et al.* (1995) described a hydrogen-dependent denitrification. In 1997, evidence was given for another mechanism to gain energy, a nitrogen dioxide-

*For correspondence email: ingo.schmidt1@uni-bayreuth.de

dependent ammonia oxidation called the NOx-cycle (Schmidt *et al.*, 1997a; 2001a; 2001c), which obviously is a connecting link between anoxic denitrification and oxic nitrification. To give a complete overview, the three pathways (aerobic ammonia oxidation in brief) will be introduced in the following sections.

2.1. Diversity

Gram-negative ammonia oxidizers belong to the genera *Nitrosomonas*, *Nitrosovibrio*, *Nitrosolobus*, *Nitrosococcus*, and *Nitrosospira* (Watson *et al.*, 1989). Several species reveal extensive intracytoplasmic membrane (ICM) systems. Recently, molecular tools to detect the presence of ammonia oxidizing bacteria in the environment have been supplemented by PCR primers for specific amplification of the ammonia monooxygenase structural gene *amoA* (Bothe *et al.*, 2000). Environmental 16S rRNA and *amoA* libraries have extended the knowledge of the natural diversity of ammonia oxidizing bacteria (Head *et al.*, 1993). Comparative 16S rRNA sequence analyses revealed that members of this physiological group are confined to two monophyletic lineages within the *Proteobacteria*. *Nitrosococcus oceanus* is affiliated with the gamma-subclass of the *Proteobacteria*, while members of the genera *Nitrosomonas* and *Nitrosospira* form a closely related group within the beta-subclass (Purkhold *et al.*, 2000). Using these molecular tools, nitrifiers can be detected even in anoxic habitats.

2.2. Aerobic Ammonia Oxidation

Ammonia oxidizers, such as members of the genera *Nitrosomonas*, *Nitrosolobus*, *Nitrosococcus* (Watson *et al.*, 1989), have been described as obligatory aerobic chemolithoautotrophic organisms that use carbon dioxide as their main carbon source (Bock *et al.*, 1991). The only known pathway to gain energy was the aerobic oxidation of ammonia to nitrite (Hooper, 1969; Hooper *et al.*, 1997).

$$NH_3 + O_2 + 2\,H^+ + 2\,e^- \rightarrow NH_2OH + H_2O \qquad (1)$$

$$NH_2OH + H_2O \rightarrow HNO_2 + 4\,H^+ + 4\,e^- \qquad (2)$$

$$0.5\,O_2 + 2\,H^+ + 2\,e^- \rightarrow H_2O \qquad (3)$$

First, ammonia is oxidized to hydroxylamine (eq. 1) by the ammonia monooxygenase (AMO) (Hollocher *et al.*, 1981; Hooper *et al.*, 1997). Second, in this metabolism (eq. 2) is the oxidation of hydroxylamine to nitrite by the enzyme hydroxylamine oxidoreductase (HAO) (Wood, 1986). Small amounts of NO and N_2O are detectable during ammonia oxidation (Poth *et al.*, 1985; Zart *et al.*, 2000). Two of the four electrons released from this reaction (Anderson *et al.*, 1983) are required for AMO-reaction (Tsang *et al.*, 1982), while the remaining ones are used for the generation of a proton motive force (eq. 3, Hollocher *et al.*, 1982) in order to regenerate ATP and, via a reverse electron transport, NADH (Wheelis, 1984; Wood, 1986). It is noteworthy that growth on hydroxylamine was difficult to demonstrate. In the presence of high ammonia concentrations, hydroxylamine stimulated ammonia oxidation activity (Boettcher *et al.*, 1994; de Bruijn *et al.*, 1995). When hydroxylamine was the only substrate, small amounts of ammonium still had to be added, possibly to serve as a nitrogen source for assimilation. Under these conditions nitrite was produced but no growth was detected. It was speculated that hydroxylamine is not a suitable substrate, because the redox status of the organism during hydroxylamine oxidation might be unbalanced in the absence of ammonium/ammonia, and the excess of electrons would cause a collapse of energy generation.

2.3. Anaerobic Metabolism

As already indicated the three different groups of nitrifying bacteria (ammonia oxidizers, anammox, and nitrite oxidizers) have different strategies to gain energy and to grow under anoxic conditions. These strategies will be discussed in the following section.

2.3.1. Anaerobic Nitrogen Dioxide-Dependent Ammonia Oxidation

Recently published data provide the first evidence for an anaerobic ammonia oxidation by *Nitrosomonas*, *Nitrosolobus* (Schmidt *et al.*, 1997a), and *Nitrosococcus* (I. Schmidt, submitted). These results indicate a complex role of nitrogen oxides (NO and NO_2) in the metabolism of 'aerobic' ammonia oxidizers. *Nitrosomonas* can oxidize ammonia in the absence of dissolved oxygen (Schmidt *et al.*, 1997a, 1998), replacing molecular oxygen by nitrogen dioxide or dinitrogen tetroxide (dimeric form of NO_2). The overall nitrogen balance shows a ratio of about 1:1:1:2 for the conversion of ammonia, dinitrogen tetroxide, nitrite, and nitric oxide (eq. 4, 5).

$$NH_3 + N_2O_4 + 2\,H^+ + 2\,e^- \rightarrow NH_2OH + H_2O + 2\,NO \qquad (4)$$

$$NH_2OH + H_2O \rightarrow HNO_2 + 4\,H^+ + 4\,e^- \qquad (5)$$

Hydroxylamine and nitric oxide are formed in this reaction. While nitric oxide is not further metabolized, hydroxylamine is oxidized to nitrite. The nitrite produced is partly used as electron acceptor leading to the formation of dinitrogen (eq. 6), while the rest of the nitrite remains as end product.

$$HNO_2 + 3\,H^+ + 3\,e^- \rightarrow 0.5\,N_2 + 2\,H_2O \qquad (6)$$

Under anoxic conditions, ammonia and NO_2 are consumed by *Nitrosomonas* in a ratio of about one to

Table 1. Anaerobic and aerobic ammonia oxidation activities, N-loss, NADH-, and ATP-production or consumption (-) depending on the NO_2 and O_2 concentrations.

	NO_2 [ppm]	O_2 [ppm]	Ammonia oxidation activity [μmol (g protein)$^{-1}$ h^{-1}]	N-loss [%]	NADH [μmol (g protein)$^{-1}$ h^{-1}]	ATP [μmol (g protein)$^{-1}$ h^{-1}]
	0	0	0.3	-	-0.2	-0.3
	10	0	65.1	17	0.2	0.1
anoxic	25	0	134.2	47	1.3	0.9
	50	0	105.7	52	1.0	0.8
	100	0	25.2	29	-0.1	-0.2
	0	10,000	255.4	15	1.7	1.3
	25	10,000	366.9	25	1.8	1.4
oxic	70	10,000	215.7	36	1.6	1.2
	0	210,000	1377.6	6	0.6	0.4
	25	210,000	1392.1	18	0.5	0.4

one (eq. 4). Concurrently, an equivalent amount of NO is produced, but the nitrite concentration is lower than would be expected from ammonia consumption. As shown above, nitrite is consumed by denitrification leading to the formation of N_2 (eq. 6) and, to a lesser extent, N_2O. The amount of nitrite plus dinitrogen is almost equivalent to the amount of consumed ammonium. Furthermore, hydroxylamine, a typical intermediate of aerobic ammonia oxidation, can be detected. Ammonia seems to be oxidized first to hydroxylamine under anoxic as well as under oxic conditions (eq. 5). Hydroxylamine is subsequently oxidized to nitrite by the hydroxylamine oxidoreductase (Hopper *et al.*, 1979). Between 40 and 60 % of the produced nitrite is denitrified to dinitrogen (N_2). Nitrous oxide (N_2O) was shown to be an intermediate of denitrification. In a pure culture of *Nitrosomonas eutropha* in an N_2 atmosphere supplied with 25 ppm NO_2 and 300 ppm CO_2, a specific anaerobic ammonia oxidizing activity of 134.4 μmol ammonia (g protein)$^{-1}$ h^{-1} can be measured. In the same experimental setup, but in an air atmosphere (21 % oxygen), 1377.6 μmol ammonia (g protein)$^{-1}$ h^{-1} are oxidized (Schmidt *et al.*, 1997a). The specific anaerobic ammonia oxidation activity is low (10-fold lower) due to the fact that only low NO_2 could be supplemented to a pure culture of *Nitrosomonas*. Higher concentrations already lead to a stagnation of ammonia oxidation activity (50 ppm), while NO_2 concentrations of more than 75 ppm are inhibitory. In a mixed population with anammox (see chapter 'Interactions between anammox and *Nitrosomonas*), NO_2 concentrations as high as 600 ppm can be applied without an inhibitory effect on *Nitrosomonas* (Schmidt *et al.*, 2002a). As a consequence of this increased substrate (oxidant) concentration, specific ammonia oxidation activities of about 1500 μmol ammonia (g protein)$^{-1}$ h^{-1}, which is similar to the aerobic activities, are detectable. During anaerobic ammonia oxidation, a growth rate of about 0.12 h^{-1} can be measured, the amount of cell protein increased by 0.87 μg protein per mmol ammonia, and the cell number of *Nitrosomonas* increased by 5.8 x 10^6 cells per mmol ammonia consumed. In addition, the ATP and NADH concentration increased by 0.9 μmol ATP (g protein)$^{-1}$ h^{-1} and 1.3 μmol NADH (g protein)$^{-1}$ h^{-1} and were similar in both anaerobically and aerobically grown cells. Without NO_2, the ATP concentration decreased by 0.1 μmol (g protein)$^{-1}$ h^{-1}, and the NADH concentration decreased by 0.2 μmol (g protein)$^{-1}$ h^{-1}. The key data characterizing anaerobic and aerobic ammonia oxidation are summarized in Table 1 (Schmidt, 1997b). It is interesting to note that the specific ammonia oxidation activities are similar in the presence of 25 ppm NO_2 and 10,000 ppm O_2, respectively, indicating a much higher affinity of the ammonia oxidizing system for the oxidizing agent NO_2 in comparison to O_2 ($K_s(NO_2)$ 52 ppm, $K_s(O_2)$ 55000 ppm). The Ks value for the substrate ammonia is the same under oxic as well as under anoxic conditions ($K_s(NH_3)$ 20 μM). *Nitrosomonas* transfers reducing equivalents to nitrite as the terminal electron acceptor, resulting in a high denitrification activity. The increasing pool sizes of intracellular ATP and NADH indicate energy conservation in the absence of oxygen. Under these conditions, reducing equivalents are also used for the reduction of CO_2, resulting in cell growth and excretion of extracellular organic compounds like glycerol into the medium. This 'overflow metabolism' has been described for different species of bacteria (Russell *et al.*, 1995).

The differences between the anaerobic, NO_2-dependent and the aerobic, O_2-dependent ammonia oxidation are summarized in Figure 1. First, under anoxic

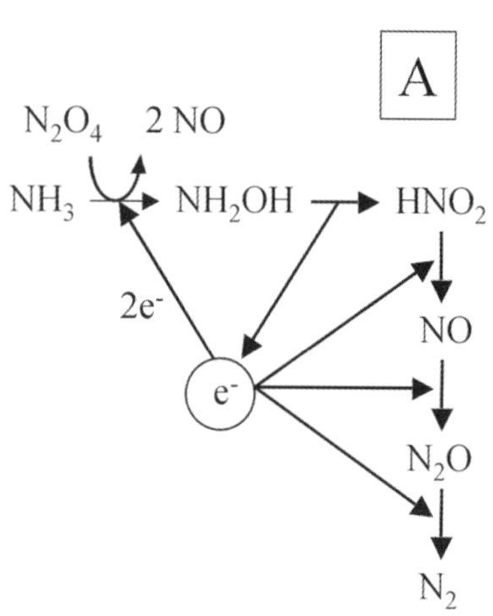

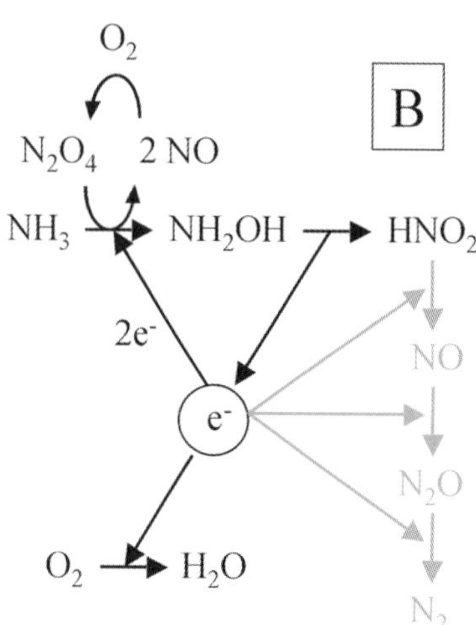

Figure 1. Anaerobic (A) and aerobic (B) ammonia oxidation of *Nitrosomonas*. Under oxic condition nitrite is the only electron acceptor available, leading to an N-loss of up to 67 % via the denitrification pathway (A). Under oxic conditions, most electrons are transferred to oxygen (respiration), and only a minor share of the electrons is discharged via denitrification (N-loss of up to 15 %). Therefore, the pathway of denitrification in (B) is shown in gray.

conditions nitrite is the only suitable electron acceptor available (Figure 1A). Two of the four electrons released during the oxidation of hydroxylamine are required for ammonia oxidation. The other two electrons, used to generate energy (e.g. proton motive force), are transferred to nitrite. To reduce nitrite to molecular dinitrogen (N_2), three electrons per N are necessary. As a consequence, up to 2/3 (the production of one nitrite gains net two electrons, but the denitrification of one nitrite needs three electrons) of the produced nitrite is converted to N_2 (N-loss). This could be shown in short term experiments in cell-free extracts of *Nitrosomonas europaea* (Schmidt *et al.*, 1998). In experiments with whole cells, less nitrite is reduced to N_2 (N-loss about 50 %), because reducing equivalents (electrons) are necessary for other purposes (e.g. CO_2 fixation). Second, NO_2 is not available in natural environments under anoxic conditions. Anaerobic ammonia oxidation is therefore dependent on the transport of NO_2 from oxic layers. Third, N_2O_4 is used as an electron acceptor and NO, an additional product, is released in anaerobic ammonia oxidation. Fourth, ammonia oxidation with N_2O_4 as oxidant is more favorable from an energetic point of view (eq. 4, $\Delta G^{0'}$ -140 kJ mol^{-1}) compared with a reaction directly with oxygen (eq. 1, $\Delta G^{0'}$ -120 kJ mol^{-1}). The first step of ammonia oxidation is not the energy- generating step of the process, but the lower reaction enthalpy might accelerate the oxidation of ammonia providing more hydroxylamine for energy generation. This fact might explain the increased ammonia oxidation activity and higher growth rate of *Nitrosomonas* with N_2O_4 as oxidant (Zart *et al.*, 1998; Schmidt *et al.*, 2002a).

NO_x also plays an important role in the aerobic metabolism of nitrifying microorganisms (Figure 1B). In the presence of O_2, the produced NO can be (re)oxidized to N_2O_4 (eq. 7) (Schmidt *et al.*, 2001c) which is also the oxidizing agent under oxic conditions (NOx-cycle). Since NO is (re)oxidized, only small amounts are detectable in the gas-phase of *Nitrosomonas* cell suspensions.

$$2\,NO + O_2 \rightarrow 2\,NO_2\,(N_2O_4) \qquad (7)$$

Since the detectable NO_x concentrations are small, the nitrogen oxides seem to cycle, possibly enzyme-bound, in the cell (Bock *et al.*, 2001; Schmidt *et al.* 2001c), and the total amount of NO_x per cells is expected to be low. This new hypothetical model of the NOx-cycle was developed on the basis of the results provided by inhibitory studies using acetylene (Schmidt *et al.*, 2001c). An important observation is that anaerobic ammonia oxidation with NO_2 (N_2O_4) as oxidant is not affected by acetylene. *Nitrosomonas eutropha* cells treated with acetylene oxidized ammonia even under oxic conditions if NO_2 is available. The specific ammonia oxidation activity of acetylene-treated cells in the presence of NO_2 is almost the same under both oxic and anoxic conditions. Ammonia oxidation is not detectable in the absence of NO_2. Acetylene does not bind to the 27 kDa polypeptide of AMO during anoxic NO_2-dependent ammonia oxidation (Schmidt *et al.*, 2001c). When oxygen is added, acetylene rapidly associates with this polypeptide. The ammonia concentration does not influence the acetylene binding reaction. It is interesting to note that not only is an active AMO (Hyman *et al.*, 1985) necessary for acetylene binding at the 27 kDa polypeptide, but the presence of oxygen is also required. Furthermore, NO protects the 27 kDa AMO subunit from acetylene binding and ammonia oxidation from the inhibitory effect of acetylene (Schmidt *et al.*, 2001c). The most plausible interpretation of these observations is that NO and acetylene compete for the same active binding site, while NO and NO_2 act at different sites. These results led to a modified model of ammonia oxidation in *Nitrosomonas*. Figure 2 shows the model of the NO_2-dependent ammonia oxidation. Anaerobic ammonia oxidation is dependent on the presence of the oxidizing agent N_2O_4 (Schmidt *et al.*, 1998). Several lines of evidence suggest that this hypothetical model (Figure 2) describes both oxic and anoxic ammonia oxidation by *Nitrosomonas*. First, NO_2 (N_2O_4) is a suitable oxidizing agent under anoxic conditions, leading to the formation of NO. Second, the addition of NO_2 or NO increases specific

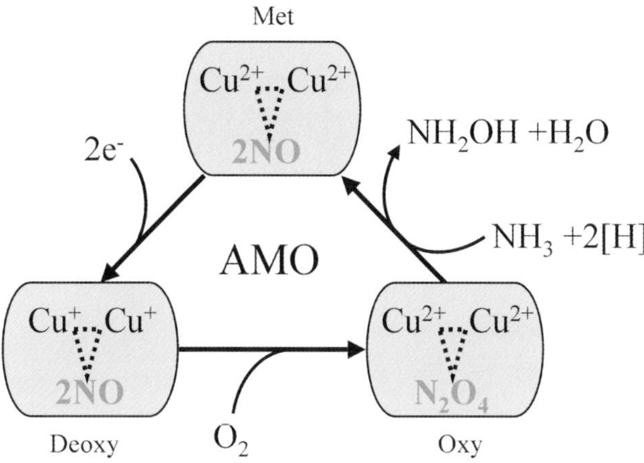

Figure 2. Hypothetical model of ammonia oxidation by *Nitrosomonas* (NO_x-cycle). The figure shows the Oxy-, Met-, and Deoxy-status of AMO. Modified version of the model by C. Pinck (doctoral thesis, University of Hamburg).

ammonia oxidation activity under oxic condition (Zart *et al.*, 1998). The addition of NO_x (NO or NO_2) may supply the NO_x-cycle with additional substrate, increasing the amount of oxidants available to the cells and thus finally increasing specific ammonia oxidation activity. Third, increasing NO concentrations protected the 27 kDa polypeptide from acetylene labeling, indicating that NO and acetylene may compete for the same binding site at the 27 kDa polypeptide. Furthermore, the 27 kDa polypeptide is only labeled in the presence of oxygen, and only O_2-dependent ammonia oxidation is inhibited, while NO_2-dependent ammonia oxidation is unaffected. If acetylene binds in the presence of O_2 at the NO-binding site, it may inactivate the NO oxidizing function of AMO as a consequence of the attempted oxidation of the acetylene triple bond.

Aerobic and anaerobic ammonia oxidation are catalyzed by the two key enzymes AMO and HAO. AMO consists of a 27 kDa membrane-associated protein (AmoA) containing its active site (Hyman *et al.*, 1992; 1995). A second iron-copper protein of 38 to 43 kDa (AmoB) was isolated and described as a probable subunit of AMO (Arciero *et al.*, 1993; Hoppert *et al.*, 1995). The sequences of two genes, *amoA* and *amoB*, encoding these subunit proteins, were determined from overlapping clones (McTavish *et al.*, 1993; Bergmann *et al.*, 1994). A 30 kDa polypeptide was discussed as being a possible precursor form of the 27 kDa polypeptide (Hyman *et al.*, 1992). HAO is a multiheme enzyme, supposedly located in the periplasmic space (Olson *et al.*, 1983; Hooper *et al.*, 1984; 1985). The enzyme complex has a relative molecular weight of about 200 kDa containing one P-460 center and seven c-type hemes per subunit (Hooper *et al.*, 1984; Igarashi *et al.*, 1997). P-460 was found to be a CO-binding heme (Hooper *et al.*, 1978; Lipscomb *et al.*, 1982). The c-type hemes of HAO could be assigned to at least four classes with different oxidation-reduction midpoint potentials (+295 to -390 mV) and associated protein environments (Lipscomb *et al.*, 1982; Prince *et al.*, 1983; Hooper, 1984). HAO constitutes about 40 % of the c-type heme of *Nitrosomonas* (Hooper *et al.*, 1978). Hooper (1989) discussed possible interactions among the described redox centers of HAO. Hydroxylamine is supposed to bind HAO near the P-460 center. Electrons are released and transferred to c-hemes (Hooper *et al.*, 1977; 1982; Olson *et al.*, 1983). The favored mechanism for the oxidation of hydroxylamine, involves the use of water as the source of the second oxygen atom of the metabolic product, nitrite (Andersson *et al.*, 1983; Hooper, 1984). Besides HAO, other soluble electron-transport components in *Nitrosomonas* include cytochromes c_{554}, c_{553}, c_{552}, P460 fragment, two CO-binding c-type cytochromes (c_{CO550}, c_{CO552}), *a*-type cytochrome, and cytochrome *b* (DiSpirito *et al.*, 1985). Currently, little is known about the fluctuation of cytochrome concentrations in connection with the growth conditions.

A rather curious finding was described by Schmidt *et al.* (2004a). Cells of *Nitrosomonas eutropha* were incubated without ammonium for more than 24 h. As soon as ammonium was added or the cells were transferred to a medium containing ammonium, the cells started a rapid uptake of ammonia. In a cell suspension with about 10^8 cell ml^{-1} more than 0.5 mM ammonium disappeared within three to five minutes, but formation of nitrite was not detected. Nitrite (5 μM) first appeared after about 10 min. Cells were harvested five minutes after ammonia uptake but before the appearance of nitrite. The disrupted cells were incubated for 10 min in 1 mM HCl. The ammonium/ammonia amount that the cells took up within the 5 min incubation was then released and detected in the cell-free extract. These results clearly demonstrate that ammonium/ammonia uptake is not coupled to an oxidative ammonia conversion to nitrite. It was calculated that about 70,000 ammonium/ammonia molecules are taken up per cell, reflecting an internal ammonium/ammonia concentration of about 1 M. There are still several unsolved questions; first, to accumulate 1 M ammonium/ammonia in the cells, an active transport mechanism for ammonium/ammonia is necessary, but ammonia is thought to diffuse passively through the membranes (Kleiner, 1981). Such an ammonium transporter has not been experimentally described but was discussed (Chain *et al.*, 2003); second, a storage system for 70,000 ammonium/ammonia molecules would have to exist somewhere in the cells.

2.3.2. Anaerobic Hydrogen-Dependent Denitrification

The apparent inflexibility of ammonium metabolism stood in sharp contrast to the diverse environments where vital ammonia oxidizers had been found (e.g. anoxic habitats). The first evidence for anaerobic metabolism of *Nitrosomonas* was given by Abeliovich *et al.* (1992) and Bock *et al.* (1995). Simple organic compounds such as pyruvate or formate might serve as a carbon source or electron donor while nitrite is used as an electron acceptor (Abeliovich *et al.*, 1992; Stüven *et al.*, 1992). In the absence of dissolved oxygen, *Nitrosomonas eutropha* and *Nitrosomonas europaea* are capable of anoxic denitrification with molecular hydrogen or acetate (Bock *et al.*, 1995). Molecular hydrogen can serve as the only electron donor for nitrite reduction provided that the apparent redox potential of the medium is low (–250 to –300 mV). Under such conditions a specific denitrification activity of about 140 µmol NO_2^- g protein^{-1} h^{-1} can be measured. During denitrification, up to 110 µmol N_2-N g protein^{-1} h^{-1}, 15 µmol N_2O-N g protein^{-1} h^{-1}, and 1.1 µmol NO-N g protein^{-1} h^{-1} are produced and released into the atmosphere (Schmidt *et al.* 2001a). The stoichiometry of this reaction has yet to be determined (eq. 8). Nitrite is consumed immediately in the presence of hydrogen while ammonium is not consumed. If nitrite is replaced by NO or N_2O as electron acceptor the metabolic activity and the growth rates are significantly reduced. Obviously, the enzymes involved in the denitrification pathway in *Nitrosomonas* are constitutively produced. Hydrogenase activity is detectable in *Nitrosomonas eutropha* as well as in *Nitrosolobus multiformis*. The cell growth of *Nitrosomonas* is directly coupled to nitrite reduction (growth rate 0.08 h^{-1}).

$$H_2 + HNO_2 \rightarrow N_2 + N_2O + NO + H_2O \qquad (8^*)$$

*: Stoichiometry unknown

Interestingly, cells growing as denitrifiers completely lose their ability to oxidize ammonia within 15 days. AMO and HAO are hardly degraded, but under these conditions, the proteins are not expressed and therefore diluted during cell growth. In addition, the cytochrome concentration decreases during anoxic denitrification.

The nitrite reductase of *Nitrosomonas europaea* has recently been described. The molecular mass is about 127 kDa consisting of three α-subunits (40 kDa). Copper is the metal center and cytochrome c_{552} is the electron donor (Dispirito *et al.*, 1985; Whittaker *et al.*, 2000).

2.4. Regulation of Metabolic Activities in Ammonia Oxidizers

Over the past 15 years, the role of NO in human biology has begun to be recognized as an environmental pollutant to an endogenously produced substance of great importance that is involved in cell communication and signal transduction. It became obvious that NO is a potent hormone-like signal molecule in human tissue, inducing various intracellular second messenger cascades (Hanafy *et al.*, 2001). Still its role in prokaryotes is poorly understood. The importance of NO and NO_2

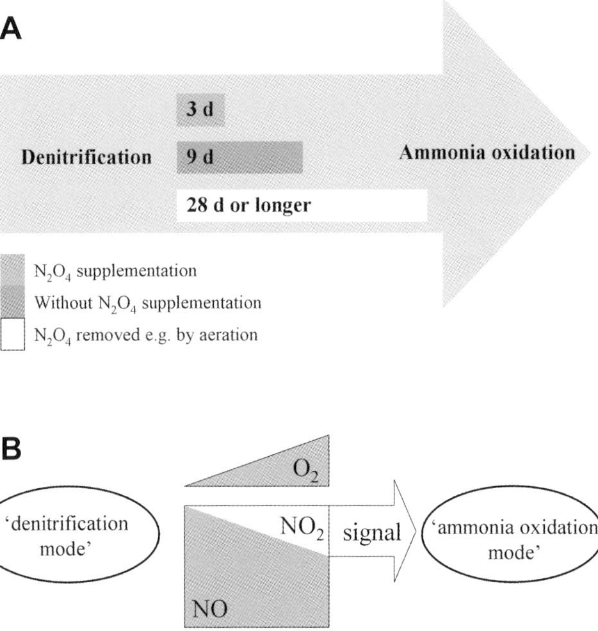

Figure 3. A) N_2O_4 dependence of ammonia oxidation recovery in *Nitrosomonas*. *Nitrosomonas* cells grown under anoxic conditions as hydrogen-dependent denitrifiers lose their ability to oxidize ammonia. The time necessary to re-establish ammonia oxidation was 3 days when N_2O_4 was supplemented, 9 days without N_2O_4 supplementation, and at least 28 days when N_2O_4 was removed by intensive aeration. **B)** Regulation of the 'denitrification mode' and 'ammonia oxidation mode'. In the absence of oxygen, NO is stable and *Nitrosomonas* cells remain in the 'denitrification mode'. In the presence of oxygen, NO is oxidized to NO_2 (N_2O_4), and the cells shift to 'ammonia oxidation mode'.

in *Nitrosomonas* as substances involved in energy generation (Schmidt *et al.*, 1997) and as signal molecules (Schmidt *et al.*, 2001a) has been recognized. Until now, NO was described as a minor product of metabolism. It is constantly produced during ammonia oxidation (Remde *et al.*, 1990; Zart *et al.*, 2000), but the mechanism of NO production (denitrification, chemodenitrification, or autooxidation of hydroxylamine) (Chalk *et al.*, 1983; Moews *et al.*, 1959; Skiba, 1993) has not been investigated in detail. When NO and N_2O_4 were discovered as key elements of ammonia oxidation by *Nitrosomonas* and other ammonia oxidizers, the interest focused on the role of NO_x as a substrate and on its regulatory functions. Several results indicated an important role of NO: First, NO is continuously produced, but the specific production rate differs depending on the growth conditions. Second, NO and NO_2 (NO_x-cycle) are directly involved in aerobic and anaerobic ammonia oxidation (Schmidt *et al.*, 2001c). Third, NO induces a high aerobic denitrification activity, leading to N-losses from soluble compounds of more than 50 % (Zart *et al.*, 1998). Fourth, ammonia oxidizers are highly NO tolerant. NO concentrations as high as 600 ppm do not significantly inhibit their activity (Schmidt *et al.*, 2002a), while 1 ppm NO has been shown to have bactericidal effects on many chemoorganotrophic microorganisms (Mancinelli *et al.*, 1983; Zumft, 1993).

A regulatory function of nitrogen oxides is observed when *Nitrosomonas eutropha* is grown under anoxic conditions as denitrifier. Here, a time-dependent complete loss of the ability to oxidize ammonia and hydroxylamine is observed. NO/N_2O_4 is the master regulatory signal for reinitiating ammonia and hydroxylamine oxidation. This process is directly connected to changes in cell structure, protein, and cytochrome concentrations (Schmidt *et al.*, 2001a; 2001b).

Evidence has emerged that recovery of ammonia oxidation activity in denitrifying *Nitrosomonas* cells strongly depends on the NO_2 concentration (Figure 3A). Upon addition of NO_2 under anoxic or oxic conditions, ammonia consumption starts after about one or three days, respectively. Without supplementing N_2O_4, the cells require nine days to resume ammonia consumption. Removing the nitrogen oxides (NO_x) produced by cells that are incubated under intensive aeration, does not result in recovery of ammonia oxidation even after four weeks. Most notably, the concentration of the 27 kDa polypeptide (AmoA) as well as the concentrations of *a*-type and *c*-type cytochromes increases with increasing ammonia oxidation activity of *Nitrosomonas eutropha*. The relative concentration of AmoA is very low in anaerobically denitrifying, whereas it is high in ammonia oxidizing cells. During ammonia oxidation, about 15 % of the cell protein is AMO (Pinck *et al.*, 2001), while during denitrification AMO is less than 1 % of the total cellular protein (Schmidt, unpublished).

In denitrifying *Nitrosomonas* cells only small amounts of cytochrome are detectable, but the cytochrome concentration of both *a*-type and *c*-type cytochromes increases simultaneously with the elevation of ammonia oxidation activity (Schmidt *et al.*, 2001a). Therefore, the concentrations of *a*-type and *c*-type cytochromes seem to be related directly to ammonia oxidation of *Nitrosomonas*. Since hydroxylamine oxidation activity also increases during the recovery period, it can be assumed that HAO (a heme protein) and electron transport components containing *a*- and *c*-type cytochromes are synthesized during recovery of ammonia oxidation activity. The addition of the translation inhibitor, chloramphenicol, renders cells unable to recover their ammonia oxidation activity. These observations indicate that recovery requires *de novo* synthesis of AmoA (27 kDa), HAO, and electron transport components. Evidence that NO_2 is necessary for regaining nitrification activity has been reported. The length of lag-phase of *Nitrosomonas eutropha* cultures seems to be linked to the concentration of gaseous NO_2. The simplest interpretation of these observations is that NO_2 is an inducer of protein synthesis required for ammonia oxidation. If this is so, then it is not surprising that the addition of NO_2 leads to a faster recovery of ammonia and hydroxylamine oxidation activity and that the removal of nitrogen oxides by intensive aeration leads to slower recovery or cell death (Schmidt *et al.*, 2001b). The nitrogen oxide couple $NO\text{-}NO_2(N_2O_4)$ might therefore function as an oxygen sensing element responsible for switching between hydrogen-dependent denitrification and ammonia oxidation (Figure 3B). Under anoxic conditions, NO is a stable compound - oxidation to NO_2 is not possible - and *Nitrosomonas* cells remain in the 'denitrification mode'. As soon as oxygen becomes available, NO is oxidized to NO_2 , which is the signal that switches metabolism to the 'ammonia oxidation mode'. Shifting back to an oxic environment, NO_2 is consumed rapidly by *Nitrosomonas*. In the absence of NO_2, the cells activate the 'denitrification mode' again. Cells also shift to the 'ammonia oxidation mode' in the absence of oxygen when NO_2 is present. The addition of oxygen is not the decisive signal, because when the nitrogen oxides are removed by intensive aeration, ammonia oxidation is not activated even though O_2 is present. Hence, the $NO\text{-}NO_2(N_2O_4)$ concentrations are the master regulatory signal for the shift between the two metabolic activities. Further investigations will be necessary to completely understand the role of NO_2 in the metabolism of ammonia oxidizers. It is still unknown if NO_2 directly influences protein biosynthesis or if it controls the production and activity of as yet unidentified second messengers.

The second example of the regulatory function of nitrogen oxides in ammonia oxidizers was found recently (Schmidt *et al.*, 2004b). The changes in growth mode (planktonic/motile and biofilm) of *Nitrosomonas europaea* had been studied by varying different parameters such as NO, ammonium, nitrite, and oxygen concentration as well as pH and temperature, then examining their effects on the protein profile of the cells. The NO concentration is the only parameter that significantly influences the growth mode and the protein pattern. By supplementing a *Nitrosomonas europaea* culture growing in a chemostat with NO, the number of planktonic/motile cells decreases while biofilm formation becomes detectable (Figure 4). About 65 % of planktonic/motile cells actively attach to the substratum, and the number of biofilm cells increases rapidly during NO supplementation. The change in growth mode is associated with a change in the protein profiles. Interestingly, the NO concentration stabilized at a high level (about 35 ppm) within twelve hours when the NO supplementation is stopped (Figure 4, day 8), and the cells remain in the biofilm growth mode. When NO is actively removed from the system by increasing the aeration rate or by addition of NO binding compounds such as 2,3-Dimercapto-1-propane-sulfonic acid (DMPS), the NO concentration is reduced to less than 5 ppm, and number of the planktonic/motile cells increases, which expresses protein pattern typical to this growth mode.

Eleven proteins are highly up- or down-regulated, depending on the growth mode. Six flagella and flagella assembly proteins are expressed in the planktonic/motile

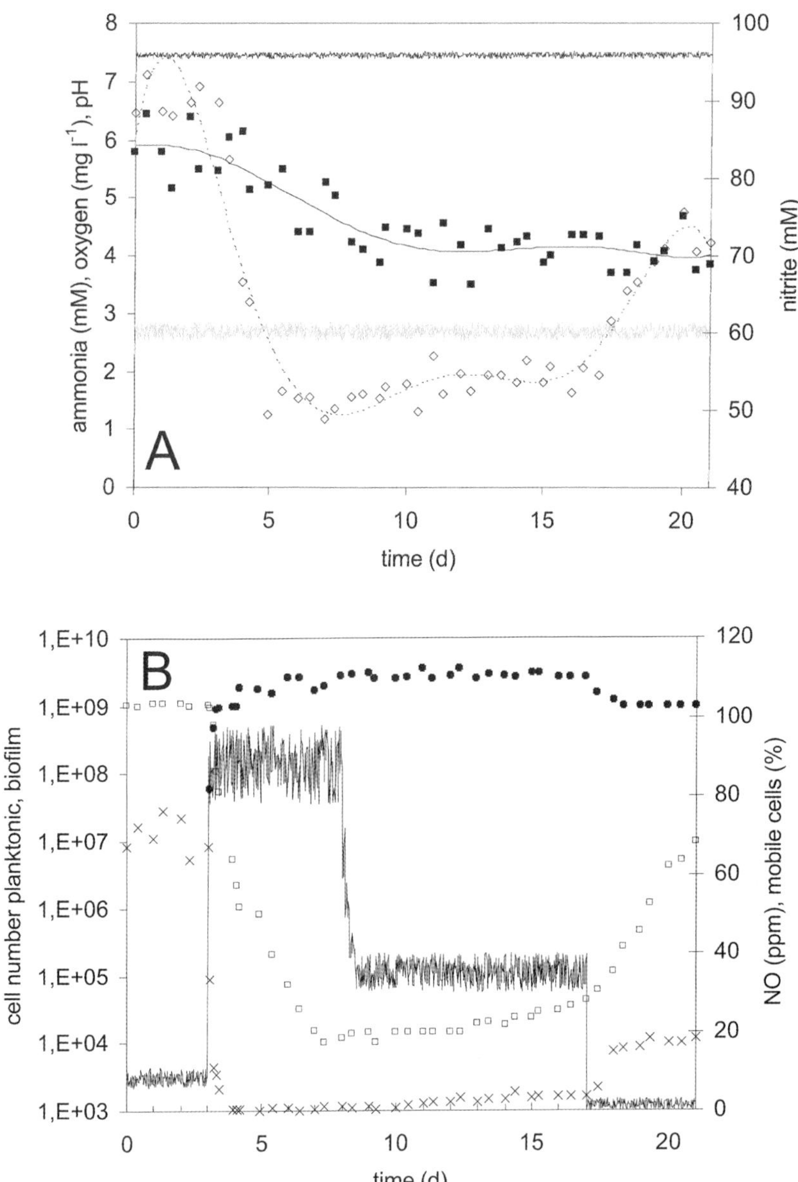

Figure 4. *Nitrosomonas europaea* grown in planktonic/motile or biofilm mode (Schmidt *et al.*, 2004b). Day 3: the aeration gas was supplemented with 200 ppm NO; day 8: NO supplementation was stopped; day 17: The aeration was increased from 0.2 to 2 l min^{-1} (stripping of NO). A: Ammonium (▯), oxygen (gray line), and nitrite (◊) concentration and pH value (black line); B: NO concentration in the off-gas (black line), cell number of planktonic/motile cells (□) and biofilm cells (●), and the ratio of motile to the planktonic/motile cells (x).

cells, but are not detectable in the immotile biofilm cells. Increasing NO concentrations (above 30 ppm) induces the change in the growth mode and flagella protein production is down regulated. The reverse process occurs when NO supplementation is stopped, and the NO concentration is adjusted to below 5 ppm; flagella protein is produced, and the number of planktonic/motile cells increase (Schmidt *et al.*, 2004b). Typical for the biofilm cells are the expression of two regulatory proteins, a CheY-like protein (response regulator consisting of a CheY-like receiver domain and a winged-helix DNA-binding domain like that of OmpR) and CheZ (chemotaxis protein). In contrast, CheW (chemotaxis protein) was detected in planktonic/motile cells. One could speculate that these regulatory proteins might be involved in NO sensing and/or signal transduction leading to regulation of flagella protein expression. Interestingly, the production of nitrosocyanin and succinyl-CoA synthetase (alpha subunit) is also modulated. Both proteins are detectable in planktonic/motile and biofilm cells, but their concentration is significantly higher in biofilm cells (by a factor of approximately 10). The succinyl-CoA synthetase catalyzes the reversible conversion of succinyl-CoA to succinate (citric acid cycle). Although the citric acid cycle is central to the energy yielding metabolism in many microorganisms, the C4 and C5 carbon intermediates serve as biosynthetic precursor for a variety of products. In view of the higher content of succinyl-CoA synthetase in biofilm cells three pathways might be discussed: first, gluconeogenesis and subsequently the production of extracellular polymer substances (EPS) via oxaloacetate as the basis of biofilm formation; second, the synthesis of glucogenic amino acids, e.g. valine, threonine, methionine, and isoleucine via succinyl-CoA; third, the synthesis of porphyrin and heme starting with succinyl-CoA, which might explain the high concentration of nitrosocyanin. Nitrosocyanin is a periplasmic protein with some homology to plastocyanin, the electron-donating copper-bearing domain of N_2O reductase (Whittaker *et al.*, 2000). The function of nitrosocyanin has not yet been determined. Evidence was reported that *Nitrosomonas europaea* produces more nitrosocyanin in the biofilm growth mode than do planktonic/motile cells (Schmidt *et al.* 2004b. Additionally, NO production increases from 110 to 450 μmol NO d^{-1}, and specific denitrification activity increases about five fold (N-loss from 9 to 50 %, Figure 4B). The results described, together with the role of plastocyanin, the homologue of nitrosocyanin, as an electron shuttle in N_2O reduction might indicate a function of nitrosocyanin in the denitrification and NO production in *Nitrosomonas europaea*. More study is needed to determine the role of succinyl-CoA synthetase and nitrosocyanin in *N. europaea* biofilms.

The third regulatory effect of NO on the metabolic activity of *Nitrosomonas* was described by Zart *et al.* (1998). Under anoxic conditions, nitrite is the only electron acceptor available (Schmidt *et al.*, 1997), but under oxic conditions, nitrite (denitrification) and oxygen (respiration) are used simultaneously (Poth *et al.*, 1985). Usually, denitrification resulting in an N-loss of soluble nitrogen compounds does not exceed 15 %. Interestingly, it was shown that the N-loss increases to about 50 % and up to 100 % (short term) in the presence of NO (Zart *et al.*, 1998). During denitrification, mainly N_2 and small amounts of N_2O are produced. The NO concentration and the N-loss in *Nitrosomonas* are directly correlated. NO is responsible for a cumulative induction of the denitrification enzymes nitrate reductase and NO reductase in *Pseudomonas fluorescens* (Ye *et al.*, 1992), leading to high denitrification rates. Furthermore, NO forms nitrosyl hemoproteins and nitrosyl complexes with other iron and copper containing proteins (Henry *et al.*, 1991) resulting in the inhibition of cytochrome oxidases (Carr *et al.*, 1990). To compensate the reduced respiration activity, ammonia oxidizers elevate their denitrification activity. These results are in accordance with the finding that NO not only induces biofilm formation, but also leads to an increased concentration of nitrosocyanin and a higher denitrification activity (N-loss) in the biofilm-cells (Figure 4, Schmidt *et al.*, 2004b).

Other compounds have been reported to influence the metabolic activity of ammonia oxidizers. After irreversible inhibition of AMO with acetylene, Hyman *et al.* (1995) reported that recovery of ammonia oxidation activity and the *de novo* synthesis of AMO (27 kDa subunit) only occurs in the presence of ammonia. This effect of ammonia might be based on a regulatory function or ammonia as the substrate might be essential for the recovery process. High nitrite concentrations are inhibitory to ammonia oxidation activity of *Nitrosomonas europaea* (Stein *et al.*, 1998). *Nitrosomonas eutropha* seems to be able to adapt to high nitrite concentrations as concentrations higher than 40 mM are required to inhibit ammonia oxidation. Slowly increasing the nitrite concentration (about 3 mM per day) up to 100 mM is not inhibitory (Zart *et al.*, 1998). By increasing nitrite concentration, the concentrations of NO and peroxynitrite also increase, in part, by chemodenitrification. Thus, nitrite might indirectly affect ammonia oxidation.

Batchelor *et al.* (1997) described stimulating effects of N-*acyl* homoserine lactones on the nitrification activity of *Nitrosomonas*. Many gram-negative bacteria, including *Nitrosomonas europaea*, produce homoserine lactones as signal molecules. When nutrients become available after a period of starvation, heterotrophs produce homoserine lactones that stimulate ammonia oxidation in *Nitrosomonas* and reduce its lag phase. The competition for ammonium is intense, and a long lag phase would seriously reduce the probability to compete for ammonium (Batchelor *et al.*, 1997). Homoserine lactones function in a signaling system designed to provide information about environmental conditions. In contrast, the effect of nitrogen oxides seems to be very specific for ammonia oxidizers, stimulating recovery of ammonia oxidation when environmental conditions change from anoxic to oxic conditions (Schmidt *et al.*, 2001b).

The metabolism of *Nitrobacteraceae* family of ammonia oxidizers is highly flexible and allows this group of organisms not only to survive, but also to grow in very diverse habitats. Although research has just started to uncover the regulation of their metabolism and the niche differentiation of different species, nothing is known about the evolutionary origins of these organisms. It had been speculated that microorganisms with intracytoplasmic membrane systems (ICM) might have evolved from a common ancester and might be closely related to phototrophic microorganisms (Watson *et al.*, 1989), but genomic data offer little support for this theory. The recently uncovered metabolic flexibility of ammonia oxidizers leads the authors to speculate about the origin of the ammonia oxidizers as denitrifying microorganisms. According to this hypothesis the exposure to oxygen exerted a selective pressure on these bacteria to develop a detoxification pathway for compounds such as NO_2. In ammonium rich environments, ammonia might have been a reasonable electron donor for detoxification. Later evolution might have established ammonia oxidation as a pathway to generate energy. Future genomic research might uncover clues to the evolutionary origins of the *Nitrobacteraceae*.

3. Anaerobic Ammonium Oxidation (Anammox)

The second metabolic pathway for anaerobic ammonia oxidation by lithotrophic microorganisms has been described by Mulder *et al.* (1995). The anaerobic ammonia oxidation by the *Planctomycete Candidatus* "Brocadia anammoxidans" is a nitrite-dependent reaction with hydrazine, a unique compound in biological systems, as intermediate.

3.1. Diversity

Although Broda (1977) predicted the existence of chemolithoautotrophic bacteria capable of anaerobic ammonium/ammonia oxidation and Abeliovich (1992) reported high cell concentrations of nitrifiers under anoxic conditions, the first experimental confirmation of anaerobic ammonia oxidation (anammox) was obtained in the early nineties (Mulder *et al.*, 1995). During experiments on a denitrifying pilot plant it was noted that ammonia and nitrate disappeared from the reactor effluent with a concomitant increase of dinitrogen gas production. The microbial nature of the process was verified, and nitrite was shown to be the preferred electron acceptor (Jetten *et al.*, 1999). Hydroxylamine and hydrazine were identified as important intermediates. Since the growth rate of the anammox biomass appeared to be very low (doubling time about 11 days), reactor systems with very efficient biomass retention were necessary for the enrichment. A sequencing batch reactor (SBR) system was chosen for the ecophysiological study of the anammox community (Strous *et al.*, 1998). The biomass in the community was dominated by more than 70 % by a morphologically conspicuous bacterium. Attempts to isolate the microorganisms with 'classical' methods failed. Therefore, the bacterium was physically purified from enrichment cultures by density gradient centrifugation (Strous *et al.*, 1999). DNA extracted from the purified cells was used as a template for PCR amplification with a universal 16S rDNA primer-set. The dominant 16S rDNA sequence obtained was planctomycete-like, branching very deep within the planctomycete lineage of descent. The anaerobic ammonium oxidizing planctomycete-like bacterium was named '*Candidatus* Brocadia anammoxidans'. The 16S rDNA sequence information was used to design specific oligonucleotide probes for application in fluorescence in situ hybridization (FISH) and to survey the presence of *B. anammoxidans* and related anammox bacteria in several wastewater treatment systems (Schmid *et al.*, 2000). Indeed, *B. anammoxidans* and the closely related '*Candidatus* Kuenenia stuttgartiensis' could be detected in many of these systems throughout the world and seem to be dominating in these microbial biofilm communities. Very recently, another new anammox bacterium "*Candidatus* Scalindua sorokinii" was discovered in the world's largest natural anoxic basin, the Black Sea (Kuypers et al 2003). Very accurate nutrient profiles, newly designed fluorescently labeled 16S rRNA gene probes specific for *S. sorokinii*, tracer experiments with ^{15}N-ammonia and ^{15}N-nitrite, and the distribution of specific anammox membrane lipids all strongly indicated that the ammonium diffusing upwards from the anoxic deep water of the Black Sea was consumed below the oxic zone by the newly discovered anammox bacteria. *S. sorokinii* was the first anammox bacteria that was directly linked to the removal of fixed inorganic nitrogen from a natural ecosystem. Nutrient profiles and ^{15}N tracer studies in other suboxic marine settings indicated that anammox bacteria might play a very important role in the oceanic nitrogen cycle (Thamdrup and Dalsgaard, 2002; Dalsgaard *et al.*, 2003).

The order *Planctomycetales*, first described in 1986 by Schlessner *et al.* (1986), includes so far only four genera (*Planctomyces, Pirellula, Gemmata*, and *Isosphaera*) with seven validly described species (Fuerst, 1995). Various environmentally derived 16S rDNA sequences (Jetten, 2001; Griepenburg, 1999) strongly indicate further planctomycete lineages (Kuenen *et al.*, 2001), including the anammox bacteria. In fact, the newly found bacterium *K. stuttgartiensis* forms a distinct branch within anammox bacteria, and the sequence similarity of less than 90 % to *B. anammoxidans* is indicative of a genus level diversity of these bacteria (Schmid *et al.*, 2000). The application of FISH probes showed the dominance of these bacteria in ecosystems with high nitrogen losses. Molecular techniques are important tools to monitor the presence and activity of microorganisms in ecosystems. For example, the growth rate of many bacteria can be deduced from their ribosome content (Poulsen *et al.*, 1993). This method, however, is not applicable to slow growing anammox- and *Nitrosomonas*-like bacteria (Schmid *et al.*, 2001) since inactive cells of both groups tend to keep their ribosome content at a high level. In such cases, the cellular concentrations of precursor rRNA might be a good indicator for physiological activity (Oerther *et al.*, 2000). Therefore, the intergenic spacer regions (ISR) between the 16S rRNA and 23S rRNA, as part of the precursor rRNA, of *B. anammoxidans* and *K. stuttgartiensis* were sequenced. Subsequently, ISR-targeted oligonucleotide probes were constructed and applied by FISH. Inhibition experiments with *B. anammoxidans* revealed a good correlation between the metabolic activity and the ISR concentrations, demonstrating the ISR targeting FISH to be a powerful method for the detection of activity changes in slow growing bacteria (Schmid *et al.*, 2001).

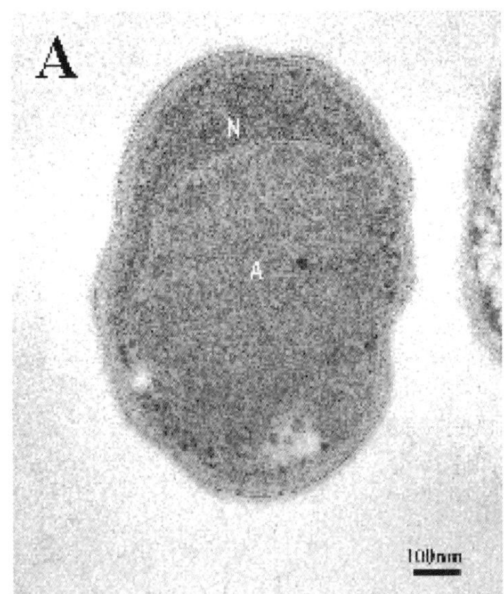

Figure 5. Ultrastructure of the anammox bacterium *Candidatus* "*Brocadia anammoxidans*". A: anammoxosome, N: bacterial nucleoid.

3.2. Physiology

The ultrastructure of *B. anammoxidans* has many features in common with previously described planctomycetes (Figure 5). These microorganisms have a proteinaceous cell wall lacking peptidoglycan and are thus insensitive to ampicillin. Anammox catabolism is at least partly located in a membrane-bound intracytoplasmic compartment, known as the anammoxosome (Schalk *et al.*, 2000; Lindsay *et al.*, 2001). All anammox cells have one anammoxosome. Anammoxosomes can be isolated intact from anammox cells. They contain little or no RNA or DNA and are surrounded by a dedicated membrane that is very impermeable because it consists of ladderane lipids (Sinninghe Damsté *et al.*, 2002). NMR analyses showed that the structure is comprised of five linearly concatenated cyclobutanes, substituted by a heptyl chain which contained a methyl ester moiety at its ultimate carbon atom. The 3D structure of the ring system is fused by *cis*-ring junctions, resulting in a staircase-like arrangement of the fused butane rings, defined as [5]-ladderane. The two ring systems are closely related because it would require only one additional C-C bond in the cyclohexyl ring to form the [5]-ladderane moiety. The bacterial nucleoid is located on the outside of the anammoxosome membrane; it is extremely condensed as is the case for the other planctomycetes. Figure 5 shows the ultrastructure of the anammox bacterium *Candidatus* "Brocadia anammoxidans". Interestingly, *B. anammoxidans* (Sinninghe Damsté *et al.*, 2002) as well as aerobic ammonia oxidizers like *Nitrosomonas* develop internal membrane systems. Whether such a membrane system is bioenergetically necessary for ammonia oxidation is still the subject of investigation since both key enzymes of *Nitrosomonas* are obviously not localized in the ICM system. According to the peptide structure, AMO was described as a membrane-bound enzyme (Suzuki *et al.*, 1981), and recent studies (Pinck, 2001) indicate localization in the cytoplasmic membrane. HAO is localized in the periplasm (Olson *et al.*, 1983).

To unravel the metabolic pathway of anaerobic ammonium oxidation in *B. anammoxidans*, a series of ^{15}N-labeling experiments were conducted. It could be shown that ammonium and nitrite are combined to yield dinitrogen gas (van de Graaf *et al.*, 1997), and radioactive bicarbonate is incorporated in the biomass. With an excess of hydroxylamine, an intermediate accumulation of hydrazine was observed, indicating that hydrazine is an intermediate of the anammox process. According to the working hypothesis, the oxidation of hydrazine to dinitrogen gas is supposed to generate four electrons for the initial reduction of nitrite to hydroxylamine. The overall nitrogen balance shows a ratio of about 1 : 1.32 : 0.26 for the conversion of ammonia, nitrite, and nitrate (eq. 9). The function of the formation of nitrate is assumed to be the generation of reducing equivalents necessary for the reduction of CO_2.

$$NH_4^+ + 1.32\ NO_2^- + 0.066\ HCO_3^- + 0.13\ H^+ \rightarrow 0.26\ NO_3^- + 1.02\ N_2 + 0.066\ CH_2O_{0.5}N_{0.15} + 2.03\ H_2O \qquad (9)$$

A high anammox activity is detectable in a pH range between 6.4 and 8.3 and a temperature range between 20°C and 43 °C (Strous *et al.*, 1999). Under optimal conditions, the specific activity is about 3.6 mmol (g protein)$^{-1}$ h^{-1}, the biomass yield about 0.066 C-mol (mol ammonium)$^{-1}$, and the specific growth rate about 0.0027 h^{-1}. Recent studies showed that *K. stuttgartiensis* is in many ways similar to *B. anammoxidans* (Egli *et al.*, 2001). *K. stuttgartiensis* cells have the same overall cell structure and also produce hydrazine from exogenously supplied hydroxylamine. Energetically favorable mechanisms with Fe^{3+}, Mn^{4+}, or even sulfate as oxidant have yet to be reported (Thamdrup *et al.*, 2000).

To assess the occurrence of the anammox reaction in both natural environments and man-made ecosystems, further data about the effect of several chemical and physical parameters are necessary. For example, the anammox bacteria are very sensitive to oxygen and nitrite. Oxygen concentrations as low as 2 μM and nitrite concentrations between 5 and 10 mM inhibit anammox activity completely, but reversibly (Jetten *et al.*, 1999).

The hydroxylamine oxidoreductase was characterized by Schalk *et al.* (2000). Many similarities to HAO of *Nitrosomonas europaea* were found. The molecular weight (about 180 kDa), the structure of the subunits ($\alpha 3$), and the heme content are quite similar, but HAO of anammox bacteria contains a special *c* type cytochrome P-468 (*Nitrosomonas europaea* P-460). Whether HAO of anammox bacteria catalyses the reduction of nitrite to hydroxylamine or more likely the oxidation of hydrazine to dinitrogen is still part of ongoing research

(I. Cirpus, 2003, unpublished). Preliminary results of the characterization of the nitrite reductase indicated that the enzyme belongs to the *c*-type heme rather than to the cd_1 heme or copper containing nitrite reductases (Schalk, 2000).

3.3. Ecology of Anammox

In various ecosystems having oxic/anoxic interfaces, growth of *B. anammoxidans* is dependent on the activity of aerobic ammonia oxidizing bacteria. Anammox biomass has already been detected in wastewater treatment plants in the Netherlands, Germany, Switzerland, England, Australia, and Japan (Jetten, 2001). Recently, anammox cells were detected in a non-artificial ecosystem, a fresh water swamp in Uganda (Jetten, 2001). Oxic/anoxic interfaces are abundant in nature, for example in biofilms and flocs. In these oxygen-limited environments, the ammonia oxidizers would oxidize ammonium to nitrite and keep the oxygen concentration low, while *B. anammoxidans* would convert the produced nitrite and the remaining ammonium to dinitrogen gas. Such conditions have been established in many different reactor systems (Kuai *et al.*, 1998; Helmer *et al.*, 2001; Third *et al.*, 2002; Sliekers *et al.*, 2003). FISH analysis and activity measurements showed that aerobic as well as anaerobic ammonia oxidizers were present and active in these oxygen-limited reactors, but aerobic nitrite oxidizers (*Nitrobacter* or *Nitrospira*) were not detected. Apparently, the aerobic nitrite oxidizers are unable to compete for oxygen with the aerobic ammonia oxidizers, or for nitrite with the anaerobic ammonia oxidizers (Helder *et al.*, 1983; Hanaki *et al.*, 1990). It seems likely that under these conditions anaerobic and aerobic ammonia oxidizers form a quite stable community. The co-operation of aerobic and anaerobic ammonium/ammonia oxidizing bacteria is not only relevant for wastewater treatment (Helmer *et al.*, 2001; Jetten *et al.*, 2001), but might play an important role in natural environments at the oxic/anoxic interface. Further interactions under anoxic conditions between both groups of ammonia oxidizers seem to be likely since an anoxic, NO_2-dependent metabolism of *Nitrosomonas*-like microorganisms was recently discovered (Schmidt *et al.*, 2001). The competition and/or co-operation between these ammonia oxidizers will be discussed later.

4. Nitrite Oxidizers

Nitrite oxidizers are a third group of nitrogen converting bacteria. Under oxic conditions, they use the oxidation of nitrite to nitrate for energy generation. In 1968, Smith and Hoare described an acetate assimilation, first indicating that these microorganisms are not obligately autotroph, and in 1976 Bock *et al.* provided evidence indicating the anaerobic heterotrophic growth of *Nitrobacter*.

4.1. Diversity

Gram-negative nitrite oxidizers belong to the genera *Nitrobacter*, *Nitrospira*, or *Nitrococcus* (Prosser, 1989; Spieck *et al.*, 1992). They are all Gram-negative, but they are obviously not closely related to each other genetically. Most of the knowledge stems from research work on *Nitrobacter* species. The results obtained cannot be generalized for all nitrite oxidizers, though, as already indicated by investigations of Watson *et al.* (1986). Several strains of *Nitrobacter* and one strain of *Nitrospira* are the only nitrite oxidizers which are not restricted to marine environments (Bock *et al.*, 1991; Ehrich *et al.*, 1995). Members of the genus *Nitrobacter* have been isolated from soils, rocks, building stones, and fresh water. Recently, the first evidence that *Nitrospira* and not *Nitrobacter* species are the predominating nitrite oxidizers in various wastewater samples. One *Nitrospira* species was also isolated from a heating system in Moscow, Russia (Ehrich *et al.*, 1995).

4.2. Nitrite Oxidation

The key enzyme of nitrite oxidizing bacteria is the membrane-bound nitrite oxidoreductase (Tanaka *et al.*, 1983) which oxidizes nitrite with water as the source of oxygen to form nitrate (Aleem *et al.*, 1965). The electrons released from this reaction are transferred via *a*- and *c*-type cytochromes to *a* cytochrome oxidase of the aa_3-type (eq. 10, 11). However, the mechanism of energy conservation in nitrite oxidizers is still unclear; NADH is thought to be produced as its first step (Sundermeyer *et al.*, 1981).

$$HNO_2 + H_2O \leftrightarrow HNO_3 + 2\,H^+ + 2\,e^- \qquad (10)$$

$$0.5\,O_2 + 2\,H^+ + 2\,e^- \rightarrow H_2O \qquad (11)$$

4.3. Anaerobic Heterotrophic Denitrification

Many strains of *Nitrobacter* were shown to grow heterotrophically, but this way of growing seems to be inefficient and slow (Smith *et al.*, 1968; Bock, 1976). In anaerobic environments, *Nitrobacter* cells are able to grow by denitrification (Freitag *et al.*, 1987; Bock *et al.*, 1988). Nitrate can be used as an acceptor for electrons derived from organic compounds. The oxidation of nitrite is a reversible process (eq. 10), and the nitrite oxidoreductase also catalyzes the reduction of nitrate to nitrite in the absence of oxygen (eq. 12, Sundermeyer-Klinger *et al.*, 1984). Thus far, data of the heterotrophic growth of *Nitrobacter* in natural and artificial ecosystems or in laboratory-scale reactor systems are missing. Whether this metabolic pathway is of importance for growth and/or maintenance of nitrite oxidizers has yet to be determined.

$$\text{Organic compounds} + HNO_3 \rightarrow HNO_2 + H_2O \qquad (12^*)$$

*: Stoichiometry unknown

The membrane-bound key enzyme nitrite oxidoreductase (NO_2^-OR) is catalyzing this reversible process. The enzyme is involved in aerobic nitrite oxidation (eq. 10), and it also catalyses the anaerobic reduction of nitrate to nitrite (eq. 12). The NO_2^-OR is an inducible membrane protein present in both lithotrophically and heterotrophically grown *Nitrobacter* cells. The purified protein of *Nitrobacter winogradskyi* is composed of three subunits of 55, 29, and 19 kDa, and, most likely, cytochrome a_1 and c_1 (Sundermeyer-Klinger *et al.*, 1984). Interestingly, the NO_2^-OR of *Nitrobacter hamburgensis* differs, consisting of three subunits of 116, 65, and 32 kDa. Cytochromes were not found (Bock *et al.*, 1990). All NO_2^-OR tested so far contained molybdenum (Mo) and iron-sulfur clusters (Ingledew *et al.*, 1976; Krüger *et al.*, 1987; Fukuoka *et al.*, 1987). The pH optimum for nitrite oxidation is about 8.0, for nitrate reduction it ranges between 6.0 to 7.0 (Tanaka *et al.*, 1983; Sundermeyer-Klinger *et al.*, 1984). Obviously, the structural grouping of NO_2^-OR on the membranes is important for the specific activity of the enzyme, since the specific activity of the isolated enzyme is 10-100 times lower than the specific activity of the NO_2^-OR located on membranes (Spieck *et al.*, 1996).

5. Anaerobic Ammonia Oxidation by Two Different Lithotrophs

Two different metabolic pathways for anaerobic ammonia oxidation by lithotrophic microorganisms have been described. The anaerobic ammonia oxidation by *Planctomycete Candidatus* "Brocadia anammoxidans" is a nitrite-dependent reaction, and anaerobic ammonia oxidation of *Nitrosomonas* is a NO_2/N_2O_4-dependent process. The influence of NO and NO_2 on *B. anammoxidans*, the microbial community structure, and specific ammonia oxidation activities of *B. anammoxidans* and *Nitrosomonas* were evaluated recently (Schmidt *et al.*, 2002a).

Both gasses, NO and NO_2, are of central importance in the metabolism of *Nitrosomonas* and the other members of this family, although the anammox process neither depends on nor produces significant amounts of both gasses. Interestingly, the anaerobic ammonia oxidizing activity of *B. anammoxidans* is not inhibited by NO concentrations up to 600 ppm and NO_2 concentrations up to 100 ppm. These high tolerances of *B. anammoxidans* to NO and NO_2 might be traced back to its capability to convert (detoxify) NO (Schmidt *et al.*, 2002a).

In the presence of NO_2, specific ammonia oxidation activity of *B. anammoxidans* increases and *Nitrosomonas*-like microorganisms recover an NO_2-dependent anaerobic ammonia oxidation activity. The addition of NO_2 to a mixed population of *B. anammoxidans* and *Nitrosomonas* induces simultaneously, a specific anaerobic ammonia oxidation activity up to 5.5 mmol NH_4^+ g protein^{-1} h^{-1} by *B. anammoxidans* and up to 1.5 mmol NH_4^+ g protein^{-1} h^{-1} by *Nitrosomonas*. *Nitrosomonas* reveals a high specific ammonia oxidation activity that is similar to the aerobic activity (see 2.3.1).

Another interesting observation was the growth features of the *B. anammoxidans* and *N. eutropha* community in the presence of ammonia and gaseous NO_2 as substrates (Schmidt *et al.*, 2002a). In this situation, *Nitrosomonas* is able to produce nitrite, N_2, and NO by combined ammonia oxidation and denitrification (eq. 4-6). *B. anammoxidans* takes advantage of the produced nitrite. Combining ammonia and nitrite yields nitrate and N_2 (equation 9). In this system, specific ammonia oxidation activity of *B. anammoxidans* is directly dependent on the nitrite production rate of *N. eutropha*. On the basis of NO_2 consumption, the specific anaerobic ammonia oxidation activity of *N. eutropha* was calculated to be about 1.6 mmol g protein^{-1} h^{-1}. The production of nitrate was used to calculate the specific activity of *B. anammoxidans* (about 0.9 mmol NH_4^+ g protein^{-1} h^{-1}). Here, it is very interesting to compare anaerobic ammonia oxidation activity of *Nitrosomonas eutropha* in pure cultures and in this mixed population. The specific anaerobic activity in pure cultures is low (about 0.14 mmol g protein^{-1} h^{-1}), which is approximately ten times lower than the specific aerobic activity (1.38 mmol g protein^{-1} h^{-1}). In contrast, the anaerobic activity in the mixed population is similar to this aerobic activity. Most likely the activity in the pure cultures are low, because the NO_2 concentration has to be adjusted to a low level (about 50 ppm), to prevent inhibitory and/or toxic effects of NO_2 (Schmidt *et al.*, 1997). This indicates an NO_2 limitation in pure cultures. In mixed populations, NO_2 concentrations as high as 250 ppm have no inhibitory effect on ammonia oxidation activity of *Nitrosomonas*, and the substrate limitation is overcome, resulting in high activities.

The results presented show that *Brocadia* and *Nitrosomonas* are able to coexist under anoxic conditions. While *B. anammoxidans* uses nitrite and *Nitrosomonas* uses gaseous NO_2 as substrate, they compete for ammonia. Key elements to analyzing the activities in co-cultures are shown in Figure 6 and in equations 13 to 15 describing anaerobic ammonia oxidation of *Brocadia* and *Nitrosomonas*. It is important to note that *Nitrosomonas* produces nitrite during anaerobic ammonia oxidation that can serve as a substrate for *Brocadia* (Figure 6). The assumption that this additional nitrite is consumed by *Brocadia* when ammonia is not the limiting substrate was confirmed. As a consequence, the NO_2^-/NH_3-ratio in such a mixed population will change depending on the metabolic activities of both groups of ammonia oxidizers. Since this is important for the interpretation

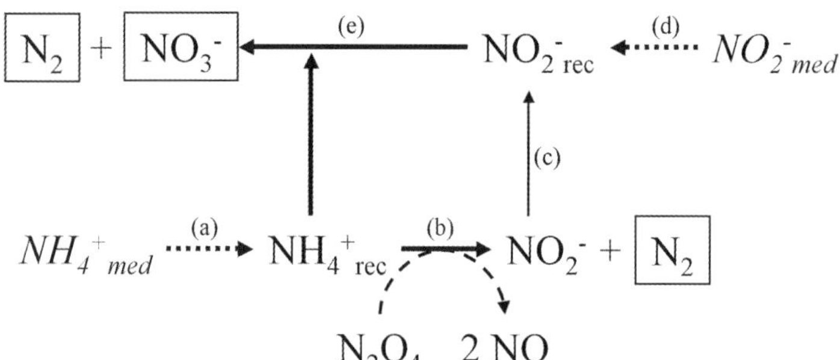

Figure 6. Schematic model describing the competition/co-operation between *B. anammoxidans* and *Nitrosomonas* under anoxic conditions (Schmidt *et al.*, 2002). (a) ammonia addition (fresh medium), (b) ammonia oxidation by *Nitrosomonas*, (c) additional nitrite in the reactor via ammonia oxidation, (d) nitrite addition (fresh medium), (e) anammox reaction, □ end products, (*med*) Medium, (rec) reactor.

$$\mathbf{1}\ NH_3 + \mathbf{1.3}\ HNO_2 \rightarrow 0.3\ HNO_3 + 1\ N_2 + 1.7\ H_2O + 0.6\ H^+ + 0.6\ e^- \quad (13)$$
$$+1\ NH_3 + 1\ N_2O_4 \rightarrow 0.5\ HNO_2 + 0.25\ N_2 + 2\ NO + 1\ H_2O + 0.5\ H^+ + 0.5\ e^- \quad (14)$$

$$2\ NH_3 + 0.8\ HNO_2 + 1\ N_2O_4 \rightarrow 0.3\ HNO_3 + 1.25\ N_2 + 2\ NO + 2.7\ H_2O + 1.1\ H^+ + 1.1\ e^- \quad (15)$$

of co-cultures, it will be illustrated with a theoretical example (equations 13-15).

In an anammox reactor-system without active *Nitrosomonas* cells (eq. 13) the NO_2^-/NH_3-ratio is about 1.3/1. If active ammonia oxidizing *Nitrosomonas* cells are present (eq. 14), the NO_2^-/NH_3-ratio of the system will decrease to about 1/2.5 (eq. 15) when both groups contribute equally to the ammonia consumption. According to this calculation, the presence of ammonia oxidizing *Nitrosomonas* cells should be indicated by a decreasing NO_2^-/NH_3-ratio. Hence, on the basis of this ratio, the contribution of *B. anammoxidans* and *Nitrosomonas* on the total ammonia conversion in a mixed population can be calculated. The results obtained from mixed populations (Schmidt *et al.*, 2002a) indicate a NO_2^-/NH_3-ratio of 1/1.2 in the presence of 10^8 cells ml^{-1} of *B. anammoxidans* and *N. eutropha*, respectively. According to equations 13-15, *B. anammoxidans* contributes about 75 % and *N. eutropha* about 25 % to the total anaerobic ammonia oxidation.

Simultaneous anaerobic ammonia oxidation activities of *B. anammoxidans* and *N. eutropha* are also detectable when ammonia is the limiting substrate. An anammox-system can be grown with low ammonium and nitrite concentrations. Nitrate is produced and accumulates in the medium. After *N. eutropha* (10^8 cells ml^{-1}) is added, the consumption of NO_2 and the production of nitrite and NO indicates an anaerobic ammonia oxidation activity by *N. eutropha*. Since the ammonia supply is not increased, *B. anammoxidans* and *N. eutropha* have to compete for this limited resource. As a result of the competition, the specific activity of *B. anammoxidans* decreases, which is indicated by a decreasing nitrate concentration. Obviously, both groups of ammonia oxidizing bacteria have similar capabilities (K_m, K_s values) to compete for the limiting substrate ammonia under these anoxic conditions.

B. anammoxidans and *Nitrosomonas* form stable co-cultures under anoxic conditions supplemented with a surplus of ammonium and gaseous NO_2 without the addition of nitrite (Schmidt *et al.*, 2002a). Under these conditions, ammonia is oxidized by *N. eutropha* to dinitrogen and nitrite. The nitrite generated is further converted by *B. anammoxidans* using ammonia as an electron donor. The model for this process is shown in Figure 6. The ammonia oxidation activity of *N. eutropha* is indicated by the NO_2 consumption and NO production profiles. The activity of *B. anammoxidans*, dependent on the nitrite production of *N. eutropha*, is indicated by the nitrate production rate. Under anoxic conditions, *N. eutropha* converts only 50-60 % of the consumed ammonia to nitrite (Schmidt *et al.*, 1997). Because of this nitrite limitation specific ammonia oxidation activity of *B. anammoxidans* is lower (0.9 mmol g $protein^{-1}$ h^{-1}) than specific ammonia oxidation activity of *N. eutropha* (1.6 mmol g $protein^{-1}$ h^{-1}). Although both groups of ammonia oxidizers compete for ammonia, a co-operation also seems to be possible when the NO_2-dependent ammonia oxidation of *Nitrosomonas* supplies *B. anammoxidans* with nitrite as oxidant. This might be of ecological relevance. On the one hand, ammonia oxidation of *B. anammoxidans* is restricted to anoxic environments. On the other hand, *Nitrosomonas*-like microorganisms need an oxidizing agent (O_2 or NO_2) that is only available under oxic conditions. The oxic/anoxic interface might be a suitable environment for both groups of ammonia oxidizers. The availability of ammonia and nitrite will be decisive for the outcome of the co-existence. At the oxic/anoxic interface, ammonia oxidation of *Nitrosomonas*-like microorganisms might be the major source of nitrite necessary for nitrite-dependent anammox metabolism. The products of this co-operation are mainly N_2 and small amounts of nitrate. When ammonia is the limiting substrate, both groups of ammonia oxidizers will compete for ammonia (Figure 6), and the specific affinities for ammonia might be decisive for the outcome of the competition.

A co-operation between anammox or anaerobically grown *Nitrosomonas* cells and nitrite oxidizers like *Nitrobacter* has yet to be documented. The cell number of nitrite oxidizers in a population under anoxic conditions dominated by anammox microorganisms was shown to be very low (Strous *et al.*, 1997). There are several explanations that might explain the existence of nitrite oxidizers in an anammox reactor system. First, these cells might have grown in the oxic activated sludge that was used to inoculate the anammox system and then remain in an inactive state as part of the anammox flocs. Second, the nitrite oxidizers might grow extremely slow under these anoxic conditions (heterotrophic nitrate reduction), and therefore a higher cell number cannot be expected. For anoxic growth, nitrite oxidizers need a suitable carbon source and nitrate. It might be speculated that, on the one hand, a favorable carbon source is not available and that, on the other hand, the nitrate concentration is too low. Even in the presence of a carbon source and sufficient nitrate, the nitrite oxidizers would have to compete with heterotrophic denitrifiers. Most likely, the nitrite oxidizers will be out-competed under these conditions by the fast growing heterotrophs.

6. Conclusions

The global nitrogen cycle is more complex that what was thought a few years ago. Beside the aerobic nitrification and the anaerobic denitrification shortcuts like anammox, anaerobic ammonia oxidation by nitrifiers, and aerobic denitrification have also been uncovered recently. The newly discovered metabolic pathways open up new possibilities for nitrogen removal from wastewater. More specifically, the paradigm that biological conversion of wastewater ammonia to dinitrogen gas necessitates complete oxidation to nitrate followed by heterotrophic denitrification, has become obsolete. New processes like Anammox, Canon (Sliekers *et al.*, 2002; 2003), and the NOx process (Schmidt *et al.*, 2002b; 2003) might contribute to meet the future challenges in wastewater treatment.

References

Abeliovich, A., and Vonhak, A. 1992. Anaerobic metabolism of *Nitrosomonas europaea*. Arch. Microbiol. 158: 267-270.

Aleem M.I.H. 1965. Path of carbon and assimilatory power in chemosynthetic bacteria. *I. Nitrobacter agilis*. Biochim. Biophys. Acta. 107: 14-28.

Allison, S.M., and Prosser J.I. 1993. Survival of ammonia oxidizing bacteria in air-dried soil. FEMS Microbiol. Lett. 79: 65-68.

Anderson, K.K., and Hooper, A.B. 1983. O_2 and H_2O are each the source of one O in NO_2^- produced from NH_3 by *Nitrosomonas*; ^{15}N-NMR evidence. FEBS Lett. 164: 236-240.

Arciero, D.M., and Hooper, A.B. 1993. Hydroxylamine oxidoreductase from *Nitrosomonas europaea* is a multimer of an octa-heme subunit. J. Biol. Chem. 268: 14645-14654.

Batchelor, S.E., Cooper, M., Chhabra, S.R., Glover, L.A., Stewart, G.S.A.B., Williams, P., and Prosser, J.I. 1997. Cell density-regulated recovery of starved biofilm populations of ammonia-oxidizing bacteria. Appl. Environ. Microbiol. 63: 2281-2286.

Bergmann, D.J., and Hooper, A.B. 1994. Sequence of the gene *amoB* for the 43-kDa polypeptide of ammonia monooxygenase of *Nitrosomonas europaea*. Biochem. Biophys. Res. Commun. 204: 759-762.

Bock, E. 1976. Growth of *Nitrobacter* in the presence of organic matter. II. Chemoorganic growth of *Nitrobacter agilis*, Arch. Microbiol. 108: 305-312.

Bock, E., Wilderer, P.A., and Freitag, A. 1988. Growth of *Nitrobacter* in the absence of dissolved oxygen. Water Res. 22: 245-250.

Bock, E., Koops, H.-P., Möller, U.C., and Rudert, M. 1990. A new facultatively nitrite oxidizing bacterium, *Nitrobacter vulgaris* sp. nov. Arch. Microbiol. 153: 105-110.

Bock, E., Koops, H.-P., Harms, H., and Ahlers, B. 1991. The biochemistry of nitrifying organisms. In: Variations of Autotrophic Life. J.M. Shively, ed. Academic Press, London, UK. p. 171-200.

Bock, E., Schmidt, I., Stüven, R., and Zart, D. 1995. Nitrogen loss caused by denitrifying *Nitrosomonas* cells using ammonia or hydrogen as electron donors and nitrite as electron acceptor. Arch. Microbiol. 163: 16-20.

Bock, E. and Wagner, M. 2002. Oxidation of inorganic nitrogen compounds as an energy source. In: The Prokaryotes, 4th Edn. M. Dworkin, S. Falkow, E. Rosenberg, K.-H. Schleifer, E. Stackebrandt, eds. Springer-Verlag, New York.

Boettcher, B., and Koops, H.-P. 1994. Growth of lithotrophic ammonium-oxidizing bacteria on hydroxylamine. FEMS Microbiol. Lett. 122: 253-266.

Bothe, H., Jost, G., Schloter, M., Ward, B.B., and Witzel, K. 2000. Molecular analysis of ammonia oxidation and denitrification in natural environments. FEMS Microbiol. Rev. 24: 673-690.

Broda, E. 1977. Two kinds of lithotrophs missing in nature. Zeitschrift für Allg. Mikrobiologie. 17: 491-493.

Burton, S.A.Q., and Prosser, J.I. 2001. Autotrophic ammonia oxidation at low pH through urea hydrolysis. Appl. Environ. Microbiol. 67: 2952-2957.

Carr, G.J., and Ferguson, S.J. 1990. Nitric oxide formed by nitrite reductase of *Paracoccus denitrificans* is sufficiently stable to inhibit cytochrome oxidase activity and is reduced by its reductase under aerobic conditions. Biochim. Biophys. Acta. 1017: 57-62.

Chain, P., Lamerdin, J., Larimer, F., Regala, W., Lao, V., Land, M., Hauser, L., Hooper, A., Klotz, M., Norton, J., Sayavedra-Soto, L., Arciero, D., Hommes, N., Whittaker, M., and Arp, D. 2003. Complete genome sequence of the ammonia-oxidizing bacterium and obligate chemolitho-autotroph *Nitrosomonas europaea*. J. Bacteriol. 185: 2759-2773. Erratum in: J. Bacteriol. 2003 185: 6496.

Chalk, P.M., and Smith, C.J. 1983. Chemodenitrification. Dev. Plant Soil Sci. 9: 65-89.

Dalsgaard, T., and Thamdrup, B. 2002. Factors controlling anaerobic ammonium oxidation with nitrite in marine sediments. Appl. Environ. Microbiol. 68:3802-3808.

Dalsgaard, T., Canfield, D.E., Petersen, J., Thamdrup, B., and Acuña-González, J. 2003. N_2 production by the anammox reaction in the anoxic water column of Golfo Dulce, Costa Rica. Nature. 422: 606-608.

De Boer, W., Klein Gunnewiek, P.J.A., Veenhuis, M., Bock, E., and Laanbroek H.J. 1991. Nitrification at low pH by aggregated chemolithotrophic bacteria. Appl. Environ. Microbiol. 57: 3600-3604.

De-Boer, W., and Kowalchuk, G.A. 2001. Nitrification in acid soils: micro-organisms and mechanisms. Soil Biol. Biochem. 33: 853-866.

De Bruijn, P., van de Graaf, A.A., Jetten, M.S.M., Robertson, L.A., and Kuenen, J.G. 1995. Growth of *Nitrosomonas europaea* on hydroxylamine. FEMS Microbiol. Lett. 125: 179-184.

DiSpirito, A.A., Taaffe, L.R., and Hooper, A.B. 1985. Localization and concentration of hydroxylamine oxidoreductase and cytochromes c-552, c-554, c_m-553, c_m-552 and a in *Nitrosomonas europaea*. Biochim. Biophys. Acta. 806: 320-330.

Egli, K., Franger, U., Alvarez, P.J.J., Siegrist, H., Vandermeer, J.R., and Zehnder, A.J.B. 2001. Enrichment and characterization of an anammox bacterium from a rotating biological contactor treating ammonium-rich leachate. Arch. Microbiol. 175: 198-207.

Egorova, L.A., and Loginova, I.G. 1975. Distribution of highly thermophilic, nonsporulating bacteria in the hot springs of Tadzhikistan. Mikrobiologiia. 44: 938-942.

Ehrich, S., Behrens, D., Lebedeva, E., Ludwig, W., and Bock, E. 1995. A new obligately chemolithoautotrophic, nitrite-oxidizing bacterium, Nitrospira moscoviensis sp. Nov. and its phylogenetic relationship, Arch. Microbiol. 164: 16-23.

Freitag, A., Rudert, M., and Bock, E. 1987. Growth of *Nitrobacter* by dissimilatoric nitrate reduction. FEMS Microbiol. Lett. 48: 105-109.

Fuerst, J.A. 1995. The planctomycetes: emerging models for microbial ecology, evolution and cell biology. Microbiology. 141: 1493-1506.

Fukuoka, M., Fukumori, Y., and Yamanaka, T. 1987. *Nitrobacter winogradskyi* cytochrome a_1c_1 is an iron-sulfur molybdo-enzyme having hemes a and c. J. Biochem. 102: 525-530.

Griepenburg, U., Ward-Rainey, N., Mohamed, S., Schlesner, H., Marxen, H., Rainey, F.A., Stackebrandt, E., and Auling, G. 1999. Phylogenetic diversity, polyamine pattern and DNA base composition of members of the order *Planctomycetales*. Int. J. Syst. Bacteriol. 49: 689-696.

Hanafy, K.A., Krumenacker, J.S., and Murad, F. 2001. NO, nitrotyrosine, and cyclic GMP in signal transduction. Med. Sci. Monit. 7: 801-819.

Hanaki, K., Wantawin, C., and Ogaki, S. 1990. Nitrification at low levels of dissolved oxygen with and without organic loading in a suspended-growth reactor. Wat. Res. 24: 297-302.

Head, I.M., Hiorns, W.D., Embley, T.M., and McCarthy, A.J. 1993. The phylogeny of autotrophic ammonia-oxidizing bacteria as determined by analysis of 16S ribosomal RNA gene sequences. J. Gen. Microbiol. 139: 1147-1153.

Helder, M.N., and de Vries, E.G. 1983 Estuarine nitrite maxima and nitrifying bacteria (Ems-Dollard estuary). Neth. J. Sea Res. 17: 1-18.

Helmer, C., Tromm, C., Hippen, A., Rosenwinkel, K.H., Seyfried, C.F., and Kunst, S. 2001. Single stage biological nitrogen removal by nitritation and anaerobic ammonium oxidation in biofilm systems. Wat. Sci. Technol. 43: 311-320.

Henry, Y., Ducrocq, C., Drapier, J.-C., Servent, D., Pellat, C., and Guissani, A. 1991. Nitric oxide, a biological effector. Electron paramagnetic resonance detection of nitrosyl-iron-protein complexes in whole cells. Eur. Biophys. J. 20: 1-15.

Hollocher, T.C., Kumar, S., and Nicholas, D.J.D. 1981. Oxidation of ammonia by *Nitrosomonas* europaea: definitive ^{18}O-tracer evidence that hydroxylamine formation involves a monooxygenase. J. Biol. Chem. 256: 10834-10836.

Hollocher, T.C., Kumar, S., and Nicholas, D.J.D. 1982. Respiration-dependent proton translocation in *Nitrosomonas europaea* and its apparent absence in *Nitrobacter agilis* during inorganic oxidations, J. Bacteriol. 149: 1013-1020.

Hooper, A.B. 1969. Biochemical basis of obligate autotrophy in *Nitrosomonas europaea*. J. Bacteriol. 97: 776-779.

Hooper, A.B., and Terry, K.R. 1977. Hydroxylamine oxidoreductase from *Nitrosomonas*: inactivation by hydrogen-peroxide. Biochemistry. 16: 455-459.

Hooper, A.B., Maxwell, P.C., and Terry, K.R. 1978. Hydroxylamine oxidoreductase from *Nitrosomonas europaea:* absorption spectra and content of heme and metal. Biochemistry. 17: 2984-2989.

Hooper, A.B., and Terry, K.R. 1979. Hydroxylamine oxidoreductase of *Nitrosomonas* - production of NO from hydroxylamine. Biochim. Biophys. Acta. 571: 12-20.

Hooper, A.B., and Balny, C. 1982. Reaction of oxygen with hydroxylamine oxidoreductase of *Nitrosomonas*: Fast kinetics. FEBS Lett. 144: 299-303.

Hooper, A.B. 1984. Ammonia oxidation and energy transduction in the nitrifying bacteria. In: Microbial Chemoautotrophy. W.R. Strohl, and O.H. Tuovinen eds. Ohio State University Press, Columbus, Ohio. p. 133-167.

Hooper, A.B., DiSpirito, A.A., Olson, T.C., Anderson, K.A., Cunningham, W., and Taaffe, L.R. 1984. Generation of the proton gradient by a periplasmic dehydrogenase. In: Microbial Growth on C_1 Compounds. R.L. Crawford, and R.S. Hanson, eds. Am. Soc. Microbiol., Washington D.C. p. 53-58.

Hooper, A.B., and DiSpirito, A.A. 1985. In bacteria which grow on simple reductants generation of a proton gradient involves extracytoplasmic oxidation of substrate. Microbiol. Rev. 49: 140-157.

Hooper, A.B. 1989. Biochemistry of the nitrifying lithoautotrophic bacteria. In: Autotrophic Bacteria. H.G. Schlegel, and B. Bowien, eds. Science Tech., Madison, Wisconsin. p. 239-265.

Hooper, A.B., Vannelli, T., Bergmann, D.J., and Arciero, D.M. 1997. Enzymology of the oxidation of ammonia to nitrite by bacteria. Antonie van Leeuwenhoek. 71: 59-67.

Hoppert, M., Mahony, T.J., Mayer, F., and Miller, D.J. 1995. Quaternary structure of the hydroxylamine oxidoreductase from *Nitrosomonas europaea*. Arch. Microbiol. 163: 300-306.

Hyman, M.R., and Wood, P.M. 1985. Suicidal inactivation and labeling of ammonia monooxygenase by acetylene. Biochem. J. 227: 719-725.

Hyman, M.R., and Arp, D.J. 1992. $^{14}C_2H_2$ and $^{14}CO_2$-labeling studies of the de novo synthesis of polypeptides by *Nitrosomonas europaea* during recovery from acetylene and light inactivation of ammonia monooxygenase. J. Biol. Chem. 267: 1534-1545.

Hyman, M.R., and Arp, D.J. 1995. Effects of ammonia on the de novo synthesis of polypeptides in cells of *Nitrosomonas europaea* denied ammonia as an energy source. J. Biol. Chem. 177: 4974-4979.

Igarashi, N., Moriyama, H., Fujiwara, T., Fukumori, Y., and Tanaka, N. 1997. The 2.8 Å structure of hydroxylamine oxidoreductase from a nitrifying chemoautotrophic bacterium. *Nitrosomonas europaea*. Nat. Struct. Biol. 4: 276-284.

Jetten, M.S.M., Strous, M., van de Pas-Schoonen, K.T., Schalk, J., van Dongen, L., van de Graaf, A.A., Logemann, S., Muyzer, G., van Loosdrecht, M.C.M., and Kuenen J.G. 1999. The anaerobic oxidation of ammonium. FEMS Microbiol. Rev. 22: 421-437.

Jetten, M.S.M. 2001. New pathways for ammonia conversion in soil and aquatic systems. Plant Soil. 230: 9-19.

Jetten, M.S.M. 2001. Adembenemende en ademloze Microbiologie. University of Nijmegen, ISBN 90-9015227-X.

Jetten, M.S.M., Wagner, M., Fuerst, J., van Loosdrecht, M., Kuenen, G., and Strous, M. 2001. Microbiology and application of the anaerobic ammonium oxidation ('anammox') process. Cur. Opinion Biotechnol. 12: 283-288.

Kleiner, D. 1981. The transport of NH_3 and NH_4^+ across biological membranes. Biochim. Biophys. Acta. 639: 41-52.

Kuenen, J.G., and Jetten, M.S.M. 2001. Extraordinary anaerobic ammonium-oxidizing bacteria. ASM News. 67: 456-463.

Kuai, L., and Verstraete, W. 1998. Ammonium removal by the oxygen-limited autotrophic nitrification-denitrification system. Appl. Environ. Microbiol. 64: 4500-4506.

Kuypers, M.M.M., Sliekers, A.O., Lavik, G., Schmid, M, Jørgensen, B.B., Kuenen, J.G., Sinninghe Damsté, J.S., Strous M, Jetten, M.S.M. 2003. Anaerobic ammonium oxidation by anammox bacteria in the Black Sea. Nature. 422: 608-611.

Lindsay, M.R., Webb, R.I., Strous, M., Jetten, M.S.M, Butler, M.K., Forde, R.J., and Fuerst, J.A. 2001. Cell compartmentalization in planctomycetes: novel types of structural organization for the bacterial cell. Arch. Microbiol. 175: 413-429.

Lipscomb, J.D., and Hooper, A.B. 1982. Resolution of multiple heme centers of hydroxylamine oxidoreductase from *Nitrosomonas:* 1. Electron paramagnetic resonance spectroscopy. Biochemistry. 21: 3965-3972.

Mancinelli, R.L., and McKay, C.P. 1983. Effects of nitric oxide and nitrogen dioxide on bacterial growth. Appl. Environ. Microbiol. 46: 198-202.

Mansch, R., and Bock, E. 1998. Biodeterioration of natural stone with special reference to nitrifying bacteria. Biodegradation. 9: 47-64.

McTavish, H., Fuchs, J.A., and Hooper, A.B. 1993. Sequence of the gene coding for ammonia monooxygenase in *Nitrosomonas europaea*. J. Bacteriol. 175: 2436-2444.

Moews, P.C., and Audrieth, L.F. 1959. The autoxidation of hydroxylamine. J. Inorg. Nucl. Chem. 11: 242-246.

Mulder, A., van de Graaf, A.A., Robertson, L.A., and Kuenen, J.G. 1995. Anaerobic ammonium oxidation discovered in a denitrifying fluidized bed reactor. FEMS Microbiol. Ecol. 16: 177-184.

Oerther, D.B., Pernthaler, J., Schramm, A., Amann, R., and Raskin, L. 2000. Monitoring precursor 16S rRNAs of *Acinetobacter* spp. in activated sludge wastewater treatment systems. Appl. Environ. Microbiol. 66: 2154-2165.

Olson, T.C., and Hooper, A.B. 1983. Energy coupling in the bacterial oxidation of small molecules: an extracytoplasmic dehydrogenase in *Nitrosomonas*. FEMS Microbiol. Lett. 19: 47-50.

Pinck, C. 2001. Immunologische Untersuchung am Schlüsselenzym der Ammoniakoxidanten. Doctoral thesis, University of Hamburg.

Pinck, C., Coeur, C., Potier, P., and Bock, E. 2001. Polyclonal antibodies recognizing the AmoB protein of ammonia oxidizers of the beta-subclass of the class Proteobacteria. Appl. Environ. Microbiol. 67: 118-124.

Poth, M., and Focht, D.D. 1985. ^{15}N kinetic analysis of N_2O production by *Nitrosomonas europaea*: an examination of nitrifier denitrification. Appl. Environ. Microbiol. 49: 1134-1141.

Poulsen, L.K., Ballard, G., and Stahl, D.A. 1993. Use of rRNA fluorescence *in situ* hybridization for measuring the activity of single cells in young and established biofilms. Appl. Environ. Microbiol. 59: 1354-1360.

Prince, R.C., Larroque, C., and Hooper, A.B. 1983. Resolution of the hemes of hydroxylamine oxidoreductase by redox potentiometry and optical spectroscopy. FEBS Lett. 163: 25-27.

Prosser, J.I. 1989. Autotrophic nitrification in bacteria, In: Advances in Microbial Physiology, Vol. 30. A.H. Rose, D.W. Tempest, eds. Academic Press, London. p. 125-181.

Purkhold, U., Pommering-Röser, A., Juretschko, S., Schmid, M.C., Koops, H.-P., and Wagner, M. 2000. Phylogeny of all recognized species of ammonia oxidizers based on comparative 16S rRNA and *amoA* sequence analysis: Implications for molecular diversity surveys Appl. Environ. Microbiol. 66: 5368-5382.

Remde, A., and Conrad, R. 1990. Production of nitric oxide in *Nitrosomonas europaea* by reduction of nitrite. Arch. Microbiol. 154: 187-191.

Russel, J.B., and Cook, G.M. 1995. Energetics of bacterial growth: Balance of anabolic and catabolic reactions. Microbiol. Rev. 59: 48-62.

Schalk, J., de Vries, S., Kuenen, J.G., and Jetten, M.S.M. 2000. A novel hydroxylamine oxidoreductase involved in the Anammox process. Biochemistry. 39: 5405-5412.

Schalk, J. 2000. A study of the metabolic pathway of anaerobic ammonium oxidation. Doctoral thesis, Technical University Delft.

Schlesner, H., and Stackebrandt, E. 1986. Assignment of the genera *Planctomyces* and *Pirella* to a new family *Planctomycetaceae* fam. nov. and description of the order *Planctomycetales* ord. nov. Syst. Appl. Microbiol. 8: 174-176.

Schmid, M., Twachtmann, U., Klein, M., Strous, M., Juretschko, S., Jetten, M., Metzger, J., Schleifer, K.H., and Wagner,

M. 2000. Molecular evidence for genus level diversity of bacteria capable of catalyzing anaerobic ammonium oxidation. Syst. Appl. Microbiol. 23: 93-106.

Schmid, M., Schmitz-Esser, S., Jetten, M., and Wagner, M. 2001. 16S-23S rDNA intergenic spacer and 23S rDNA of anaerobic ammonium oxidizing bacteria: implications for phylogeny and *in situ* detection. Environ. Microbiol. 3: 450-459.

Schmidt, I., and Bock, E. 1997a. Anaerobic ammonia oxidation with nitrogen dioxide by *Nitrosomonas eutropha*. Arch. Microbiol. 167: 106-111.

Schmidt, I. 1997b. Anaerobe Ammoniakoxidation von *Nitrosomonas eutropha*. Doctoral thesis, Universität Hamburg.

Schmidt, I., and Bock, E. 1998. Anaerobic ammonia oxidation by cell free extracts of *Nitrosomonas eutropha*. Antonie van Leeuwenhoek. 73: 271-278.

Schmidt, I., Look, C., Bock, .E, and Jetten, M.S. 2004a. Ammonium and hydroxylamine uptake and accumulation in *Nitrosomonas*. Microbiology. 150:1405-1412.

Schmidt, I., Steenbakkers, P.J., op den Camp, H.J., Schmidt, K., and Jetten, M.S.2004b. Physiologic and proteomic evidence for a role of nitric oxide in biofilm formation by *Nitrosomonas europaea* and other ammonia oxidizers. J. Bacteriol. 186:2781-2788.

Schmidt, I., Zart, D., and Bock, E. 2001a. Effects of gaseous NO_2 on cells of *Nitrosomonas eutropha* previously incapable of using ammonia as an energy source. Antonie van Leeuwenhoek. 79: 39-47.

Schmidt, I., Zart, D., and Bock, E. 2001b. Gaseous NO_2 as a regulator for ammonia oxidation of *Nitrosomonas eutropha*. Antonie van Leeuwenhoek. 79: 311-318.

Schmidt, I., Bock, E., and Jetten, M.S.M. 2001c. Ammonia oxidation by *Nitrosomonas eutropha* with NO_2 as oxidant is not inhibited by acetylene. Microbiology. 147: 2247-2253.

Schmidt, I., Hermelink, C., van de Pas-Schoonen, K., Strous, M., op den Camp, H.J., Kuenen, J.G., and Jetten, M.S.M. 2002a. Anaerobic ammonia oxidation in the presence of nitrogen oxides (NOx) by two different lithotrophs. Appl. Environ. Microbiol. 68: 5351-5357.

Schmidt, I., Zart, D., Stüven, R., Bock, E. 2002b. A new process for ammonia removal from waste water. Chemical & Engineering Technology 25, Eng. Life Sci. 2: 59-62.

Schmidt, I., Sliekers, O., Schmid, M., Bock, E., Kuenen, J.G., Jetten, M.S.M., and Strous, M. 2003. New concepts of microbial treatment processes for the nitrogen removal in wastewater. FEMS Microbiol. Rev. 27: 481-492.

Sinninghe Damsté., J.S., Strous, M., Rijpstra, W.I.C., Hopmans, E.C., Geenevasen, J.A.J., Van Duin, A.C.T., Van Niftrik, L.A., and Jetten, M.S.M. 2002. Linearly concatenated cyclobutane lipids form a dense bacterial membrane. Nature. 419: 708-712.

Skiba, U., Smith, K.A., and Fowler, D. 1993. Nitrification and denitrification as sources of nitric oxide and nitrous oxide in a sandy loam soil. Soil. Biol. Biochem. 25: 1527-1536.

Sliekers, A.O., Derwort, N., Campos Gomez, J.L., Strous, M., Kuenen, J.G., and Jetten M.S.M. 2002. Completely autotrophic nitrogen removal over nitrite in one single reactor. Wat. Res. 36: 2475-2482.

Sliekers, A.O., Third, K.A., Abma, W., Kuenen, J.G., and Jetten, M.S. 2003. CANON and Anammox in a gas-lift reactor. FEMS Microbiol Lett. 218:339-344.

Smith, A.J., and Hoare, D.S. 1968. Acetate assimilation by *Nitrobacter agilis* in relation to its "obligate autotrophy." J. Bacteriol. 95: 844-855.

Sorokin, D., Muyzer, G., Brinkhoff, T., Kuenen, J.G., and Jetten M.S.M. 1998. Isolation and characterization of a novel facultatively alkaliphilic *Nitrobacter* species, *N. alkalicus* sp. nov. Arch. Microbiol. 170: 345-352.

Sorokin, D., Tourova, T., Schmid, M., Wagner, M., Koops, H.-P., Kuenen, J.G., and Jetten M.S.M 2001. Isolation and properties of obligately chemolithotrophic and extremely alkali-tolerant ammonia-oxidizing bacteria from Mongolian soda lakes. Arch. Microbiol. 176: 170-177.

Spieck, E., Meinecke, M., and Bock, E. 1992. Taxonomic diversity of Nitrosovibrio strains isolated from building sandstones. FEMS Microbiol. Ecol. 102: 21-26.

Spieck, E., Müller, S., Engel, A., Mandelkow, E., Patel, H., and Bock, E. 1996. Two-dimensional structure of membrane-bound nitrite-oxidoreductase from Nitrobacter hamburgensis. J. Struct. Biol. 117: 117-123.

Stein, L.Y., and Arp, D.J. 1998. Loss of ammonia monooxygenase activity in *Nitrosomonas europaea* upon exposure to nitrite. Appl. Environ. Microbiol. 64: 4098-4102.

Steinmüller, W., and Bock, E. 1976. Growth of *Nitrobacter* in the presence of organic matter. I. Mixotrophic growth, Arch. Microbiol. 108: 299-304.

Strous, M., van Gerven, E., Kuenen, J.G., and Jetten, M.S.M. 1997. Effects of aerobic and microaerobic conditions on anaerobic ammonium-oxidation (Anammox) sludge. Appl. Environ. Microbiol. 63: 2446-2448.

Strous, M., Heijnen, J.J., Kuenen, J.G., and Jetten, M.S.M. 1998. The sequencing batch reactor as a powerful tool to study very slowly growing micro-organisms. Appl. Microbiol. Biotechnol. 50: 589-596.

Strous, M., Fuerst, J., Kramer, E., Logemann, S., Muyzer, G., van de Pas, K., Webb, R., Kuenen, J.G., and Jetten M.S.M. 1999. Missing lithotroph identified as new planctomycete. Nature. 400: 446-449.

Strous, M., Kuenen, J.G., and Jetten, M.S.M. 1999. Key physiology of anaerobic ammonium oxidation. Appl. Environ. Microbiol. 65: 3248-3250.

Stüven, R., Vollmer, M., and Bock, E. 1992. The impact of organic matter on nitric oxide formation by *Nitrosomonas europaea*. Arch. Microbiol. 158: 439-443.

Sundermeyer, H., and Bock, E. 1981. Energy metabolism of autotrophically and heterotrophically grown cells of *Nitrobacter winogradskyi*. Arch. Microbiol. 130: 250-254.

Sundermeyer-Klinger, H., Meyer, W., Warninghoff, B., and Bock, E. 1984. Membrane-bound nitrite oxidoreductase of *Nitrobacter*: evidence for a nitrate reductase system, Arch. Microbiol. 140: 153-158.

Suzuki, I., Kwok, S.-C., Dular. U., and Tsang, D.C.Y. 1981. Cell-free ammonia-oxidizing system of *Nitrosomonas europaea*: general conditions and properties. Can. J. Biochem. 59: 477-483.

Tanaka Y., Fukumori, Y., and Yamanaka, T. 1983. Purification of cytochrome a_1c_1 from *Nitrobacter agilis* and characterization of nitrite oxidation system of the bacterium. Arch. Microbiol. 135: 265-271.

Thamdrup, B., and Dalsgaard, T. 2000. The fate of ammonium in anoxic manganese oxide-rich marine sediment. Geochim. et Cosmochim. Acta. 64: 4157-4164.

Thamdrup. B., and Dalsgaard, T. 2002. Production of N_2 through anaerobic ammonium oxidation coupled to nitrate reduction in marine sediments. Appl. Environ. Microbiol. 68: 1312-1318.

Third, K.A., Sliekers, A.O., Kuenen, J.G., and Jetten, M.S.M. 2002. The CANON System (Completely Autotrophic Nitrogen-removal Over Nitrite) under Ammonia Limitation: Interaction and Competition between Three Groups of Bacteria. Sys. Appl. Microbiol. 24: 588-596.

Tsang, D.C.Y., and Suzuki, I. 1982. Cytochrome c554 as a possible electron donor in the hydroxylation of ammonia and carbon monoxide in *Nitrosomonas europaea*. Can. J. Biochem. 60: 1018-1024.

Van de Graaf, A.A., De Bruijn, P., Robertson, L.A., Jetten, M.S.M., and Kuenen, J.G. 1997. Metabolic pathway of anaerobic ammonium oxidation on the basis of N-15 studies in a fluidized bed reactor. Microbiology. 143: 2415-2421.

Watson, S.W., Bock, E., Valois, F.W., Waterbury, J.B., and Schlosser, U. 1986. *Nitrospira marina* gen. nov. sp. nov.: a chemolithotrophic nitrite oxidizing bacterium. Arch. Microbiol. 144: 1-7.

Watson, S.W., Bock, E., Harms, H., Koops, H.-P., and Hooper, A.B. 1989. Genera of ammonia-oxidizing bacteria, In: Bergey's Manual of Systematic Bacteriology. J.T. Staley, M.P. Bryant, N. Pfennig, and J.G. Holt, eds. Williams & Wilkins, Baltimore, Maryland. p. 1822-1834.

Weber, S., Stubner, S., and Conrad, R. 2001, Bacterial populations colonizing and degrading rice straw in anoxic paddy soil. Appl. Environ. Microbiol. 67: 1318-1327.

Wheelis, M. 1984. Energy conservation and pyridine nucleotide reduction in chemoautotrophic bacteria: a thermodynamic analysis, Arch. Microbiol. 138: 166-169.

Whittaker, M., Bergmann, D., Arciero, D., and Hooper, A.B. 2000. Electron transfer during the oxidation of ammonia by the chemolithotrophic bacterium *Nitrosomonas europaea*. Biochim. Biophys. Acta 1459: 346-355.

Wilson, K., Sprent, J.I., and Hopkins, D.W. 1997. Nitrification in aquatic soils. Nature. 385: 404.

Winogradsky, S. 1949. Principes de la microbiologie oecologique. In: Microbiologie du Sol - Problèmes et Méthodes - Cinquante ans de Recherches. Masson et cie éditeurs, Paris.

Wood, P.M. 1986. Nitrification as a bacterial energy source. In: Nitrification. J.I. Prosser, ed. Soc. Gen. Microbiol. (IRL Press), Washington, D.C. p. 39-62.

Ye, R.W., Arunakumari, A., Averill, B.A., and Tiedje, J.M. 1992. Mutants of *Pseudomonas fluorescens* deficient in dissimilatory nitrite reduction are also altered in nitric oxide reduction. J. Bacteriol. 174: 2560-2564.

Zart, D., and Bock, E. 1998. High rate of aerobic nitrification and denitrification by *Nitrosomonas eutropha* grown in a fermentor with complete biomass retention in the presence of gaseous NO_2 or NO. Arch. Microbiol. 169: 282-286.

Zart, D., Schmidt, I., and Bock, E. 2000. Significance of gaseous NO for ammonia oxidation by *Nitrosomonas eutropha*. Antonie van Leeuwenhoek 77: 49-55.

Zumft, W.G. 1993. The biological role of nitric oxide in bacteria. Arch. Microbiol. 160: 253-264.

Chapter 16

Reductive Dehalogenation of Polychlorinated Benzenes and Dioxins

Lorenz Adrian* and Ute Lechner

Abstract

Highly chlorinated aromatic compounds like chlorobenzenes, polychlorinated dibenzo-*p*-dioxins, dibenzofurans and biphenyls are toxic, poorly water-soluble compounds, which are persistent in the environment and tend to accumulate in food chains. They are resistant to most aerobic biodegradation processes. Under anaerobic conditions, however, they can serve as electron acceptors in a respiratory process called dehalorespiration, thereby undergoing a reductive dehalogenation. In this review we summarize the current knowledge about the reductive dechlorination of chlorobenzenes and polychlorinated dioxins. A variety of pathways of microbial reductive dechlorination of dioxins and chlorobenzenes are described suggesting the involvement of different microbes in the observed dechlorination processes. The first described anaerobic bacterium able to respire with chlorinated benzenes and dioxins belongs to the genus *Dehalococcoides*, forming an isolated cluster within the green non-sulfur bacteria. Therefore, this review also summarizes physiological and biochemical properties of *Dehalococcoides*, including the remarkable specialization to dehalorespiration as a lifestyle and the capability to use a broad spectrum of halogenated compounds as electron acceptors. Biostimulation of dehalogenating bacteria present in contaminated anaerobic habitats can be a potentially useful approach to enhance natural attenuation processes and is discussed with respect to the versatility of *Dehalococcoides* species.

1. Introduction

Substrate level phosphorylation (fermentation) and electron transport coupled phosphorylation (respiration) have been identified as the two processes involved in energy fixation by microorganisms. For respiration, the difference in the redox potentials of two redox pairs is exploited to generate a proton gradient across a membrane which can be used for ATP regeneration. Under anaerobic conditions the available electron acceptors often limit respiratory growth. Different terminal reductases have evolved that reduce nitrate, sulfate, sulfite, Fe(III), Mn(IV), fumarate or other electron acceptors. Few bacteria are able to use halogenated compounds as electron acceptors in an anaerobic respiration, a process designated dehalorespiration (Holliger *et al.*, 1999). By transferring two electrons on a halogenated compound a halogen substituent is cleaved off as anion (Figure 1). The electrons stem from the oxidation of an electron donor such as hydrogen, formate, pyruvate or lactate, and flow across the membrane to reduce the halogenated molecule by the activity of a membrane-bound dehalogenase. A proton-gradient across the membrane generated by components of this electron transport chain drives a membrane-bound ATPase (Mohn and Tiedje, 1992; Schumacher and Holliger, 1996; Miller *et al.*, 1997a). Reductive dehalogenation has attracted tremendous interest during the last decade because it can lead to the efficient detoxification of highly recalcitrant chlorinated compounds including toxic congeners of polychlorinated dibenzo-*p*-dioxins, dibenzofurans, and biphenyls.

Several bacteria have been identified as being able to use the energy generated by reductive dehalogenation for growth. Reductive dechlorination of halogenated phenols, chlorinated benzoates, trichloroethane and trichloroacetic acid, as well as reductive dechlorination of trichloroethene and tetrachloroethene to *cis*-dichloroethene has been shown with pure cultures of the low-G+C Gram-positive genera *Desulfitobacterium* and *Dehalobacter* and δ- and ε-*Proteobacteria* (*Desulfomonile*, *Desulfovibrio*, *Anaeromyxobacter*, *Desulfuromonas*, *Trichlorobacter* and *Sulfurospirillum*) (formerly *Dehalospirillum*, Luijten *et al.*, 2003) (for reviews see El Fantroussi *et al.*, 1988; van Agteren *et al.*, 1998; Holliger *et al.*, 1999, see also Boyle *et al.*,

*For correspondence email: lorenz.adrian@tu-berlin.de

Abbreviations:
DiCDD, dichlorodibenzo-*p*-dioxin,
MCDD, monochlorodibenzo-*p*-dioxin,
TeCDD, tetrachlorodibenzo-*p*-dioxin,
TrCDD, trichlorodibenzo-*p*-dioxin,
PCB, polychlorinated biphenyl,
PCDD, polychlorinated dibenzo-*p*-dioxin,
PeCDD, pentachlorodibenzo-*p*-dioxin

$$R - Cl + 2e^- + H^+ \longrightarrow R - H + Cl^-$$

Figure 1 Scheme of reductive dehalogenation reactions

1999; Sun *et al.*, 2000; De Wever *et al.*, 2000; Sanford *et al.*, 2002; Sun *et al.*, 2002). However, reductive dechlorination of chlorinated benzenes, chlorinated dioxins, chlorinated furans, chlorinated biphenyls, dichloroethene or vinyl chloride has been assigned up to now only to bacterial strains with a phylogenetic affiliation to the green non-sulfur bacteria, in particular to the genus *Dehalococcoides*. In this chapter we focus on the reductive dechlorination of chlorinated aromatics especially dioxins and benzenes. However, we will also refer to recent results in the field of reductive dehalogenation of chloroaliphatic compounds and PCBs that highlight the remarkable substrate spectrum of the genus *Dehalococcoides*. A short introduction to biochemistry and genetics of dehalorespiration is given.

1.1. Toxicity

Chlorinated benzenes, dioxins and biphenyls (Figure 2) are similar in many chemical characteristics: A common structural characteristic of the compounds is that they do not contain any hydroxyl, carbonyl, keto or carboxyl groups. All three substance classes are aromatic molecules which are highly hydrophobic leading to their accumulation in the food chain. Each group consists of many congeners substituted with chlorine atoms at different positions. The various congeners are different with respect to their physico-chemical characteristics but also with respect to their toxicity and accessibility to microbial degradation. As a rule of thumb, the solubility in water decreases with increasing numbers of chlorine substituents. Also the degradability by aerobic microorganisms decreases drastically with increasing numbers of chlorine substituents. Highly chlorinated benzenes, biphenyls and dioxins are not degraded under aerobic conditions. In contrast, the more highly chlorinated congeners are amenable to reductive dehalogenation under anaerobic conditions. The toxicity, however, is not directly related to the number of chlorine substituents. The dioxins, especially the laterally substituted congeners, are toxic to humans. The 2,3,7,8-tetrachlorodibenzo-*p*-dioxin with its four lateral substitutions is the most toxic congener of the polychlorinated dibenzo-*p*-dioxins due to the planar conformation of the molecule. Planar structured dioxins (and also other chemicals such as biphenyls) act as ligands of the cytosolic arylhydrocarbon (Ah) receptor, which then forms a complex with the nuclear translocator protein Arnt. The complex binds to so called xenobiotic response elements in DNA and induces expression of a series of genes, e.g. those that encode detoxifying enzymes. In addition, this complex can also mimic the effect of oestrogen, because it binds to oestrogen receptors, leading to activation of transcription of oestrogen-regulated genes and respective oestrogenic effects (Ohtake *et al.*, 2003). This gives rise to adverse, endocrine-disrupting actions of dioxin-type environmental contaminants (Geyer *et al.*, 2000). Chlorobenzenes show mostly sub acute and chronic toxicity with trichlorobenzenes being the most toxic congeners (Koch, 1991).

1.2. Natural Formation of Halogenated Benzenes and Dioxins

Although alarming concentrations of halogenated benzenes and dioxins in contaminated environments are clearly of anthropogenic origin, it is now widely accepted that they are also formed by natural processes. For instance, 1,2,3,4-tetrachlorobenzene was found as a component of the oil of the needle rush *Juncus*

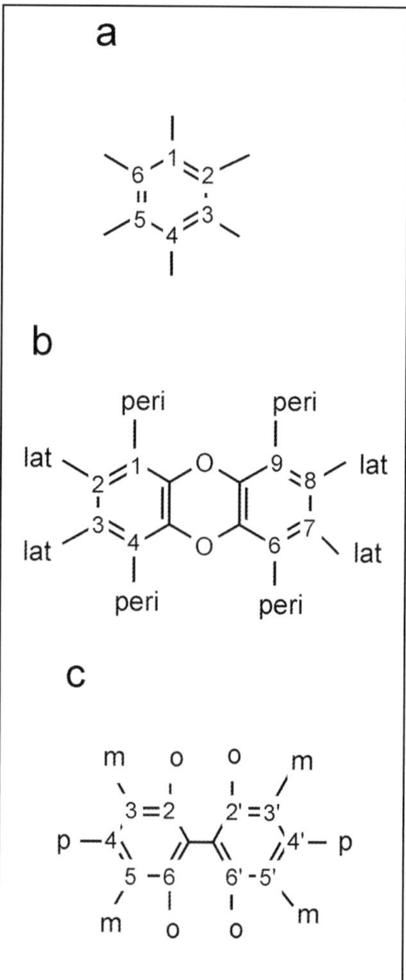

Figure 2 Nomenclature for chlorobenzenes (a) polychlorinated dibenzo-*p*-dioxins (b) and polychlorinated biphenyls (c). lat - lateral, peri - peripheral, p - para, m - meta, o – ortho.

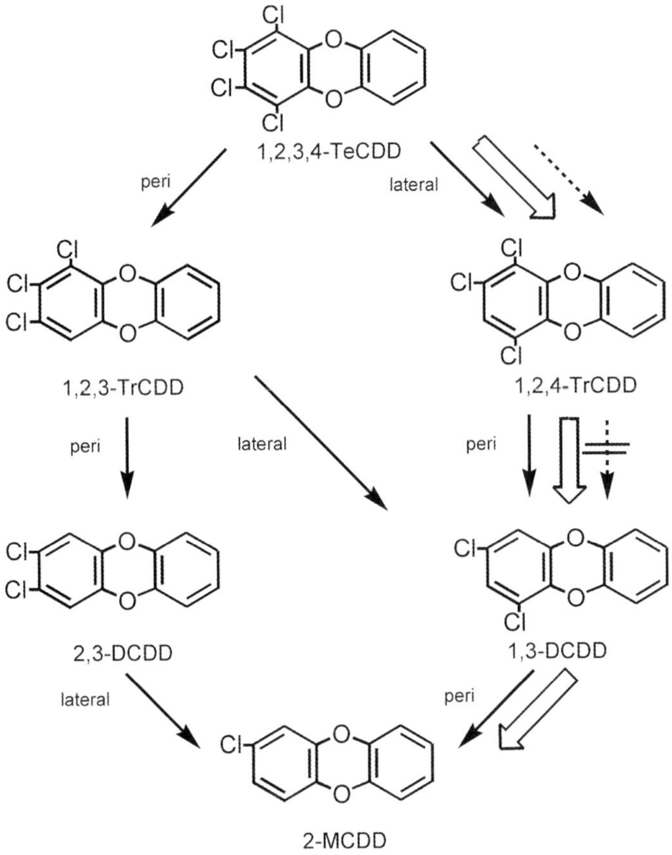

Figure 3 Pathways of reductive dehalogenation of 1,2,3,4-TeCDD by enrichment cultures from different aquatic sediments. Pathways "M" (solid arrows), "S" (open arrows) and "SP" (broken arrows). See text for the origin of sediment samples.

roemerianus (Gribble, 1998). Natural formation of chlorophenols and dioxins in soil of a Douglas fir forest has been demonstrated by Hoekstra *et al.* (1999). Peroxidase-catalyzed reactions with chlorophenols as precursors have been indicated as a possible source of dioxins (Öberg and Rappe, 1992). Some ancient sediments and clays contain high concentrations of dioxins (Ferrario *et al.*, 2000; Gauss *et al.*, 2001). Marine organisms, particularly plants and sponges, are known to produce a wide range of organohalogens (Gribble, 1998; 2003), including polybrominated dibenzo-*p*-dioxins (Utkina *et al.*, 2001), although mostly in only low quantities. Halogenated bipyrroles, which are similar to PCBs with respect to their bioaccumulative behaviour, have been detected in marine environments (Tittlemier *et al.*, 2002; Vetter *et al.*, 2003). The ubiquitous presence of naturally formed halogenated organic molecules might have been the driving force for the evolution of bacteria growing by dehalorespiration. A close association of dehalogenating bacteria with a marine sponge containing 7-12 % of brominated metabolites of its dry weight has recently been described (Ahn *et al.*, 2003).

2. Dioxin Dehalogenation

A first indication for the occurrence of *in situ* reductive dehalogenation of PCDDs was obtained from the comparison of the congener pattern within dated sediment cores of Lake Ketelmeer, The Netherlands, with an archived top-layer sample taken 18 years before from the same location (Beurskens *et al.*, 1993). In another study a relative decrease of OCDD and a relative increase of lower chlorinated dioxin congeners were found in sediment layers containing a natural dioxin source also indicating slow reductive dechlorination of OCDD in natural habitats (Gaus *et al.*, 2002). Abiotic dechlorination of dioxins was shown to be mediated by metals (e.g. zero-valent Zn), vitamin B_{12} (Adriaens *et al.*, 1996) or model humic constituents like resorcinol and catechol (Barkovskii and Adriaens, 1998) or humic acids (Fu *et al.*, 1999) and, therefore, might also considerably influence the congener profile of a deposited dioxin contaminant under anaerobic conditions. The endpoint of abiotic dechlorination of 1,2,3,4,6,7,9-heptachlorodibenzo-*p*-dioxin by model humic acids was DiCDD (Fu *et al.*, 1999), and OCDD was abiotically dechlorinated to tetrachlorinated dioxins in the presence of model humic constituents (Barkovskii and Adriaens, 1998). In the latter case, the addition of anaerobic bacteria changed the profile of dechlorination products, leading to an increase of tetrachlorinated congeners (among them also the most toxic 2,3,7,8-TeCDD isomer) and to the formation of tri- to monochlorodioxins (Barkovskii and Adriaens, 1998).

Clear evidence for the involvement of microbes in the reductive dechlorination of hexa- and heptachlorinated dioxins and hepta- and pentachlorinated furans came from the comparison of live and autoclaved microcosms established with Hudson River sediment and a creosote contaminated aquifer (Adriaens and Grbic-Galic, 1994, Adriaens *et al.*, 1995). Bacteria eluted from PCDD-contaminated Passaic River sediment dechlorinated freshly spiked OCDD to hepta-, hexa-, penta-, tetra-, tri-, di- and monochlorinated congeners in a dichotomous dechlorination pathway characterized by (i) a mixed *peri*-lateral dechlorination of non-2,3,7,8-substituted hepta- to tetrachlorinated congeners and (ii) *peri*-dechlorination of 2,3,7,8-substituted hepta- through pentachlorinated congeners (Barkovskii and Adriaens, 1996). The *peri*-lateral dechlorination was assigned to subpopulations of thermoresistant anaerobic sporeformers in combination with abiotic dechlorination reactions. Dechlorination stopped here at the level of trichlorinated congeners.

In contrast, non-methanogenic, non-thermoresistant, non-spore-forming microorganisms were responsible for the distinct *peri*-dechlorination and for the final formation of monochlorinated dioxin. Beurskens *et al.* (1995)

reported for the first time the reductive dechlorination of 1,2,3,4-TeCDD to tri- and dichlorinated congeners in a sediment-free enrichment culture obtained after several successive transfers from Lake Ketelmeer sediment with hexachlorobenzene as an electron acceptor. Subsequently, 1,2,3,4-TeCDD was used as a model substrate also in other studies using inocula from different river and estuarine sediments (Ballerstedt *et al.*, 1997; Bunge *et al.*, 2001; Vargas *et al.*, 2001).

Three different pathways of chlorine removal from 1,2,3,4-TeCDD were proposed (Figure 3). The most complex pathway was characterized by the simultaneous removal of chlorines from lateral and peripheral positions and was designated pathway "M" according to the origin of the respective sediment from the river Mulde, Germany (Bunge and Lechner, 2001). This dechlorination route was originally described for the enrichment culture obtained from Lake Ketelmeer sediment (Beurskens *et al.*, 1995) and was also found in cultures inoculated with surface sediment samples of Spittelwasser creek, Germany (Bunge *et al.*, 2001). Pathway "SP" was described only for enrichment cultures inoculated with selected samples of Spittelwasser sediment (Bunge *et al.*, 2001). In this case, dechlorination of a chlorine atom strictly depended on the presence of two flanking chlorine atoms. Thus, the final product of the lateral dechlorination of 1,2,3,4-TeCDD was 1,2,4-TrCDD, which was not further converted. In accordance, 1,2,3-TrCDD was converted to 1,3-DiCDD as the final product. A similar specificity for the removal of doubly-flanked chlorines is known for PCB dehalogenating communities (pattern "H", Wiegel and Wu, 2000) and the PCB-dehalogenating bacterium DF-1 (Wu *et al.*, 2002a).

Another pathway, designated pathway "S", was first observed in enrichment cultures from the Saale River, Germany (Ballerstedt *et al.*, 1997), which was characterized by the removal of a lateral chlorine atom from 1,2,3,4-TeCDD leading to the formation of 1,2,4-TrCDD. 1,2,4-TrCDD was then further converted by *peri*-dechlorination to 1,3-DiCDD and finally to 2-MCDD. One culture obtained from Mulde sediment (Bunge and Lechner, 2001) and the bromophenol-amended culture from Arthur Kill estuarine sediment (Vargas *et al.*, 2001) exhibited almost the same pathway. Except the last step, the formation of 2-MCDD, this dechlorination sequence followed the thermodynamically most favorable reactions predicted by the respective redox potentials according to Huang *et al.* (1996). In general, dechlorination of lateral chlorine substituents from partially positively charged carbon atoms is highly favored thermodynamically over the dechlorination of peripheral substituents (Damborsky *et al.*, 1998). Therefore, the formation of the most toxic congener 2,3,7,8-TeCDD by reductive dechlorination is thermodynamically unfavorable.

However, the transient formation of 2,3,7,8-TeCDD has been found in a PCDD contaminated estuarine sediment (Albrecht *et al.*, 1999) and in cultures spiked with heptachlorinated congeners (Fu *et al.*, 2001). The fraction of 2,3,7,8-TeCDD formed among the TeCDD homologue group seemed to depend on the culture conditions. Stimulation of fermenting and methanogenic microorganisms in the community led to a proportional increase of 2,3,7,8-TeCDD formation whereas stimulation of sulfate-reducing bacteria led to a decrease of 2,3,7,8-TeCDD formation (Fu *et al.*, 2001). However, addition of hydrogen to methanogenic cultures reduced the accumulation of 2,3,7,8-TeCDD and supported further dechlorination to 2-MCDD, which finally represented a considerable molar fraction of total dioxin congeners (Albrecht *et al.*, 1999).

Summarizing these observations, dioxin-dehalogenating bacteria of some methanogenic communities (Barkovskii and Adriaens, 1996; Albrecht *et al.*, 1999) seem to possess dehalogenase enzymes specific for the removal of peripheral chlorine atoms from highly chlorinated PCDDs. Reductive dechlorination therefore results in the formation of toxicologically critical 2,3,7,8-substituted dioxins. On the other hand, 2-MCDD is the final main dechlorination product in most dioxin-dechlorinating communities studied so far, a congener with low toxicity compared to other PCDDs.

Direct isolation of dioxin-dehalogenating bacteria from enrichment cultures has not yet been achieved. Most-probable-number analysis indicated 10^4 dioxin-dechlorinating bacteria ml^{-1} in enrichment cultures from Spittelwasser sediment compared to a total cell number of more than 10^7 ml^{-1} (Bunge *et al.*, 2003). This low number of dioxin dechlorinators might be due to the low water solubility of dioxins. First insights into the bacterial composition of dioxin-dehalogenating communities came from the analysis of 16S rRNA genes present in dehalogenating mixed cultures. Total community DNA obtained from different trichlorodibenzo-*p*-dioxin-dehalogenating enrichment cultures from freshwater sediments of rivers in Middle Germany was used as template for a 16S rDNA directed nested PCR approach with universal bacterial primers and primers specific for different dehalogenating organisms (Holliger *et al.*, 1999). A surprising diversity of bacteria with a known potential to dehalogenate different chlorinated aliphatic and aromatic compounds was detected in each of the anaerobic communities (Table 1). The spectrum of organisms differed, but two organisms were present in each of the cultures and were represented by almost identical 16S rDNA sequences throughout the different cultures. Sequence analysis revealed a *Desulfitobacterium* strain as most similar to *D. hafniense* and a *Dehalococcoides* strain, which shared an identical sequence with *Dehalococcoides* sp. strain CBDB1, a chlorobenzene respiring bacterium (Adrian *et al.*, 2000). Further properties of the mixed cultures such as sensitivity to pasteurization, resistance

Table 1. Detection of bacteria with a general dechlorination potential in enrichment cultures from river sediments in Middle Germany by nested PCR with universal and specific primers complementary to the 16S rDNA of the target organisms. DNA was extracted from almost sediment-free subcultures transferred five to seven times (10 % vol/vol each) from the primary enrichment culture (adapted from Bunge and Lechner 2001). +: PCR product of the expected size (confirmed by sequencing); -: no PCR product; data from enrichment cultures obtained from duplicate sediment samples are given if available

Target organisms	16S rDNA-PCR products obtained from enrichment cultures inoculated with sediment from:						
	Saale	Mulde		Spittelwasser		Leine	
	Incubation with TrCDDs substituted at positions						
	1,2,4	1,2,4	1,2,3	1,2,4	1,2,3	1,2,4	1,2,3
Dehalobacter restrictus	-	-/-	+	-/-	-/-	-	+
Sulfurospirillum multivorans	-	+/-	-	+/+	+/+	-	-
Desulfomonile tiedjei	-	-/-	-	-/-	-/-	-	-
Desulfuromonas chloroethenica	+	+/-	-	+/+	+/-	-	-
Desulfitobacterium spp.	+	+/+	+	+/+	+/+	+	+
Dehalococcoides spp.	+	+/+	+	+/+	+/+	+	+

to vancomycin (an inhibitor of Gram-positive bacteria) and the ability to dechlorinate 1,2,3-trichlorobenzene (Ballerstedt *et al.,* 2004) suggested an involvement of *Dehalococcoides* rather than *Desulfitobacterium* in reductive dehalogenation of spiked dioxins. To test this hypothesis, the pure culture of *Dehalococcoides* sp. strain CBDB1 was tested for its ability to dechlorinate several different dioxin congeners. The results show that strain CBDB1 is the first known bacterium able to grow by dehalorespiration with dioxins (Bunge *et al.*, 2003).

3. Reductive Dehalogenation of Chlorinated Benzenes and Other Chlorinated Compounds and the Involvement of *Dehalococcoides* Species

3.1. Chlorinated Ethenes

The best-investigated chlorinated substrate that is transformed by reductive dehalogenation is tetrachloroethene. Tetrachloroethene is dechlorinated by many different mixed and pure cultures to *cis*-dichloroethene (Holliger *et al.,* 1999). While most described pure cultures cannot dechlorinate chlorinated ethenes beyond *cis*-dichloroethene, several mixed cultures have been described and enriched that are able to reductively dechlorinate tetrachloroethene to vinyl chloride and ethene. From one of those enrichments Maymó-Gatell *et al.* (1997) isolated a pure bacterial culture and proposed the name *Dehalococcoides ethenogenes* for the strain referring to the coccus-like morphology, the dehalogenating activity and the capacity to produce ethene from tetrachloroethene. However, vinyl chloride was dechlorinated only slowly and the reduction of vinyl chloride did not contribute to the energy metabolism of the organism (Maymó-Gatell *et al.,* 2001). The bacterium was the first representative of a new taxonomic group of organisms loosely affiliated with the *Chloroflexi.* Other mixed and pure cultures were successively described that dechlorinated chlorinated ethenes past *cis*-dichloroethene with different specificities: they catabolically dechlorinated trichloroethene and *cis*-dichloroethene (Löffler *et al.,* 2000), mainly *cis*-dichloroethene and vinyl chloride (Rosner *et al.,* 1997; Cupples *et al.,* 2003), or *cis*-dichloroethene, *trans*-dichloroethene and vinyl chloride (He *et al.,* 2003a). In all these cultures *Dehalococcoides* species have been detected and the capacity to dechlorinate past *cis*-dichloroethene was correlated to the presence of these species (Löffler *et al.,* 2000; Hendrickson *et al.,* 2002; He *et al.,* 2003a; 2003b; Cupples *et al.,* 2003).

3.2. PCB Dechlorination

Polychlorinated biphenyls can be reductively dechlorinated by different pathways in mixed cultures (Bedard and Quensen, 1995; Wiegel and Wu, 2000). These processes are characterized by the position in the molecule from which chlorine substituents are removed and by the number of chlorine substituents that flank the removed chlorine substituents. For example in some cultures predominantly flanked substituents in meta position are dechlorinated (process "N"), whereas in other cultures predominantly flanked substituents in *para* position are dechlorinated (process "P") (Wiegel and Wu, 2000). It has been shown that environmental conditions e.g. temperature or pH and also the supply of electron donors have an influence on the type and extent of the dechlorinating reactions that occur in a culture (Wu *et al.,* 1997a; Wiegel and Wu, 2000). It was possible to establish sediment-free cultures in synthetic media that dechlorinated chlorine substituents specifically and exclusively in the *ortho* position (Cutter *et al.*, 1998) or doubly flanked substituents in the *para* or *meta* position (Wu *et al.*, 2000). Recently, dechlorinating bacteria were identified in both cultures (Cutter *et al.*, 2001; Wu *et al.*, 2002a). The identified bacteria, designated strains *o*-17 and DF-1 for the *ortho* and the doubly flanked *para* and *meta* dechlorinating strain, respectively, are affiliated with each other and cluster in a subgroup of the *Chloroflexi* in which also the *Dehalococcoides* can be found. Pure PCB dechlorinating strains have not yet been obtained. The experiments, however, demonstrate that the properties of a culture are dependent on the source of the inoculum used and suggest that different dechlorination patterns are catalyzed by different populations.

3.3. Chlorobenzene Dechlorination

Different types of cultures reductively dechlorinating chlorinated benzenes are known: many mixed cultures from different sources catalyze the reductive dechlorination of doubly flanked chlorine substituents from chlorobenzenes (Tiedje *et al.*, 1987; Fathepure *et al.*, 1988; Holliger *et al.*, 1992; Beurskens *et al.*, 1994; Chang *et al.*, 1997; Susarla *et al.*, 1997; Adrian *et al.*, 1998; Pavlostathis and Prytula 2000). Characteristic reactions for this pattern are the reductive dechlorination of 1,2,3-trichlorobenzene to 1,3-dichlorobenzene, the dechlorination of 1,2,3,4-tetrachlorobenzene to 1,2,4-trichlorobenzene, the dechlorination of 1,2,3,5-tetrachlorobenzene to 1,3,5-trichlorobenzene, and the dechlorination of pentachlorobenzene to 1,2,4,5- or 1,2,3,5-tetrachlorobenzene.

However, another pathway of chlorobenzene dechlorination has been identified in some mixed cultures in which predominantly singly flanked chlorine substituents are dechlorinated (Ramanand *et al.*, 1993; Middeldorp *et al.*, 1997). Characteristic products of this pathway are 1,2,3,4-tetrachlorobenzene, 1,2,3-trichlorobenzene and 1,2-dichlorobenzene.

As described for cultures dechlorinating chlorinated dioxins, ethenes or biphenyls the chlorobenzene dechlorinating cultures also retain their characteristic dechlorination pattern when they are subcultured over many transfers, indicating that the pattern is characteristic to a specific type of organism.

Two strains have now been identified to reductively dechlorinate chlorinated benzenes. Both are affiliated with the *Dehalococcoides* spp. The first identified bacterium, *Dehalococcoides* sp. strain CBDB1, was enriched and isolated from Saale river sediment (Sachsen-Anhalt, Germany) on a mixture of 1,2,3- and 1,2,4-trichlorobenzene with a procedure in which colonies grew in an agarose-solidified medium (Adrian *et al.*, 2000b). Strain CBDB1 dechlorinates hexachlorobenzene, pentachlorobenzene (Jayachandran *et al.*, 2003), all three tetrachlorobenzene congeners, 1,2,3-trichlorobenzene and 1,2,4-trichlorobenzene (Adrian *et al.*, 2000b). The strain predominately dechlorinates doubly flanked chlorine substituents. For example 1,2,3-trichlorobenzene is completely converted to 1,3-dichlorobenzene and 1,2,3,5-tetrachlorobenzene is converted to 1,3,5-trichlorobenzene which is not further dechlorinated by CBDB1. From 1,2,4,5-tetrachlorobenzene and 1,2,4-trichlorobenzene singly flanked chlorine substituents were also removed by strain CBDB1 leading to the formation of 1,2,4-trichlorobenzene and 1,3- plus 1,4-dichlorobenzene, respectively. Both hexa- and pentachlorobenzene were dechlorinated to a mixture of 1,3,5-trichlorobenzene, 1,3-dichlorobenzene, and 1,4-dichlorobenzene (Jayachandran *et al.*, 2003). For 1,2,3- and 1,2,4-trichlorobenzenes, pentachlorobenzene and hexachlorobenzene it has been shown that dechlorination is coupled to growth of the organism. In addition to chlorobenzenes and hydrogen that are used as electron acceptor and electron donor for energy generation, respectively, acetate is needed as a source of carbon. No fermentative growth or growth with non-chlorinated electron acceptors has been found. The strain is strictly anaerobic and requires highly reduced medium. The strain grows in medium containing high concentrations of the cell wall antibiotic, vancomycin or ampicillin, and electron microscopy indicates that the bacterium does not contain a peptidoglycan cell wall. Electron-microscopic pictures suggest that the cell wall is formed by a crystalline-like structure known as an S-layer e.g. in *Archaea.*

The second strain identified to reductively dechlorinate chlorobenzenes is strain DF-1. This strain is only loosely affiliated to the genus *Dehalococcoides* with approximately 89% identity to the 16S rDNA sequence of *Dehalococcoides*. Strain DF-1 was primarily identified as the PCB dechlorinating organism in a PCB dechlorinating mixed culture (Wu *et al.*, 2002a). In the same mixed culture strain DF-1 was also identified as the bacterium responsible for hexachlorobenzene, pentachlorobenzene, and 1,2,3,5-tetrachlorobenzene dechlorination (60 days for the almost complete reduction of 176 μM pentachlorobenzene to 1,3,5-tetrachlorobenzene) (Wu *et al.*, 2002b). Growth of the strain only occurred when chlorobenzenes or PCBs were reductively dechlorinated. The chlorobenzene dechlorination pathway catalyzed by strain DF-1 differed from that catalyzed by strain CBDB1 as DF-1 dechlorinates only doubly flanked chlorine substituents. Pentachlorobenzene is exclusively dechlorinated to 1,2,3,5-tetrachlorobenzene so that the only final product from hexa-, penta-, and 1,2,3,5-tetrachlorobenzene is 1,3,5-trichlorobenzene. This dechlorination pattern has previously been shown for other mixed cultures as well (e.g. Tiedje *et al.*, 1987; Holliger *et al.*, 1992) and the work of Wu *et al.* now suggests that similar *Dehalococcoides*-like bacteria were present also in those previously described mixed cultures.

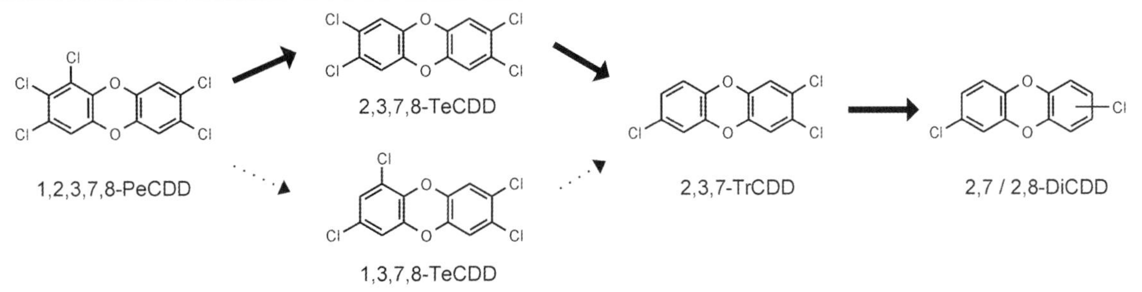

Figure 4 Proposed pathway of chlorine removal from 1,2,3,7,8-PeCDD by a pure culture of *Dehalococcoides* sp. strain CBDB1. Main route (solid arrows) and side route (dashed arrows) of reductive dechlorination as suggested by the ratio of products formed (adapted from Bunge *et al.*, 2003).

While doubly flanked chlorine substituents are preferentially or exclusively removed by strains CBDB1 and DF-1, respectively, other cultures preferentially remove singly flanked chlorine substituents (Ramanand *et al.*, 1993; Middeldorp *et al.*, 1997). This dechlorination pathway could not yet be attributed to a single strain and therefore it is not clear if *Dehalococcoides*-like populations are involved. Bacteria catalyzing such a preferential dechlorination of singly flanked chlorine substituents are potentially of more practical importance than those dechlorinating doubly flanked chlorobenzenes because the formation of the highly persistent "dead end" product 1,3,5-trichlorobenzene is avoided. Therefore, further studies are needed to investigate the bacteria catalyzing dechlorination of singly flanked substituents in more detail.

3.4. Dechlorination of Dioxins by *Dehalococcoides* sp. Strain CBDB1

The pathways of dioxin dehalogenation by strain CBDB1 were studied using five different dioxin congeners, four of which were substituted at only one ring so that none of the extremely toxic 2,3,7,8-substituted congeners could be produced by reductive dechlorination (Bunge *et al.*, 2003). In contrast to the preferred removal of doubly flanked chlorines from chlorinated benzenes, the removal of chlorine substituents from dioxins occurred preferentially at singly flanked *peri*-positions. Thus, 1,2,4-TrCDD was transformed to 1,3-DiCDD, and 1,2,3-TrCDD was transformed to a mixture containing more 2,3-DiCDD than 1,3-DiCDD. 1,2,3,4-TeCDD was dechlorinated to 2,3-DiCDD and 2-MCDD, also indicative of this preferential dechlorination from the flanked *peri*-positions. However, unflanked peripheral chlorine substituents and singly flanked lateral chlorine substituents were also dechlorinated as shown in the production of 2-MCDD from 1,3-DiCDD and 2,3-DiCDD, respectively. Unflanked substituents from the lateral positions, however, were not dechlorinated by strain CBDB1, so that 2-MCDD was not further dechlorinated. To test if the results obtained with congeners that were chlorinated at only one ring were transferable also to dioxin congeners substituted on both rings, the environmentally significant congener 1,2,3,7,8-PeCDD was added as an electron acceptor to cultures of strain CBDB1 (Figure 4). Here, the same patterns as previously found with dioxins chlorinated only on one ring were obtained: peripheral chlorine substituents were removed first, leading to formation of the highly toxic 2,3,7,8-TeCDD. However, as observed with 2,3-DiCDD, singly flanked lateral chlorine substituents were also dechlorinated leading to the formation of 2,7- and/or 2,8-DiCDD. These results obtained with a pure culture are in good agreement with the previous observations of Barkovskii and Adriaens (1996) and Albrecht *et al.* (1999) who reported the *peri*-dechlorination of higher chlorinated 2,3,7,8-substituted dioxins and the further dechlorination of tetra- and trichlorinated congeners by anaerobic, non-thermoresistant mixed cultures.

4. Biochemistry and Genetics of Reductive Dehalogenation

Very little is known about the biochemistry and the genetic background of reductive dehalogenation of chlorinated dioxins, biphenyls and benzenes. As described above, all chlorobenzene dechlorinating cultures available do not grow by fermentation and do not use electron acceptors for respiration other than chlorinated compounds. Because the chlorinated dioxins, biphenyls and benzenes transformable by reductive dechlorination have very low solubility in water and are often toxic to the dechlorinating bacteria, electron acceptor concentrations of a few μmolar or below have to be used for cultivation. However, a high amount of electron acceptors has to be reduced because the chlorinated compounds are used for energy regeneration by the bacteria. Molar growth yields between 0.2 and 5 g cell protein per mole of chloride released by dehalorespiration have been reported (reviewed by Adrian *et al.*, 2000a; Cupples *et al.*, 2003; He *et al.*, 2003b). This means that a concentration of 1 μM of a chlorinated compound can provide a sufficient amount of energy for the production of only 0.2 to 5 μg of cell protein per liter, which corresponds to the growth of about 10^5 bacteria per mL. This cultivation problem significantly complicates biochemical studies so that no dehalogenases reductively dechlorinating chlorinated benzenes, dioxins or biphenyls have yet been isolated.

However, some enzymatic characteristics could be determined even without the isolation of reductive dehalogenases. The fact that chlorobenzene dechlorinating bacteria were able to grow with hydrogen as an electron donor and the chlorinated compound as an electron acceptor in the absence of fermentative processes was taken as evidence of a respiratory process for energy fixation (Adrian *et al.*, 2000b). A similar assumption was drawn much earlier from studies of bacteria growing with different chlorinated compounds such as tetrachloroethene or chlorophenols as the electron acceptor. Supporting evidence for a respiratory process involved in these dechlorination reactions has accumulated over the past 10 years (Mohn and Tiedje, 1991).

Detailed enzymatic studies of the reductive dehalogenases from *Sulfurospirillum multivorans, Dehalobacter restrictus,* different *Desulfitobacterium* species, *Desulfomonile tiedjei* and *Dehalococcoides ethenogenes* (summarized by El Fantroussi *et al.*, 1998; Holliger *et al.*, 1999; see also Magnuson *et al.*, 2000; Krasotkina *et al.*, 2001; Suyama *et al.*, 2002; Maillard *et al.*, 2003; Boyer *et al.*, 2003) have revealed important characteristics of reductive dehalogenases. The general picture is; i) these enzymes are associated with the membrane, ii) they contain two iron-sulfur clusters, and iii) a corrinoid cofactor is present in the active site. However, some of the reductive dehalogenases described differ from this scheme by being localized in the cytosol (*S. multivorans*, Neumann *et al.*, 1995), by containing a heme as a cofactor instead of a corrinoid (*Desulfomonile tiedjei*, Ni *et al.*, 1995) or by missing the iron-sulfur clusters (Boyer *et al.*, 2003).

Recently, a biochemical characterization of chlorobenzene dechlorinating activities in crude extracts of *Dehalococcoides* strain CBDB1 has been published (Hölscher *et al.*, 2003). Many characteristics of chlorobenzene reductive dehalogenases were found to be similar to those of chloroethene reductive dehalogenases: i) the chlorobenzene dehalogenase activity accepts methyl and ethyl viologen as an artificial electron donor, ii) the activity is light-reversibly inhibited by propyl or ethyl iodide, indicating the presence of a corrinoid cofactor, and iii) it is loosely associated to the cell membrane. However, in contrast to *Desulfitobacterium* strain PCE-S, which showed higher activity with cell extracts than with whole cells (Miller *et al.*, 1997b), *Dehalococcoides* strain CBDB1 did not show an increased dehalogenase activity after cell disruption, indicating an orientation of the reductive dehalogenase to the outside of the cell. The specific reductive dehalogenase activity in crude extracts of CBDB1 was very high, amounting to 11 nkat/mg of protein for 1,2,3-trichlorobenzene, 76 nkat/mg for 1,2,3,5-tetrachlorobenzene, 171 nkat/mg for pentachlorobenzene and 355 nkat/mg for 1,2,3,4-tetrachlorobenzene. Dechlorination activity was also detected for hexachlorobenzene, 1,2,4,5-tetrachlorobenzene, and 1,2,4-trichlorobenzene (Hölscher *et al.*, 2003).

No data are available yet for the dioxin or PCB dechlorinating enzymes. However, the results found with *Dehalococcoides* strain CBDB1, which dechlorinates chlorobenzenes and dioxins, as well as many more studies showing priming effects might indicate that dioxin and PCB dehalogenases have similar characteristics.

4.1. Genetics

Nothing is yet known on the genetics of chlorodioxin, PCB and chlorobenzene reductive dehalogenation. Due to the similarities in the biochemical characteristics of chlorobenzene and PCE dehalogenation it can only be hypothesized that similar genes found in tetrachloroethene dehalogenating organisms are present in dioxin, PCB or chlorobenzene dechlorinating organisms. Reductive dehalogenase genes have been described in several organisms including *Desulfitobacterium dehalogenans, Desulfitobacterium* strain PCE-S, *Desulfitobacterium chlororespirans, Desulfitobacterium* strain *Y51, Sulfurospirillum multivorans*, and *Dehalococcoides ethenogenes* (summarized by Smidt *et al.*, 2000; Villemur *et al.*, 2002) and many more putative reductive dehalogenase sequences are available now in public databases (www.ncbi.nlm.nih.gov) which have been obtained by PCR from mixed or pure cultures.

The genes encode reductive dehalogenases that transform a wide range of chlorinated substrates such as tetrachloroethene, trichloroethene, 3-chloro-4-hydroxy-phenylacetate, 3-chloro-4-hydroxybenzoat, 3-chlorobenzoate, or *ortho*-chlorophenol. All these genes have similarities in their organization. In summary, reductive dehalogenases are encoded by operons consisting of two genes. One gene encodes the catalytic subunit of the reductive dehalogenase, which carries the corrinoid cofactor and two iron-sulfur clusters. While the binding motif for the two iron-sulfur clusters can be found in the encoded amino acid sequence, no corrinoid binding motif has been described yet in reductive dehalogenase sequences. The catalytic subunit has a leader peptide that presumably targets the protein for transport to or across the cell membrane via the twin arginine transport (TAT) pathway (Berks *et al.*, 2000) but does not appear to be a membrane-integrated enzyme, because it contains no hydrophobic membrane-spanning helices. However, the second gene in the operon, which is much smaller than that of the catalytic subunit, encodes a small integral membrane protein which presumably serves as a membrane anchor for the catalytic subunit. The organization of the genes indicates that both are co-expressed. No studies are available yet describing the interaction between the two gene products.

5. Use of Multiple Halogenated Electron Acceptors by Dehalogenating Bacteria and Practical Implications

Because there are several putative dehalogenase genes in genomes of dehalogenating species such as *Dehalococcoides ethenogenes* or *Desulfitobacterium hafniense* (S. Zinder, personal communication; Villemur *et al.*, 2002) it is not surprising that pure cultures are able to use a variety of chlorinated compounds as electron acceptors. Biochemical evidence is available for the presence of different dehalogenases in *Desulfitobacterium* sp. strain PCE1 (tetrachloroethene and *ortho*-chlorophenol reductive dehalogenase; van de Pas *et al.*, 2001) and in *D. ethenogenes* strain 195 (tetra- and trichloroethene reductive dehalogenase; Magnuson *et al.*, 1998). Versatility in the use of different electron acceptors has also been observed with chlorobenzene dehalogenating bacteria. *Dehalococcoides* sp. strain CBDB1 is able to use chlorobenzenes and dioxins. Strain DF-1 has been identified as a PCB- and chlorobenzene-dechlorinating bacterium (Wu *et al.*, 2002a; Wu *et al.*, 2002b). Different chlorobenzene-dehalogenating mixed cultures were shown to also dechlorinate PCBs (Middeldorp *et al.*, 1997; Chang *et al.*, 1999), PCBs and dioxins (Beurskens *et al.*, 1995) or chlorotoluenes (Ramanand *et al.*, 1993). It is not clear yet, if this diversity is based on different populations with different dechlorination activities or on the broad substrate spectrum of one bacterium as seen with strains CBDB1 and DF-1.

For bioremediation, microbial-catalyzed dehalogenation can be useful in directing dechlorination activities towards a selected compound or to support a particular dechlorination pathway that leads to benign end products and avoids the accumulation of toxic chlorinated intermediates. Reductively dehalogenating bacteria might be present at a contaminated habitat, but exhibit almost no dechlorination activity due to limiting concentrations of electron donors or to adsorption of chlorinated pollutants to the environmental matrix. The latter effect has been observed by Prytula and Pavlostathis (1996) for hexachlorobenzene bound tightly to sediment samples. Although reductive dehalogenation of additionally amended hexachlorobenzene was catalyzed by the autochthonous bacterial community, the hexachlorobenzene that was initially bound to the sediment material was only dehalogenated after grinding the sediment. The low accessibility in turn should result in a low population density of dehalogenating bacteria.

In PCB dechlorination studies, Bedard *et al.* (1996) observed that the addition of 2,5,3',4'-tetrachlorobiphenyl to PCB contaminated sediments enhanced the dechlorination of PCBs, an effect described as "priming". The priming substrate was applied at a 7 to 14 times higher concentration compared to the background of PCBs. Both the priming substrate and the contaminating PCBs were dechlorinated at the *para*-position (process "P"). This suggests that a specific population of dechlorinating bacteria was enriched growing with 2,5,3',4'-tetrachlorobiphenyl as an electron acceptor. Using a sediment sample from the same location but with other PCB congeners as "haloprimers" (particularly those containing a 2,3,6-substitution pattern), reductive dechlorination was stimulated at *meta*-positions (process "N") (Van Dort *et al.*, 1997). However, additional factors such as the temperature strongly influenced which pathway was stimulated (Wu *et al.*, 1997b). Stimulation of process "N" occurred only between 8°C and 30°C, whereas other pathways occurred at temperatures of up to 60 °C. Bromobiphenyls primed twice as effectively as chlorobiphenyls (Bedard *et al.*, 1998). Wu *et al.* (1999) demonstrated that the addition of 2,6-dibromobiphenyl to a dechlorinating mixed culture resulted in an increase of the number of PCB dechlorinating bacteria and also in an enhanced dechlorination of PCBs. This shows that growth stimulation of a particular dehalogenating population can be a valuable tool for bioremediation.

The addition of PCBs or brominated biphenyls to the environment is clearly not acceptable. A more feasible approach is the addition of naturally occurring non-persistent compounds as "haloprimers" for the bioremediation of highly persistent contaminants. However, the stimulating effect of such naturally occurring halogenated compounds has yet to be studied. Another approach would be the application of halogenated priming compounds, which are known to be completely dehalogenated and mineralized under anaerobic conditions. For example, DeWeerd and Bedard (1999) demonstrated that specific congeners of brominated and iodated (but not chlorinated or fluorinated) benzoates were effective primers for PCB dehalogenation, and undergo complete dehalogenation to benzoate, which was mineralized by the cultures. Halobenzoate dehalogenation always preceded PCB dehalogenation and PCB dehalogenation rates were highest during subsequent benzoate mineralization. Here, a possible function of benzoate as an additional source of reducing equivalents for dehalogenation can be considered (Mohn and Tiedje, 1992). Cho *et al.* (2002) found indications that chlorinated phenols, chlorinated benzoates and chlorinated benzenes enhanced the number of PCB dechlorinating bacteria in cultures enriched from river sediments.

Also the stimulation of dioxin dechlorination has been observed using brominated compounds. Albrecht *et al.* (1999) found enhanced dechlorination of sediment-associated dioxins in the presence of 2-monobromodibenzo-*p*-dioxin, whereas Vargas *et al.* (2001) described dechlorination of 1,2,3,4-TeCDD by cultures enriched from estuarine sediments on three monobromophenols.

6. Summary

Reductive dechlorination of chlorinated aromatics including benzenes and dioxins occurs in the environment, but mostly at very low rates. Despite the obstacles, which hinder dioxin dechlorination in the environment such as adsorption processes and low water solubility, reductive dechlorination not only of higher chlorinated, but also of 2,3,7,8-tetra- and lower chlorinated congeners has been clearly demonstrated. Evidence for the involvement of microbes in this process came from studies of microcosms and enrichment cultures and recently from the isolation and investigation of a *Dehalococcoides* strain, which grows by dehalorespiration with chlorinated benzenes and dioxins. Also other *Dehalococcoides* strains seem to be able to reductively dechlorinate various halogenated aromatics. These organisms are well adapted to a life with halogenated compounds as the electron acceptor in anaerobic environments, however, they also seem to be restricted to this energy gaining process. Further physiological, biochemical and genetic studies of *Dehalococcoides* species are needed to allow for a better prediction and selective transformation of chlorinated aromatics in the environment.

References

Adriaens, P., Chang, P.R., and Barkovskii, A.L. 1996. Dechlorination of PCDD/F by organic and inorganic electron transfer molecules in reduced environments. Chemosphere. 32: 433-441.

Adriaens, P., Fu, Q., and Grbic-Galic, D. 1995. Bioavailability and transformation of highly chlorinated dibenzo-*p*-dioxins and dibenzofurans in anaerobic soils and sediments. Environ. Sci. Technol. 29: 2252-2260.

Adriaens, P., and Grbic-Galic, D. 1994. Reductive dechlorination of PCDD/F by anaerobic cultures and sediments. Chemosphere. 29: 2253-2259.

Adrian, L., and Görisch, H. 2002. Microbial transformation of chlorinated benzenes under anaerobic conditions. Res. Microbiol. 153: 131-137.

Adrian, L., Manz, W., Szewzyk, U., and Görisch, H. 1998. Physiological characterization of a bacterial consortium reductively dechlorinating 1,2,3- and 1,2,4-trichlorobenzene. Appl. Environ. Microbiol. 64: 496-503.

Adrian, L., Szewzyk, U., and Görisch, H. 2000a. Bacterial growth linked to reductive dechlorination of trichlorobenzenes. Biodegradation. 11: 73-81.

Adrian, L., Szewzyk, U., Wecke, J., and Görisch, H. 2000b. Bacterial dehalorespiration with chlorinated benzenes. Nature. 408: 580-583.

Ahn, Y.-B., Rhee, S.-K., Fennell, D.E., Kerkhof, L.J., Hentschel, U., and Häggblom, M.M. 2003. Reductive dehalogenation of brominated phenolic compounds by microorganisms associated with the marine sponge *Aplysina aerophoba*. Appl. Environ. Microbiol. 69: 4159-4166.

Albrecht, I.D., Barkovskii, A.L., and Adriaens, P. 1999. Production and dechlorination of 2,3,7,8-tetrachlorodibenzo-*p*-dioxin in historically contaminated estuarine sediments. Environ. Sci. Technol. 33: 737-744.

Ballerstedt, H., Hantke, J., Bunge, M., Werner, B., Gerritse, J., Andreesen, J.R., and Lechner, U. 2004. Properties of a trichlorodibenzo-*p*-dioxin dechlorinating mixed culture with a *Dehalococcoides* as putative dechlorinating species. FEMS Microbial. Ecol. 47: 223-234.

Ballerstedt, H., Kraus, A., and Lechner, U. 1997. Reductive dechlorination of 1,2,3,4-tetrachlorodibenzo-*p*-dioxin and its products by anaerobic mixed cultures from Saale river sediment. Environ. Sci. Technol. 31: 1749-1753.

Barkovskii, A.L., and Adriaens, P. 1996. Microbial dechlorination of historically present and freshly spiked chlorinated dioxins and diversity of dioxin-dechlorinating populations. Appl. Environ. Microbiol. 62: 4556-4562.

Barkovskii, A.L., and Adriaens, P. 1998. Impact of humic constituents on microbial dechlorination of polychlorinated dioxins. Environ. Toxicol. Chem. 17: 1013-1020.

Bedard, D.L., and Quensen, J.F. 1995. Microbial reductive dechlorination of polychlorinated biphenyls. In: Microbial Transformation and Degradation of Toxic Organic Chemicals. L.Y. Young and C. E. Cerniglia, eds. John Wiley & Sons, New York, New York. p. 127-216.

Bedard, D.L., Bunnell, S.C., and Smullen, L.A. 1996. Stimulation of microbial *para*-dechlorination of polychlorinated biphenyls that have persisted in Housatonic River sediment for decades. Environ. Sci. Technol. 30: 687-694.

Bedard, D.L., van Dort, H., and DeWeerd, K.A. 1998. Brominated biphenyls prime extensive microbial reductive dehalogenation of aroclor 1260 in Housatonic River sediment. Appl. Environ. Microbiol. 64: 1786-1795.

Berks, B.C., Sargent, F., and Palmer, T. 2000. The TAT protein export pathway. Mol. Microbiol. 35: 260-274.

Beurskens, J.E.M., Mol, G.A.J., Barreveld, H.L., van Munster, B., and Winkels, H.J. 1993. Geochronology of priority pollutants in a sedimentation area of the Rhine River. Environ. Toxicol. Chem. 12: 1549-1566.

Beurskens, J.E.M., Dekker, C.G.C., van den Heuvel, H., Swart, M., de Wolf, J., and Dolfing, J. 1994. Dechlorination of chlorinated benzenes by an anaerobic microbial consortium that selectively mediates the thermodynamic most favorable reactions. Environ. Sci. Technol. 28: 701-706.

Beurskens, J.E.M., Toussaint, M., de Wolf, J., van der Steen, J.M.D., Slot, P.C., Commandeur, L.C.M., and Parsons, J.R. 1995. Dehalogenation of chlorinated dioxins by an anaerobic microbial consortium from sediment. Environ. Toxicol. Chem. 14: 939-943.

Boyer, A., Page-Belanger, R., Saucier, M., Villemur, R., Lepine, F., Juteau, P., and Beaudet, R. 2003. Purification, cloning and sequencing of an enzyme mediating the reductive dechlorination of 2,4,6-trichlorophenol from *Desulfitobacterium frappieri* PCP-1. Biochem. J. 373: 297-303.

Boyle, A.W., Phelps, C.D. and Young, L.Y. 1999. Isolation from estuarine sediments of a *Desulfovibrio* strain which can grow on lactate coupled to the reductive dehalogenation of 2,4,6-tribromophenol. Appl. Environ. Microbiol. 65: 1133-1140.

Bunge, M., Adrian, L., Kraus, A., Opel, M., Lorenz, W.G., Andreesen, J.R., Görisch, H., and Lechner, U. 2003. Reductive dehalogenation of chlorinated dioxins by an anaerobic bacterium. Nature. 421: 357-360.

Bunge, M., Ballerstedt, H., and Lechner, U. 2001. Regiospecific dechlorination of spiked tetra- and trichlorodibenzo-*p*-dioxins by anaerobic bacteria from PCDD/F- contaminated Spittelwasser sediments. Chemosphere. 43: 675-681.

Bunge, M., and Lechner, U. 2001. Anaerobic transformation of dioxins by bacteria from contaminated sediments: diversity of the dehalogenating community. In: Schriftenreihe Biologische Abwasserreinigung, Technische Universität Berlin, Vol. 15: Anaerobic dehalogenation. p. 69-81.

Chang, B.V., Chen, Y.M., Yuan, S.Y., and Wang, Y.S. 1997. Reductive dechlorination of hexachlorobenzene by an anaerobic mixed culture. Wat. Air Soil Poll. 100: 25-32.

Chang, B.V., Chou, S.W., and Yuan, S.Y. 1999. Microbial dechlorination of polychlorinated biphenyls in anaerobic sewage sludge. Chemosphere. 39: 45-54.

Cho, Y.C., Ostrofsky, E.B., Sokol, R.C., Frohnhoefer, R.C., and Rhee, G.Y. 2002. Enhancement of microbial PCB dechlorination by chlorobenzoates, chlorophenols and chlorobenzenes. FEMS Microbiol. Ecol. 42: 51-58.

Cupples, A.M., Spormann, A.M., and McCarty, P.L. 2003. Growth of a *Dehalococcoides*-like microorganism on vinyl chloride and cis-dichloroethene as electron acceptors as determined by competitive PCR. Appl. Environ. Microbiol. 69: 953-959.

Cutter, L., Sowers, K.R., and May, H.D. 1998. Microbial dechlorination of 2,3,5,6-tetrachlorobiphenyl under anaerobic conditions in the absence of soil or sediment. Appl. Environ. Microbiol. 64: 2966-2969.

Cutter, L.A., Watts, J.M., Sowers, K.R., and May, H.D. 2001. Identification of a microorganism that links its growth to the reductive dechlorination of 2,3,5,6-chlorobiphenyl. Environ. Microbiol. 3: 699-709.

Damborsky, J., Lynam, M., and Kuty, M. 1998. Structure-biodegradability relationships for chlorinated dibenzo-*p*-dioxins and dibenzofurans. In: Biodegradation of Dioxins and Furans, R.-M. Wittich, ed. Springer, Berlin. p. 165-228.

De Wever, H., Cole, J.R., Fettig, M.R., Hogan, D.A., and Tiedje, J.M. 2000. Reductive dehalogenation of trichloroacetic acid by *Trichlorobacter thiogenes* gen. nov., sp. nov. Appl. Environ. Microbiol. 66: 2297-2301.

DeWeerd, K.A., and Bedard, D.L. 1999. Use of halogenated benzoates and other halogenated aromatic compounds to stimulate the microbial dechlorination of PCBs. Environ. Sci. Technol. 33: 2057-2063.

El Fantroussi, S., Naveau, H., and Agathos, S.N. 1998. Anaerobic dechlorinating bacteria. Biotechnol. Prog. 14: 167-188.

Fathepure, B.Z., Tiedje, J.M., and Boyd, S.A. 1988. Reductive dechlorination of hexachlorobenzene to tri- and dichlorobenzenes in anaerobic sewage sludge. Appl. Environ. Microbiol. 54: 327-330.

Ferrario, J.B., Byrne, C.J., and Cleverly, D.H. 2000. 2,3,7,8-Dibenzo-*p*-dioxins in mined clay products from the United States: evidence for possible natural origin. Environ. Sci. Technol. 34: 4524-4532.

Fu, Q.S., Barkovskii, A.L., and Adriaens, P. 1999. Reductive transformation of dioxins: An assessment of the contribution of dissoved organic matter to dechlorination reactions. Environ. Sci. Technol. 33: 3837-3842.

Fu, Q.S., Barkovskii, A.L., and Adriaens, P. 2001. Dioxin cycling in aquatic sediments: the Passaic River Estuary. Chemosphere. 43: 643-648.

Gaus, C., Päpke, O., Dennison, N., Haynes, D., Shaw, G.R., Connell, D.W., and Müller, J.F. 2001. Evidence for a widespread PCDD source in coastal sediments and soils from Queensland, Australia. Chemosphere. 43: 549-558.

Gaus, C., Brunskill, G.J., Connell, D.W., Prange, J., Müller, J.F., Päpke, O., and Weber, R. 2002. Transformation processes, pathways, and possible sources of distinctive polychlorinated dibenzo-*p*-dioxin signatures in sink environments. Environ. Sci. Technol. 36: 3542-3549.

Geyer, H.J., Rimkus, G.G., Scheunert, I., Kaune, A., Schramm, K.-W., Kettrup, A., Zeemann, M., Muir, D.C.G., Hansen, L.G., and Mackay, D. 2000. Bioaccumulation and occurrence of endocrine-disrupting chemicals (ECDs), persistent organic pollutants (POPs), and other organic compounds in fish and other organisms including humans. In: The Handbook of Environmental Chemistry. B. Beek, ed. Springer, Berlin. vol. 2, p. 1-178.

Gribble, G. W. 1998. Chlorinated compounds in the biosphere, natural production. In: Encyclopedia of Environmental Analysis and Remediation. R.A. Meyers, ed. John Wiley & Sons, Hoboken, New Jersey. p. 972-1035.

Gribble, G. W. 2003. The diversity of naturally produced organohalogens. Chemosphere. 52, 289-297.

He, J., Ritalahti, K.M., Aiello, M.R., and Löffler, F.E. 2003a. Complete detoxification of vinyl chloride by an anaerobic enrichment culture and identification of the reductively dechlorinating population as a *Dehalococcoides* species. Appl. Environ. Microbiol. 69: 996-1003.

He, J., Ritalahti, K.M., Yang, K.L., Koenigsberg, S.S., and Löffler, F.E. 2003b. Detoxification of vinyl chloride to ethene coupled to growth of an anaerobic bacterium. Nature. 424: 62-65.

Hendrickson, E.R., Payne, J.A., Young, R.M., Starr, M.G., Perry, M.P., Fahnestock, S., Ellis, D.E., and Ebersole, R.C. 2002. Molecular analysis of *Dehalococcoides* 16S ribosomal DNA from chloroethene-contaminated sites throughout North America and Europe. Appl. Environ. Microbiol. 68: 485-495.

Hoekstra, E.J., De Weerd, H., De Leer, E.W.B., and Brinkman, U.A.T. 1999. Natural formation of chlorinated phenols, dibenzo-p-dioxins, and dibenzofurans in soil of a Douglas fir forest. Environ. Sci. Technol. 33: 2543-2549.

Holliger, C., Schraa, G., Stams, A.J.M., and Zehnder, A.J.B. 1992. Enrichment and properties of an anaerobic mixed culture reductively dechlorinating 1,2,3-trichlorobenzene to 1,3-dichlorobenzene. Appl. Environ. Microbiol. 58: 1636-1644.

Holliger, C., Wohlfarth, G., and Diekert, G. 1999. Reductive dechlorination in the energy metabolism of anaerobic bacteria. FEMS Microbiol. Rev. 22: 383-398.

Hölscher, T., Görisch, H., and Adrian, L. 2003. Reductive dehalogenation of chlorobenzene congeners in cell extracts of *Dehalococcoides* sp. strain CBDB1. Appl. Environ. Microbiol. 69: 2999-3001.

Huang, C.-L., Harrison, B.K., Madura, J., and Dolfing, J. 1996. Gibbs free energies of formation of PCDDs: evaluation of estimation methods and application for predicting dehalogenation pathways. Environ. Toxicol. Chem. 15: 824-836.

Jayachandran, G., Görisch, H., and Adrian, L. 2003. Dehalorespiration with hexachlorobenzene and pentachlorobenzene by *Dehalococcoides* sp. strain CBDB1. Arch. Microbiol. 180: 411-416.

Koch, R. 1991. Umweltchemikalien. 2. Auflage. Verlag Chemie, Weinheim.

Krasotkina, J., Walters, T., Maruya, K.A., and Ragsdale, S.W. 2001. Characterization of the B_{12}- and iron-sulfur-containing reductive dehalogenase from *Desulfitobacterium chlororespirans*. J. Biol. Chem. 276: 40991-40997.

Löffler, F.E., Sun, Q., Li, S., and Tiedje, J.M. 2000. 16S rRNA gene-based detection of tetrachloroethene-dechlorinating *Desulfuromonas* and *Dehalococcoides* species. Appl. Environ. Microbiol. 66: 1369-1374.

Luijten, M.L.G.C., de Weert, J., Smidt, H., Boschker, H.T.S., de Vos, W.M., Schraa, G., and Stams, A.J.M. 2003. Description of *Sulfurospirillum halorespirans* sp. nov., an anaerobic, tetrachloroethene-respiring bacterium, and transfer of *Dehalospirillum multivorans* to the genus *Sulfurospirillum* as *Sulfurospirillum multivorans* comb. nov. Int. J. Sys. Evol. Microbiol. 53: 787-793.

Magnusen, J.K., Romine, M.F., Burris, D.R., and Kingsley, M.T. 2000. Trichloroethene reductive dehalogenase from *Dehalococcoides ethenogenes*: Sequence of *tceA* and substrate range characterization. Appl. Environ. Microbiol. 66: 5141-5147.

Magnusen, J.K., Stern, R.V., Gossett, J.M., Zinder, S.H., and Burris, D.R. 1998. Reductive dechlorination of tetrachloroethene to ethene by a two-component enzyme pathway. Appl. Environ. Microbiol. 64: 1270-1275.

Maillard, J., Schumacher, W., Vazquez, F., Regeard, C., Hagen, W.R., and Holliger, C. 2003. Characterization of the corrinoid iron-sulfur protein tetrachloroethene reductive dehalogenase of *Dehalobacter restrictus*. Appl. Environ. Microbiol. 69: 4628-4638.

Maymó-Gatell, X., Chien, Y.T., Gossett, J.M., and Zinder, S.H. 1997. Isolation of a bacterium that reductively dechlorinates tetrachloroethene to ethene. Science. 276: 1568-1571.

Maymó-Gatell, X., Nijenhuis, I., and Zinder, S.H. 2001. Reductive dechlorination of cis-1,2-dichloroethene and vinyl chloride by "*Dehalococcoides ethenogenes*". Environ. Sci. Technol. 35: 516-521.

Middeldorp, P.J.M., de Wolf, J., Zehnder, A.J.B., and Schraa, G. 1997. Enrichment and properties of a 1,2,4-trichlorobenzene-dechlorinating methanogenic microbial consortium. Appl. Environ. Microbiol. 63: 1225-1229.

Miller, E., Wohlfarth, G., and Diekert, G. 1997a. Studies on tetrachloroethene respiration in *Dehalospirillum multivorans*. Arch. Microbiol. 166: 379-387.

Miller, E., Wohlfarth, G., and Diekert, G. 1997b. Comparative studies on tetrachloroethene reductive dechlorination mediated by *Desulfitobacterium* sp. strain PCE-S. Arch. Microbiol. 168: 513-519.

Mohn, W.W., and Tiedje, J.M. 1992. Microbial reductive dehalogenation. Microbiol. Rev. 56: 482-507.

Mohn, W.W., and Tiedje, J.M. 1991. Evidence for chemiosmotic coupling of reductive dechlorination and ATP synthesis in *Desulfomonile tiedjei*. Arch. Microbiol. 157: 1-6.

Neumann, A., Wohlfarth, G., and Diekert, G. 1995. Properties of tetrachloroethene and trichloroethene dehalogenase of *Dehalospirillum multivorans*. Arch. Microbiol. 163: 276-281.

Ni, S., Fredrickson, J.K., and Xun, L. 1995. Purification and characterization of a novel 3-chlorobenzoate-reductive dehalogenase from the cytoplasmic membrane of *Desulfomonile tiedjei* DCB-1. J. Bacteriol. 177: 5135-5139.

Öberg, L.G., and Rappe, C. 1992. Biochemical formation of PCDD/Fs from chlorophenols. Chemosphere. 25: 49-52.

Ohtake, F., Takeyama, K., Matsumoto, T., Kitagawa, H., Yamamoto, Y., Nohara, K., Tohyama, C., Krust, A., Mimura, J., Chambon, P., Yanagisawa, J., Fujii-Kuriyama, Y., and Kato, S. 2003. Modulation of oestrogen receptor signalling by association with the activated dioxin receptor. Nature. 423: 545-550.

Pavlostathis, S.G., and Prytula, M.T. 2000. Kinetics of the sequential microbial reductive dechlorination of hexachlorobenzene. Environ. Sci. Technol. 34: 4001-4009.

Prytula, M.T., and Pavlostathis, S.G. 1996. Effect of contaminant and organic matter bioavailability on the microbial dehalogenation of sediment-bound chlorobenzenes. Water Research. 30: 2669-2680.

Ramanand, K., Balba, M.T., and Duffy, J. 1993. Reductive dehalogenation of chlorinated benzenes and toluenes under methanogenic conditions. Appl. Environ. Microbiol. 59: 3266-3272.

Rosner, B.M., McCarty, P.L., and Spormann, A.M. 1997. *In vitro* studies on reductive vinyl chloride dehalogenation by an anaerobic mixed culture. Appl. Environ. Microbiol. 63: 4139-4144.

Sanford, R.A., Cole, J.R., and Tiedje, J.M. 2002. Characterization and description of *Anaeromyxobacter dehalogenans* gen. nov., sp. nov., an aryl-halorespiring facultative anaerobic Myxobacterium. Appl. Environ. Microbiol. 68: 893-900.

Schumacher, W., and Holliger, C. 1996. The proton/electron ratio of the menaquinone-dependent electron tansport from dihydrogen to tetrachloroethene in "*Dehalobacter restrictus*". J. Bacteriol. 178: 2328-2333.

Smidt, H., Akkermans, A.D.L., van der Oost, J., and de Vos, W.M. 2000. Halorespiring bacteria - molecular characterization and detection. Enzyme Microb. Technol. 27: 812-820.

Sun, B., Cole, J.R., Sanford, R.A., and Tiedje, J.M. 2000. Isolation and characterization of *Desulfovibrio dechloracetivorans* sp. nov., a marine dechlorinating bacterium growing by coupling the oxidation of acetate to the reductive dechlorination of 2-chlorophenol. Appl. Environ. Microbiol. 66: 2408-2413.

Sun, B.L., Griffin, B.M., Ayala-del-Rio, H.L., Hashsham, S.A., and Tiedje, J.M. 2002. Microbial dehalorespiration with 1,1,1-trichloroethane. Science. 298: 1023-1025.

Susarla, S., Yonezawa, Y., and Masunaga, S. 1997. Transformation kinetics and pathways of chlorophenols and hexachlorobenzene in fresh water lake sediment under anaerobic conditions. Environ. Technol. 18: 903-911.

Suyama, A., Yamashita, M., Yoshino, S., and Furukawa, K. 2002. Molecular characterization of the PceA reductive dehalogenase of *Desulfitobacterium* sp strain Y51. J. Bacteriol. 184: 3419-3425.

Tiedje, J.M., Boyd, S.A., and Fathepure, B.Z. 1987. Anaerobic degradation of chlorinated aromatic hydrocarbons. J. Ind. Microbiol. Suppl. 27: 117-127.

Tittlemier, S.A., Fisk, A.T., Hobson, K.H., and Norstrom, R.J. 2002. Examination of the bioaccumulation of halogenated dimethyl bipyrroles in an arctic food web using stable nitrogen isotope analysis. Environ. Pollut. 116: 85-93.

Utkina, N.K., Denisenko, V.A., Scholokova, O.V., Virovaya, M.V., Gerasimenko, A.V., Popov, D.Y., Krasokhin, V.B., and Popov, A.M. 2001. Spongiodioxins A and B, two new polybrominated dibenzo-*p*-dioxins from an Australien marine sponge *Dysidea dendyi*. J. Nat. Prod. 64: 151-153.

Van Agteren, M.H., Keuning, S., and Janssen, D.B. 1998. Chlorinated aromatic compounds. In: Handbook on Biodegradation and Biological Treatment of Hazardous Organic Compounds, M.H. van Agteren, and S. Keuning, eds. Kluwer Academic Publishers, Dordrecht. p. 351-474.

Van de Pas, B.A., Gerritse, J., de Vos, W.M., Schraa, G., and Stams, A.J.M. 2001. Two distinct enzyme systems are responsible for tetrachloroethene and chlorophenol reductive dehalogenation in *Desulfitobacterium* strain PCE1. Arch. Microbiol. 176: 165-169.

Van Dort, H.M., Smullen, L.A., May, R.J., and Bedard, D.L. 1997. Priming microbial *meta*-dechlorination of polychlorinated biphenyls that have persisted in Housatonic River sediments for decades. Environ. Sci. Technol. 31: 3300-3307.

Vargas, C., Fennell, D., and Häggblom, M.M. 2001. Anaerobic reductive dechlorination of chlorinated dioxins in estuarine sediments. Appl. Microbiol. Biotechnol. 57: 786-790.

Vetter, W., Jun, W., and Althoff, G. 2003. Non-polar halogenated natural products bioaccumulated in marine samples. I. 2,3,3´,4,4´, 5,5´-heptachloro-1´-methyl-1,2´-bipyrrole (Q1). Chemosphere. 52: 415-422.

Villemur, R., Saucier, M., Gauthier, A., and Beaudet, R. 2002. Occurrence of several genes encoding putative reductive dehalogenases in *Desulfitobacterium hafniense/frappieri* and *Dehalococcoides ethenogenes*. Can. J. Microbiol. 48: 697-706.

Wiegel, J., and Wu, Q.Z. 2000. Microbial reductive dehalogenation of polychlorinated biphenyls. FEMS Microbiol. Ecol. 32: 1-15.

Wu, Q., Bedard, D.L., and Wiegel, J. 1999. 2,6-Dibromobiphenyl primes extensive dechlorination of Aroclor 1260 in contaminated sediment at 8-30EC by stimulating growth of PCB-dehalogenating microorganisms. Environ. Sci. Technol. 33: 595-602.

Wu, Q., Bedard, D.L., and Wiegel, J. 1997a. Effect of incubation temperature on the route of microbial reductive dechlorination of 2,3,4,6-tetrachlorobiphenyl in polychlorinated biphenyl (PCB)-contaminated and PCB-free freshwater sediments. Appl. Environ. Microbiol. 63: 2836-2843.

Wu, Q., Bedard, D.L., and Wiegel, J. 1997b. Temperature determines the pattern of anaerobic microbial dechlorination of Aroclor 1260 primed by 2,3,4,6-tetrachlorobiphenyl in Woods Pond sediment. Appl. Environ. Microbiol. 63: 4818-4825.

Wu, Q., Sowers, K.R., and May, H.D. 2000. Establishment of a polychlorinated biphenyl-dechlorinating microbial consortium, specific for doubly flanked chlorinated, in a defined, sediment-free medium. Appl. Environ. Microbiol. 66: 49-53.

Wu, Q., Watts, J.E.M., Sowers, K.R., and May, H.D. 2002a. Identification of a bacterium that specifically catalyzes the reductive dechlorination of polychlorinated biphenyls with doubly flanked chlorines. Appl. Environ. Microbiol. 68: 807-812.

Wu, Q.Z., Milliken, C.E., Meier, G.P., Watts, J.E.M., Sowers, K.R., and May, H.D. 2002b. Dechlorination of chlorobenzenes by a culture containing bacterium DF-1, a PCB dechlorinating microorganism. Environ. Sci. Technol. 36: 3290-3294.

Chapter 17

Biotransformation of Carbon Tetrachloride by the Facultative Anaerobic Bacterium *Pseudomonas stutzeri*

Andrzej Paszczynski*, Jonathan Sebat, Daniel Erwin, and Ronald L. Crawford

Abstract

Carbon tetrachloride (CCl_4) and its dechlorination products are toxic and/or carcinogenic in mammals. Thus, their environmental fate is of concern to both environmental scientists and government regulators. A unique CCl_4 dechlorination mechanism has been discovered in cultures of iron-limited *Pseudomonas stutzeri* strain KC. This mechanism is characterized by extensive hydrolysis to give CO_2 as a major product, with low or undetectable levels of chloroform. A low-molecular-weight metal chelator promotes this route of CCl_4 decomposition. That agent is pyridine-2,6-bis(thiocarboxylic acid) (pdtc). Although pdtc synthesis is regulated by iron stress, preliminary results from ^{59}Fe uptake experiments suggest that pdtc is not the bacterium's primary siderophore. Unlike other siderophores, pdtc chelates copper(II), cobalt(III), and iron(III) with similar, very high affinity (Kd = 10^{34}). The affinity decreases dramatically (10^{12}) when iron(III) is reduced to iron(II), suggesting that pdtc functions as a siderophore towards iron in *P. stutzeri* but is not required for growth of the bacterium in low-iron media. Pdtc chelates many transition metals, some heavy metals, and also some lanthanides. Pdtc has antimicrobial activity, and this may be its main physiological function in nature. Pdtc is produced by *P. stutzeri* during both anaerobic and aerobic growth. *P. stutzeri* is a classic facultative anaerobic bacterium (denitrifier) and can be used for CCl_4 degradation in either the presence or absence of oxygen. The reaction of pdtc with CCl_4 does not require oxygen since it forms thiophosgene and this intermediate hydrolyzes to CO_2, H_2S, and HCl. Current research is elucidating the genetic basis of pdtc biosynthesis and resistance to it by its producers. This knowledge should allow the ultimate use of pdtc or pdtc-producing microorganisms for *in situ* bioremediation of CCl_4 and/or heavy metals and radionuclides in locations such as anaerobic aquifers.

*For correspondence email: andrzej@uidaho.edu

1. Introduction

Carbon tetrachloride (CCl_4) and its dechlorination products such as chloroform ($CHCl_3$) and dichloromethane (CH_2Cl_2) are carcinogenic in mammals. Thus, their environmental fate is of concern to both environmental scientists and government regulators. In the liver, CCl_4 is transformed by cytochrome P450 through reductive mechanisms, forming reactive species such as trichloromethyl radical and dichlorocarbene (Recknagel *et al.*, 1973; Pohl *et al.*, 1983) that are thought to be responsible for toxicity after CCl_4 exposure. Microbial transformations of CCl_4 have also been investigated as a means to develop active bioremediation processes or monitor *in situ* natural attenuation of CCl_4 in contaminated environments (Egli *et al.*, 1987; 1988; Galli *et al.*, 1989). Initial investigations showed that the primary biochemical agents from microbial sources that dehalogenate CCl_4 are tetrapyrrole-based cofactors, such as cobalamins, porphyrins, and factor F_{430} (Wood *et al.*, 1968; Castro *et al.*, 1985; Krone *et al.*, 1989a; 1989b; Gantzer *et al.*, 1991). These agents also transform CCl_4 through reductive mechanisms.

Mechanisms of CCl_4 dehalogenation mediated by transition metal cofactors appear to involve an initial one-electron reduction to give radical species (Assaf-Anid *et al.*, 1994; Chiu and Reinhard, 1995; Lewis *et al.*, 1995; Glod *et al.*, 1997). These carbon-centered radicals then undergo a second one-electron reduction by an available reductant, which is usually already present to regenerate the active form of the cofactor. This net two-electron reduction yields chloroform via replacement of one chlorine atom by one hydrogen atom. Other products may include carbon monoxide and formate, which arise through hydrolysis of dichlorocarbene. Products such as chloroform are not desirable end products from an environmental health perspective since they are also toxic.

An alternative type of CCl_4 dechlorination mechanism has been discovered in cultures of iron-limited *P. stutzeri* strain KC. This mechanism is characterized by extensive hydrolysis to CO_2 as a major product, with low or undetectable levels of chloroform (Criddle *et al.*,

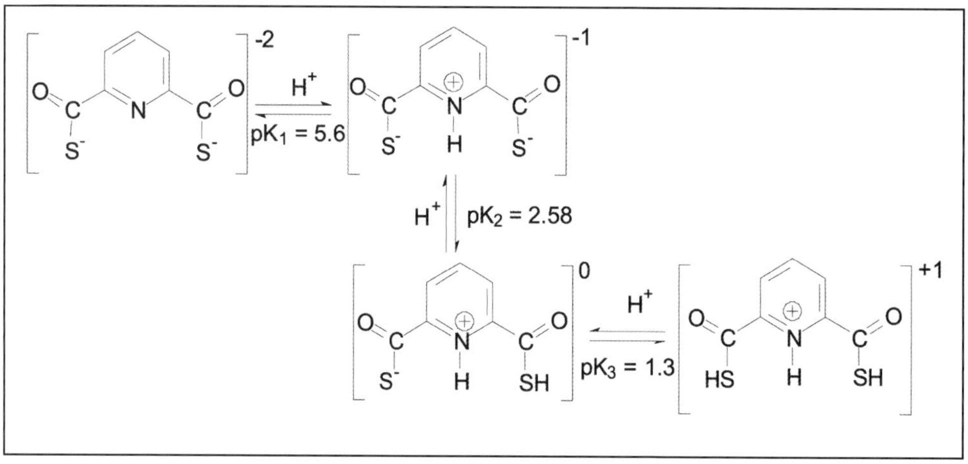

Figure 1. Pyridine-2,6-bis(thiocarboxylic acid) (pdtc) stepwise protonation. This amphoteric biochelator has three distinctive pKas. Depending on pH, the net charge of pdtc can change from negative to positive. Protonation constants were determined by potentiometric titration as described in Stolworthy *et al.*, 2001.

1990; Lewis *et al.*, 1996). This route of CCl_4 destruction is more environmentally friendly than the reductive processes seen previously since toxic products are not produced. Details of the process by which this facultative anaerobic *P. stutzeri* strain degrades CCl_4 are discussed in the following sections.

2. Carbon Tetrachloride is Dehalogenated by *P. stutzeri* Strain KC

Pseudomonas stutzeri strain KC represents a new genomovar (genomovar 9) within the species *P. stutzeri* (Sepulveda-Torres *et al.*, 2001). Early work with *P. stutzeri* strain KC showed that this strain mineralizes CCl_4 to CO_2 but does not dehalogenate chloroform. This observation indicated that CCl_4 is degraded via a novel pathway that does not involve step-wise dehalogenation. Details of this new degradative route became clearer when thiophosgene ($CSCl_2$) was identified as an intermediate of CCl_4 degradation by strain KC, indicating a net hydrolysis of the solvent molecule (Lewis and Crawford, 1995). Further study confirmed that the pathway involving thiophosgene accounts for most of the CCl_4 transformation observed in strain KC cultures under anoxic conditions (Lewis and Crawford, 1999). Oxygen substitution at the carbon atom of CCl_4 was also observed in the form of carbonyl-containing products, and the amounts of these products increased in the presence of O_2 (Lewis and Crawford, 1995). Assuming the trichloromethyl radical was involved, this oxygen substitution could be explained as a result of the phosgene intermediate reacting under aerobic conditions with O_2 (Asmus *et al.*, 1985). Under anoxic conditions carbonyl sulfide (COS) was observed, likely arising from thiophosgene hydrolysis by sulfur nucleophiles common in the cultures. These data suggested that a radical substitution mechanism initiated by one-electron reduction of CCl_4 at a metal center and followed by reaction of a trichloromethyl radical with a sulfur species could be involved in CCl_4 transformation by strain KC.

3. The Agent Involved in CCl_4 Dehalogenation by *P. stutzeri* Strain KC is Pyridine-2,6-Bis(Thiocarboxylic Acid)

In a breakthrough study of this novel pathway, a low-molecular-weight metabolite produced by *P. stutzeri* strain KC, pyridine-2,6-bis(thiocarboxylic acid) (pdtc; Figure 1), was found to be the extracellular agent responsible for CCl_4 dechlorination activity (Lee *et al.*, 1999). This compound was previously identified as a metal-chelating agent from iron-limited cultures of a strain of *Pseudomonas putida* (Ockels *et al.*, 1978). *P. putida* and *P. stutzeri* strain KC thus far are the only two species of bacteria known to produce pdtc. The occurrence of two thiocarboxylic groups in pdtc and its ability to coordinate transition metals suggested a potential mechanism for reaction with CCl_4. Specifically, this would involve the reduction of CCl_4 at the metal center to produce trichloromethyl radical followed by the condensation of the radical with one of the pdtc sulfur atoms.

The addition of transition metals to cultures had previously been observed to strongly affect CCl_4 transformation activity of strain KC (Criddle *et al.*, 1990; Lewis and Crawford, 1993; Tatara *et al.*, 1993). Iron as Fe(II) or Fe(III) inhibited CCl_4 transformation when present initially in cultures at 10–100 μM; however, no inhibition was seen when iron was added to cultures already showing dehalogenation activity. Co(II) inhibited CCl_4 transformation in low micromolar concentrations and inhibited growth of strain KC at higher concentrations (Lewis and Crawford, 1993). $CuCl_2$ stimulated CCl_4 transformation activity at very low concentrations (5 nM) (Criddle *et al.*, 1990; Lewis and Crawford, 1993; Tatara *et al.*, 1995). Data obtained from transposon mutants derived from strain KC indicated the effects of iron were due to repression of genes necessary for CCl_4 transformation in response to iron supplementation (Sepulveda-Torres *et al.*, 1999; Lewis *et al.*, 2000; Sepulveda-Torres *et al.*, 2002). The effects of copper and cobalt were explained later during observations of biophysical chemistry of pure pdtc-metal

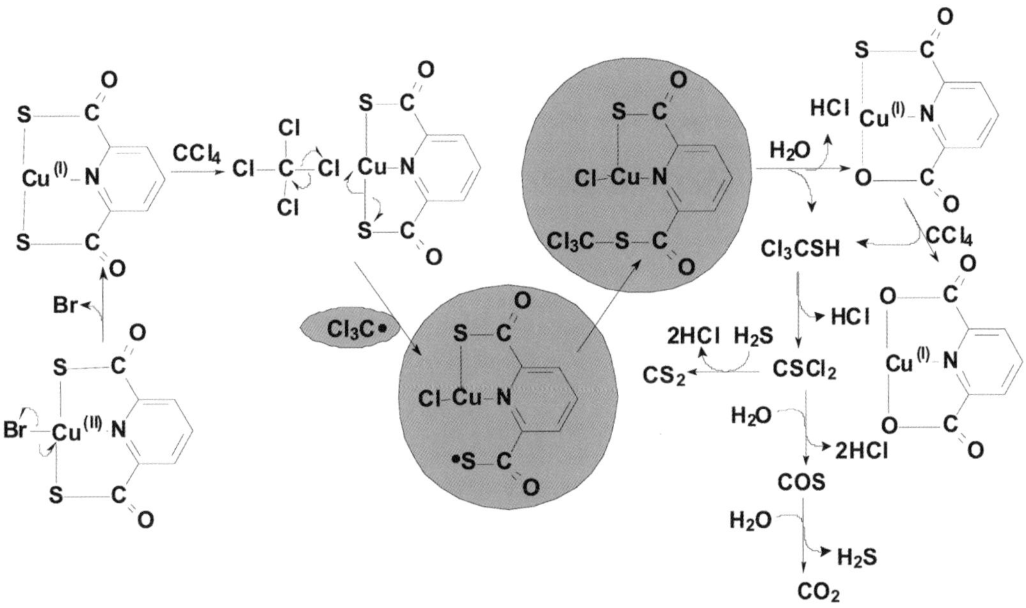

Figure 2. Mechanism of CCl_4 dehalogenation by pdtc produced by *Pseudomonas stutzeri* strain KC. (Modified from Lewis *et al.*, 2001). The pathway involves the stoichiometric reaction of the Cu(I):pdtc molecule with two successive molecules of CCl_4. The essential reactions in the pathway are formation of the trichloromethyl and pdtc-carbonylthiyl radicals followed by coupling of the radicals to form the unstable trichloromethylthioester of Cu:pdtc (grey background). The trichloromethylthioester of Cu:pdtc, trichloromethanethiol, and thiophosgene are unstable intermediates. The hydrolysis of these intermediates to HCl, CO_2, and H_2S results in complete mineralization of two moles of CCl_4 by one mole of Cu(I):pdtc. Byproducts of this pathway are two moles of the copper complex of pyridine-2,6-dicarboxylic acid (dipicolinic acid - dpa).

complexes. Cu is required for CCl_4 dehalogenation, and Co probably out competes Cu for binding by pdtc. The Co: pdtc complex is not active toward CCl_4 dehalogenation. The identification of pdtc as the active agent and the availability of chemically synthesized pdtc made it possible to resolve the mechanism of CCl_4 degradation by *P. stutzeri* strain KC. More information on pyridine derivatives substituted with monothiocarboxylic acid groups, the unique metabolites of certain *Pseudomonas* species, can be found in a recent review (Budzikiewicz, 2003).

3. The Reaction Pathway for CCl_4 Dehalogenation by *P. stutzeri* strain KC is Known

Lewis *et al.* (2001) proposed a pathway to explain the formation of end products during defined chemical reactions between carbon tetrachloride (CCl_4) and either metal complexes of pdtc or pure cultures of *P. stutzeri* KC. The pathway (Figure 2) involves chloride abstraction from CCl_4 by a Cu(I):pdtc complex followed by condensation of the formed trichloromethyl and thiyl radicals, with hydrolysis of a thioester intermediate. Products detected during *in vitro* reactions included carbon dioxide, chloride, carbonyl sulfide, carbon disulfide, and dipicolinic acid. Spin-trapping and electrospray MS/MS experiments gave evidence for trichloromethyl and thiyl radicals generated by reaction of CCl_4 with pdtc:copper complex. A subsequent experiment showed that dechlorination of CCl_4 by pdtc required copper and is inhibited by cobalt but not by iron or nickel. Pdtc was shown to react stoichiometrically (2 CCl_4 per 1 pdtc:Cu) rather than catalytically without added reducing equivalents.

4. Current Research is Elucidating the Genetic Basis of CCl_4 Dehalogenation by *P. stutzeri* Strain KC

Lewis *et al.* (2000) observed a spontaneous mutant of *P. stutzeri* strain KC that lacked the ability to transform CCl_4. Restriction digests separated by pulsed-field gel electrophoresis showed that the mutant strain (CTN1) differed from strain KC by deletion of approximately 170 kb of chromosomal DNA. CTN1 did not produce pdtc and did not degrade CCl_4. Cosmids from a genomic library of strain KC containing DNA from within the deleted region were identified by hybridization with a 148-kb genomic *Spe*I fragment absent in strain CTN1. One of these cosmids (pT31) complemented the pdt- phenotype of CTN1 and conferred CCl_4 transformation activity and pdtc production upon other pseudomonads, including *P. aeruginosa, P. putida,* and *P. fluorescens.* Southern analysis showed that none of three other *P. stutzeri* strains representing three genomovars contained DNA that would hybridize with the 25,746 bp insert of pT31. Sequencing of this DNA placed a previously known 8.27-kb *Eco*RI fragment containing genes involved in CCl_4 transformation within this region (Sepulveda-Torres *et al.*, 1999). Transposon mutagenesis

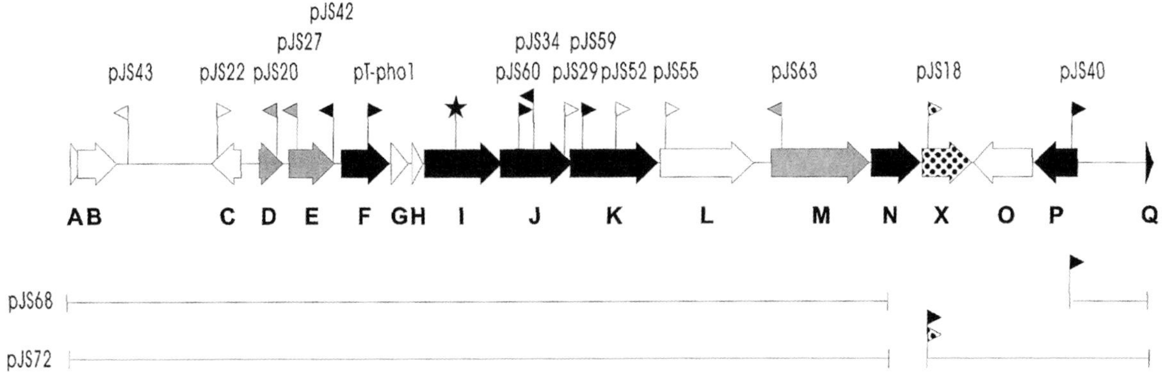

Figure 3. Phenotypes of *pdt* locus mutants. Shown is a map of the insert of cosmid pT31. It has been revised since previous publication (Lewis *et al.*, 2000) to include a new gene *pdtX*. Flags indicate the location and orientation of each unique transposon insertion. Each flag and open reading frame is colored to indicate the phenotype abolished by the mutation: (black) mutants that fail to confer pdtc production to *P. stutzeri* CTN1; (gray) mutants that fail to enhance the pdtc resistance of *E. coli* DH5a; (polka-dot) mutants that enhance the pdtc resistance of *E. coli*; (white) unaffected mutants; (★) a mutation by Sepulveda-Torres *et al.* (1999) that abolishes pdtc production and is not characterized in this review.

of pT31 identified open reading frames the disruption of which affected the ability to make pdtc in the strain CTN1 background. Thus, the *pdt* locus of strain KC resides in a non-essential region of the chromosome and is subject to spontaneous deletion.

The *pdt* locus encodes much of the pdtc biosynthesis pathway of strain KC and is sufficient for pdtc biosynthesis by other pseudomonads. Mutational studies have identified six genes that are essential for pdtc production (Sepulveda-Torres *et al.*, 1999; Lewis *et al.*, 2000; Figure 3). Several of the *pdt* gene products are similar to the enzymes involved in the biosynthesis of siderophores and antibiotics, and a mechanism of pdtc biosynthesis has been proposed (Sepulveda-Torres *et al.*, 2002).

Although pdtc synthesis by *P. stutzeri* strain KC is regulated by iron stress (Criddle *et al.*, 1990; Lewis and Crawford, 1993; Tatara *et al.*, 1993), preliminary results from ^{59}Fe uptake experiments suggest that pdtc is not the bacterium's primary siderophore (Cortese *et al.*, 2002). Unlike other siderophores, pdtc chelates copper as Cu(II), cobalt as Co(III), and iron as Fe(III) with similar, very high affinity (Kd = 10^{34}; Stolworthy *et al.*, 2001). Also, pdtc is not required for growth of the bacterium in low-iron medium. *P. stutzeri* strain KC appears to produce other siderophores, including a hydroxamate (Dybas *et al.*, 1995). We determined that the stability constant (Kd) of Fe(II):pdtc is 10^{12}. We used potentiometric and spectrophotometric titration and ligand-ligand competition studies using 2,6-pyridinedicarboxylic acid (dipicolinic acid – dpa; Brandon *et al.*, 2003). Dpa was chosen because it has the same iron-binding geometry as pdtc, and the binding constant of dpa for iron was known. Comparing the pdtc Kd of Fe(II) to the Kd for Fe(III) shows that the stability constant of FeII(pdtc) is approximately 21 orders of magnitude smaller. This represents a very significant decrease in the binding strength of pdtc toward iron, suggesting a

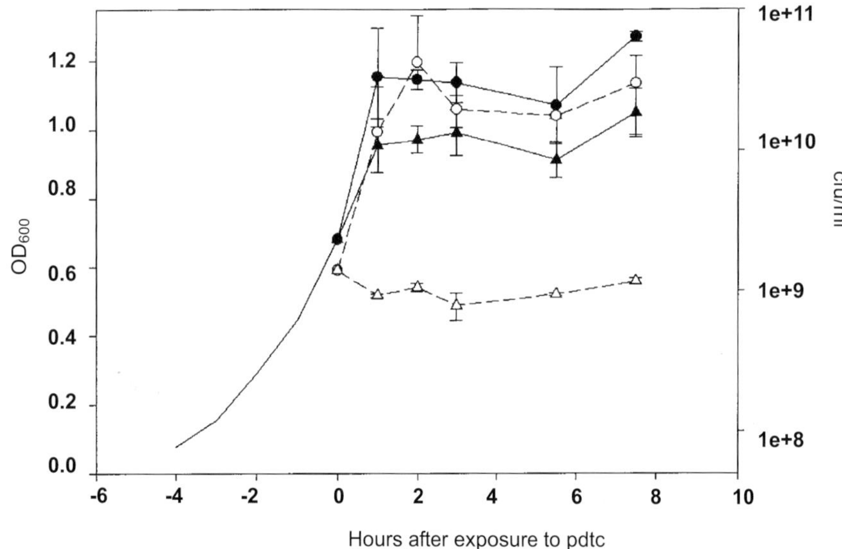

Figure 4. Growth and viability of *E. coli* after exposure to pdtc. A single culture of *E. coli* was grown to an OD ~ 0.7. At time = 0, the culture was split into six flasks (25 mL each). 30 μM pdtc was added to three flasks, and the remaining three flasks received an equivalent amount of solvent (DMF). Viable plate counts and measurements of optical density were performed at hourly intervals. (●) OD_{600}, no pdtc; (▲) OD_{600}, 30 μM pdtc; (○) cfu/mL, no pdtc; (Δ) cfu/mL, 30 μM pdtc. (Sebat *et al.*, 2001).

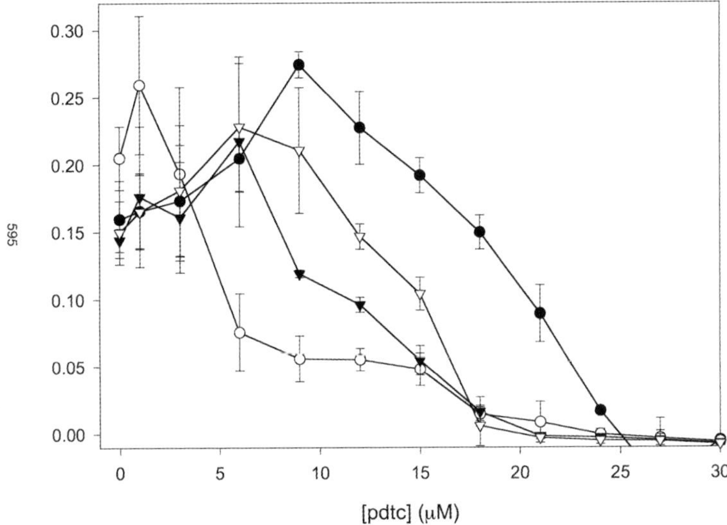

Figure 5. Plasmid pT31 (*pdt* locus) enhances the pdtc resistance of *pdt⁻* mutant strain *P. stutzeri* CTN1. *P. stutzeri* strain KC or its derivatives were grown in microtiter plates containing succinate media (SM) supplemented with various concentrations of pdtc. Measurements of optical density were performed after 48 hours. Values shown are the mean and standard deviation of three replicates. (●) KC/pRK311; (○) CTN1/pRK311; (▲) CTN1/pT31; (Δ) CTN1 pJS60.

siderophore-like function for pdtc in pseudomonad cells. Additionally, iron-specificity is typical of siderophores, a group of natural compounds produced by *Pseudomonas* spp. and other aerobic microorganisms in order to acquire and transport ferric ion under low iron conditions. The two most common types of iron(III)-chelating siderophores, the hydroxamate and catecholate chelators, are both highly specific for ferric iron. Less-specific chelators are seldom observed, and sulfur-containing siderophores like pdtc are very rare. Examples of sulfur-containing siderophores are: o-*N*-hydroxy-o-*N*-[2'-(2",3"-dihydroxy-phenyl)thiazolin-4'-yl]histamine anguibactin (Jalal *et al.*, 1989) isolated from the fish pathogen *Vibrio anguillarum*; pyochelin, 2-(2-*o*-hydroxyphenyl-2-thiazoline-4-yl)-3-methylthiazolidine-4-carboxylic acid isolated from a low-iron culture of *P. aeruginosa* (Cox *et al.*, 1981); and ferrithiocin, an unusual orange-brown ferric ion organic complex isolated from the culture supernatant of *Streptomyces antibioticus* (Naegeli *et al.*, 1980).

Our knowledge of the genetics and physiological properties of pdtc provide insight into the function of pdtc. Pdtc is inhibitory to numerous species of bacteria (Sebat *et al.*, 2001), and its antimicrobial activity is eliminated by titration with iron, copper, or cobalt. Furthermore, the effect of pdtc on *E. coli* is bacteriostatic, suggesting that the mechanism of inhibition is the sequestration of metals. The effect of pdtc on *E. coli* is illustrated in Figure 4. Pdtc was added to log-phase cultures of *E. coli*. Two hours after exposure to pdtc, cultures appeared to have reached stationary phase at a significantly lower density than untreated controls (Figure 4). The number of viable cells in the presence of pdtc remained constant for 7.5 hours.

The *pdt* locus of *P. stutzeri* KC also encodes an unidentified mechanism that confers pdtc resistance. Plasmid pT31 enhances the resistance of *E. coli* to pdtc (Figure 5). Mutational studies have shown that three of the genes within the *pdt* locus (Figure 3) appear to be required for pdtc resistance including the putative transporter gene *pdt*E. A mutation in a second transporter gene *pdt*N eliminated the ability to produce pdtc but did not affect pdtc resistance. Examination of the derived amino acid sequences of these transporters indicated that PdtE is homologous to transporters YbtX and Irp8 that are encoded by the yersiniabactin uptake operon of *Yersinia pestis* and *Y. entercolitica*. PdtN is homologous to a putative hexuronate permease and numerous multidrug efflux pumps.

In order to assess the potential for PdtE to bind and transport metals and/or pdtc, we constructed a model for the membrane topology of this protein. The topology model of PdtE revealed characteristics common to siderophore transporters and a pocket of acidic residues that may serve as a metal-binding site. We also modeled PdtN. PdtN is hydrophobic and contains five cysteines, which is consistent with the transport of an aromatic, sulfur-containing compound (Figure 6). These findings support the proposal that PdtN is the primary mechanism of pdtc secretion, and PdtE is involved in the utilization of trace-metal:pdtc complexes.

5. The Metal-Chelating Properties of pdtc and the Biological Activities of the Formed Chelates are Complex

Using several analytical techniques, our group observed that either one or two pdtc molecules would bind one metal ion, depending on the metal, forming a tri-dentate or tetra-dentate planar complex (Figure 7A) or a hexa-dentate octahedral complex with the central metal ion (Figure 7B). The sulfur and nitrogen atoms are metal-coordinating atoms in all cases, with the exception of UO_2:(pdtc)$_2$, where carbonyl oxygen substitutes for sulfur in this complex formation (Table 1). The spectra presented in

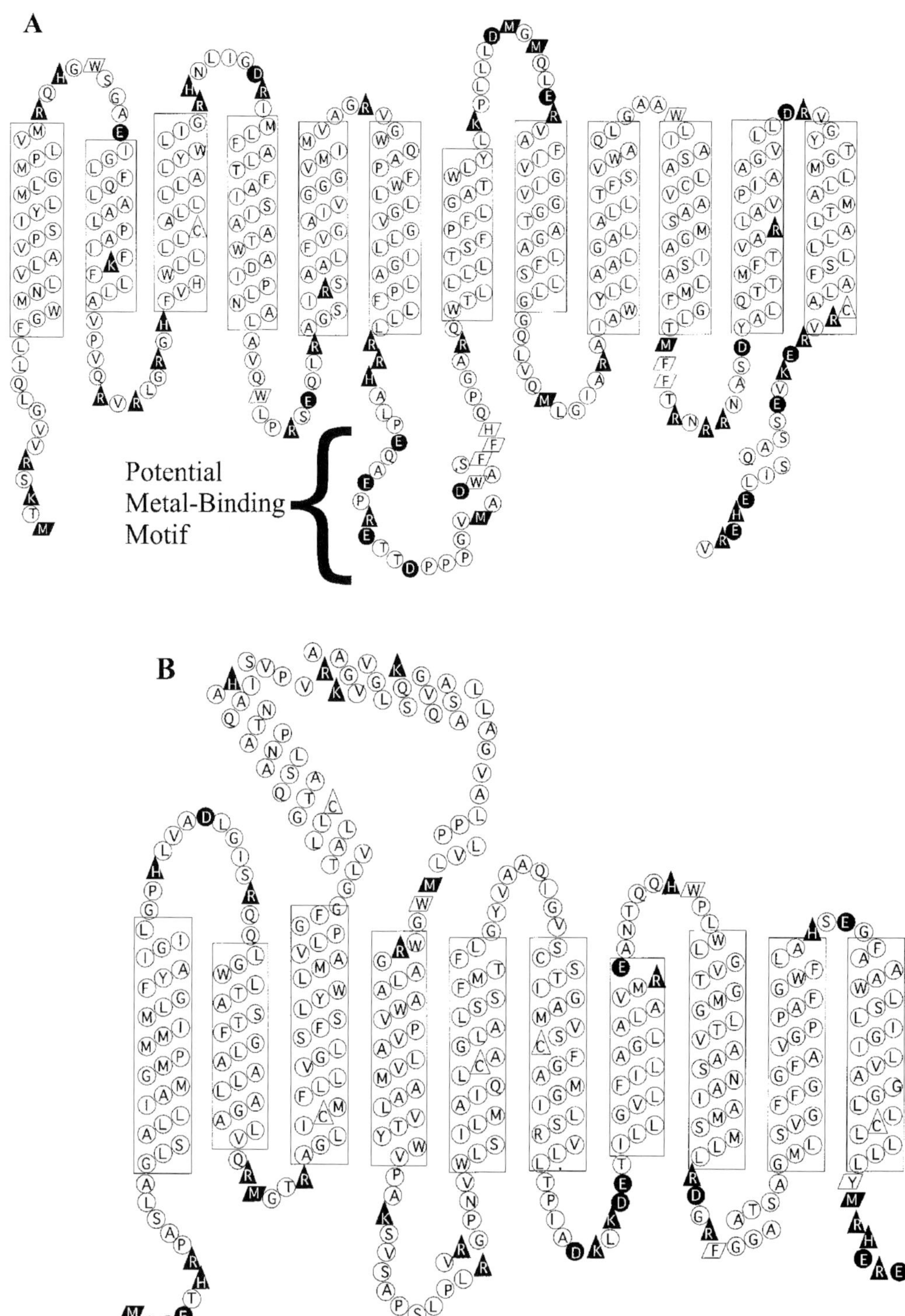

Figure 6. Predicted topology of hypothetical pdt transporters. (A) PdtE; (B) PdtN. Transmembrane regions were determined by the TMHMM program and alignment of protein families (Pfam B). Additional helical folds were identified with the secondary structure predictor 3D-pssm. All cysteine residues (open triangles) as well as all acidic (filled circles) and basic (filled triangle) residues are highlighted. In addition, methionine (filled prism) and aromatic residues (open prisms) are indicated at positions outside of the transmembrane regions. The region of acidic residues resembles a repeating iron binding motif similar to the repeating EXXE motif present in the yeast iron transporter FTR1 (Stearman *et al.*, 1996) and the bacterial response regulator PmrB (Wosten *et al.*, 2000). This region of the loop is predicted to remain unfolded (determined by analysis with the software program PONDR), while the hydrophobic region is predicted to have an ordered helical conformation.

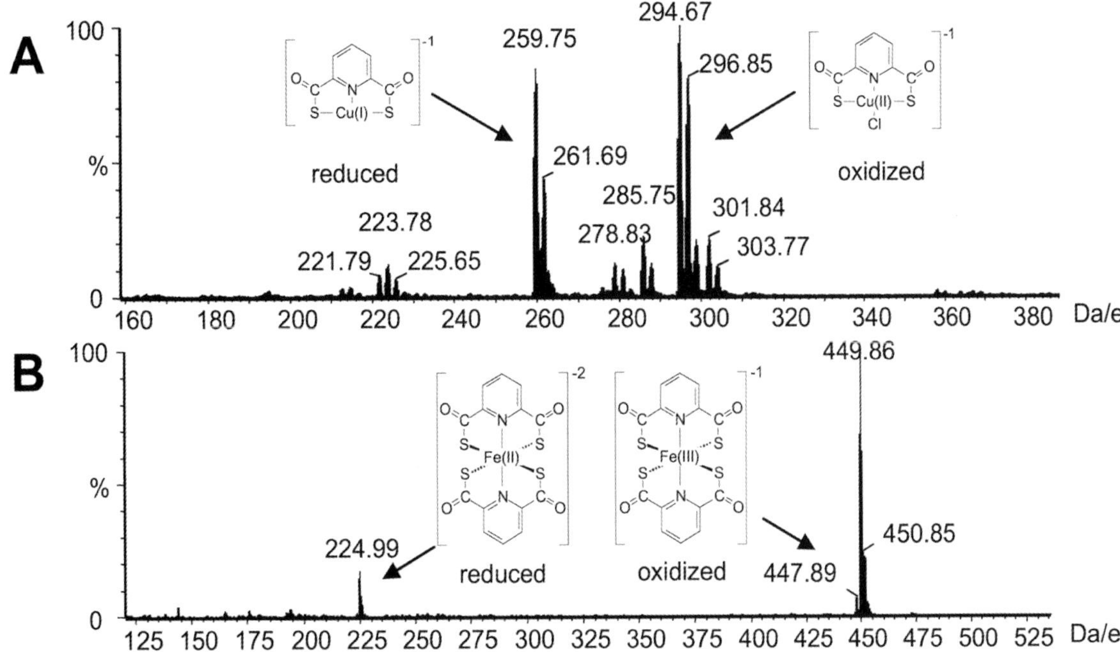

Figure 7. Electrospray mass-spectra of pdtc:metal comlexes (Cortese *et al.* 2002). **A.** Example of electrospray mass spectrum of the Cu(II/I): pdtc couples. The solution color changes from green to brown as cupric:pdtc is reduced to cuprous:pdtc. The Cu(II) complex shows a large parent +2 peak because of the natural isotope distribution patterns of copper and chloride, ^{63}Cu = 69.1%; ^{65}Cu = 30.9%; ^{35}Cl = 75.5%; ^{37}Cl = 24.5%. The chloride ion in the complex can be replaced by other halides, by inorganic ions such as nitrate, or by chelating organic molecules. **B.** Example of an electrospray mass spectrum of Fe(II/III):$(pdtc)_2$ complexes. Ferric/ferrous iron redox cycling changes the charge of the molecule from –1 to –2, which is reflected in a solution color change from brown to blue and apparent mass change on the spectrum from 450 Da/e to 225 Da/e.

Figures 4 and 5 show that electrospray mass spectroscopy (ESMS), with its soft ionization, is particularly well suited for analysis of these types of molecules, providing information not only about structure but also about the redox state of the metal ion.

Our group also observed that heavy metals such as Pb(II), Hg(II), and Cd(II) form 1:1 complexes with pdtc. The mercury:pdtc complex has very low solubility in aqueous solutions. The structure of this complex has not yet been determined. Some metals will catalyze acceleration of pdtc hydrolysis. We found that the hydrolysis product of pdtc will also bind these metals, forming characteristic mass patterns on ESMS (Figure 8). The mass spectra peaks for pdtc, lead, bismuth, and chromium (Figure 8A, B, C) are separated by 16 Daltons (Da), confirming that oxygen atoms replace sulfur atoms in the complexes.

Cortese *et al.* (2002) evaluated the ability of pdtc to form complexes with 19 metals and three metalloids; it formed complexes with 14 of the metals. Examples of some pdtc:metal complexes are shown in Table 1. Four of these complexes, Fe:$(pdtc)_2$, Co:$(pdtc)_2$, Mn:$(pdtc)_2$, and Cu:pdtc were found to cycle between redox states. Precipitates formed when pdtc was added to solutions of As, Cd, Hg, Mn, Pb, and Se. Additionally, 14 of 16 microbial strains tested were protected from Hg toxicity when pdtc was present. Pdtc also mediated protection from the toxic effects of Cd and Te for some strains. Pdtc by itself did not greatly facilitate iron uptake but modestly increased the overall level of iron uptake of *P. stutzeri* strain KC and *P. putida* DSM301. Both these bacteria could reduce amorphous Fe(III) oxyhydroxide in culture. *In vitro* reactions showed that copper and pdtc were required for this activity. Thus, pdtc appears to have some activity as a siderophore to facilitate iron uptake, but it is not the primary function of this metabolite.

Stolworthy *et al.* (2001) studied the metal-binding properties of pdtc by potentiometric and spectrophotometric techniques. They described the first two stepwise protonation constants (pK) for successive proton addition to pdtc; these were 5.48 and 2.58, respectively (Figure 1). The third stepwise protonation constant was estimated to be 1.3. The stability (affinity) constants for iron(III), nickel(II), and cobalt(III) were determined and were found to be exceptionally high. The stability constants (log K) were 33.93 for $(Co{:}pdtc_2)^{-1}$, 33.36 for $(Fe{:}pdtc_2)^{-1}$ and 33.28 for $(Ni{:}pdtc_2)^{-2}$. These protonation constants and high affinity constants show that over a physiological pH range, the ferric pdtc complex has one of the highest effective stability constants for iron binding among known bacterial chelators.

Sebat *et al.* (2001) examined the antimicrobial properties of pdtc and the mechanism of this antibiotic activity. The growth of *P. stutzeri* strain KC was

Table 1. Molecular ion peaks from electrospray mass spectra and associated structures for complexes formed by pdtc (Cortese *et al.* 2002)

Metal	Mol. Ion (*m/z*)	Structure	Metal	Mol. Ion (*m/z*)	Structure	Metal	Mol. Ion (*m/z*)	Structure
Au	464	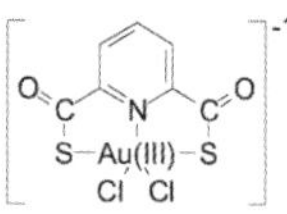 S–Au(III)–S, Cl Cl; -1	Mn	450	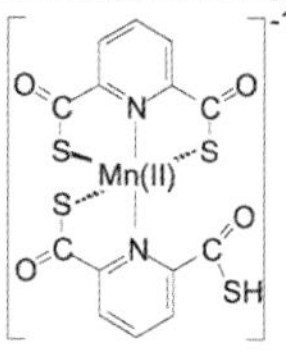 Mn(II), SH; -1	Sc	439	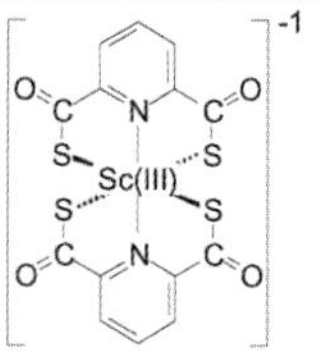 Sc(III); -1
Cd	373	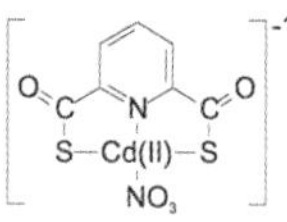 S–Cd(II)–S, NO_3; -1	Nd	538	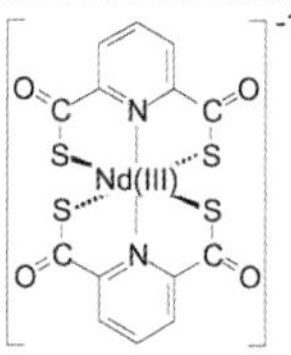 Nd(III); -1	UO_2[4]	N.D.	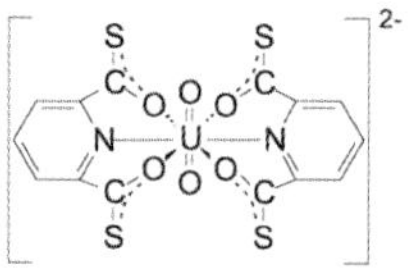 U; 2-
Co[1]	453	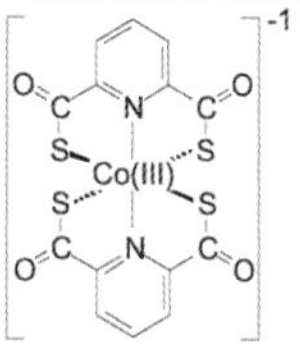 Co(III); -1	Ni[1,3]	226	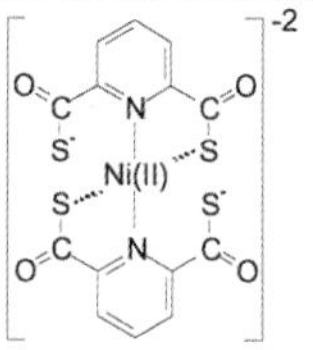 Ni(II); -2	Zn	298	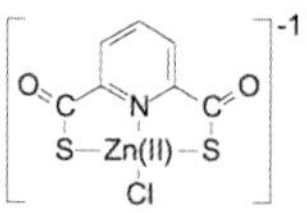 S–Zn(II)–S, Cl; -1
Mn	224.5	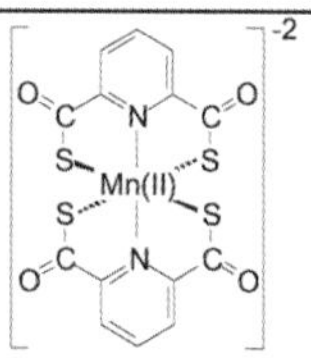 Mn(II); -2	Pd[2]	N.D.	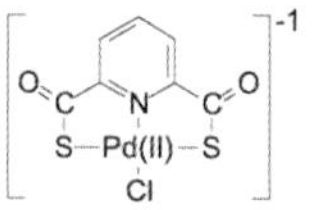 S–Pd(II)–S, Cl; -1	Zn	229	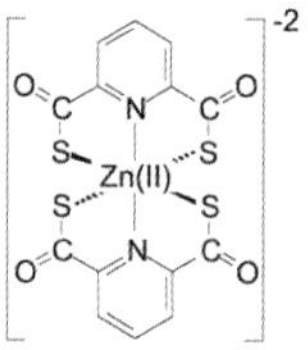 Zn(II); -2

[1](Hildebrand 1988) ; [2](Espinet *et al.*, 1994); [3](Krüger *et al.*, 1990); [4](Neu *et al.*, 2001)

significantly enhanced by 32 μM pdtc; however, all non-pseudomonads and two strains of *P. stutzeri* were sensitive to 16 to 32 μM pdtc. Fluorescent pseudomonads were generally resistant to all concentrations tested. In competition experiments, strain KC demonstrated antagonism toward *Escherichia coli*, and this effect was partially alleviated by 100 μM $FeCl_3$. Less antagonism was observed in mutant derivatives of strain KC (Lewis *et al.*, 2000) that lack the ability to produce pdtc. A competitive advantage was restored to these strains by a cosmid (pT31) that confers pdtc production. This cosmid also enhanced pdtc resistance of all pdtc-sensitive strains, indicating that this plasmid contains elements responsible for resistance to pdtc. The antimicrobial effect of pdtc was reduced by the addition of Fe(III), Co(III), and Cu(II) and enhanced by Zn(II). Analyses by mass spectrometry determined that Cu(I):pdtc and Co(III):$pdtc_2$ formed immediately. Overall, these results suggested that pdtc is an antagonist and that metal sequestration is the primary mechanism of its antimicrobial activity.

6. Mechanism of pdtc:copper Complex Interaction with CT

Our recent research has focused on probing the molecular mechanism of the pdtc reaction with CCl_4. We used crystallographic, electrospray mass spectrometry methods, and computer simulations (Figure 9) to further refine our understanding of the interaction of pdtc: copper with CCl_4. We observed by ESMS (Figure 10) the tetradentate Cu(II):pdtc:halide and a tridentate Cu(I): pdtc. The amount of Cu(I):pdtc in solution is proportional to the size of the halide atom (affinity for valence electrons decreases as the atomic number of the halide atom increases). Electron transfer from the carbonyl sulfide (-COS^-) group or halide ligands to form Cu(I) could not be ruled out. Consequently, our data suggest that the electron transfer to the copper from –COS^- could be the main mechanism of the carbonyl-thiyl radical and trichloromethyl radical formations (Figure 2). In each halide solution, a large peak of Cu(I):pdtc complex was observed (Figure 10) though the relative abundance of these peaks does not correspond to the observed rates

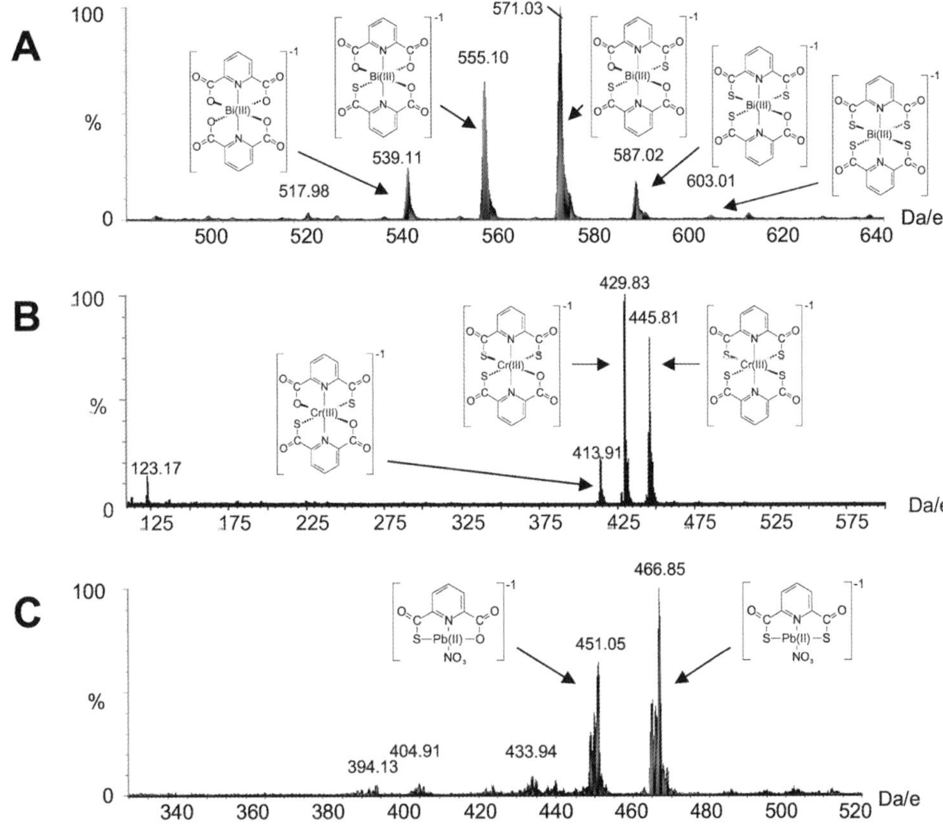

Figure 8. ESMS spectra of A. Bi(III):$(pdtc)_2$; B. Cr(III):$(pdtc)_2$; and C. Pb(II):pdtc complexes (Cortese *et al.* 2002). A 100-ppm solution of these metals in DMF was used for EMS analysis. The spectra show additional complexes formed between the metal ions and pdtc hydrolysis products. In the case of bismuth, the complex with two dipicolinic acids (product of complete hydrolysis of pdtc) was observed (Da/e = 539.11). Low solubility of lead and bismuth pdtc complexes in water was observed.

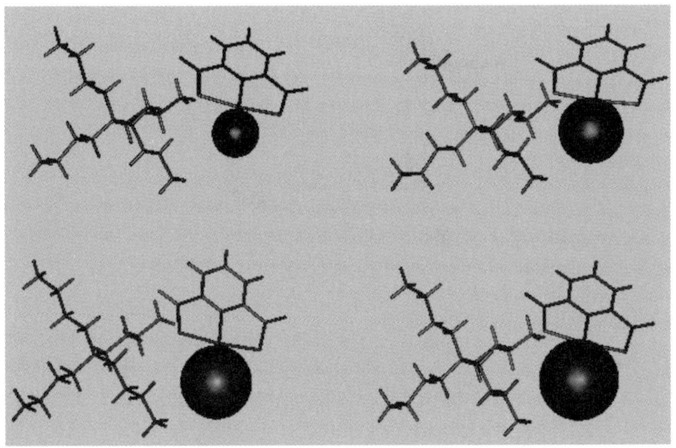

Figure 9. Computer-generated models of tetrabutylammonium: Cu(II):pdtc:fluoride (upper left), Cu(II):pdtc:chloride (upper right), Cu(II):pdtc:bromide (lower left), and pdtc:iodine (lower right) complexes. The chloride, bromide, and iodine complexes were chemically synthesized and crystallized from acetone:diethyl ether (1:2) solutions (Hildebrand *et al.*, 1983; Espinet *et al.*, 1994). The atom positions were measured using X-ray diffraction methods. The computer model of the pdtc: Cu:fluoride complex was generated using ECCE (Pacific Northwest National Laboratory) and InsightII (Accelrys Inc.,) software. The diameters of fluoride, chloride, bromide, and iodine atoms represent relative difference in van der Waals radii of the respective halides. These differences reflected decreasing affinity for valence electrons as the halide atom radius increases, generating more Cu(I):pdtc in solution (see Figure 10).

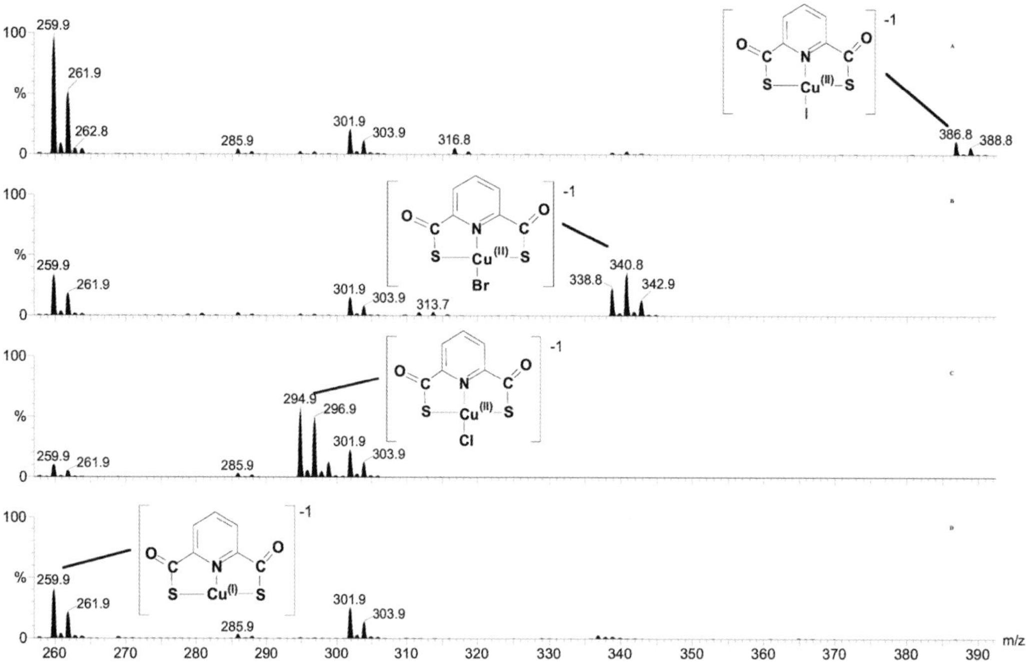

Figure 10. Electrospray mass spectra of pdtc:copper halide complexes. 2-mM solutions of copper halide pdtc complex in 50:50 water - DMF were analyzed by ESMS. (A) Cu:pdtc:I; (B) Cu:pdtc:Br; (C) Cu:pdtc:Cl; (D) Cu:pdtc:F. The characteristic ion 259.85 representing the pdtc: Cu(I) complex was observed in each halide solution. We were not able to observe the Cu(II):pdtc:F complex using ESMS. The concentration of $Cu^{(I)}$:pdtc in solution (m/z = 259.9) decreased as the halide atom radius was decreased (A>B>C).

of CCl_4 degradation by the respective complexes (data not shown). We also used the Cu(I) specific chelator, bathocuproine disulfonate (BC), to confirm the presence of pdtc:Cu(I) in the reaction mixture. All copper was sequestered from pdtc by the addition of BC to the solution, inhibiting CCl_4 degradation (data not shown) and confirming the existence of a Cu(I):pdtc complex in the presence and in the absence of CCl_4 in the reaction mixture. From this work we conclude that carbon tetrachloride transformation by pdtc involves Cu(II)/Cu(I) redox cycling and formation of radicals on CCl_4 (the substrate) and Cu:pdtc (the catalyst). Formation of this radical and consequent dechlorination activity is attributable to three factors: (1) the redox activity of copper, (2) the strong affinity of copper for the halide ligand, and (3) the proximity of a sulfide nucleophile and copper electrophile in the Cu:pdtc catalytic complex (Figure 2).

7. *Pseudomonas stutzeri* Strain KC and pdtc have Great Potential for Use in *In Situ* Bioremediation of Carbon Tetrachloride and Metals

Dybas *et al.* (2002) have demonstrated that pdtc-producing cells of *P. stutzeri* KC can be employed for *in situ* bioremediation of a groundwater carbon tetrachloride plume. A full-scale field evaluation of bioaugmentation using *P. stutzeri* KC was conducted in a carbon tetrachloride (CT) and nitrate impacted aquifer at Schoolcraft, MI. Strain KC was introduced into the aquifer using closely spaced injection/extraction wells installed normal to the direction of groundwater flow and near the leading edge of the CT plume. This system produced a "biocurtain" for promoting CT degradation as the contaminant plume passed through the treatment zone. Intermittent addition of base was used to create favorable pH conditions (alkaline) that decreased the availability of iron that might repress KC's pdtc production pathway. Weekly additions of acetate (electron donor), alkali, and phosphorus were employed. CT removal efficiencies were 98-99.9%, and KC colonized the aquifer at $>10^5$ cells KC/g aquifer material. Sustained and efficient (98%) removal of CT was observed over the duration of the four-year study. Transient, low levels of chloroform (CF; <20 ppb) and H_2S (<2 ppm) were observed, but these compounds disappeared when the concentration of acetate in the weekly feed was reduced. Strain KC is a denitrifier (Criddle *et al.*, 1990), and nitrate removal efficiencies ranged from 60% to nearly 100% depending on acetate concentrations. About 18,600 m^3 of CT-contaminated groundwater was treated during the project. The authors concluded that closely spaced wells and intermittent substrate additions were an effective technique for delivering both bacterial cells and substrates to this subsurface environment.

There are other potential options for employing pdtc-based systems for bioremediation purposes. Since pdtc can be prepared synthetically (Hildebrand, *et al.*, 1983), in theory, it could be prepared commercially for introduction into carbon tetrachloride-contaminated waters. However, chemical synthesis of pdtc requires a hydrogen sulfide-saturated pyridine solution, which is extremely toxic and volatile and renders the synthetic procedure environmentally unfriendly. Therefore, production using the microbial systems available would probably be preferable. *Pseudomonas stutzeri* KC and *P. putida* naturally produce pdtc at a concentration of 30-40 ppm in iron-limited cultures. It is possible that this level might be increased by genetic manipulation of the strains, since much is already known of the genetic controls over the biosynthetic pathway for pdtc (strain KC), and producer organisms' pdtc self-resistance mechanisms are being unraveled. Cell-free broths from fermenters could be produced on the surface at a site of CT contamination and pumped into appropriate zones of the contaminant plume. The use of this small molecule (<500 Daltons) would directly avoid the well-known problems associated with transport of whole bacterial cells in aquifers where they are often immediately trapped within biofilms close to their injection points within groundwater matrixes (Fuller *et al.*, 2000). Whole microbial cells thus often do not reach the areas of an aquifer where much of the contaminant resides, necessitating a biocurtain-based approach such as described by Dybas *et al.* (2002).

There should be little concern about the fate of the pdtc itself once it is introduced into the environment. Its decomposition product is dipicolinic acid (dpa), a natural product that makes up a large fraction of the mass of common bacterial endospores. Preparing ^{14}C-labeled pdtc and confirming its biodegradability in microcosm tests could easily assess the likely fate of pdtc in a CT plume environment.

Pdtc could also be used in treatment of heavy metal-contaminated aquifers. As discussed in preceding sections, pdtc forms insoluble complexes with many of the most toxic metals (e.g., Pb and Hg). Thus, one can envision scenarios where pdtc might be used to immobilize heavy metals or radionuclides *in situ*. In some locations where carbon tetrachloride and heavy metals are co-contaminants, it might be possible to treat both pollutants simultaneously. Similar methods as described directly above would be appropriate here as well. These are hypothetical approaches where little research has been performed, so additional applied research in these areas is warranted.

Acknowledgements

This work of the authors reported here was supported by the US DOD NABIR Program under grant # DEFG03-96ER62273 and the Inland Northwest Research Alliance under grant # UI002.

References

Asmus, K.-D., Bahnemann, D., Krischer, K., Lal, M., and Mönig, G. 1985. One-electron induced degradation of halogenated methanes and ethanes in oxygenated and anoxic aqueous solutions. Life Chem. Rep. 3: 1–15.

Assaf-Anid, N., Hayes, K.F., and Vogel, T.M. 1994. Reduction dechlorination of carbon tetrachloride by cobalamin(II) in the presence of dithiothreitol: mechanistic study, effect of redox potential and pH. Environ. Sci. Technol. 28: 246–252.

Brandon, M.S., Paszczynski, A.J., Korus, R., and Crawford, R.L. 2003. The determination of the stability constant for the iron(II) complex of the biochelator pyridine-2,6-bis(monothiocarboxylic acid). Biodegrad. 14: 73-82.

Budzikiewicz, H. 2003. Heteroaromatic monothiocarboxylic acids from *Pseudomonas* spp. Biodegrad. 14: 65-72

Castro, C.E., Wade, R.S., and Belser, N.O. 1985. Biodehalogenation: reactions of cytochrome P-450 with polyhalomethanes. Biochemistry. 24: 204–210.

Chiu, P.-C., and Reinhard, M. 1995. Metallocoenzyme-mediated reductive transformation of carbon tetrachloride in titanium(III) citrate aqueous solution. Environ. Sci. Technol. 29: 595–603.

Cortese, M.S., Paszczynski, A., Lewis, T.A., Sebat, J. L., Borek, V., and Crawford, R.L. 2002. Metal chelating properties of pyridine-2,6-bis(thiocarboxylic acid) produced by *Pseudomonas* spp. and the biological activities of the formed complexes. Biometals. 15: 103-120.

Cox, C.D., Rinehart Jr., K.L., Moore M.L., and Cook Jr., J.C. 1981. Pyochelin: novel structure of an iron-chelating growth promoter for *Pseudomonas aeruginosa*. Proc. Natl. Acad. Sci. USA. 78: 4256-4260.

Criddle, C.S., DeWitt, J.T., Grbic-Galic, D., and McCarty, P.L. 1990. Transformation of carbon tetrachloride by *Pseudomonas* sp. strain KC under denitrification conditions. Appl. Environ. Microbiol. 56: 3240–3246.

Dybas, M.J., Tatara, G.M., and Criddle, C.S. 1995. Localization and characterization of the carbon tetrachloride transformation activity of *Pseudomonas* sp. strain KC. Appl. Environ. Microbiol. 61: 758-762.

Dybas, M.J., Hyndman, D.W., Heine, R., Tiedje, J., Linning, K., Wiggert, D., Voice, T., Zhao, X., Dybas, L., and Criddle, C.S. 2002. Development, operation, and long-term performance of a full-scale biocurtain utilizing bioaugmentation. Environ. Sci. Technol. 36: 3635-3644.

Egli, C., Tschan, T., Scholtz, R., Cook, A.M., and Leisinger, T. 1988. Transformation of tetrachloromethane to dichloromethane and carbon dioxide by *Acetobacterium woodii*. Appl. Environ. Microbiol. 54: 2819-2824.

Egli, C., Scholtz, R., Cook, A.M., and Leisinger, T. 1987. Anaerobic dechlorination of tetrachloromethane and 1,2-dichloroethane to degradable products by pure cultures of *Desulfobacterium* sp. and *Methanobacterium* sp. FEMS Microbiol. Lett. 43: 257-261.

Espinet, P., Lorenzo, C., and Miguel, J.A. 1994. Palladium complexes with the tridentate dianionic ligand pyridine-2,6-bis(thiocarboxylate), ptdc. Crystal structure of (n-Bu_4N)[Pd(pdtc)Br]. Inorg. Chem. 33: 2052-2055.

Fuller, M.E., Streger, S.H., Rothmel, R.K., Mailloux, B.J., Hall, J.A., Onstott, T.C., Fredrickson, J.K., Balkwill, D.L., and DeFlaun, M.F. 2000. Development of a vital fluorescent staining method for monitoring bacterial transport in subsurface environments. Appl. Environ. Microbiol. 66: 4486-4496.

Galli, R., and McCarty, P.L. 1989. Biotransformation of 1,1,1-trichloroethane, trichloromethane, and tetrachloromethane by a *Clostridium* sp. Appl. Environ. Microbiol. 55: 837–844.

Gantzer, C.J., and Wackett, L.P. 1991. Reductive dechlorination catalyzed by bacterial transition-metal coenzymes. Environ. Sci. Technol. 25: 715-722.

Glod, G., Angst, W., Holliger, C., and Schwarzenbach, R.P. 1997. Corrinoid-mediated reduction of tetrachloroethene, trichloroethene, and trichlorofluoroethene in homogeneous aqueous solution: reaction kinetics and reaction mechanisms. Environ. Sci. Technol. 31: 253–260.

Hildebrand, U., Ockels, W., Lex, J., and Budzikiewicz, H. 1983. Zur Struktur eines 1:1-Adduktes von Pyridin-2,6-dicabothiosaure und Pyridin. Phosphorus Sulfur. 16: 361-364.

Jalal, M.A.F., Hossain, M.B., van der Helm, D., Sanders-Loehr, J., Actis, L.A., and Crosa, J.H. 1989. Structure of anguibactin, a unique plasmid related bacterial siderophore from the fish pathogen *Vibrio anquillarum*. J. Am. Chem. Soc. 111: 292-296.

Krone, U.E., Laufer, K., and Thauer, R.K. 1989b. Coenzyme F430 as a possible catalyst for the reductive dehalogenation of chlorinated C1 hydrocarbons in methanogenic bacteria. Biochemistry. 28: 10061–10065.

Krone, U.E., Thauer, R.K., and Hogenkamp, H.P.C. 1989a. Coenzyme F430 as a possible catalyst for the reductive dehalogenation of chlorinated C1 hydrocarbons in methanogenic bacteria. Biochemistry. 28: 4908–4914.

Krüger H-J., and Holm, R.H. 1990. Stabilization of trivalent nickel in tetragonal NiS_4N_2 and NiN_6 environments: Synthesis, structures, redox potentials, and observations related to [NiFe]-hydrogenases. J. Am. Chem. Soc. 112: 2955-2963.

Lee, C.-H., Lewis, T.A., Paszczynski, A.J., and Crawford R.L. 1999. Identification of an extracellular catalyst of carbon tetrachloride dehalogenation from *Pseudomonas stutzeri* strain KC as pyridine-2,6-bis(thiocarboxylate). Biochem. Biophys. Res. Commun. 261: 562-566.

Lewis, T.A., and Crawford, R.L. 1993. Physiological factors affecting carbon tetrachloride dehalogenation by the denitrifying bacterium *Pseudomonas* sp. Strain KC. Appl. Environ. Microbiol. 59: 1635-1641.

Lewis, T.A., and Crawford, R.L. 1995. Transformation of carbon tetrachloride via sulfur and oxygen substitution by *Pseudomonas* sp. Strain KC. J. Bacteriol. 177: 2204-2208.

Lewis, T.A., and Crawford, R.L. 1999. Chemical studies of carbon tetrachloride transformation by *Pseudomonas stutzeri* strain KC. In: Novel Approaches for Bioremediation of Organic Pollution: Proceedings of the 42nd OHOLO Conference. R. Fass,, Y. Flashner, and S. Reuveny, eds. Plenum Publishing, New York, New York. p. 1-11.

Lewis, T.A., Cortese, M.S., Sebat, J.L., Green, T.L., and Crawford, R.L. 2000. A *Pseudomonas stutzeri* gene cluster encoding the biosynthesis of the CCl_4-dechlorination agent pyridine-2,6-bis(thiocarboxylic acid). Environ. Microbiol. 2: 407-416.

Lewis, T.A., Morra, M.J., and Brown, P.D. 1996. Comparative product analysis of carbon tetrachloride dehalogenation catalyzed by cobalt corrins in the presence of thiol or titanium(III) reducing agents. Environ. Sci. Technol. 30: 292–300.

Lewis, T.A., Paszczynski, A., Gordon-Wylie, S.W., Jeedigunta, S., Lee, C.H., and Crawford, R.L. 2001. Carbon tetrachloride dechlorination by the bacterial transition metal chelator pyridine-2,6-bis(thiocarboxylic acid). Environ. Sci. Technol. 35: 552-559.

Naegeli, H.-U., and Zähner, H. 1980. Stoffwechselprodukte von Microorganismen. Ferrithiocin. Helv. Chem. Acta. 63: 1400.

Neu, M.P., Johnson, M.T, Matonic, J.H., and Scott, B.L. 2001. Actinide interactions with microbial chelators: the dioxobis[pyridine-2,6-bis(monothiocarboxylato)]uranium (VI) ion. Acta Crystallogr. C. 57: 240-242.

Ockels, W., Römer, A., and Budzikiewicz, H. 1978. An Fe(III) complex of pyridine-2,6-di-(monothiocarboxylic acid)—A novel bacterial metabolic product. Tetrahed. Lett. 1978: 3341–3342.

Pohl, L.R., and George, J.W., 1983. Identification of dichloromethyl carbene as a metabolite of carbon tetrachloride. Biochem. Biophys. Res. Commun. 117: 367–371.

Recknagel, R.O., and Glende, E. A. 1973. Carbon tetrachloride hepatotoxicity: an example of lethal cleavage. Crit. Rev. Toxicol. 2: 263-297.

Sebat, J.L., Paszczynski, A.J., Cortese, M.S., and Crawford, R.L. 2001. Antimicrobial properties of pyridine-2,6-dithiocarboxylic acid, a metal chelator produced by *Pseudomonas* spp. Appl. Environ. Microbiol. 67: 3934-3942.

Sepulveda-Torres, L.D.C., Rajendran, N., Dybas, M.J., and Criddle, C.S. 1999. Generation and initial characterization of *Pseudomonas stutzeri* KC mutants with impaired ability to degrade carbon tetrachloride. Arch. Microbiol. 171: 424–429.

Sepulveda-Torres, L.C., Zhou, J., Guasp, C., Lalucat, J., Knaebel, D., Plank, J.L., and Criddle, C.S. 2001. *Pseudomonas* sp. strain KC represents a new genomovar within *Pseudomonas stutzeri*. Int. J. Syst. Evol. Microbiol. 51: 2013-2019.

Sepulveda-Torres, L., Huang, A., Kim, H., and Criddle, C.S. 2002. Analysis of regulatory elements and genes required for carbon tetrachloride degradation in *Pseudomonas stutzeri* strain KC. J. Mol. Microbiol. Biotechnol. 4: 151-161.

Stolworthy, J.C., Paszczynski, A.J., Korus, R.A., and Crawford, R.L. 2001. Metal binding by pyridine-2,6-bis(monothiocarboxylic acid), a biochelator produced by *Pseudomonas stutzeri* and *Pseudomonas putida*. Biodegrad. 12: 411-418.

Tatara, G.M., Dybas, M.J., and Criddle, C.S. 1993. Effects of medium and trace metals on kinetics of carbon tetrachloride transformation by *Pseudomonas* sp. strain KC. Appl. Environ. Microbiol. 59: 2126–2131.

Tatara, G.M., Dybas, M.J., and Criddle, C.S. 1995. Biofactor-mediated transformation of carbon tetrachloride by diverse cell types. In: Bioremediation of Chlorinated Solvents. R.E. Hinchee, A. Leeson, and L. Semprini, eds. Battelle Press, Columbus, Ohio. p 69–76.

Wood, J.M., Kennedy, J.S., and Wolfe, R.S. 1968. The reaction of multihalogenated hydrocarbons with free and bound reduced vitamin B 12. Biochemistry. 7: 1707–1713.

Chapter 18

Solventogenesis by Clostridia

Peter Dürre*

Abstract
Clostridium acetobutylicum and *C. beijerinckii* are able to produce industrially important solvents such as acetone, butanol, and 2-propanol. The respective enzymes are induced shortly before the transition from exponential to stationary growth phase. The bacteria thus counteract the deleterious effects of butyric and acetic acids that had been synthesized during active growth. Regulation of solventogenesis is closely coupled to that of sporulation, a developmental program that guarantees long-time survival. Five operons are meanwhile known that are essential for acetone and butanol synthesis. Their regulation is complex, involving probably several transcription factors (among them Spo0A, the master regulator of sporulation), RNA processing, and co- or posttranslational modification of the gene product. DNA supercoiling plays an important role in signal transduction. Solventogenesis is also coupled to the stress response and a number of other metabolic reactions, as revealed by RNA analyses, two-dimensional gel electrophoresis, and DNA microarrays. Characterization of the genes and the still growing understanding of their regulation has allowed the metabolic engineering of recombinant strains with improved solvent formation ability and with clostridial genes for the production of commercially important polyesters.

1. Introduction

Acetone, 2-propanol (formerly isopropanol), and 1-butanol (formerly n-butanol) are solvents of considerable importance in the chemical industry. Acetone is used for the production of methacrylates and methyl isobutyl ketone (about 39 % of the worldwide production) as well as smokeless powder (cordite). It also represents a solvent for adhesives, cellulose acetate, fats, lacquers, oils, paints, plastics, printing inks, resins, rubber cements, varnishes, and waxes. It is a component of nail polish remover as well as paint and varnish removers. It is used in wet and dry spinning, for extraction of compounds from animal and plant substances, and the preparation of vitamin intermediates. Commercial synthesis is mainly performed by catalytic dehydrogenation of 2-propanol. 2-Propanol is a component in antifreeze mixtures, in quick-drying oils and ink, in hand and after-shave lotions, and in some healthcare personnel antiseptic handwash mixtures. It is used as a solvent in perfumes and oils, in extraction processes, as an intermediate or processing solvent in the synthesis of other chemicals (about 40 % of the worldwide production), as a coating solvent, and as a deicing agent for liquid fuels (gasoline additive). It is an ingredient in liquid soaps and window cleaners. 2-Propanol is commercially made by dissolving propylene gas in sulfuric acid and then hydrolyzing the sulfate ester that is formed. It is one of the cheapest alcohols and has replaced ethanol for many uses. 1-Butanol is converted to a large part into derivatives for use in the coating industry. 42 % of the worldwide production (2,533 millions of pounds, data of July 1999 from http://www.chemexpo.com/news/PROFILE990705.cfm) are needed for butyl acrylate and methacrylate, 25 % for butyl glycol ethers, and 15 % for butyl acetate. 1-Butanol is an eminently suitable solvent for acid-curing lacquers and baking finishes. About 9 % of the worldwide production are directly used as solvent. Some butyl esters are established plasticizers for dispersions, plastics, and rubber mixes (such as butyl acrylate, benzylbutyl phthalate, and dibutyl phthalate). Continued use of esters for latex architectural paints in the housing market is driving this demand. 1-Butanol is also used as solvent for dyes, e. g. in printing inks, as an extractant in the production of drugs and natural substances such as alkaloids, antibiotics, hormones, and vitamins, as an additive in polishes and cleaners, e. g. floor cleaners and stain removers, as a solubilizer in the textile industry, e. g. additive in spinning baths or carrier for coloring plastics, as an additive in deicing fluids, as an antiicing additive of gasoline, as a mobile phase in paper and thinlayer chromatography, as a humectant for cellulose nitrate, in the manufacture of hydraulic fluids, as a diluent for formulating brake fluids suitable for use in passenger cars, and as a feedstock for the production of the ore flotation agent butyl xanthate. Most commercial production processes are now based on an oxo reaction

*For correspondenc email: peter.duerre@biologie.uni-ulm.de

with propylene, yielding butyraldehyde, which then is converted into 1-butanol by hydrogenation. The largest producers worldwide are Union Carbide, European Oxo GmbH, and BASF.

Historically, about two thirds of world's butanol production stemmed from fermentation with several clostridia, now recognized as *Clostridium acetobutylicum*, *C. beijerinckii*, and *C. saccharobutylicum* (Keis *et al.*, 1995; Johnson *et al.*, 1997; Jones, 2001; Keis *et al.*, 2001). The first report of biological butanol formation was published by the French microbiologist Louis Pasteur in 1862. He found that a culture of "vibrion butyrique" (no pure culture, but probably dominated by *C. butyricum*, which is also able to form butanol (Zoutberg *et al.*, 1989; Sauer *et al.*, 1993), produced butanol in addition to butyrate under anaerobic conditions. Pasteur's pioneering studies were followed by work of Fitz, Beijerinck, Winogradsky, and other famous microbiologists (for a detailed historical treatise see Dürre and Bahl, 1996). Fitz was probably the first who obtained a pure culture of a butanol-producing *Clostridium* (named *Bacillus butylicus* at that time) (Fitz, 1876; 1877; 1878; 1882). Biological acetone production was first discovered with *Bacillus macerans* (Schardinger, 1904; 1905) and that of 2-propanol again with an anaerobic bacterium, *Clostridium americanum* (Pringsheim, 1906a; 1906b).

Synthesis of acetone by clostridial fermentation in the United Kingdom was a requirement during World War I, when this compound was needed for production of ammunition (cordite) and the starting material for chemical synthesis, calcium acetate, was no longer available in large quantities (Gibbs, 1983; Jones and Woods, 1986). The armistice in 1918 resulted in a massive drop of acetone demand and a closure of the fermentation plants. However, soon after the end of the war, prohibition was introduced in the United States, which also meant that the industrial solvent amyl alcohol was no longer available. Butanol (in the form of its acetate ester) proved to be an excellent replacement. Consequently, the fermentation plants were reopened and even new ones were erected. Starting in about 1950, the biotechnological process declined due to increasing substrate prices (molasses) and cheap crude oil avalability (for reviews see Jones and Woods, 1986; Dürre and Bahl, 1996; Santangelo and Dürre, 1996; Dürre, 1998).

The oil crisis in the 1970s and political turmoils in the Middle East, followed by the First Gulf War, prompted new research activities with solventogenic clostridia, with a major focus on molecular biology. Gene cloning from these bacteria started 1986, in David Woods' laboratory at the University of Cape Town, South Africa (reviewed in Young *et al.*, 1989). The first genes encoding enzymes involved in solvent formation, namely acetoacetyl-coenzyme A: acetate/butyrate: coenzyme A-transferase and acetoacetate decarboxylase, were cloned and sequenced in 1990 (Cary *et al.*, 1990; Gerischer and Dürre, 1990; Petersen and Bennett, 1990). Since then, genome sequencing of *C. acetobutylicum*, RNA analyses, reporter gene constructions, gene knock-outs, and DNA microarrays provided a wealth of information on the regulation of solvent formation, which will be summarized in this chapter.

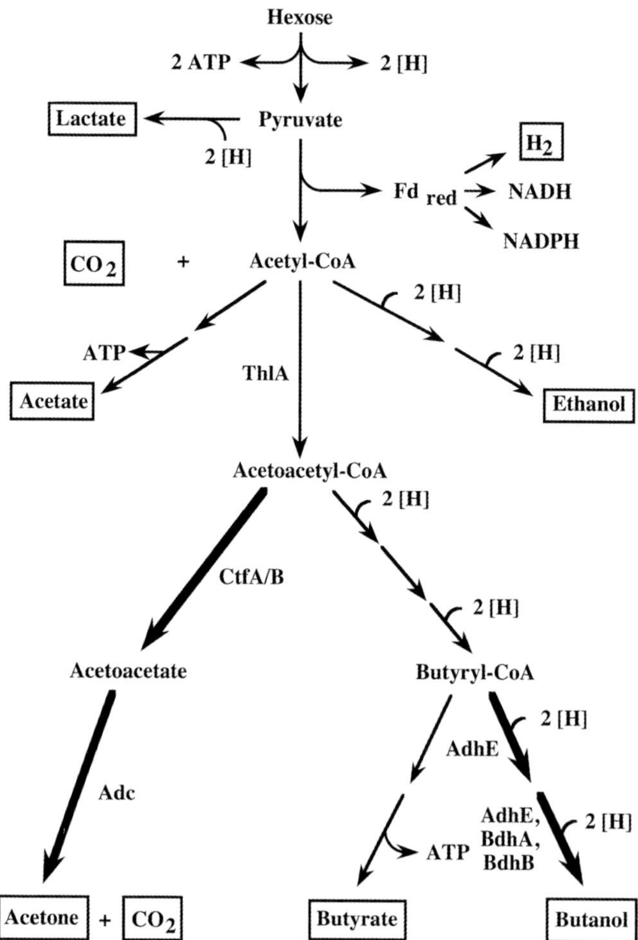

Figure 1. Catabolic pathways of acid and solvent formation in *Clostridium acetobutylicum*. Gene products mentioned in the text are listed aside the arrows indicating the catalyzed reaction(s). Bold arrows indicate acetone and butanol production pathways. Adc, acetoacetate decarboxylase; AdhE, bifunctional butyraldehyde/butanol dehydrogenase; BdhA, butanol dehydrogenase; BdhB, butanol dehydrogenase; CtfA/B, acetoacetyl-CoA: acetate/butyrate-coenzyme A transferase; ThlA, thiolase.

2. Clostridial Solvent Formation

A number of clostridia produce the aforementioned solvents. *C. acetobutylicum* [and also *C. saccharobutylicum* and *Clostridium saccharoperbu tylacetonicum* (Shaheen *et al.*, 2000; Keis *et al.*, 2001] forms acetone, butanol, and small amounts of ethanol, while *C. beijerinckii* strains usually produce butanol and 2-propanol (a notable exception is NRRL B592, producing acetone, butanol, and ethanol). These are species in which solvents represent major fermentation

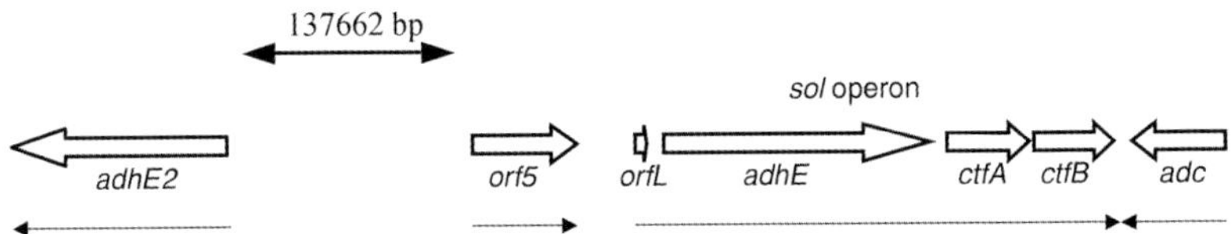

Figure 2. Schematic map of *C. acetobutylicum* DNA regions, encoding essential enzymes for solventogenesis. Arrows and arrowheads represent lengths, locations, and orientation of genes; lines with arrows indicate primary mRNA transcripts. Distance between *orf5* and *adhE2* genes is indicated in bp.

products. Other solventogenic clostridia comprise *C. aurantibutyricum, C. butyricum, C. cadaveris, C. pasteurianum, C. puniceum, C. sporogenes, C. tetanomorphum*, and *C. thermosaccharolyticum* (for reviews see Sauer *et al.*, 1993, Dürre and Bahl, 1996).

Typically, acetone, butanol, and 2-propanol are not formed during the exponential growth stage, but at the transition to stationary phase. The organisms first perform a typical butyrate fermentation (Figure 1) and induce enzymes required for solventogenesis concomitantly with those required for endospore formation. These two overlapping regulatory networks are very complex. Crucial parameters include pH, limitation of some salts, surplus of substrate, and redox control. The coupling of the two processes provides solventogenic clostridia with a distinct advantage: Due to the massive production of butyric and acetic acid, the pH of the environment decreases to between 4.5 and 4. Like other anaerobic bacteria (for a review see Dürre *et al.*, 1988), *C. acetobutylicum* is unable to maintain cytoplasmic pH homeostasis, but keeps a relatively constant transmembrane pH gradient (Gottwald and Gottschalk, 1985; Huang *et al.*, 1985). That means that a decreasing external pH will lead to a decreasing internal pH, which is about 1 pH unit more alkaline. At low external pH, the organic acids produced will become protonated and can thus enter the cell again by diffusion. Inside, they will dissociate, releasing protons. As a consequence, the proton gradient over the cytoplasmic membrane will collapse and the cell will die. The only means for survival is endospore formation. However, if initiated too late, the cell will not have enough time to complete the spores. Solventogenesis represents a kind of emergency reaction that allows the organism to buy time. The acids are taken up and converted into solvents, thus raising the external (and also the internal) pH. In the long run, the solvents also will exert toxic effects on the membrane, with butanol being the most dangerous compound (Moreira, 1983; Bowles and Ellefson, 1985), but by that time spore synthesis will have been completed.

Formation of acetone and butanol in *C. acetobutylicum* is catalyzed by the activities of acetoacetyl CoA: acetate/butyrate-coenzyme A transferase (short: CoA transferase), acetoacetate decarboxylase, butyraldehyde/butanol dehydrogenase E, and butanol dehydrogenase B (Figure 1). Another butanol dehydrogenase (A or Bdh I, Welch *et al.*, 1989) rather seems to serve as an electron sink in low butanol and possibly ethanol formation (Petersen *et al.*, 1991; Sauer and Dürre, 1995; Dürre *et al.*, 1995). In addition, an $NADP^+$-dependent alcohol dehydrogenase has been purified from *C. acetobutylicum* and characterized (summarized in Sauer *et al.*, 1993). It could be separated from the NAD^+-dependent activity by ultracentrifugation (Dürre *et al.*, 1987). Transposon-generated mutants indicated a role of this enzyme in ethanol formation (Bertram *et al.*, 1990). From *C. saccharobutylicum* [previously *C. acetobutylicum* NRRL B643 (Jones, 2001)] a butyraldehyde dehydrogenase has been isolated and characterized (Palosaari and Rogers, 1988). For 2-propanol formation in *C. beijerinckii*, another primary/secondary alcohol dehydrogenase is required for reduction of acetone (Chen, 1995). Several of these enzymes have been purified and their properties have been summarized in a number of reviews (Chen, 1993; Dürre and Bahl, 1996; Girbal and Soucaille, 1998, Mitchell, 1998).

3. Essential Genes Required for Solventogenesis and Their Transcriptional Regulation

The genes encoding enzymes required for solventogenesis in *C. acetobutylicum* are organized in a number of operons. The genome sequencing project revealed that the organism contained a chromosome of 3.94 Mbp and a 192-kbp megaplasmid, designated pSOL1 (Cornillot and Soucaille, 1996; Cornillot *et al.*, 1997; Nölling *et al.*, 2001). So far, five operons have been identified to encode enzymes catalyzing acetone and butanol formation (Figure 2).

3.1. The *bdhA* and *bdhB* Operons

On the chromosome, the two butanol dehydrogenase genes *bdhA* and *bdhB* are located in two contiguous, monocistronic transcription units, each controlled by a σ^A-consensus type promoter (deduced from primer extension experiments and carrying 4 and 3 mismatches, respectively) and a rho-independent terminator (Petersen *et al.*, 1991; Walter *et al.*, 1992). Sequence analysis indicated the presence of two or one, respectively, Spo0A-binding motifs upstream of *bdhA* and *bdhB* (Ravagnani *et al.*, 2000). Spo0A is the cardinal sporulation-specific transcription factor and obviously the essential regulatory link between the solventogenesis and sporulation networks (Ravagnani *et al.*, 2000; Harris *et al.*, 2002). RNA analyses showed that peak induction of *bdhB* occurred before massive butanol production starts, indicating an essential role of its gene product in solventogenesis (Sauer and Dürre, 1995). Maximal expression of *bdhB* was observed about 2 hours after *sol* operon expression peaked (see below). *bdhA* expression, on the other hand, could be observed from the beginning of growth, corresponding to its assumed role as an electron sink (Sauer and Dürre, 1995). Recent studies with two different reporter gene sytems corroborated these findings. The pattern of induction as determined by measuring activities of β-galactosidase and luciferase matched exactly the results from the Northern blots. The promoter of *bdhB* was found to be 17-fold stronger than that of *bdhA* (Feustel *et al.*, 2004). However, these data also raise a question with respect to the Spo0A-binding motifs. While *bdhA* is constitutively expressed, Spo0A (in its phosphorylated form) is active only later or late in the exponential growth phase. These two features are clearly contradictory. It might be that additional regulatory factors are involved or that consensus sequences for Spo0A-binding motifs do not represent real binding sites in every case. Evidence for the latter assumption has been obtained in recent gel retardation studies of the *sigH* gene of *C. acetobutylicum* with phosphorylated Spo0A from the same organism (Hollergschwandner, 2003). No butyraldehyde dehydrogenase genes have so far been identified in *C. acetobutylicum*. A short report on cloning such a gene from strain NRRL B643 [now recognized as *C. saccharobutylicum* (Keis *et al.*, 2001)] (Contag and Rogers, 1988) later turned out as cloning of a lactate dehydrogenase gene (Contag *et al.*, 1990). However, in *C. beijerinckii* NRRL B593 a gene encoding a coenzyme A-acylating aldehyde dehydrogenase (*ald*) has been cloned and characterized. This gene was found to be present in all strains tested of *C. beijerinckii, C. saccharobutylicum*, and *C. saccharoperbutylacetonicum*, but not in *C. acetobutylicum* (Toth *et al.*, 1999).

3.2. The *sol* Operon

The other known genes directly required for solventogenesis reside on the megaplasmid pSOL1. The so-called *sol* operon comprises *orfL, adhE, ctfA*, and *ctfB* (Fischer *et al.*, 1993). These genes encode a small peptide of so far unkown function, a bifunctional aldehyde/alcohol dehydrogenase, and the two subunits of the CoA transferase. *adhE* shows significant homology to analogous genes from *Escherichia coli* (Goodlove *et al.*, 1989), *Salmonella typhimurium* (Dailly *et al.*, 2000), *Lactococcus lactis* (Arnau *et al.*, 1998), as well as from the eukaryotic *Entamoeba histolytica* (Yang *et al.*, 1994) and *Giardia lamblia* (Sánchez, 1998).

An interesting case involves the genes of succinate-semialdehyde dehydrogenase and 4-hydroxybutyrate dehydrogenase from *C. kluyveri*. *sucD* and *4hbD* are homologous to the 5′ and 3′ parts of the *adhE* gene, respectively, are tandemly arranged, and are separated by just 36 nucleotides (Söhling and Gottschalk, 1996). This could indicate that the multifunctional aldehyde/alcohol dehydrogenases resulted from a gene fusion. Support for this assumption might come from the finding that the *adhE*-containing transcript of *C. acetobutylicum* yields two different translation products, the mature enzyme and separately the C-terminal alcohol dehydrogenase domain. The latter is probably due to the presence of a ribosome binding site sequence within the structural gene in the correct distance to an ATG codon, representing the translational start of the alcohol dehydrogenase domain (Thormann *et al.*, 2002). Another indication that this bifunctional enzyme is probably an evolutionary product of a gene fusion is the high homology of the N terminus to the family of aldehyde: NAD^+ oxidoreductases, whereas the C terminus shows homology to group III alcohol dehydrogenases, which mostly are iron-activated (Reid and Fewson, 1994; Membrillo-Hernández *et al.*, 2000). The *E. coli* protein, in addition to its acetyl-CoA and acetaldehyde reduction activity, also catalyzes the deactivation of pyruvate: formate-lyase (Kessler *et al.*, 1991). The *C. acetobutylicum* AdhE represents a butyraldehyde/butanol dehydrogenase, as shown by induction of its gene just prior to butanol synthesis (Sauer and Dürre, 1995) [ethanol is constitutively produced by the $NADP^+$-dependent ethanol dehydrogenase (Bertram *et al.*, 1990)], successful restoration of butanol production in a solvent-negative mutant upon transformation of the *adhE* [also called *aad* (Nair *et al.*, 1994)] gene (Nair and Papoutsakis, 1994), and the drastic decrease of the amount of butanol formed when the gene was insertionally inactivated (Green and Bennett, 1996).

The composition of the *sol* operon of *C. acetobutylicum* (consisting of *orfL, adhE, ctfA*, and *ctfB*) was found to be unique within the solventogenic clostridia. *C. beijerinckii, C. saccharobutylicum*, and *C. saccharoperbutylacetonicum* do not possess an *adhE* gene, but carry the *ald* gene in the respective location.

In addition, the *adc* gene is enclosed in the *sol* operon as well (Toth *et al.*, 1999; D.T. Jones, presentation at the 9th International Symposium on the Genetics of Industrial Microorganisms (GIM), July 1-5, 2002, Gyeongju, Korea).

Transcription of the *sol* operon is induced several hours before massive solvent production can be observed (Sauer and Dürre, 1995; Feustel *et al.*, 2004). Primer extension experiments revealed two transcription startpoints, a distal P_1 (or S_2) and a proximal P_2 (or S_1) (Fischer *et al.*, 1993; Nair *et al.*, 1994). While the deduced promoter structure of P_1 represented an almost perfect σ^A-type consensus sequence (just one mismatch in the –35 region), in P_2 only the first 3 bases of both hexamer motifs were identical to the consensus, with an additional unusual spacing of the two boxes (Fischer *et al.*, 1993; Nair *et al.*, 1994; Dürre *et al.*, 1995). However, signal intensities indicated that most transcripts originated from P_2 (Fischer *et al.*, 1993). This somewhat contradictory finding could be resolved by recent reporter gene studies (Thormann *et al.*, 2002). Only the distal P_1 motif acts as a promoter for the *sol* operon. P_2 rather represents an mRNA processing site. Interestingly, the *adhE* transcript of *E. coli* is processed by RNase G, which probably recognizes the secondary structure rather than the primary sequence (Wachi *et al.*, 2001; Kaga *et al.*, 2002). This fits perfectly with the results obtained from analysis of the clostridial transcript, where alterations in secondary structure led to loss of processing (Thormann *et al.*, 2002).

Reporter gene studies revealed that *sol* expression was much lower than that of *bdhB*, as also found for *bdhA* expression (Feustel *et al.*, 2004). It has been proposed that *sol* is regulated by a transcriptional repressor, encoded by the adjacent upstream gene *orf5* (Nair *et al.*, 1999). However, a thorough investigation of the *orf5* gene product revealed that it is involved in glycosylation/deglycosylation reactions and is located at the outside of the cytoplasmic membrane. Furthermore, it contains a motif required for protein-protein interactions (tetratrico peptide repeat) rather than a protein-DNA interaction motif such as HTH (Thormann and Dürre, 2001). Finally, overexpression of *orf5* did not result in the solvent-negative phenotype reported by Nair *et al.* (1999). This contradiction could be resolved by a detailed analysis of the regulatory regions of the *sol* promoter, P_1. Upstream of P_1, a Spo0A-binding site and two out of three imperfect sequence repeats (R1 and R3) were found to be essential for induction of *sol*. This part had erroneously been subcloned together with the *orf5* gene by Nair *et al.* (1999) and was in fact responsible for the effects reported (Thormann *et al.*, 2002). The binding of Spo0A~P to a DNA fragment carrying P_1 and its regulatory region could recently be demonstrated in gel retardation assays (Hollergschwandner, 2003). This clearly represents the molecular basis for the link between butanol formation and sporulation. However, mutation of the 0A box (Thormann *et al.*, 2002) did not completely abolish regulation. This result, in combination with the relatively long distance between R1 and the 0A box, suggests the involvement of an additional transcription factor, probably acting in concert with Spo0A~P (Thormann *et al.*, 2002). In addition, inactivation of the *spo0A* gene did reduce, but not completely prevent expression of the *sol* operon (and also the *adc* operon) (Harris *et al.*, 2002).

The signal for induction of *sol* operon expression has not yet been unequivocally identified. Physiological prerequisites are a surplus of substrate, pH below 4.3, limiting phosphate or sulfate concentrations, and high concentrations of acetate and butyrate (for reviews see Dürre and Bahl, 1996; Dürre, 1998; Dürre *et al.*, 2002). The effects are obviously additive and interchangeable, as e. g. addition of larger amounts of acetate and butyrate could compensate for raising the pH to about 7 (Holt *et al.*, 1984). In addition, temperature was found to be another parameter that influences induction of butanol production (Bahl, 1983). All these conditions are known to influence DNA topology directly. Thus, it was tempting to speculate that changes in DNA supercoiling affect gene expression, providing the direct connection between alterations in the environment and the cellular response at the molecular level. There are two reports providing data in accordance with this assumption. Transcription of the *sol* operon was found to be immediately increased to about 300 % by inhibition of DNA gyrase by novobiocin (resulting in relaxation or less negative supercoiling) (Ullmann *et al.*, 1996) and DNA isolated from *C. acetobutylicum* at various stages of growth was indeed shown to become relaxed during the switch from acidogenesis to solventogenesis (Wong and Bennett, 1996). Thus, it seems that changes in DNA topology serve as a transcriptional sensor and are at least participating in regulating the onset of solventogenesis (Ullmann and Dürre, 1998).

3.3. The *adc* Operon

Adjacent to the *sol* operon, but convergently orientated is the monocistronic *adc* operon, encoding acetoacetate decarboxylase of *C. acetobutylicum* (Gerischer and Dürre, 1990, Petersen and Bennett, 1990). This transcription unit shares the same rho-independent transcriptional terminator with *sol* (which is functional in both directions) and is controlled by a single promoter (Gerischer and Dürre, 1992). The *adc* gene is induced early in the exponential growth phase, but reaches maximal expression in stationary phase. Its promoter allows 1.3-fold higher expression than that of *bdhB* and 84-fold higher expression than that of *sol* (Feustel *et al.*, 2004). Three 0A boxes are located upstream of the *adc* promoter (one in reverse orientation).

In gel retardation assays, band shifts of the *adc* promoter fragment could be observed using phosphorylated Spo0A of *C. beijerinckii* (Ravagnani *et al.*, 2000) and *C. acetobutylicum* (Hollergschwandner, 2003). Again, this obviously represents the regulatory link between the metabolic networks of solventogenesis and sporulation. However, as in the case of the *sol* operon, Spo0A~P does not seem to be the only regulator. Mutation of the 0A boxes (Ravagnani *et al.*, 2000) and even their complete removal (Böhringer, 2002) reduced the level of induction, but did not completely abolish it. No sequences with homology to the imperfect repeats upstream of *sol* are present in the vicinity of the *adc* promoter. Thus, it is very likely that different additional transcription factors are involved in the regulation of the *adc* and *sol* operons. This assumption is supported by the finding that the ratio of acetone and butanol formation can vary dramatically (Bahl *et al.*, 1986), indicating different regulatory systems. The cell's need to adjust respective syntheses of the fermentation products to internal redox conditions (i.e. availability of reducing equivalents) also provides the explanation for why the *adc* gene is organized in a separate transcription unit. Its gene product, acetoacetate decarboxylase, is only needed to remove the end product of the CoA transferase reaction (acetoacetate), since this conversion is thermodynamically less favorable, thus forcing butyryl-CoA formation. This is only necessary when reducing equivalents are limiting and butyryl-CoA cannot be completely converted into butanol (Dürre *et al.*, 1995; Dürre *et al.*, 2002). Relaxation (i.e. less negative supercoiling of DNA) also had a stimulatory effect on *adc* expression. However, it was not as pronounced as in the case of the *sol* operon (only about 30 % increase, Ullmann *et al.*, 1996), again indicating different regulatory systems for the two operons.

Acetoacetate decarboxylase was found to occur in different forms in solventogenic *C. acetobutylicum* cells by two-dimensional protein gel electrophoresis (Schaffer *et al.*, 2001). N-terminal amino acid sequencing revealed two different spots for this enzyme, with only one copy of the gene being present in the genome according to the sequencing data (Nölling *et al.*, 2001). This finding indicates a post-translational modification of Adc. Both, the nature and physiological relevance of this modification are still unknown. In contrast to other solventogenic enzymes, the activity of Adc was found to be very stable throughout the solventogenic phase (Gerischer and Dürre, 1992). Thus, it is tempting to speculate that the modification is involved in stabilization of the protein.

3.4. The *adhE2* Operon

The fifth operon known to be directly required for butanol formation (however, only under specific conditions) encodes another aldehyde/alcohol dehydrogenase gene (*adhE2*) and is also located on the megaplasmid (Fontaine *et al.*, 2002). This gene is only induced when the overall degree of substrate reduction is increased, e.g. by substitution of part of the glucose by glycerol. Under such conditions, *C. acetobutylicum* no longer forms acetone, but only butanol and increased amounts of ethanol. This type of fermentation has been designated "alcohologenic" to differentiate it from "solventogenic", which includes acetone production (though in varying amounts) (Girbal *et al.*, 1995; Girbal and Soucaille, 1998). Instead of using a substrate mixture, such a fermentation can also be induced artificially by adding methyl viologen to actively growing cultures (Rao and Mutharasan, 1986; 1987). RNA analysis of cells treated with methyl viologen, however, revealed that neither the *sol* nor the *bdhB* operon was expressed. Weak hybridization could only be detected with an *adhE* probe, raising the assumption that another butyraldehyde/butanol dehydrogenase gene was present in *C. acetobutylicum* (Dürre *et al.*, 1995). This assumption was indeed verified by the genome sequencing project (Nölling *et al.*, 2001). The *adhE2* gene is monocistronic and encodes a product that shares 66.1 % identity with AdhE from *C. acetobutylicum* (Fontaine *et al.*, 2002).

The regulatory region seems to be very similar to that of the *sol* operon. Primer extension analysis revealed two transcription start points, with most of the signal intensity stemming from the proximal one. The deduced distal promoter (S_2) showed perfect homology to the consensus motif, while the proximal one (S_1) had two mismatches in either hexamer motif and a very unusual spacing of the two (20 nucleotides) (Fontaine *et al.*, 2002). Thus, it is very tempting to speculate that the situation resembles the *sol* operon promoter, with only S_2 being a proper promoter and S_1 being a processing site. Further support for this assumption is a 0A box, located 35 nucleotides upstream of S_2 [5′-TGGCAAT-3′, consensus motif 5′-TGNCGAA-3′ (Ravagnani *et al.*, 2000)] (in case of the *sol* operon, the 0A box is reversed and located 62 nucleotides upstream of P_1) and providing the link to the sporulation regulatory network. An additional feature of the *adhE2* gene is the presence of a putative FNR (fumarate and nitrate reduction) binding site, located in a region of dyad symmetry and just 19 nucleotides upstream of the predicted ribosome binding site (Fontaine *et al.*, 2002). This motif might be responsible for the redox state-dependent regulation.

4. Genes Related to Solventogenesis and DNA Microarrays

In addition to genes, whose products directly catalyze reactions leading to acetone or butanol, several other operons were found to change expression at the onset of solventogenesis.

4.1. The *thlA* and *thlRBC* Operons

In *C. acetobutylicum*, two thiolase-encoding genes have been identified and characterized (Stim-Herndon *et al.*, 1995; Winzer *et al.*, 2000). Thiolase catalyzes the (reversible) formation of acetoacetyl-CoA from 2 acetyl-CoA (Figure 1) and is thus essential for the formation of C4 compounds in both acidogenic butyrate and solventogenic butanol production pathways. This enzyme is encoded by the monocistronic *thlA* operon (Stim-Herndon *et al.*, 1995). The physiological role of the protein would suggest a constituitive expression of its gene. So, it came as a surprise when RNA analyses revealed high expression during the acidogenic stage, followed by repression, and again by an induction at the onset of solventogenesis (together with the *sol* and *adc* operons) (Winzer *et al.*, 2000). So far, no other gene is known in *C. acetobutylicum* that shows a similar regulatory pattern. Primer extension experiments revealed a putative promoter upstream of *thlA*, which shows high homology to consensus motifs recognized by σ^A (Winzer *et al.*, 2000). Further investigations are required to understand the regulatory mechanisms controlling *thlA* expression.

The second thiolase gene, *thlB*, is part of an operon that contains the genes *thlR-thlB-thlC*. ThlR shows significant similarity to transcriptional regulators of the TetR/AcrR family, whereas ThlC represents a protein with still unknown function (Winzer *et al.*, 2000). Expression of the operon is low, with a transient, threefold increase at the onset of solventogenesis (Winzer *et al.*, 2000). The physiological function of ThlB and ThlC still remain to be elucidated.

4.2. Stress Proteins

Proteomic analysis that compared extracts from acidogenic and solventogenic cells was performed in order to further identify proteins synthesized preferentially or exclusively during solventogenesis (Schaffer *et al.*, 2002). Low pH and high concentrations of acids certainly represent environmental stress, so it was not surprising to find a number of stress proteins being induced under these conditions. Among them were DnaK, GroEL, and Hsp18, which also had been found earlier to be induced at the transcriptional and translational level when cells started to produce butanol or were exposed to solvents, air, and heat (Terraciano *et al.*, 1988; Pich *et al.*, 1990; Sauer and Dürre, 1993; 1995; Bahl *et al.*, 1995). Hsp18 was also found to be modified co- or posttranslationally (Schaffer *et al.*, 2002). Another protein with increased synthesis rate, PdxY (for pyridoxine biosynthetic protein), might be a stress protein as well, as inferred from its homology to respective proteins from eukaryotes and *Bacillus*. Increasing pyridoxine synthesis might be a means for *C. acetobutylicum* to quench reactive oxygen species (Schaffer *et al.*, 2002).

4.3. The *ser* Operon

An unexpected finding in the above mentioned proteomic analysis was the increased synthesis of enzymes catalyzing serine biosynthesis at the onset of solventogenesis (Schaffer *et al.*, 2002). The respective spots were cut out of the 2D-gels and identified by N-terminal amino acid sequencing and BLAST comparison to deduced proteins from the known *C. acetobutylicum* genome sequence as well as those of other bacteria. The proteins showed between 40 and 63 % identity to serine aminotransferase (SerC), 3-phosphoglycerate dehydrogenase (SerA), and a seryl-tRNA synthetase (SerS), enzymes needed for biosynthesis of serine and its incorporation into proteins. They were all induced about threefold at the onset of solventogenesis, which already suggested a common organization of the respective genes. Indeed, inspection of the genome sequence revealed such a clustering in the succession *serC-serA-serX-serS*. *serX* encodes a protein of yet unknown function (SerX), which did not show up during the proteome analysis, presumably due to its larger size. It is tempting to speculate that SerX might express 3-phosphoserine phosphatase activity, the only missing enzyme of the serine biosynthesis pathway in *C. acetobutylicum*. If so, it would represent an unsual phosphatase, since it does not share significant homology with respective known enzymes. RNA analyses revealed that the *ser* genes indeed form an operon, which is induced concomitantly with the *sol* operon (Schaffer *et al.*, 2002). Primer extension experiments uncovered a promoter consisting of motifs typical for σ^A-dependent transcription. No 0A boxes could be detected upstream or downstream of this regulatory region, and no information is available on its control so far. There are also no proteins known that play a role in solventogenesis or sporulation and possess an unusually high serine content. So, the physiological reason for induction of the *ser* operon at the onset of solventogenesis still remains to be resolved.

4.4. The *gap* Operon

Another protein with increased abundance during solventogenesis (about twofold) was identified as glyceraldehyde-3-phosphate dehydrogenase (Gap) (Schaffer *et al.*, 2002). However, contrary to the aforementioned proteins, its production is not induced during the metabolic shift to acetone and butanol formation (synthesis rate was determined by incorporation of labelled methionine). The respective gene is located within an operon, encoding other glycolytic enzymes. The order of genes was found to be *gap-pgk-tpi*. *pgk* encodes phosphoglycerate kinase and *tpi* triosephosphate isomerase (Schreiber and Dürre, 1999). Downstream of this operon, the gene encoding the 2,3-bisphosphoglycerate-independent phosphoglycerate mutase (*pgm*(i)) is located (Schreiber and Dürre, 1999). Primer extension experiments revealed transcription start points upstream of *gap* (probably the natural promoter),

Table 1. Metabolic engineering of *Clostridium acetobutylicum* in order to enhance solvent production

Feature of strain	Acetate	Butyrate	Acetone [mM]	Butanol	Ethanol	Reference
wild type (w.t.)	112 71	108 76	79 85	131 158	11 16	Green *et al.*, 1996 Harris *et al.*, 2001
mutant M5 (no AdhE and CoA transferase activities) plus overexpression of *adhE* gene (encoding butyraldehyde/butanol dehydrogenase)	101	99	-	84	8	Nair and Papoutsakis, 1994
inactivation of *pta* gene (encoding phosphotransacetylase	87	159	72	133	13	Green *et al.*, 1996
inactivation of *buk* gene (encoding butyrate kinase)	149 (111	37 18	39 76	146 225	16 57	Green *et al.*, 1996 Harris *et al.*, 2000)
inactivation of *buk* gene plus over-expression of *adhE* gene	113	18	66	226	98	Harris *et al.*, 2000
inactivation of *orf5* gene (encoding a deglycosylase)	84	74	97	197	29	Harris *et al.*, 2001
inactivation of *orf5* gene plus overexpression of *adhE* gene	87	97	141	238	47	Harris *et al.*, 2001
repression of *ctfB* (encoding CoA transferase subunit B) plus overexpression of *adhE*	125	52	26	130	190	Tummala *et al.*, 2003
overexpression of *groESL* genes	80	70	148	231	21	Tomas *et al.*, 2003b
overexpression of *adhE* gene			Same as w.t.			Nair *et al.*, 1994
overexpression of *buk* and *ptb* genes	60	30	80 (plasmid effect reported)	140	10	Walter *et al.*, 1994a (Walter *et al.*, 1994b)

upstream of *tpi* (possibly representing another promoter), and upstream of *pgm*(i), which probably stems from the natural promoter of that operon (Schreiber and Dürre, 2000). RNA analyses showed a most abundant transcript encoded by the *gap* gene, indicating that the transcript with a prolonged half-life is generated by processing of the full-length message. The *gap* promoter proved to be constitutively active (Schreiber and Dürre, 2000), which is in accordance with the finding that although the protein level was higher in solventogenic cells, its observed synthesis rate remained unchanged (Schaffer *et al.*, 2002). However, the *tpi* transcript was found to be slightly enhanced at the onset of solventogenesis (by app. 15 %) (Schreiber and Dürre, 2000), the physiological relevance of which is unknown.

4.5. DNA Microarrays

A recent new approach to study regulatory phenomena in *C. acetobutylicum* is the use of DNA microarray-based transcriptional analysis. Such cDNA arrays included 1,019 genes from *C. acetobutylicum* (representing app. 25 % of the whole genome), among them all 178 pSOL1 open reading frames (orfs), 123 DNA replication and repair genes [90 % of total such genes as identified by the genome annotation (Nölling *et al.*, 2001)], 97 cell division- and sporulation-related genes (92 %), 85 carbohydrate and/or primary metabolism genes (31 %), 67 energy conservation genes (52 %), 63 membrane and cell envelope genes (36 %), 48 lipid metabolism genes (80 %), and 42 motility and chemotaxis genes (39 %) (http://www.chem-eng.northwestern.edu/Faculty/papou.html). Conditions in which transcription was examined were those that induce solvent production (Tummala *et al.*, 2003), sporulation (Tomas *et al.*, 2003a), and stress response (Tomas *et al.*, 2003b). Simultaneous overexpression of *adhE* and downregulation of *ctfB* [by antisense RNA (asRNA)] resulted in massive ethanol production (190 mM). While *adhE* was highly expressed throughout, *adhE2* expression level decreased when ethanol formation started. Interestingly, overexpression of *adhE* (as compared to the asRNA construct alone) led to differential expression of 273 genes out of the 1,019 tested. This drastic change of the transcriptional profile of *C. acetobutylicum* came as a surprise and might be caused by titration of transcription factors and/or altered product concentrations that initiate a stress response (Tummala *et al.*, 2003).

Table 2. Synthetic production operons composed of genes from *Clostridium acetobutylicum*

Recombinant genes (host)	Acetate	Butyrate	Acetone	Butanol	Ethanol [mM]	PHA[a]	Reference
adc, ctfA, ctfB encoding acetoacetate decarboxlase and CoA transferase (*C. acetobutylicum*)	10	0.5	149	177	31	n.d.	Mermelstein *et al.*, 1993
thlA, adc, ctfA, ctfB encoding thiolase, acetoacetate decarboxylase and CoA transferase (*Escherichia coli*)	23	-	93	-	3	n.d.	Bermejo *et al.*, 1998
adc, ctfA, ctfB encoding acetoacetate decarboxylase and CoA transferase (*Escherichia coli*, acetoacetate added)	n.d.	-	8	-	n.d.	n.d.	Guillot, 1997
buk, ptb encoding butyrate kinase and phosphotransbutyrylase from *C. acetobutylicum* plus *phaE, phaC* encoding PHA synthase from *Thio-capsa pfennigii* (*Escherichia coli*)						68.7	Liu and Steinbüchel, 2000a

[a] PHA, poly(hydroxyalkanoic acid); amount given in % of cell dry weight
n. d., not determined

With respect to sporulation, the wild-type strain was compared to a *spo0A* knock-out construct (SKO1) and a strain that had lost the megaplasmid pSOL1 (M5). SKO1 showed downregulation of solvent formation genes, *sigF*, carbohydrate metabolism genes, and several electron transport genes. Upregulation was observed for *abrB* (encoding a transitional state gene regulator), major chemotaxis and motility operon genes, as well as glycosylation genes. In M5, several other electron transport, motility, and chemotaxis genes were more highly expressed, while *sigF*, several two-component histidine kinase genes, *spo0A*, *cheA*, *cheC*, many stress response, *fts* family, DNA topoisomerase, and central-carbon metabolism genes were downregulated. The expression pattern of gyrase and DNA helicase genes appeared to be closely linked to expression of genes involved in acid formation and glycolysis (Tomas *et al.*, 2003a). This finding further supports the predicted role of DNA supercoiling in regulation of solvent formation (Ullmann and Dürre, 1998).

The third analysis was prompted by the rationale that metabolic engineering of stress genes might lead to better suited strains for bioprocessing. The objective of this research is to identify genes that contribute to solvent tolerance and to use genetic modifications (involving these genes) to generate solvent tolerant constructs. Therefore, the *groES* and *groEL* genes (Narberhaus and Bahl, 1992) were overexpressed in *C. acetobutylicum* and the large-scale transcriptional program of the cells was examined in response to various levels of butanol challenge (Tomas *et al.*, 2003b). 174 Genes (out of 1,019 tested) were found to be differentially expressed. Transcript levels of motility and chemotaxis genes increased, while those of other major stress response genes (e.g. *dnaK* and *dnaJ*) decreased. Acetone and butanol production were also increased, despite the fact that enzyme levels of acetoacetate decarboxylase and CoA transferase were not significantly higher compared to the control strain. Possibly, overexpression of the GroESL machinery led to a broader stabilization of the biosynthetic machinery. This construct also proved to be more resistant towards externally added butanol (Tomas *et al.*, 2003b). Evidence was also obtained for complex host-plasmid interactions. The presence of the control plasmid represented a cellular stress for *C. acetobutylicum*, as indicated by altered expression of stress proteins, motility, chemotaxis, sporulation, and DNA topology-related genes (Tomas *et al.*, 2003b).

5. Metabolic Engineering

During the last decade, many attempts have been made to improve solvent formation of *C. acetobutylicum* by metabolic engineering (Table 1). These included overexpression of *adhE* (with no success, Nair and Papoutsakis, 1994; Nair *et al.*, 1994) and inactivation of acid formation genes (also without success, Green *et al.*, 1996). However, another report showed that inactivation of the butyrate kinase gene (*buk*) led to solvent superproduction, when fermentation was carried out at pH 5 or below (Harris *et al.*, 2000). This was the first time that the assumed barrier of biological solvent production of 200 mM was exceeded by forming 76 mM acetone and 225 mM butanol. Virtually identical amounts were obtained (66 mM acetone and 226 mM butanol) if, in addition to *buk* inactivation, the *adhE* gene was overexpressed (Harris *et al.*, 2000). Similar effects were observed in the case of *orf5* inactivation (encoding a deglycosylase/glycosylase, Thormann and Dürre, 2001) without and with *adhE* overexpression.

The 238 mM butanol found with the latter construct represent the highest value ever reported (Harris *et al.*, 2001). As already described in the preceding paragraph, overexpression of the *groES* and *groEL* genes also led to massive overproduction of acetone (148 mM) and butanol (231 mM), associated with enhanced resistance towards externally added butanol (Tomas *et al.*, 2003b). Another approach to increase solvent tolerance was an artificial change in lipid composition by overexpression of the cyclopropane fatty acid synthase gene (*cfa*). Indeed, cyclopropane fatty acid content (17cyc and especially 19cyc) increased significantly, while the amount of unsaturated fatty acids (16:1 and especially 18:1) decreased. Initial acid and butanol resistance was also increased. However, solvent production was considerably decreased (Zhao *et al.*, 2003). An interesting case relates to a recombinant *C. acetobutylicum* strain, in which *ctfB* gene expression was reduced by production of respective asRNA and simultaneous *adhE* overexpression. This led to a dramatic increase in ethanol production (190 mM), while concentrations of acetone (26 mM) and butanol (130 mM) remained unaffected (Tummala *et al.*, 2003). It must be kept in mind, however, that the presence of plasmids in *C. acetobutylicum* cells causes a complex host response (Walter *et al.*, 1994b; Tomas *et al.*, 2003b). For industrial production, the presence of small plasmids is less desirable, especially if associated with antibiotic selection or maintenance. Thus, it will be necessary to transfer the most promising gene overexpression/inactivation combinations to the chromosome.

In addition to alterations in expression of one or two genes, several artificial operons have been constructed using genes from *C. acetobutylicum* (Table 2). Most of the attempts focused on building a transcription unit for acetone production in either *E. coli* or *C. acetobutylicum*. A combination of *adc*, *ctfA*, and *ctfB* genes worked very well in the natural host (Mermelstein *et al.*, 1993), but only with limited success in the heterologous *E. coli* (Guillot, 1997). Performance in *E. coli* could be significantly improved by addition of a thiolase gene (*thlA*), whose gene product increases carbon flux towards C4 compounds (Bermejo *et al.*, 1998).

A completely different rationale prompted the construction of an operon consisting of the butyrate kinase (*buk*) and phosphotransbutyrylase (*ptb*) genes from *C. acetobutylicum* as well as the poly(hydroxyalkanoic acid) synthase genes (*phaE* and *phaC*) from *Thiocapsa pfennigii*. This transcription unit allowed synthesis of different poly(hydroxyalkanoic acids) from added hydroxyfatty acids in *E. coli* (Liu and Steinbüchel, 2000a). For example, 3-hydroxybutyrate was phosphorylated to 3-hydroxybutyryl phosphate by butyrate kinase (Buk), further converted to 3-hydroxybutyryl-CoA by phosphotransbutyrylase (Ptb), and then polymerized into poly(3-hydroxybutyrate) by poly(hydroxyalkanoic acid) synthase (PHA synthase). Similarly, 4-hydroxybutyrate and 4-hydroxyvalerate were used. Pure polyesters as well as mixtures could be obtained in amounts of up to 68.7 % of cell dry weight (Liu and Steinbüchel, 2000a). *In vitro* biosynthesis was also possible by using purified Buk, Ptb, and PHA synthase. A yield of up to 2.8 mg polymers per ml could be obtained (Liu and Steinbüchel, 2000b).

6. Outlook

C. acetobutylicum currently represents the model organism of non-pathogenic clostridia. Recombinant DNA methods, reporter genes, proteome analysis, transcriptional profiling, and knock-out strategies are well-established (e. g. Nakotte *et al.*, 1998; Tummala *et al.*, 1999; Young *et al.*, 1999; Harris *et al.*, 2002; Schaffer *et al.*, 2002; Girbal *et al.*, 2003; Tomas *et al.*, 2003a; Feustel *et al.*, 2004). The genome sequence has been determined (Nölling *et al.*, 2001). This will allow further research that forcuses on regulatory mechanisms in this bacterium. The primary goal will be comprehension of the regulatory networks governing solventogenesis and sporulation, thus forming the basis for new industrial and medical applications such as solvent formation as reviewed above, removal of toxic products (Watrous *et al.*, 2003; Zhang and Hughes, 2003), and cancer therapy by recombinant clostridial spores (Minton *et al.*, 2001) (see Chapter 11). Thus, the enormous metabolic potential of *C. acetobutylicum* still awaits application, further elucidation, and exploitation.

Acknowledgements

Work in my laboratory was supported by grants from the Deutsche Forschungsgemeinschaft within the priority program "Molekulare Analyse von Regulationsnetzwerken in Bakterien", the BMBF GenoMik project (Competence Network Göttingen), and the European Community (contract No. QLK3-CT-2001-01737).

References

Arnau, J., Jørgensen, F., Madsen, S.M., Vrang, A., and Israelsen, H. 1998. Cloning of the *Lactococcus lactis adhE* gene, encoding a multifunctional alcohol dehydrogenase, by complementation of a fermentative mutant of *Escherichia coli*. J. Bacteriol. 180: 3049-3055.

Bahl, H. 1983. Kontinuierliche Aceton-Butanol-Gärung durch *Clostridium acetobutylicum*. Ph. D. thesis, University of Göttingen, Germany.

Bahl, H., Gottwald, M., Kuhn, A., Rale, V., Andersch, W., and Gottschalk, G. 1986. Nutritional factors affecting the ratio of solvents produced by *Clostridium acetobutylicum*. Appl. Environ. Microbiol. 52: 169-172.

Bahl, H., Müller, H., Behrens, S., Joseph, H., and Narberhaus, F. 1995. Expression of heat shock genes in *Clostridium acetobutylicum*. FEMS Microbiol. Rev. 17: 341-348.

Bermejo, L.L., Welker, N.E., and Papoutsakis, E.T. 1998. Expression of *Clostridium acetobutylicum* ATCC 824 genes in *Escherichia coli* for acetone production and acetate detoxification. Appl. Environ. Microbiol. 64: 1079-1085.

Bertram, J., Kuhn, A., and Dürre, P. 1990. Tn*916*-induced mutants of *Clostridium acetobutylicum* defective in regulation of solvent formation. Arch. Microbiol. 153: 373-377.

Böhringer, M. 2002. Molekularbiologische und enzymatische Untersuchungen zur Regulation des Gens der Acetacetat-Decarboxylase von *Clostridium acetobutylicum*. Ph. D. thesis, University of Ulm, Germany.

Bowles, L.K., and Ellefson, W.L. 1985. Effects of butanol on *Clostridium acetobutylicum*. Appl. Environ. Microbiol. 50: 1165-1170.

Chen, J.-S. 1993. Properties of acid- and solvent-forming enzymes of clostridia. In: The Clostridia and Biotechnology. D.R. Woods, ed. Butterworth-Heinemann, Stoneham. p. 51-76.

Chen, J.-S. 1995. Alcohol dehydrogenase: multiplicity and relatedness in the solvent-producing clostridia. FEMS Microbiol. Rev. 17: 263-273.

Contag, P.R., and Rogers, P. 1988. The cloning and expression of the *Clostridium acetobutylicum* B643 butyraldehyde dehydrogenase by complementation of an *Escherichia coli* aldehyde dehydrogenase deficient mutant. Abstracts 88th Ann. Meeting ASM, Miami Beach, May 8-13, 1988, p. 169, H-146.

Contag, P.R., Williams, M.G., and Rogers, P. 1990. Cloning of a lactate dehydrogenase gene from *Clostridium acetobutylicum* B643 and expression in *Escherichia coli*. Appl. Environ. Microbiol. 56: 3760-3765.

Cornillot, E., and Soucaille, P. 1996. Solvent-forming genes in clostridia. Nature. 380: 489.

Cornillot, E., Nair, R., Papoutsakis, E. T., and Soucaille, P. 1997. The genes for butanol and acetone formation in *Clostridium acetobutylicum* ATCC 824 reside on a large plasmid whose loss leads to degeneration of the strain. J. Bacteriol. 179: 5442-5447.

Dailly, Y.P., Bunch, P., and Clark, D.P. 2000. Comparison of the fermentative alcohol dehydrogenases of *Salmonella typhimurium* and *Escherichia coli*. Microbios. 103: 179-196.

Dürre, P., Kuhn, A., Gottwald, M., and Gottschalk, G. 1987. Enzymatic investigations on butanol dehydrogenase and butyraldehyde dehydrogenase in extracts of *Clostridium acetobutylicum*. Appl. Microbiol. Biotechnol. 26: 268-272.

Dürre, P., Bahl, H., and Gottschalk, G. 1988. Membrane processes and product formation in anaerobes. In: Handbook on Anaerobic Fermentations. L.E. Erickson and D.Y.-C. Fung, eds. Marcel Dekker, Inc., New York, New York. p. 187-206.

Dürre, P., Fischer, R.-J., Kuhn, A., Lorenz, K., Schreiber, W., Stürzenhofecker, B., Ullmann. S., Winzer, K., and Sauer, U. 1995. Solventogenic enzymes of *Clostridium acetobutylicum*: catalytic properties, genetic organization, and transcriptional regulation. FEMS Microbiol. Rev. 17: 251-262.

Dürre, P., and Bahl, H. 1996. Microbial production of acetone/butanol/isopropanol. In: Biotechnology, vol. 6, 2nd ed. M. Roehr, ed. VCH Verlagsgesellschaft mbH, Weinheim. p. 229-268.

Dürre, P. 1998. New insights and novel developments in clostridial acetone/butanol/isopropanol fermentation. Appl. Microbiol. Biotechnol. 49: 639-648.

Dürre, P., Böhringer, M., Nakotte, S., Schaffer, S., Thormann, K., and Zickner, B. 2002. Transcriptional regulation of solventogenesis in *Clostridium acetobutylicum*. J. Mol. Microbiol. Biotechnol. 4: 295-300.

Feustel, L., Nakotte, S., and Dürre, P. 2004. Characterization and development of two reporter gene systems for *Clostridium acetobutylicum*. Appl. Environ. Microbiol. 70: 798-803.

Fischer, R. J., Helms, J., and Dürre, P. 1993. Cloning, sequencing, and molecular analysis of the *sol* operon of *Clostridium acetobutylicum*, a chromosomal locus involved in solventogenesis. J. Bacteriol. 175: 6959 - 6969.

Fitz, A. 1876. Ueber die Gährung des Glycerins. Ber. Dtsch. Chem. Ges. 9: 1348-1352.

Fitz, A. 1877. Ueber Schizomyceten-Gährungen II [Glycerin, Mannit, Stärke, Dextrin]. Ber. Dtsch. Chem. Ges. 10: 276-283.

Fitz, A. 1878. Ueber Schizomyceten-Gährungen III. Ber. Dtsch. Chem. Ges. 11: 42-55.

Fitz, A. 1882. Ueber Spaltpilzgährungen. VII. Mittheilung. Ber. Dtsch. Chem. Ges. 15: 867-880.

Fontaine, L., Meynial-Salles, I., Girbal, L., Yang, X., Croux, C., and Soucaille, P. 2002. Molecular characterization and transcriptional analysis of *adhE2*, the gene encoding the NADH-dependent aldehyde/alcohol dehydrogenase responsible for butanol production in alcohologenic cultures of *Clostridium acetobutylicum* ATCC 824. J. Bacteriol. 184: 821-830.

Gerischer, U., and Dürre, P. 1990. Cloning, sequencing, and molecular analysis of the acetoacetate decarboxylase gene region from *Clostridium acetobutylicum*. J. Bacteriol. 172: 6907- 6918.

Gerischer, U., and Dürre, P. 1992. mRNA analysis of the *adc* gene region of *Clostridium acetobutylicum* during the shift to solventogenesis. J. Bacteriol. 174: 426 - 433.

Gibbs, D.F. 1983. The rise and fall (... and rise?) of acetone/butanol fermentations. Trends Biotechnol. 1: 12-15.

Girbal, L., Croux, C., Vasconcelos, I., and Soucaille, P. 1995. Regulation of metabolic shifts in *Clostridium acetobutylicum* ATCC 824. FEMS Microbiol. Rev. 17: 287-297.

Girbal, L., and Soucaille, P. 1998. Regulation of solvent production in *Clostridium acetobutylicum*. Trends Biotechnol. 16: 11-16.

Girbal, L., Mortier-Barriere, I., Raynaud, F., Rouanet, C., Croux, C., and Soucaille, P. 2003. Development of a sensitive gene expression reporter system and an inducible promoter-repressor system for *Clostridium acetobutylicum*. Appl. Environ. Microbiol. 69: 4985-4988.

Goodlove, P.E., Cunnigham, P.R., Parker, J., and Clark, D.P. 1989. Cloning and sequence analysis of the fermentative alcohol-dehydrogenase-encoding gene of *Escherichia coli*. Gene. 85: 209-214.

Gottwald, M., and Gottschalk, G. 1985. The internal pH of *Clostridium acetobutylicum* and its effect on the shift from acid to solvent formation. Arch. Microbiol. 143: 42-46.

Green, E. M, and Bennett, G. N. 1996. Inactivation of an aldehyde/alcohol dehydrogenase gene from *Clostridium acetobutylicum* ATCC 824. Appl. Biochem. Biotechnol. 57/58: 213-221.

Green, E.M., Boynton, Z.L., Harris, L.M., Rudolph, F.B., Papoutsakis, E.T., and Bennett, G.N. 1996. Genetic manipulation of acid formation pathways by gene inactivation in *Clostridium acetobutylicum* ATCC 824. Microbiology. 142: 2079-2086.

Guillot, J.-P.K.M. 1997. Konstruktion und Expression artifizieller Transkriptionseinheiten zur Lösungsmittelsynthese in *Escherichia coli*. Diploma thesis, University of Ulm, Germany.

Harris, L.M., Desai, R.P., Welker, N.E., and Papoutsakis, E.T. 2000. Characterization of recombinant strains of the *Clostridium acetobutylicum* butyrate kinase inactivation mutant: need for new phenomenological models for solventogenesis and butanol inhibition? Biotechnol. Bioeng. 67: 1-11.

Harris, L.M., Blank, L., Desai, R.P., Welker, N.E., and Papoutsakis, E.T. 2001. Fermentation characterization and flux analysis of recombinant strains of *Clostridium acetobutylicum* with an inactivated *solR* gene. J. Ind. Microbiol. Biotechnol. 27: 322-328.

Harris, L.M., Welker, N.E., and Papoutsakis, E.T. 2002. Northern, morphological, and fermentation analysis of *spo0A* inactivation and overexpression in *Clostridium acetobutylicum* ATCC 824. J. Bacteriol. 184: 3586-3597.

Hollergschwandner, C. 2003. Unterschiede in der Regulation der Sporulation von *Clostridium acetobutylicum* und *Bacillus subtilis*. Ph.D. Thesis, University of Ulm.

Holt, R.A., Stephens, G.M., and Morris, J.G. 1984. Production of solvents by *Clostridium acetobutylicum* cultures maintained at neutral pH. Appl. Environ. Microbiol. 48: 1166-1170.

Huang, L., Gibbins, L.N., and Forsberg, C.W. 1985. Transmembrane pH gradient and membrane potential in *Clostridium acetobutylicum* during growth under acetogenic and solventogenic conditions. Appl. Environ. Microbiol. 50: 1043-1047.

Johnson, J.L., Toth, J. Santiwatanakul, S., and Chen, J.-S. 1997. Cultures of „*Clostridium acetobutylicum*" from various collections comprise *Clostridium acetobutylicum*, *Clostridium beijerinckii*, and two other distinct types based on DNA-DNA reassociation. Int. J. Syst. Bacteriol. 45: 420-424.

Jones, D.T., and Woods, D.R. 1986. Acetone-butanol fermentation revisited. Microbiol. Rev. 50: 484-524.

Jones, D.T. 2001. Applied acetone-butanol fermentation. In: *Clostridia*. Biotechnological and Medical Applications. H. Bahl and P. Dürre, eds. Wiley-VCH Verlag GmbH, Weinheim. p. 125-168.

Kaga, N., Umitsuki, G., Clark, D.P., Nagai, K., and Wachi, M. 2002. Extensive overproduction of the AdhE protein by *rng* mutations depends on mutations in the *cra* gene or in the Cra-box of the *adhE* promoter. Biochem. Biophys. Res. Commun. 295: 92-97.

Keis, S., Bennett, C.F., Ward, V.K., and Jones, D.T. 1995. Taxonomy and phylogeny of industrial solvent-producing clostridia. Int. J. Syst. Bacteriol. 45: 693-705.

Keis, S., Shaheen, R., Jones, D.T. 2001. Emended descriptions of *Clostridium acetobutylicum* and *Clostridium beijerinckii*, and descriptions of *Clostridium saccharoperbutylacetonicum* sp. nov. and *Clostridium saccharobutylicum* sp. nov. Int. J. Syst. Evol. Microbiol. 51: 2095-2103.

Kessler, D., Leibrecht, I., and Knappe, J. 1991. Pyruvate: formate-lyase-deactivase and acetyl-CoA reductase activities of *Escherichia coli* reside on a polymeric operon particle encoded by *adhE*. FEBS Lett. 281: 59-63.

Liu, S.-J., and Steinbüchel, A. 2000a. A novel genetically engineered pathway for synthesis of poly(hydroxyalkanoic acids) in *Escherichia coli*. Appl. Environ. Microbiol. 66: 739-743.

Liu, S.-J., and Steinbüchel, A. 2000b. Exploitation of butyrate kinase and phosphotransbutyrylase from *Clostridium acetobutylicum* for the *in vitro* biosynthesis of poly(hydroxyalkanoic acid). Appl. Microbiol. Biotechnol. 53: 545-552.

Membrillo-Hernández, J., Echave, P., Cabiscol, E., Tamarit, J., Ros, J., and Lin, E.C.C. 2000. Evolution of the *adhE* gene product of *Escherichia coli* from a functional reductase to a dehydrogenase. J. Biol. Chem. 275: 33869-33875.

Mermelstein, L.D., Papoutsakis, E.T., Petersen, D.J., and Bennett, G.N. 1993. Metabolic engineering of *Clostridium acetobutylicum* ATCC 824 for increased solvent production by enhancement of acetone formation enzyme activities using a synthetic acetone operon. Biotechnol. Bioeng. 42: 1053-1060.

Minton, N.P., Brown, J.M., Lambin, P., and Anné, J. 2001. Clostridia in cancer therapy. In: *Clostridia*: Biotechnology and Medical Applications. H. Bahl and P. Dürre, eds. Wiley-VCH Verlag GmbH, Weinheim. p. 251-270.

Mitchell, W.J. 1998. Physiology of carbohydrate to solvent conversion by clostridia. Adv. Microbial Physiol. 39: 31-130.

Moreira, A.R. 1983. Acetone-butanol fermentation. In: Organic Chemicals from Biomass. D.L. Wise, ed. Benjamin/Cummins Publ. Co. Inc., Menlo Park. p. 385-406.

Nair, R. V., Bennett, G. N., and Papoutsakis, E. T. 1994. Molecular characterization of an aldehyde/alcohol dehydrogenase gene from *Clostridium acetobutylicum* ATCC 824. J. Bacteriol. 176: 871 - 885.

Nair, R. V., and Papoutsakis, E. T. 1994. Expression of plasmid-encoded *aad* in *Clostridium acetobutylicum* M5 restores vigorous butanol production. J. Bacteriol. 176: 5843-5846.

Nair, R. V., Green, E. M., Watson, D. E., Bennett, G. N., and Papoutsakis, E. T. 1999. Regulation of the *sol* locus genes for butanol and acetone formation in *Clostridium acetobutylicum* ATCC 824 by a putative transcriptional repressor. J. Bacteriol. 181: 319-330.

Nakotte, S., Schaffer, S., Böhringer, M., and Dürre, P. 1998. Electroporation of, plasmid isolation from and plasmid conservation in *Clostridium acetobutylicum* DSM 792. Appl. Microbiol. Biotechnol. 50: 564-567.

Narberhaus, F., and Bahl, H. 1992. Cloning, sequencing, and molecular analysis of the *groESL* operon of *Clostridium acetobutylicum*. J. Bacteriol. 174: 3282-3289.

Nölling, J., Breton, G., Omelchenko, M. V., Makarova, K. S., Zeng, Q., Gibson, R., Lee, H. M., Dubois, J., Qiu, D., Hitti, J., GTC Sequencing Center Production, Finishing, and Bioinformatics Teams, Wolf, Y. I., Tatusov, R. L., Sabathe, F., Doucette-Stamm, L., Soucaille, P., Daly, M. J., Bennett, G. N., Koonin, E. V., and Smith, D. R.

2001. Genome sequence and comparative analysis of the solvent-producing bacterium *Clostridium acetobutylicum*. J. Bacteriol. 183: 4823-4838.

Palosaari, N., and Rogers, P. 1988. Purification and properties of the inducible coenzyme A-linked butyraldehyde dehydrogenase from *Clostridium acetobutylicum*. J. Bacteriol. 170: 2971-2976.

Pasteur, L. 1862. Quelques résultats nouveaux relatifs aux fermentations acétique et butyrique. Bull. Soc. Chim. Paris May 1862: 52-53.

Petersen, D. J., and Bennett, G. N. 1990. Purification of acetoacetate decarboxylase from *Clostridium acetobutylicum* ATCC 824 and cloning of the acetoacetate decarboxylase gene in *Escherichia coli*. Appl. Environ. Microbiol. 56: 3491 - 3498.

Petersen, D.J., Welch, R.W., Rudolph, F.B., and Bennett, G.N. 1991. Molecular cloning of an alcohol (butanol) dehydrogenase gene cluster from *Clostridium acetobutylicum* ATCC 824. J. Bacteriol. 173: 1831-1834.

Pich, A., Narberhaus, F., and Bahl, H. 1990. Induction of heat shock proteins during initiation of solvent formation in *Clostridium acetobutylicum*. Appl. Microbiol. Biotechnol. 33: 697-704.

Pringsheim, H.H. 1906a. Ueber den Ursprung des Fuselöls und eine Alkohole bildende Bakterienform. Cbl. Bakteriol. Parasitenkd. Infektionskrankh., II. Abt. 15: 300-321.

Pringsheim, H. 1906b. Ueber ein Stickstoff assimilierendes *Clostridium*. Cbl. Bakteriol. Parasitenkd. Infektionskrankh., II. Abt. 15: 795-800.

Rao, G., and Mutharasan, R. 1986. Alcohol production by *Clostridium acetobutylicum* induced by methyl viologen. Biotechnol. Lett. 8: 893-896.

Rao, G., and Mutharasan, R. 1987. Altered electron flow in continuous cultures of *Clostridium acetobutylicum* induced by viologen dyes. Appl. Environ. Microbiol. 53: 1232-1235.

Ravagnani, A., Jennert, K. C. B., Steiner, E., Grünberg, R., Jefferies, J. R., Wilkinson, S. R., Young, D. I., Tidswell, E. C., Brown, D. P., Youngman, P., Morris, J. G., and Young, M. 2000. Spo0A directly controls the switch from acid to solvent production in solvent-forming clostridia. Mol. Microbiol. 37: 1172-1185.

Reid, M.F., and Fewson, C.A. 1994. Molecular characterization of microbial alcohol dehydrogenases. CRC Crit. Rev. Microbiol. 20: 13-56.

Sánchez, L.B. 1998. Aldehyde dehydrogenase (CoA-acetylating) and the mechanism of ethanol formation in the amitochondriate protist, *Giardia lamblia*. Arch. Biochem. Biophys. 354: 57-64.

Santangelo, J.D., and Dürre, P. 1996. Microbial production of acetone and butanol: Can history be repeated? Chimica oggi/Chemistry today. 14:29-35.

Sauer, U., and Dürre, P. 1993. Sequence and molecular characterization of a DNA region encoding a small heat shock protein of *Clostridium acetobutylicum*. J. Bacteriol. 175: 3394-3400.

Sauer, U., Fischer, R., and Dürre, P. 1993. Solvent formation and its regulation in strictly anaerobic bacteria. Curr. Top. Mol. Genet. 1: 337-351.

Sauer, U., and Dürre, P. 1995. Differential induction of genes related to solvent formation during the shift from acidogenesis to solventogenesis in continuous culture of *Clostridium acetobutylicum*. FEMS Microbiol. Lett. 125: 115-120.

Schaffer, S., Isci, N., Zickner, B., and Dürre, P. 2002. Changes in protein synthesis and identification of proteins specifically induced during solventogenesis in *Clostridium acetobutylicum*. Electrophoresis. 23: 110-121.

Schardinger, F. 1904. Azetongärung. Wiener Klin. Wochenschr. 17: 207-209.

Schardinger, F. 1905. *Bacillus macerans*, ein Aceton bildender Rottebacillus. Zbl. Bakteriol. Parasitenkd. Infektionskrankh., II. Abt. 4: 772-781.

Schreiber, W., and Dürre, P. 1999. The glyceraldehyde-3-phosphate dehydrogenase of *Clostridium acetobutylicum*: isolation and purification of the enzyme, and sequencing and localization of the *gap* gene within a cluster of other glycolytic genes. Microbiology. 145: 1839-1847.

Schreiber, W., and Dürre, P. 2000. Differential expression of genes within the *gap* operon of *Clostridium acetobutylicum*. Anaerobe 6: 291-297.

Shaheen, R., Shirley M. and Jones, D.T. 2000. Comparative fermentation studies of industrial strains belonging to four species of solvent-producing clostridia. J. Mol. Microbiol. Biotechnol. 2: 115-124.

Söhling, B., and Gottschalk, G. 1996. Molecular analysis of the anaerobic succinate degradation pathway in *Clostridium kluyveri*. J. Bacteriol. 178: 871-880.

Stim-Herndon, K. P., Petersen, D. J., and Bennett, G. N. 1995. Characterization of an acetyl-CoA acetyltransferase (thiolase) gene from *Clostridium acetobutylicum* ATCC 824. Gene. 154: 81-85.

Terracciano, J.S., Rapoport, E., and Kashket, E.R. 1988. Stress- and growth phase-associated proteins of *Clostridium acetobutylicum*. Appl. Environ. Microbiol. 54: 1989-1995.

Thormann, K., and Dürre, P. 2001. Orf5/SolR: a transcriptional repressor of the *sol* operon of *Clostridium acetobutylicum*? J. Ind. Microbiol. Biotechnol. 27: 307-313.

Thormann, K., Feustel, L., Lorenz, K., Nakotte, S., and Dürre, P. 2002. Control of butanol formation in *Clostridium acetobutylicum* by transcriptional activation. J. Bacteriol. 184: 1966-1973.

Tomas, C.A., Alsaker, K.V., Bonarius, H.P.J., Hendriksen, W.T., Yang, H., Beamish, J.A., Paredes, C.J., and Papoutsakis, E.T. 2003a. DNA array-based transcriptional analysis of asporogenous, nonsolventogenic *Clostridium acetobutylicum* strains SKO1 and M5. J. Bacteriol. 185: 4539-4547.

Tomas, C.A., Welker, N.E., and Papoutsakis, E.T. 2003b. Overexpression of *groESL* in *Clostridium acetobutylicum* results in increased solvent production and tolerance, prolonged metabolism, and changes in the cell's transcriptional program. Appl. Environ. Microbiol. 69: 4951-4965.

Tummala, S.B., Welker, N.E., and Papoutsakis, E.T. 1999. Development and characterization of a gene expression reporter system for *Clostridium acetobutylicum* ATCC 824. Appl. Environ. Microbiol. 65: 3793-3799.

Tummala, S.B., Junne, S.G., and Papoutsakis, E.T. 2003. Antisense RNA downregulation of coenzyme A transferase combined with alcohol-aldehyde dehydrogenase overexpression leads to predominantly alcohologenic *Clostridium acetobutylicum* fermentations. J. Bacteriol. 185: 3644-3653.

Ullmann, S., Kuhn, A., and Dürre, P. 1996. DNA topology and gene expression in *Clostridium acetobutylicum*: implications for the regulation of solventogenesis. Biotechnol. Lett. 18: 1413-1418.

Ullmann, S., and Dürre, P. 1998. Changes in DNA topology are involved in triggering the onset of solventogenesis in *Clostridium acetobutylicum*. Recent Res. Dev. Microbiol. 2: 281-294.

Wachi, M., Kaga, N., Umitsuki, G., Clark, D.P., and Nagai, K. 2001. A novel RNase G mutant that is defective in degradation of *adhE* mRNA but proficient in the processing of 16S rRNA precursor. Biochem. Biophys. Res. Commun. 289: 1301-1306.

Walter, K.A., Bennett, G.N., and Papoutsakis, E.T. 1992. Molecular characterization of two *Clostridium acetobutylicum* ATCC 824 butanol dehydrogenase isozyme genes. J. Bacteriol. 174: 7149-7158.

Walter, K.A., Mermelstein, L.D., and Papoutsakis, E.T. 1994a. Studies of recombinant *Clostridium acetobutylicum* with increased dosages of butyrate formation genes. Ann. New York Acad. Sci. 721: 69-72.

Walter, K.A., Mermelstein, L.D., and Papoutsakis, E.T. 1994b. Host-plasmid interactions in recombinant strains of Clostridium acetobutylicum ATCC 824. FEMS Microbiol. Lett. 123: 335-342.

Watrous, M.M., Clark, S., Kutty, R., Huang, S., Rudolph, F.B., Hughes, J.B., and Bennett, G.N. 2003. 2,4,6-Trinitrotoluene reduction by an Fe-only hydrogenase in *Clostridium acetobutylicum*. Appl. Environ. Microbiol. 69: 1542-1547.

Welch, R.W., Rudolph, F.B., and Papoutsakis, E.T. 1989. Purification and characterization of the NADH-dependent butanol dehydrogenase from *Clostridium acetobutylicum* (ATCC 824). Arch. Biochem. Biophys. 273: 309-318.

Winzer, K., Lorenz, K., Zickner, B., and Dürre, P. 2000. Differential regulation of two thiolase genes from *Clostridium acetobutylicum* DSM 792. J. Mol. Microbiol. Biotechnol. 2: 531-541.

Wong, J., and Bennett, G. N. 1996. The effect of novobiocin on solvent production by *Clostridium acetobutylicum*. J. Ind. Microbiol. 16: 354-359.

Yang, W., Li, E., Kairong, T., and Stanley, S.L.; Jr. 1994. *Entamoeba histolytica* has an alcohol dehydrogenase homologous to the multifunctional *adhE* gene product of *Escherichia coli*. Mol. Biochem. Parasitol. 64: 253-260.

Young, M., Minton, N.P., and Staudenbauer, W.L. 1989. Recent advances in the genetics of the clostridia. FEMS Microbiol. Rev. 63: 301-326.

Young, D. I., Evans, V. J., Jefferies, J. R., Jennert, K. C. B., Phillips, Z. E. V., Ravagnani, A., and Young, M. 1999. Genetic methods in clostridia. Methods Microbiol. 29: 191-207.

Zhang, C., and Hughes, J.B. 2003. Biodegradation pathways of hexahydro-1,3,5-trinitro-1,3,5-triazine (RDX) by *Clostridium acetobutylicum* cell-free extract. Chemosphere, 50: 665-671.

Zhao, Y., Hindorff, L.A., Chuang, A., Monroe-Augustus, M., Lyristis, M., Harrison, M.L., Rudolph, F.B., and Bennett, G.N. 2003. Expression of a cloned cyclopropane fatty acid synthase gene reduces solvent formation in *Clostridium acetobutylicum* ATCC 824. Appl. Environ. Microbiol. 69: 2831-2841.

Zoutberg, G.R., Willemsberg, R., Smit, G., Texeira de Mattos, M.J., and Neussel, O.M. 1989. Solvent production by an aggregate-forming variant of *Clostridium butyricum*. Appl. Microbiol. Biotechnol. 32: 22-26.

Chapter 19

The Clostridial Cellulosome

Anne Belaich*, Chantal Tardif, Henri-Pierre Fierobe, Sandrine Pagès, and Jean-Pierre Belaich

Abstract

The cellulosomes of five cellulolytic *Clostridia* have been investigated. Cellulosomes are large molecular complexes which efficiently hydrolyze crystalline cellulose and where the catalytic subunits are anchored onto a non-enzymatic protein called scaffoldin. Many components of the various cellulosomes, produced as recombinant proteins in *Escherichia coli,* have been studied from a biochemical and structural point of view. Special attention was devoted to the cellulosome of a sixth *Clostridium*: *Clostridium acetobutylicum.* This bacterium is non-cellulolytic but recent sequencing of its genome revealed that this species harbours all the genes coding for cellulosomal components. In four of the six *Clostridia*, the genes encoding the major components of the cellulosome are organized in large clusters on the chromosome. The study of a spontaneous mutant of *Clostridium cellulolyticum*, affected in cellulolysis, provided interesting information on the regulation of the genes belonging to these clusters. Previously established genetic tools, now available for the two mesophilic *Clostridia*, *C. cellulolyticum* and *C. acetobutylicum*, will allow new approaches for studying the cellulosomes "*in clostridio*"

1. Introduction

Cellulose is the most important biopolymer on earth and constitutes the major component of plant cell wall where it is associated with other structural polymers: hemicellulose, pectin and lignin. In nature, cellulose is degraded by a wide variety of microorganisms and this degradation plays a fundamental role in carbon recycling on the planet. The major part of the cellulosic material is degraded in aerobiosis by white-rot fungi, soft-rot fungi and aerobic bacteria. About 10% of the cellulosic material is degraded in anaerobiosis essentially by bacteria. In anaerobic biotopes, these bacteria are often accompanied by non-cellulolytic bacteria, such as acetogens and methanogens; these consortia allow the entire mineralization of the biomass to methane and carbon dioxide. Despite numerous studies, attempts to enhance the rate of biomass fermentation for chemical or biofuel production had only limited success as hydrolysis of its polymeric components, especially cellulose, confers limits upon the rate of the entire process.

Cellulose is a linear polymer of β-1,4-linked glucose units. The individual cellulose chain can contain up to 10,000 glucose units. The chains are packed in insoluble microfibrils by means of intra- and inter-molecular hydrogen bonds making them insoluble and chemically resistant. The organization of the macromolecules in a cellulose fiber is not uniform. The fiber of cellulose is mainly composed of crystalline regions that are rich in hydrogen bonds and difficult to hydrolyze, and less ordered regions, called amorphous or paracrystalline regions, less rich in hydrogen bonds and more easily hydrolyzed. The degree of crystallinity of a given cellulose reflects crystalline /amorphous region ratios. Furthermore, cellulose fibers are embedded in a matrix of hemicellulose, pectin and lignin; the proportion of these polymers in the plant cell wall depends on the tissue, the species and the age of the vegetal. All these properties explain why natural cellulose is so recalcitrant to enzymatic degradation. Extensive studies on the various forms of cellulose have been published in specialized reviews (Atalla *et al.*, 1984; Chanzy, 1990; Atalla, 1999). To complete these processes, aerobic and anaerobic microorganisms secrete various specific enzymes that act in concert to hydrolyze the various types of plant cell walls.

Anaerobic cellulolytic microorganisms (essentially bacteria and a few fungi) have been isolated in a variety of ecological niches including animal digestive tracts, soil, sediments, composts and sewages. They have been found in thermophilic, mesophilic and even in psychrophilic environments. Bacteria belonging to the genus *Clostridium* are widespread in almost all these ecosystems. A brief non exhaustive list could include: *C. thermocellum*, *C. papyrosolvens*, *C. cellulovorans*, *C. herbivorans*, *C. cellulofermentans*, *C. cellulosi*,

*For correspondence email: abelaich@ibsm.cnrs-mrs.fr

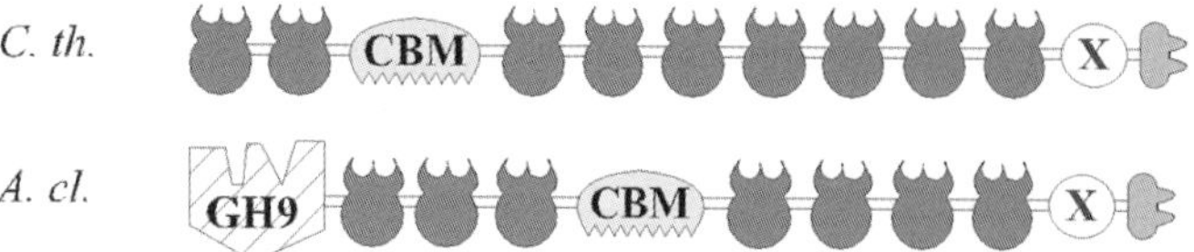

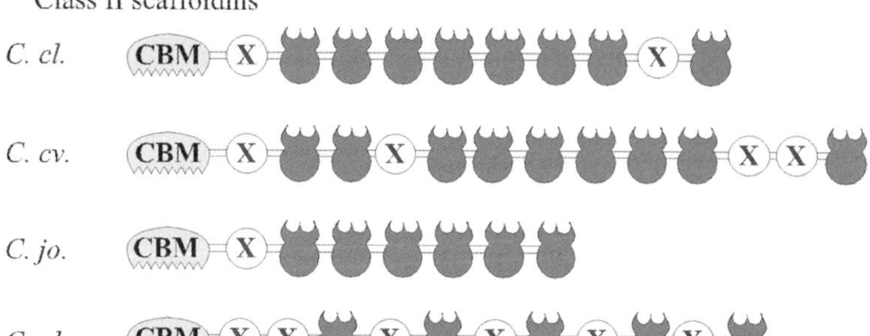

Figure 1. Modular organization of the scaffoldins. *C.th., Clostridium thermocellum*; *A.cl., Acetivibrio cellulolyticus*; *C.cl., Clostridium cellulolyticum*; *C.cv., Clostridium cellulovorans*; *C. jo., Clostridium josui*; *C.ab., Clostridium acetobutylicum*. CBM, carbohydrate-binding module; X, module of unknown function; GH9, glycosyl hydrolase family 9 catalytic domain; dark grey, type I cohesin module; light grey, type II dockerin module.

C. aldrichii, *C. cellulolyticum*, *Acetivibrio cellulolyticus*, *C. josui* , *C. phytofermentans* and *Clostridium* strain PXYL1, a novel cold-tolerant *Clostridium*. For some of them, only a brief description has been given. However others have been the subject of extensive biochemical and molecular studies.

A characteristic of the anaerobic cellulolytic microorganisms is that their cellulolytic enzymes are, conversely to cellulolytic systems of aerobic microorganisms, generally associated in large well-structured complexes that efficiently hydrolyze crystalline cellulose. These complexes are present in bacteria belonging to the families of *clostridiacae* and *bacteroidacae*.

The first complex was discovered in the thermophilic *C. thermocellum* and named the cellulosome (Bayer *et al.*, 1983; 1985). Biochemical analysis showed that the cellulosome is a complex with an apparent size of 1 Mda composed of numerous enzymes anchored on a non-cellulolytic proteinaceous skeleton called scaffoldin (Lamed *et al.*, 1988). Cellulosomes were subsequently found in other cellulolytic anaerobic bacteria (Gal *et al.*, 1997; Murashima *et al.*, 2002). Attempts to dissociate the cellulosomes in order to purify separate active proteins were unsuccessful; however, thanks to a molecular biological approach and genetic engineering over the last 15 years, information was obtained concerning the cellulosome components and many reviews have been published to describe their role in hydrolysis of crystalline cellulose (Béguin, 1990; Lamed *et al.*, 1988; Bayer *et al.*, 1998; 2000; Schwartz, 2001; Lynd *et al.*, 2002; Doi *et al.*, 2003)

This chapter is focused on the cellulosomes of six *Clostridia* belonging to family I of *clostridiacae* in the order of Clostridiales. They are: *C. thermocellum*, *C. cellulovorans*, *C. cellulolyticum*, *C. josui*, *Acetivibrio cellulolyticus* and *C. acetobutylicum*. This last *Clostridium* is not a cellulolytic bacterium, but recent sequencing of its genome revealed that this species harboured all the genes coding for cellulosome components (Nolling *et al.*, 2001). However this cellulosome is inactive and very poorly produced (Sabathé *et al.*, 2002). It was hypothesized that *C. acetobutylicum* was at one time a cellulolytic bacterium but it has lost its cellulytic activity (Sabathé, 2002).

2. Organization of the Cellulosomes

The clostridial cellulosomes are made up of two kinds of components, the scaffoldins and the catalytic units.

2.1. Scaffoldins

The scaffoldins are skeleton proteins on which the catalytic units are anchored (Shoseyov *et al.*, 1992; Gerngross *et al.*, 1993; Pagès *et al.*, 1996, 1999; Kakiuchi *et al.*, 1998; Ding *et al.*, 1999). The most striking feature of this architecture is that it is built by using two complementary affinity modules called cohesins, located on the scaffoldin, and dockerins, borne by enzymatic subunits. The scaffoldins are made up of three kinds of modules.

2.1.1. Cohesins

The scaffoldin cohesin modules (cohesins I) contain about 150 residues and each cohesin can be produced in *E. coli* by genetic engineering. The 3D structures of three different recombinant cohesin modules, two from *C. thermocellum* and one from *C. cellulolyticum*, have been determined (Shimon *et al.*, 1997; Tavares *et al.*, 1997; Spinelli *et al.*, 2000). The overall folds of these cohesins are very similar showing a nine-stranded β–sandwich with a jellyroll topology with a dimension of 40X31X26 Å. All the *clostridiaceae* scaffoldins contain a set of type I cohesins, which are involved in the binding of the catalytic units by means of the complementary dockerins (Tokatlidis *et al.*, 1991; Salamitou *et al.*, 1994; Pagès *et al.*, 1996; Park *et al.*, 2001). A schematic view of the clostridial scaffoldins is shown in Figure 1. The number of cohesins ranges from 6 to 9.

2.1.2. Cellulose Binding Modules

CBMs are found in all cellulase systems and play a prominent role in cellulose breakdown (Tomme *et al.*, 1995). 33 families of CBM are listed on the CAZy website (http://afmb.cnrs-mrs.fr/CAZY/acc.html). All the scaffoldins of the *Clostridiaceae* species contain a family-3a cellulose-binding module (CBM), (about 160 residues), which is located in an internal position or at the N-terminal in scaffoldins. These CBMs bind strongly to crystalline cellulose. The binding of the *C. cellulolyticum* 3a-CBM onto different crystalline celluloses has been studied. The Kd's were found to range from $4x10^{-7}$ to $3.5x10^{-8}$ M (Pagès *et al.*, 1997b). The 3-D structure of the CBMs from *C. thermocellum* and *C. cellulolyticum* were determined by X-ray crystallography (Tormo *et al.*, 1996; Shimon *et al.*, 2000). Both revealed a nine-stranded jellyroll topology, a well-conserved calcium-binding site, a putative cellulose binding surface and a conserved shallow groove of unknown function. Only minor differences were observed between these two structures. 3a-CBM, which is only found in scaffoldins, anchors the cellulosomes onto the crystalline cellulose. This module can be produced separately without any decrease of its activity towards cellulose. By studying chimeric cellulases harbouring a 3a-CBM fused to the catalytic domain of a *C. cellulolyticum* cellulase, it was shown that this CBM enhanced the accessibilitiy of the enzyme to the substrate by modifying the structure of the bound cellulose and creating new hydrolysable sites (Pagès *et al.*, 1997b; Belaich *et al.*, 1998). The same results were obtained using a *C. thermocellum* cellulase associated with the 3a-CBM of scaffoldin by means of cohesin-dockerin interaction (Garcia-Campayo *et al.*, 1997; Kataeva *et al.*, 1997). Finally, these results were confirmed by demonstrating that 3a-CBM was the most efficient in promoting the cellulose hydrolysis when compared to CBMs borne by *Trichoderma reesei* or *Cellulomonas fimi* cellulases (Carrard *et al.*, 2000).

2.1.3. X Modules

The third kind of module which is found in all the scaffoldins, is called the X-module, thus named because its function is unknown. In *C. thermocellum* and *A. cellulolyticus* only one X module is present and is located just before the last domain of the protein. The scaffoldins of four other mesophylic *Clostridia* contain one or more X2 modules. All the scaffoldins from these *Clostridia* contain an X2 module located just behind the N-terminal CBM. The *C. cellulolyticum* scaffoldin contains an extra X2 module located just before the last cohesin and the *C. cellulovorans* scaffoldin contains four X2 domains (Figure 1). In CipA from *C. acetobutylicum*, two X2 domains are located between the CBM and the cohesin 1, and one X2 domain is inserted between each of the following cohesins. The NMR structure of the first X2 domain from *C. cellulolyticum* scaffoldin showed an immunoglobulin-like fold with two β-sheets packed against each other (Mosbah *et al.*, 2000). The X2 module is smaller than the 3a-CBM and cohesin modules.

In two cases (*C. thermocellum* and *A. cellulolyticus*) a dockerin module is located at the C-terminus of scaffoldin. These dockerins (type II) have no affinity for the type I cohesins of scaffoldin and recognize cohesins of class-II located in the ancillary proteins which play a role in the attachment of cellulosomes to the cell surface within these species (see section 3). The position of the CBM in scaffoldin and the presence of a type II dockerin at the C-terminus have been used to classify the scaffoldins (Bayer *et al.*, 2000). Scaffoldins of class I contain an internal CBM and a C-terminal type II dockerin, while scaffoldins of class II contain an N-terminal family-3a CBM and are dockerinless

Acetivibrio cellulolyticus produces a class I scaffoldin (Cip V), although it containins a family 9 glycosyl hydrolase catalytic domain at its N-terminus. To date it is the only scaffoldin found to contain a catalytic domain (Ding *et al.*, 1999).

2.2. Catalytic Units

2.2.1. Cellulases and Cellulose Hydrolysis

Both bacteria and fungi hydrolyze crystalline cellulose is performed by a set of glycosyl hydrolases that act synergistically. These enzymes cleave the β-1,4 glycosidic bonds with either inversion or retention of the of anomeric carbon configuration (Mc Carter *et al.*, 1994; White *et al.*, 1997). In the classification of glycosyl hydrolases that are based on hydrophobic cluster analysis and sequence alignments (Henrissat *et al.*, 1997; Couthino *et al.*, 1999; website: http://afmb.cnrs-mrs.fr/CAZy) 13 families out of 91 contain cellulases. Roughly, three categories of cellulases can be defined, the endo-cellulases which hydrolyze the glycosidic bonds located in the interior of the cellulose chain, the processive endo-cellulases which act initially as endo-cellulases by cleaving in the middle of the chain but then glide along the chain to cleave glycosidic bonds every two glucose units; and the exo-cellulases acting at the extremities. Exo-cellulases were found that were specific for each end (reducing or not reducing) of the cellulose

Table 1. Actual or putative cellulosomal glysoside hydrolases from the best known four clostridia species

Family Number	Bacterial Species			
	C.cellulolyticum	*C.cellulovorans*	*C.thermocellum*	*C.acetobutylicum*
5	**Cel5A** :GH5-Doc, **Cel5D** :GH5-CBM11-Doc, **Cel5N** : GH5-Doc, **Man5K :** Doc-GH5	**EngB** : GH5-Doc, **EngE** : SLH-GH5-X-Doc, **ManA** : Doc-GH5	**Cel5B** : GH5-Doc, **Cel5E** :GH5-Doc-CE2, **Cel5G** : GH5-Doc, **Cel5H** : G26-GH5-CBM11-Doc, **Cel5O** : CBM3-GH5-Doc,	**Cel5B** :GH5-Doc, **Cel5D** :GH5-Doc, **Man5G** : GH5-Doc **Cel5Y** :(SLH)3- GH5-Doc
8	**Cel8C** : GH8-Doc	Not found	**Cel8A** : GH8-Doc	Not found
9	**Cel9E** :CBM4-Ig-GH9-Doc, **Cel9G** :GH9-CBM3c-Doc, **Cel9H** : GH9-CBM3c-Doc, **Cel9J** : GH9-CBM3c-Doc, **Cel9M** : GH9-Doc	**EngH** :GH9-CBM3c-Doc, **EngK** :CBM4-Ig-GH9-Doc, **EngL**: GH9-Doc **EngM** :CBM4-Ig-GH9-Doc, **EngY** : X7-GH9-CBM3c-Doc.	**CbhA** :CBM4-Ig-GH9-X1-X1-CBM3-Doc, **Cel9D** : Ig-GH9-Doc, **Cel9F** : GH9-CBM3c-Doc, **Cel9J** : X7-Ig- GH9-GH44-Doc-X, **Cel9K** : CBM4-Ig-GH9-Doc, **Cel9N** : GH9-CBM3c-Doc, **Cel9Q** : GH9-CBM3c-Doc, **Cel9T** : GH9-Doc	**Cel9C** :GH9-CBM3c-Doc, **Cel9E** :GH9-CBM3c-Doc, **Cel9F** : GH9-Doc, **Cel9H** : GH9-CBM3-Doc, **Cel9X** : CBM4-Ig-GH9-Doc
10	Not found	Not found	**Xyn10C** :CBM22-GH10-Doc, **Xyn10Y** :CBM22-GH10-CBM22-Doc-CE1, **Xyn10Z** : CE1-CBM6-Doc-GH10.	2 genes found
11	Not found	**Xyn11A** : GH11-doc-CE4	**Xyn11A** : GH11-CBM6-Doc-CE4, **Xyn11B** : GH11-CBM6-Doc	Not found
26	Not found	Not found	**CelH** : G26-GH5-CBM11-Doc, **ManA** : G26-Doc, **ManB** : G26-Doc	5 genes found
44	Not found	Not found	**CelJ** : X7-Ig- GH9-GH44-Doc-X,	1 gene found
48	**Cel48F** : GH48-Doc	**ExgS** : GH48-Doc	**CelS** : GH48-Doc	**Cel48A** : GH48-Doc

Name of cellulases in bold.

chain. The 3D structures of the catalytic modules of these three enzymes have been reported. The endocellulases are characterized by a long open cleft on the surface of the protein into which the cellulose chain can fit (Juy *et al.*, 1992; Ducros *et al.*, 1995; Dominguez *et al.*, 1996; Guerin *et al.*, 2002). The exo-cellulases possess a tunnel into which the extremity of the cellulose chain can enter and they act processively (Barr *et al.*, 1996). The structure of the endo-processive enzymes is more complex. It was observed that processivity was obtained in two very different ways. In one way, a family 9 endo-catalytic domain is fused with a family-3c CBM that extends the active site cleft by increasing the number of glucose-binding sub-sites, thus enhancing interaction of the cellulose chain with the enzyme (Sakon *et al.*, 1997; Mandelman *et al.*, 2003). Whereas in family 48 enzymes, the cellulose chain was found to interact with aromatic residues present in a long tunnel formed by several loops followed by an open cleft (Reverbel-Leroy *et al.*, 1997; Parsiegla *et al.*, 1998; 2000; Guimaraes *et al.*, 2002).

2.2.2. Cellulases from Clostridiaceae Species

All the cellulases, which are part of the cellulosomes, contain at least two modules: a catalytic module and a type I dockerin module. The dockerins are composed of a duplicated sequence of about 22 to 25 residues, the first 12 of which are homologous to the well-known EF-hand motif of calcium binding proteins (Chauvaux *et al.*, 1990). In the case of *C. thermocellum* and *C. cellulolyticum*, the interaction of type I cohesin and type I dockerin was strongly species-specific (Pagès *et al.*, 1997a; Mechaly *et al.*, 2000). In both cases the cohesins of one bacterium interact with dockerin borne by enzymes from its own cellulosomes but fail to recognize dockerins of enzymes from the other species. It is thought that four residues are responsible for this specificity. The 3D structure of the dockerin of Cel48S from *C. thermocellum* was obtained from NMR studies and is in accordance with a theoretical prediction derived from sequence analysis. The structure consists of two Ca^{2+} binding loop-helix motifs connected by a linker (Lytle *et al.*, 2001). In species containing the two types of cohesins and dockerins, cohesins of type I only recognize dockerins belonging to the same type and vice versa. In all cases, the Kd of the dockerin-cohesin complex is about 10^{-9} M, corresponding to a value of –52 Kj for the free enthalpy accompanying the reaction of complex formation at 30°C.

Three *Clostridia*, *C. thermocellum, C. cellulolyticum* and *C. cellulovorans,* have been extensively studied from a biochemical point of view, whereas few components of *C. josui* cellulosome have been studied. Their glycosyl hydrolases were found to belong to five families (Table 1): families 5, 8, 9, 44 and 48. Only one GH48 cellulase is produced in all species, while cellulases belonging to families 5 and 9 are present in many copies in the cellulosomes. Although the genome sequence of the three cellulolytic clostridia (*C. thermocellum, C. cellulolyticum, C. cellulovorans)* are not yet available, the on-going genome sequencing of *C. thermocellum* will uncover the number of genes encoding cellulosomal proteins.

Only a few cellulosomal cellulases contain only 2 modules, a catalytic one and a dockerin one. This is the case for enzymes belonging to families 48 and 8 and for some enzymes belonging to families 5, 9 and 44. Most of cellulases contain one or two extra domains belonging to the following list: CBM, immunoglobulin-like domain (Ig), S-layer homology (SLH) domain, X module (of unknown function) and a second catalytic domain. The majority of the family 9 catalytic domains are fused with extra-domains. To date, only four enzymes were found to be composed of a GH9 catalytic domain and a dockerin. One of them was found in each of the well-known *Clostridium species*: *C. cellulolyticum* (Cel9M) (Belaich *et al.*, 2002), *C. thermocellum* (Cel9T) (Kurokawa *et al.*, 2002), *C. cellulovorans* (Cel9L) and *C. acetobutylicum* (Cel9L). Cellulases containing a GH9 fused to a 3c-CBM domain are frequently found in all species. In contrast to 3a-CBM that is found exclusively in the scaffoldins, 3c-CBM binds cellulose with a low affinity. The 3D structures were determined for a fragment of the cellulase E4 from *Thermomonospora fusca* (Sakon *et al.*, 1997) and for CelG from *C. cellulolyticum* (Mandelman *et al.*, 2003). It was shown that 3c-CBM can be considered as a helper CBM (Gal *et al.*, 1997; Irwin *et al.*, 1998) which allows the enzyme to be processive (see above). This conclusion was recently reinforced by comparing the structural and catalytic properties of cellulases Cel9M and Cel9G from *C. cellulolyticum* (Parseglia *et al.*, 2002; Belaich *et al.*, 2002). It was observed that Cel9M is a genuine endo-glucanase, whereas the 3c-CBM of Cel9G confers a processive activity to this cellulase. A second type of domain fusion frequently observed is family-4 CBM-Ig-GH9. The catalytic properties of Cel9K from *C. thermocellum* (Kataeva *et al.*, 1999) and Cel9E from *C. cellulolyticum* (Gaudin *et al.*, 2000), both enzymes harbouring fusion of these three domains, have been reported. Cel9K and Cel9E are exocellulases (cellobiohydrolase) that have a certain capacity for a random mode of attack. The characteristic properties of these proteins seem to be due to the presence of the family-4 CBM; moreover, the deletion of this CBM induces a total inactivation of Cel9E. It is thus clear that the type of modules associated with GH9 catalytic domains have an important impact on the activity profile of such enzymes.

2.2.3. Helper Enzymes

Plant cell walls contain several polymeric compounds (cellulose, hemicellulose, pectin and lignin) which are tightly packed together. The cellulolytic bacteria synthesize a set of enzymes that degrade the accompanying polymers of cellulose. Hemicellulases, esterases and lyases, by their action, allow the cellulases to reach the cellulose. All these enzymes, which facilitate access to the cellulose substrate can be defined as helper enzymes. Many helper enzymes have been found to be present in cellulosomes (Mohand-Oussaid *et al.*, 1999; Kosugi *et al.*, 2001).

Hemicellulases can be divided into two categories: the xylanases and mannanases that hydrolyze the main chain backbone of the hemicelluloses and the enzymes that hydrolyze the side chain substituents. The xylanases are the best-known hemicellulases. These glysosyl hydrolases all belong to families 10 and 11. Among the five xylanases found in *C. thermocellum,* three of them have been classified in family 10 and two in family 11. Only one xylanase belonging to family 11 has been reported in *C. cellulovorans* (Kosugi *et al.*, 2001; 2002b). Although a β-xylosidase was purified from this *Clostridium* eight years ago (Saxena *et al.*, 1995), no xylanase has yet been isolated from *C. cellulolyticum*. Nevertheless this species grows very well on xylan suggesting that the bacterium probably produces xylanases which could either function independently or as part of a cellulosome. SDS-PAGE and zymogram analyses of the cellulosomes purified from cultures grown with xylan showed the existence of two proteins harbouring xylanase activity with an apparent molecular masses of 54 and 31 kDa (Mohand-Oussaid *et al.*, 1999). Family 5 mannanases have been found in *C. cellulolyticum* (Man5K), *C. cellulovorans* (Man5A) (Tamaru *et al.*, 2000a), and *C. acetobutylicum* (Man5A). In all these cases, the enzymes only contain a GH5 and a dockerin module located at the N-terminus of the protein. In the case of *C. thermocellum* three GH26 mannanases have been reported (Kurokawa *et al.*, 2001).

The second category of hemicellulases hydrolyzes the side chain susbtituents and contains one carbohydrate esterase catalytic module belonging to families CE1, CE2 or CE4 (CAZy website). Both families contain acetyl xylan esterases, whereas cynnamoyl esterases and feruloyl esterases belong to family 1. Such domains have been found in Xyn10Y (Xyn10B), Xyn10Z and Xyn11A produced by *C. thermocellum* and Xyn11A produced by *C. cellulovorans.* A cellulosomal α-galactosidase, Aga27A, which contributes to the degradation of galactomannan in the plant cell wall, has been found in the cellulosomes of *C. josui* (Jindou *et al.*, 2001). In addition to their catalytic module, the helper enzymes contain several types of CBMs, generaly belonging to families 6 (which binds xylane and amorphous cellulose), 11 and 22.

Pectins are the most complex plant cell wall polysaccharides. Only two cellulosomal polysaccharide lyases (PL) have been described to date. A pectate lyase (PelA) belonging to family 4 produced by *C. cellulovorans* was shown to degrade polygalacturonic acid (Tamaru and Doi 2001) and the rhamnogalacturonase Rgl11Y from *C. cellulolyticum* cleaves glycosidic bonds in the backbone of rhamnogalaturonan I via a β-elimination mechanism (Pagès *et al.*, 2003).

Finally, the cellulosomes from *C. thermocellum* contain a lichenase belonging to family 16 (Fuchs *et al.*, 2003) and a chitinase belonging to family 18 (Zverlov *et al.*, 2002).

2.2.4. Bifunctional Enzymes

In some cases, found in *C. thermocellum* and in *C. cellulovorans,* a GH domain is fused to another catalytic domain giving rise to a bifunctional enzyme. *C. thermocellum* CelH is composed of a GH5 cellulase and a mannanase (GH26), CelJ contains a GH9 and a GH44, and CelE bears a GH5 domain fused to a carbohydrate esterase (CE2). The three xylanases Xyn10Y (Prates *et al.*, 2001), Xyn10Z (Blum *et al.*, 2000), Xyn11A from *C. thermocellum* and Xyn11A produced by *C. cellulovorans* are bifunctional enzymes that contain a xylanase module and a carbohydrate esterase module (GH10 with CE1 and GH11 with CE4). In the last five cases, the dockerin is located between the xylanase module and the carbohydrate esterase module (see Table 1).

The association of two catalytic domains generating bifunctional enzymes is also frequently observed in cellulolytic bacteria which do not form cellulosomes such as *Anaerocellum thermophylum, Caldicellulosiruptor* and *Cellulomonas fimi.* Such a device, namely a pair of catalytic modules fused in one protein, may facilitate synergistic actions on complex substrates such as plant cell wall. In cellulosomes, the bifunctional enzymes might increase the number of catalytic domains contained therein and thus enhancing the catalytic power of these cellulolytic machines.

2.3. Non Cellulosomal Enzymes

It is noteworthy that *Clostridia* that produce cellulosomes also produce non-cellulosomal enzymes which are devoid of dockerin. However the catalytic domains of these enzymes are similar to those of cellulosomal enzymes. *Clostridium thermocellum,* for example, produces a family 5 endoglucanase (Cel5C), a family 9 cellulase (Cel9I), a GH10 xylanase (Xyn10X) and a lichenase (LicA). Both Xyn10X and LicA contain three SLH modules indicating that these enzymes are bound to the cell surface. *C. josui* produces a GH10 xylanase which contains also two SLH modules. *Clostridium cellulovorans* synthesizes three endoglucanases, two belonging to family 5 (EngA and EngF) and one belonging to family 9 (EngD), one arabinofuranosidase (ArfA) and one beta-galactosidase (BgaA). It was shown that ArfA and BgaA act in synergy with the cellulosomal bifunctionnal XynA for degradation of plant cell wall polymers containing galactose residues (Kosugi *et al.*, 2002c). Therefore, from these two examples, it appears that the efficient degradation of plant cell wall could be the result of sophisticated synergistic actions between cellulosomes, non-cellulosomal enzymes bound to the cell surface and free enzymes.

One of the major questions raised by the existence of the cellulosome is whether the cellulosome structure is a parietal docking device and/or if this complex machinery provides a catalytic advantage. Some recent findings are in favour of the second hypothesis. Using the specificity of the interaction of cohesins and dockerins belonging to *C. thermocellum* and *C. cellulolyticum* it was possible to engineer *in vitro* chimeric mini-cellulosomes containing, stoichiometrically, two cellulases and one family-3a CBM located in appropriate positions (Fierobe *et al.*, 2001). 75 cellulosome chimeras, each containing a pair of different cellulases belonging to *C. cellulolyticum*, were designed and assayed on different types of cellulose substrates. It was observed that the synergy between cellulases incorporated into minicellulomes was significantly enhanced when compared to the mixture of the same but free cellulase pair (Fierobe *et al.*, 2002). The same kind of results were also obtained with mini-cellulosomes engineered with *C. cellulovorans* components (Murashima *et al.*, 2003).

2.4. Diversity of Cellulosomes

SDS-PAGE analysis of cellulosomes purified from cultures grown in the presence of cellulose, cellobiose or xylan showed diverse gel profiles. Cellulolytic *Clostridia* are able to adapt the composition of their cellulosomes according to the available substrate. However, in the case of a culture that was grown on one defined substrate (Avicel), various complexes could be present. Indeed the number of dockerin-harbouring enzymes is always higher (at least twice) than the number of cohesin domains on the scaffoldin. Only three or four proteins are visualized as major components in every cellulosomal preparation. As individual complexes have never been isolated, it is impossible to predict, at the present time, if each cellulosome harbours one copy of the major enzymes. Early experiments were performed using *C. papyrosolvens* cellulosomal preparations (Pohlschroder *et al.*, 1994). The cellulosomes of *C. papyrosolvens* were fractionated by exchange chromatography into at least seven high-molecular-weight complex fractions which displayed different enzymatic properties. These complexes were investigated by transmission electron microscopy and the ultrastructural observations showed a distinct morphology for each biochemical fractions (Pohlschroder *et al.*, 1995). These experiments, performed eight years ago, could not be pursued since no molecular tools from *C. papyrosolvens* were available. This approach, combining biochemical separation of cellulosomes and electron microscopy studies of the different fractions obtained could now be undertaken using cellulosome preparations of the well-studied *C. thermocellum, C. cellulovorans* or *C. cellulolyticum*. In this way, preliminary studies on structural organization of intact bacterial cellulosomes using electron microscopy have recently been published (Madour *et al.*, 2003).

3. Attachment of the Cellulosomes on the Cell Surface

Visualization of polycellulosomal protuberances on the cell surface of *C. thermocellum* was reported about twenty years ago using immunochemical labelling and electron microscopy (TEM and SEM). These protuberances were not detected in an adherence defective mutant, AD2. Upon contact with cellulose, the protuberances protract to form a corridor so that the cellulosomes become attached to the cell surface and to the cellulosic substrate (Bayer *et al.*, 1985; Bayer *et al.*, 1986). Such protuberances were visualized later on other cellulolytic bacteria, such as *A. cellulolyticus*, *C. cellulovorans* and *C. cellobioparum* (Lamed *et al.*, 1987). The molecular basis of the cellulosome attachment to the cell surface of *C. thermocellum* was discovered in the last decade. As mentioned above the scaffoldin CipA contains a C-terminal type II dockerin module which does not recognize its own type I cohesin domains. The discovery of three anchoring proteins, SdbA, Orf2p and OlpB, each containing one SLH module and one or more type II cohesins led to the elucidation of the the cellulosome attachment mechanism (Lemaire *et al.*, 1995; Leibovitz and Béguin, 1996; Leibovitz *et al.*, 1997). The SLH module anchors the proteins on the cell surface and the type II cohesins (one for SdbA, two for Orf2p and four for OlpB) bind selectively to the type II dockerin of CipA, anchoring the entire cellulosome to the cell surface. The regulation of the genes coding for the anchoring proteins will be examined in the next section.

The ScaA scaffoldin of *A. cellulolyticus* (previously called CipV) also bears a C-terminal dockerin which does not recognize its own cohesin domains. The anchoring protein ScaB is composed of four cohesin domains (Noach *et al.*, 2003) and one dockerin domain whereas ScaC is composed of three cohesin domains and one SLH domain. A schematic representation of the anchorage has recently been proposed (Xu *et al.*, 2003). In this model, ScaB plays the role of an adaptor protein between ScaA and ScaC. The enzymatic units are bound onto ScaA *via* the specific type I cohesin-dockerin interaction. The C-terminal dockerin of ScaA binds onto one cohesin domain of ScaB which binds, *via* its own dockerin domain, onto one cohesin domain of ScaC which is anchored onto the cell surface *via* its SLH domain. Some additional experiments are still necessary to corroborate

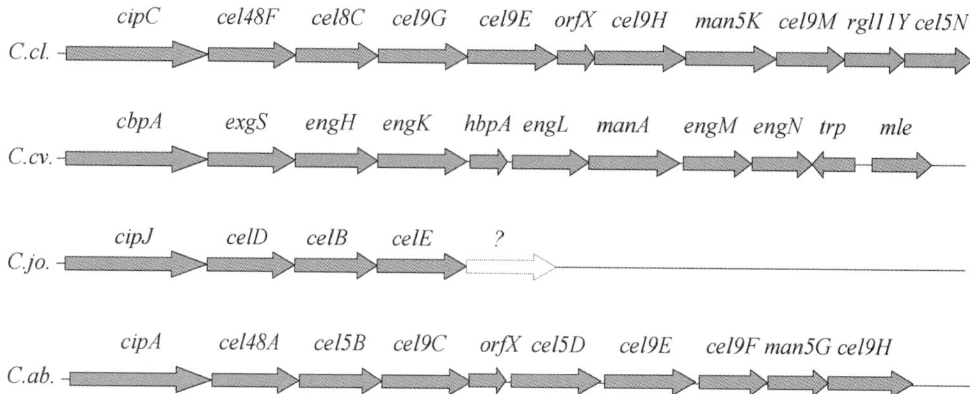

Figure 2. Organization of genes in "*cel* clusters" found in mesophilic *Clostridia. C.cl, Clostridium cellulolyticum*; *C.cv, Clostridium cellulovorans*; *C.jo, Clostridium josui*; *C.ab. Clostridium acetobutylicum*. For *C. cellulolyticum* and *C. acetobutylicum*, the genes harbour the number of the corresponding glycosyl hydrolases. For *C. cellulovorans* and *C. josui*, the genes encode the glycosyl hydrolases (in brackets). *C.cv*: *exgS* (GH48), *engH* (GH9), *engK* (GH9), *engL* (GH9), *ManA* (GH5), *engM* (GH9), *engN* (GH5). *C.jo*: *celD* (GH48), *celB* (GH8), *celE* (GH9). *orfXp*, *hbpA* are coding for small non-enzymatic proteins harbouring a type I cohesin module.

this hypothesis, but if this model is confirmed, twelve different cellulosomes could be anchored on each ScaC protein.

So far, the mechanism by which the cellulosome(s) of other *Clostridia* (*C. cellulovorans*, *C. josui*, *C. acetobutylicum* and *C. cellulolyticum*) are anchored to the cell surface has not yet been established; their anchoring is accomplished using a different molecular mechanism, since no C-terminal dockerin domain has been found in their scaffoldins. A hypothesis has been proposed in the case of the attachment of the cellulosomes of *C. cellulovorans* cell surface (Tamaru *et al.*, 1998). The authors found slight similarities between the four X2 modules of CbpA and S-layers proteins from *Mycoplasma hyorhinis* and *Plasmodium reichenowi* and suggested that these modules could be involved in the attachment of the cellulosome to the cell surface. This hypothesis was however not supported by subsequent experiments. Another hypothesis was proposed later (Tamaru *et al.*, 1999) that the cellulosome of this *Clostridium* contains the unusual endoglucanase EngE which is composed of three highly conserved repeats of 163, 126, and 163 amino acids, a GH5 catalytic domain and a dockerin domain. The three N-terminal repeats showed moderate identities with known SLH domains. EngE binds to CbpA through the specific dockerin-cohesin interaction and could anchor CbpA (and consequently the cellulosome) on the cell surface through its SLH modules. This model was investigated recently (Kosugi *et al.*, 2002a) using recombinant EngE produced with and without the SLH modules. Only the first form of the recombinant protein was able to bind cell wall preparations of *C. cellulovorans*. If this model is accurate, it implies that each cellulosome contains at least one EngE subunit. This situation is plausible since EngE is one of the three major subunits of the *C. cellulovorans* cellulosome. This proposal is not exclusive; *C. cellulovorans* may possess other systems, not known to date, for anchoring the cellulosomes.

Sequencing of *C. acetobutylicum* chromosome DNA revealed the presence of a gene *CAC03469* encoding a protein called Cel5Y (Sabathe *et al.*, 2002) that was very similar to EngE (about 58% of identity) and possessed similar domains, *ie* three conserved repeats, a GH5 catalytic domain and a dockerin domain. The three repeats have significant similarities with the three SLH domains of EngE. Further studies of this protein could provide more information concerning its function. Until now, such proteins have not been identified either in *C. cellulolyticum* or in *C. josui*. The SLH-GH5 containing-proteins identified to date in these *Clostridia* (Cel5I of *C. cellulolyticum,* C. Tardif, unpublished, and Cel5A in *C. josui,* Fujino *et al.*, 1996, Genbank n° D85526) did not contain a dockerin domain and, evidently, are not components of the cellulosomes.

4. Genetic Organization of *cel* Genes

4.1. Identification of Gene Clusters

The organization of the *cel* genes differs greatly depending on the *Clostridium* strain. In the thermophilic *C. thermocellum* most of the genes coding for the known cellulosomal components are generally scattered over the chromosome. Few gene clusters containing only two or three genes have been identified. The largest known cluster (about 14 kb long) contains *cipA*, *olpB*, *orf2p* and *olpA,* which code for the cellulosome scaffoldin, the proteins OlpB and Orf2p involved in the cell surface attachment of the cellulosome and the cell surface protein OlpA harbouring a type I cohesin domain, respectively (Fujino *et al.*, 1993a).

Conversely, the main feature of the genetic organization in mesophilic *Clostridia* C. *cellulolyticum*, *C. cellulovorans*, *C. josui* and *C. acetobutylicum* is that numerous genes are clustered into one locus. This was first found in *C. cellulolyticum* and then in *C. josui* and in *C. cellulovorans*. Recently, the sequencing of the *C. acetobutylicum* genome also revealed the presence of a similar cluster. The large (26 kb) " *cel* cluster" of *C. cellulolyticum* (Belaich *et al.*, 1993; 1997; 2002; Pagès *et al.*, 2003; Maamar, 2004) encodes twelve proteins. In *C. cellulovorans* nine genes reside in a 22-kb locus (Tamaru *et al.*, 2000b). The cluster of *C. acetobutylicum* contains ten genes (Nolling *et al.*, 2001; Sabathé *et al.*, 2002) and that of *C. josui,* only partially sequenced at the present time, contains at least four genes (Fujino *et al.*, 1993b; Kakiuchi *et al.*, 1998). A comparison of these clusters is shown in Figure 2. The first gene in each cluster is the gene coding for the scaffoldin, followed by the gene coding for the processive GH48 endocellulase, onc of the major proteins of the cellulosomes. All clusters contain a large majority of genes encoding cellulases which belong to families 5 and 9. There is significant similarity in the positioning of the genes in all clusters. For example, a gene (*orfX* in *C. cellulolyticum* and *C. acetobutylicum* and *hbpA* in *C. cellulovorans*) coding for a non-hydrolytic protein was always found downstream of a gene coding for a GH9. These genes code for a small protein containing a type I cohesin. OrfX is attached to the membrane of *C. cellulolyticum* but its function remains unknown. The gene coding for a GH5 mannanase (*manA* or *manG* or *manK*) is located between two genes coding for GH9 cellulases. The main differences between these clusters are a gene encoding a GH8 found only in *C. cellulolyticum* and *C. josui* clusters and *CAC0915* coding for a putative GH44 in *C. acetobutylicum,* and *rglY* coding for a rhamnogalacturonan lyase only found in *C. cellulolyticum*. The similarities in the organization of the gene clusters suggest horizontal genetic transfers from a common ancestor.

Only a few data are available concerning the genetic organization of the *cel* genes of *A. cellulolyticus*. A scaffoldin gene cluster was recently identified (Xu *et al.*, 2003). This cluster of about 12.5 kbp, contains the *scaA* gene, coding for scaffoldin, followed by *scaB* and *scaC* encoding the anchoring proteins ScaB and ScaC described in the above section. The beginning of a new *orf* was identified downstream of *scaC* but it is not known yet whether this gene is a part of the gene cluster. In the first place, it seems that *A. cellulolyticus* adopted the organization that is similar to that described for the thermophilic *C. thermocellum*, however, the presence of a large cluster of *cel* genes in *A. cellulolyticus*, as was described in the mesophilic *Clostrida*, cannot be completely excluded. The cohesin 3 domain of ScaC primraily recognizes, in the cellulose-adsorbed supernatant fraction, a 170 kDa protein, which could be a new scaffoldin. The corresponding gene has not been identified to date and consequently its genetic origin remains unknown.

4.2. Localization of *cel* Genes in the Chromosome

Where do the clusters map on the bacterial chromosome? In the sequenced genome of *C. acetobutylicum*, *CAC 0909* encoding a methyl accepting chemotaxis protein was identified upstream of the cluster and *CAC* 0920, coding for a metal dependent amido hydrolase, was located downstream of the cluster. Walking on the chromosome of *C. cellulolyticum* identified three orfs at the 5′ end of *cipC*. The nearest, orf3, is located 1076 bp upstream of the translational start site of the *cipC* gene and codes for a protein similar to a hypothetical protein of *C. perfringens*. Downstream of *celN*, the last gene of the cluster, two orfs were identified: *orf4* (966 bp long), which is transcribed in the opposite direction, codes for a transcriptional regulator belonging to the LysR family and *orf 5* (1098 b long) transcribed in the same direction as *celN*, codes for a protein similar to Cap 64 from *Pleurotus oestreatus* (Maamar, 2003b). The 5′-end of the gene cluster of *C. cellulovorans* is flanked at its 5′ end by the gene cluster *nifV-orf1-sigX-regA* (Liu *et al.*, 1998; Tamaru *et al.*, 2000). The last gene of the cluster, *engN*, is a truncated gene; it is disrupted by *trn,* coding for a transposase, followed by *mle* that codes for a malate permease (Tamaru *et al.*, 2000). In conclusion, it seems that the three gene clusters from *C. acetobutylicum*, *C. cellulolyticum*, and *C. cellulovorans* are located in very different genomic contexts.

Some of genes encoding the cellulosomal protein reside outside of the cluster; they are *cel5A* (Faure *et al.*, 1989), *cel5D* (Shima *et al.*, 1991), and *cel44O* (Maamar, 2003b) from *C. cellulolyticum* and *engE*, *xynA* and *pelA* coding for *C. cellulovorans* EngE, which seems to play a major role in the attachment of the cellulosome to the cell surface, and for two hemicellulases, XynA and PelA, respectively. In *C. acetobutylicum*, two cellulosomal genes, *celX* and *celY*, have also been identified. In *C. josui*, *aga27A*, coding for a cellulosomal α-galactosidase, was located upstream of *celA*, which codes for a non-cellulosomal family 5 cellulase. The direction of transcription of *aga27A* is opposite to that of *cipA*. A gene encoding the dockerin-GH9B protein of *A. cellulolyticus* has recently been identified (Xu *et al.*, 2003, sequence unpublished). In addition to the cellulosomal genes, numerous genes coding for dokerinless cellulases and hemicellulases have been identified in these *Clostridia*.

5. Regulation of *cel* Genes

5.1. Variations in Cellulosome Composition Depending on the Growth Substrate

Only a few molecular data concerning the regulation of cellulosomal genes are available. Several experiments showed that the composition of the cellulosomes is dependant on the nature of the substrate and on the growth conditions. For example, a comparison of *C. cellulovorans* cellulose, cellobiose and xylan-grown cultures showed that two bands, corresponding to the major xylanases were significantly increased in the presence of xylan. Similar results were obtained with *C. cellulolyticum*. When subjected to SDS-PAGE, the cellulosomes derived from cellulose-grown cells and xylan-grown cells exhibited very different profiles (Mohand-oussaid *et al.*, 1999). In particular, the 94-kDa band corresponding to Cel9E, which is one of the major components of the cellulosomes of cellulose-grown cells, decreased significantly when cells were grown on xylan whereas a 31-kDa band, corresponding to a xylanase, increased dramatically.

In *C. cellulovorans*, the three major components of the cellulosome were expressed at the same level when the bacterium was grown in the presence of cellulose or cellobiose. Different results were obtained with *C. thermocellum*. This *Clostridium* synthesizes cellulosomes when the substrate is cellulose or cellobiose; however, depending on the substrate, the proportion of the compounds varied. In particular, it has been shown that the synthesis of Cel48S, which is a major protein of the cellulosome and is essential for crystalline cellulose hydrolysis, is affected in cellobiose-grown cells. The decrease in the quantity of Cel48S, which is observed when *C. thermocellum* is grown on cellobiose medium, was confirmed by the recent and elegant experiments performed by Dror *et al.* (Dror *et al.*, 2003a). Using continuous cultures under cellobiose limiting conditions (by measuring the transcript level of *cel48S* in batch cultures and in chemostat), the authors observed a connection between the growth rate and the expression of *cel48S*. During exponential growth (at a growth rate of 0.35 h^{-1}) the level of mRNA was evaluated at 28 transcripts per cell. In chemostat experiments, a low dilution rate was accompanied by an increase in the level of *cel48S* transcript until a critical rate of 0.21 h^{-1} (170 transcripts per cell) was reached. To determine if this connection was due to an effect of the variation of cellobiose concentration at each steady state of the continuous cultures or to the dilution rate *per se*, chemostat experiments with limited nitrogen were performed confirming that the expression of *cel48S* is modulated by growth rate both under conditions of cellobiose and of nitrogen limitation. Such conditions exist in biotopes containing cellulose which dictates slow rates of growth (the maximum growth rate in cellulose culture conditions is 0.23 h^{-1}).

Another set of experiments was recently published (Dror *et al.*, 2003b) where the mRNA level of *cipA* and of *olpB*, *orf2* and *sdbA* was determined quantitatively in batch cultures and in continuous cultures under cellobiose or nitrogen limitation. The mRNA level of *cipA*, *olpB* and *orf2* varied from 40 to 60 transcripts per cell in chemostat experiments at a dilution rate of 0.04 h^{-1} under conditions of cellobiose limitation to 2 to 10 transcripts per cell in batch experiments during exponential growth phase (μ = 0.35 h^{-1}). On the contrary, the mRNA level of *sdbA*, which is not located in the cluster, was not growth rate-dependent and was evaluated at 3 transcripts per cell. Simultaneous transcription of *cbpA* (coding for the scaffoldin of *C. cellulovorans* cellulosomes) and *engE* (coding for the "cell anchorage protein" of the cellulosomes) has been observed (Han *et al.*, 2003), suggesting that the expression of these two genes is coordinately regulated.

5.2. Identification of Transcription Units

Predictions of monocistronic or polycistronic transcription units have been made by analysing the sequences and the stabilities of the putative intergenic hairpins. On the other hand, few data were published concerning the experimental determination of the transcript size. It seems that in *C. thermocellum* the genes are generally transcribed as monocistronic units. Northern data showed that *cel8C* and *cel9G* from *C. cellulolyticum* appeared to be present in a polycistronic transcriptional unit which probably contains another gene, since two mRNA of 5-kb and 6-kb have been detected (Bagnara-Tardif *et al.*, 1992). Two major mRNA species were detected using a *cipC* antisense-RNA probe on a Northern blot. The first one of 4.9-kb was large enough to carry the *cipC* gene. The second one, of 7.5-kb, was long enough to carry *cipC* and *cel48F*. These two genes appear to be transcriptionally coupled (Maamar *et al.*, 2004). The study of the other transcripts encoded by genes within the cluster is ongoing. The *cel5A* gene, which resides outside the cluster, is probably transcribed as a monocistronic unit as sequences similar to factor-independent transcription terminators are located both 5′ and 3′ of the gene; more, the first *orf* identified upstream *cel5A* was located 658-bp upstream of the translation start codon of the gene and no *orf* was identified by sequencing 1000-bp downstream of the terminator. Nothing can be proposed to date for the open reading frame of *cel5D* because only the DNA region at the 5′ end of *cel5D* has been sequenced. One *orf* 903-bp long has been identified, located at 250-bp upstream of the translation initiation site of *cel5D* and transcribed on the opposite strand. This gene codes for a protein similar to a transcriptional regulator of the AraC family (Maamar, 2003b).

Recently, a study on the transcription of *C. cellulovorans* cellulosomal cellulase and hemicellulase genes was carried out (Han *et al.*, 2003). Northern hybridization analyses showed two polycistronic transcripts of 8- and 12-kb for *cbpA* and a 12-kb transcript for *engH*. It was demonstrated that the smaller 8-kb mRNA from *cbpA-engS* was not a processed product of the larger 12-kb mRNA. A putative terminator located in the intergenic region between *exgS* and *engH* may be bypassed by an antitermination mechanism allowing synthesis of the 12-kb transcript encompassing *cbpA-exgS-engH*. RT-PCR experiments indicated that the *engK* and *hbpA* genes were linked and constituted an operon with a promoter in front of the *engK* gene, with an internal promoter proximal to the *hbpA* gene. The two other genes of the cluster, *engM* and *manA* are transcribed as monocistronic units. The genes *engE* and *xynA*, not included in the cluster, are monocistronic. The 4-kb transcript detected for *pelA* is larger than the size of the gene (2.7-kb), suggesting that *pelA* could be a part of an operon. The major cellulase genes, such as *cbpA-exgS*, and *engE* are constitutively expressed under all conditions of growth substrate while *pelA* is only expressed in the presence of pectin.

5.3. Identification of Promoters

No direct experiments using reporter genes have been performed to locate the promoters of the *cel* genes. Potential promoters were identified by analysis of the DNA sequences located upstream of the transcriptional start sites determined for various *cel* genes. The first experiment was carried out to map the mRNA encoding the endoglucanase A from *C. thermocellum* (Béguin *et al.*, 1986). Two potential promoters similar to those recognized by *B. subtilis* σ^D and *B. subtilis* σ^A were identified upstream of *celA*. A study of the transcription of the *celD* and *celF* genes of the same bacterium then revealed two potential promoters similar to those recognized by *B. subtilis* σ^A and *B. subtilis* σ^D upstream of *celD* while DNA sequences upstream of the *celF* start site did not reveal any homology with known *B. subtilis* or *E. coli* promoters (Mishra *et al.*, 1991). In a recent study (Dror *et al.*, 2003b) two promoter sequences were identified in the region upstream of *cipA*. The majority of the *cipA* mRNA could be transcribed from a promoter resembling the σ^L-type promoter of *B. subtilis*. An additional transcript was observed in continuous culture initiated at a site similar to the *B. subtilis* σ^A-type promoter. Transcriptional analysis of *celS* showed two putative promoters homologous to σ^A and σ^B sequence promoters of *B. subtilis* although the space between the –10 and –35 regions was not the typical 17 bp.

Alignment of the putative promoter regions of *cbpA*, *engE*, *manA* and *hbpA* of *C. cellulovorans* revealed a consensus sequence similar to the *B. subtilis* σ^A promoter (Han *et al.*, 2003). The distance between the –35 and –10 regions was exactly 17 bp. The promoter regions of *cel* genes of the other cellulolytic *Clostridia* are not known.

6. In *Clostridio* Studies

Considerable information concerning various protein subunits of the cellulosome in terms of structure, individual activities, subunit interaction and synergistic activity within minicellulosomes were provided by *in vitro* studies. Nevertheless, the contribution of each component to the process of cellulolysis *in vivo* cannot be assessed from these studies. It was necessary to develop gene transfer techniques to modify the composition of cellulosomes. Attempts to efficiently transfer genetic material to *C. thermocellum* were unsuccessful; results were only obtained with *C. acetobutylicum* and with *C. cellulolyticum* using electrotransformation technique and/or conjugation.

6.1. Transfer by Conjugation in *C. cellulolyticum*

The pCTC1 plasmid containing the replication machinery of the pAMβ1 and the oriT from plasmid RK2 of *Enterococcus faecalis* was transferred at a low frequency to *C. cellulolyticum* (Jennert, 2000). Tn*1545* was also transferred from *E. faecalis* BM4110, which contains multiple copies of the transposon, to the *Clostridium* strain. However, reproducibility of experiments was poor and this approach cannot be used routinely to transfer genetic material in *C. cellulolyticum*.

6.2. Electrotransformation

Electrotransformation is an efficient method for obtaining gene transfer, but only once the incoming DNA has been protected from the restriction barrier of the host. The restriction endonuclease *Cac*8241 has been shown to constitute a barrier to transformation of *C. acetobutylicum* ATCC 824. Methylated plasmid can be obtained *in vivo* in a strain of *E. coli* harbouring the gene coding for the methyltransferase *B. subtilis* Φ3TI phage (Mermelstein and Papoutsakis, 1993). An active restriction system, responsible for degrading incoming DNA has been discovered in *C. cellulolyticum*. Incubation of the pJIR418 plasmid with a *C. cellulolyticum* cell extract resulted in a pattern of DNA fragments, the number and the size of which indicated that this extract contained an isoschizomer of *Msp*I that has been named *Cce*I. For transfer in *C. cellulolyticum*, the recognition sequence of *Cce*I may be modified *in vitro* by the commercial methylase M. *Msp*I (Jennert *et al.*, 2000). Optimized electrotransformation procedure has been obtained for *C. acetobutylicum* (Nakotte *et al.*, 1998) and for *C. cellulolyticum* (Tardif *et al.*, 2001). Nevertheless, the transformation efficiencies for the various plasmids used were rather low (especially for *C. cellulolyticum,* when transformants were selected on solid medium). The transformation efficiencies were reproducible and the plasmids were maintained in the host when the appropriate selection factor was added to the medium (generally erythromycin or thiamphenicol). This approach is now used routinely to transform the two *Clostridium* species.

6.3. Production of a Minicellulosome in *C. acetobutylicum*

A minicellulosome is a complex obtained using a truncated form of the scaffoldin as the skeleton protein. To produce a minicellulosome, a DNA fragment from *C. acetobutylicum* chromosome, coding for a miniscaffoldin (miniCipA) composed of the CBD, the first two X2 domains, the cohesin 1, the third X2 domain, the cohesin 2 and the fourth X2 domain (CBD-HD1-HD2-Coh1-HD3-Coh2-HD4) was introduced into an expression vector (pSOS95) under the control of the constitutive thiolase (*thl*) promoter. After electrotransformation of *C. acetobutylicum*, a complex with an apparent molecular weight of 250 kDa was detected in addition to the large native complex (665 kDa). This minicomplex consisted of three major proteins and two of them were identified as the miniCipA and Cel48A. Unfortunately, the strain harbouring this minicellulosome was still unable to grow on crystalline cellulose. It was nevertheless observed that, when miniCipA was over-expressed, sedimentation of cellulose after mixing occurred at a slower rate, probably due to the effect of the increased amounts of 3a-CBD bound to cellulose (Sabathe *et al.*, 2003).

6.4. Modification of the Ratio of Cel48F in *C. cellulolyticum* by Using the Antisense RNA Strategy

Several genes of the cellulosome are clustered within a broad 26-kb cluster and some of them are probably co-transcribed. The use of antisense technology could represent an effective means of reducing gene expression without any strong effect on adjacent genes. In order to regulate *cel48F* encoding the major cellulase of the cellulosomes, the strain ATCC 35319 (pSOS*asrF*) was developed. This strain synthesized a 469-base antisense RNA (asRNA) covering the translation initiation site and the beginning of the coding region of *cel48F* from a plasmid construct bearing the *thl* promoter. A control strain, ATCC 35319 (pSOSzero), containing the vector plasmid, was also constructed (Perret *et al.*, 2004).

The main results obtained after studying the growth of the control strain and the asRNA-producing strain are enumerated as follows. The vector plasmid affects the growth rate of *C. cellulolyticum* and affects the production of cellulosome. A similar effect has already been observed with other *Clostridia* transformed with a vector plasmid (Tumala *et al.*, 2003). The growth of the asRNA producing strain in cellulose-containing medium was not severely impaired when compared to the control strain, in spite of a 3- to 4-fold decrease of the Cel48F content in cellulosomes. This means that the other processive cellulases present in the cellulosome can compensate for the reduced amount of Cel48F. Nevertheless, a 30% decrease of the specific activity of the cellulolytic system from the pSOS*asrF*-containing strain was found compared to the control strain. These results highlight the important role of Cel48F in the cellulosome activity. The production of CipC was affected by asRNAs targeted against *cel48F*, depending on the culture samples. This effect is probably due to the fact that *cipC* and *cel48F* might be partly translated from a common mRNA (see the following section). In the asRNA-producing strain, two non-cellulosomal proteins of unknown function, P105 and P98, were found to be the major components of the cellulolytic system. These proteins, which were present at lower amounts in the wild type system, were also encountered as major components of the cellulolytic system of a mutant greatly affected in cellulolysis (described below).

6.5. Study of Spontaneous Mutants of *C. cellulolyticum* Affected in Cellulolysis

Spontaneous mutant strains of *C. cellulolyticum* affected in cellulolysis have been isolated. It was found that these mutants were only affected in cellulose hydrolysis since the mutant and wild type strains showed the same growth parameters when the carbon source was cellobiose. Analysis of several mutant colonies showed that two insertion sequences, ISCce*1* and ISCce*2*, were inserted in the *cipC* gene. *cipC* was disrupted in the region encoding the cohesin number 7 (Maamar *et al.*, 2003a) by ISCce*1* alone in the first type of mutant, and by a ISC*ce1* derivative that was interrupted by ISC*ce2* in the second type of mutant. ISCce*1* is also present as at least 20 copies in the genome of the wild-type strain. Both mutant strains shared the same phenotype and displayed a very similar SDS-PAGE pattern from the cellulose-absorbed protein fraction (Maamar *et al.*, 2004). In the mutants, only small amounts of a truncated form of CipC (which exhibits a size of 120 kDa) were found in the fraction adsorbed on the cellulosic substrate. Molecular filtration analysis of this fraction revealed the absence of typical WT cellulosome complexes. Interestingly the analysis of the proteins complexed by the truncated scaffoldin revealed that none of the cellulosomal enzymes encoded by the genes located downstream of *cipC*, even the proteins encoded by distant genes such as *man5K* and *rgl11Y*, were immunodetected (see Figure 2). The major components of the cellulosomes, Cel48F and Cel9E, were clearly lacking in the mutants. Nevertheless, the endoglucanase Cel5A, which is encoded by a gene located outside the "*cel* cluster", was immunodetected in the cellulose-adsorbed proteins produced by the mutant. As the low quantity of truncated CipC produced by the mutants might significantly contribute to the loss of the cellulolytic phenotype, a trans-complemented strain, where the entire *cipC* gene was expressed under the control of the strong *thl* promoter, was obtained. The cellulose-adsorbed protein fraction purified from a culture of the trans-complemented strain was shown to contain large amounts of intact CipC, but none of the enzymatic proteins encoded by the "*cel* cluster" were (once again) produced (Maamar *et al.*, 2004). It was concluded that the disruption of the chromosomal *cipC* led to a strong polar effect preventing the expression of the downstream " *cel* " genes. A very large mRNA covering the major part of the cluster might be synthesized as a primary transcript. As mentioned in an earlier section, a messenger 12 kb-long, covering four genes of the *C. cellulovorans* "*cel* cluster" was detected. A complete study of the regulation of the *cel* genes and the identification of potential promoters in the "*cel* cluster" will provide more information concerning this strong polar effect.

Another major result was obtained through analysis of the cellulose-bound proteins of the trans-complemented mutant. The activity of this fraction was tested on various substrates. Hydrolysis of crystalline cellulose was the most affected with a decrease of 80% when compared to the WT; about 30-40% of the WT activities on carboxymethylcellulose and phosphoric acid swollen cellulose were retained, indicating that the amorphous cellulose degradation was less affected than the degradation of the crystalline cellulose. The analysis of this cellulose-bound protein fraction revealed twelve dockerin-containing proteins. One of them has been identified as Cel5A and three others might be Cel5D and the two 54kDa and 31kDa xylanases detected by zymograms (Mohand-Oussaid *et al.*, 1999). Taking into account the proteins encoded by the "*cel* cluster," these results indicate that at least 22 dockerin-containing proteins could be produced by *C. cellulolyticum* WT and probably inserted into the cellulosomes. Nevertheless, it can be assumed that the enzymes encoded by the "*cel* cluster" are required for crystalline cellulose hydrolysis Finally, three non-cellulosomal proteins, P125, P105 and P98, were found to be the major components of the mutant and of the complemented mutant cellulolytic system. Two of them, P105 and P98, have already been identified as the major components of the cellulolytic system produced by the strain ATCC35319 (pSOS*asrF*) (which overproduces asRNAs targeted against *cel48F*). These proteins have not been characterized to date.

7. Conclusions

The numerous studies performed over these last twenty years have significantly increased our knowledge of cellulosome components. The study of the various enzymes and different scaffoldin modules obtained by recombinant DNA expression have provided fresh insights into the biochemical and structural properties of the cellulosome components. Notably, the modules involved in the building of the cellulosomes have been clearly identified as well as the proteins necessary for hydrolysis of crystalline cellulose.

Numerous helper enzymes belonging to the cellulosomes have also been identified. The ratio of these enzymes in the composition of the cellulosomes is dependent on the nature of the substrate and may dictate how the cellulolytic *Clostridia* are able to adapt to different ecological niches. Studies on the regulation of cel gene expression that have recently been initiated in several laboratories will provide new insights concerning the strategy used by the various *Clostridia* to efficiently hydrolyze plant cell wall polymers. From the preliminary studies, it appears that some genes coding for anchoring proteins are genetically linked. It also seems that the genes belonging to the large "*cel* clusters" identified in mesophilic *Clostridia* are co-regulated.

Despite the numerous data obtained over the past two decades, several questions remain unsolved. Most notable among these: when and where are the cellulosomes assembled? The new genetic tools now available, and the recently determined genomic nucleotide sequence of the model *Clostridia*, will certainly help to answer these questions.

The complexity of the hydrolytic machinery, containing cellulosomal and non-cellulosomal enzymes acting in synergy, squares with that of the entire plant cell wall. Moreover the fact that plant cell wall structure varies according to the different species, as well as within a given species, this being dependent on the tissue and age of the plant, has probably contributed to the selection of increasing complex hydrolytic machinery and regulation of cellulosomal component synthesis. Careful studies of the specificity of some enzymes, as for example bifunctional enzymes, and of their targeting CBM modules could provide some useful indications on the structure of the polymers on which they act.

Finally the potential applications of the cellulosomes in molecular and cellular engineering are considerable. The cellulosome has been considered as " a treasure trove for technology " (Bayer *et al.*, 1994). The possibility of using the affinity modules of the cellulosomes in association with the SLH modules could allow the engineering of nanoreactors containing proteins (enzymatic or not) of industrial interest in many fields of application. It is reasonable to envisage further biotechnological development of nanoreactors that resemble the cellulosomes.

References

Atalla, R.H., and Vanderhart, D.L. 1984. Native cellulose: a composite of two distinct crystalline cellulose forms. Science. 223:283-285.

Atalla, R.H. 1999. Celluloses. In: Comprehensive Natural Products Chemistry. B.M. Pinto, ed. Elsevier, Cambridge. p. 529-598.

Bagnara-Tardif, C., Gaudin, C., Belaich, A., Hoest, P., Citard, T., and Belaich, J.P. 1992. Sequence analysis of a gene cluster encoding cellulases from *Clostridium cellulolyticum*. Gene. 119: 17-28.

Barr, B.K., Hsieh, Y.L., Ganem, B., and Wilson, D.B. 1996. Identification of two functionally different classes of exocellulases. Biochemistry. 35: 586-592.

Bayer, E.A., Kenig, R., and Lamed, R. 1983. Adherence of *Clostridium thermocellum* to cellulose. J. Bacteriol. 156: 818-827.

Bayer, E.A., Setter, E., and lamed, R. 1985. Organization and distribution of the cellulosome in *Clostridium thermocellum*. J. Bacteriol. 163: 552-559.

Bayer, E.A., and lamed, R. 1986. Ultrastructure of the cell surface cellulosome of *Clostridium thermocellum* and its interaction with cellulose. J. Bacteriol. 167: 828-836.

Bayer, E.A., Morag, E., and Lamed, R. 1994. The cellulosome-A treasure-trove for biotechnology. Trends Biotechnol. 12: 378-386.

Bayer, E.A., Chanzy, H., Lamed, R., and Shoham, Y. 1998. Cellulose, cellulases and cellulosomes. Curr. Opin. Struct. Biol. 8: 548-557.

Bayer, E.A., Shoham, Y., and Lamed, R. 2000. The cellulosome: an exocellular organelle for degrading plant cell wall polysaccharides. In: Glycomicrobiology. R.J. Doyle, ed. Kluver academic/ Plenum publishers, New York. p. 387-439.

Béguin, P., Rocancourt, M., Chebrou, M.C., and Aubert, J.P. 1986. Mapping of mRNA encoding endocellulase CelA from *Clostridium thermocellum*. Mol. Gen. Genet. 202: 251-254.

Béguin, P. 1990. Molecular biology of cellulose degradation. Annu. Rev. Microbiol. 44: 219-248.

Belaich, J.P., Gaudin, C., Belaich, A., Bagnara-Tardif, C., Fierobe, H.P., and Reverbel, C. 1993. Molecular organization and biochemistry of the cellulolytic system of *Clostridium cellulolyticum*. In: Genetics, Biochemistry and Ecology of Lignocellulose Degradation. K. Shimada, K. Ohmiya, Y. Kobayashi, S. Hoshino, K. Sakka, and S. Karita, eds. Uni Publisher CO., LTD. p. 53-62.

Belaich, J.P., Tardif, C., Belaich, A., and Gaudin, C. 1997. The cellulolytic system of *Clostridium cellulolyticum*. J. Biotechnol. 57: 3-14.

Belaich, A., Belaich, J.P., Fierobe, H.P., Gal, L., Gaudin, C., Pagès, S., Reverbel-Leroy, C., and Tardif, C. 1998. Cellulosome analysis and cellulases CelF and CelG from *Clostridium cellulolyticum*. In: Carbohydrates from *Trichoderma reesei* and Other Microorganisms. M. Claeyssens, W. Nerinckx and K. Piens, eds. The Royal Society of Chemistry, Cambridge. p. 73-86.

Belaich, A., Parsiegla, G., Gal, L., Villard, C., Haser, R., and Belaich, J.P. 2002. Cel9M, a new family 9 cellulase of the Clost*ridium cellulolyticum* cellulosome. J. Bacteriol. 184: 1378-1384.

Blum, D.L., Kataeva, I.A., Li, X.L., and Ljungdahl, L.G. 2000. Feruloyl esterase activity of the *Clostridium thermocellum* cellulosome can be attributed to previously unknown of Xyn Y and Xyn Z. J. Bacteriol. 182: 1346-1351.

Carrard, G., Koivula, A., Soderlund, H., and Béguin, P. 2000. Cellulose-binding domain promote hydrolysis of different sites on crystalline cellulose. Proc. Natl. Acad. Sci. USA. 97: 10342-10347.

Chanzy, H. 1990. Aspects of cellulose srructure. In: Cellulose Sources and Exploitation. J.F. Kennedy, G.O. Philips and P.A. Williams, eds. Ellis Horwood limited, New York, New York. p. 3-12.

Chauvaux, S., Béguin, P., Aubert, J.P., Bath, K.M., Gow, L.A., Wood, T.M., and Bairoch, A. 1990. Calcium-binding affinity and calcium-enhanced activity of *Clostridium thermocellum* endoglucanase D. Biochem. J. 265: 261-265.

Coutinho, P.M., and Henrissat, B. 1999. Carbohydrate-active enzymes: an integrated database approach. In: Recent Advances in Carbohydrate Bioengineering. H.J. Gilbert, G.J. Davies, B. Henrissat, and B. Svensson, eds. The Royal Society of Chemistry, Cambridge. p. 3-12.

Ding, S.Y., Bayer, E.A., Steiner, D., Shoham, Y., and Lamed, R. 1999. A novel cellulosomal scaffoldin from *Acetivibrio cellulolyticus* that contains a family 9 glycosyl hydrolase. J. Bacteriol. 181: 6720-6729.

Doi, R.H., Kosugi, A., Murashima, K., Tamaru, Y., and Ok Han, S. 2003. Cellulosomes from mesophilic bacteria. J. Bacteriol. 185: 5907-5947.

Dominguez, R., Souchon, H., Lascombe, M., and Alzari, P.M. 1996. The crystal structure of a family 5 endoglucanase mutant in complexed and uncomplexed forms reveals an induced fit activation mechanism. 1996. J. Mol. Biol. 257: 1042-1051.

Dror, T.W., Morag, E., Rolider, A., Bayer, E.A., Lamed, R., and Shoham, Y. 2003a. Regulation of the cellulosomal *celS* (*cel48A*) gene of *Clostridium thermocellum* is growth rate dependant. J. Bacteriol. 185: 3042-3048.

Dror, T.W., Rolider, A., Bayer, E.A., Lamed, R., and Shoham, Y. 2003b. Regulation of expression of scaffoldin-related genes in *Clostridium thermocellum*. J. Bacteriol. 185: 5109-5116.

Ducros, V., Czjzek, M., Belaich, A., Gaudin, C., Fierobe, H.P., Belaich, J.P., Davies, G.J., and Haser, R. 1995. Crystal structure of the catalytic domain of a bacterial cellulase belonging to family 5. Structure. 3: 939-949.

Faure, E., Belaich, A., Bagnara, C., Gaudin, C., and Belaich, J.P. 1989. Sequence analysis of the *Clostridium cellulolyticum* endoglucanase-A-encoding gene *celccA*. Gene. 84: 39-46

Fierobe, H.P., Mechaly, A., Tardif, C., Belaich, A., Lamed, R., Shoham, Y., Belaich, J.P., and Bayer, E.A. 2001. Design and production of active cellulosome chimeras. Selective incorporation of dockerin-containing enzymes into defined functional complexes. J. Biol. Chem. 276: 21257-21261.

Fierobe, H.P., Bayer, E.A., Tardif, C., Czjzek, M., Mechaly, A., Belaich, A., Lamed, R., Shoham, Y., and Belaich, J.P. 2002. Degradation of cellulose substrates by cellulosome chimeras. J. Biol. Chem. 277: 49621-49630.

Fuchs, K.P., Zverlov, V.V., Velikodvorskaya, G.A., Lottspeich, F., and Schwartz, W.H. 2003. Lic16A of *Clostridium thermocellum*, a non-cellulosomal, highly complex endo-beta-1,3-glucanase bound to the outer cell surface. Microbiology. 149: 1021-1031.

Fujino, T., Béguin, P., and Aubert, J.P. 1993. Organization of a *Clostridium thermocellum* gene cluster encoding the cellulosomal scaffolding protein CipA and a protein possibly involved in attachment of the cellulosome to the cell surface. J. Bacteriol. 175: 1891-1899.

Fujino, T., Karita, S., and Ohmiya, K. 1993. Cloning of the gene cluster encoding putative cellulosomal components from *Clostridium josui*. In: Genetics, Biochemistry and Ecology of Lignocellulose Degradation. K. Shimada, K. Ohmiya, Y. Kobayashi, S. Hoshino, K. Sakka, and S. Karita, eds. Uni Publisher CO., Ltd. p. 67-75.

Gal, L., Gaudin, C., Belaich, A., Pagès, S., Tardif, C., and Belaich, J.P. 1997. CelG from *Clostridium cellulolyticum*: a multidomain endoglucanase acting efficiently on crystalline cellulose. J. Bacteriol. 179: 6595-6601.

Gal, L., Pagès, S., Gaudin, C., Belaich, A., Reverbel-Leroy, C., Tarfif, C., and Belaich, J.P. 1997. Characterization of the cellulolytic complex (cellulosome) produced by *Clostridium cellulolyticum*. Appl. Environ. Microbiol. 63: 903-909.

Garcia-Campayo, V., and Béguin, P. 1997. Synergism between the cellulosome-integrating protein CipA and endoglucanase D of *Clostridium thermocellum*. J. Biotechnol. 57: 39-47.

Gaudin, C., Belaich, A., Champ, S., and Belaich, J.P. 2000. CelE, a multidomain cellulase from *Clostridium cellulolyticum*: a key enzyme in the cellulosome? J Bacteriol. 182: 1910-1915.

Gerngross, U.T., Romaniec, M.P., Kobayashi, T., Huskisson, N.S., and Demain, A.L. 1993. Sequencing of a *Clostridium thermocellum* gene (cipA) encoding the cellulosomal SL-protein reveals an unusual degree of internal homology. Mol. Microbiol. 8: 325-334. Erratum in Mol. Microbiol. 10: 1155.

Guerin, D.M., Lascombe, M.B., Costabel, M., Souchon, H., Lamzin, V., Béguin, P., and Alzari, P.M. 2002. Atomic (0.94Å) resolution structure of an inverting glycosidase in complex with substrate. J. Mol. Biol. 316: 1061-1069.

Guimaraes, B.G., Souchon, H., Lytle, B.L., David Wu, J.H., Alzari, P.M. 2002. The crystal structure and catalytic mechanism of cellobiohydrolase CelS, the major enzymatic component of the *Clostridium thermocellum* cellulosome. J. Mol. Biol. 320: 587-596.

Han, S.O., Yukawa, H., Inui, M., and Doi, R.H. 2003. Transcription of *Clostridium cellulovorans* cellulosomal cellulase and hemicellulase genes. J. Bacteriol. 185: 2520-2527.

Henrissat, B., and Davies, G. 1997. Structural and sequence-based classification of glycoside hydrolases. Curr. Opin. Struct. Biol. 7: 637-644.

Irwin, D., Shin, D.H., Zhang, S., Barr, B.K., Sakon, J., Karplus, P.A., and Wilson, D.B. 1998. Roles of the catalytic domain of two cellulose binding domains of *Thermomonospora fusca* E4 in cellulose hydrolysis. J. Bacteriol. 180: 1709-1714.

Jennert, K., Tardif, C., Young, D., and Young, M. 2000. Gene transfer to *Clostridium cellulolyticum* ATCC 35319. Microbiology. 146: 3071-3080.

Jindou, S., Karita, S., Fujino, T., Hayashi, H., Kimura, T., Sakka, K., and Ohmiya, K. 2002. Alpha-galactosidase Aga27A, an enzymatic component of the *Clostridium josui* cellulosome. J. Bacteriol. 184: 600-604.

Juy, M., Amit, A.G., Alzari, P.M., Poljak, R.J., Claeyssens, M., Béguin, P., and Aubert, J.P. 1992. Three-dimensional structure of a thermostable bacterial cellulase. Nature. 357: 89-91.

Kakiuchi, M., Isui, A., Suzuki, K., Fujino, T., Fujino, E., Kimura, T., Karita, S., Sakka, K., and Ohmiya, K. 1998. Cloning and DNA sequencing of the genes encoding *Clostridium josui* scaffolding protein CipA and cellulase CelD and identification of their gene products as major components of the cellulosome. J. Bacteriol. 180: 4303-4308.

Kataeva, I., Gugliemi, G., and Béguin, P. 1997. Interaction between *Clostridium thermocellum* endoglucanase CelD and polypeptides derived from the cellulosome-integrating protein CipA: stoichiometry and cellulolytic activity of the complexes. Biochem. J. 326: 617-624.

Kataeva, I., Li, X.L., Chen, H., Choi, S.K., and Ljungdahl, L.G. 1999. Cloning and sequence analysis of a new cellulase gene encoding CelK, a major cellulosome component of *Clostridium thermocellum*: Evidence for gene duplication and recombination. J. Bacteriol. 181: 5288-5295.

Kosugi, A., Murashima, K., and Doi, R.H. 2001. Characterization of xylanolytic enzymes in *Clostridium cellulovorans*: Expression of xylanase activity dependent on growth substrates. J. Bacteriol. 183: 7037-7043.

Kosugi, A., Murashima, K., Tamaru, Y., and Doi, R.H. 2002a. Cell-surface anchoring role of N-terminal surface layer homology domains of *Clostridium cellulovorans* EngE. J. Bacteriol. 184: 884-888.

Kosugi, A., Murashima, K., and Doi, R.H. 2002b. Xylanase and acetyl xylan esterase activity of Xyn A, a key subunit of the *Clostridium cellulovorans* cellulosome for xylan degradation. Appl. Environ. Microbiol. 68: 6399-6402.

Kosugi, A., Murashima, K., and Doi, R.H.2002c. Characterization of two noncellulosomal subunits, ArfA and BgaA, from *Clostridium cellulovorans* that cooperate with the cellulosome plant cell wall degradation. J. Bacteriol. 184: 6859-6865.

Kurokawa, J., Hemjinda, E., Arai, T., Karita, S., Kimura, T., Sakka, K., and Ohmiya, K. 2001. Sequence of the *Clostridium thermocellum* mannanase gene *man26B* and characterization of the translated product. Biosci. Biotechnol. Biochem. 65: 548-554.

Kurokawa, J., Hemjinda, E., Arai, T., Kimura, T., Sakka, K., and Ohmiya, K. 2002. *Clostridium thermocellum* cellulase CelT, a family 9 endoglucanase without an Ig-like domain or family 3c carbohydrate-binding module. Appl. Microbiol. Biotechnol. 59: 455-461.

Lamed, R., Naimark, J., Morgenstern, E., and Bayer, E.A. 1987. Specialized cell surface structures in cellulolytic bacteria. J. Bacteriol. 169: 3792-3800.

Lamed, R., and Bayer, E.A. 1988. The cellulosome concept: exocellular/extracellular enzyme reactor centers for efficient binding and cellulolysis. In: Biochemistry and Genetics of Cellulose Degradation. J.P. Aubert, P. Béguin and J. Millet, eds. Academic Press, London. p. 101-116.

Leibovitz, E., and Béguin, P. 1996. A new type of cohesin domain that specifically binds the dockerin domain of the *Clostridium thermocellum* cellulosome-integrating protein CipA. J. Bacteriol. 178: 3077-3084.

Leibovitz, E., Ohayon, H., Gounon, P., and Béguin, P. 1997. Characterization and subcellular localization of the *Clostridium thermocellum* scaffoldin dockerin binding protein SdbA. J. Bacteriol. 179: 2519-2523.

Lemaire, M., Ohayon, H., Gounon, P., Fujino, T., and Béguin, P. 1995. OlpB, a new outer layer protein of *Clostridium thermocellum*, and binding of its S-layer-like domains to components of the cell enveloppe. J. Bacteriol. 177: 2451-2459.

Liu, C.C., and Doi, R.H. 1998. Properties of *exgS*, a gene for a major subunit of the *Clostridium cellulovorans* cellulosome. Gene. 211: 39-47.

Lynd, L.R., Weimer, P.J., van Zyl, W.H., and Pretorius, I.S. 2002. Microbial cellulose utilization: fundamentals and biotechnology. Microbiol. Mol. Biol. Rev. 66: 506-577.

Lytle, B.L., Volkman, B.F., Westler, W.M., Heckman, M.P., and Wu, J.H. 2001. Solution structure of a type I dockerin domain, a novel prokaryotic, extracellular calcium-binding domain. J. Mol. Biol. 307: 745-753.

Maamar, H., de Philip, P., Belaich, J.P., and Tardif, C. 2003a. ISCce1 and ISCce2, two novel insertion sequences in *Clostridium cellulolyticum*. J. Bacteriol. 185: 714-725.

Maamar, H. 2003b. Etude *in vivo* du système cellulolytique de *Clostridium cellulolyticum*: Caracterisation du premier mutant d'insertion *cipC*. PHD thesis, Université Marseille I, France.

Maamar, H., Valette, O., Fierobe, H.P., Belaich, A., Belaich, J.P., and Tardif, C. 2004.Cellulolysis is severely affected in *Clostridium cellulolyticum* strain cipCMut1. Mol. Microbiol.51: 589-592.

Madkour, M., and Mayer, F. 2003. Structural organization of the intact bacterial cellulosome as revealed by electron microscopy. Cell. Biol. Int. 27: 831-836.

Mandelman, D., Belaich, A., Belaich, J.P., Aghajari, N., Driguez, H., and Haser, R. 2003. X-ray crystal structure of the multidomain endoglucanase cel9G from *Clostridium cellulolyticum* complexed with natural and synthetic cello-oligosaccharides. J. Bacteriol. 185: 4127-4135.

McCarter, J.D., and Withers, S.G. 1994. Mechanisms of enzymatic glycoside hydrolyze. Curr. Opin. Struct. Biol. 4: 885-892.

Mechaly, A., Yaron, S., Lamed, R., Fierobe, H.P., Belaich, A., Belaich, J.P., Shoham, Y., and Bayer, E.A. 2000. Cohesin-dockerin recognition in cellulosome assembly: Experiment versus hypothesis. Proteins. 39: 170-177.

Mermelstein, L.D., and Papoutsakis, E.T. 1993. *In vivo* methylation in *Escherichia coli* by the B*acillus subtilis* phage Φ3T I methyltransferase to protect plasmids from restriction upon transformation of *Clostridium acetobutylicum* ATCC 824. Appl. Environ. Microbiol. 59: 1077-1081.

Mishra, S., Béguin, P., and Aubert, P. 1991. Transcription of *Clostridium thermocellum* endoglucanase genes *celF* and *celD*. J. Bacteriol. 173: 80-85.

Mohand-Oussaid, O., Payot, S., Guedon, E., Gelhaye, E., Youyou, A., and Petitdemange, H. 1999. The extracellular xylan degradative system in *Clostridium cellulolyticum* cultivated on xylan: Evidence for cell-free cellulosome production. J. Bacteriol. 181: 4035-4040.

Mosbah, A., Belaich, A., Bornet, O., Belaich, J.P., Henrissat, B., and Darbon, H. 2000. Solution structure of the module X2 1 of unknown function of the cellulosomal scaffolding protein CipC of *Clostridium cellulolyticum*. J. Mol. Biol. 304: 201-217.

Murashima, K., Kosugi, A., and Doi, R.H. 2002. Determination of subunit composition of *Clostridium cellulovorans* cellulosomes that degrade plant cell walls. Appl. Environ. Microbiol. 68: 1610-1615.

Murashimaa, K., Kosugi, A., and Doi, R.H. 2003. Synergistic effects of cellulosomal xylanase and cellulases from *Clostridium cellulovorans* on plant cell wall degradation. J. Bacteriol. 158: 1518-1524.

Nakotte, S., Schaffer, S., Bohringer, M., and Durre, P. 1998. Electroporation of, plasmid isolation from and plasmid conservation in *Clostridium acetobutylicum* DSM 792. Appl. Microbiol. Biotechnol. 50: 564-567.

Noach, I., Lamed, R., Xu, Q., Rosenheck, S., Shimon, L.J., Bayer, E.A., and Frolow, F. 2003. Preliminary X-ray characterization and phasing of a type II cohesin domain from the cellulosome of *Acetivibrio cellulolyticus*. Acta Crystallogr. D Biol. Crystallogr. 59: 1670-1673.

Nolling, J., Breton, G., Omelchenko, M.V., Makarova, K.S., Zeng, Q., Gibson, R., Lee, H.M., Dubois, J., Qiu, D., Hitti, J., Wolf, Y.I., Tatusof, R.L., Sabath, F., Doucette-Stamm, L., Soucaille, P., Daly, M.J., Bennett, G.N., Koonin, E.V., and Smith, D.R. 2001. Genome sequence and comparative analysis of the solvent-producing bacterium *Clostridium acetobutylicum*. J. Bacteriol. 183: 4823-4838.

Pagès, S., Belaich, A., Tardif, C., Reverbel-Leroy, C., Gaudin, C., and Belaich, J.P. 1996. Interaction between the endoglucanase CelA and the sacffolding protein CipC of the *Clostridium cellulolyticum* cellulosome. J. Bacteriol. 178: 2279-2286.

Pagès, S., Belaich, A., Belaich, J.P., Morag, E., Lamed, R., Shoham, Y., and Bayer, E.A. 1997a. Species-specificity of the cohesin-dockerin interaction between *Clostridium thermocellum* and Clostridium *cellulolyticum*: Prediction of specificity determinants of the dockerin domain. Proteins. 29: 517-527.

Pagès, S., Gal, L., Belaich, A., Gaudin, C., Tardif, C., and Belaich, J.P. 1997b. Role of the scaffolding protein CipC of *Clostridium cellulolyticum* in cellulose degradation. J. Bacteriol. 179: 2810-2816.

Pagès, S., Belaich, A., Fierobe, H.P., Tardif, C., Gaudin, C., and Belaich, J.P. 1999. Sequence analysis of scaffolding protein CipC and ORFXp, a new cohesin-containing protein in *Clostridium cellulolyticum*: comparison of various cohesin domains and subcellular localization of ORFXp. J. Bacteriol. 181: 1801-1810.

Pagès, S., Valette, O., Abdou, L., Belaich, A., and Belaich, J.P. 2003. A rhamnogalacturonan lyase in the *Clostridium cellulolyticum* cellulosome. J. Bacteriol. 185: 4727-4733.

Park, J.S., Matano, Y., and Doi, R.H. 2001. Cohesin-dockerin interactions of cellulosomal subunits of *Clostridium cellulovorans*. J. Bacteriol. 183: 5431-5435.

Parsiegla, G., Juy, M., Reverbel-Leroy, C., Tardif, C., Belaich, J.P., Driguez, H., and Haser, R. 1998. The crystal structure of the processive endocellulase CelF of *Clostridium cellulolyticum* in complex with a thiooligosaccharide inhibitor at 2 resolution. EMBO J. 17: 5551-5562.

Parsiegla, G., Reverbel-Leroy, C., Tardif, C., Belaich, J.P., Driguez, H., and Haser, R. 2000. Crystal structures of the cellulase Cel48F in complex with inhibitor and substrates give insights into processive action. Biochemistry. 39: 11238-11246.

Parsiegla, G., Belaich, A., Belaich, J.P., and Haser, R. 2002. Crystal structure of the cellulase Cel9M enlightens structure/function relationships of the variable catalytic modules in glycosyl hydrolases. Biochemistry. 41: 11134-11142.

Perret, S., Maamar, H., Belaich, J.P., and Tardif, C. 2004. Use of antisense RNA to modify the composition of cellulosomes produced by *Clostridium cellulolyticum*. Mol. Microbiol. 51: 599-607.

Pohlschroder, M., Leschine, S.B., and Canale-Parola, E. 1994. Multicomplex cellulase-xylanase system of *Clostridium papyrosolvens* C7. J. Bacteriol. 176: 70-76.

Pohlschroder, M., Canale-Parola, E., and Leschine, S. 1995. Ultrastructural diversity of the cellulase complexes of *Clostridium papyrosolvens* C7. J. Bacteriol. 177: 6625-6629.

Prates, J.A., Tarbouriech, N., Charnock, S.J., Fontes, C.M., Ferreira, L.M., and Davies, G.J. 2001. The structure of the feruloyl esterase module of xylanase 10B from *Clostridium thermocellum* provides insights into substrate recognition. Structure. 9: 1183-1190.

Reverbel-Leroy, C., Pagès, S., Belaich, A., Belaich, J.P., and Tardif, C. 1997. The processive endocellulase CelF, a major component of the *Clostridium cellulolyticum* cellulosome: purification and characterization of the recombinant form. J. Bacteriol. 179: 46-52.

Sabathé, F., Belaich, A., and Soucaille, P. 2002. Characterization of the cellulolytic complex (cellulosome) of *Clostridium acetobutylicum*. FEMS. Microbiol. Lett. 217: 15-22.

Sabathé, F., and Soucaille, P. 2003. Characterization of the CipA scaffolding protein and in vivo production of minicellulosome in *Clostridium acetobutylicum*. J. Bacteriol. 185: 1092-1096.

Sakon, J., Irwin, D., Wilson, D.B., and Karplus, P.A. 1997. Structure and mechanism of endo/exocellulase E4 from *Thermomonospora fusca*. Nature Struct. Biol. 4: 810-818.

Salamitou, S., Raynaud, O., Lemaire, M., Coughlan, M., Béguin, P., and Aubert, J.P. 1994. Recognition specificity of the duplicated segments present in *Clostridium thermocellum* endoglucanase CelD and in the cellulose-integrating protein CipA. J. Bacteriol. 176: 2822-2827.

Saxena, S., Fierobe, H.P., Gaudin, C., Guerlesquin, F., and Belaich, J.P. 1995. Biochemical properties of a beta-xylosidase from *Clostridium cellulolyticum*. Appl. Environ. Microbiol. 61: 3509-3512.

Shima, S., Igarashi, Y., and Kodama, T. 1991. Nucleotide sequence analysis of the endoglucanase-encoding gene, *celCCD*, of *Clostridium cellulolyticum*. Gene. 104: 33-38.

Shimon, L., Bayer, E.A., Morag, E., Lamed, R., Yaron, S., Shoham, Y., and Frolow, F. 1997. The crystal structure at 2.15 resolution of a cohesin domain of the cellulosome from *Clostridium thermocellum*. Structure. 5: 381-390.

Shimon, L., Pagès, S., Belaich, A., Belaich, J.P., Bayer, E.A., Lamed, R., Shoham, Y., and Frolow, F. 2000. Structure of a family IIIa scaffoldin CBD from the cellulosome of *Clostridium cellulolyticum* at 2.2Å resolution. Acta. Crystallogr. D. Biol. Crystallogr. 12: 1560-1568.

Shoseyov, O., Tagaki, M., Goldstein, M.A., and Doi, R.H. 1992. Primary sequence analysis of *Clostridium cellulovorans* binding protein A. Proc. Natl. Acad. Sci. USA. 89: 3483-3487.

Spinelli, S., Fierobe, H.P., Belaich, A., Belaich, J.P., Henrissat, B., and Cambillau, C. 2000. Crystal structure of a cohesin module from *Clostridium cellulolyticum*: Implications for dockerin recognition. J. Mol. Biol. 304: 189-200.

Schwarz, W.H. 2001. The cellulosome and cellulose degradation by anaerobic bacteria. Appl. Microbiol. Biotechnol. 56: 634-649.

Tamaru, Y., Liu, C.C., Ichi-ishi, A., Malburg, L., and Doi, R. 1998. The *Clostridium cellulovorans* cellulosome and non-cellulosomal cellulases. In: Genetics, Biochemistry and Ecology of Cellulose Degradation. K. Ohmiya, K. Hayashi, K. Sakka, Y. Kobayashi, S. Karita, T. Kimura, eds. Uni publishers Co., LTD, Tokyo. p. 488-494.

Tamaru, Y., and Doi, R.H. 1999. Three surface layer homology domains at the N terminus of the *Clostridium cellulovorans* major cellulosomal subunit EngE. J. Bacteriol. 181: 3270-3276.

Tamaru, Y., and Doi, R.H. 2000a. The *engL* gene cluster of *Clostridium cellulovorans* contains a gene for cellulosomal ManA. J. Bacteriol. 182: 244-247.

Tamaru, Y., Karita, S., Ibrahim, A., Chan, H., and Doi, R.H. 2000b. A large gene cluster for the *Clostridium cellulovorans* cellulosome. J. Bacteriol. 182: 5906-5910.

Tamaru, Y., and Doi, R.H. 2001. Pectate lyase A, an enzymatic subunit of the *Clostridium cellulovorans* cellulosome. Proc. Natl. Acad. Sci. USA. 98: 4125-4129.

Tardif, C., Maamar, H., Balfin, M., and Belaich, J.P. 2001. Electrotransformation studies in *Clostridium cellulolyticum*. J. Ind. Microbiol. Biotechnol. 27: 271-274.

Tavares, G.A., Béguin, P., and Alzari, P.M. 1997. The crystal structure of a type I cohesin domain at 1.7A resolution. J. Mol. Biol. 273: 701-713.

Tokatlidis, K., Salamitou, S., Béguin, P., Dhurjati, P., and Aubert, J.P. 1991. Interaction of the duplicated segment carried by *Clostridium thermocellum* cellulases with cellulosome components. FEBS. Lett. 291: 185-188.

Tomme, P., Warren, R.A., Miller, R.C., Kilburn, D.G., and Gilkes, N.R. 1995. Cellulose-binding domains: classification and properties. In: Enzymatic Degradation of Insoluble Polysaccharides. J.N. Saddler and M. Penner, eds. American Chemical Society, Washington, D.C. p. 142-163.

Tormo, J., Lamed, R., Chirino, A.J., Morag, E., Bayer, E.A., Shoham, Y., and Steitz, T.A. 1996. Crystal structure of a bacterial family-III cellulose-binding domain: a general mechanism for attachment to cellulose. EMBO. J. 15: 5739-5751.

Tummala, S.B., Welker, N.E., and Papoutsakis, E.T. 2003. Design of antisense RNA constrcuts for downregulation of the acetone pathway of *Clostridium acetobutylicum*. J. Bacteriol. 185: 1923-1964.

White, A., and Rose, D.R. 1997. Mechanism of catalysis by retaining β-glycosyl hydrolases. Curr. Opin. Struct. Biol. 7: 645-651.

Xu, Q., Gao, W., Ding, S.Y., Kenig, R., Shoham, Y., Bayer, E.A., and Lamed, R. 2003. The cellulosome system of *Acetivibrio cellulolyticus* includes a novel type of adaptator protein and a cell surface anchoring. J. Bacteriol. 185: 4548-4557.

Zverlov, V.V., Fuchs, K.P., and Schwarz, W.H. 2002. Chitin18A, the endochitinase in the cellulosome of the thermophilic, cellulolytic bacterium *Clostridium thermocellum*. Appl. Environ. Microbiol. 68: 3176-3179.

Chapter 20

Microbial Community Structure and Functions in Methane Fermentation Technology for Wastewater Treatment

Yuji Sekiguchi* and Yoichi Kamagata

Abstract

To date, anaerobic (methanogenic) fermentation technology has been widely applied to the treatment of municipal and industrial wastes and wastewaters. A number of anaerobic processes have been intensively developed, and the application of these processes is being expanded to low-strength wastewaters, wastes and wastewaters under extreme temperature conditions, and more complex wastewaters containing anthropogenic compounds and/or compounds that are recalcitrant to biodegradation. The recent development of molecular techniques in the field of microbial ecology has allowed us to explore the microbial diversity and community structure of those anaerobic processes. As a result of the development and application of these techniques, we now have better insight into the community composition and architecture of anaerobic sludge, which can be adapted to treat a variety of waste/wastewaters under different operation conditions. Importantly, the community was found to be composed in large part of various yet-to-be cultured microorganisms, some of which were often found to play significant roles in those anaerobic processes. To reveal the function of the community constituents, numerous efforts have been made to isolate relevant microbes in the anaerobic processes, and the information on the functions of the microbes in anaerobic sludge is accumulating at an encouraging rate. In this chapter, the state-of-the-art anaerobic waste/wastewater treatment technologies, the microbial community structure in anaerobic sludge, and the functions of individual populations are summarized.

1. Introduction

The anaerobic microbial degradation of organic compounds takes place naturally in a variety of habitats on the earth. Such ecosystems are everywhere; examples of them include sediments, sub-surfaces, and intestinal tracts in animals. It was roughly estimated that the number of prokaryotes and the total amount of their cellular carbon on the earth are 4-6×10^{30} cells and 350-550 Pg of carbon, respectively, and this estimate suggested that the total cellular carbon of prokaryotes would be equivalent to 60-100% of the total carbon found in plants (Whitman *et al.*, 1998). Importantly, a large fraction of the total number of prokaryotes is thought to inhabit anaerobic environments: >90% of the total prokaryotic cells live under anoxic conditions (Whitman *et al.*, 1998). These estimations imply that a large part of all the biota on earth is anoxic, and that anaerobic microorganisms are abundantly present as one of the major natural populations on the earth. Furthermore, it can also be speculated that such a large population of microbes under anoxic conditions has an enormous capacity for harboring genetically diverse anaerobes. In fact, anaerobic microorganisms enjoy various forms of life under those conditions (see other chapters).

Anaerobic waste/wastewater treatment technology artificially employs a vast array of such diverse anaerobes mainly for the mineralization of organic matter in waste and wastewaters. Anaerobic (methanogenic) fermentation technology is enormously important because it has the following advantages: (1) anaerobic processes do not require aeration (energy-input) as do conventionally applied aerobic biological treatment processes, (2) anaerobic processes produce less excess sludge biomass (i.e., wastes) when compared to aerobic processes, and (3) anaerobic processes produce biogas (methane) that can be used as an energy source (Lettinga, 1995; Speece, 1996; Ahring, 2003). Those advantages are conferred by the anaerobes that participate in these processes. In particular, if efficient terminal electron acceptors such as sulfate, nitrate, and oxidized forms of metals are not

*For correspondence email: y.sekiguchi@aist.go.jp

Abbreviations

COD, chemical oxygen demand;
CLSM, confocal laser scanning microscopy;
EGSB, expanded granular sludge bed;
FISH, fluorescence *in situ* hybridization;
PLFA, polar-lipid fatty acid;
SSCP, single-strand conformation polymorphism;
UASB, upflow anaerobic sludge blanket.

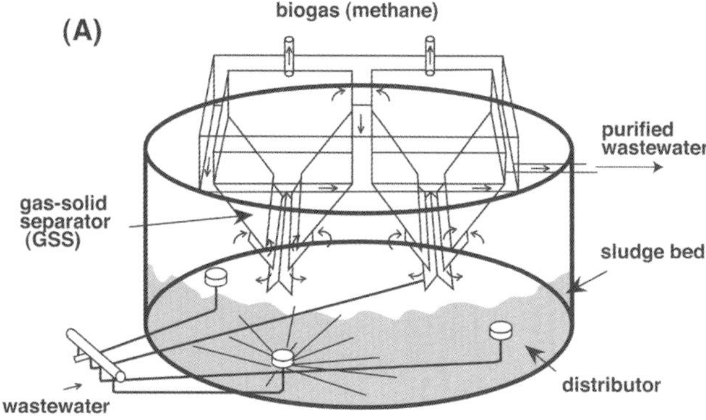

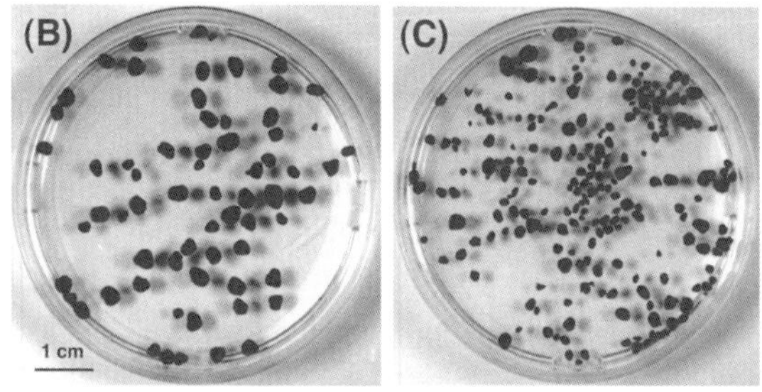

Figure 1. Schematic representation of a UASB process [A, modified based Qict Brocades Co. Ltd. UASB BIOTHANE PROCESS Pamphlet (1986)] and granular sludge retained in a mesophilic (37°C) UASB process (B) and in a thermophilic (55°C) UASB process (C).

abundantly present in waste/wastewater, methanogenesis occurs as the primary degradation pathway of the organic substances (Schink, 1997; Stams *et al.*, 2003). Since the conversion of complex organic material to methane and carbon dioxide is the least exergonic microbial reaction if compared to other aerobic and anaerobic respiratory processes, microorganisms gain only a small amount of energy through methanogenic degradation. This trait of anaerobes leads to lower excess sludge production during anaerobic processes (5-10 times lower than that of typical aerobic processes in general). Instead, the end product (i.e., methane) still retains a large portion of the available energy, and thus can be further utilized for various purposes. Owing to these advantages, anaerobic biotechnology provides an attractive means not only of providing the need for efficient and cost-effective treatment of waste/wastewater, but also of generating energy from organic waste/wastewater.

To date, anaerobic (methanogenic) biotechnology has been widely applied for the treatment of municipal and industrial wastes and wastewaters (Lettinga, 1995; Angelidaki *et al.*, 2003; Kleerebezem and Macarie, 2003). Decomposition of organic wastes, such as sewage sludge, municipal solid wastes and manures, has been the major application of this technology (this application is often referred to as "anaerobic digestion"). Furthermore, the treatment of high-strength organic wastewaters (containing high concentrations of organic compounds typically above 1,000 mg-COD [chemical oxygen demand] per liter) is also successfully performed by anaerobic biotechnology worldwide (Kleerebezem and Macarie, 2003). By contrast, in certain cases these anaerobic treatment processes are often associated with process failures, such as (1) unstable treatment efficiency, (2) slow start-up, and (3) sudden washout of the sludge. These negative features are, again, often associated with traits of the microbial populations in sludge. For example, reduced quality of the treated water was found in certain cases in which high concentrations of volatile fatty acids, particularly propionate, accumulated as the main organic fraction in the effluent (e.g., see van Lier *et al.*, 1992; 1993). This finding is considered to be due to the unstable and sensitive nature of the microbes responsible for propionate degradation in these processes.

Methanogenic degradation of complex organic material requires different trophic groups of anaerobes that perform different metabolic tasks to complete degradation. Therefore, it is important for anaerobic treatment processes to develop a well-balanced consortium, containing all of the active trophic groups of anaerobes necessary for treatment. In addition, since the growth of anaerobes is generally slow, retaining a large amount of active cells within the process is a crucial factor for all anaerobic wastewater treatment processes for a high-rate digestion. To accomplish this, anaerobic wastewater treatment processes often need well-settled sludge aggregates (or stable biofilms). For the formation of such sludge aggregates/biofilms, it is recognized that certain microbial species play important roles (e.g., Wiegant, 1987). In contrast, other (or often the same) microbial species can also be detrimental to the settling properties of sludge (e.g., Sekiguchi *et al.*, 2001b). In this manner, the community of anaerobes in sludge represents the core component of the treatment processes, and can determine all process performance associated with the treatment. Consequently, a thorough knowledge of the microbial community ecology is required in order to elucidate the factors affecting the stability, development, and disintegration of anaerobic sludge, as well as to develop promising strategies for reactor operation.

In this chapter, we review state-of-the-art anaerobic waste/wastewater treatment technology and the microbial ecology of anaerobic processes. Special emphasis is placed on the upflow anaerobic sludge blanket (UASB) reactor for wastewater treatment and the community structures of its granular sludge.

2. Recent Developments in Anaerobic Wastewater Treatment Biotechnology

Anaerobic biological treatment technology is presently accepted as a proven technology for the treatment of medium- and high-strength organic wastewaters (such

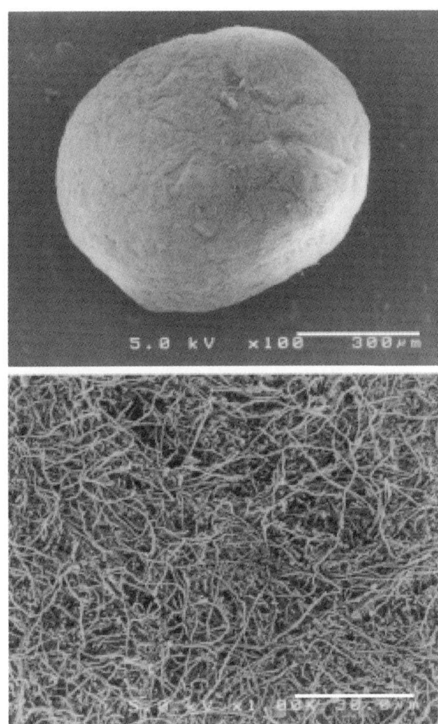

Figure 2. Scanning electron micrographs of a sludge granule. (A) Whole view of a sludge granule taken from a UASB reactor treating food-processing wastewater under mesophilic (35-40°C) conditions. (B) Magnified view of the surface of the granule showing that certain filamentous microbial populations entirely cover the surface of the granule.

as those from breweries, distilleries, and food-processing manufacturers) and solid wastes (Lettinga, 1995). However, anaerobic biotechnology has been considered for a long time to be slow, and not applicable for the treatment of low-strength organic wastewaters (such as domestic sewage), wastes and wastewaters under certain temperature conditions (such as low temperatures; i.e., <20°C), and more complex wastewaters containing anthropogenic compounds and/or compounds recalcitrant to biodegradation. As a consequence, most of these situations require aerobic biological systems. It is now recognized, however, that most biodegradable organic pollutants can be metabolized by either aerobic or anaerobic processes at about the same rate per unit of biomass, and the only fundamental problem associated with anaerobic biotechnology is thought to be the slow growth rate of most anaerobes (Speece, 1996). In fact, a variety of compounds have been found to be degradable in methanogenic environments, and the number of such chemicals has increased enormously during the past decades (Kleerebezem and Macarie, 2003). For example, under methanogenic conditions most aliphatic and homocyclic aromatic compounds are degradable (Alexander, 1998; Kleerebezem and Macarie, 2003). Additionally, some aromatic hydrocarbons, such as toluene and xylene, can also be completely mineralized even under methanogenic conditions (Edwards and Grbic-Galic, 1994). However, depending on the inoculum used, it is a general observation that anaerobic (in particular methanogenic) degradation of those anthropogenic compounds requires long lag periods (generally months to a year, depending on the chemical and inoculum used) prior to the degradation, mainly due to the slow growth of anaerobes. As mentioned above, since anaerobes can gain only a limited amount of energy through the degradation of organic compounds, and because most anaerobes are capable of metabolizing a limited range of substances, the growth of certain trophic groups of anaerobes is often the major limiting factor in anaerobic biotechnology, and is responsible for the slow start-up of treatment processes. In addition, restoring anaerobic processes is often difficult (i.e. slow) after a sudden process failure (such as a sudden reduction in pollutant removal efficiency due to the loss of certain microbial activities), which is a problem inherent in processes involving anaerobes. To overcome these disadvantages, various reactor configurations and operation strategies have been proposed.

Among a number of the reactor technologies developed, the upflow anaerobic sludge blanket (UASB) reactor came of age as the simplest and most innovative reactor technology for efficient wastewater treatment (Lettinga, 1995; Speece, 1996). The UASB process can perform anaerobic degradation of organic material and the subsequent separation in a single vessel (Figure 1). The wastewater is supplied from the bottom of the vessel, and is evenly dispersed by distributing devices on the bottom. Within the vessel, a high concentration of biomass settles down, forming a dense sludge layer (referred to as the "sludge bed" or "sludge blanket"). In the sludge blanket, the organic matter in wastewater is anaerobically digested as the wastewater flows up to the top. During digestion, biogas bubbles (i.e., methane and carbon dioxide in general) are produced; both the upflowing wastewater and biogas carry the sludge up, resulting in the spontaneous formation of dense sludge granules (so called "granulation"; see Figures 1B and 1C). Sludge granules are primarily composed of complex microbial populations that are necessary for the anaerobic degradation of pollutants (Figure 2). In each pellet, heterogeneous microorganisms are packed as a spherical biofilm, forming an interesting microbial ecosystem with a characteristic internal architecture (this is discussed in more detail below). Biogas, purified water, and sludge granules are individually separated at the top of the vessel by the gas-solid separator, by which the effluent is discharged and the produced biogas is collected (Figure 1).

The striking advantages of the UASB process are (1) a massive amount of biomass can be retained in vessels as well-settled granules, which leads to the high-rate conversion of organic matter at a high organic loading rate, and (2) the reactor configuration is simple and economical because it requires neither an added substratum as is needed in anaerobic filters, nor effluent recirculation as in fluidized bed reactors. This process and the number of modified processes that appeared later, such as the expanded granular sludge bed (EGSB) reactor (Kato *et al.*, 1994), anaerobic baffled (AB) reactor (Barber and Stuckey, 2000), internal circulation (IC) reactor (Habets *et al.*, 1997), and upflow stage sludge bed (USSB) reactor (Lettinga *et al.*, 1997), have been successfully used worldwide. These new technologies have now been expanding the feasibility of anaerobic biotechnology, and have overcome several drawbacks that characterized the original process (i.e., the drawbacks related to using anaerobes). Over 1,000 full-scale, UASB-type reactors are running worldwide, and the UASB-based reactor is now recognized as one of the well-established biogas reactors (the following web site is useful for gaining more information on UASB processes: http://www.uasb.org/index.htm). Other reactor types have also been investigated which enhance the retention of biomass by immobilizing sludge on support materials in the reactors, such as fluidized-bed reactors (AB), anaerobic filter (AF) reactors, and hybrid reactors composed of a combination of, for example, a UASB-type reactor and an AF-type reactor (Speece, 1996).

As mentioned above, all of these new generation reactor technologies have the capacity to harbor a high amount of biomass by means of their unique reactor configurations. By combining the power of anaerobes and advanced reactor technology, the anaerobic waste/wastewater treatment processes now offer a powerful alternative for the treatment of many other wastewater streams that have been considered untreatable by anaerobic processes. Examples of such wastewater are low-strength wastewater and complex wastewater containing compounds recalcitrant to biodegradation. Over several decades, special attention has been paid to the anaerobic treatment of low-strength wastewater (<1,000 mg COD per liter) such as municipal sewage. Recent studies demonstrated the enhanced capabilities of anaerobic low-strength wastewater treatment systems under sub-tropical and even moderate climate conditions by incorporating proper reactor technologies (such as EGSB and two-phase separation) into reactor systems (Lettinga *et al.*, 1999).

The recent advances in anaerobic treatment technology have also allowed the processes to treat various types of complex organic wastewaters discharged in the chemical, petrochemical, textile, and pulp/paper industries (Macarie, 2000). Phthalate-containing wastewater is among the best examples of such waste streams. Phthalate isomers, which are primarily anthropogenic compounds, have been produced in massive amounts for use in the manufacturing of polyester resins, plastic bottles, plasticizers, polyester fibers, and other petroleum-based products in the world, and are consequently eluted in the waste streams generated in the corresponding industries, such as purified terephthalic acid (PTA) manufacturing (Macarie *et al.*, 1992). Generally, those waste streams contain high concentrations of phthalate isomers, benzoate, acetate, and 4-methylbenzoic acid, but it has been demonstrated that these streams can be successfully treated by UASB-based systems (Macarie *et al.*, 1992; Cheng *et al.*, 1997). Now, anaerobic (methanogenic) processes have increasingly been introduced to treat such wastewaters, and more than 10 full-scale anaerobic bioreactors are currently in operation for the treatment of phthalate isomer-containing wastewaters (Macarie, 2000) (the number of such plants is increasing).

High-rate anaerobic wastewater treatment technology under psychrophilic (4°C to 20°C) and thermophilic (above 45°C) conditions has also been developed (van Lier *et al.*, 1994; van Lier *et al.*, 1997; Lettinga *et al.*, 1999). Wastewaters, such as municipal sewage, are discharged at low temperature; sometimes such wastewater holds a temperature below 10°C. Recent studies have clearly demonstrated that anaerobic treatment under 4°C to 10°C is also possible with a sufficient organic removal efficiency at a high loading rate by employing advanced anaerobic technologies like the EGSB systems (Lettinga *et al.*, 1999). Thermophilic anaerobic digestion (normally operated between 50°C and 60°C) is expected to be a promising approach for high temperature wastewaters (such as those from food, chemical and petrochemical industries) due to the fact that the metabolic and growth rate of thermophilic anaerobes are higher than those of mesophiles, and many investigators have attempted to establish stable processes using these organisms (van Lier, 1996; van Lier *et al.*, 1997). Traditionally, anaerobic digestion of solid wastes has been successfully performed under two major temperature ranges: mesophilic (25-40°C) and thermophilic (50-65°C). In addition to solid waste digestion, recent reports indicate that thermophilic wastewater treatment processes can accommodate a very high loading rate at a feasible removal efficiency (van Lier *et al.*, 1997).

Thanks to the development of those anaerobic reactor technologies, and owing to the anaerobes that naturally occur in these processes, anaerobic treatment technology fundamentally contributes to the sustainable development of our society.

3. Community Structure and Microbiology of Anaerobic Sludge

To achieve stable and efficient operation of those anaerobic processes, knowledge of the microbial ecology of anaerobic sludge is fundamental in addition to the development of the reactor configuration, since the microbial communities in these processes play central roles in the decomposition/removal of pollutants in waste/wastewater. The recent development of molecular techniques in the field of microbial ecology has allowed us to explore the microbial diversity and community structure of these anaerobic processes. Owing to the application of these techniques (often in combination with classical methodologies), we now have a much clearer insight into community composition and architecture of anaerobic sludge, which are adapted to the treatment of a variety of wastes/wastewaters under different operation conditions.

3.1. Development and Application of Modern Molecular Techniques for Analyzing Community Composition and Functions in Anaerobic Processes

Traditionally, microbial communities in anaerobic sludge were analyzed either by light, fluorescent and electron microscopic observation, or by culture-dependent techniques such as most-probable number (MPN) counting and activity measurements (see a review by Schmidt and Ahring, 1996). Methanogenic archaea have a unique fluorescent deazaflavin-like cofactor (F_{420}) as a potential biomaker, and hence epifluorescence microscopy has also been widely used to detect and enumerate methanogens in complex anaerobic ecosystems by direct counting. However, detecting methanogen-cells based solely on F_{420} signals under microscopy has limitations in the precise counting of methanogens; i.e., it is recognized that intercellular concentrations of this cofactor vary up to 100-fold depending on the species and cultivation conditions (Lin and White, 1986; DiMarco *et al.*, 1990; Kamagata and Mikami, 1991); particularly cells of most *Methanosaeta* species have low levels of the cofactor which are not sufficient for identifying these cells based on the F_{420} fluorescent signals (Kamagata and Mikami, 1991). In addition to this feature, the cofactor has also been found in other non-methanogenic *Archaea* (Lin and White, 1986; DiMarco *et al.*, 1990).

Electron microscopy has also been widely used to view the spatial distribution of anaerobes within sludge granules, and a number of important findings have been made (MacLeod *et al.*, 1990; Grotenhuis *et al.*, 1991a; Guiot *et al.*, 1992; Fang *et al.*, 1994; 1995; MacLeod *et al.*, 1995). Scanning and transmission electron micrographs of sections of sludge granules showed that a variety of morphologically diverse microbes are present. These granules often exhibit a layered structure, in which each layer contains a different microbial community than those of other layers. Additionally, close juxtaposition of syntrophic bacteria-like and methanogenic archaea-like cells was found to occur within sludge granules. However, it should be noted that all of the descriptions that were made based solely on morphological observations were not sufficiently reliable for the precise identification of individual cells.

To overcome such drawbacks of morphological identification, an immunohistochemical technique has been used to unambiguously describe anaerobes in bioreactors (Macario and Conway de Macario, 1988; Macario *et al.*, 1989; Visser *et al.*, 1991; Zellner *et al.*, 1997). A number of calibrated antibody probes have been developed for various anaerobes and used to detect and quantify the targeted cells in various sludge samples. For example, 17 different antibody probes for 15 species of typical methanogenic archaea and 6 different antibody probes for 6 species of fermentative, syntrophic, and sulfate-reducing bacteria were used to quantitatively detect these populations within three different anaerobic sludge communities (Zellner *et al.*, 1997). In addition, these probes have been used to detect the targeted cells by microscopy; using probes for methanogens and syntrophic bacteria, the proximity of those microbes has been clearly demonstrated with thin sections of anaerobic sludge granules (Grotenhuis *et al.*, 1991b). However, since pure cultures of the targeted organisms are required to create antibody probes, this method is applicable only for previously cultured microbes. In fact, quantitative microbial community analysis of anaerobic sludge with these calibrated antibody probes revealed that the reactive cells accounted for only a very limited fraction of the total population of the anaerobic communities (Macario and Conway de Macario, 1988). These facts suggest that a large part of the community constituents in anaerobic sludge is still composed of unknown (yet-to-be-cultured) microbes.

Cultivation-dependent methods have also been widely used for describing the community structure of anaerobic bioreactors. Methanogenic degradation of complex organic substances involves at least three different trophic groups of anaerobes, namely fermentative heterotrophs, proton-reducing syntrophic bacteria, and methanogenic archaea, and each group contains diverse microorganisms that undertake different metabolic tasks. The enumeration of respective trophic groups of anaerobes has been done using MPN counting using selective media containing specific substrates for each trophic group (see a review of Schmidt and Ahring, 1996). However, the data obtained by means of cultivation-based methods are only relevant for microbial populations that are able to grow in artificial media, and it should be noted that a certain fraction of community members might not be able to grow under those conditions. Activity measurements (such as the measurement of methane-producing activity; i.e., maximum methane-producing rate per unit of biomass) using selective media have been widely used to evaluate the potential metabolic activity of respective trophic groups (Schmidt and Ahring, 1996), and they continue to provide valuable information on the metabolic potential of each sludge, in particular for the selection of good inoculum for the start-up of the process. However, this

Table 1. Community structure analyses of anaerobic waste/wastewater treatment sludges by retrieval of 16S rRNA genes

	No. 1	No. 4	No. 2	No. 3	No. 5	No. 6
Operation temperature (°C)	35	28	37	55	35	35
Reactor type	AnFB[a]	AnFB	UASB[b]	UASB	UASB	UASB
	wine distillation wastewater	synthetic wastewater containing acetate, methanol, and trichlorobenzene	synthetic organic wastewater composed of sucrose, propionate, acetate, and yeast extract	synthetic organic wastewater composed of sucrose, propionate, acetate, and peptone	synthetic terephthalate wastewater	synthetic 4-methylbenzoate wastewater
Target range of analysis	Bacteria	Bacteria	Prokaryotes	Prokaryotes	Bacteria	Bacteria
Number of sequences analyzed	369	145	114	110	106	139
Reference	Godon *et al.*, 1997	von Wintzingerode *et al.*, 1999	Sekiguchi *et al.*, 1998	Sekiguchi *et al.*, 1998	Wu *et al.*, 2001	Wu *et al.*, 2001
Archaea[c]						
Euryarchaeota						
Methanomicrobia	-	-	●●[d]	●●	-	-
Methanobacteria	-	-	○	●	-	-
Thermoplasmata	-	-	○	○	-	-
WSA2 (clone cluster)	-	-	○○	○	-	-
Crenarchaeota						
C2 (clone cluster)[e]	-	-	○	○	-	-
Bacteria						
Proteobacteria						
Alphaproteobacteria	●	●	○	○	○	○
Betaproteobacteria	●	●	○	○	○	○
Gammaproteobacteria	●	●	○	○	○	●
Deltaproteobacteria	●●	●	●●	○	●●●	●●●
Chloroflexi	●	●●	●●	●●	●	●
Firmicutes	●●●	●●	●	●	○	●
Spirochaetes	●	●●	●	○	○	○
Bacteroides	●●	○	●	○	○	●
Actinobacteria	●	○○	●	○	○	○
Synergistes	●	○	○○	●	○○	○
Planctomycetes	●	○	○○	●	○	●

Chlorobi	○	○○	●	●	○	○
Nitrospira	○	○	○○	●●	○	○
Acidobacteria	○	●●	○	○	○	○
Verrucomicrobia	●	○	○	○	○	○
Bacterial candidate phyla (clone clusters)[e]						
OP10	○	●	○	●	○	●
BA024 group	○	○	○	○	●●	●
OP8	○	○	●	●	○	○○
TM6	●	●	○	○	○	○
OP11	○	○	○	○	○○	○○
EM3	○	○	●	●	○	○
OP9	○	○	○	●	○	○
Termite group I	○○	○	○	○	○	○
OP3	○	○	●	○	○	○
OS-K	○	○	●	○	○	○
TM7	○	○	○○	○	○	○
others	○○	○○	○○	○○	○	○

[a] AnFB, anaerobic fluidized bed reactor. [b] UASB, upflow anaerobic sludge blanket reactor.

[c] Archaeal and bacetrial phyla and orders are named according to the taxonomic outline of Bergey's Manual.

[d] Frequency of sequences assigned with a phylogenetic group in percentage of the total number of sequences analyzed: ○ , 0%; ○○, 0.1-0.9%; ●, 1-9%; ●●, 10-29%; ●●●, >29%.

[e] Candidate phyla (order) are named based on a review reported by Hugenholtz (2002).

Table 2. Community analyses of archaeal populations in various anaerobic wastewater treatment systems by retrieval of 16S rRNA genes

	No. 1	No. 4	No. 9	No. 2	No. 3	No. 5	No. 6	No. 7	No. 8
Operation temperature (°C)	35	35	55	37	30	37	37	55	10-14
Reactor type[a]	FB	UASB	ABR	FB	DF	IC	HR	HR	HR
Wastewater type	wine distillation wastewater	synthetic terephthalate wastewater	industrial dye wastewater	citric acid production wastewater	milk processing wastewater	potato processing wastewater	synthetic wastewater containing acetate, propionate, and butyrate	molasses wastewater	synthetic wastewater composed of glucose and peptone
Target range of analysis	Archaea	Archaea	Archaea	Archaea	Archaea	Archaea	Archaea	Archaea	Archaea
Number of sequences analyzed	74	72	98	33	37	26	30	34	40
Reference	Godon *et al.*, 1997	Wu *et al.*, 2001	Plumb *et al.*, 2001	McHugh *et al.*, 2003	McHugh *et al.*, 2003	McHugh *et al.*, 2003	McHugh *et al.*, 2003	McHugh *et al.*, 2003	McHugh *et al.*, 2003
Archaea[b]									
Euryarchaeota									
Methanomicrobia									
Methanosarcinales	●●●[c]	●●●	●●●	●●●	●●●	●●●	●●●	●●●	●●●
Methanomicrobiales	○	●●	●	○	○	○	○	○	●
Methanobacteria	●●●	●●	●	●●●	●●	●●	○	●●●	●●
Thermoplasmata	●	○	○	○	○	○	○	○	○
WSA2 (clone cluster)	○	○	○	●	○	●●●	●●●	○	●
Crenarchaeota									
C2 (clone cluster)[d]	●	○	○	○	○	○	○	○	○

[a] FB, fluidized bed reactor; UASB, upflow anaerobic sludge blanket reactor; ABR, anaerobic buffled reactor; DF, downflow filter reactor; IC, internal circulation reactor primarily based on UASB-type system;
[a] HR, hybrid reactor primarily based on UASB-type system.
[b] Archaeal phyla and orders are named according to the taxonomic outline of Bergey's Manual.
[d] Frequency of sequences assigned with a phylogenetic group in percentage of the total number of sequences analyzed: ○, 0%; ○○, 0.1-0.9%; ●, 1-9%; ●●, 10-29%; ●●●, >29%.
[e] Candidate phyla (order) are named based on a review reported by Hugenholtz (2002).

method provides less information about which organism is responsible for the metabolic activities of interest.

So far, a number of microorganisms from anaerobic sludge samples have been isolated and characterized (Harmsen, 1996; Stams *et al.* 1992; Schink. 1997). In fact, microbiologists have spent a great deal of time isolating functionally important microorganisms from methanogenic communities due to the fastidious characteristics of anaerobes (slow growth rate, syntrophic association, or low growth yield). However, only a few of the microorganisms and their functions in the degradation processes have been characterized due to the limitations of these cultivation-dependent methods (Amann, 1995; Hugenholtz *et al.*, 1998a; Hugenholtz, 2002). To compensate for these limitations, cultivation-independent molecular methods have been developed and intensively utilized for the study of microbial communities within anaerobic sludge. These molecular techniques have substantially expanded our knowledge of community composition and structure of anaerobic sludge (Stams and Oude Elferink, 1997; Oude Elferink *et al.*, 1998b; Sekiguchi *et al.*, 2001a; Hofman-Bang *et al.*, 2003). In particular, rRNA-based detection and identification methods have become indispensable tools for precisely determining the characteristics of anaerobic sludge communities. Several rRNA- and rRNA gene-based molecular methods have been developed, such as rRNA gene-clone library analysis, rRNA-targeted membrane hybridization with oligonucleotide probes, and fluorescence *in situ* hybridization (FISH). Studies of anaerobic sludge employing such methods (in combination with other methods described above) have become standard.

It should also be noted that microbiologists are still interested in isolating and characterizing the anaerobes in the processes, since a pure culture provides much more information than culture-independent molecular community analyses can ever provide. Therefore, if we could combine the data from pure cultures with molecular analyses, we could obtain much more precise information on the microbes of interest in conjunction with their abundance and spatial distribution, and changes in their population over time.

3.2. Microbial Community Composition in Anaerobic Processes

Sequence analysis of rRNA genes has become a most powerful tool to determine the composition of entire microbial populations inhabiting an ecosystem (Amann *et al.*, 1995). By 16S rRNA gene cloning and subsequent sequencing of each clone, it is possible to uncover each species residing within a complex microbial community, and a rough estimate of the phylogenetic diversity (species richness and evenness) of anaerobes in the ecosystems can be obtained. Studies in which 16S rRNA gene-cloning analysis was conducted on anaerobic sludge adapted to waste/wastewater treatment under various operation conditions have been reported (Tables 1 and 2); these rRNA gene-based analyses have uncovered a vast diversity of microorganisms in these ecosystems (Godon *et al.*, 1997; Sekiguchi *et al.*, 1998; von Wintzingerode *et al.*, 1999; Wu *et al.*, 2001a; 2001b; Liu *et al.*, 2002a). 16S rRNA cloning analyses were performed on digested sewage sludge as well (Ng *et al.*, 1994; Moffett *et al.*, 2000). As can be seen in Figure 3, diverse microbial constituents belonging to various bacterial phyla (divisions) have been detected in anaerobic sludge used to treat various wastewaters under different conditions. Typical 16S rRNA genes frequently detected in these anaerobic sludges are those of the bacterial phyla *Proteobacteria* (in particular *Deltaproteobacteria*), *Firmicutes*, and *Chloroflexi* (see Table 1). Some examples of commonly observed 16S rRNA gene clones, which may represent the principal constituents of anaerobic (methanogenic) reactors that treat waste/wastewater are briefly summarized below. In addition, representatives of the important anaerobes isolated and characterized so far are also indicated, along with their closely related clones found in sludge.

3.2.1. Deltaproteobacteria

Important Gram-negative sulfate-reducers and syntrophic bacteria belong to the delta-subphylum of *Proteobacteria* [the class *Deltaproteobacteria* (Garrity, 2001)]. Among these, species of the genus *Syntrophobacter* are particularly well recognized as syntrophic bacteria capable of degrading propionate in association with hydrogenotrophs (Boone and Bryant, 1980; Wallrabenstein *et al.*, 1995a; Harmsen *et al.*, 1998), because propionate is one of the most important intermediates in the anaerobic degradation of organic compounds. Recently, *Smithella propionica* has been isolated by Liu *et al.* as a new propionate-oxidizing syntroph (Liu *et al.*, 1999), although an organism very closely related to *S. propionica* (in terms of phylogeny) was previously enriched on propionate medium and identified phylogenetically (formerly known as strain SYN7) (Harmsen, 1996; Harmsen *et al.*, 1996a). Species of the genus *Syntrophus* are known as well-characterized benzoate-degrading syntrophs (Mountfort *et al.*, 1984; Wallrabenstein *et al.*, 1995b). Among these species, *Syntrophus aciditrophicus* isolated by Jackson *et al.* is known to be a versatile syntroph that is capable of oxidizing not only fatty acids but also benzoate (Jackson *et al.*, 1999).

16S rRNA gene sequences affiliated with this group were frequently retrieved from mesophilic (30-40°C) anaerobic wastewater sludge. In particular, clones phylogenetically belonging to the genera *Desulfovibrio*, *Desulfobulbus*, *Syntrophobacter*, *Syntrophus* and

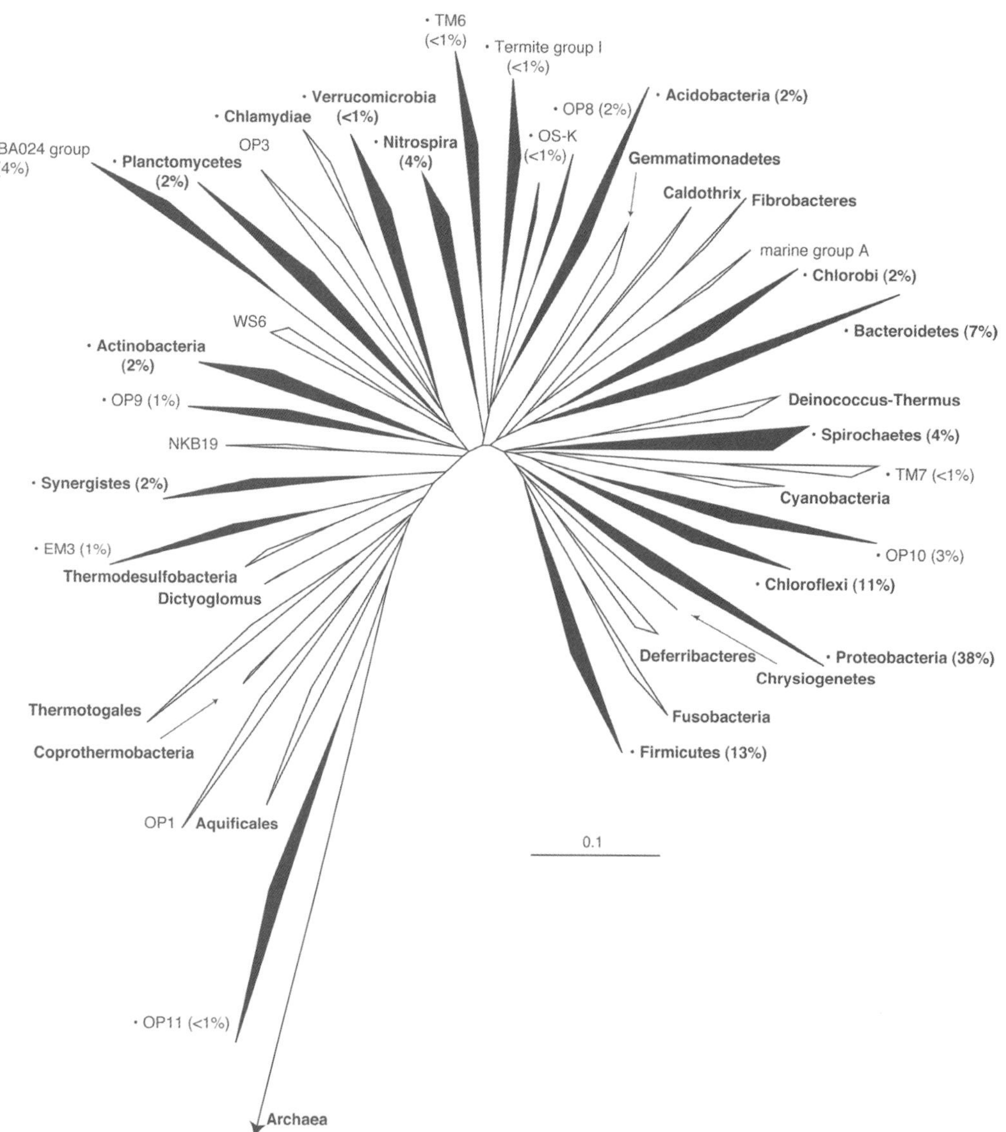

Figure 3. Evolutionary distance dendrogram of bacterial diversity derived from comparative analysis of 16S rRNA gene sequences. The tree was constructed using the ARB software package based on a sequence database deposited by Hugenholtz (2002) in the Ribosomal Database Project user-submitted alignments download site (http://rdp.cme.msu.edu/html/alignments.html). Major lineages (phyla) are shown as wedges. Phyla with cultivated representatives are in gray, and are named according to the taxonomic outline of Bergey's Manual. Phyla known only from environmental sequences are in white, and are named based on a review by Hugenholtz (2002). The dots shown at the phylum names indicate that 16S rRNA gene clones belonging to respective phyla have been detected in anaerobic processes. The numbers after the dots show the abundance of these clones (as percentages) in the total bacterial clones retrieved from such bioprocesses. The scale bar represents 0.1 changes per nucleotide.

Smithella have been frequently retrieved from methanogenic wastewater sludge. For example, Godon *et al.* (1997) reported the microbial community structure of methanogenic sludge in a fluidized bed reactor treating vinasses (wine distillation waste) based on a 16S rRNA gene clone library analysis, in which a number of clones closely related with species of the genera *Desulfovibrio*, *Syntrophus* and *Smithella* were detected (Godon *et al.*, 1997). Clones related with these anaerobes have also been detected at high frequencies in methanogenic sludge used in the treatment of actual and artificial wastewater (Sekiguchi *et al.*, 1998; Wu *et al.*, 2001a, b; Liu *et al.*, 2002a). The findings strongly imply that these populations are fundamental constituents of reactors for treating high-strength organic wastewater.

The interpretation of the physiological functions of microbial constituents based solely on the phylogenetic affiliation determined by 16S rRNA gene sequences should be carefully considered, since the genetic distances (i.e., physiological properties) between two organisms do not always correlate with the difference between their 16S rRNA genes (Stackebrandt and Goebel, 1994). However, we can roughly suppose the physiological functions of microbes identified by sequencing of 16S rRNA gene clones that are closely related to organisms that have been successfully cultivated. Within this subphylum, however, several uncultured (and hence functionally unknown) clusters have also been detected in anaerobic processes in addition to those recognizable clones. Wu *et al.* (2001) reported the community structure of methanogenic, synthetic terephthalate-wastewater treating sludge based on 16S rRNA gene cloning analysis (Wu *et al.*, 2001a). Notably, they reported that a large fraction of the retrieved clones formed a phylogenetically independent, yet-to-be-cultured bacterial cluster (referred to as "group TA") within *Deltaproteobacteria*; the clones belonging to this cluster accounted for approximately 64% of the total bacterial clones analyzed. These populations were found to be abundant (more than 87% of total bacterial cells) in the original sludge based on direct counting with FISH using an oligonucleotide probe specific for the 16S rRNA of group TA (Wu *et al.*, 2001a). Due to the abundance of these populations, this new, yet-to-be-cultured group was presumed to be involved in the methanogenic terephthalate degradation. However, because no cultured representatives are involved in the group, we cannot precisely know the physiological traits of these populations at the moment.

3.2.2. Firmicutes

The majority of well-characterized Gram-positive heterotrophs, acetogens, and sulfate-reducing bacteria are known to be part of the bacterial phylum *Firmicutes* (previously known as "low G+C Gram-positive bacteria") (Garrity, 2001). These organisms include those of the genera *Clostridium*, *Thermoanaerobacter*, and *Desulfotomaculum*. Members of the family *Syntrophomonadaceae* (Zhao *et al.*, 1993), such as those of the genera *Syntrophomonas* (McInerney *et al.*, 1981), *Syntrophospora* (Zhao *et al.*, 1990), *Thermosyntropha* (Svetlitshnyi *et al.*, 1996), and *Syntrophothermus* (Sekiguchi *et al.*, 2000), belong to this phylum. All of these are well-known Gram-positive syntrophs capable of degrading butyrate and its analogues in syntrophic association with hydrogenotrophic methanogens. Acetate-oxidizing syntrophs such as *Clostridum ulutnense* (Schnürer *et al.*, 1999) and *Thermoacetogenium phaeum* (Hattori *et al.*, 2000), and amino acid-degrading syntrophs such as *Caloramator coolhaasii* (Plugge *et al.*, 2000) and *Gelria glutamica* (Plugge *et al.*, 2002b) also belong to this phylum.

A number of diverse 16S rRNA gene clones within this phylum have been found in anaerobic processes. Clones that are relatively close to previously cultured species such as those of the genera *Clostridium*, *Eubacterium*, *Streptococcus*, and members of the family *Syntrophomonadaceae* were frequently found within anaerobic sludge. However, it should be noted that various 16S rRNA gene clones that are distantly related to cultured microbes were found in addition to the recognizable clones.

One of the important features of this phylum is that a relatively high number of 16S rRNA gene clones can be retrieved from thermophilic (55°C) UASB sludge. We compared the microbial community structure and diversity of mesophilic (37°C) and thermophilic (55°C) UASB sludge from the same wastewater treatment by 16S rRNA gene cloning analysis (Sekiguchi *et al.*, 1998). rRNA gene clones affiliated with *Deltaproteobacteria*, which are the most frequently retrieved clones among all bacterial phyla from the mesophilic anaerobic sludge, were not found in the clone library from the thermophilic sludge. Instead, clones belonging to *Firmicutes* are more abundant in this clone library. In particular, clones closely related with members of the genus *Pelotomaculum* were found within the sludge. A thermophilic propionate-oxidizing bacterium *Pelotomaculum thermopropionicus* was isolated and described by Imachi *et al.* (Imachi *et al.*, 2000; Imachi *et al.*, 2002). The isolation of thermophilic propionate-oxidizing syntrophic bacteria had long been difficult in spite of their importance in thermophilic anaerobic digesters (Stams *et al.*, 1992). However, these researchers accomplished the isolation using a molecular probe-mediated cultivation and isolation strategy; i.e., they employed a full-cycle rRNA approach for the enrichment culture that utilized propionate under methanogenic conditions, and used the designed probe to identify the targeted cells in cultures. Very interestingly, the isolate (strain SI) was affiliated with *Desulfotomaculum* clade ['*Desulfotomaculum* lineage I' (Stackebrandt *et al.*, 1997)], although it was not able to reduce sulfate. The researchers also demonstrated, using fluorescence *in situ* hybridization direct count, that a species of the genus *Pelotomaculum* (*P. thermoproionicus*) was present in thermophilic UASB sludge granules as a significant population (Imachi *et al.*, 2000). It is, therefore, likely that such populations compensate for the absence of *Syntrophobacter* and *Smithella* species in thermophilic sludges, playing a role in methanogenic propionate degradation within these ecosystems. Furthermore, *Desulfotomaculum thermobenzoicum* subsp. *thermosyntrophicum,* also of the phylum *Firmicutes,* has been isolated as another strain capable of degrading propionate in syntrophic association with methanogens under thermophilic conditions (Plugge *et al.*, 2002a).

Recently, phthalate-degrading syntrophs were found to be within the same clade of the genus *Pelotomaculum* in '*Desulfotomaculum* lineage I' (Qiu *et al.*, 2004). In addition, a new species of the genus *Sporotomaculum* (*S. syntrophicum*), which is also within '*Desulfotomaculum* lineage I', was identified as a syntrophic benzoate-degrading bacterium (Qiu *et al.*, 2003). However, none of the 16S rRNA gene clones closely related to these syntrophic bacteria have been found in the community analyses of the sludge used to treat terephthalate-containing wastewater, and, hence, the distribution and abundance of these populations within anaerobic sludge remain unknown.

It was also discovered that some of the clones retrieved from thermophilic anaerobic sludge are related to the known thermophilic, acetate-oxidizing syntrophic bacterium, *Thermoacetogenium phaeum* (Hattori *et al.*, 2000), although their actual physiological functions remain to be characterized. All of the known acetate-oxidizing syntrophs, except a newly isolated acetate-oxidizing bacterium *Thermus lettingae* (Balk *et al.*, 2002), are known to belong to the phylum *Firmicutes*. So far, these include one mesophilic and three thermophilic strains that are known to perform syntrophic acetate oxidation under methanogenic conditions (Zinder and Koch, 1984; Stieb and Schink, 1985; Lee and Zinder, 1988; Hattori *et al.*, 2000). A recent study showed that some cultures of acetate-oxidizing iron-reducers (such as the genus *Geobacter*) could oxidize acetate in co-culture with hydrogen-utilizing sulfate- or nitrate-reducers in the absence of oxidized forms of iron (Cord-Ruwisch *et al.*, 1998). In general, acetate is converted to methane by the acetate-cleaving methanogens, *Methanosarcina* and *Methanosaeta*, and we still believe that they are the primary acetate consumers in reactors. However, these findings together with previous descriptions and other studies of sludges (Petersen and Ahring, 1991; Ahring *et al.*, 1993; Uemura and Harada, 1993, 1995; Schnürer *et al.*, 1999) strongly indicate that methane reactors harbor diverse organisms even when they metabolize such a simple substrate. It will be of great interest to determine the populations of these organisms and their competition with the acetate-cleaving methanogens in the sludge communities.

3.2.3. Chloroflexi

To date, a number of clone clusters which are only distantly related to other known phyla and classes in the domain *Bacteria* but containing few or no cultured microorganisms have been recognized (Hugenholtz *et al.*, 1998a; Hugenholtz *et al.*, 1998b; Hugenholtz, 2002). Among these uncultured clades, the bacterial phylum *Chloroflexi* (formerly known as "green non-sulfur bacteria") (Garrity and Holt, 2001a) has been recognized as a typical bacterial phylum represented by a number of diverse environmental 16S rRNA gene clones with only a few cultured representatives (Hugenholtz *et al.*, 1998a). The phylum *Chloroflexi* can be divided into more than four major subphyla (subphyla I, II, II, and IV) on the basis of 16S rRNA gene sequences (Hugenholtz *et al.*, 1998a; Garrity and Holt, 2001a). The phylum contains cultured microbes belonging to the genera *Chloroflexus*, *Oscillochloris*, *Herpetosiphon*, *Sphaerobacter*, and *Roseiflexus*, which belong to subphylum III. The other three subphyla (I, II, and IV) are composed almost solely of a variety of environmental clones.

Subphylum I contains the most diverse environmental clones among the four subphyla of *Chloroflexi*, e.g., clones from hot spring environments, sediments, subsurface environments, aerobic and anaerobic wastewater treatment sludge, and contaminated aquifers (Hugenholtz *et al.*, 1998b; Sekiguchi *et al.*, 1998; Björnsson *et al.*, 2002; Juretschko *et al.*, 2002). Importantly, members of subphylum I are known to be widespread in anaerobic ecosystems and were frequently retrieved from various anoxic environments. Anaerobic wastewater treatment sludge is also one of the typical habitats of these microbes (Sekiguchi *et al.*, 1998; von Wintzingerode *et al.*, 1999; Wu *et al.*, 2001a; 2001b). These findings highlight the ecological and physiological breadth of these microbes in anaerobic processes. In fact, it was shown that these bacteria are numerically abundant in anaerobic, thermophilic granular sludge for wastewater treatment, and that they are important constituents in sludge granules, in that they maintain the granule structure and also trigger filamentous bulking [(Sekiguchi *et al.*, 1998; 1999; 2001b), see below for details]. Studies employing whole cell *in situ* hybridization analyses have suggested that members of subphylum I have a filamentous morphotype with a wide range of thickness (Sekiguchi *et al.*, 1999; 2001b; Björnsson *et al.*, 2002; Juretschko *et al.*, 2002).

For a long time, there had been no cultured representatives of the subphylum, and we were, therefore, not able to determine the physiological traits of the filamentous organisms. However, very recently, pure cultures of two representative filaments belonging to this subphylum, i.e., *Anaerolinea thermophila* and *Caldilinea aerophila*, have been obtained from thermophilic UASB sludge and a hot spring (Sekiguchi *et al.*, 2003). *A. thermophila* was isolated from a UASB sludge as a thermophilic, anaerobic, filamentous heterotroph. The organism can slowly grow anaerobically on a limited number of carbohydrates in the presence of yeast extract. Interestingly, it grew more rapidly when it was co-cultivated with hydrogenotrophic methanogens; this implies that the organism is a "semi-syntrophic" bacterium that requires interspecies hydrogen transfer for efficient growth (Sekiguchi *et al.*, 2001b). *Chloroflexi*-type clones belonging to subphyla other than subphylum I have not been found to be abundant in typical anaerobic waste/wastewater treatment processes.

3.2.4. Other Bacterial Phyla

In addition to the groups mentioned above, a variety of 16S rRNA genes belonging to other bacterial phyla have been found to be present in anaerobic processes, suggesting that anaerobic sludge harbors greater microbial diversity than we ever imagined (Figure 3, Table 1). Importantly, 16S rRNA gene-based analysis has shown that a large fraction of these clone sequences do

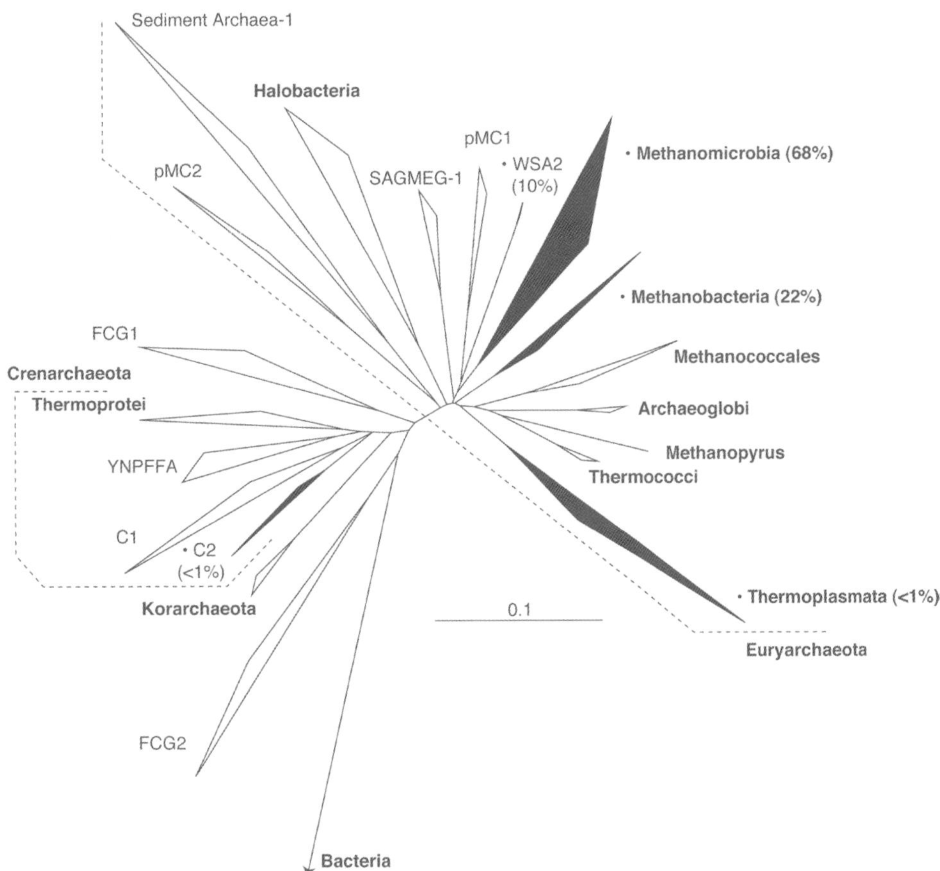

Figure 4. Evolutionary distance dendrogram of archaeal diversity derived from comparative analysis of 16S rRNA gene sequences. The tree was constructed using the ARB software package as in Figure 1. Major lineages (class) are shown as wedges. Classes with cultivated representatives are in gray, and are named according to the taxonomic outline of Bergey's Manual. Putative classes known only from environmental sequences are in white, and are named based on a review by Hugenholtz (2002). The dots shown at the class names indicate that 16S rRNA gene clones belonging to the respective class have been detected in anaerobic processes. The numbers after the dots show the abundance of these clones (as percentages) in the total archaeal clones retrieved from such bioprocesses. The scale bar represents 0.1 changes per nucleotide.

not match with 16S rRNA genes of previously cultured bacteria. Typical examples of these clones in recognized bacterial phyla are those of the phyla *Synergistes*, *Chlorobi*, and *Nitrospira*. In addition, a number of clones have been found to fall into putative (candidate) phyla, in which various environmental 16S rRNA gene sequences are recognized with no cultured representatives. Table 1 lists the possible candidate phyla that have been detected in anaerobic processes. Interestingly, clones belonging to the candidate phylum BA024 group were frequently found in methanogenic sludge used to treat terephthalate and 4-methylbenzoate-containing wastewaters (Wu *et al.*, 2001a; 2001b). Because these clones have only been found in sludge that degrades these aromatic compounds under methanogenic conditions, these microbes probably participate, directly or indirectly, in the degradation of those compounds. However, due to the lack of pure cultures representing the candidate phylum, we do not have accurate information pertaining to these populations.

3.2.5. *Archaea*

From the microbiological point of view, methanogens compared to other heterotrophic bacteria are relatively well-understood microbes in anaerobic reactors. To date, a large number of methanogens have been isolated and described (Garcia *et al.*, 2000; Garrity and Holt, 2001b). All of the methanogens recognized so far belong to the phylum *Euryarchaeota* of the domain *Archaea*. In general, almost all of archaeal 16S rRNA genes retrieved from anaerobic sludge were found to be those of *Euryarchaeota*. By 16S rRNA gene cloning analysis of *Archaea* within methanogenic sludge, several clones closely related with those of methanogens are found (Figure 4, Table 2). Most of the studies based on 16S rDNA cloning analyses of methanogens suggest that the number of genera of predominant methanogens in biogas reactors is limited (Hiraishi *et al.*, 1995; Plumb *et al.*, 2001; McHugh, 2003). The most frequently retrieved archaeal clones are closely related with members of the genera *Methanosaeta* [formerly *Methanothrix* (Huser *et al.*, 1982; Patel and Sprott, 1990;

Kamagata and Mikami, 1991; Kamagata *et al.*, 1992)], *Methanosarcina*, *Methanospirillum*, *Methanobacterium*, and *Methanothermobacter* [formerly known as *Methanobacterium* (Wasserfallen *et al.*, 2000)]. Among these, 16S rRNA gene clones closely related with *Methanosaeta* spp. are the clones most frequently retrieved from methanogenic processes.

However, we still have difficulty in precisely characterizing the community structures and functions of archaeal populations within these ecosystems since some of the 16S rRNA clones retrieved show no similarity to 16S rRNA of any microbes that were cultured and characterized. Examples are those of the phylum *Creanerchaeota*, class *Thermoplasmata*, and a putative clone cluster at the class level (tentatively called "WSA2") (Figure 4, Table 2). Interestingly, clones belonging to the candidate class WSA2 are abundant in certain methanogenic sludges; this may imply that these are significant populations that play roles in the treatment processes. In addition, there are a number of clones that are distantly related to those of the cultured methanogens. Such clones correspond to the order *Methanomicrobiales*.

3.3. Phylogenetic Diversity

One of the important aspects of 16S rRNA gene clone library analysis is that we can roughly view the phylogenetic diversity (species richness and evenness) of anaerobes in these ecosystems through analyzing the cumulative number of different phylotypes along with the number of clones analyzed. Godon *et al.* estimated the bacterial diversity of methanogenic sludge fed with vinasses by depicting the cumulative number of different phylotypes against the number of bacterial clones analyzed (Godon *et al.*, 1997). They found 139 different phylotypes (in which clones having more than 96% sequence similarity were grouped into one phylotype) with 460 clones analyzed (Godon *et al.*, 1997). This suggests that at least 140 different bacterial "species" were present within the community. We also estimated the difference in microbial diversity between mesophilic and thermophilic UASB sludges treating identical artificial wastewater (Sekiguchi *et al.*, 1998). From this estimation, it was suggested that the microbial contents in the mesophilic granules were almost twice as diverse as that of the thermophilic granule, although both of the granules had received almost the same substrates over two years of operation. In addition, by comparing the estimations obtained in the above two studies, the estimated diversity of the sludges used to treat artificial waste streams was suggested to be much lower than that of the sludge used to treat actual wastes (i.e., vinasses).

3.4. Quantitative Detection of Individual Populations in Anaerobic Sludges

As mentioned above, the retrieval of 16S rRNA gene sequences provides a more comprehensive view of the microbial community structure within anaerobic biological processes. However, the results should be interpreted with caution with respect to the limitations associated with the technique. Several steps, such as DNA extraction, PCR amplification, and cloning steps, are known to introduce possible biases in qualitative and quantitative determinations of microbial communities (Cariello *et al.*, 1990; Brakenhoff *et al.*, 1991; Kopczynski *et al.*, 1994; Suzuki and Giovannoni, 1996; Wintzingerode *et al.*, 1997; Qiu *et al.*, 2001; Speksnijder *et al.*, 2001; Kanagawa, 2003). In particular, 16S rRNA gene cloning analysis provides less (or to some extent biased) information on the quantitative composition of a microbial community. Therefore, results from the gene retrieval should be supported by other molecular or conventional analyses to precisely and quantitatively determine population structure.

The relative abundance and population dynamics of several important groups of anaerobes in anaerobic reactors have thus been studied mainly by 16S rRNA-targeted oligonucleotide hybridization (Raskin *et al.*, 1994b; 1995; Hansen *et al.*, 1999; Angenent *et al.*, 2002; Casserly and Erijman, 2003). For this purpose, group-specific oligonucleotide probes targeting 16S rRNA of various sulfate-reducing bacteria (Devereux *et al.*, 1992; Daly *et al.*, 2000; Hristova *et al.*, 2000), syntrophic bacteria (Harmsen, 1996; Hansen *et al.*, 1999), fermentative heterotrophs (Liu *et al.*, 2002a), and methanogenic archaea (Raskin *et al.*, 1994a; Zheng and Raskin, 2000) have been extensively developed (for more details, see a review by Hofman-Bang *et al.*, 2003). For example, Raskin *et al.* performed 16S rRNA-based community analyses of methanogens and Gram-negative sulfate-reducers in 22 different types of anaerobic digester sludges including one thermophilic (52°C) process (Raskin *et al.*, 1995). Based on the 16S rRNA-targeted quantitative membrane hybridization using probes for various phylogenetic groups of methanogens, they found that nearly all of active archaeal populations could be considered methanogens. Importantly, they indicated that *Archaea* (i.e., methanogens) in "healthy" sewage sludge digesters accounted for 8-12% of the total microbial population, based on rRNA sequence analysis. Furthermore, it was found that members of the orders *Methanosarcinales* and *Methanomicrobiales* were the major populations constituting the total archaeal populations. Similarly, fluorescence *in situ* hybridization direct count has been used for the same purpose (Sekiguchi *et al.*, 1998; Tagawa *et al.*, 2000; Gonzalez-Gil *et al.*, 2001). These methods provide precise information on the actual species abundance.

In addition, for particular groups of microbes, cellular contents other than RNA and DNA have used for the quantitative population measurement. For example, polar-lipid fatty acid (PLFA) analysis has been used to detect certain species of sulfate reducers and *Syntrophobacter* sp. (Moffett *et al.*, 2000). In this study, the polar-lipid fatty acid profiles of three species of the genus *Syntrophobacter*, which are recognized as representative propionate-oxidizing syntrophic bacteria, as well as that of *Desulforhabdas amnigenus* were examined to find possible biomakers. In addition, both 16S rRNA-targeting dot-blot hybridization and PLFA profiling were applied to characterize populations within anaerobic sludge used to treat different types of wastewater. The analyses showed that PLFA analysis of the whole microbial community of anaerobic sludge was useful for obtaining an overall impression of the microbial sludge composition and a quick comparison of different sludge types, although the PLFA profiling did not seem very useful for estimating the relative amount of sulfate reducers and syntrophs in sludge, and that the 16S rRNA dot-blot hybridization method provides a means for more sensitive and precise quantification of populations (Oude Elferink *et al.*, 1998a).

3.5. Spatial Distribution and Organization of Different Trophic Groups of Anaerobes in UASB Sludge Granules

Granulation of sludge is the most characteristic phenomenon associated with UASB reactors. In this phenomenon, microorganisms form dense aggregates as a result of their penchant for self-immobilization. This has been one of the most intriguing topics in the fields of microbial ecology and wastewater treatment engineering. This unique biofilm has been intensively studied, and a number of unique characteristics have been reported (Schmidt and Ahring, 1996). As mentioned above, one feature of the granules is the spatial organization of microorganisms. The spatial distributions of anaerobes within granular sludges have been visualized in thin-sections of granules by 16S rRNA-targeted fluorescence *in situ* hybridization analysis with confocal laser scanning microscopy (CLSM) (Harmsen, 1996; Harmsen *et al.*, 1996a; Rocheleau *et al.*, 1999; Santegoeds *et al.*, 1999; Sekiguchi *et al.*, 1999; Plumb *et al.*, 2001) as well as with immunological techniques (Macario *et al.*, 1991; Schmidt and Ahring, 1999). These studies have revealed detailed community structures that we had never been able to see by the conventional techniques. Generally, the inner layer consists mostly of aceticlastic methanogens, and the outer layer is composed of fermentative bacteria (MacLeod *et al.*, 1990; Guiot *et al.*, 1992; Sekiguchi *et al.*, 1999). One of the most fascinating findings is that a number of microcolonies of different microbes have been found to be present inside the granule, and they form a well-organized architecture performing interspecies hydrogen and substrate transfer between microcolonies (Stams *et al.*, 1989).

For example, Hermsen *et al.* (1996b) investigated the microbial composition and spatial orientation of microbes in the sludge granules of two UASB reactors used to treat either sucrose or a mixture of volatile fatty acids using oligonucleotide probes for *Bacteria*, *Archaea*, several phylogenetic groups of methanogens, and two syntrophic propionate-oxidizing strains, i.e., *Syntrophobacter fumaroxidans* (formerly known as MPOB; probe, MPOB1) (Harmsen *et al.*, 1995; 1998) and *Syntrophobacter pfennigii* (previously known as KoProp1; probe, KOP1) (Wallrabenstein *et al.*, 1995a). They found that both types of sludge granules had layered structures, in which the outer layer consisted mainly of bacterial microcolonies. In both sludge granules, large microcolonies were found internally within sludge granules, which consisted mainly of MPOB1 probe-positive, *S. fumaroxidans*-like cells. The microcolonies were also found to contain archaeal cells, which reacted with the ARC915 probe (a probe specific for the domain *Archaea* (Stahl *et al.*, 1988)). Based on *in situ* hybridization with various group specific probes for methanogens in thin-sections of the granules using CLSM, they found that the archaeal cells were very closely juxtaposed with *S. fumaroxidans*-like cells, and that the archaeal cells belonged to the family *Methanobacteriaceae*, indicating that the populations residing within syntrophic microcolonies perform interspecies hydrogen (and/or formate) transfer. This work demonstrated convincingly the power of molecular techniques to study the microbial architecture of sludge granules.

Using similar techniques, the location of members of the genera *Desulfobulbus*, *Syntrophobacter* and *Smithella* was visualized within sludge granules used to treat artificial wastewater containing propionate alone (methanogenic conditions) or with sulfate (sulfidogenic conditions) (Harmsen *et al.*, 1996a). The researchers found that *Smithella*-type cells formed large microcolonies in methanogenic sludge granules, in which *Methanospirillum*-like methanogens and *Smithella*-type cells were closely juxtaposed with each other. Similarly, Imachi *et al.* (2000) found that a thermophilic propionate-oxidizing syntroph (*Pelotomaculum thermopropionicus*) also forms microcolonies that are surrounded by hydrogenotrophic thermophilic methanogens (*Methanothemrobacter*), and their proximity is close enough to make propionate oxidation via interspecies hydrogen transfer energetically feasible.

Microsensors have also been used for estimating the distribution of particular metabolic activities within sludge granules. Microsensors for several parameters, such as pH, glucose, H_2S, H_2, and CH_4, have been developed so far, and applied to various types of granular sludge (or anaerobic biofilms). Additionally, fluorescence *in situ* hybridization has been performed in combination with microsensors to view the population structure and the activity distribution in sludge granules simultaneously. For example, Santegoeds *et al.* (1999) employed microsonsors for H_2S and CH_4 with molecular techniques such as FISH for methanogenic, methanogenic-sulfidogenic, and sulfidogenic sludges to view the profiles of activities and locations of sulfate-reducers and methanogens within the three different types of anaerobic sludge.

16S rRNA cloning analysis and subsequent design of fluorescence *in situ* hybridization probes [called "full-cycle rRNA approach" (Amann *et al.*, 1995)] allowed us to visualize several yet-to-be cultured populations within granular sludges and to speculate on their function (Sekiguchi *et al.*, 1999; Liu *et al.*, 2002a). We used the *in situ* hybridization technique combined with CSLM to visualize the location of several microorganisms of particular interest in both mesophilic and thermophilic granular sludges, which were detected in our previous 16S rRNA cloning analysis (Sekiguchi *et al.*, 1999). In this study, some unidentifiable bacteria, which were considered to be major bacterial components in the granules by the previous 16S rRNA cloning analysis (Sekiguchi *et al.*, 1998), were visualized by specifically designed and fluorescently labeled probes to reveal their *in situ* morphology and locations in the granule. One of the interesting findings was that the subsequent fluorescence *in situ* hybridization using an oligonucleotide probe targeting 16S rRNAs of the uncultured *Chloroflexi*-type clones (subphylum I) revealed that those microbes were thin filamentous cells that constituted a major population within the outer most layers of thermophilic sludge granules. Because they inhabit only the outer most layer of the sludge granules, they were thought to be fermentative heterotrophs that might degrade some primary substrates in the anaerobic degradation of organic compounds, such as certain carbohydrates. Later, this physiological trait was confirmed by a pure culture of one of these previously uncultured filaments (Sekiguchi *et al.*, 2001b). This work also demonstrated that granular sludge is an ideal ecosystem for study in the microbial ecology of uncultured microbial phylotypes, since the sludge granules harbor a vast array of diverse uncultured microbes, and because these populations can be easily visualized within sludge granules for the purpose of speculating on their physiological functions.

These results using molecular cloning of rRNA genes, 16S rRNA-based quantitative measurements, and fluorescence *in situ* hybridization detection of microbial constituents within anaerobic bioprocesses indicate that molecular techniques are important approaches in the exploration of the anaerobes that inhabit such ecosystems.

3.6. Microbial Causative Agents for Granulation and Bulking Phenomena in Methane Fermentation Processes

Besides the internal architecture of sludge aggregates, other important issues include the mechanisms of granular sludge formation (granulation) and disintegration (e.g. bulking). Because the granulation of seed sludge (dispersed sludge like sewage sludge) is the major prerequisite for the successful start-up of UASB reactors, enhancing and controlling the granulation phenomenon in reactors are the fundamental engineering needs in reactor operations, and thorough knowledge of the mechanisms of these phenomena is required. In contrast, the deterioration of the settling properties of granular sludge is also one of the important problems associated with UASB technology. Once the sludge granules within a reactor become fluffy (bulking) or disintegrated (dispersed type), they do not properly settle down, and consequently the reactor cannot further accept high loading rates and needs to be inoculated with additional seed sludge transferred from other plants. Molecular techniques reveal the important constituents of the granulation and bulking of sludge granules and provide valuable information for the stable operation of these processes.

Several factors are known to affect the granulation phenomenon, i.e., (i) the physico-chemical properties of constituents in sludge such as the surface hydrophobicity of cells (Grotenhuis *et al.*, 1992; Daffonchio *et al.*, 1995; Schmidt and Ahring, 1996), (ii) the presence of extracellular polymers (Schmidt and Ahring, 1996; Veiga *et al.*, 1997), and (iii) the composition of microorganisms (Wiegant and de Man, 1986; Wiegant, 1987; Hulshoff Pol *et al.*, 1988; Schmidt and Ahring, 1996). Among these, the microbial constituents are thought to be the decisive factor in many cases. In general, filamentous and aggregating types of *Methanosaeta* cells are commonly observed in mesophilic UASB granules (Dubourguier *et al.*, 1987; Wiegant, 1987; Kamagata and Mikami, 1990; MacLeod *et al.*, 1990; Grotenhuis *et al.*, 1991b; Macario *et al.*, 1991; Sekiguchi *et al.*, 1999). For a long time, they have been considered important for making the cores of sludge granules by constructing web-like structures. In fact, the high abundance of *Methanosaeta*-like archaea has been frequently confirmed in various UASB sludge aggregates using molecular and conventional techniques. In contrast, *Methanosarcina*-like methanogens have been found to be less abundant in such aggregates (Sekiguchi *et al.*, 1999), although they share the same substrate for methanogenesis, i.e., acetate.

For the formation of well-settled thermophilic (55-60°C) sludge, thin filamentous microorganisms, which are morphologically different from *Methanosaeta*, have been considered essential for granulation (Uemura and Harada, 1993; 1995; Syutsubo *et al.*, 1997; Sekiguchi *et al.*, 1999). In fact, long filament-type *Methanosaeta* cells are rarely observed in thermophilic UASB granules, whereas short filament-type (dispersing-type) *Methanosaeta* cells can be frequently found. The thin filamentous microbes appear to entirely cover the thermophilic granules forming a web-like coating. As mentioned above, the full-cycle rRNA approach for analyzing thermophilic granular sludge provided solid evidence that this type of filament phylogenetically belongs to *Chloroflexi* subphylum I (Sekiguchi *et al.*, 1999). In addition, we have shown that at least one of these filamentous populations on the surface of the thermophilic granules was a slow growing bacterium [*Anaerolinea thermophila* (Sekiguchi *et al.*, 2003)] that is able to utilize only a few types of carbohydrates in the presence of yeast extract (see above). It has been reported that it was difficult for the granulation of thermophilic sludges to occur when only volatile fatty-acid mixtures were used as the sole substrate, while the addition of sucrose or glucose to the influent wastewater resulted in the formation of granular sludge that settled properly (Wiegant *et al.*, 1985; van Lier *et al.*, 1992; Uemura and Harada, 1993; van Lier, 1996; van Lier *et al.*, 1997). Considering these findings together with the physiological properties of a pure representative strain (*A. thermophila*), this type of filamentous organism is likely one of the microbial species that is indispensable for the granulation of thermophilic UASB sludges.

The bulking phenomenon of granular sludge is also an important matter that warrants further investigation (Speece, 1996). We found that a thermophilic (55°C) UASB reactor used to treat actual organic wastewater from a fried soybean curd-manufacturing factory formed fluffy sludge granules, although the process involved well-settled granules in the earlier period of the operation (Sekiguchi *et al.*, 2001b). The fluffy sludge retained the granular structure, but a number of filaments were found to be sticking out of the surface of the granules. The fluffy granules had far less settling ability than the normal granules, indicating an anaerobic bulking phenomenon. Scanning electron microscopy observations of the spine-like projections on the granules showed that the projections were comprised mainly of filamentous organisms, which were identified as members of *Chloroflexi* subphylum I (i.e., *Anaerolinea thermophila*). Through the study, *A. thermophila* was also found to be a causative agent for the bulking of thermophilic UASB sludge. Considering the fact that this type of filamentous populations was essential for the formation and preservation of the granular structure of thermophilic UASB sludge, these observations suggest that controlling the growth of these filamentous cells is important for thermophilic UASB processes, not only to enhance the granule formation but also to prevent the bulking of sludge granules. We still do not know exactly what triggers the outbreak of the filamentous cells, but more detailed physiological studies of pure strains as well as molecular ecological studies of these populations will elucidate the mechanism of filament outbreak.

3.7. Changes in Population Over Time

The analysis of population changes over the duration of reactor operation is also important for determining how community populations are affected by changes in the reactor environment and/or operation. Molecular techniques provide a promising method for the rapid and quantitative monitoring of microbes in the communities. Those techniques also bridge the gaps between engineer (reactor operation) and microbiologist (culture-based study) and lead to an interactive communication between them to improve reactor performance.

rRNA-based quantitative membrane hybridization and fluorescence *in situ* hybridization direct count are generally employed to determine the dynamics of particular populations of interest. For example, the population dynamics of members of the genus *Smithella* (formerly known as strain SYN7 in earlier studies as mentioned above) were examined under methanogenic and sulfidogenic conditions in anaerobic granular sludge using quantitative membrane hybridization (Harmsen *et al.*, 1996a). Changes in the populations of methanogens and some important sulfate-reducers and syntrophs (such as members of *Desulfovibrionaceae*, *Desulfobacterium*, *Syntrophobacter*, *Smithella*, and *Syntrophomonadaceae*) within anaerobic systems decomposing municipal solid wastes were monitored using the same method. Griffin *et al.* (1998) evaluated the population dynamics of methanogens and sulfate-reducers in mesophilic (37°C) and thermophilic (55°C) digesters during the start-up period of operation (Griffin *et al.*, 1998). Similarly, McMahon *et al.* (2001) determined the population dynamics of important anaerobes in digesters using rRNA-based membrane hybridization to link the community dynamics with various operational parameters. Importantly, they found that stable digesters contained *Methanosaeta*-type archaea as the predominant aceticlastic methanogens, while unstable digesters exhibited higher levels of *Methanosarcina*-type archaea in the communities. For similar purposes, immunochemical antibody probes were used. For example, population changes of methanogens in a UASB reactor shifting from mesophilic (37°C) to thermophilic (55°C) conditions were evaluated using antibody probes for a variety of methanogens. The population dynamics of the aceticlastic methanogens *Methanosaeta concilii* and *Methanosarcina mazei* were also monitored in granular sludge in UASB reactors using antibody probes specific for each organism (Schmidt and Ahring, 1999).

Rather than focusing only on changes in particular populations, changes in overall community composition during the operation have been detected using 16S rRNA gene-based DGGE (denatured gradient gel electrophoresis), SSCP (single strand conformation polymorphism), and cloning analysis. Liu *et al.* (2002) evaluated the bacterial and archaeal community shifts during the start-up of mesophilic (37°C) and thermophilic (55°C) anaerobic reactors using DGGE analyses with PCR-amplified 16S rRNA genes, and subsequently quantified the respective populations using 16S rRNA-based membrane hybridization (Liu *et al.*, 2002b). Similarly, Pereira *et al.* (2002) determined the microbial diversity and population shifts of bacterial and archaeal cells in two EGSB reactors fed with oleic acid-containing wastewater using 16S rRNA gene-based DGGE analysis. In their study, bacterial and archaeal population shifts with increasing organic loading rates were clearly observed. Subsequent sequencing of DGGE bands uncovered the phylogenetic position of microbes representing each band; particularly members of the family *Syntrophomonadaceae* were found to be present in stable reactors. SSCP analysis of PCR-amplified community 16S rRNA genes has been also used for the same purpose as DGGE analysis. The laboratory of Godon *et al.* used the SSCP method to roughly view the population shifts of bacterial and archaeal cells in various anaerobic digesters (Delbes *et al.* 2000; Zumstein *et al.*, 2000; Delbes *et al.* 2001; Leclerc *et al.*, 2001). Although the resolution of SSCP patterns seems less than that of DGGE profiles, the method provides a useful means for the study of microbial community dynamics. The overall dynamics of anaerobic communities in bioreactors were also estimated by 16S rRNA gene clone library analysis (Fernandez *et al.*, 1999). One of the interesting findings of these studies by Fernandez *et al.* was that dynamic changes in community structure were occurring even within the functionally stable reactor (Fernandez *et al.*, 1999; 2000). This finding raises the question of whether stable operation actually does not signify the stability of community structure.

3.8. Yet-to-be Cultured but Important Anaerobes in Anaerobic Biotechnology

Through the application of molecular techniques in the field of microbial ecology to anaerobic biotechnology systems, we could obtain a more precise view of the world of microbes in anaerobic waste/wastewater treatment systems. At present, it may be too early to gain general knowledge of this black box, which links process parameters (i.e., reactor performance) and microbial community dynamics based solely on the above studies. However, information on the essential constituents, the functions of individual microbes, and their population shifts along with the operation of various anaerobic processes has been accumulating at an encouraging rate. The further application of these molecular techniques to various anaerobic processes as well as continuing attempts to cultivate relevant (but often recalcitrant) anaerobes in the community is still needed. In particular, the cultivation of yet-to-be-cultured but relevant anaerobes from these populations is important if the functions of bacteria that inhabit the ecosystems are to be understood. Even at the phylum level, there are a number of uncultured (putative) clone clusters (Tables 1 and 2, such as BA024, OS-K, OP9, OP8, and TM7), some of which have been suggested to be significant populations that may play roles in the anaerobic processes. Although several groups are trying to cultivate these populations, there remain technical obstacles to accomplishing this task. Recently, several techniques such as fluorescence *in situ* hybridization combined with microautoradiography (MAR-FISH) (Lee *et al.*, 1999; Ouverney and Fuhrman, 1999), DNA-based stable isotope probing (SIP) (Radajewski *et al.*, 2000), and RNA-SIP (MacGregor *et al.*, 2002; Manefield *et al.*, 2002) have been developed to estimate the potential metabolic activities of these uncultured microorganisms; these provide a powerful means for indirectly evaluating the substrate uptake capabilities of those microbes by combining molecular techniques with cultivation-mediated processes. These techniques will undoubtedly help us to further understand the nature of the principal but uncultured microbial components. In addition to these newly developed techniques, conventional cultivation techniques still have value for providing unambiguous knowledge of the physiological traits of individual populations. Since most of the microorganisms in anaerobic environments (not only anaerobic bioreactors but also other anaerobic habitats) are yet-to-be cultured and their *in situ* metabolic functions still remain to be characterized, methane reactors also provide microbiologists a fruitful research field. Such information will continue to provide the basis for better controlling the processes, as well as for further expanding the application of waste treatment technology.

4. Conclusions

In summary, anaerobic wastewater treatment technology is an attractive alternative for several types of organic waste/wastewater. The practical application of this technology will increase in the future due not only to its economical and environmental advantages but also to the recent advancement of the reactor technology for a variety of types of waste/wastewater. The identification and quantification of microbes in such reactors is essential for optimizing reactor operation, and precise data have been accumulating through the application of recently introduced modern molecular techniques to anaerobic microbial ecosystems. In particular, rRNA and rRNA gene-based analyses provide a powerful means for elucidating the community composition and species abundance. One of the important findings made recently through such analysis of anaerobic processes is that a large portion of the microbial community is found to be made up of uncultured organisms. This finding has led to the conclusion that more cultivation-dependent isolation attempts should be performed to uncover the functions of these populations. In addition, new technology to cultivate as yet uncultured microbes may be needed.

To further understand the microbial ecology of anaerobic sludge, the following two research strategies should be further explored: (1) more detailed physiological and functional studies on principal (or key) constituents of anaerobic processes, particularly for yet-to-be cultured microbes, should be performed, and (2) the community dynamics in conjunction with process performance and operation parameters (such as hydraulic retention time, organic loading rate, wastewater compositions, etc.) should be studied. For the purpose of advancing anaerobic biotechnology and expanding the types of waste/wastewater to which it applies, much further information on these microbial aspects of anaerobic processes should be forthcoming.

References

Ahring, B. 2003. Perspectives for anaerobic digestion. In: Biomethanation I.B. Ahring, ed. Springer-Verlag, Berlin Heidelberg. p. 1-30.

Ahring, B.K., Schmidt, J.E., Winther-Nielsen, M., Macario, A.J.L., and Conway de Macario, E. 1993. Effect of medium composition and sludge removal on the production, composition, and architecture of thermophilic (55°C) acetate-utilizing granules from an upflow anaerobic sludge blanket reactor. Appl. Environ. Microbiol. 59: 2538-2545.

Alexander, M. 1998. Biodegradation and Bioremediation 2nd edn. Academic Press, San Diego.

Amann, R.I. 1995. *In situ* identification of micro-organisms by whole cell hybridization with rRNA-targeted nucleic acid probes. In: Molecular Microbial Ecology Manual. A.D.L. Akkermans, and J.D. van Elsas, eds. Kluwer Academic Publishers, London. p. 3.3.6/1-15.

Amann, R.I., Ludwig, W., and Schleifer, K.-H. 1995. Phylogenetic identification and *in situ* detection of individual microbial cells without cultivation. Microbiol. Rev. 59: 143-169.

Angelidaki, I., Ellegaard, L., and Ahring, B. 2003. Applications of the anaerobic digestion process. In: Biomethanation II. B. Ahring, ed. Springer-Verlag, Berlin Heidelberg. p. 1-34.

Angenent, L.T., Zheng, D., Sung, S., and Raskin, L. 2002. Microbial community structure and activity in a compartmentalized, anaerobic bioreactor. Wat. Env. Res. 74: 450-461.

Balk, M., Weijma, J., and Stams, A.J.M. 2002. *Thermotoga lettingae* sp. nov., a novel thermophilic, methanol-degrading bacterium isolated from a thermophilic anaerobic reactor. Int. J. Syst. Evol. Micribiol. 52: 1361-1368.

Barber, W.P., and Stuckey, D.C. 2000. Nitrogen removal in a modified anaerobic baffled reactor (ABR): 1, denitrification. Wat. Res. 15: 2413-2422.

Björnsson, L., Hugenholtz, P., Tyson, G.W., and Blackall, L.L. 2002. Filamentous *Chloroflexi* (green non-sulfur bacteria) are abundant in wastewater treatment processes with biological nutrient removal. Microbiology. 148: 2309-2318.

Boone, D.R., and Bryant, M.P. 1980. Propionate-degrading bacterium, *Syntrophobacter wolinii* sp. nov. gen. nov., from methanogenic ecosystems. Appl. Environ. Microbiol. 40: 626-632.

Brakenhoff, R.H., Schoenmarkers, J.G.G., and Lubsen, N.H. 1991. Chimeric cDNA clones: a novel PCR artifact. Nucleic Acids Res. 19: 1949-1950.

Cariello, N.F., Thilly, W.G., Swenberg, J.A., and Skopek, T.R. 1990. Deletion mutagenesis during polymerase chain reaction: dependence on DNA polymerase. Gene. 99: 105-108.

Casserly, C., and Erijman, L. 2003. Molecular monitoring of microbial diversity in an UASB reactor. Int. Biodeterio. Biodeg. 52: 7-12.

Cheng, S.-S., Ho, C.-Y., and Wu, J.-H. 1997. Pilot study of UASB process treating PTA manufacturing wastewater. Wat. Sci. Tech. 36: 73-82.

Cord-Ruwisch, R., Lovley, D.R., and Schink, B. 1998. Growth of *Geobacter sulfurreducens* with acetate in syntrophic cooperation with hydrogen-oxidizing anaerobic partners. Appl. Environ. Microbiol. 64: 2232-2236.

Daffonchio, D.D., Thaveesri, J., and Verstraete, W. 1995. Contact angle measurement and cell hydrophobicity of granular sludge from upflow anaerobic sludge bed reactors. Appl. Environ. Microbiol. 61: 3676-3680.

Daly, K., Sharp, R.J., and McCarthy, A.J. 2000. Development of oligonucleotide probes and PCR primers for detecting phyogenetic sybgroups of sulfate-reducing bacteria. Microbiology. 146: 1693-1705.

Delbes, C., Moletta, R., and Godon, J.-J. 2001. Bacterial and archaeal 16S rDNA and 16S rRNA dynamics during an acetate crisis in an anaerobic digestor ecosystem. FEMS Microbiol. Ecol. 35: 19-26.

Delbes, C., Moletta, R., and Godon, J.J. 2000. Monitoring of activity dynamics of an anaerobic digester bacterial community using 16S rRNA polymerase chain reaction-single-strand conformation polymorphism analysis. Environ. Microbiol. 2: 506-515.

Devereux, R., Kane, M.D., Winfrey, J., and Stahl, D.A. 1992. Genus- and group specific hybridization probes for determinative and environmental studies of sulfate-reducing bacteria. System. Appl. Microbiol. 15: 601-609.

DiMarco, A.A., Bobik, T.A., and Wolfe, R.S. 1990. Unusual coenzymes of methanogenesis. Annu. Rev. Biochem. 59: 355-394.

Dubourguier, H.C., Prensier, G., and Albagnac, G. 1987. Structure and microbial activities of granular anaerobic sludge. In: Granular Anaerobic Sludge; Microbiology and Technology. G. Lettinga, A.J.B. Zehnder, J.T.C. Grotenhuis, and L.W. Hulshoff Pol, eds. Centre for agricultural publishing and documentation, Wageningen. p. 18-41.

Edwards, E.A., and Grbic-Galic, D. 1994. Anaerobic degradation of toluene and o-xylene by a methanogenic consortium. Appl. Environ. Microbiol. 60: 313-322.

Fang, H.H.P., Chui, H.K., and Li, Y.Y. 1994. Microbial structure and activity of UASB granules treating different wastewater. Wat. Sci. Tech. 30: 87-96.

Fang, H.H.P., Chui, H.K., and Li, Y.Y. 1995. Microstructural analysis of UASB granules treating brewery wastewater. Wat. Sci. Tech. 31: 129-135.

Fernandez, A., Huang, S., Seston, S., Xing, J., Hickey, R., Criddle, C., and Tiedje, J. 1999. How stable is stable? Function versus community composition. Appl. Environ. Microbiol. 65: 3697-3704.

Fernandez, A.S., Hashsham, S.A., Dollhopf, S.L., Raskin, L., Glagoleva, O., Dazzo, F.B., Hickey, R.F., Criddle, C.S., and Tiedje, J.M. 2000. Flexible community structure correlates with stable community function in methanogenic bioreactor communities perturbed by glucose. Appl. Environ. Microbiol. 66: 4058-4067.

Garcia, J.-L., Patel, B.K.C., and Ollivier, B. 2000. Taxonomic, phylogenetic, and ecological diversity of methanogenic *Archaea*. Anaerobe. 6: 205-226.

Garrity, G.E., and Holt, J.G. 2001a. Phylum BVI. *Chloroflexi* phy. nov. In: Bergey's Manual of Systematic Bacteriology. G.M. Garrity, ed. Springer-Verlag, New York, New York. p. 427-446.

Garrity, G.E., and Holt, J.G. 2001b. Phylum AII. *Euryarchaeota* phy. nov. In: Bergey's Manual of Systematic Bacteriology. D.R. Boone and R.W. Castenholz, eds. Springer-Verlag, New York, New York. p. 211-354.

Garrity, G.M. 2001. Bergey's Manual of Systematic Bacteriology. Springer-Verlag, New York, New York.

Godon, J.-J., Zumstein, E., Dabert, P., Habouzit, F., and Moletta, R. 1997. Molecular microbial diversity of an anaerobic digestor as determined by small-subunit rDNA sequence analysis. Appl. Environ. Microbiol. 63: 2802-2813.

Gonzalez-Gil, G., Lens, P.N.L., van Aelst, A., van As, H., Versprille, A.I., and Lettinga, G. 2001. Cluster structure of anaerobic aggregates of an expanded granular sludge bed reactor. Appl. Environ. Microbiol. 67: 3683-3692.

Griffin, M.E., McMahon, K.D., Mackie, R.I., and Raskin, L. 1998. Methanogenic population dynamics during start-up of anaerobic digesters treating municipal solid waste and biosolids. Biotech. Bioeng. 57: 342-355.

Grotenhuis, J.T.C., Plugge, C.M., Stams, A.J.M., and Zehnder, A.J.B. 1992. Hydrophobicities and electrophoretic mobilities of anaerobic bacterial isolates from methanogenic granular sludge. Appl. Environ. Microbiol. 58: 1054-1056.

Grotenhuis, J.T.C., Smit, M., van Lammeren, A.A.M., Stams, A.J.M., and Zehnder, A.J.B. 1991a. Localization and quantification of extracellular polymers in methanogenic granular sludge. Appl. Microbiol. Biotechnol. 36: 115-119.

Grotenhuis, J.T.C., Smit, M., Plugge, C.M., Yuansheng, X., van Lammeren, A.A.M., Stams, A.J.M., and Zehnder, A.J.B. 1991b. Bacteriological composition and structure of granular sludge adapted to different substrate. Appl. Environ. Microbiol. 57: 1942-1949.

Guiot, S.R., Pauss, A., and Costerton, J.W. 1992. A structured model of the anaerobic granule consortium. Wat. Sci. Tech. 25: 1-10.

Habets, L.H., Engelaar, A.J.H.H., and Groeneveld, N. 1997. Anaerobic treatment of inuline effluent in an internal circulation reactor. Wat. Sci. Tech. 35: 189-197.

Hansen, K.H., Ahring, B.K., and Raskin, L. 1999. Quantification of syntrophic fatty acid-β-oxidizing bacteria in a mesophilic biogas reactor by oligonucleotide probe hybridization. Appl. Environ. Microbiol. 65: 4767-4774.

Harmsen, H.J.M. 1996. Detection, phylogeny and population dynamics of syntrophic propionate-oxidizing bacteria in anaerobic granular sludge. Wageningen Agricultural University.

Harmsen, H.J.M., Akkermans, A.D.L., Stams, A.J., and de Vos, W.M. 1996a. Population dynamics of propionate-oxidizing bacteria under methanogenic and sulfidogenic conditions in anaerobic granular sludge. Appl. Environ. Microbiol. 62: 2163-2168.

Harmsen, H.J.M., Kengen, K.M.P., Akkermans, A.D.L., Stams, A.J.M., and de Vos, W.M. 1995. Phylogenetic analysis of two syntrophic propionate-oxidizing bacteria in enrichments cultures. System. Appl. Microbiol. 18: 67-73.

Harmsen, H.J.M., Kengen, H.M.P., Akkermans, A.D.L., Stams, A.J.M., and de Vos, W.M. 1996b. Detection and localization of syntrophic propionate-oxidizing bacteria in granular sludge by *in situ* hybridization using 16S rRNA-based oligonucleotide probes. Appl. Environ. Microbiol. 62: 1656-1663.

Harmsen, H.J.M., van Kuijk, B.L.M., Plugge, C.M., Akkermans, A.D.L., de Vos, W.M., and Stams, A.J.M. 1998. *Syntrophobacter fumaroxidans* sp.nov., a syntrophic propionate-degrading sulfate-reducing bacterium. Int. J. Syst. Bacteriol. 48: 1383-1387.

Hattori, S., Kamagata, Y., Hanada, S., and Shoun, H. 2000. *Thermacetogenium phaeum* gen. nov., sp. nov., a strictly anaerobic, thermophilic, syntrophic acetate-oxidizing bacterium. Int. J. Syst. Evol. Microbiol. 50: 1601-1610.

Hiraishi, A., Kamagata, Y., and Nakamura, K. 1995. Polymerase chain reaction amplification and restriction fragment length polymorphism analysis of 16S rRNA genes from methanogens. J. Ferment. Bioeng. 79: 523-529.

Hofman-Bang, J., Zheng, D., Westermann, P., Ahring, B., and Raskin, L. 2003. Molecular ecology of anaerobic reactor systems. In: Biomethanation. I.B. Ahring, ed. Springer-Verlag, Berlin Heidelberg. p. 151-204.

Hristova, K.R., Mau, M., Zheng, D., Aminov, R.I., Mackie, R.I., Gaskins, H.R., and Raskin, L. 2000. *Desulfotomaculum* genus- and subgenus-specific 16S rRNA hybridization probes for environmental studies. Environ. Microbiol. 2: 143-160.

Hugenholtz, P. 2002. Exploring prokaryotic diversity in the genomic era. Genome Biol. 3: reviews0003 1-008.8.

Hugenholtz, P., Goebel, B.M., and Pace, N.R. 1998a. Impact of culture-independent studies on the emerging phylogenetic view of bacterial diversity. J. Bacteriol. 180: 4765-4774.

Hugenholtz, P., Pitulle, C., Hershberger, K.L., and Pace, N.R. 1998b. Novel division level bacterial diversity in a yellowstone hot spring. J. Bacteriol. 180: 366-376.

Hulshoff Pol, L.W., Heijnekamp, K., and Lettinga, G. 1988. The selection pressure as a driving force behind the granulation of anaerobic sludge. In: Granular Anaerobic Sludge; Microbiology and Technology. G. Lettinga, A.J.B. Zehnder, J.T.C. Grotenhuis, and L.W. Hulshoff Pol, eds. Centre for agricultural publishing and documentation, Wageningen. p. 153-161.

Huser, B.A., Wuhrmann, K., and Zehnder, J.B. 1982. *Methanothrix soehngenii* gen. nov. sp. nov., a new acetotrophic non-hydrogen-oxidizing methane bacterium. Arch. Microbiol. 132: 1-9.

Imachi, H., Sekiguchi, Y., Kamagata, Y., Ohashi, A., and Harada, H. 2000. Cultivation and *in situ* detection of a thermophilic bacterium capable of oxidizing propionate in syntrophic association with hydrogenotrophic methanogens in a thermophilic methanogenic granular sludge. Appl. Environ. Microbiol. 66: 3608-3615.

Imachi, H., Sekiguchi, Y., Kamagata, Y., Hanada, S., Ohashi, A., and Harada, H. 2002. *Pelotomaculum thermopropionicum* gen. nov., sp. nov., an anaerobic, thermophilic syntrophic propionate-oxidizing bacterium. Int. J. Syst. Evol. Micribiol. 52: 1729-1735.

Jackson, B.E., Bhupathiraju, V.K., Tanner, R.S., Woese, C.R., and McInerney, M.J. 1999. *Syntrophus aciditrophicus* sp. nov., a new anaerobic bacterium that degrades fatty acids and benzoate in syntrophic association with hydrogen-using microorganisms. Arch. Microbiol. 171: 107-114.

Juretschko, J., Loy, A., Lehner, A., and Wagner, M. 2002. The microbial community composition of a nitrifying-denitrifying activated sludge from an industrial sewage treatment plant analyzed by the full-cycle rRNA approach. System. Appl. Microbiol. 25: 84-99.

Kamagata, Y., and Mikami, E. 1990. Some characteristics of 2 morphotypes of *Methanothrix soehngenii* from mesophilic anaerobic digesters. J. Ferment. Bioeng. 70: 272-274.

Kamagata, Y., and Mikami, E. 1991. Isolation and characterization of a novel thermophilic methanosaeta strain. Int. J. Syst. Bacteriol. 41: 191-196.

Kamagata, Y., Kawasaki, H., Oyaizu, H., Nakamura, K., Mikami, E., Endo, G., Koga, Y., and Yamasato, K. 1992. Characterization of three thermophilic strains of *Methanothrix* ("*Methanosaeta*") *thermophila* sp. nov. and rejection of *Methanothrix* ("*Methanosaeta*") *thermoacetophila*. Int. J. Syst. Bacteriol. 42: 463-468.

Kanagawa, T. 2003. Bias and artifacts in multitemplate polymerase chain reaction (PCR). J. Biosci. Bioeng. 96: 317-323.

Kato, M., Field, J.A., Versteeg, P., and Lettinga, G. 1994. Feasibility of the expanded granular sludge bed (EGSB) reactors for the anaerobic treatment of low strength soluble wastewaters. Biotechnol. Bioeng. 44: 469-479.

Kleerebezem, R., and Macarie, H. 2003. Treating industrial wastewater: Anaerobic digestion comes of age. Chemical Engineering. April 01: 56-64.

Kopczynski, E.D., Bateson, M.M., and Ward, D.M. 1994. Recognition of chimeric small-subunit ribosomal DNAs composed of genes from uncultivated microorganisms. Appl. Environ. Microbiol. 60: 746-748.

Leclerc, M., Delbes, C., Moletta, R., and Godon, J.-J. 2001. Single strand conformation polymorphism monitoring of 16S rDNA Archaea during start-up of an anaerobic digester. FEMS Microbiol. Ecol. 34: 213-220.

Lee, M.J., and Zinder, S.H. 1988. Isolation and characterization of a thermophilic bacterium which oxidizes acetate in syntrophic association with a methanogen and which grows acetogenically on H_2-CO_2. Appl. Environ. Microbiol. 54: 124-129.

Lee, N., Nielsen, P.H., Andreasen, K.H., Juretschko, S., Nielsen, J.L., Schleifer, K.-H., and Wagner, M. 1999. Combination of fluorescent *in situ* hybridization and microautoradiography - a new tool for structure-function analysis in microbial ecology. Appl. Environ. Microbiol. 65: 1289-1297.

Lettinga, G. 1995. Anaerobic digestion and wastewater treatment systems. Antonie van Leeuwenhoek. 67: 97.93.28.

Lettinga, G., Field, J., van Lier, J., Zeeman, G., and Hulshoff Pol, L.W. 1997. Advanced anaerobic wastewater treatment In the near future. Wat. Sci. Tech. 35: 5-12.

Lettinga, G., Rebac, S., Parshina, S., Nozhevnikova, A., van Lier, J.B., and Stams, A.J.M. 1999. High-rate anaerobic treatment of wastewater at low temperatures. Appl. Environ. Microbiol. 65: 1696-1702.

Lin, X., and White, R.H. 1986. Occurrence of coenzyme F_{420} and its gamma-monoglutamyl derivative in nonmethanogenic archaebacteria. J. Bacteriol. 168: 2317-2324.

Liu, W.-T., Chan, O.-C., and Fang, H.H.P. 2002a. Characterization of microbial community in granular sludge treating brewery wastewater. Wat. Res. 36: 1767-1775.

Liu, W.-T., Chan, O.-C., and Fang, H.H.P. 2002b. Microbial community dynamics during start-up of acidogenic anaerobic reactors. Wat. Res. 36: 3203-3210.

Liu, Y., Balkwill, D.L., Aldrich, H.C., Drake, G.R., and Boone, D.R. 1999. Characterization of the anaerobic propionate-degrading syntrophs *Smithella propionica* gen. nov., sp. nov. and *Syntrophobacter wolinii*. Int. J. Syst. Bacteriol. 49: 545-556.

Macarie, H. 2000. Overview of the application of anaerobic treatment to chemical and petrochemical wastewater. Wat. Sci. Tech. 42: 201-214.

Macarie, H., Noyola, A., and Guyot, J.P. 1992. Anaerobic treatment of a petrochemical wastewater from a terephthalic acid plant. Wat. Sci. Tech. 25: 223-235.

Macario, A.J.L., and Conway de Macario, E. 1988. Quantitative immunologic analysis of the methanogenic flora of digesters reveals a considerable diversity. Appl. Environ. Microbiol. 54: 79-86.

Macario, A.J.L., Visser, F.A., van Lier, J.B., and Conway de Macario, E. 1991. Topography of methanogenic subpopulations in a microbial consortium adapting to thermophilic conditions. J. Gen. Microbiol. 137: 2179-2189.

Macario, A.J.L., Conway de Macario, E., Ney, U., Schoberth, S.M., and Sahm, H. 1989. Shifts in methanogenic subpopulations measured with antibody probes in a fixed-bed loop anaerobic bioreactor treating sulfite evaporator condensate. Appl. Environ. Microbiol. 55: 1996-2001.

MacGregor, B.J., Bruchert, V., Fleischer, S., and Amann, R. 2002. Isolation of small-subunit rRNA for stable isotopic characterization. Environ. Microbiol. 4: 451-464.

MacLeod, F., Guiot, S.R., and Costerton, J.W. 1990. Layered structure of bacterial aggregates produced in an upflow anaerobic sludge bed and filter reactor. Appl. Environ. Microbiol. 56: 1598-1607.

MacLeod, F., Guiot, S.R., and Costerton, J.W. 1995. Electron microscopic examination of the extracellular polymeric substances in anaerobic granular biofilms. World J. Microb. Biotech. 11: 481-485.

Manefield, M., Whiteley, A.S., Griffiths, R.I., and Bailey, M.J. 2002. RNA stable isotope probing, a novel means of linking microbial community function to phylogeny. Appl. Environ. Microbiol. 68: 5367-5373.

McInerney, M.J., Bryant, M.P., Hespell, R.B., and Costerton, J.W. 1981. *Syntrophomonas wolfei* gen. nov. sp. nov., an anaerobic, syntrophic, fatty acid-oxidizing bacterium. Appl. Environ. Microbiol. 41: 1029-1039.

Moffett, B.F., Walsh, K.A., Harris, J.A., and Hill, T.C.J. 2000. Analysis of bacterial community structure using 16 rDNA analysis. Anaerobe. 6: 129-131.

Mountfort, D.O., Brulla, W.J., Krumholz, L.R., and Bryant, M.P. 1984. *Syntrophus buswellii* gen. nov., sp. nov.: a benzoate catabolizer from methanogenic ecosystems. Int. J. Syst. Bact. 34: 216-217.

Ng, A., Melvin, W.T., and Hobson, P.N. 1994. Identification of anaerobic digester bacteria using a polymerase chain reaction method. Biores. Technol. 47: 73-80.

Oude Elferink, S.J.W.H., Boschker, H.T.S., and Stams, A.J.M. 1998a. Identification of sulfate reducers and *Syntrophobacter* sp. in anaerobic granular sludge by fatty-acid biomarker and 16S rRNA probing. Geomicrobiology J. 15: 3-17.

Oude Elferink, S.J.W.H., van Lis, R., Heilig, H.G.H.J., Akkermans, A.D.L., and Stams, A.J.M. 1998b. Detection and quantification of microorganisms in anaerobic bioreactors. Biodegradation. 9: 169-177.

Ouverney, C.C., and Fuhrman, J.A. 1999. Combined microautoradiography-16S rRNA probe technique for determination of radioisotope uptake by specific microbial cell types *in situ*. Appl. Environ. Microbiol. 65: 1746-1752.

Patel, G.B., and Sprott, G.D. 1990. *Methanosaeta concilii* gen. nov., sp. nov. ("*Methanothrix concilii*") and *Methanosaeta thermoacetophila* nom. rev., comb. nov. Int. J. Syst. Bacteriol. 40: 79-82.

Pereira, M.A., Roest, K., Stams, A.J.M., Mota, M., Alves, M., and Akkermans, A.D.L. 2002. Molecular monitoring of microbial diversity in expanded granular sludge bed (EGSB) reactors treating oleic acid. FEMS Microbiol. Ecol. 41: 95-103.

Petersen, S.P., and Ahring, B. 1991. Acetate oxidation in a thermophilic anaerobic sewage-sludge digestor: the importance of non-aceticlastic methanogenesis from acetate. FEMS Microbiol. Ecol. 86: 149-158.

Plugge, C., Zoetendal, E.G., and Stams, A.J. 2000. *Caloramator coolhaasii* sp. nov., a glutamate-degrading, moderately thermophilic anaerobe. Int. J. Syst. Evol. Micribiol. 50: 1155-1162.

Plugge, C., Balk, M., and Stams, A.J. 2002a. *Desulfotomaculum thermobenzoicum* subsp. *thermosyntrophicum* subsp. nov., a thermophilic, syntrophic, propionate-oxidizing, spore-forming bacterium. Int. J. Syst. Evol. Micribiol. 52: 391-399.

Plugge, C.M., Balk, M., Zoetendal, E.G., and Stams, A.J. 2002b. *Gelria glutamica* gen. nov., sp nov., a thermophilic, obligately syntrophic, glutamate-degrading anaerobe. Int. J. Syst. Evol. Micribiol. 52: 401-407.

Plumb, J.J., Bell, J., and Stuckey, D.C. 2001. Microbial populations associated with treatment of an industrial dye effluent in an anaerobic baffled reactor. Appl. Environ. Microbiol. 67: 3226-3235.

Qiu, X., Wu, L., Huang, H., McMDonel, P.E., Palumbo, A.V., Tiedje, J.M., and Zhou, J. 2001. Evaluation of PCR-generated chimeras, mutations, and heteroduplexes with 16S rRNA gene-based cloning. Appl. Environ. Microbiol. 67: 880-887.

Qiu, Y.-L., Sekiguchi, Y., Imachi, H., Kamagata, Y., Tseng, I.-C., Cheng, S.-S., Ohashi, A., and Hanada, S. 2003. *Sporotomaculum syntrophicum* sp. nov., a novel anaerobic, syntrophic benzoate-degrading bacterium isolated from methanogenic sludge treating wastewater from terephthalate manufacturing. Arch. Microbiol. 179: 242-249.

Qiu, Y.-L., Sekiguchi, Y., Imachi, H., Kamagata, Y., Tseng, I.-C., Cheng, S.-S., Ohashi, A., and Hanada, S. 2004. Identification and isolation of anaerobic, syntrophic phthalate isomers-degrading microbes from methanogenic sludges treating wastewater from terephthalate manufacturing. Appl. Environ. Microbiol. 70: 1617-1626.

Radajewski, S., Ineson, P., Parekh, N.R., and Murrell, J.C. 2000. Stable-isotope probing as a tool in microbial ecology. Nature. 403: 646-649.

Raskin, L., Stromley, J.M., Rittmann, B.E., and Stahl, D.A. 1994a. Group-specific 16S rRNA hybridization probes to describe natural communities of methanogens. Appl. Environ. Microbiol. 60: 1232-1240.

Raskin, L., Poulsen, L.K., Noguera, D.R., Rittmann, B.E., and Stahl, D.A. 1994b. Quantification of methanogenic groups in anaerobic biological reactors by oligonucleotide probe hybridization. Appl. Environ. Microbiol. 60: 1241-1248.

Raskin, L., Zheng, D., Griffin, M.E., Stroot, P.G., and Misra, P. 1995. Characterization of microbial communities in anaerobic bioreactors using molecular probes. Antonie van Leeuwenhoek. 68: 297-308.

Rocheleau, S., Greer, C.W., Lawrence, J.R., Cantin, C., Laramee, L., and Guiot, S.R. 1999. Differentiation of *Methanosaeta concilii* and *Methanosarcina barkeri* in anaerobic mesophilic granular sludge by fluorescent *in situ* hybridization and confocal scanning laser microscopy. Appl. Environ. Microbiol. 65: 2222-2229.

Santegoeds, C.M., Damgaard, L.R., Hesselink, G., Zopfi, J., Lens, P., Muyzer, G., and de Beer, D. 1999. Distribution of sulfate-reducing and methanogenic bacteria in anaerobic aggregates determined by microsensor and molecular analyses. Appl. Environ. Microbiol. 65: 4618-4629.

Schink, B. 1997. Energetics of syntrophic cooperation in methanogenic degradation. Microbiol. Mol. Biol. Rev.: 262-280.

Schmidt, J.E., and Ahring, B.K. 1996. Granular sludge formation in upflow anaerobic sludge blanket (UASB) reactors. Biotech. Bioeng. 49: 229-246.

Schmidt, J.E., and Ahring, B.K. 1999. Immobilization patterns and dynamics of acetate-utilizing methanogens immobilized in sterile granular sludge in upflow anaerobic sludge blanket reactors. Appl. Environ. Microbiol. 65: 1050-1054.

Schnürer, A., Zellner, G., and Svensson, B.H. 1999. Mesophilic syntrophic acetate oxidation during methane formation in biogas reactors. FEMS Microbiolo. Ecol. 29: 249-261.

Sekiguchi, Y., Kamagata, Y., and Harada, H. 2001a. Recent advances in methane fermentation technology. Curr. Opin. Biotechnol. 12: 277-282.

Sekiguchi, Y., Kamagata, Y., Nakamura, K., Ohashi, A., and Harada, H. 1999. Fluorescence *in situ* hybridization using 16S rRNA-targeted oligonucleotides reveals localization of methanogens and selected uncultured bacteria in mesophilic and thermophilic sludge granules. Appl. Environ. Microbiol. 65: 1280-1288.

Sekiguchi, Y., Kamagata, Y., Nakamura, K., Ohashi, A., and Harada, H. 2000. *Syntrophothermus lipocalidus* gen. nov., sp. nov., a novel thermophilic, syntrophic, fatty-acid-oxidizing anaerobe which utilizes isobutyrate. Int. J. Syst. Evol. Microbiol. 50: 771-779.

Sekiguchi, Y., Takahashi, H., Kamagata, Y., Ohashi, A., and Harada, H. 2001b. *In situ* detection, isolation, and physiological properties of a thin filamentous microorganism abundant in methanogenic granular sludges: a novel isolate affiliated with a clone cluster, the green non-sulfur bacteria, subdivision I. Appl. Environ. Microbiol. 67: 5740-5749.

Sekiguchi, Y., Kamagata, Y., Syutsubo, K., Ohashi, A., Harada, H., and Nakamura, K. 1998. Phylogenetic diversity of mesophilic and thermophilic granular sludges determined by 16S rRNA gene analysis. Microbiology. 144: 2655-2665.

Sekiguchi, Y., Yamada, T., Hanada, S., Ohashi, A., Harada, H., and Kamagata, Y. 2003. *Anaerolinea thermophila* gen. nov., sp. nov. and *Caldilinea aerophila* gen. nov., sp. nov., novel filamentous thermophiles that represent a previously uncultured lineage of the domain Bacteria at the subphylum level. Int. J. Syst. Evol. Micribiol. 53: 1843-1851.

Speece, R.E. 1996. Anaerobic Biotechnology for Industrial Wastewaters. Archae Press, Nashville, Tennessee.

Speksnijder, A., Kowalchuk, G.A., De Jong, S., Kline, E., Stephan, J.R., and Laanbroek, H.J. 2001. Microvariation artifacts intriduced by PCR and cloning of closely related 16S rRNA gene sequences. Appl. Environ. Microbiol. 67: 469-472.

Stackebrandt, E., and Goebel, B.M. 1994. Taxonomic note: a place for DNA-DNA reassociation and 16S rRNA sequence analysis in the present species definition in bacteriology. Int. J. Syst. Bacteriol. 44: 846-849.

Stackebrandt, E., Sproer, C., Rainey, F.A., Burghardt, J., Pauker, O., and Hippe, H. 1997. Phylogenetic analysis of the genus *Desulfotomaculum*: evidence for the misclassification of *Desulfotomaculum guttoideum* and description of *Desulfotomaculum orientis* as *Desulfosporosinus orientis* ge.mov., comb. nov. Int. J. Syst. Bacteriol. 47: 1134-1139.

Stahl, D.A., Flesher, B., Mansfield, H.R., and Montgomery, L. 1988. Use of phylogenetically based hybridization probes for studies of ruminal microbial ecology. Appl. Environ. Microbiol. 54: 1079-1084.

Stams, A.J., Oude Elferink, S.J.W.H., and Westermann, P. 2003. Metabolic interaction between methanogenic consortia and anaerobic respiring bacteria. In: Biomethanation. I. B. Ahring, ed. Springer-Verlag, Berlin Heidelberg. p. 31-56.

Stams, A.J.M., and Oude Elferink, S.J.W.H. 1997. Understanding and advancing wastewater treatment. Curr. Opin. Biotechnol. 8: 328-334.

Stams, A.J.M., Grotenhuis, J.T.C., and Zhender, A.J.B. 1989. Structure-function relationship in granular sludge. In: Recent Advances in Microbial Ecology. T. Hattori, ed. Japan Scientific Societies Press, Tokyo. p. 440-445.

Stams, A.J.M., Grolle, K.C.F., Frijters, C.T.M.J., and van Lier, J.B. 1992. Enrichment of thermophilic propionate-oxidizing bacteria in syntrophy with *Methanobacterium thermoautotrophicum* or *Methanobacterium thermoformicicum*. Appl. Environ. Microbiol. 58: 346-352.

Stieb, M., and Schink, B. 1985. Anaerobic oxidation of fatty acids by *Clostridium bryantii* sp. nov., a sporefprming, obligately syntrophic bacterium. Arch Microbiol. 140: 387-390.

Suzuki, M.T., and Giovannoni, S.J. 1996. Bias caused by template annealing in the amplification of mixtures of 16S rRNA genes by PCR. Appl. Environ. Microbiol. 62: 625-630.

Svetlitshnyi, V., Rainey, F., and Wiegel, J. 1996. *Thermosyntropha lipolytica* gen. nov., sp. nov., a lipolytic, anaerobic, alkalitolerant, thermophilic bacterium utilizing short- and long-chain fatty acids in syntrophic coculture with a methanogenic archaeum. Int. J. Syst. Bacteriol. 46: 1131-1137.

Syutsubo, K., Harada, H., Ohashi, A., and Suzuki, H. 1997. An effective start-up of thermophilic UASB reactor by seeding mesophilically-grown granular sludge. Wat. Sci. Tech. 36: 391-398.

Tagawa, T., Syutsubo, K., Sekiguchil, Y., Ohashi, A., and Harada, H. 2000. Quantification of methanogen cell density in anaerobic granular sludge consortia by fluorescence in-situ hybridization. Water Sci. Technol. 42: 77-82.

Uemura, S., and Harada, H. 1993. Microbial characteristics of methanogenic sludge consortia developed in thermophilic UASB reactors. Appl. Microbiol. Biotechnol. 39: 654-660.

Uemura, S., and Harada, H. 1995. Inorganic composition and microbial characteristics of methanogenic granular sludge frown in a thermophilic upflow anaerobic sludge blanket reactor. Appl. Microbiol. Biotechnol. 43: 358-364.

van Lier, J.B. 1996. Limitations of thermophilic anaerobic wastewater treatment and the consequences for process design. Antonie van Leeuwenhoek. 69: 1-14.

van Lier, J.B., Rebac, S., and Lettinga, G. 1997. High-rate anaerobic wastewater treatment under psychrophilic and thermophilic conditions. Wat. Sci. Tech. 35: 199-206.

van Lier, J.B., Boersma, F., Debets, M.M.W.H., and Lettinga, G. 1994. High-rate thermophilic anaerobic wastewater treatment in compartmentalized upflow reactors. Wat. Sci. Tech. 30: 251-261.

van Lier, J.B., Grolle, K.C.F., Stams, A.J.M., de Macario, E.C., and Lettinga, G. 1992. Start-up of a thermophilic upflow anaerobic sludge bed (UASB) reactor with mesophilic granular sludge. Appl. Microbiol. Biotechnol. 37: 130-135.

van Lier, J.B., Grolle, K.C.F., Frijters, C.T.M.J., Stams, A.J.M., and Lettinga, G. 1993. Effects of acetate, propionate, and butyrate on the thermophilc anaaerobic degradation of propionate by methanogenic sludge and defined cultures. Appl. Environ. Microbiol. 59: 1003-1011.

Veiga, M.C., Jain, M.K., Wu, W.-M., Hollingsworth, R.I., and Zeikus, J.G. 1997. Composition and role of extracellular polymers in methanogenic granules. Appl. Environ. Microbiol. 63: 403-407.

Visser, F.A., van Lier, J.B., Macario, A.J.L., and de Macario, E.C. 1991. Diversity and population dynamics of methanogenic bacteria in a granular consortium. Appl. Environ. Microbiol. 57: 1728-1734.

von Wintzingerode, F., Selent, B., Hegemann, W., and G¨obel, U.B. 1999. Phylogenetic analysis of an anaerobic, trichlorobenzene-transforming microbial consortium. Appl. Environ. Microbiol. 65: 283-286.

Wallrabenstein, C., Hauschild, E., and Schink, B. 1995a. *Syntrophobacter pfennigii* sp. nov., new syntrophically propionate-oxidizing anaerobe growing in pure culture with propionate sulfate. Arch. Microbiol. 164: 346-352.

Wallrabenstein, C., Gorny, N., Springer, N., Ludwig, W., and Schink, B. 1995b. Pure culture of *Syntrophus buswellii*, definition of its phylogenetic status, and description of *Syntrophus gentianae* sp. nov. System. Appl. Microbiol. 18: 62-66.

Wasserfallen, A., Nölling, J., Pfister, P., Reeve, J., and Coway de Macario, E. 2000. Phylogenetic analysis of 18 thermophilic *Methanobacterium* isolates supports the proposals to createa new genus, *Methanothermobacter* gen., nov., and to reclassify several isolates in three species, *Methanothermobacterthermoautotrophicus* comb. nov., *Methanothermobacter wolfeii* comb. nov., and *Methanothermobacter marburgensis* sp. nov. Int. J. Syst. Evol. Microbiol. 50: 43-53.

Whitman, W.B., Coleman, D.C., and Wiebe, W.J. 1998. Prokaryotes: The unseen majority. Proc. Natl. Acad. Sci. USA. 95: 6578-6583.

Wiegant, W.M., and de Man, A.W.A. 1986. Granulation of biomass in thermophilic upflow anaerobic sludge blanket reactors treating acidified wastewaters. Biotechnol. Bioeng. 28: 718-727.

Wiegant, W.M., Claassen, J.A., and Lettinga, G. 1985. Thermophilic anaerobic digestion of high strength wastewaters. Biotechnol. Bioeng. 27: 1374-1381.

Wiegant, W.W. 1987. The "spaghetti theory" on anaerobic granular sludge formation, or the inevitabulity of granulation. In: Granular Anaerobic Sludge; Microbiology and Technology. G. Lettinga, A.J.B. Zehnder, J.T.C. Grotenhuis, and L.W. Hulshoff Pol, eds. Centre for Agricultural Publishing and Documentation, Wageningen. p. 146-152.

Wintzingerode, F.V., Gobel, U.B., and Stackebrandt, E. 1997. Determination of microbial diversity in environmental samples: pitfalls of PCR-based rRNA analysis. FEMS Microbiol. Rev. 21: 213-229.

Wu, J.-H., Liu, W.-T., Tseng, I.-C., and Cheng, S.-S. 2001a. Characterization of microbial consortia in a terephthalate-degrading anaerobic granular sludge system. Microbiology 147: 373-382.

Wu, J.-H., Liu, W.-T., Tseng, I.-C., and Cheng, S.-S. 2001b. Characterization of a 4-methylbenzoate-degrading methanogenic consortium as determined by small-subunit rDNA sequence analysis. J. Biosci. Bioeng. 91: 449-455.

Zellner, G., Macario, A.J.L., and Conway de Macario, E. 1997. A study of three anaerobic methanogenic bioreactors reveals that syntrophs are diverse and different from reference organisms. FEMS Microbiol. Ecol. 22: 295-301.

Zhao, H., Yang, D., Woese, C.R., and Bryant, M.P. 1990. Assignment of *Clostridium bryantii* to *Syntrophospora bryantii* gen. nov., comb. nov. on the basis of a 16S rRNA sequence analysis of its crotonate-grown pure culture. Int. J. Syst. Bacteriol. 40: 40-44.

Zhao, H., Yang, D., Woese, C.R., and Bryant, M.P. 1993. Assignment of fatty acid-beta-oxidizing syntrophic bacteria to *Syntrophomonadaceae* fam. nov. on the basis of 16S rRNA sequence analyses. Int. J. Syst. Bacteriol. 43: 278-286.

Zheng, D., and Raskin, L. 2000. Quantification of *Methanosaeta* species in anaerobic bioreactors using genus- and species-specific hybridization probes. Microb. Ecol. 39: 246-262.

Zinder, S.H., and Koch, M. 1984. Non-acetoclastic methanogenesis from acetate: acetate-oxidation by a thermophilic syntrophic coculture. Arch. Microbiol. 138: 263-272.

Zumstein, E., Moletta, R., and Godon, J.-J. 2000. Examination of two years of community dynamics in an anaerobic bioreactor using fluorescence polymerase chain reaction (PCR) single-strand conformation polymorphism analysis. Environ. Microbiol. 2: 69-78.

Index

1-Butanol (n-Butanol) 212, 266, 329-338
2-Propanol (isopropanol) 266, 269, 329-331
3-Hydroxypropionate cycle 41
5-Fluorocytosine (5-FC) 215, 217
5-Fluorouracil (5-FU) 215, 217
16S Ribosomal RNA gene 4-19,126-130, 226, 242, 253, 268, 284, 292, 306-308, 366-378
  PCR of 128, 292, 306-307, 374, 378
  phylogeny of 9-10, 241, 253, 370-374

A

Acetivibrio cellulolyticus 344-345, 349, 351
Acetobacterium carbinolicum 265
Acetobacterium dehalogenans 264-265
Acetobacterium woodii 252, 256, 258, 260-268
  ATPase of 260
Acetobacter xylinum 73
Acetogen 29, 38, 39, 53, 251-271, 343, 371
  in gastrointestinal 270
  in terrestrial, subsurface, and aquatic 270-271
  phylogeny of 252-254
Acetogenesis 31, 35, 37-38, 55, 251-259, 263, 267-271
  electron donors 263-265
  in situ acetogenesis 268
Acetone 212, 325, 329-338
Acetyl-CoA 29-30, 37-41, 47-50, 251-252, 259-264, 267-270, 332, 335
Acetyl-CoA pathway 254-255, 268. *See also* Wood-Ljungdahl pathway
  carbonyl branch of 254-256
  methyl branch of 254-256
Acetyl-CoA synthase 256-258, 267-268
  Cu in 257
  iron sulfur in 257
  Ni in 257
Acidaminococcus fermentans 31, 45, 51-52
Acidobacteria 7, 15, 19-20, 225-226, 367
Acr (α-crystallin) 150-153
Actinobacillus actinomycetemcomitans 4, 110-117
Actinomyces naeslundii 110, 114, 118
Acylated homoserine lactone 98, 116-117, 136
adc operon 333-335
ADEPT 212, 215-216
adhE 330, 332-338
ADP-ribosylation 174, 182-184
Aer 73
Aerobic respiration 33, 76, 78, 90-94, 151, 164
Aerolysin 171, 175, 177
Aeromonas 177, 225, 240,
Aeromonas hydrophila 177, 240
Aerotolerance 55, 161-167. *See also* Oxygen
AerR 75, 77. *See also* PpaA
Agrobacterium tumefaciens 68
AI-2 117, 119, 202-203
Alginate production 87-88, 101-102, 104. *See also* Mucoid
AlgR 102-104, 202
AlgT 102-104
α-Crystallin 150-153. *See also* Acr
α-Toxin 171, 173-180, 200-203. *See also* Phospholipase C of *Staphylococcus aureus* 176
Alternative sigma 102, 150, 199
  BotR 203-206
  SigE 203-204
  SigK 203-204
  TetR 203-206
  TxeR 203-206
Ammonia oxidation 283-285, 289-293, 295-297
  aerobic 283
  anaerobic 295
  anaerobic nitrogen dioxide-dependent 284
  Anammox 283, 292-294
Anaerobe
  classification of 3-20
  evolution of 1-2
  facultative 3
  obligate 2-3
  strict 2
Anaerobic baffled (AB) reactor 364
Anaerobic degradation
  of aromatic compounds 44-45
  of aromatic hydorocarbon 231
  of hydrocarbon 41-44
  of polycyclic aromatic hydrocarbons (PAHs) 44
  of polymer 28
Anaerobic heterotrophic denitrification 294
Anaerobic metabolism 27
Anaerobic nutrition 35-36
Anaerobic photosynthetic bacteria 10, 74. *See also* Purple nonsulfur bacteria
Anaerocellum thermophylum 348
Anammox (anaerobic nitrite-dependent ammonia oxidizer) 283, 292-294. *See also* Ammonia oxidation
  ecology of 294
ANR 68, 95, 99, 104. *See also* FNR/CRP homologue
ANR-binding domain 92-93, 104
Antibody-directed enzyme prodrug therapy See ADEPT
Antisense RNA 205, 336, 352, 354
AppA 76-77
Aquificales 7-8
Arc 70-73
  controlled genes 70-73
  homologues of 72
ArcA 70-73
ArcA modulon 70-73. *See also* Arc-controlled genes
ArcB 70-73
  crossregulation by 71
arcDABC 95-96, 99, 103-105
Archaea 2, 6-9, 36, 54, 68, 225, 241, 259, 264, 365, 373-375
ArfM 70, 77-78
Arginine deiminase 95, 99, 103
ArgR 95
Aromatic hydrocarbon oxidation. *See also* Anaerobic degradation
  coupled to electrode reduction 230
  coupled to Fe(III) reduction 227
Arsenate reductase (Arr) 239, 242-245. *See also* Electron acceptor
  phylogeny of ArrA 246
Arsenate respiration 9, 240-247
arsRDABC 239-240

As-yet uncultured bacteria 6
ATPase of *Acetobacterium woodii* 260
Azoarcus sp. 12, 41-45, 245
Azotobacter vinelandii 68-69, 72, 79, 164

B

Bacillus anthracis 68, 182-185
Bacillus cereus 179, 182, 184
Bacillus infernos 3
Bacillus macerans 330
Bacillus selenitireducens 241-242
Bacillus subtilis 37, 54-55, 68, 70, 77-79, 112, 203, 245
  redox regulation 77-79
Bacteroidetes 7, 18-19
Bacteroides fragilis 55, 132, 161-167, 171, 173, 181-182
Bacteroides-Prevotella 127
Bacteriodes ruminicola 35-36
Bacteroides thetaiotaomicron 54, 127, 131-132, 161-162, 165-166
badDEFG 45
bdhA 330-333
bdhB 330-334
Benzoyl-CoA 36, 43, 45, 52-53
  benzoyl-CoA reductase 45
Beta-toxin 176
Bifidobacterium longum 131, 217
Biofilm 290
  extracellular polymer in 113
  of cystic fibrosis lung 88-91, 104
  of oral bacteria 109-115, 119-120
  wasterwater treatment for 362-365, 376
Bioremediation 317
  coupled to reduction of Fe(III) oxide 221-232
  gemome-enabled 231
  in situ 229-230
    of CCl_4 and metals 326-327
  of dehalogenation 311
Biotin-dependent decarboxylase 51
BoNT 174, 187-189, 205-206
BotR 203-206
Bradyrhizobium japonicum 68-69, 76, 80
Brocadia anammoxidans 292-296
brpA 114
Brucella spp. 68
1-Butanol (n-Butanol) 212, 266, 329-338

C

Caldicellulosiruptor 348
Campylobacter rectus 110
Cancer therapy 211-217, 338
Candida albicans 135
Carbon flow 28-31
  regulation of 29
Carbon metabolism 37-41, 77, 149
Carbon tetrachloride (CCl_4) 317-327
Carboxydothermus hydorgenoformans 257
Catenabacterium catenaforme 129
CCl_4 317-327
CcpA 205-206
CDEPT 211, 215-217
cel genes 350-351, 355
  promoters of 353
Cell-cell communication. *See also* Quorum sensing
  between dental bacteria 116-118
  in *Pseudomonas aeruginosa* 98-101
  in toxin production 202
Cellulase 345-355
  hemicellulase 348
Cellulomonas fimi 345, 348-352
Cellulose binding module (CBM) 345-355
Cellulosome 343-355
  minicellulosome 354
Chemiosmotic potential 27, 48-49, 52, 55
Chitinase 348
Chlorobenzene 303-311
  nomenclature of 304
  reductive dehalogenation 307-309
Chlorobi 5, 7, 14-15, 373
Chlorobium 40
Chloroflexi 2, 5-7, 14, 307-308, 366-377
Chloroflexus aurantiacus 41
Cholorinated ethene 307
Chrysiogenetes 7-9, 240
Chrysinogenes arsenatis 9, 242-243
Citric acid cycle 7, 37, 41, 45, 149, 290
  modified 31
  oxidative 40
  reductive 39- 40
Clostridia-directed Enzyme Prodrug Therapy *See* CDEPT
Clostridium absonum 173
Clostridium aceticum 251-252, 266-269
Clostridium acetobutylicum 8, 212, 215, 329-338, 343-354. *See also Clostridium beijerinckii*
  electrotransformation of 353
Clostridium aldrichii 344
Clostridium americanum 330
Clostridium aminobutyricum 46
Clostridium baratii 173-174, 179, 187
Clostridium beijerinckii 212-213, 329-338. *See also Clostridium acetobutylicum*
Clostridium bifermentans 173, 179
Clostridium botulinum 36, 173-174, 181-184, 187, 199, 203-206
Clostridium butyricum 46, 211-212, 329. *See also Clostridium oncolyticum*
Clostridium cellobioparum 349
Clostridium cellulofermentans 343
Clostridium cellulolytcum 343-344, 347-351
  conjugation of 353
  electrotransformation of 353
Clostridium cellulose 343
Clostridium cellulovorans 343-355
Clostridium chauvoei 173
Clostridium coccoides 127, 254
Clostridium difficile 185-186, 199, 204-206
Clostridium felsineum 253
Clostridium formicoaceticum 253, 256, 265-267
Clostridium glycolicum 253, 270
Clostridium herbivorans 343
Clostridium histolyticium 36, 173, 181
Clostridium josui 344, 347-351
Clostridium leptum 127
Clostridium limosum 174, 184
Clostridium ljundgahlii 253
Clostridium novyi 173-174, 179, 185-186

Clostridium oncolyticum 212, 216. *See also Clostridium butyricum*
Clostridium pasteurianum 246, 331
Clostridium papyrosolvens 343, 349
Clostridium perfringens 55, 69, 171-184, 199-206, 351
Clostridium phytofermentans 344
Clostridium saccharobutylicum 330-332
Clostridium saccharoperbutylacetonicum 330, 332
Clostridium scatologenes 254
Clostridium septicum 171, 173, 175
Clostridium sordellii 173-174, 185
Clostridium spiroforme 174, 182, 184
Clostridium sporogenes 36, 212-216, 331
  transformation of 212-214
Clostridium sporosphaeroides 45
Clostridium stricklandii 36
Clostridium symbiosum 45
Clostridium tetani 171-173, 181, 187, 199, 204-206, 211
Clostridium thermoaceticum 5, 36, 252
  See also Morella thermoacetica
Clostridium thermoautotrophicum 258.
  See also Moorella thermoautotrophica
Clostridium thermocellum 343-353
Clostridium viride 46-47
CLSM 111, 116, 375
CO_2 1-2, 6-7, 13, 28-40, 48, 50-55, 96, 206, 255-259, 263-270, 285-286, 293, 317-319
  fixation of 76, 251-252
  reduction of 254
CodY 205-206
Cohesins 344-354
Collagenase 171, 173, 181, 200-203. *See also* Kappa-toxin
com 101, 114, 118
Confocal laser scanning microscopy See CLSM
Conjugation 118-119, 199
  of *Clostridium cellulolyticum* 353
CooA 68. *See also* FNR/CRP homologue
Corrinoid 37-39, 77, 254-257, 260-265, 310
cpe 203-204
Crenarchaeota 37, 41, 240, 366, 368
CRP 67-69, 73-76, 95
CrtJ 76-77. *See also* PpsR
ctfA 330, 332, 337-338
ctfB 332, 336-338
Cultivation-independent molecular method 369
CydR 69. *See also* FNR/CRP homologue
Cystic fibrosis 87-104
Cytosine deaminase 212-217

D

Deferribacteres 7, 9, 240
Dehalobacter 16, 229, 303, 307, 310
Dehalococcoides 229, 303-312
Dehalococcoides ethenogenes 307-311
Dehalogenation of CCl_4 317-327
  biochemistry 309-310
  dioxin 305-309
  genetics 319-321
  reaction pathway 319
  reductive 303
  under Fe(III)-reducing conditions 229
Dehalospirillum 229, 303
Dehydration 27, 41, 45-47, 54-55
Delta-toxin 173, 176
Denaturing gradient gel electrophoresis *See* DGGE
Denitrification 33
  anaerobic hydrogen-dependent 287
Dental plaque 109-120
  gene transfer between dental plaque bacteria 118-119
  subgingival 110-111
  supragingival 110-111
Desulfitobacterium 229, 307
Desulfitobacterium hafniense 68, 69, 243-245, 306, 311
Desulfobacterium autotrophicum 38
Desulfobulbus 13, 369, 375
Desulforhabdas amnigenus 375
Desulfotomaculum acetoxidans 38
Desulfovibrio desulfuricans 54, 245
Desulfovibrio termitidis 54
Desulfovibrio 7, 13, 54, 165, 175, 369
Desulfovibrio termitidis 54
Desulfovibrio vulgaris 50
Desulfuromonas acetoxidans 31, 40
Desulfuromonas 13, 225-226, 229, 303
Desulfuromusa 13, 225
DevR 151-154. *See also* DosR
DGGE 126-132, 378
Dictyoglomus 6-7, 14-15
Dioxin 303-312
  dehalogenation 305-309. *See also* Dehalogenation
Dissimilatory reduction of Fe(III), Mn(IV) 32, 225. *See also* Electron acceptor
DMSO reductase 76, 244
  family of 239, 242-246
DNR 92, 95, 103-104. *See also* FNR/CRP homologue
Dos 71,73
DosR 151-154. *See also* DevR
Dps 55, 161-164

E

Eikenella corrodens 110
Electrode reduction 230-231
Electron acceptor. *See also* Terminal electron acceptor
  arsenate, As(V) 9, 32-33, 227, 239-245, 247
  carbon dioxide 2, 32-33, 266
  chlorate, perchlorate 32-33
  Co(III) 32-33
  Cr(VI) 9, 32-33, 230
  Fe(III) 2, 9, 12, 32-33, 303
  Fe(III)oxide 222-232
  fumarate 2, 32-33, 48-50, 70, 74, 162, 164, 227, 242, 247, 266-268, 303, 334
  heterodisulfide, CoM-S-S-CoB 50
  Mn (III) (IV) 2, 9, 32-33, 223, 303
  multiple halogenated electron acceptors 311
  nitrate 2, 32-33, 67-79, 87, 104, 150, 221-227, 242, 247, 266-269, 283, 294-295, 303, 361,
  oxidized metals 2
  selenate 32-33, 227, 239-247
  sulfate, sulfite, sulfur, thiosulfate 2, 9, 32-35, 38, 44, 51, 55, 247, 266-268, 303

trimethylamine-N-oxide 32-33
U(V) (VI) 9, 32-33, 229
End-product efflux 52
Energy conservation in acetogens 258-260, 267-268
Energy generation in ammonia oxidation 286
EngE 346, 350-353
Enterotoxin 171, 173, 175, 181
of *Bacteroides fragilis* 181-182
of *Clostridium perfringens* 175, 199, 203-204
Epsilon toxin 171-178
Escherichia coli 46- 55, 67-74, 76-79, 80, 94-96, 103, 133-135, 145, 151, 163-165, 203-205, 212-215, 244-246, 256-262, 267-268, 320-321, 332-333, 338, 344, 353
Esterase 348
EtrA 70, 74. *See also* FNR/CRP homologue
Expanded granular sludge bed (EGSB) 364, 378

F

Facultative anaerobe 3, 12, 27, 67-80, 109, 127, 163, 317-327
Fe(III) reduction 13, 221-232. See also Electron acceptor
enzymatic 226
humics 227
in sediments 223-225
Fermentation. 1, 3, 7-19, 28, 32-36, 47-48, 70-77, 129, 132, 161, 199, 254, 263-270, 303, 309, 330-337, 343
See also Substrate-level phosphorylation
fermentative bacteria 29-31
gastrointestinal 32
methane fermentation 361-379
substrate-level phosphorylation 47
Ferrimonas 225
FhlA 33
FhlC/FhlD 71, 73, 80
Fibrobacteres 7, 14-15, 225-226
Firmicutes 7, 15-19, 225, 369-372
FISH 126, 292, 294, 375-378
FixK 69
Flagella, flagellar 114, 290
flagella and type IV pili in biofilm 88, 90
FLP 69. See also FNR/CRP homologue
FlpA 68-70. See also FNR/CRP homologue
Fluorescent *in situ* hybridisation *See* FISH
fluorescently labeled RNA probes 6
5-Fluorocytosine (5-FC) 215, 217
5-Fluorouracil (5-FU) 215, 217
FNR 33, 67-78, 95-96, 334
FNR/CRP homologue 67-70
FnrL 69, 77. *See also* FNR/CRP homologue
FnrN 68-69. *See also* FNR/CRP homologue
FnrP 69. *See also* FNR/CRP homologue
Formate dehydrogenase 30, 36-38, 49, 69, 244, 255, 258
Fumarate reduction 48-50
fumarate reductase 49-50, 70-74, 162-164, 267
Fusobacteria 7, 19
Fusobacterium nucleatum 45, 110, 112-113, 117-118

G

gap operon 335
Gene directed enzyme drug therapy (GDEPT) 215-216
Gene transfer between dental plaque bacteria 118-119.
See also Oral communities
Genomic sequence tags 6
Geobacter 13, 221, 221-232, 372
Geobacter metallireducens 228, 231
Geobacter sulfurreducens 165
Geobacteraceae 13, 221-232
Gradient gel electrophoresis 6, 126. *See also* DGGE
Gut microflora 125-138
gangrenes or gastrointestinal disease 171-187
gastrointestinal tract 15, 19, 28, 32, 36, 161, 171-187, 241

H

Haemophilus influenzae 72-73
Halogenated benzenes and dioxins 303-312
natural formation of 304
toxicity of 304
multiple halogenated electron acceptors 311
HbaR 68. *See also* FNR/CRP homologue
Helicobacter pylori 132
Hemiceullulase 348-352. *See also* Cellulase
Heterodisulfide reductase 50
Holophaga foetida 20, 252-253, 264
Hydrogen peroxide 53-55, 102, 162
Hydrogenase 2, 29-39, 48-52, 70, 76, 246, 258, 267, 288
3-Hydroxypropionate cycle 37, 41
hyp7 200-201. *See also* VirR-regulated RNA (VR-RNA)

I

Immune system 125, 127, 132-138
In Clostridio 353
In situ acetogenesis 268
In situ bioremediation 221-232, 317, 326-327
In vivo expression technology See IVET
Internal circulation (IC) reactor 364
Iron reduction 33 *See also* Electron acceptor
Iron stress 317, 320
Iron-sulfur clusters (centers) 14-15, 36-38, 43, 48-54, 68-70, 162-165, 243-244, 256-258, 310
IVET 115, 131-132

J

Juncus roemerianus 305

K

Kappa-toxin 171, 181, 200-203. *See also* Collagenase
Klebsiella pneumoniae 51, 69, 245
Kuenenia stuttgartiensis 292-293

L

Lactobacillus 18, 69, 125-132
Lactobacillus acidophilus 128-132
Lactobacillus brevis 128-130
Lactobacillus casei 69, 128-129
Lactobacillus crispatus 128-133
Lactobacillus curvatus 128-129
Lactobacillus delbrueckii 128-129
Lactobacillus fermentum 128-129
Lactobacillus gasseri 128-130
Lactobacillus johnsonii 128-129
Lactobacillus paracasei 128-130
Lactobacillus plantarum 128-130
Lactobacillus reuteri 128-132

Lactobacillus rhamnosus 128-130
Lactobacillus ruminis 128-129, 131
Lactobacillus sakei 128-129
Lactobacillus salivarius 128-130
Lactococcus lactis 68, 332
Laser capture microdissection See LCM
LasR 88, 98-99, 117
LCM 135
Leuconostoc 48
Leuconostoc argentinum 129
Leuconostoc mesenteroides 129
Lichenase 348
Listeria monocytogenes 173
Lithotroph 283, 295
LuxS 117-119, 200-203. *See also* Quorum sensing
Lyase 47, 54, 72, 149, 152, 332, 348, 351

M

Malonomonas rubra 13, 52
MAR-FISH 378. See also FISH
Medicargo truncatula 136
Methanomicrobiales 374
Mesophilic 31, 226, 343, 350-355, 362-364, 369-378
Metabolic engineering 329, 337
Metal 32, 36
  cobalamin 47, 317
  for catalysis 27, 36-39, 45-55, 94, 239-247, 255-257, 295, 319, 323
  metal chelator 317-327
Methane fermentation 361-379
Methanobrevibacter ruminantium 36
Methanogenesis 27-40, 52, 55, 221-225, 259, 270-271, 362, 376
  methanogen 1-3, 7, 29-39, 55, 227, 256-260, 268-271, 343, 365, 371-377
  methanogenic ecosystems 28-31
Methanomicrobium mobile 36
Methanosarcinales 374
Methanoseata thermophila 39
Methyltransferase 37-39, 52, 256-267, 353
Microarray analysis 127, 132-136, 329
  in *Bacillus subtilis* 77
  in *Mycobacterium tuberculosis* 150-151
  in *Porphyromonas gingivalis* 115
  in *Pseudomonas aeruginosa* 99-101, 115
  related to solventogenesis 334-337
Microautoradiography-fluorescence *in situ* hybridization *See* MAR-FISH
Microbial community 27, 32-35, 306, 311
  in anaerobic sludge 364-378
  of dental bacteria 109-120
  in gut microflora 125-135
  in sediment 221-231
  of ammonia oxidation 292-295
Minicellulosome 353-354. *See also* Cellulosome
Mn (IV) reduction 32-33, 223. 303 *See also* Electron acceptor
Moorella thermoacetica 36, 253-269. *See also Clostridium thermoautotrophicum*
Moorella thermoautotrophica 258-261, 266-267. *See also Clostridium thermoautotrophicum*
MucA 102-104
MucB 102-104
Mucoid 88, 101-104. S*ee also* Alginate production
Mycobacterium bovis BCG 145-154
Mycobacterium marinum 152
Mycobacterium tuberculosis 145-154

N

NADH/NAD$^+$ ratio 72, 80
narGHJI 70, 77-80, 94, 246-247
NarL 33, 74, 93-95, 103-104
NarX 33, 74, 93-95, 151, 153
Neurotoxin 74, 171, 174, 181, 187-188, 205-206. *See also* Toxin
Ni/Fe hydrogenase 70
NifA 69
NifL 69
Nitrate reduction 33, 49, 67, 94, 223-225, 295, 334. *See also* Electron acceptor
Nitric oxide 78-79, 87, 94-98, 102-104, 132, 146, 151-154, 284-297
  nitric oxide reductase 90, 92, 94, 97, 103
Nitrification 283
Nitrite oxidation 294-295
Nitrite reductase 78-79, 92, 94, 99, 104, 288, 294
Nitrobacter 7-11, 283, 294-297
Nitrobacteraceae 283, 291-292
Nitrococcus 294
Nitrogen oxide 87-99, 284-291
  transport of 94
Nitrogenase 36-37, 45, 69, 72
Nitrosococcus 284
Nitrosolobus 284, 288
Nitrosomonas 284-296
Nitrosomonas europaea 284-296
Nitrosospira 284
Nitrosovibrio 284
Nitrospira 7, 20, 294, 367, 373
Nitrous oxide reductase 90, 94, 99, 104
NnrR 76. *See also* FNR/CRP homologue
NtcA 68. *See also* FNR/CRP homologue

O

Obligate anaerobe 35, 112, 161-167, 211, 217
Oral communities 109-120. *See also* Microbial community
  gene transfer 118-119
Oxalate metabolism 49, 52
Oxalobacter formigenes 52
Oxidative stress response 55, 161-164
Oxygen
  aerotolerance 55, 161-167
  protective enzymes 54
  sensitivity to 54
OxyR 55, 161, 163-164

P

Paracoccus denitrificans 69, 166
PAS domain 70, 76
  in ArcB 71, 72
  in CrtJ 76
  in Dos 73
  in ResE 78-79

PCB 304-311
PCR
  of 16S rRNA gene 128
  PCR-DGGE 126, 128-129, 132. *See also* DGGE
Pdtc 317-327. *See also* Pyridine-2,6,-bis(thiocarboxylic acid)
  Pdt transporter 322
  Pdtc-metal complex 323-326
Pediococcus acidilactici 129
Pediococcus pentosaceus 129
Pelobacter 13, 36, 225-226
Pelobacter acetylenicus 36
Perfringolysin171-175, 200-201. *See also* Theta-toxin
Peroxidases 2, 53-54, 69, 93, 161-166, 269, 305
Phospholipase C 92, 95, 171, 175, 178-180, 200-203. *See also* α-Toxin
Photosynthetic genes 74-77
Phylogenic tree 9, 18, 253
  by16S rRNA gene sequence 370-374
  of arsenate/selenate respiring bacteria 241
Planctomycetales 6, 292
Polar-lipid fatty acid (PLFA) analysis 375
Polychlorinated benzene 303
Polymerase chain reaction *See* PCR
Polymorphism 4, 135, 146. *See also* Single strand polymorphism
Polyphosphate kinase. *See ppk*
Polysaccharide 29. *See also* Cellulose
Porphyromonas gingivalis 55, 110-115, 117-118, 134, 161-163, 166
PpaA 77. *See also* AerR
ppk 88, 114-115, 118-119
PpsR 76,77. *See also* CtrJ
Prevotella intermedia 110, 161
Prevotella melaninogenica 166
Prevotella nigrescens 112-113
2-Propanol (isopropanol) 329-331
Propionate 35, 46-55, 362
  degradation of 29-32, 362, 369-371
  oxidation of 31, 375, 371
Propionigenium modestum 51
Protective enzymes to oxygen 54. *See also* Oxygen
Proteobacteria 2-19, 225, 366-371
PrrA-PrrB 74-77. *See also* RegA-RegB
PrrC 74-75. *See also* SenC
Pseudomonas aeruginosa 76, 87-104, 114-117, 132, 136, 175, 202, 319, 321
Pseudomonas fluorescens 291, 319
Pseudomonas putida 245, 318-319, 323, 327
Pseudomonas spp. 68
Pseudomonas stutzeri 317-327
Psychrophilic 269, 343, 364
Purple nonsulfur bacteria 10-11, 74. *See also* Anaerobic photosynthetic bacteria
Pyridine-2,6,-bis(thiocarboxylic acid) 317-327. *See also* Pdtc
Pyrobaculum islandicum 40
Pyrococcus furiosus 48

Q

Quorum sensing 98-101, 109-119. 136, 202. *See also* Cell-cell communication

R

Reactive oxygen species *See* ROS
Real time PCR *See* RT-PCR
Redox regulation 67-80
Redox-active(sensitive) cysteine 76
  in RegB 75
  in ResD 79
RegA-RegB 74-77. See also PrrA-PrrB
  homologues of 76
ResD-ResE 70, 77-79
ResE, autophosphorylation 78
Reverse electron transport 31, 284
Rex 80
Rhizobium spp. 68-69
Rhl system 88, 98-99, 101, 116-117
Rhodobacter spp. 11, 68-69
Rhodobacter capsulatus 74-77
Rhodobacter sphaeroides 74-77
Rhodopseudomonas palustris 45, 68
Rhodospirillum rubrum 257
Rhodovulum robiginosum 2
Rhodovulum iodosum 2
Rhodovulum sulfidophilum 76, 245
16S Ribosomal RNA gene 4-19,126-130, 226, 242, 253, 268, 284, 292, 306-308, 366-378
  PCR of 128
  phylogeny of 9-10, 241, 253, 370-374
RNA polymerase 68, 74, 79, 96, 98, 103, 164, 203-206
RNA-SIP 378. *See also* SIP
ROS 53-55, 161-167, 335
  damage by 162-163
  detoxification of 165
  prevention of 164-165
  sensitive targets of 166
  tolernance 163-167
Roseobacter denitrificans 76
RT-PCR 73, 80, 104-105, 132, 153, 353
Rubrerythrin (Rbr) 54, 166, 269
Ruminococcus albus 36
Ruminococcus productus 252, 256, 259, 266-267

S

Salmonella enterica 72.
Salmonella typhimurium 131-133, 243-245, 332
Scaffoldins 344-350
Selective-capture-of-transcribed sequences (SCOTS) 132
Selenate respiration 240-247. See also Electron acceptor
  selenate reductase (Ser) 245-247
Selenomonas ruminantium 52
SenC 74-75. *See also* PrrC
ser (serine biosynthesis) operon 335
ser (selenate reduction) operon 247
Shewanella alga 32
Shewanella oneidensis 2, 70-74, 243-244
Shewanella spp. 68, 70, 225, 240-244
Siderophore 317, 320-323
SigE 203-204
SigH 332
SigK 203-204
Single strand polymorphism 6, 378. *See also* SSCP
Sinorhizobium meliloti 76
SIP (stable isotope probing) 378. *See also* RNA-SIP

Sludge aggregates 362-365, 376
Smithella 13, 370-377
Smithella propionica 369
SNARE 188-189
Sodium pump 38, 51-52
sol operon 332-335
Soluble NSF attachment protein receptor *See* SNARE
Solventogenesis 329-338
 gene regulation of 331-334
Spirochaetes 5-7, 18-19, 366
Spo0A 112, 329-334, 337
Sporomusa malonica 52
Sporomusa ovata 252, 256
SSCP (single strand conformation polymorphism) 378
Stable isotope 44
 SIP 378
Staphylococcus aureus 87, 175-177, 184, 201-202, 239-244
Streptococcus cremoris 52
Streptococcus gordonii 110-118
Streptococcus milleri 110
Streptococcus mitior 110
Streptococcus mutans 68,110-120
Streptococcus oralis 110, 114
Streptococcus pneumoniae 118, 173
Streptococcus pyogenes 118, 173
Streptococcus sanguinis 110, 119
Streptococcus sobrinus 110
Streptomyces coelicolor 79
Stress proteins 95, 115, 151, 335-337
Subgingival 109-119. *See also* Dental plaque
Substrate-level phosphorylation 27, 40, 47-55, 254-255, 303. *See also* Fermentation
Succinivibrio dextrinosolvens 36
Succinyl-CoA synthetase 290-291
Sulfate reduction 27-33, 50-51, 223-225, 271
 sulfate reducer 1-13, 306, 365, 369-377
 sulfate-reducing ecosystems 28
Sulfurospirillum barnesii 240, 242, 247
Superoxide 2, 53-54, 95, 162
 superoxide dismutase 2, 53-54, 161-165, 269
 superoxide reductase 54, 269
Supragingival 109-111. *See also* Dental plaque
Syntrophic metabolism 30-34
 of benzoate 52-53
Syntrophobacter 13, 369-377
Syntrophobacter fumaroxidans 30, 31, 375
Syntrophobacter wolinii 31
Syntrophococcus sucromutans 36
Syntrophomonas wolfei 31
Syntrophospora bryantii 5, 30
Syntrophus 369
Syntrophus aciditrophicus 53, 369
Syntrophus buswellii 31
Syntrophus gentianae 53

T

Tannerella forsythensis 110-114, 161
TCA cycle *See* Citric acid cycle
tcbR 114
tdc operon 73-74
TdcA 74
TdcR 74
Terminal electron
 accepting process (TEAP) 222-225
 acceptor 29-33, 50, 70, 74, 94, 150, 228, 239, 241, 247, 251, 263, 266-269, 285, 302, 361 *See also* Electron acceptor
TetR 204-206
TeTx 174, 181, 187-188, 205, 210
Thauera 12
 Thauera aromatica 41-43, 45, 53
 Thauera selenatis 240, 244-247
Thermicanus aegyptius 269
Thermodesulfobacteria 6-9
Thermodynamic 27-35, 53-55, 222-224, 260-262, 268, 306, 334
Thermophilic 6-9, 18-20, 34, 40, 225, 252, 343-344, 350-351, 362-364, 371-378
Thermoproteus tenax 40
Thermotogae 6-8
Theta-toxin 172, 200-203. *See also* Perfringolysin
Thiolase
 thl promoter 354
 thlA 330, 335, 338
 thlRBC 335
Toll-like receptor (TLR) 125, 133-137
ToxA 174, 185-186, 204-206
ToxB 174, 185-186, 204-206
Toxin
 α-toxin 171, 175-180, 200-203 *See also* Phopholipase C
 beta-toxin 176-177, 201-201
 delta-toxin 176
 enterotoxin 175, 181, 203
 epsilon-toxin 177-178
 gene regulation of 199-206
 iota-toxin 181-183, 202
 kappa-toxin 173, 181, 200-203. *See also* Collagenase
 mode of action 171-189
 modifying cell membranes 178
 modifying actin 182
 neurotoxin 74, 171, 174, 181, 187-188, 205-206
 nutritional regulation of toxin genes 206
 of *Bacteroides* 171-189, 199-206
 of *Clostridia* 171-189, 199-206
 pore-forming 172-178, 183
 therapeutic use 189, 205
 theta-toxin 200-203. *See also* Perfringolysin
ToxT 73
Transformation
 of *Clostridum beijerinckii* 212-214. See also *Clostridum beijerinckii*
 of *Clostridium sporogenes* 212-214. See also *Clostridum sporogenes*
Treponema denticola 110-114, 117-119
TspO 77
Tuberculosis (TB) 145-154
 TB and hypoxia 147-154
 TB latency 146-154. *See also* Wayne model
TxeR 204-206
Type I pili swarming 114

U

Upflow anaerobic sludge blanket (UASB) 361-363, 376-377
UASB sludge 371-375
Upflow state sludge bed (USSB) reactor 364

V

Vascular-targeting agents 216
Veillonella spp. 16, 110
Vibrio cholerae 72-73,114
VirR-VirS 200-203
VirR-regulated RNA (VR-RNA) 200-201. *See also hyp7*
Virulence factors 136, 171-186, 199-206
controlled by
Arc 72
las 98
luxS 117

W

Wastewater treatment 361-379
mesophilic 362-364, 369-378
psychrophilic 364
thermophilic 362, 364, 371-378
Wayne model 149-152. *See also* TB and hypoxia
Wolinella succinogenes 14, 49, 243-245
Wood-Ljungdahl pathway 37, 254-255. *See also* Acetyl-CoA pathway

X

X module 344-347
Xylanase 348, 352, 355

Y

YeiL 68, 69. *See also* FNR/CRP homologue
Yersinia pestis 72, 321